太陽活動 1992-2003
──フレア監視望遠鏡が捉えたサイクル23──

柴田一成
Kazunari Shibata
北井礼三郎
Reizaburo Kitai
門田三和子
Miwako Kateda
他著
et al.

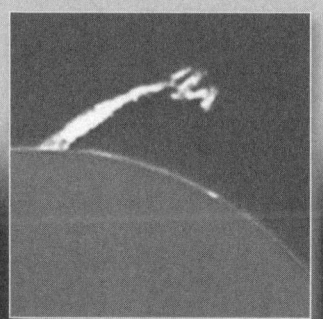

Solar Activity in 1992-2003
Solar Cycle 23 Observed by Flare Monitoring Telescope

京都大学
学術出版会

Solar Activity in 1992-2003

Solar Cycle 23 Observed by Flare Monitoring Telescope

*

Kazunari Shibata, Reizaburo Kitai and Miwako Katoda et al.

Kyoto University Press, 2011

ISBN978-4-87698-987-4

京都大学飛騨天文台の全体写真.
Hida Observatory.

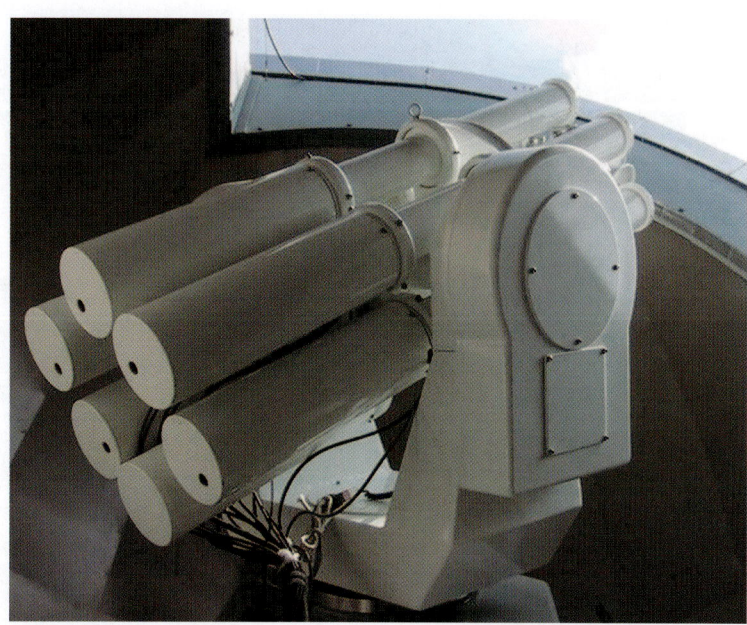

フレア監視望遠鏡（FMT）の外観.
Flare monitoring telescope at the Hida Observatory.

ペルーのイカ大学に設置されたフレア監視望遠鏡．奥に見える小さな小屋は2本のレールに沿って移動できるスライド式の格納庫である．
The FMT after its relocation to Ica Univ., Peru. The small cabin seen just behind the telescope is a sliding hangar that can move along the two rails.

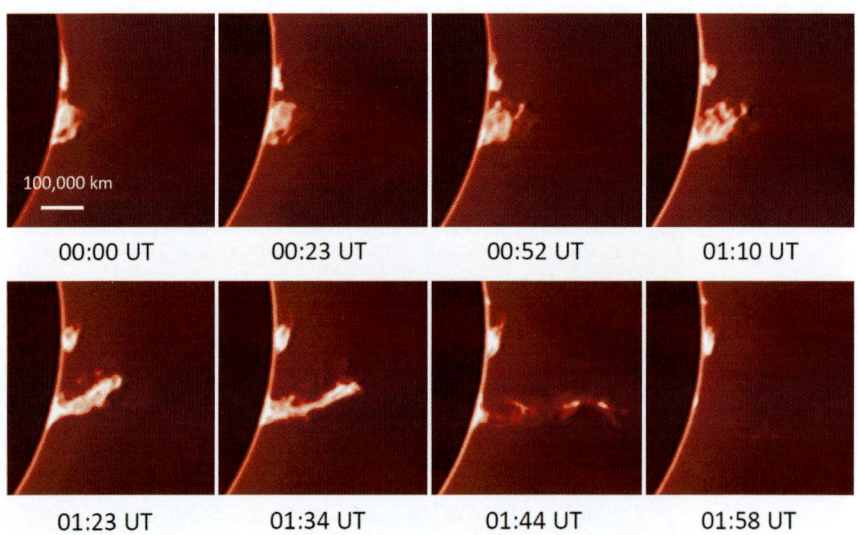

FMTで2000年11月4日に観測されたプロミネンス
Temporal development of a prominence observed with the FMT on November 4, 2000.

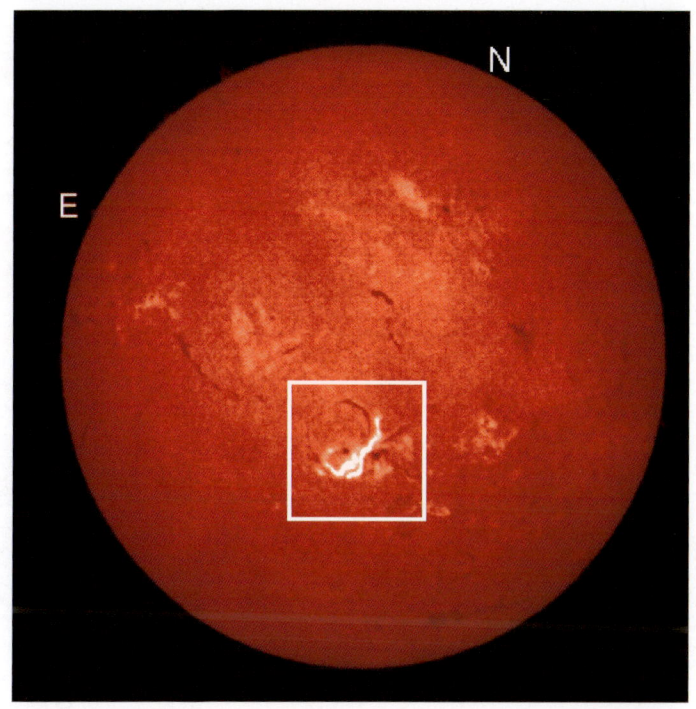

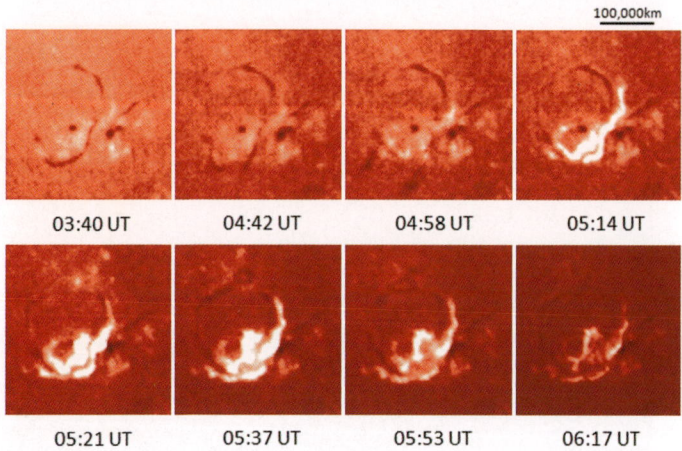

2001年4月10日に観測されたツーリボンフレア．上段：太陽全面Hα線中心像，下段：フレアの時間発展を示す時系列画像（Hα線中心像）．
A two-ribbon flare observed on April 10, 2001. Upper panel: solar full-disk image in Hα center; lower panels: sequential images showing the temporal development of the flare in Hα center.

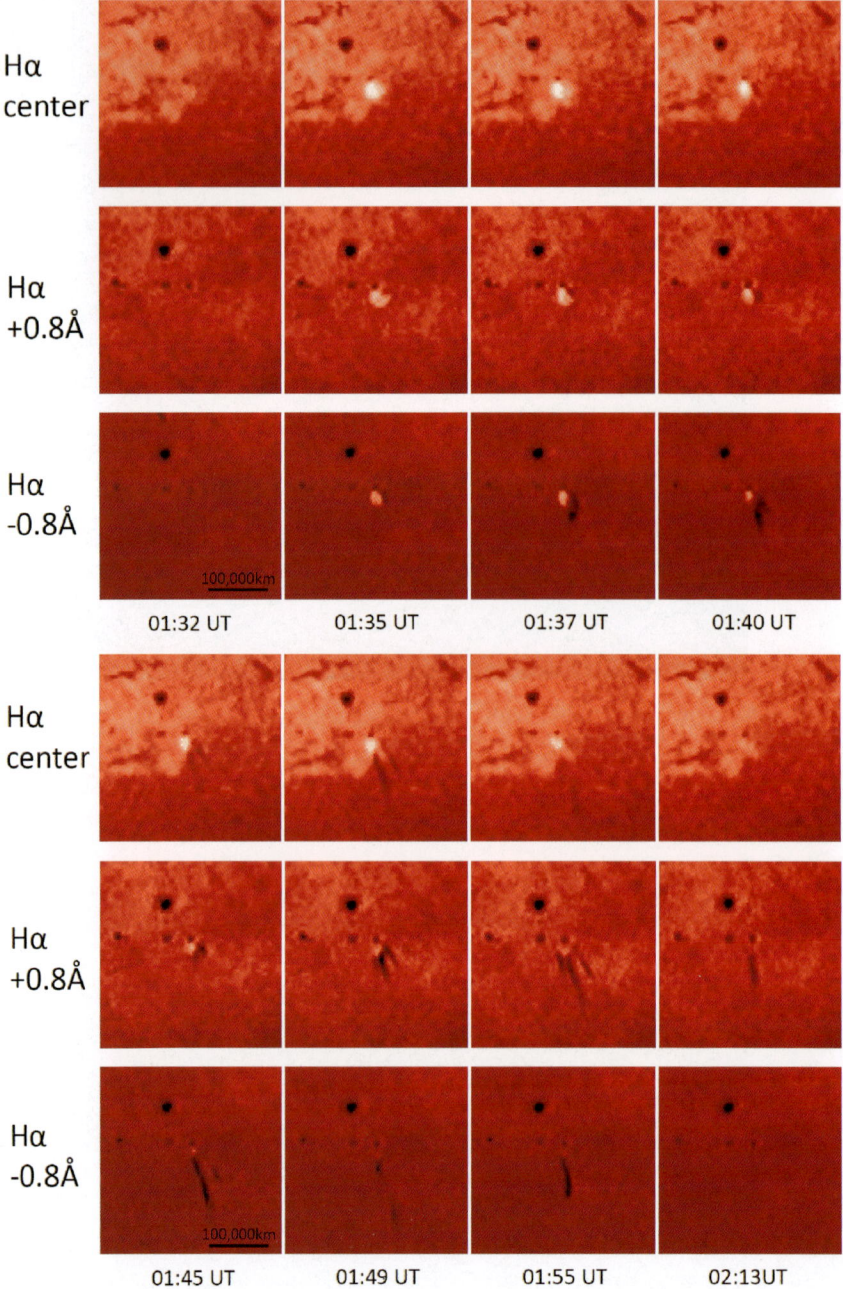

1999 年 8 月 25 日に観測されたサージ.
A surge observed through the FMT on August 25, 1999.

はじめに

太陽は，地球上の諸現象や生命のエネルギーの源であるだけでなく，人類文明の未来の究極のエネルギー源でもある．このように太陽は我々人類や地球にとって恵みの星であるが，その正体は実は爆発だらけの恐ろしい天体であることが，近年の太陽観測の発展によってわかってきた．さらに，太陽の爆発現象は，地球の磁気圏や超高層大気に甚大な影響を及ぼしていることもわかってきた．

1957年，人類は人工衛星を打ち上げることに成功し，以来，半世紀余にわたって多くの科学衛星や実用衛星を打ち上げてきた．今や，宇宙ステーションに宇宙飛行士が常時滞在して宇宙開発を進める時代となった．宇宙の利用なしに人類は現代文明を維持するのが困難になりつつある．ところが，ひとたび太陽で爆発（フレア）が起きると，大量のX線や紫外線，放射線粒子が太陽フレアから放射され，人工衛星を故障させたり，宇宙飛行士を被ばくの危険に陥れる．実はフレアの影響は「宇宙（スペース）」だけにとどまらない．地球の磁気圏や電離圏で磁気嵐や電離圏嵐が起こり，通信障害，地上電力網寸断，石油パイプライン腐食などの被害が起きることもわかってきた．このような被害を未然に防ぐためには，太陽の爆発によって地球周辺の宇宙や超高層大気でどのような現象が起こるのか事前に予報する必要がある．このような予報は，地上の天気予報になぞらえて，「宇宙天気予報」と呼ばれている．宇宙天気予報の実現が現代の人類文明にとって緊急の課題といえる．

太陽観測は宇宙天気予報にとって不可欠の観測である．太陽面のどこでどれくらいの規模のフレアが起きたか，どの方向にどれくらいの速度でプロミネンス噴出が起きたか，というような情報が，宇宙天気予報の基礎データとなる．

京都大学飛騨天文台フレア監視望遠鏡（Flare Monitoring Telescope＝FMT）は，1992年STEP（Solar Terrestrial Energy Program）プロジェクト予算により太陽フレアやプロミネンス噴出の監視用望遠鏡として建設された．目的は，地球の磁気嵐や電離圏嵐の元となる太陽面爆発現象の始まりをとらえること，すなわち，宇宙天気予報のためである．

フレア監視望遠鏡は口径6.4cmの屈折望遠鏡6本から成る6連式望遠鏡であり，Hα中心波長だけでなく，Hα＋0.8Å，Hα－0.8Å，プロミネンスモード（太陽そのものはマスクで隠してしまい，太陽リム外のプロミネンスだけを観測可能にした状態），白色光の5種類の太陽全面像を同時に連続観測しているのが特徴である．これにより，プロミネンス噴出の速度とその飛び出す方向を求めることが可能となる．このような太陽全面に渡る彩層活動の現象の速度場の連続観測を行なっているのは，世界広しといえどもこ

の飛騨天文台のフレア監視望遠鏡だけであった．

　フレア監視望遠鏡は，1992年以来，11年以上の長きにわたってフレアを始めとする様々な太陽活動現象の観測で活躍してきた．上記の特色のおかげで，フレア監視望遠鏡はモートン波と呼ばれるフレアから発生する衝撃波を20例以上発見した．これは，これまでに世界で観測されたモートン波の1/3以上に相当する．また，フレア監視望遠鏡は，「ようこう」衛星などのスペースからの太陽観測を補助する役割も果たした．現代はスペース太陽観測の黄金時代である．素晴らしい高空間分解能の多波長太陽観測データが続々と得られる時代である．そのような折り，フレア監視望遠鏡によるHα速度場データは貴重な速度情報を与えてくれる．

　さて，以上のようにフレア監視望遠鏡のHα多波長連続太陽全面観測は大変有効であることが判明したので，同じような観測システムを，国際プログラムCAWSES（Climate And Weather of the Sun and Earth System, 2004–2008）とIHY（International Heliophysical Year）の一環として，世界各国に広げようという国際協力プロジェクト，CHAIN（＝Continuous H-Alpha Imaging Network）プロジェクトを開始した．国際会議などで呼びかけたところ発展途上国からの反響は大きく，各国との国際共同研究の模索が始まっている．

　一方，飛騨天文台では，2003年に太陽磁場活動望遠鏡（Solar Magnetic Activity Research Telescope＝SMART）が完成し，太陽全面をさらに高精度で観測することが可能となった．SMARTは口径25cm望遠鏡2本と口径20cm望遠鏡2本からなる太陽望遠鏡であり，口径20cm望遠鏡で太陽全面を観測した場合，フレア監視望遠鏡より3倍以上，空間分解能がいい．Hα速度場観測の場合，1本の望遠鏡で時間差をつけて複数の波長で太陽全面を観測するので，フレア監視望遠鏡と違って「同時」でないという欠点があるが，SMARTはその欠点を上まわる精度の向上が実証されている．2005年には1つのフレアから3連発モートン波の発生を世界で初めて発見するなど，早くもその特色を発揮している．

　したがって，SMARTはフレア監視望遠鏡の役割を完全にカバーできることがわかり，フレア監視望遠鏡を世界の適地に移設する計画の検討が始まった．そうした折り，ペルー地球物理学研究所の石塚睦先生がペルーにおける太陽物理学・天文学の教育・普及活動のために非常な苦労と努力をされていることを聞き及び，フレア監視望遠鏡をペルーに移設する可能性を検討し始めた．2007年夏にペルーの候補地を視察した結果，ペルーのイカ大学の構内が最適地であることが判明し，イカ大学に移設することが2007年秋に決定された．この移設プロジェクトは，石塚先生支援，発展途上国支援というだけでなく，学問的にも意義が大きい．というのは，ペルーは日本から見て地球の裏側にあり，日本の夜の時間帯で観測が可能なため，日本では決して観測できない太陽観測データを供給してくれるからである．また，イカ大学周辺はペルーの砂漠地帯であるため晴天率が高く，CHAINプロジェクトを推進する上で，最も重要な場所に位置している．以上のことに基づいて，フレア監視望遠鏡を2010年3月にペルー・イカ大学に移設し，設

置が完了した．移設にあたっては，石塚睦先生，ご子息のホセ・イシツカ博士には，大変お世話になった．これを機に，両国の協力関係が更に実りのあるものになるよう力を尽くしたいと考えている．

　本書出版の目的は，以下のようになる．
(1) フレア監視望遠鏡によって観測された1992年5月～2003年4月の太陽活動現象データの総合報告をすること．
(2) フレア監視望遠鏡をペルーに移設するにあたり，ペルーでのフレア監視望遠鏡の活用にとってテキストの役割を果たせるような書物をまとめること．

　この目的を満たすように本書の編集に努め，上梓することができた．本書がペルーだけでなく，日本や世界の関連研究者や学生にとっても役に立つ貴重な文献になることを期待している．

　本書の読者は，太陽物理学者，関連研究者（天文学，地球物理学，地球環境学，物理学），アマチュア天文家，大学院生，大学生，教育普及関係者（学校教員，博物館解説員など）を想定している．本書が，これらの読者に広く読まれ活用されることを，京都大学附属天文台一同，心より願うものである．

2010年10月1日
京都大学大学院理学研究科附属天文台 台長 柴田一成

Introduction

In addition to being a source of energy for various phenomena and living creatures on the Earth, the Sun is also expected to be the ultimate source of energy for the future of human civilization. With this in mind, the Sun can be said to be a blessing for the Earth, including mankind. However, recent solar observations have revealed that the Sun is actually a horrifying heavenly body, full of explosions. Solar research has also demonstrated the powerful effects of these explosive activities on the magnetosphere and upper atmosphere of the Earth.

A large number of artificial satellites have been launched for scientific or commercial purposes over the last 50 years, since the first successful launch in 1957. We are now in the age of space development, with a group of astronauts staying on the International Space Station on a continuous basis. It is becoming more and more difficult for mankind to maintain its modern civilization without using resources from beyond the Earth's atmosphere. However, when an explosion (flare) occurs on the Sun, enormous amounts of X-ray radiation, ultraviolet light, and radiation particles are emitted, causing artificial satellites to become disordered, or putting astronauts in danger of exposure. In fact, the influences of these flares are not limited to space. They can produce magnetic or ionospheric storms in the magnetosphere or ionosphere of the Earth, which can cause various problems such as communication failures, disruption to power supply networks, or corrosion of oil pipelines. To prevent such damage, it is necessary to predict what kinds of phenomena occur in space near the Earth or in the upper atmosphere of the Earth because of the Sun's explosions. Such a prediction is called "space weather prediction", a name coined from the weather forecasts on Earth. The realization of such space weather predictions is an emergent issue that must be addressed for modern human civilization.

Solar observations are indispensable for space weather predictions. The size of a flare and its location on the solar surface, the direction and velocity of a prominence eruption, etc., are the kinds of information used as fundamental data for space weather predictions.

The FMT (Flare Monitoring Telescope) of Hida Observatory, Kyoto University, was constructed in 1992 as a telescope system for monitoring solar flares and prominence eruptions; it was constructed using the budget for the STEP (Solar Terrestrial Energy

Program) project. The purpose was to detect the beginnings of explosive phenomena on the Sun that can cause magnetic or ionospheric storms on Earth. In other words, the FMT was specifically designed for space weather predictions.

The FMT is a telescope system with six 6.4-cm refractors. This system is characterized by its continuous and simultaneous observations of the solar full disk in five different modes: Hα center, Hα+0.8Å, Hα-0.8Å, prominence mode, and continuum light. (The prominence mode uses a mask that occults the solar disk so as to observe only the prominences rising from the solar limb). Using this system, the velocities of prominence eruptions and their eruption directions can be determined. Among the large number of solar telescopes in the world, the FMT operated by Hida Observatory was the only instrument continuously observing the velocity fields of active phenomena in the chromosphere over the entire solar full disk.

Since 1992, the FMT has made great contributions to the observation of various solar activities, including flares, over long periods of time (exceeding 11 years). Thanks to the above mentioned features, the FMT has succeeded in detecting more than 20 cases of Moreton waves (a type of shock wave generated from flares), which is more than one third of all the Moreton waves observed to date worldwide. The telescope has also played an important role in assisting space-based solar observations by the Yohkoh and other artificial satellites. In this golden age of space-based solar observations, multi-wavelength solar observation data of high spatial resolution are produced every day. The velocity field data obtained by the Hα observations by the FMT provide valuable information on velocities.

After confirming the significant effectiveness of the Hα multi-wavelength solar full disk observation by the FMT, we launched an international collaboration project, named the CHAIN (Continuous H-Alpha Imaging Network) project, which is aimed at setting up similar observation systems in other countries. This project was initiated as a part of two international programs: CAWSES (Climate And Weather of the Sun and Earth System, 2004–2008) and IHY (International Heliophysical Year). We used international conferences and various other opportunities to appeal for support of and participation in our project, and received many affirmative responses. Thus, efforts to begin new international collaborative research have been initiated.

In the meantime, a new solar telescope was installed at Hida Observatory in 2003. This telescope, named SMART (Solar Magnetic Activity Research Telescope), has enabled us to observe the solar full disk with even higher accuracy and resolution. SMART is a telescope system with two 25-cm refractors and two 20-cm refractors. One of the two 20-cm refractors is designed for solar full disk observation, and has a spatial resolution over three times higher than that of the FMT. Unlike the FMT, the

Hα velocity field observation by this 20-cm refractor cannot simultaneously capture solar full disk images at different wavelengths, because it uses only one telescope operated at different observation wavelengths on a time-division basis. However, it has been demonstrated that the extremely high performance of SMART easily compensates for this "drawback". For example, in 2005, SMART became the first telescope in the world to discover three successive Moreton waves generated by a single solar flare.

Thus, it has been proved that SMART can substitute the FMT completely. This allowed us to embark on a project to move the FMT to an appropriate overseas location. We soon came across timely information about Dr. Mutsumi Ishitsuka at the Geophysical Institute of Peru, who has been intensively involved with educational and promotional activities in solar physics and astronomy in Peru. Driven by this news, we began to consider the possibility of moving the FMT to Peru. In the summer of 2007, we visited candidate sites in Peru and found that a location on the campus of Ica University was the most suitable. Ica University was officially designated as the destination for the FMT in the fall of the same year. This project is not only aimed at supporting Dr. Ishitsuka and assisting a developing country, but is also expected to make significant academic contributions. Firstly, because Peru is on the opposite side of the Earth from Japan, having a telescope there enables us to monitor the Sun when it is nighttime in Japan and to acquire solar observation data that cannot be obtained from any telescope located in Japan. Another critical point is that Ica University is located in a desert area and therefore enjoys a high probability of clear skies. These features of Ica University match the most important conditions for the promotion of the CHAIN project. For all these reasons, the FMT was moved to Ica University in March 2010. We sincerely thank Dr. Mutsumi Ishitsuka and his son Dr. Jose Ishitsuka for their cooperation in moving the FMT to Peru. We would like to take this opportunity to make an effort to enrich the cooperative relationships between our two countries.

This book has been published for the following reasons:
(1) To give a general report of data on the solar activity phenomena observed through the FMT from May 1992 through April 2003.
(2) To provide a guide for Peruvian students and researchers using the FMT for solar observation.

This book was compiled to achieve these goals. We expect this book to be an informative reference for solar researchers and students, not only in Peru but also in Japan and other countries.

This book is targeted at researchers in solar physics as well as in other related fields (such as astronomy, geophysics, environmental science, and physics), amateur astronomers, graduate students, undergraduate students, and people involved in educational activities (such as school teachers and museum curators). We sincerely hope that this book will be a useful resource for many readers in these fields.

Kazunari Shibata
Director of Kwasan and Hida Observatories, Graduate School of Science, Kyoto University

目　次
CONTENTS

目次

はじめに i
- 第1章 太陽研究と地球環境 2
  - 1.1 太陽観測に関する基礎知識 2
  - 1.2 太陽活動の地球への影響 16
- 第2章 フレア監視望遠鏡 27
  - 2.1 望遠鏡観測システムとデータ公開 27
  - 2.2 ペルー国移設の写真記録 37
- 第3章 世界の太陽全面観測 41
  - 3.1 太陽全面Hα観測 42
  - 3.2 その他の波長での太陽全面パトロール（Hα線以外） 45
- 第4章 11年間の観測のまとめ 50
  - 4.1 フレア 52
  - 4.2 サージ 57
  - 4.3 プロミネンス 60
  - 4.4 フィラメント活動 61
  - 4.5 モートン波 64
- 第5章 イベントリスト 67
  - 5.1 Hα-0.8Åイベントリスト（ディスクイベント） 68
  - 5.2 Hαプロミネンスイベントリスト（リムイベント） 70
- 第6章 代表的イベント 73
  - 6.1 イベント図版に関する全体的な説明 73
  - 6.2 代表的ディスクイベント 80
  - 6.3 代表的リムイベント（プロミネンス） 306
- 付録A FMT関係論文 330
  - A-1 森本、黒河「太陽フィラメント消失現象の三次元速度場決定方法」 330
  - A-2 森本、黒河「噴出型および準噴出型太陽フィラメント消失現象とコロナ活動との関係」 371
  - A-3 衛藤、他「1997年11月4日に観測されたモートン波とEIT波の関係」 403
  - A-4 成影、他「1997年11月3日のモートン波におけるHαと軟X線の同時観測」 425
  - A-5 成影、他「モートン波に伴うX線で見られた爆発現象」 438
  - A-6 森本、他「太陽フィラメント消失とそれに随伴するフレアアーケードの間のエネルギー関係について」 450
- 付録B フレア監視望遠鏡関連論文リスト 476
- 付録C 太陽地球系エネルギー国際共同研究（STEP）シンポジウム報告よりFMT関連報告集 480
- あとがき 521
- 索引 525

CONTENTS

Introduction iv
Chapter 1 Solar Research and the Terrestrial Environment 2
  1.1 Basics of Solar Observation 2
  1.2 Impacts of the Sun on the Earth 16
Chapter 2 Flare Monitoring Telescope 27
  2.1 Telescope Observation System and Public Data Release 27
  2.2 Photographic Records of the FMT Before and After its Relocation to Peru 37
Chapter 3 Solar Full-disk Hα Observations in the World 41
  3.1 Solar Hα Full-disk Observation 42
  3.2 Solar Full-disk Observation in Other Wavelengths 45
Chapter 4 Summary of the Observations Carried Out over a Period of 11 Years 50
  4.1 Flare 52
  4.2 Surge 57
  4.3 Prominence 60
  4.4 Filament Activity 61
  4.5 Moreton Waves 64
Chapter 5 Event Lists 67
  5.1 Active Phenomena in Hα $-$ 0.8 Å (Disk Events) 68
  5.2 Active Phenomena in Prominence Images (Limb Events) 70
Chapter 6 Representative Events 73
  6.1 General Descriptions of the Illustrations 73
  6.2 Representative Disk Events 80
  6.3 Representative Limb Events (Prominences) 306
Appendix A Selected Papers Relating to FMT 330
  A-1 Morimoto & Kurokawa, "A Method for the Determination of 3-D Velocity Fields of Disappearing Solar Filaments," 2003 330
  A-2 Morimoto & Kurokawa, "Eruptive and Quasi-Eruptive Disappearing Solar Filaments and Their Relationship with Coronal Activities," 2003 371
  A-3 Eto et al., "Relation between a Moreton Wave and an EIT Wave Observed on 1997 November 4," 2002 403
  A-4 Narukage et al., "SIMULTANEOUS OBSERVATION OF A MORETON WAVE ON 1997 NOVEMBER 3 IN Hα AND SOFT X-RAYS," 2002 425
  A-5 Narukage et al., "X-Ray Expanding Features Associated with a Moreton Wave," 2004 438
  A-6 Morimoto et al., "Energetic Relations between the Disappearing Solar Filaments and the Associated Flare Arcades," 2010 450
Appendix B FMT-Related Papers 476
Appendix C FMT-Related Reports in Proceedings of the Symposiums of Solar-Terrestrial Energy Program (STEP) 480
Afterword 522
INDEX 527

太陽活動 1992-2003
Solar Activity in 1992-2003

第 1 章　太陽研究と地球環境
Chapter 1　Solar Research and the Terrestrial Environment

第 2 章　フレア監視望遠鏡
Chapter 2　Flare Monitoring Telescope

第 3 章　世界の太陽全面観測
Chapter 3　Solar Fulldisk

第 4 章　11年間の観測のまとめ
Chapter 4　Summary of the Observations Carried Out over a Period of 11 Years

第 5 章　イベントリストの説明と定義
Chapter 5　Explanation and Definition of the Events Mentioned in the Event Lists

第 6 章　代表的イベント図版
Chapter 6　Illustrations of the Representative Events

第1章
太陽研究と地球環境

Chapter 1
Solar Research and the Terrestrial Environment

遠く離れたところにある太陽で起こる爆発的な活動が，地球環境に大きな影響をもつことが近年明らかになってきた．では，太陽での爆発の影響が，太陽系空間をどのように伝わり，それが地球にどのような現象を引き起こすのか．この章では，このような太陽－地球間の相互作用という最先端の研究について解説する．

章の前半では，フレア監視望遠鏡（FMT）で行われているHα線による太陽活動の観測について基礎的な事柄を簡単に解説する．後半では，地球環境への影響や宇宙天気予報などの話題を，具体例に即してわかりやすく説明する．

In recent years, studies have shown that the explosions occurring on the Sun have a significant impact on the terrestrial environment, even though the Sun is far away from the Earth. This finding has posed the question of how the impact of such explosions can spread through the interplanetary space of the solar system and what the consequences of such an impact would be. This chapter illustrates some of the most advanced research achievements relating to Sun-Earth interactions.

The first section of this chapter briefly describes the basics of solar-activity observation in Hα light carried out by using the Flare Monitoring Telescope (FMT). The second section presents illustrative examples relating to the Sun's impact on the terrestrial environment and discusses space weather prediction.

1.1 太陽観測に関する基礎知識

なぜ太陽を観測するのか

太陽は45億年以上もの長い間輝き続

1.1 Basics of Solar Observation

Why do we observe the Sun?

The Sun has been shining for more than 4.5 billion years. Its source of energy

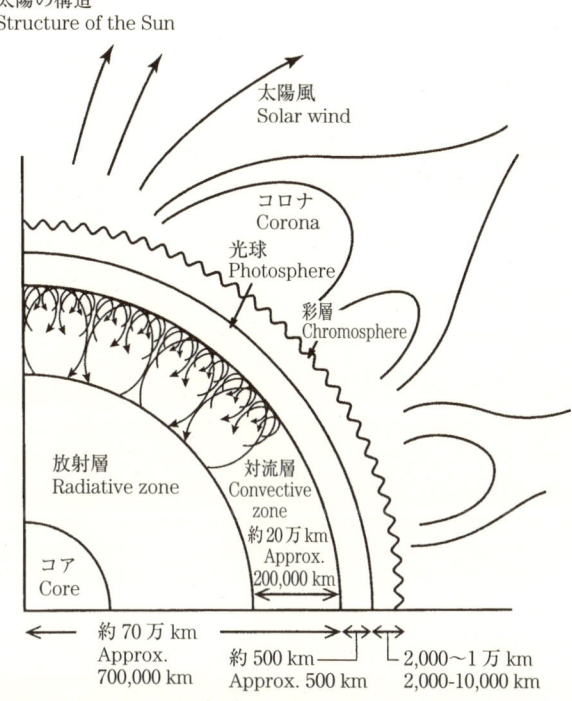

図 1.1 太陽の構造（柴田・大山「写真集太陽」より）．

Fig. 1.1 Structure of the Sun (illustration cited from Shibata and Ohyama 2002).

けている．そのエネルギー源となっているのは中心部で起きている水素の核融合反応である（図1.1に太陽の内部構造と大気の説明図を示す）．この核融合により発生するエネルギーは桁違いに大きい．発生したエネルギーは，可視光線を始めとする電磁波，粒子線など，様々な形で宇宙空間に放出され，その影響は広く太陽系全体に及ぶ．地球もこのような太陽の勢力圏の中にあることはいうまでもない．

太陽の地球側の表面で大きな活動現象が起きると，強烈なX線，紫外線，高エネルギー粒子線が放出されるほか，秒

is the nuclear fusion of hydrogen that occurs at the core of the Sun (see Fig. 1.1, which illustrates the Sun's internal structure and atmosphere). This nuclear fusion produces an enormous amount of energy, which is released into space in various forms such as electromagnetic waves and particle beams. These forms of energy can have a wide-ranging impact on the entire solar system, including our planet Earth.

If a large-scale active phenomenon occurs on the side of the Sun facing the Earth, large amounts of radiation, including, but not limited to, X-rays and ultraviolet light and high-energy particle beams, are released. This is followed by a

速数100km以上にもなるプラズマの雲が地球周辺に押し寄せてくる．この影響で，地球上では地磁気嵐，オーロラなどの現象が発生する．しかし，その影響は自然現象にとどまらない．人間活動においても，通信の乱れ，高圧電力網の破壊などの障害が発生することが確かめられている．

大気圏の外に出れば太陽の影響はより直接的なものとなる．例えば，人工衛星は太陽からくる高エネルギーの電磁波や粒子線を直接受けながら地球のまわりを周回している．したがって，太陽から異常に高いエネルギーをもつ電磁波や粒子線が到来すると，人工衛星の動作に不具合が生じたり，衛星が故障したりする可能性がある．また，宇宙ステーションに

surge of plasma through space toward the Earth at speeds higher than several hundreds of kilometers per second. This surge of plasma results in the occurrence of geomagnetic storms, aurora, and other phenomena on the Earth. However, the impact is not limited to causing these natural phenomena. It has been confirmed that telecommunication systems and high-voltage power supply systems suffer damages due to this surge of plasma.

Solar activities have a more direct impact outside the Earth's atmosphere. For example, artificial satellites orbiting the Earth directly receive high-energy electromagnetic waves and particle beams from the Sun; as a result, they may suffer from functional failure or serious damages. Another serious problem is the exposure of the

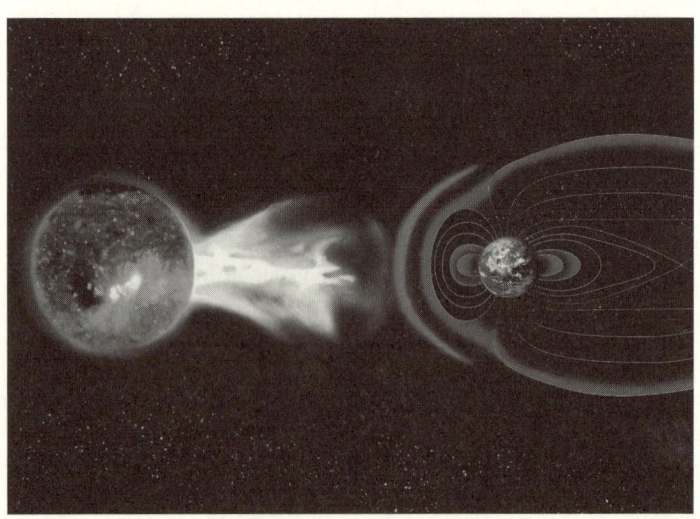

図1.2　太陽—地球環境の想像図．太陽面爆発（フレア）から噴出したコロナ質量放出（CME）が地球磁気圏に衝突すると地磁気嵐が発生する．

Fig. 1.2 Conceptional picture of the Sun-Earth environment. When the coronal mass ejection (CME) erupted from an explosion on the Sun's surface (a flare) collides with the magnetosphere of the Earth, a geomagnetic storm occurs.

滞在する宇宙飛行士は船外活動を含む様々な活動を行っているが，これには放射線被曝の危険性が常に伴う．

このような宇宙空間における人間の活動や人工的システムの動作を安全かつ効率的に進めるには，太陽の様々な活動現象を研究し，その活動が地球にどのような影響を及ぼすかを知っておく必要がある．そのため，現在も世界中で数多くの研究者が日々太陽を観測し，それにより得られたデータの解析に励んでいる．

太陽の何を観測するのか

では，研究者たちは太陽の何を観測しているのだろうか．

最も有名な観測対象は黒点であろう．中国や日本の古代の文献にも登場するように，黒点の存在は古くから知られている．望遠鏡を使った黒点観測は1611年にガリレオらによって始められた．そして，長年の観測の結果，その数が約11年の周期で増減することがわかった．黒点は比較的小さな望遠鏡でも観察できるため，現在でも数多くの研究者やアマチュア天文家により観測され，貴重なデータが集められている．図1.3（上）は多数の黒点が発生した太陽像（白色光像）の例である．

プロミネンスも比較的よく知られている現象だろう．皆既日食の際，太陽の縁（リムと呼ばれる）から赤い炎のようなものが立ちのぼって見えることがある．これがプロミネンスである（図1.3および1.4）．紅炎とも呼ばれるこの現象の正体は，周囲よりも低温のプラズマである（ただし，それでもその温度は数千〜1万度に達する）．プロミネンスが太陽の

International Space Station astronauts to such radiations when performing various tasks, including extravehicular activities.

In order to ensure the safety and efficiency of human activities and artificial system operations in space, it is necessary to study various kinds of solar activities and understand their impacts on the Earth. For this purpose, many researchers all over the world are currently observing the Sun on a daily basis and are analyzing the obtained data.

What are the aspects of the Sun that are being observed?

What is the specific target of solar observational research?

The most well-known target is probably the sunspot. The existence of sunspots has been known for a long time, and references to sunspots can be found in the ancient literature of China and Japan. Galileo Galilei was among the first to observe sunspots through a telescope in 1611. Observations conducted by many people over the course of time have revealed that the number of sunspots varies cyclically with a period of approximately 11 years. Sunspots can also be observed through small telescopes, and important data are currently being collected by many amateur astronomers as well as by professional researchers. The upper panel of Fig. 1.3 shows an example of the solar image (in white light) in which a large number of sunspots can be seen on the Sun's surface.

Prominence may also be a relatively well-known solar activity. On a total solar eclipse, a red, flame-like structure can be often seen outside the edge (or "limb") of the Sun. This is called a prominence (see Figs. 1.3 and 1.4). A prominence, which has a high temperature on the

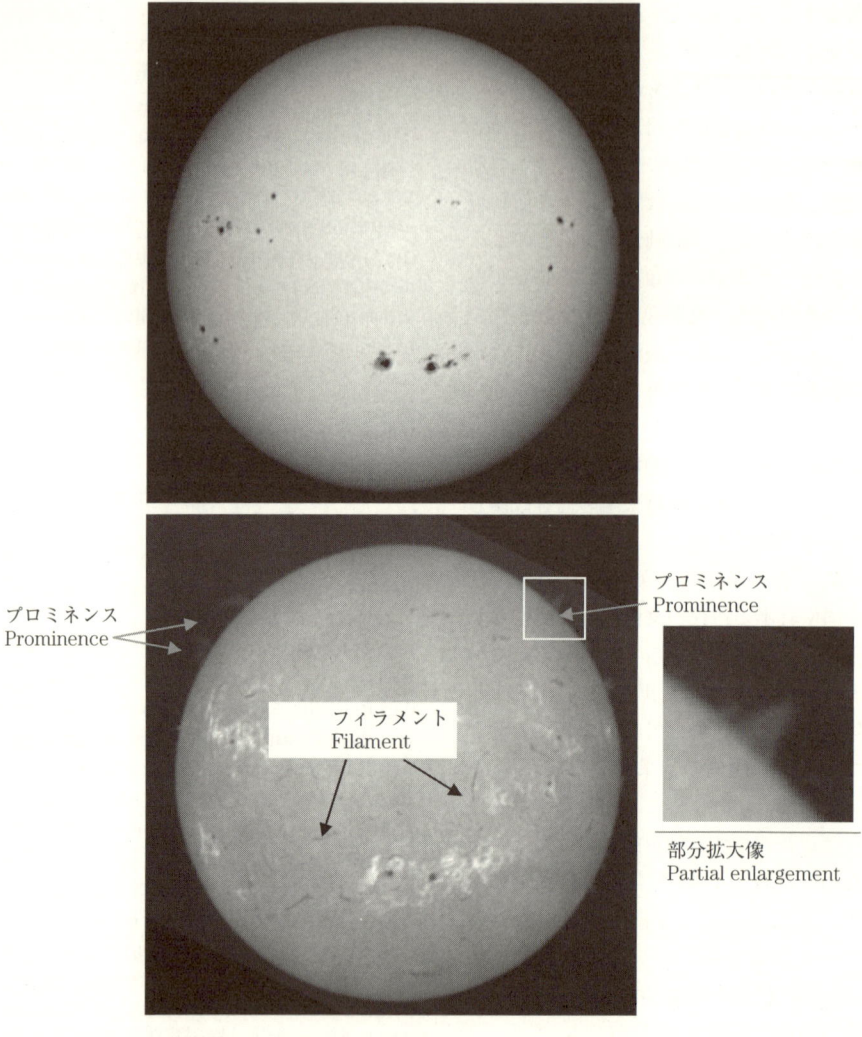

図 1.3 上：FMT の連続光モードで撮影された黒点（2001 年 9 月 26 日 04:16 UT）．下：同じ日時に撮影された太陽 Hα 線中心像．

Fig. 1.3 Sunspots observed with the FMT on Sep 16, 2001 at 04:16 UT. Upper panel: under continuum light. Lower panel: under Hα center light.

縁ではなくその内側の太陽面（ディスクと呼ばれる）で発生すると黒い筋状に見える．この場合はプロミネンスではなくフィラメントと呼ばれる．これらは同じ order of 1,000–10,000 K, is actually a low-temperature plasma that is cooler than the surrounding coronal space. When such a cool gas is observed on the Sun's bright disk, it appears as a dark

現象を違う角度から見たものである．例えば，図1.3（下）の太陽像の右上には，同じ筋状の構造物がディスクの上ではフィラメントとして，またディスクの外側ではプロミネンスとして見えている（図1.3下段の部分拡大像を参照）．

filamentary structure; this is called a filament, not a prominence. These are two aspects of the same structure. For example, the filamentary structure located in the upper-right quadrant of the solar image in the lower panel of Fig. 1.3 appears as a filament on the solar disk and as a prominence outside the disk (see the partial enlargement in the lower panel of Fig. 1.3).

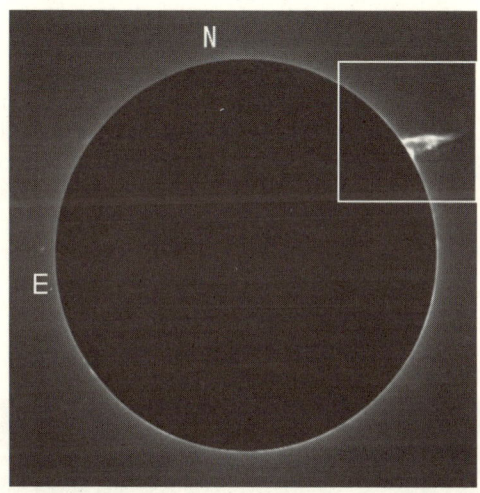

図1.4　FMTで撮影された噴出型プロミネンス（2000年8月3日 08:14 UT）．

Fig. 1.4 Prominence observed with the FMT (Aug 3, 2000, 08:14 UT).

しかし，太陽の表面で起きる現象は黒点やプロミネンスだけではない．フレア，サージ，粒状斑，白斑など，一般にはあまりなじみのない名前の現象や構造も数多く存在する．もちろん，これらの現象や構造はずっと昔から存在していたのであって，科学技術の進歩のおかげで初めて発見されたものである．これは科学一般にいえることだが，太陽の研究においても，観測方法や観測装置の進歩が新たな現象の発見につながることが少なくな

Sunspots and prominences are not the only phenomena that occur on the Sun. There are many other types of specific phenomena and structures on the Sun, such as flares, surges, granulations, and plages. These solar structures and phenomena have been discovered thanks to the progress and development in the instruments and methods used for making observations. The FMT was planned and built in 1992 with a view to developing a novel high-performance telescope and promoting research on

い．1992年に設置されたFMTもそのような科学的使命を担って誕生したものと言える．

一方で，太陽表面で突発的に生じる動的な現象のメカニズムの解明も進んだ．中でも重要なのがフレアである（図1.5）．フレアとは，太陽の磁気エネルギーが太陽外層大気（彩層，コロナなど）において熱や電磁波，粒子線などのエネルギーに変換される現象である．太陽系内最大の爆発現象といわれる通り，そのエネルギー解放量は莫大であり（最大級のフレアでは水爆一億個に相当するエネルギーが放出される），地球環境への影響もそれだけ大きい．

さらに，より直接的に地球に影響を与えるものとしては，コロナ質量放出（Coronal Mass Ejection = CME）（図1.6

solar activity.

The mechanisms underlying the occurrence of dynamic phenomena on the solar surface have theoretically been elucidated through observations using advanced methods. The most important explosive phenomenon is a solar flare (Fig. 1.5). A flare is a phenomenon in which the magnetic energy of the Sun is converted into different forms of energy such as heat, electromagnetic waves, and particle beams in the outer atmosphere of the Sun (e.g., in the chromosphere or corona). The amount of energy released by a flare is enormous, and hence, it is often called the largest explosive phenomenon in the solar system (a very large flare can release an amount of energy that is comparable to 100 million H-bombs.) Such solar flares can have a strong impact on the terrestrial environment.

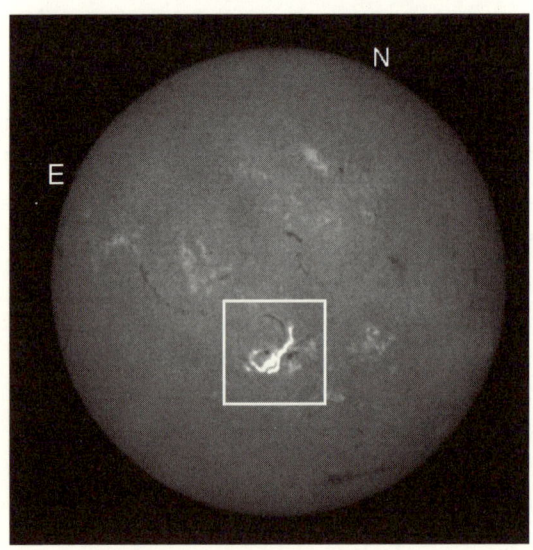

図1.5　FMTで撮影されたフレア（2001年4月10日05:14 UT）．

Fig. 1.5 A flare observed with the FMT (Apr 10, 2001, 05:14 UT).

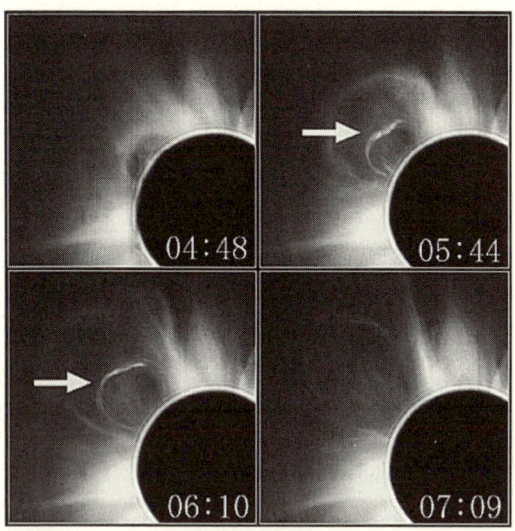

図 1.6 コロナ質量放出（CME）．矢印は噴出型プロミネンス．

Fig. 1.6 A coronal mass ejection (CME); the arrows show an eruptive prominence.

参照）がある．CME はフレアやプロミネンス噴出の発生に伴って起こるもので，磁場を含んだ大量のプラズマが太陽コロナから放出される現象である（図 1.6）．磁場の南向き成分を含んだ CME の磁気プラズマが地球磁気圏に衝突すると磁気リコネクションを引き起こし，エネルギーが地球磁気圏に侵入可能となる．このようにして磁気嵐が発生する．

フレアやプロミネンス爆発のような突発的現象は，「突発的」という言葉が示す通り，その発生を予測することが難しい．一方で，これらの現象は黒点近傍の領域で見られることが多い．これは，黒点の周辺に強い磁場が存在することと関係がある．このように，強い磁場の存在によりフレアやプロミネンス噴出などの大規模な爆発現象が発生する領域を活動

Another phenomenon that can have a more direct impact on the Earth is a coronal mass ejection (CME; see Fig. 1.6). A CME is a phenomenon that is associated with a flare or a prominence eruption in which an enormous amount of plasma, which contains magnetic fields, is released from the solar corona. When the magnetic plasma of a CME comprising southward-oriented magnetic fields collides with the magnetosphere of the Earth, a magnetic reconnection occurs, allowing an inflow of energy into the magnetosphere and resulting in the occurrence of a strong magnetic storm.

Since the phenomena such as flares and prominence eruptions are unpredictable and occur randomly, it is difficult to forecast their occurrences. However, observations have revealed that they often occur in a region close to a sunspot, where there exists a strong

領域 (active region = AR) と呼ぶ．太陽観測では，その時々の活動領域を見出し，それを継続的に監視して，突発的現象の発生を検出することが重要な仕事となる．

さらには，活動領域以外の場所においても，何らかの弱い磁場の影響で突発的現象が発生することもあり，これらのすべての「突発的現象」を本書では，「イベント」と呼ぶことにする．

次章で述べるように，FMTはこのようなイベントの観測を効率よく，高精度に行うことができるように設計された望遠鏡である．1992年から2003年までの11年の間にFMTにより観測されたイベントは膨大な数にのぼる．本書の第6章ではその中からごく一部のイベントを厳選し，画像や数値データを用いて紹介している．

なぜHα線を使うのか

先に述べたような太陽の活動現象を地上から観測する場合，Hα線と呼ばれる光を使うことが多い．このHα線とはどういう光なのか．そして，なぜこの光が太陽観測に適しているのか．

Hα線とは水素原子 (H) のスペクトル線の一種であって，水素原子において電子がエネルギー準位の3番目から2番目へ遷移する際に発する光である（図1.7）．その波長は656.3 nm (= 6563Å) で，可視光のうち赤色の領域に属する．逆に，水素原子においてエネルギー準位の2番目にある電子が，波長656.3 nmの光を

magnetic field. Such a region where large-scale explosive phenomena such as flares or prominence eruptions occur is called an active region (AR). In solar activity observations, it is important to focus on the active regions on the sun and monitor them in order to detect sudden phenomena.

It is known that sudden phenomena also occasionally occur at regions of weak magnetic fields, remote from the active region. In this book, we will refer to both types of sudden phenomena—the ones that occur within and the ones that occur outside the active region—as "events".

The FMT, which is explained in detail in the next chapter, is a telescope system that is specially designed to efficiently detect events with a high level of accuracy. From 1992 to 2003, we observed an enormous number of events by using this telescope. In Chapter 6, brief descriptions, images, graphical plots, and numerical text data of some of the important events observed with the FMT are provided.

Why is Hα light preferred?

For most of the ground-based observations of solar activities, a specific kind of light called Hα light is used. This section explains what Hα light is and why it is suitable for making solar observations.

Hα light is emitted when a bound electron in a hydrogen atom (H) falls from the third to the second energy level (Fig. 1.7). This light has a wavelength of 656.3 nm (6563Å) and belongs to the red part of the visible spectrum. When an electron in the second level in a hydrogen atom absorbs light of wavelength 656.3 nm and jumps to the third energy level, the Hα light will appear as an absorption line in the solar

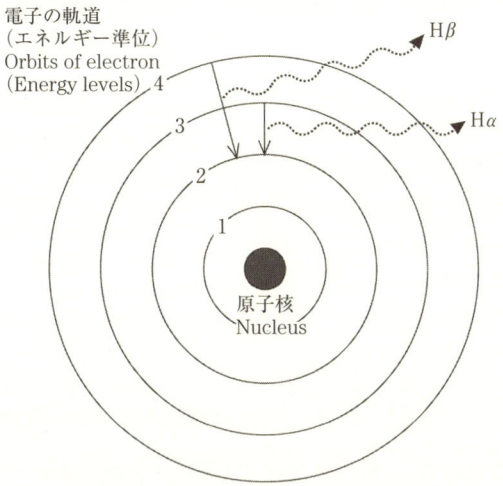

図 1.7 水素原子における電子の遷移のイメージ図.

Fig. 1.7 Transition of an electron in a hydrogen atom.

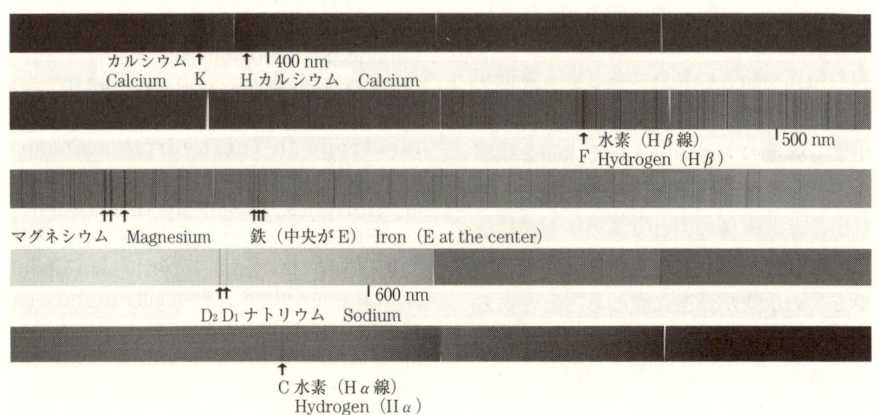

図 1.8 様々な吸収線を含む太陽スペクトル.

Fig. 1.8 Solar spectrum with various absorption lines.

吸収して 3 番目のエネルギー準位に遷移することもある. このとき, Hα 線は連続スペクトルの中に吸収線として現れる (図 1.8).

spectrum (Fig. 1.8).

先に述べたように，太陽は水素の核融合反応をエネルギー源とする恒星であり，そこには大量の水素原子が存在している。太陽の光球から発せられた光のうち，Hα線を始めとする水素のスペクトル線は，光球の上層にある大量の水素原子による吸収を受ける。こうして吸収された光のエネルギーにより水素原子の内部の電子が高いエネルギー準位へ押し上げられ，ついには原子の外へと追い出される（つまり，水素原子が電離する）。放出された電子は，再び水素の原子核（陽子）に捕えられ，Hα線などのスペクトル線を出しながら低いエネルギー準位へと遷移するようになる。このような変化は光球の外側にある大気の層で主に生じている。この層を彩層という。

このように彩層では，水素原子における電子の遷移と，それに伴うスペクトル線の発光が顕著に見られるという特徴がある。したがって，水素のスペクトル線，中でも強度の大きいHα線で太陽を観察すると，彩層の様子がよくわかる。一方，彩層には，磁場の力とガスの圧力が拮抗し，両者のバランスの変化に応じて磁場やガスの状態が急激に変わることがあるという特徴もある。以上のようなことから，地上からの太陽活動の観測ではHα線による彩層の観測が中心となっている。

As stated earlier, the Sun is a star whose source of energy is the nuclear fusion of hydrogen. There are an enormous number of hydrogen atoms on the Sun. When light is emitted from the photosphere, the components of light corresponding to the hydrogen spectral lines, including Hα light, are absorbed by the large number of hydrogen atoms present in the layer above the photosphere. The energy of the absorbed light results in the excitation of the electrons in the hydrogen atom to higher energy levels, and these excited electrons are eventually expelled from the atom (i.e., the hydrogen atom is ionized). The expelled electrons are recaptured by the nuclei of hydrogens (protons) and fall to lower energy levels, thereby emitting Hα light or other spectral lines. Such changes in hydrogen atoms mainly take place within the chromosphere, which is the atmospheric layer immediately outside the photosphere.

As explained above, the chromosphere is marked by the frequent transitions of electrons in the hydrogen atoms, accompanied by a strong emission of spectral lines. This means that observing the Sun in a spectral line of hydrogen selectively provides information about the chromosphere, especially in the case of Hα light, which is the strongest spectral line. Additionally, the gas pressure of the chromosphere is nearly balanced with the force of a magnetic field, and the state of the magnetic field or the gas can suddenly change with a change in the balance between the two forces. Given these factors, most of the ground-based solar observations of chromospheric activity are performed in Hα light.

Hα線を中心とする多波長観測

Hα線による観測では，中心波長（656.3 nm）での観測のほか，それよりわずかに長波長側（＋方向，赤色側）あるいは短波長側（－方向，青色側）に波長をずらした観測もよく行われる．このようにHα線の中心波長（これを単にHα線中心と呼ぶこともある）から少しずれた波長をHα線のウィング（翼）と呼ぶ（図1.9）．

このように波長をずらした観測を行うと，彩層におけるガスの運動状態に関する情報を得ることができる．例えば，彩層の中で観測者（つまり地球）に向かって高速で動いている水素原子を考えてみよう．この水素原子がHα線を発すると，その波長は中心波長（656.3 nm）よりわ

Observations at multiple wavelengths around the wavelength of Hα light

In many cases, Hα observation is performed not only at the central wavelength (656.3 nm) but also at other wavelengths that are either longer (+ve direction; red shift) or shorter (-ve direction; blue shift) than the central wavelength. A wavelength that is slightly shifted from the central wavelength of Hα light (hereafter referred to as the "Hα center") is called Hα wing wavelength (Fig. 1.9).

Observations made at shifted wavelengths provide information relating to the movement of the gas in the chromosphere. For example, imagine a hydrogen atom moving at a high velocity in the chromosphere toward the observer (or toward the Earth). When this atom emits Hα light, the wavelength

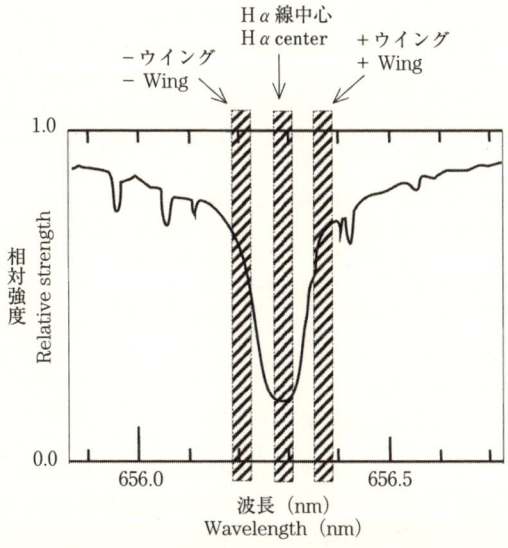

図 1.9 Hα線中心と両ウィングの関係．

Fig. 1.9 Hα center with wings on both sides.

ずかに短くなる（ドップラー効果）．したがって，このHα線は中心波長よりも少し短い波長（Hα線の－ウィング）で観測した方がよく見える．逆にいえば，Hα線の－ウィングで太陽の画像を撮影すれば，観測者の方向へ運動している水素原子の分布を示す画像が得られることになる．同様に，Hα線の＋ウィングでの観測では，観測者から離れる方向へ動いている水素原子の分布がわかる．さらに，ウィングとして見る波長のずれの大きさを変えれば，異なる運動速度（正確には，観測者から見た視線方向速度）を持つ水素原子の分布状態を知ることができる．また，フィラメントなどの太陽画像上で見えるものの伸びたり縮んだりする速さを測定すると，視線に対して垂直な方向の速度を求めることができる．この視線方向速度と視線に垂直な方向の速度を合成することで，彩層における水素原子の運動状態を三次元的に捕えることが可能となる．

1つ例をあげてみよう．図1.10はスピキュールをHα線中心とその両ウィング（＋および－）でそれぞれ撮影した画像である．Hα線中心で撮影した画像ではスピキュールの針状の構造があまりはっきり見えないが，両ウィングで撮影した画像ではスピキュールの構造がよく見える．先に述べたように，観測波長をずらすと視線方向に運動するガスから出る光がよく見えるようになる．スピキュールは高速で運動するガスのジェットであるから，Hαの両ウィングで観測する方がその姿をよく捕えることができるのである．またスピキュールの長さの時間変化から視線に垂直な方向の速度がわかり，

of the emitted light will be shorter than the central wavelength (656.3 nm) as a result of the Doppler effect. This light can be efficiently captured when the observation is performed at a wavelength slightly shorter than the Hα central wavelength (i.e., in the minus-wing of the Hα line). In other words, if an image of the Sun is taken in the minus-wing of the Hα line, the resulting image will strongly reflect the distribution of the hydrogen atoms moving toward the observer. Similarly, observations in the plus-wing of the Hα line provide information about the distribution of the hydrogen atoms moving away from the observer. By varying the amount of shift of the wing, an observer can determine the distribution of hydrogen atoms that are moving with different velocities (more precisely, the line-of-sight velocities relative to the observer). It is also possible to determine the velocity perpendicular to the direction of the line-of-sight by measuring the expanding or contracting velocity of a filament or a similar structure from the solar image. By combining the line-of-sight velocity and the velocity perpendicular to the direction of the line of sight, we can determine the three-dimensional movements of hydrogen atoms in the chromosphere.

Figure 1.10 shows three images of the same spicules taken in the Hα center and in both wings (+ve and -ve). In these images, the spike-like structure of the spicules is clearly seen in the images taken in the two wings, whereas it is rather blurred in the image taken in the Hα center. As stated previously, by shifting the observation wavelength, the components of light emitted from a gas moving in the line-of-sight direction will be stronger, thereby resulting in a clear image. Since a spicule is a jet of gas

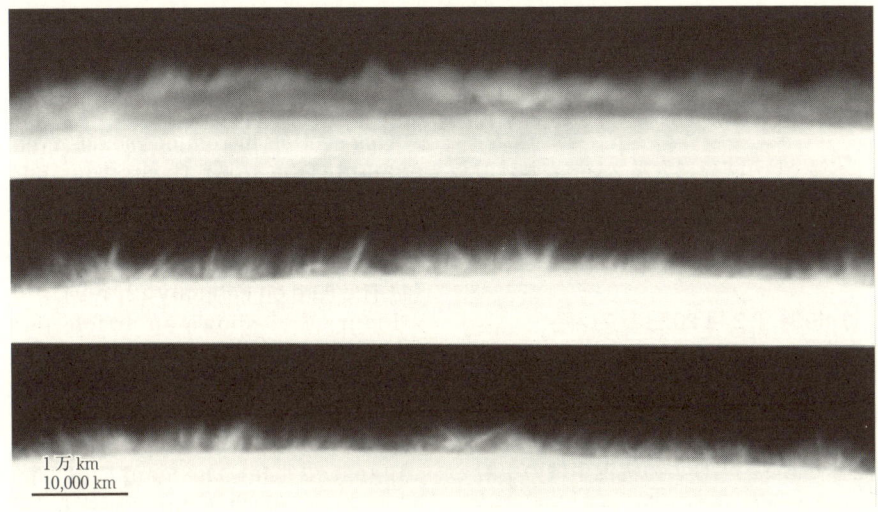

図1.10 スピキュール（1985年8月15日）．上段：Hα線中心（21:48:06 UT）．中段：Hα－0.9Å（21:48:14 UT）．下段：Hα＋0.9Å（21:48:25 UT）．

Fig. 1.10 Images showing spicules (Aug 18, 1985). The upper image was taken in the Hα center light (21:48:06 UT), the middle image in Hα-0.9Å (21:48:14 UT), and the lower image in Hα +0.9Å (21:48:14 UT).

2つの速度を合成すると三次元的な速度場が得られるというわけである．

　後で説明するように，FMTはHα線中心だけでなくその両ウィングの波長でも同時に太陽面の観測を行うことができる．さらには，連続光による太陽全面の観測とプロミネンスの観測も同時に実行可能である．このような5波長同時観測により，FMTは数多くの観測成果を上げてきた．これについては次章以降で詳しく説明する．

moving at a high velocity, its structure is more clearly recognized when observed in both wings of the Hα center. Furthermore, its velocity perpendicular to the line-of-sight can also be determined by measuring the change in the length of the spicule. By combining the two kinds of velocities, a three-dimensional velocity field can be obtained.

　By using the FMT, it is possible to simultaneously observe the solar disk in the Hα center as well as in both its wings. The telescope can also perform prominence observation and full solar disk observation in continuum light simultaneously. The five-wavelength observations have enabled us to achieve a large number of significant observational results, which will be described in detail in the following

1.2 太陽活動の地球への影響

本節では,実際に太陽フレアが地球環境に影響を及ぼした例をいくつか紹介しよう.

(1) 1994年2月20日のフレア

図1.11は,1994年2月22日にリレハンメルで行われていた冬季オリンピックの実況中継の中断を報じる北海道新聞の記事である.実はこの実況中継の中断(21:15 JST = 12:15 UT)は,元をたどると太陽フレアが原因だったと考えられている.

原因となったフレアは2月20日に太陽の真ん中から少し北に位置する場所で発生した.図1.12左にFMTで観測されたフレア,図右にGOES衛星で観測された軟X線強度の時間変化を示す.フ

1.2 Impacts of the Sun on the Earth

This section introduces some actual examples of solar flares that had significant impacts on the terrestrial environment.

(1) The flare on February 22, 1994

Figure 1.11 shows an article in a Japanese newspaper reporting a sudden interruption in the live coverage of the Winter Olympic Games in Lillehammer, Norway, on February 22, 1994. It is suggested that the cause of this interruption, which occurred at 12:15 UT, was a solar flare.

The flare that caused this interruption occurred slightly north of the center of the Sun on February 20, 1994. Figure 1.12 shows an image of this flare obtained by using the FMT (left) and the temporal change in the soft X-ray intensity as observed by the Geostationary Operational Environmental Satellite

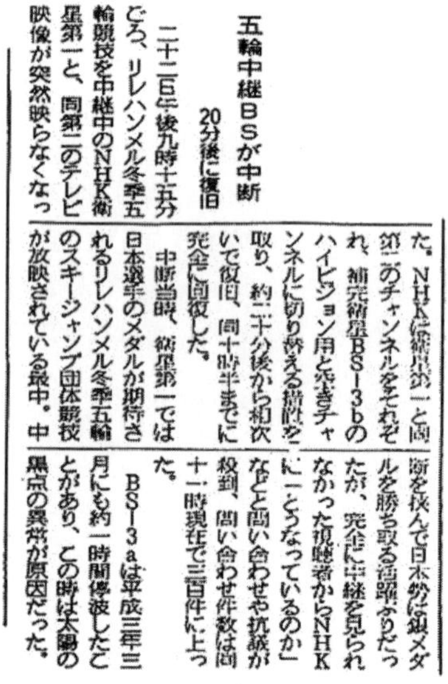

図1.11 リレハンメルの五輪中継の中断を伝える新聞記事(1994年2月23日,北海道新聞朝刊,共同通信社配信).太陽フレアによって引き起こされた磁気嵐が原因だったと考えられている.

Fig. 1.11 A newspaper article reporting a sudden interruption in the live coverage of the Lillehammer Olympic Games (Hokkaido Shimbun, dated February 23, 1994). The trouble is believed to be due to a magnetic storm resulting from a solar flare.

レアは太陽全体に比べると小さいが典型的なツーリボンフレアであり，軟X線強度（GOESクラス）はM4.0と中規模のフレアである．FMTのHα画像を詳しく解析すると，図1.13(a)-(c)のHα中心像にあるように，フレア前（00:53 UT）に黒点の近くに存在したダーク・フィラメントが次第に消失し，その後フレアが起きていることがわかる（フレアの軟X線強度のピーク時刻は01:38 UT頃）．フィラメントが消失したのは，噴出したためである．実際，FMTのHα−0.8Åの画像(d)-(f)で見ると，消失途中（00:59 UT）によく見えている．青色側（ブルーシフト）で見えているので，地球に向かって飛び出していることがわ

(GOES; right). The flare was rather small in size as compared to the size of the Sun, and it had a typical two-ribbon structure. Its soft X-ray intensity (GOES class) was M4.0, which can be classified as a middle-class event. A detailed analysis of the Hα images taken by the FMT has shown the gradual disappearance of a dark filament lying near a sunspot before the onset of the flare (00:53 UT), as shown in the Hα images in Figs. 1.13(a)-(c). Subsequently, the flare occurred, and the soft X-ray intensity reached its peak at around 01:38 UT. The disappearance of the filament was due to its eruption. This can be confirmed from the images taken with the FMT in Hα -0.8 Å (Fig. 13 (d)-(f)), which clearly show the filament in the middle of the disappearing phase

Feb 20, 1994 00:17−02:47 UT Flare, Filament Eruption

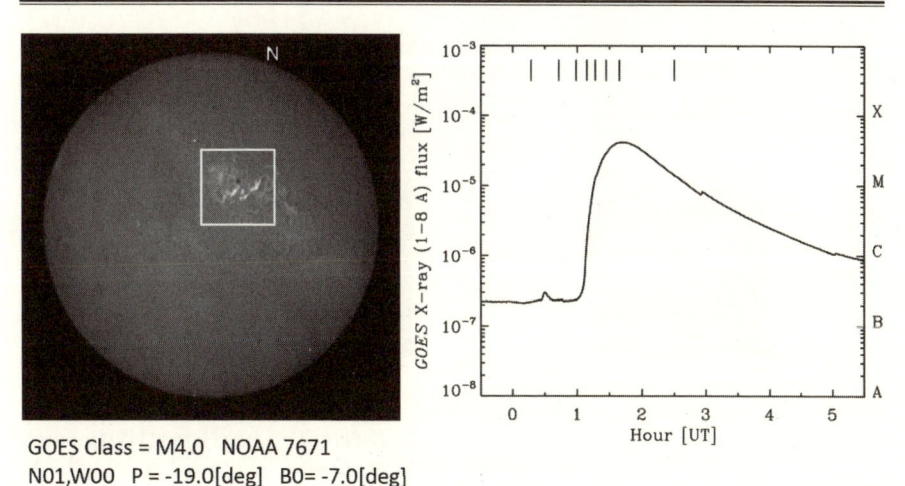

GOES Class = M4.0   NOAA 7671
N01,W00   P = -19.0[deg]   B0= -7.0[deg]

図1.12　左：FMTで観測された1994年2月20日の太陽フレアのHα像．右：同フレアの軟X線強度の時間変化（GOES衛星による）．

Fig. 1.12 Hα image of the solar flare taken using the FMT on February 20, 1994 (left) and the temporal change in the soft X-ray intensity observed by the GOES satellite (right).

かる．0.8 Å を速度に換算すると 36.5 km/s だから，その程度の噴出速度はあると考えられる．フィラメント消失は野辺山電波観測所の電波ヘリオグラフでも観測されている（Hanaoka and Shinkawa 1999）．また，「ようこう」軟 X 線望遠鏡では，明るい X 線アーケードが見られた［図 1.13(h)］．

フレア開始後 1 日あまり経った 2 月

(00:59 UT). The presence of this filament in the blue-shift images proves that the filament was ejected toward the Earth. A wavelength shift of 0.8 Å corresponds to a velocity of 36.5 km/s, which can be inferred to be the minimum velocity of the eruption. The disappearance of the filament was also observed with the Nobeyama radioheliograph (Hanaoka and Shinkawa 1999). In addition, the soft X-ray telescope (SXT) of the *Yohkoh* satellite found a bright X-ray arcade [Fig. 1.13(h)].

Approximately one day after the onset

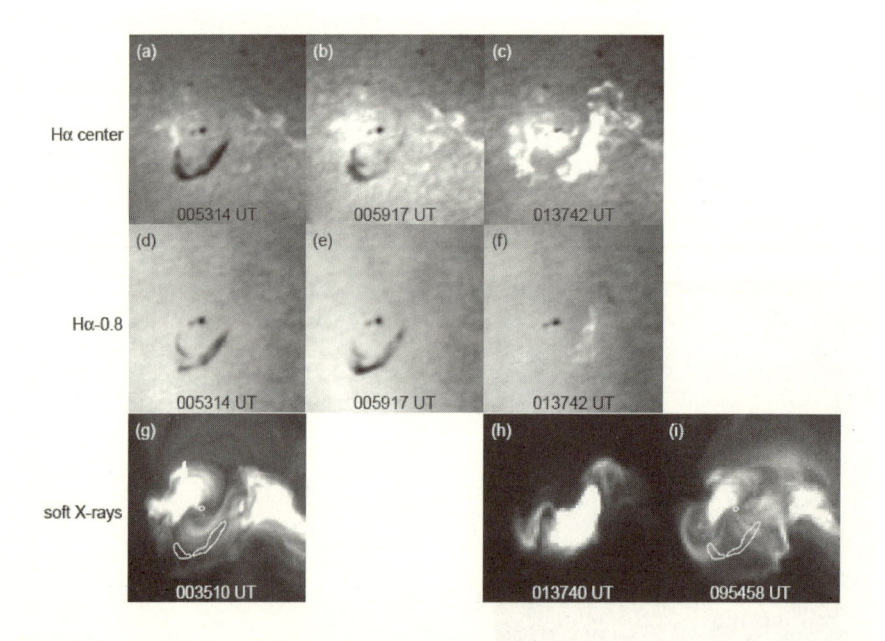

図 1.13 （a）-（c）：FMT の Hα 中心波長で見た 1994 年 2 月 20 日のフレア．00:53-01:37 UT．フィラメント消失にともなって，ツーリボンフレアが発生したのがわかる．(d)-(f)：FMT の Hα - 0.8Å で見た同フレア．フィラメントが地球方向に向かって噴出している．(g)-(i)：「ようこう」軟 X 線望遠鏡で見た同フレア（以上は，Hanaoka and Shinkawa (1999) より）．

Fig. 13 (a)-(c): Flare observed with the FMT in the Hα center on February 20, 1994, at 00:53-01:37 UT. The occurrence of a filament eruption accompanied by a two-ribbon flare can be seen. (d)-(f): The same flare observed with the FMT in Hα -0.8Å, showing the eruption of the filament toward the Earth. (g)-(i): The same flare observed with the *Yohkoh* soft-X-ray telescope (All images cited from Hanaoka and Shinkawa 1999).

21日09:00 UT頃，GOES衛星で観測された太陽風速度が500 km/sから900 km/sへと急に増大した（図1.14参照）．これはフレア（およびフィラメント噴出／CME）から惑星間空間に向けて放出された衝撃波の到来を意味する．そのとき同時に太陽風中の磁場強度も5 nTから

of the flare, the velocity of the solar wind as observed by the GOES suddenly increased from 500 km/s to 900 km/s at 09:00 UT on February 21, 1994 (Fig. 1.14). This suggests that the shock waves released from the flare (including the filament eruption and the CME) spread through interplanetary space. Simultaneously, the magnetic field

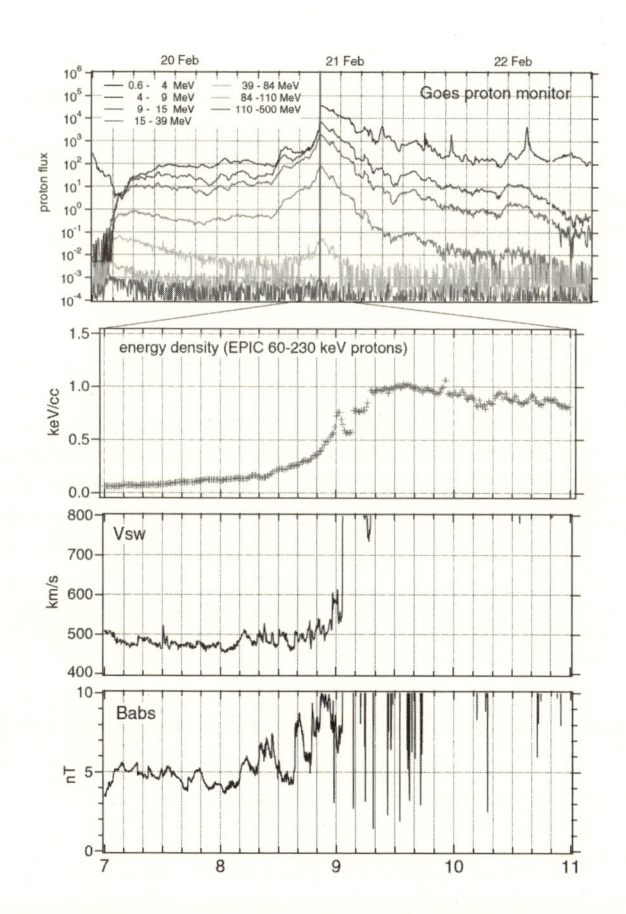

図1.14　1994年2月20日のフレアにともなって発生した高エネルギー陽子フラックス，太陽風速度（Vsw），太陽風中の磁場強度の時間変化（Babs）．（Terasawa et al. 2006 より）

Fig. 1.14 Temporal changes in the high-energy proton flux, the velocity of solar wind (Vsw), and the magnetic field strength in the solar wind (Babs) associated with the solar flare that occurred on February 20, 1994. (Cited from Terasawa et al. 2006)

20 nT 以上に増大した（衝撃波到来時間から，太陽地球間の平均の衝撃波伝播速度を見積もると 1,500 km/s となる．したがって，実際はかなり減速されたことがわかる）．さらに興味深いのは 0.4～84 MeV のプロトン（陽子）の数が衝撃波到来の 6 時間前から徐々に増大していることである．これは衝撃波で加速された高エネルギー陽子（Solar Energetic Particle ＝ SEP）が，磁力線に沿って衝撃波より速く地球に到来していることを意味する．陽子数フラックスの変化は，衝撃波粒子加速理論の予言とよく一致する（Terasawa et al. 2006，寺沢敏夫 2002）．これらの高エネルギー粒子が人工衛星を故障させたり，宇宙飛行士の被ばくの原因となる．

図 1.15 には地磁気変動を表す Dst の時間変化を示す．衝撃波の地球への到来に合わせて 2 月 21 日 09:00 UT 頃，地磁気が突然増大しているのがわかる．これは衝撃波によって地球の太陽側の磁場が突然圧縮されて地磁気が強くなったことを表す（Sudden Commencement と呼ばれる）．しばらくすると，地磁気は減少を始め，最大で 140 nT も弱くなった．

strength in the solar wind increased from 5 nT to 20 nT or even higher levels (on the basis of the arrival time of the shock waves, the average propagation speed of the shock waves in the space between the Sun and the Earth can be estimated to be 1,500 km/s. This result suggests that the shock waves were considerably decelerated). Another interesting phenomenon was the increase in the number of 0.4-84 MeV protons, which are called solar energetic particles (SEP); their numbers began to gradually increase six hours before the arrival of the shock waves. This means that the high-energy protons that are accelerated by the shock waves can travel faster than the shock waves along the lines of magnetic force and arrive at the Earth earlier. The change in the proton number flux agrees well with the theory of shock wave particle acceleration (Terasawa et al. 2006, Toshio Terasawa 2002). These high-energy particles can damage artificial satellites and cause severe radiation exposure to astronauts.

Figure 1.15 shows the temporal change in the geomagnetic fluctuation (Dst). From the figure, it can be seen that there was a sudden increase in the geomagnetism on February 21, 1994, at around 09:00 UT, with the arrival of the

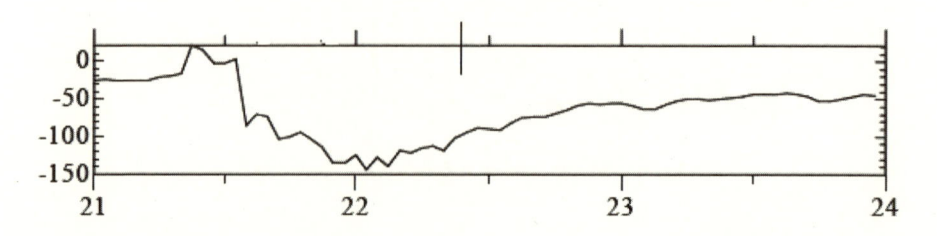

図 1.15　地磁気変動（Dst）の時間変化．縦軸の単位は nT（ナノテスラ）．横軸の単位は日であり，左端が 1994 年 2 月 21 日 00:00 UT.

Fig. 1.15 Temporal change in geomagnetic fluctuation (Dst) in nT. The unit of horizontal axis is day. The stant time is 00:00 UT of Feb 21, 1994.

このように地磁気強度が突然変動する現象を磁気嵐と呼ぶ。この磁気嵐はそれほど大きな磁気嵐というわけではなかったが、23日頃まで続き、放射線帯粒子数を増大させた結果、放送衛星を故障させ、上記のリレハンメル五輪中継の中断を引き起こしたのではないかと考えられている（片岡, private communication 2009）.

(2) 2003年10月28日のフレア

太陽サイクル極大期第23期で最大級の地磁気嵐を引き起こしたのが、2003年10月28日 09:51 UT のフレアであった。図1.16にフレアのピーク時における極端紫外線像（SOHO衛星）と、軟X線強度の時間変化（GOES衛星）を示す。X線強度（GOESクラス）はX17.2. この時点で28年間のフレアX線観測史上、3番目の大きさだった。

図1.17に、このフレアの地球への影響を報じる新聞記事を示す。「太陽の嵐、地球直撃. 14年ぶり大爆発. 微粒子、磁場乱す. テキサスでオーロラ.」などと、夕刊のトップを飾ったほどだから、大ニュースであった。記事を読むと、「データ中継衛星『こだま』が一時的に機能を停止」、「宇宙ステーションでは20分間ほど飛行士2人がステーション内の安全

shock waves on the Earth. This implies that the magnetic field of the Earth on the side facing the Sun was suddenly compressed by the shock waves, thereby strengthening the geomagnetism (this phenomenon is called sudden commencement). After a while, the geomagnetism began to decrease, recording a maximum fall of 140 nT. Such a sudden change in the geomagnetic intensity is called a geomagnetic storm. Although the magnetic storm in question was not very large in scale, it continued until around February 23, causing an increase in the number of radiation belt particles. It is thought that this increase in the radiation belt particles damaged the broadcasting satellite and caused the aforementioned sudden interruption of the live coverage of the Lillehammer Olympic Games (Kataoka, private communication, 2009).

(2) The flare on October 28, 2003

The flare that occurred on October 28, 2003, at 09:51 UT caused one of the most powerful geomagnetic storms in the maximum phase of the 23rd solar cycle. Figure 1.16 shows an extreme ultra-violet (EUV) image taken by the SOHO satellite at the peak time of the flare (left) and the temporal change in the soft X-ray intensity observed by the GOES (right). The X-ray intensity recorded X17.2, which, until then, was the third-highest recorded value in 28 years.

Figure 1.17 shows a Japanese newspaper article reporting the impact of this flare on Earth. The news made headlines in the paper, reading "Direct Attack of Solar 'Storm' on Earth / Largest Explosion in 14 Years / Particles Disturbing Magnetic Fields / Aurora Observed in Texas". The article reported many terrestrial events such as the

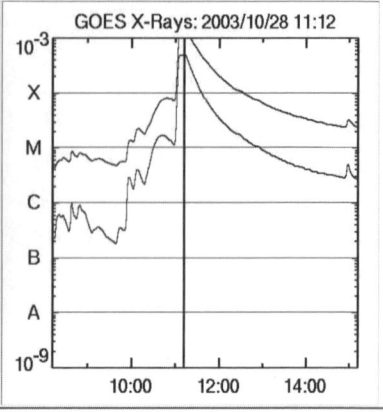

図1.16 2003年10月28日09:51 UTのフレアの極端紫外線像（SOHO衛星：撮影時刻は11:12 UTでX線のピーク時）（左図）と，軟X線強度の時間変化（GOES衛星）（右図）．

Fig. 1.16 EUV image of the flare that occurred on October 28, 2003, at 09:51 UT (left; taken by the SOHO satellite at 11:12 UT, when the X-ray intensity peaked), and temporal change in the soft X-ray intensity (right; taken by the GOES satellite).

な場所に避難」，「カナダでは航空機と通信の一部に障害が出た」，などとある．

　図1.18には，このフレアから発生したコロナ質量放出（CME）の画像を示す．画像全体が乱れているのは，フレア・CMEで発生した高エネルギー粒子（後述，図1.19）がCCDカメラを被ばくさせているためである．CMEのスピードは2,459 km/sときわめて高速だった．また，CMEの形が太陽を中心に同心円状に広がっていることもわかる．このような形のCMEのことをハローCMEと呼んでおり，地球に向かっていることを示す．つまり，ハローCMEが発生したら，地球に衝突する可能性が高く要注意，というわけである．

temporary breakdown of the Kodama data relay satellite and the partial communication failure between airplanes and ground stations in Canada. It also reported that two astronauts inside a space station had to take refuge in a safe zone for approximately 20 minutes.

　Figure 1.18 shows a sequence of images of the CME generated from this flare. The noisy appearance of the images is caused by the strong impingement of high-energy particles produced by the flare and the CME on the CCD camera (to be described later with reference to Fig. 1.19). The CME had an extremely high velocity of 2,459 km/s. The sequential images also demonstrate that the CME expanded concentrically around the Sun. Such a CME is called a halo CME. Its shape suggests that the CME is heading for the Earth. It is therefore important to issue a warning about terrestrial disturbance when a halo CME occurs, because it has a high probability of

Chapter 1 Solar Research and the Terrestrial Environment | 23

図 1.17 2003 年 10 月 29 日の磁気嵐を報じる朝日新聞記事（新聞は 30 日の夕刊）．10 月 28 日のフレアの結果，磁気嵐が 29 日に発生した．

Fig. 1.17 A newspaper article (Asahi Shimbun, dated October 30, 2003) reporting the occurrence of a geomagnetic storm on October 29, 2003. This geomagnetic storm was caused by a flare, which occurred on Oct 28, 2003.

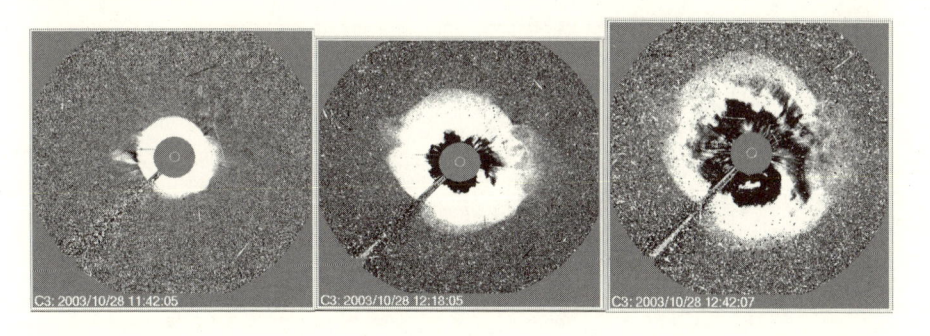

図 1.18 2003 年 10 月 28 日のフレア（X17.2）から発生した CME の時系列画像（撮影時刻は左から 11:42 UT，12:18 UT，12:42 UT）．SOHO/LASCO による可視光時系列像の差分画像（差分画像については第 6 章の 6.1.3 を参照）．中心の白丸が太陽．

Fig. 1.18 Sequential images of the CME produced by the flare that occurred on October 28, 2003 (X17.2), taken at 11:42 UT, 12:18 UT, and 12:42 UT (from left to right), respectively. These are the running difference images created from a series of visible-light images taken by the SOHO/LASCO (see Section 6.1.3 of Chapter 6 for an explanation of the running difference images). The white circle at the center is the Sun.

図 1.19 には，地球近傍の GOES 衛星で観測された高エネルギー陽子フラックスの時間変化を示す．10 MeV 以上の高エネルギー陽子フラックスは，10 月 28 日 12:00 UT（フレア発生から 2 時間後）には 1,000 pfu (proton flux unit) を超え，29 日 00:00 UT には 1 万 pfu に達した．最終的には 29 日 06:15 UT に，2 万 9,500 pfu となり，10 MeV の陽子フラックスとしては，史上 5 番目の強さとなった．

図 1.20 には，地磁気変動 (Dst) の時間変化を示す．磁気嵐は 29 日 06:00 UT 頃始まり，30 日になると地磁気変動量は −350 nT 以上という大磁気嵐となった．日本でも北海道でオーロラが見えた．

collision with the Earth.

Figure 1.19 shows the temporal change in the high-energy proton flux observed by the GOES near the Earth. The high-energy proton flux above 10 MeV exceeded a level of 1,000 proton flux units (pfu) on October 28, 2003, at 12:00 UT (two hours after the onset of the flare) and reached 10,000 pfu at 00:00 UT, the next day. The flux eventually reached a level of 29,500 pfu on October 29, 2003, at 06:15 UT, the fifth highest recorded value for the 10 MeV proton flux.

Figure 1.20 shows the temporal change in the geomagnetic fluctuation (Dst). The magnetic storm began at around 06:00 UT on October 29, 2003, and developed into a large-scale magnetic storm on October 30, 2003, with a

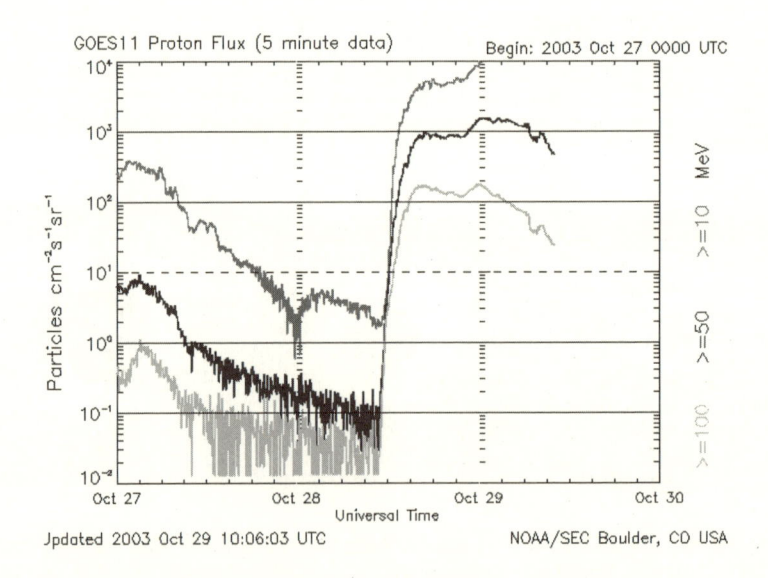

図 1.19　太陽風中の高エネルギー陽子フラックスの時間変化（GOES 衛星による）．2003 年 10 月 28 日 09:51 UT のフレアおよび CME によって加速された陽子のフラックスが 12:00 UT 頃から急に増大した．

Fig. 1.19 Temporal change in the high-energy proton flux in the solar wind (recorded by the GOES satellite). The flux of protons, which were accelerated by a flare (which occurred on October 28, 2003, at 09:51 UT) and CME, began to rapidly increase around 12:00 UT.

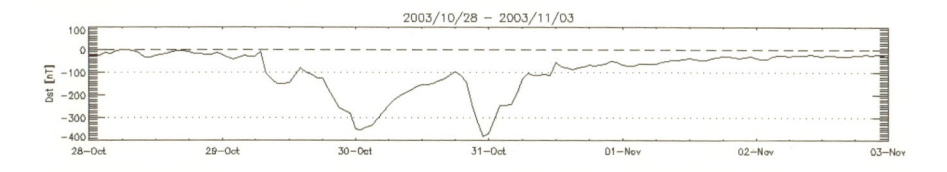

図 1.20 2003 年 10 月 28 日 09:15 UT のフレア（X17.2）にともなう地磁気変動（Dst）の時間変化. 31 日の磁気嵐は，29 日 21:00 UT 頃に起きた別のフレアによる.

Fig. 1.20 Temporal change of in the geomagnetic fluctuation (Dst) associated with the flare (X17.2) that occurred on October 28, 2003, at 09:15 UT. The other geomagnetic storm of October 31 was caused by another flare, which had occurred on October 29 at around 21:00 UT.

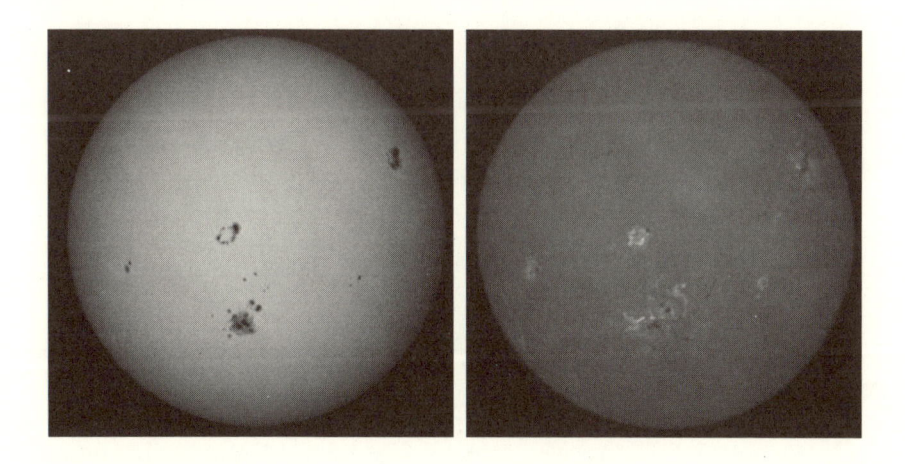

図 1.21 FMT で観測された 2003 年 10 月 28 日の太陽全面の連続光像（左図）と，Hα（中心）像（右図）.

Fig. 1.21 Full-disk solar images in continuum light (left) and Hα center light (right) taken with the FMT on October 28, 2003.

このフレアが発生したときは残念ながら，日本の夜の時間帯だったので，FMT によるフレアそのものの観測データはない．しかし，フレアの少し前の，太陽全面像が FMT によって撮られているので，それを図 1.21 に示そう．

フレアが発生したのは南半球の黒点群（活動領域）である．黒点群は大きく，

geomagnetic fluctuation of greater than -350 nT. On that day, the aurora could be seen even at such low-latitude areas as Hokkaido, Japan.

When this flare occurred, it was nighttime in Japan and thus no FMT observation data for this flare could be obtained. However, full-disk solar images were taken by the FMT slightly before the occurrence of the flare. Figure 1.21

形がかなり複雑である．このような観測から，複雑で大きな黒点群が現れたら大フレア発生の可能性が高い，ということがわかってきた．その意味で，フレアが起きていないときの太陽全面の観測データも宇宙天気予報にとって重要である．

shows two of these images.

The flare occurred in a sunspot group (active region) in the southern hemisphere. This sunspot group was large in size and had a complex configuration. From such observations, it has been found that the formation of a large-sized, complex sunspot group can possibly produce a large-scale flare. Therefore, for space weather prediction, it is important to collect full-disk solar observation data even when no flare occurs.

参考文献／References

Hanaoka, Y. and Shinkawa, T. (1999), ApJ 510, 466.

シリーズ現代の天文学第 10 巻太陽（2009），日本評論社
"Series Modern Astronomy Vol 10. The Sun," (2009), Nippon hyouronsya, in Japanese

篠原学（2009），宇宙天気，誠文堂新光社
Shinohara, M. "Space Weather," (2009), Seibundou shinkousya, in Japanese

Terasawa, T., et al. (2006), Adv Space Res. 37, 1408.

寺沢敏夫（2002），太陽圏の物理，岩波書店
Terasawa, T. "Physics of Heliosphere," (2002), Iwanami syoten, in Japanese

http://www2.nict.go.jp/pub/whatsnew/press/031105/031105.html

柴田一成，大山真満（2004），写真集太陽，裳華房
Shibata, K. and Ohyama M., "Scientific Photo Collection: The Sun," (2004), Shokabo, in Japanese

第 2 章
フレア監視望遠鏡

Chapter 2
Flare Monitoring Telescope

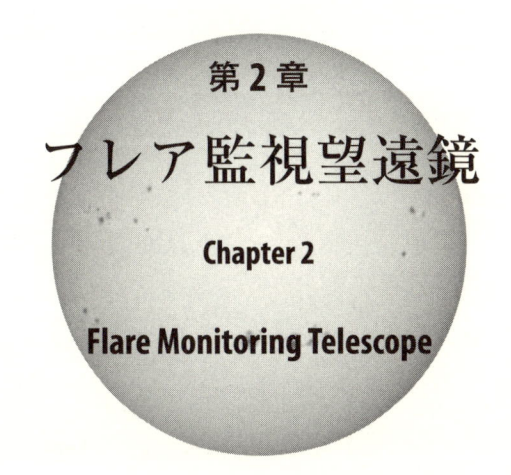

　この章では，フレア監視望遠鏡の機器構成，観測システムの基本仕様，および観測データの公開システムを紹介する．また，望遠鏡設置以来，順次行われてきた高性能化の道筋の技術的側面をこの章でまとめておく．さらに，太陽観測世界ネットワーク展開の第一歩として，ペルー国に移設されたフレア監視望遠鏡が，新天地で活躍し始めた姿を紹介する．

2.1 望遠鏡観測システムとデータ公開

　フレア監視望遠鏡（FMT）は，太陽活動の長期的変動や爆発現象の調査を目的として，1992年に京都大学飛騨天文台に設置され（Kurokawa et al. 1995, 図 2.1），1991年から始まった国際協同観測プログラム（STEP）の一翼を担う役割を果たしてきた（付録 C 参照）．
　FMTは，5本の太陽全面撮像用鏡筒と，

　This chapter introduces the system configuration of the Flare Monitoring Telescope (FMT) and the basic specifications of its observation system and WWW data release system. Technical improvements to enhance the system performance are also briefly summarized in this chapter. The FMT has been relocated to Peru, as the first step for developing a worldwide solar observation network. The last section of this chapter shows images of the FMT in operation at the new site of observation.

2.1 Telescope Observation System and Public Data Release

　The FMT was constructed in 1992 at the Hida Observatory in Japan (Fig. 2.1) to investigate the long-term variations in solar activity and explosive events (Kurokawa et al. 1995). The construction of the FMT has been one of the many successful projects launched by the international coordinated observations program (STEP), which was started in 1991 (see Appendix C).

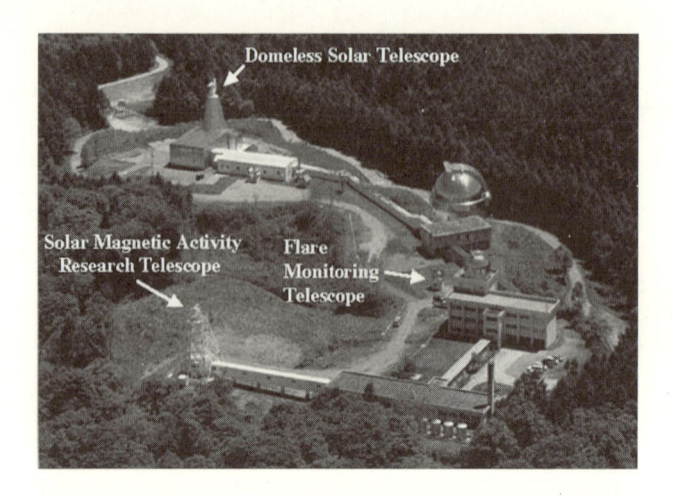

図 2.1 京都大学飛騨天文台の全体写真と 3 種の太陽観測用望遠鏡．口絵参照．

Fig. 2.1 Three solar telescopes at the Hida Observatory. Refer to the frontspiece.

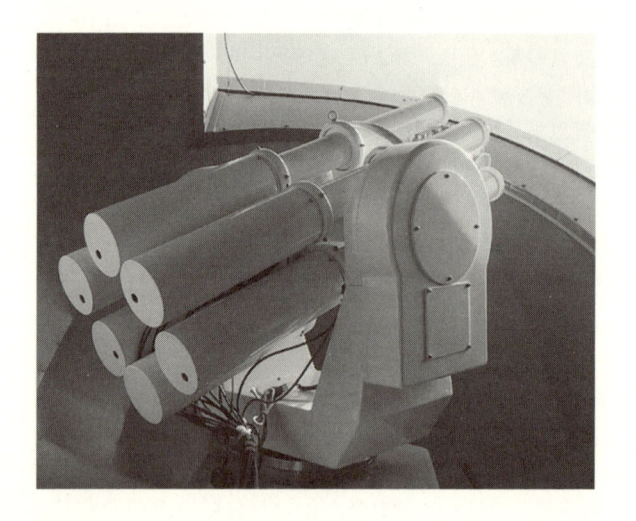

図 2.2 フレア監視望遠鏡（FMT）の外観．口絵参照．

Fig. 2.2 An image picture of the FMT (Flare Monitoring Telescope). Refer to the frontspiece.

1 本の光電ガイド用鏡筒から構成された，シンプルでコンパクトな構造を成している（図 2.2）．その光学系と主たる仕様は，

The FMT consists of five solar imaging telescopes and one photoelectric guide telescope, all of which have a simple and compact design (Fig.2.2). The optical

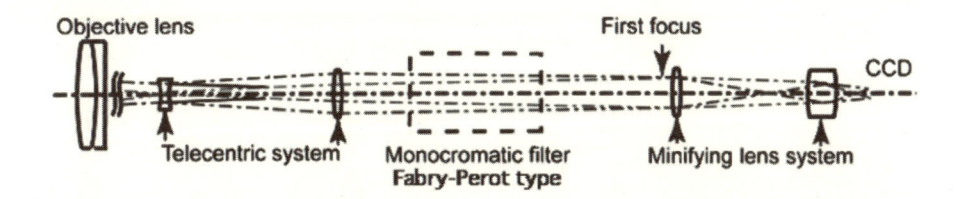

図 2.3 FMT のある 1 本の鏡筒内の光学系概略図（Kurokawa et al, 1995 より）. 5 本の撮像用望遠鏡は，全てすべてほぼ同様の構造をしている．

Fig. 2.3 Schematic drawing of the optical system of one of the five imaging telescopes in the FMT. All of the five telescopes have almost the same structure.

表 2.1 飛騨天文台 FMT の仕様（Kurokawa et al, 1995 より）.

Table 2.1 Specification of the FMT at the Hida Observatory.

Optics		
Aperture 64 mm		
F-ratio 30		
Focal length 1920 mm		
Spatial resolution 2.4 arcsec		
Filters		
Telescope name	Center wavelength	Passband
Hα center	6562.8Å	0.42Å
Hα +0.8Å	6563.6Å	0.5Å
Hα -0.8Å	6562.0Å	0.5Å
Continuum	6100Å	60Å
Prominence	6562.8Å	3Å

図 2.3，表 2.1 に示した．光学系は，フィルタ（Daystar 社製）を除き，5 本の撮像用鏡筒とも，ほぼ共通の構成となっている．この 5 本の撮像用鏡筒を用いることにより，FMT は Hα 吸収線周辺の異なる波長での，もしくは異なるモードでの太陽全面像をモニター観測することが可能である（図 2.4）．

これらから得られる複数種の画像に，「クラウドモデルフィッティング」と呼ばれる手法（図 2.5）を適用させれば，

system and the specifications of the telescopes are shown in Fig. 2.3 and Table 2.1, respectively. The optical layout of all the five imaging telescopes is almost similar, except for the filters used in them (supplied by Daystar Company). By using the five telescopes, we make simultaneous full-disk solar observations at different wavelengths around the Hα absorption line or in different modes (Hα center, Hα-0.8Å, Hα+0.8Å, continuum light, and prominence mode with an occulting disk; see Fig. 2.4).

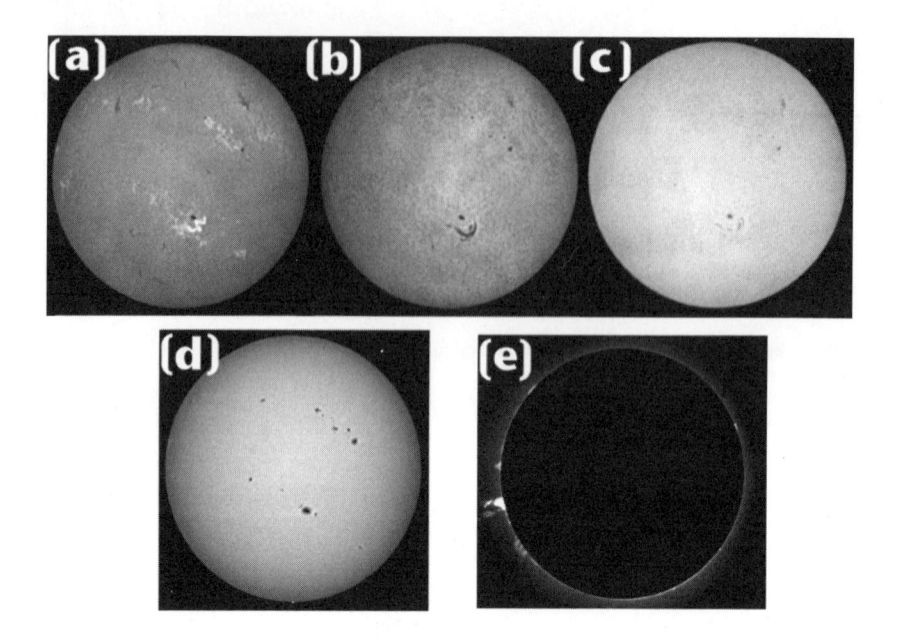

図 2.4 FMT にて撮影された 5 種類の太陽像のサンプル．(a)：Hα 線中心像，(b)：Hα 線 -0.8 Å 像，(c)：Hα 線 $+0.8$ Å 像，(d)：連続光画像，(e)：プロミネンスモード画像．

Fig. 2.4 Sample images obtained by using the five telescopes of the FMT. (a) Image in Hα line center, (b) image in H$\alpha-0.8$Å, (c) image in H$\alpha+0.8$Å, (d) continuum-light image, (e) prominence-mode image.

太陽彩層面上ガスの視線方向速度（ドップラー速度）を算定することができる．特にこの FMT は，複数波長の画像をほぼ同時に撮影できるため，地球大気による太陽画像の歪みの時間変化による影響を，他の地上望遠鏡に比べて格段に抑えることができるので，より精度のいいドップラー速度を算出することができるという点が，大きな特長である．そういう点において，FMT は宇宙天気環境に直接的に影響を及ぼしうる太陽面上の大規模な活動現象の運動状態を研究する目的に大変適した装置であるといえる．

具体的な撮像システムとして，1992年5月～2006年4月までは，まず CCD

If we apply cloud model fitting (Fig. 2.5) to these multi-wavelength images, we can also measure the Doppler velocities of the structures moving in the chromosphere. In particular, since the FMT can almost simultaneously observe the Sun at different wavelengths, the effect of the terrestrial atmosphere on the time variation of image distortion can be greatly suppressed in comparison with other ground-based telescopes. Further, the FMT can accurately measure the Doppler velocity field on the full-disk Sun; hence, it is particularly suitable for studying motion of large-scale active phenomena on the whole solar-disk, which are directly connected with the space weather environment.

Cloud Model Fitting

Observed contrast: $C(x, \Delta\lambda) = [I_P(x, \Delta\lambda) - I_{R0}(\Delta\lambda)]/I_{R0}(\Delta\lambda)$

$$C(\Delta\lambda) = \left[\frac{S}{I_{R0}(\Delta\lambda)} - 1\right]\{1 - \exp[-\tau(\Delta\lambda)]\}$$

$$\tau(\Delta\lambda) = \tau_0 \exp\{-[(\Delta\lambda - \Delta\lambda_S)/\Delta\lambda_D]^2\}$$

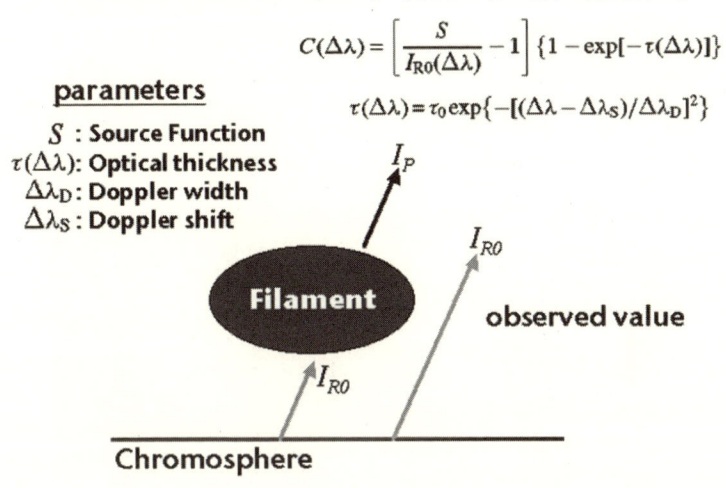

parameters
- S : Source Function
- $\tau(\Delta\lambda)$: Optical thickness
- $\Delta\lambda_D$: Doppler width
- $\Delta\lambda_S$: Doppler shift

図 2.5 クラウドモデルフィッティングの原理.この手法は,フィラメント(Filament)が,彩層 (Chromosphere)上空に雲(クラウド)のように浮かんでいることを利用する解析手法である.フィラメントが彩層から来る光をさえぎるため,その場所のHα中心像,ウィング像の明るさがフィラメントのない場所と異なって観測される.特に,フィラメントが視線方向に動いているときには,Hαウィング画像の見え方が大きく異なってくる(1.1節 Hα線を中心とする多波長観測を参照).Hα多波長での見え方(コントラスト)を利用するこの方法では,視線速度ばかりではなく,フィラメントの不透明度やフィラメント内のガスの乱流速度も求めることができる.

Fig. 2.5 Principle of the cloud model fitting. This analysis technique is based on the model that considers filaments as being like clouds above the chromosphere. The filament absorbs or scatters the light radiated from the chromosphere and thus appears darker in Hα light when compared to the surrounding area. In particular, when the filament is moving in the line-of-sight direction, it appears as a dark feature that stands in strong contrast to its surroundings, in the Hα wing images (see "Observations at multiple wavelengths around the wavelength of Hα light" in Section 1.1). By analyzing the appearance of a filament (or its contrast with the surroundings) observed at multiple wavelengths in the Hα line, this technique enables us to determine not only the line-of-sight velocity of the filament but also the opacity of the filament and the velocity of a turbulent gas inside the filament.

ビデオカメラ(SONY AVC-D7, 640×480 pix^2, 4.2 arcsec/pix)とタイムラプスビデオレコーダー(Panasonic AG-6570A)を用い,2秒毎に全種類の画像を同時にアナログ的に記録していたが,1996年9月からは,それと併行にA/D変換装置

The images were taken every two seconds and recorded using a CCD video camera (SONY AVC-D7, 640 × 480 pix^2, 4.2 arcsec/pix) and a time-lapse video camera (Panasonic AG-6570A) from 1992 to May 2006. In September 1996, we added a digital data recording system

(NEC PC-9821 & PHOTRON FRM2-512) を導入することにより，デジタル画像データ（512×512 pix^2, 4.2 arcsec/pix, 8bit）を1分毎に記録するシステムを追加した．さらに，2006年5月には，新たにデジタルCCDカメラ（竹中システム，デジタルフルフレームシャッターカメラ FC1500CL, 1392×1040 pix^2, 2.1 arcsec/pix, 10bit）を導入し，20秒毎に全種類の画像を同時にデジタルデータとして直接取得できるシステムへの更新を行なった．

このようなFMTの特徴をいかして，Morimoto & Kurokawa (2003a, 2003b) は，フィラメント消失現象の三次元速度場と太陽周辺現象との関連についての研究を行なった．彼らは，消失したフィラメントの太陽表面垂直方向の速度成分をFMTデータから求め，その時間変化を調べることにより（図2.6），それらフィラメントが本当に宇宙空間に噴出したのか，消滅しただけなのかを各々区別し，それらとコロナ中の構造やコロナ質量放出（CME）発生の有無との関係を調べた．その結果，彩層のHαフィラメントが実際に宇宙空間に噴出した場合は，ほぼ間違いなくコロナ中にアーケード型構造ができ，さらに上空にCMEが発生するということを，観測的に明確に示すに至った．

一方，Narukage et al. (2002) は，太陽フレア発生時にFMTで検出された，モートン波と呼ばれる太陽彩層中の波状現象の伝播の特徴と，ようこう衛星によりX線で観測されたその上空のコロナ中の波状現象の特徴とを比較し（図2.7），それらが，共にフレアによって引き起こ

(NEC PC-9821 & PHOTRON FRM2-512, 512×512 pix^2, 4.2 arcsec/pix, 8 bits) of 1-min cadence. Moreover, in May 2006, we introduced a new digital CCD camera system (Takenaka System Co. Ltd., digital full-frame shutter camera FC1500CL, 1392×1040 pix^2, 2.1 arcsec/pix, 10 bits) in order to improve the pixel resolutions and digital bit depths and reduce the exposure time (4 ms). In this new system, the time cadence for obtaining one data set of digital images has been set at 20 s.

Morimoto & Kurokawa (2003a, 2003b) used the FMT to measure the three-dimensional velocity fields of disappearing solar filaments. They then distinguished between the erupting and non-erupting active filaments by analyzing the time-variations of their radial upward velocities (Fig. 2.6) and investigated the relationship between the filaments and the coronal structures or coronal mass ejections (CMEs). On the basis of the results obtained in their study, they showed that the erupting filaments, which escaped from the Sun, almost perfectly corresponded to coronal arcade structures and CMEs.

Narukage et al. (2002) investigated the relationship between the Moreton wave observed using the FMT and the coronal wave as observed by the *Yohkoh* satellite in X-ray light (Fig. 2.7 (A)). The Moreton wave has been theoretically considered as the front intersection in the chromosphere of a shockwave generated by a strong solar flare. They actually found the consistency among speeds, timings and directions (Figs. 2.7 (A) & (B)). Moreover, they confirmed that the speed of disturbance propagation corresponds to that of the expected shockwave that is generated by a flare and propagates through space.

In the future, we intend to investigate

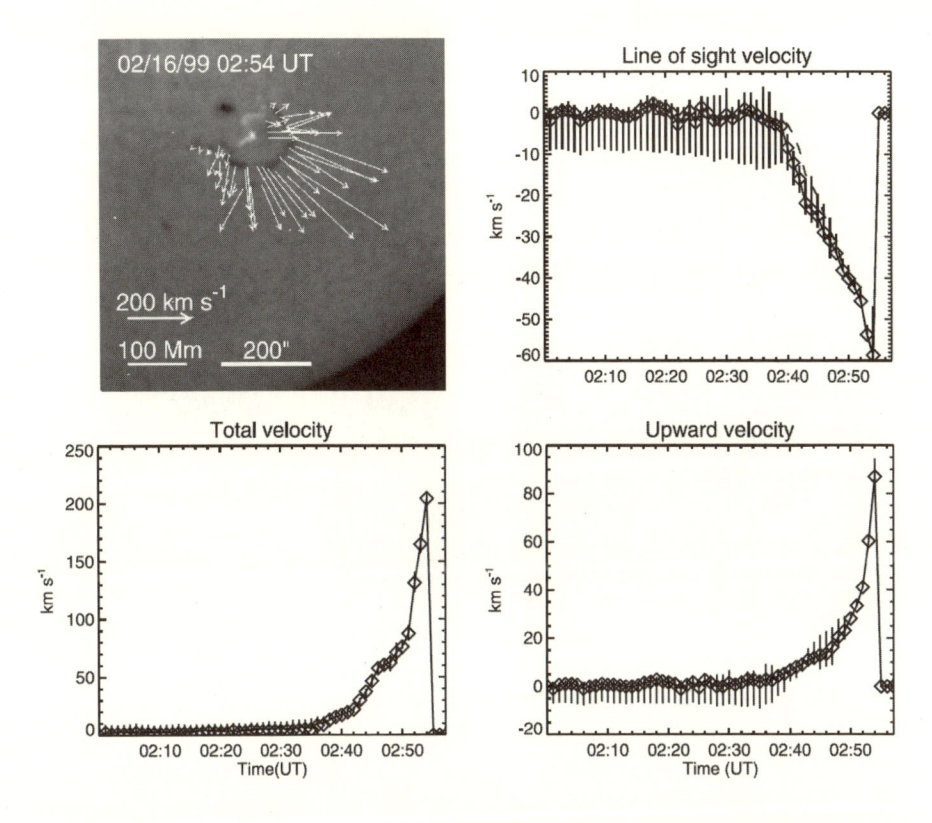

図 2.6 FMT によって観測されたフィラメント消失現象における，視線垂直方向の速度成分マップと，フィラメント中心部のドップラー速度，速度絶対値，太陽面に垂直方向の速度成分の時間発展を示すグラフ（Morimoto & Kurokawa 2003a の fig. 2 & fig. 7 より）．

Fig. 2.6 Transversal velocity field distribution and time-variation of the Doppler velocity, total velocity, and upward velocity of the disappearing filament observed with the FMT (cited from Figs. 2 & 7 in Morimoto & Kurokawa 2003a).

された，惑星間空間に伝播していく衝撃波の異なる一面を見ているものであろうことを示すに至った．

当天文台では，今後，さらにこれら噴出フィラメントや衝撃波の特徴と，惑星間空間 CME の三次元的特徴との関係，地球磁気圏への影響（磁気嵐など）の大きさとの相関などについて，研究を進展させる予定である．

the correlation between the velocity, strength and direction of an erupting filament or a shockwave and the geo-effectiveness of the corresponding CMEs (intensity of geomagnetic storm etc.).

The characteristic data obtained using the FMT, spanning a period of approximately 15 years, are now available for public access on the Internet (http://www.kwasan.kyoto-u.ac.jp/general/facilities/fmt/database_

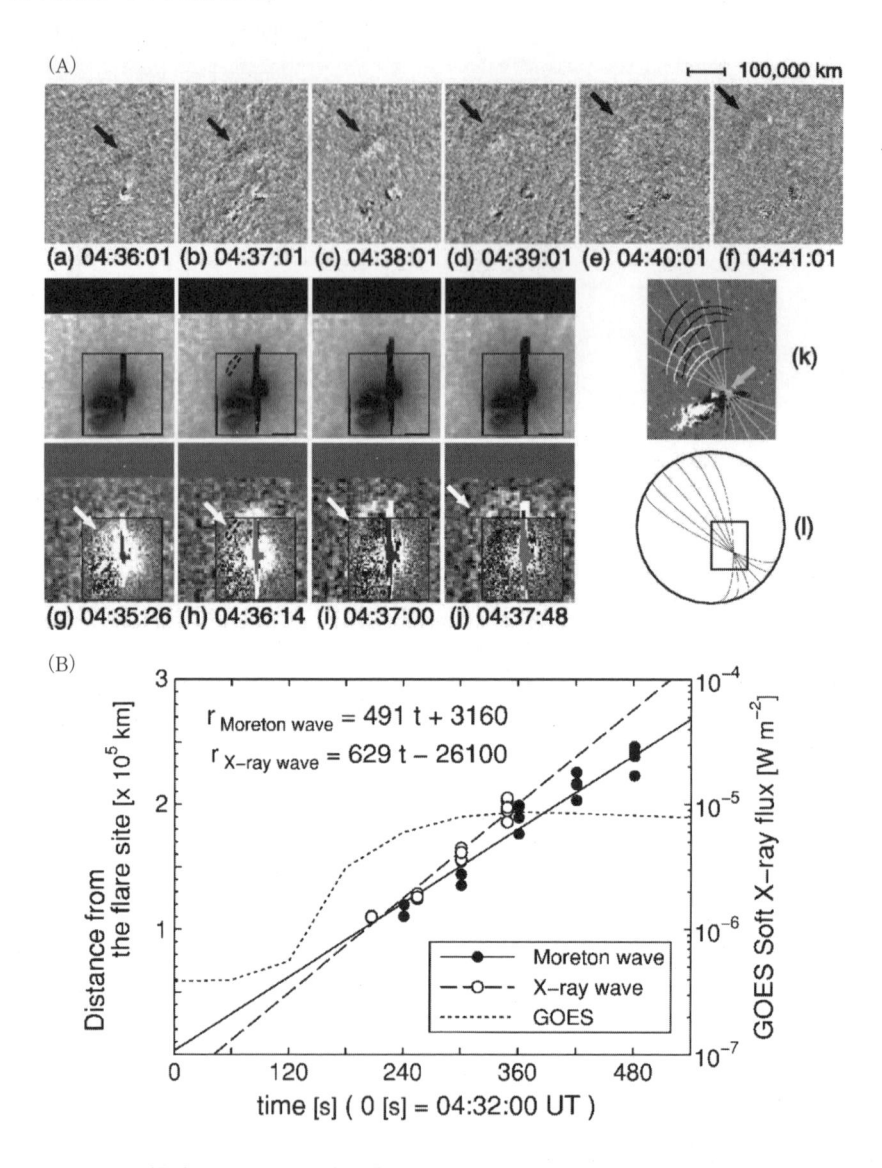

図 2.7 FMT で観測されたモートン波と「ようこう」衛星の X 線望遠鏡で観測されたコロナ中の波状現象との関係．(A)：各々の波状現象の時間発展の様子と伝播方向．(B)：各々の波面位置の時間発展．発生タイミングや伝播速度が読み取れる（Narukage et al. 2002 の fig.1 & fig.2 より）．

Fig. 2.7 Relationship between the Moretonwave observed using the FMT and the coronal wave detected with the *Yohkoh* Soft X-ray telescope. (A) Time-evolution and direction of each wave. (B) Timing and propagating speed of each wave (from Figs. 1 & 2 in Narukage et al. 2002).

このような研究分野で大変有用なFMTデータは，現在当天文台ホームページ上でも，以下のような4種類の形で公開を行なっている．
URL: http://www.kwasan.kyoto-u.ac.jp/general/facilities/fmt/database_en.html

en.html). At present, the archive consists of the following categories of data:

(1) イベントリスト

FMTで検出された2003年4月までの太陽円盤上で発生した全活動現象と，2004年8月までに太陽の縁外で発生した全活動現象の特徴や分類を示したリスト．代表的な現象については，その時間帯のGIF画像も閲覧，ダウンロード可能．

(1) Event lists

Monthly compiled lists comprising the physical characteristics and classification for each detected event. GIF images of major events can also be seen by clicking the record field.

(2) 代表的現象のMPEGムービー集

フレア，フィラメント噴出，サージ，プロミネンス噴出の4カテゴリー別に，過去の代表的な現象について，ムービーを収集して掲載．

(2) MPEG movies of significant events

The movies are classified into categories of flares, filament eruptions, surges, and prominence eruptions.

(3) リアルタイム画像中継

FMTで観測されている画像データのうち，連続光画像を除く4種類の最新画像をリアルタイムでインターネット上に自動更新．

(3) Real-time images

Four kinds of the latest images (except continuum images) are available.

(4) 全デジタル生データダウンロードサイト

本書に添付のDVDには，これらの内の「イベントリスト」を，インターネット上で公開しているすべてのデータに，更に追加情報を加えた上で収録している．

(4) Downloadable raw digital images

The DVD provided along with this book contains data of the event lists, along with supplemental information.

参考文献／References

Kurokawa, H., Ishiura, K., Kimura, G., Nakai, Y., Kitai, R., Funakoshi, Y. and Shinkawa, T.,1995, J.Geomag.Geoelectr., 47, 1043-1052

Morimoto, T. & Kurokawa, H.,2003a, PASJ 55, 503-518

Morimoto, T. & Kurokawa, H.,2003b, PASJ 55, 1141-1151

Narukage, N., Hudson, H.S., Morimoto, T., Akiyama, S., Kitai, R., Kurokawa, H. and Shibata, K.,2002, ApJ 572, L109-L112

2.2 ペルー国移設の写真記録

フレア監視望遠鏡は CHAIN プロジェクト（http://www.kwasan.kyoto-u.ac.jp/CHAIN/）の主要な観測装置の1つとして2010年3月にペルーのイカ大学へ移設された．本節では，移設前（2009年5月）に撮影されたフレア監視望遠鏡の写真とペルーへの移設後（2010年6月）に撮影された写真を紹介する．

2.2 Photographic Records of the FMT Before and After its Relocation to Peru

The FMT was moved to Ica University, Peru, in March 2010 as a key instrument for the CHAIN-Project (http://www.kwasan.kyoto-u.ac.jp/CHAIN/). This section presents some photographs of the FMT taken in May 2009, before its relocation to Peru, as well as those taken in June 2010, after its relocation.

フレア監視望遠鏡のドーム．開いたスリットから望遠鏡の本体がのぞいている．ドームの直径は3m．奥に見える塔状の建物はドームレス太陽望遠鏡（DST）である．

The dome of the Flare Monitoring Telescope (FMT). The main body of the telescope is visible through the opened slit. The dome is 3 m in diameter. The tower-like building seen in background is the DST (Domeless Solar Telescope).

ドーム内部でフレア監視望遠鏡の本体を側面から見たところ．各望遠鏡の口径は 64 mm，焦点距離は 1,920 mm である．

A side view of the main body of the FMT inside the dome. Each telescope has an aperture of 64 mm and a focal length of 1,920 mm.

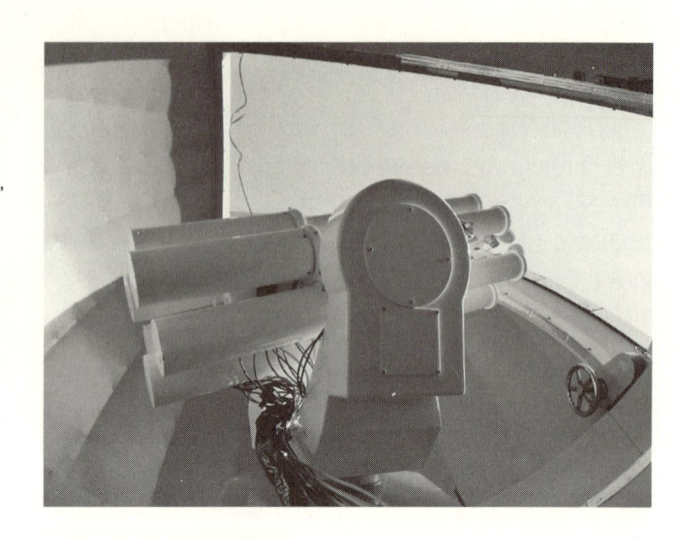

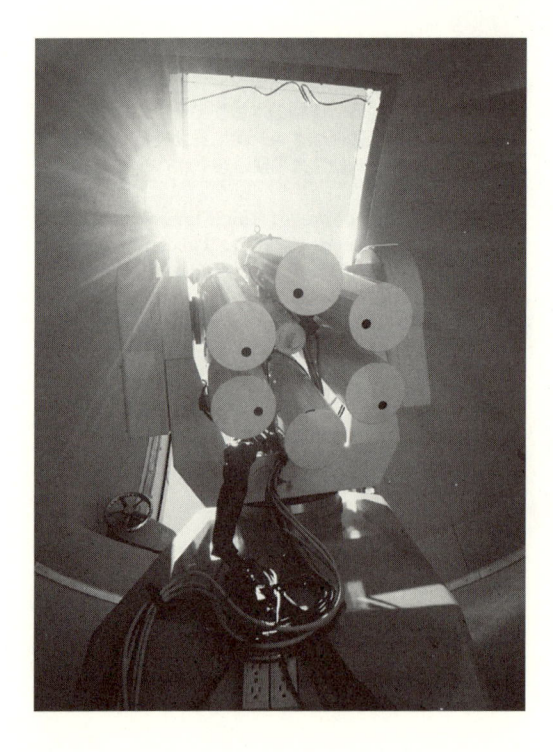

ドーム内部でフレア監視望遠鏡の本体を後方から見たところ．1つの架台に6本の望遠鏡が搭載されているのがわかる．うち1本は太陽追尾用で，残りの5本の望遠鏡は各々異なるモード（Hα 線中心，Hα ± 0.8Å，連続光，プロミネンス）で太陽を観測するためのものである．

A rear view of the FMT inside the dome. As can be seen in the photograph, the six telescopes are mounted on a single equatorial drive. One of the telescopes is used for the photoelectrical tracking of the Sun, while the other five are designed for observing the Sun in different modes (Hα center, Hα ± 0.8Å, continuum, and prominence).

Chapter 2　Flare Monitoring Telescope | 39

フレア監視望遠鏡の観測用コンピュータシステム．5本の観測用望遠鏡で撮影された太陽全面像が5つのモニターにそれぞれ表示されている．

The FMT observation control system. The five monitors show the full-disk solar images observed with the five telescopes, respectively.

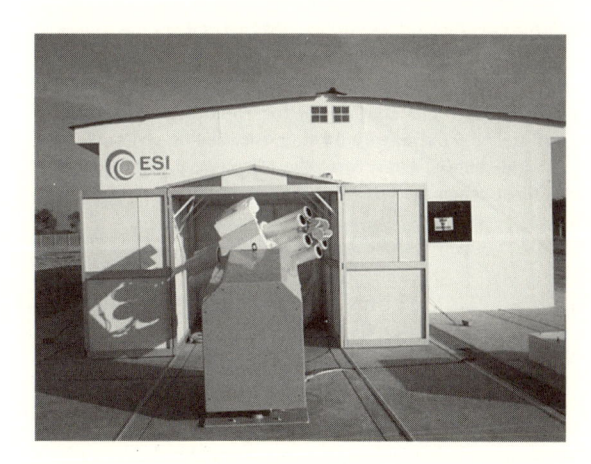

ペルーのイカ大学に設置されたフレア監視望遠鏡．奥に見える小さな小屋は2本のレールに沿って移動できるスライド式の格納庫である．口絵参照．

The FMT after its relocation to Ica Univ., Peru. The small cabin seen just behind the telescope is a sliding hangar that can move along the two rails. Refer to the frontspiece.

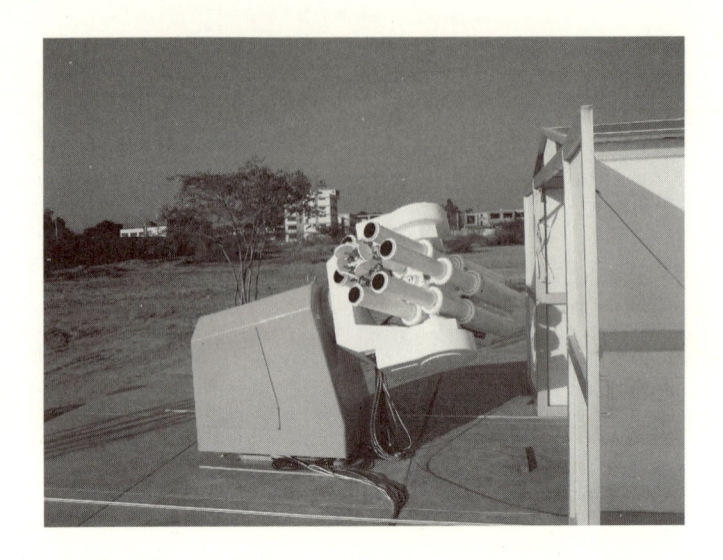

望遠鏡と架台を側面から見たところ．向かって右側が南．極軸の仰角は約 14 度である．イカは南半球の低緯度に位置しているため，望遠鏡の姿勢や極軸の向きが日本に設置されていたころ頃とはかなり違っている．

A side view of the telescope with a new mount. The south is to the right side of the photo. The elevation angle of the polar axis is approximately 14 degrees. Ica is situated at a relatively low latitude in the southern hemisphere. Therefore, the telescope position and the axis direction are quite different from those when the telescope was in Japan.

なお，現在イカ大学では同太陽ステーション内に，高さ 1 m の望遠鏡設置用コンクリート土台や制御室，スライディングルーフを含む，より本格的な格納庫の建設を進めており，2010 年度中にその中に望遠鏡を設置し，より好条件で太陽を観測できるようにする予定である．

Ica University is currently constructing a new hangar at the site of its Solar Station. The hangar will have a 1-m-high concrete base for the telescope, along with a control room and a sliding roof. The telescope is scheduled to be set in this hangar by the end of the fiscal year 2010. The new, fully equipped hanger will be better suited for solar observation.

第3章
世界の太陽全面観測

Chapter 3
Solar Full-disk Hα Observations in the World

　京都大学花山・飛騨天文台ではフレア監視望遠鏡（FMT）を飛騨天文台に設置し，彩層観測のためにHα太陽全面単色像を連続観測してきた．世界の天文台でも，それぞれ特色のある望遠鏡を備えて太陽活動の全面観測を行っている．この章では，太陽全面Hα観測（3-1節），その他の波長での太陽全面観測（3-2節）の2つにわけて世界の観測状況を簡単に見てみる．

　なお，世界の太陽全面観測網についての詳細は次のURLが有益である．

Global High-Resolution Hα Network
http://www.bbso.njit.edu/Research/Halpha/
Max Millenium Program
http://mithra.physics.montana.edu/max_millennium/

　From 1992, when the Flare Monitoring Telescope (FMT) was installed at the Hida Observatory, until the present time, we have continued to carry out Hα full disk observation of the solar chromosphere. On the other hand, solar activities have also been observed through different telescopes by various other observatories in the world. In this section, we summarize the observational systems of solar observatories in the world by classifying them into two categories: (1) solar Hα full-disk observation and (2) solar full-disk observation in other wavelengths.

　You can get more information on the details of world-wide network of solar full disk observations from the following URLs.

Global High-Resolution Hα Network
http://www.bbso.njit.edu/Research/Halpha/
Max Millenium Program
http://mithra.physics.montana.edu/max_millennium/

3.1 太陽全面 Hα 観測

フレア爆発が太陽で発生するとHα線の明るさがわずか数十秒の間に普段の10倍にもなる．それに伴って彩層のガスが噴出する．このガスの噴出もHα線での単色像観測で捉えることができる．また，200万度のコロナ中に浮かんでいる1万度のプロミネンスの複雑な運動やその爆発現象もHα線での観測で詳しく見ることができる．これらのことから，太陽活動現象はHα線の単色像観測を行なうのが，従来から基本的な研究方法であったし，現在もその重要性は変わらない．表3.1は，太陽活動Hα線単色像観測を行っている世界の地上太陽観測所の主なものの観測の特徴をまとめたものである．

これら諸外国のHα線の太陽全面観測に比べて，飛騨天文台のFMTで観測されてきたものは(1) 2秒に1枚という高い時間分解能，(2) 彩層全面をHα線の中心および±0.8Åの3波長同時観測という大きな特徴をもつ．この高い時間分解能があるため，FMTでの観測は急激に変化する活動現象の詳しい時間変化を捉えることができるという大きな利点がある．また，Hα線の中心，短波長側および長波長側の3波長同時観測は，爆発現象が発生したときの彩層ガスおよびフィラメントガスの運動状態（速度，向き）を正確に捉えることができるものである．ガスの三次元的な動きの様子を導くことができるこのデータは，太陽活動の振る舞いを研究するために貴重なものである．この特徴は世界でも他に例を見ないものとなっている．

3.1 Solar Hα Full-disk Observation

When flare explosions occur on the Sun, there are sudden increases in the brightness of the Hα line, up to 10 times the initial value in 10 s. Sometimes a tremendous quantity of chromospheric gases are ejected out from the flare sites. Complex motions of ejected gases can be observed with monochromatic imaging observation in Hα. Prominence activations and filament eruptions can also be studied closely in Hα observations. Thus, monochromatic imaging observation in the Hα line has been the most useful method of observing solar activities. In Table 3.1, a summary of the representative institutes where Hα monochromatic imaging observations are carried out is provided.

In comparison with the other observatories, Hida observatory, which uses the FMT, has two distinct features: (1) it has a high time resolution of 1 frame/2 s and (2) it enables simultaneous observation in three wavelengths (Hα line center and Hα±0.8Å). The first feature enables the FMT to detect very rapid temporal variations in active phenomena. The second feature enables us to determine the line-of-sight velocities of moving gases. By combining the line-of-site velocity with the transversal velocities derived from the temporal movement of the absorbing gases, we can obtain the three-dimensional vector velocities of the ejected gases. These merits of the FMT observation are distinctive ones among the world-wide network of full disk Hα observations.

表 3.1　世界の主な太陽全面 Hα 観測所.

天文台名	国名	望遠鏡	単色像装置	観測波長	経度	緯度	時間分解能	備考
京都大学飛騨天文台 FMT	日本	6.4cm 屈折望遠鏡×6	ファブリペロー	Hα 中心, Hα+0.8Å, Hα−0.8Å, 連続光, プロミネンス	東経137度	北緯36度	1枚/2秒	6連速（うち1本は太陽追尾用). 2010年3月 ペルーへ移設.
http://www.kwasan.kyoto-u.ac.jp/general/facilities/fmt/index.html								
京都大学飛騨天文台 SMART	日本	20cm 屈折望遠鏡×2 (T1, T2) 25cm 屈折望遠鏡×2 (T3, T4)	T1-T3: リオフィルター T4: ファブリペロー	T1: Hα中心, Hα+0.5Å, Hα−0.5Å, Hα+0.8Å, Hα−0.8Å T2: 6302.50Å T3: Hα中心 T4: 6302.50Å / Hα中心	東経137度	北緯36度	1枚/2秒	T1: Hα 全体像望遠鏡 T2: 磁場望遠鏡 T3: Hα 部分像望遠鏡 T4: 多目的望遠鏡
http://www.kwasan.kyoto-u.ac.jp/general/facilities/smart/index.html								
ビッグ・ベア太陽天文台	アメリカ	15cm 屈折望遠鏡	リオフィルター	Hα中心	西経117度	北緯34度	1枚/1分	
http://www.bbso.njit.edu/								
カンツェルヘーエ太陽天文台	オーストリア	10cm 屈折望遠鏡	リオフィルター	Hα中心	東経14度	北緯46度	1枚/1分	
http://www.kso.ac.at/index_en.php								
カターニア天文台	イタリア	15cm 屈折望遠鏡	リオフィルター	Hα中心	東経15度	北緯38度	1枚/5分	
http://web.ct.astro.it/index_en.html								
ムードン天文台	フランス	25cm シーロスタット	分光単色太陽儀	Hα中心	東経2度	北緯49度	1枚/1分	
http://bass2000.obspm.fr/instru_guide.php								
ピック・デュ・ミディ天文台	フランス	9cm 屈折望遠鏡	コロナド0.6Åフィルター	Hα中心	東経0度	北緯43度	1枚/(9秒〜1分)	CLIMSO
http://www.climso.fr/en/pages/project/CLIMSO-WhatWhoHowHowmuch.html								
ホワイロウ太陽観測所	中国	14cm 屈折望遠鏡	リオフィルター	Hα中心	東経117度	北緯40度	1枚/10分	
http://sun10.bao.ac.cn/smct/intro_smct2_e.html								
ユンナン天文台	中国	18cm 屈折望遠鏡	リオフィルター	Hα中心	東経103度	北緯25度	1枚/(15秒〜1分)	
http://www.ynao.ac.cn/en/index.html								
マウナロア太陽天文台	アメリカ	6cm 屈折望遠鏡	コロナド0.7Åフィルター	Hα中心	西経155度	北緯19度	1枚/3分	
http://mlso.hao.ucar.edu/mlso_about.html								
国立天文台（三鷹）	日本	4cm 屈折望遠鏡	リオフィルター	Hα中心	東経139度	北緯35度	1枚/1分	
http://solarwww.mtk.nao.ac.jp/jp/monochro.html								
情報通信研究機構・平磯太陽観測センター	日本	15cm 屈折望遠鏡	リオフィルター	Hα中心 連続光	東経140度	北緯36度	1枚/10分 1枚/20分	Hα+0.8Å, −0.8Å で部分像も観測（1セット/2分）
http://sunbase.nict.go.jp/solar/system/halpha/SAISO.pdf								

Table 3.1: Representative institutes in which Hα imaging observations are being carried out.

Observatory	Country	Telescope	Monochromatic device	Wavelength	Longitude	Latitude	Time Cadence	Remark
Kyoto Univ. Hida Observatory, FMT	Japan	6.4cm Refractor × 6	Fabry-Perot	Hα line center, Hα+0.8Å, Hα−0.8Å, Continuum, and Prominence	E137	N36	2 sec	Consists of 6 telescopes (including the one for tracking the Sun). Moved to Peru in March 2010.
			http://www.kwasan.kyoto-u.ac.jp/general/facilities/fmt/index.html					
Kyoto Univ. Hida Observatory, SMART	Japan	20cm Refractor × 2 (T1, T2) 25cm Refractor × 2 (T3, T4)	T1-T3: Lyot Filter T4: Fabry-Perot	T1: Hα line center, Hα+0.5Å, Hα−0.5Å, Hα+0.8Å, Hα−0.8Å T2: 6302.50Å T3: Hα line center T4: 6302.50Å/Hα line center	E137	N36	2 sec	T1: For Hα full-disk observation T2: For vector magnetic field measurement T3: For Hα partial frame observation T4: Multipurpose telescope
			http://www.kwasan.kyoto-u.ac.jp/general/facilities/smart/index.html					
Big Bear Solar Observatory	USA	15cm Refractor	Lyot Filter	Hα line center	W117	N34	1 min	
	http://www.bbso.njit.edu/							
Kanzelhoehe Solar Observatory	Austria	10cm Refractor	Lyot Filter	Hα line center	E14	N34	1 min	
	http://www.kso.ac.at/index_en.php							
Catania Astrophysical Observatory	Italy	15cm Refractor	Lyot Filter	Hα line center	E15	N38	5 min	
	http://web.ct.astro.it/index_en.html							
Meudon Observatory	France	25cm Coelostat	Spectroheliograph	Hα line center	E2	N49	1 min	
	http://bass2000.obspm.fr/instru_guide.php							
Pic-du-Midi Observatory	France	9cm Refractor	Coronado 0.6Å filter	Hα line center	E0	N43		CLIMSO
	http://www.climso.fr/en/pages/project/CLIMSO-WhatWhoHowHowmuch.html							
Huairou Solar Station	China	14cm Refractor	Lyot Filter	Hα line center	E117	N40	10 min	
	http://sun10.bao.ac.cn/smct/intro_smct2_e.html							
Yunnan Observatory	China	18cm Refractor	Lyot Filter	Hα line center	E103	N25		
	http://www.ynao.ac.cn/en/index.html							
Mauna-Loa Solar Observatory	USA	6cm Refractor	Coronado 0.7Å filter	Hα line center	W155	N19	3 min	
	http://mlso.hao.ucar.edu/mlso_about.html							
National Astronomical Observatory of Japan	Japan	4cm Refractor	Lyot Filter	Hα line center	E139	N35	1 min	
	http://solarwww.mtk.nao.ac.jp/jp/monochro.html							
National Institute of Information and Communications Technology (NICT), Hiraiso Solar Observatory	Japan	15cm Refractor	Lyot Filter	Hα line center Continuum	E140	N36	10min 20min	Partial disk observation in Hα+0.8Å and −0.8Å (1 set per 2 min)
	http://sunbase.nict.go.jp/home.html							

3.2 その他の波長での太陽全面パトロール（Hα線以外）

太陽大気の中でも外層の遷移層やコロナは，太陽表面に比べ希薄で高温（10万度〜1000万度）となっており，磁場が支配的な物理状況の下，太陽表面の活動現象が盛んに生じている現場である．一方で，そこから放射される紫外線（特に，極端紫外線）やX線が地球大気の吸収により地表面には届かないため，人工衛星による観測が必須となっている．太陽観測に用いられている主な衛星とその観測装置を表3.2に示す．

太陽コロナから放射される軟X線は，熱的な制動放射（電子がイオンに衝突して減速するときに出る放射）であり，コロナ中のエネルギー解放現象に伴って生じた高温プラズマから強く放射される．1991年に打ち上げられ2001年に観測を終了した日本の太陽観測衛星「ようこう」には，軟X線望遠鏡（SXT）が搭載され，ほぼ1太陽活動周期にわたって，コロナの全面画像が定常的に観測された．そしてSXTは，太陽コロナが実にダイナミックであり，規模（サイズ）においても形状においても様々な観測現象が絶え間なく起きていることを示した．その観測結果から，太陽フレアによるエネルギー解放現象が磁気リコネクションモデルで矛盾なく説明できることが，強く裏付けられた．また活動領域以外でも，極域でのジェット現象やコロナホールの構造など，宇宙天気研究に大きな功績を残し，長期観測という点では，コロナの活動度が太陽周期活動に同期して，大きく変動することも明らかにした．また，「ようこう」

3.2 Solar Full-disk Observation in Other Wavelengths

The transition region (TR) and corona, which lie outside the chromosphere, consist of hot and rarefied plasmas of temperature $10^5 \sim 10^7$ K. The magnetic field in regions having such conditions tends to be the dominant force, owing to which, various kinds of active phenomena related to magnetic plasmas constantly occur. However, the solar radiations in X-rays and/or in extreme ultraviolets (EUVs), both of which are mainly emitted from the TR and corona, are blocked by the Earth's atmosphere, and therefore, cannot come down to the ground. The use of satellites for making solar observations is therefore important. Some of these satellites and the key instruments used by them for making solar observations are listed in Table 3.2.

In the solar corona, the generation of hot plasmas, which effectively emit the thermal bremsstrahlung radiation (a radiation that is emitted from an electron when the electron is decelerated by collision with ions), is accompanied by a release of energy. The Japanese satellite *Yohkoh*, which was launched in 1991, used an on-board Soft X-ray Telescope (SXT) to obtain the full-disk solar images in X-rays at regular intervals spanning an entire solar cycle, until the completion of its mission in 2001. The solar corona, when observed through the SXT, appears dynamic and spectacular. Observations through the SXT also show that many active phenomena, which vary both in scale and in morphological form, constantly occur. Further, observations through the SXT have strongly supported the validity of the magnetic reconnection model in explaining the energy-release mechanism in solar flares. The observations of solar polar jets, coronal

表 3.2 Hα以外の波長を用いた太陽全面観測衛星及び観測所.

衛星名 (または天文台名)	国名	望遠鏡	観測波長 (またはエネルギー帯)	運用期間	備考
ようこう	日本, アメリカ	SXT (Soft X-ray Telescope)	軟X線	1991 ～ 2001	
		http://www.isas.jaxa.jp/home/solar/yohkoh/			
ひので	日本, アメリカ, 英国	XRT (X-Ray Telescope)	軟X線	2006 ～ 現在	
		http://solar-b.nao.ac.jp/index_e.shtml			
GOES (Geostationary Operational Environmental Satellites)	アメリカ	SXI (Solar X-ray Imager)	軟X線	2001 ～ 現在	
		http://sxi.ngdc.noaa.gov/			
SOHO (Solar and Heliospheric Observatory)	アメリカ, 欧州	EIT (Extreme ultraviolet Imaging Telescope)	17.1, 19.5, 28.4 および 30.4 nm	1995 ～ 現在	
		http://umbra.nascom.nasa.gov/eit/			
STEREO (Solar TErrestrial RElations Observatory)	アメリカ	SECCHI (Sun Earth Connection Coronal and Heliospheric Investigation) /EUVI (Extreme UltraViolet Imager)	17.1, 19.5, 28.4 および 30.4 nm	2006 ～ 現在	
		http://secchi.lmsal.com/EUVI/			
RHESSI (Reuven Ramaty High Energy Solar Spectroscopic Imager)	アメリカ	衛星名と同じ	3 keV ～ 20 MeV	2002 ～ 現在	
		http://hesperia.gsfc.nasa.gov/hessi/			
SDO (Solar Dynamic Observatory)	アメリカ	AIA (Atmospheric Imaging Assembly)	9.4, 13.1, 17.1, 19.3, 21.1, 30.4, 33.5, 160.0, 170.0 nm および 白色光	2010 ～ 現在	
		http://aia.lmsal.com/			
TRACE (Transition Region and Coronal Explorer)	アメリカ	衛星名と同じ	17.1, 19.5, 28.4, 121.6, 155.0, 160.0 nm および 白色光	1998 ～ 2010	部分画像観測 (最大で 8.5× 8.5 分角 = 約 510×510 秒角)
		http://trace.lmsal.com/			
CORONAS- Photon	ロシア	TESIS (Solar telescope/imaging spectrometer) MISH (MgXII Imaging Spectroheliometer) FET (Full-disk EUV Telescopes) SEC (Solar EUV Coronograph)	X線 (0.842 nm) 極端紫外線 (13.4, 30.4 nm) 極端紫外線コロナグラフ (30.4 nm)	2009 ～ 現在	
		http://iaf.mephi.ru/coronos-photon_main_e.htm			
マウナロア 太陽天文台	アメリカ	Mark-IV K-coronameter	可視連続光 (700 ～ 900 nm) のストークス I,Q,U 強度	1998 ～ 現在	
		http://mlso.hao.ucar.edu/mlso_about.html			
野辺山 太陽電波観測所	日本	野辺山電波ヘリオグラフ	17 GHz および 34 GHz	1992 ～ 現在	
		http://solar.nro.nao.ac.jp/norh/html/introduction.html			

Table 3.2 Satellites and representative instruments in which solar full-disk observation is carried out in other wavelengths.

Satellite (or Observatory)	Country	Telescope	Wavelength	Operation period	Remark
Yohkoh	Japan, USA	SXT (Soft X-ray Telescope)	Soft X-ray	1991 - 2001	
	http://www.isas.jaxa.jp/home/solar/yohkoh/				
Hinode	Japan, USA, UK	XRT (X-Ray Telescope)	Soft X-ray	2006 - current	
	http://solar-b.nao.ac.jp/index_e.shtml				
GOES (Geostationary Operational Environmental Satellites)	USA	SXI (Solar X-ray Imager)	Soft X-ray	2001 - current	
	http://sxi.ngdc.noaa.gov/				
SOHO (Solar and Heliospheric Observatory)	USA, Europe	EIT (Extreme ultraviolet Imaging Telescope)	17.1, 19.5, 28.4 and 30.4 nm	1995 - current	
	http://umbra.nascom.nasa.gov/eit/				
STEREO (Solar TErrestrial RElations Observatory)	USA	SECCHI (Sun Earth Connection Coronal and Heliospheric Investigation)/ EUVI (Extreme UltraViolet Imager)	17.1, 19.5, 28.4 and 30.4 nm	2006 - current	
	http://secchi.lmsal.com/EUVI/				
RHESSI (Reuven Ramaty High Energy Solar Spectroscopic Imager)	USA	HESSI (High Energy Solar Spectroscopic Imager)	3 keV – 20 MeV	2002 - current	
	http://hesperia.gsfc.nasa.gov/hessi/				
SDO (Solar Dynamic Observatory)	USA	AIA (Atmospheric Imaging Assembly)	9.4, 13.1, 17.1, 19.3, 21.1, 30.4, 33.5, 160.0, 170.0 nm and white light	2010 - current	
	http://aia.lmsal.com/				
TRACE (Transition Region and Coronal Explorer)	USA	TRACE	17.1, 19.5, 28.4, 121.6, 155.0, 160.0 nm and white light	1998 - 2010	Partial disk observation (max 8.5×8.5 arcmin = approx. 510×510 arcsec)
	http://trace.lmsal.com/				
CORONAS-Photon	Russia	TESIS (Solar telescope/imaging spectrometer) MISH (MgXII Imaging Spectroheliometer) FET (Full-disk EUV Telescopes) SEC (Solar EUV Coronograph)	X-Ray (0.842 nm) EUV (13.4, 30.4 nm) EUV coronagraph (30.4 nm)	2009 - current	
	http://iaf.mephi.ru/coronos-photon_main_e.htm				
Mauna-Loa Solar Observatory	USA	Mark-IV K-coronameter	Stokes intensities I, Q and U in visible continuum (700 ~ 900 nm)	1998 - current	
	http://mlso.hao.ucar.edu/mlso_about.html				
Nobeyama Solar Radio Observatory	Japan	Nobeyama Radioheliograph	17 GHz and 34 GHz	1992 - current	
	http://solar.nro.nao.ac.jp/norh/html/introduction.html				

衛星の長期間にわたる軟X線での観測は，この波長における太陽全面パトロールの重要性を強く印象付けるものとなった．現在でも，2006年に打ち上げられた，「ひので」衛星搭載のX線望遠鏡（XRT）では，より高空間分解能で，低温プラズマへの感度ももった太陽全面観測として引き継がれる一方，パトロールの観点からは，GOES衛星にSXI（Solar X-ray Imager）が搭載され，常時軟X線での太陽の全面像が取得されている．

極端紫外線域での太陽全面観測も，太陽遷移層・太陽コロナの活動現象を知る上で，重要である．極端紫外線域には数多くのイオン輝線があり，狭帯域フィルターを用いてそれらの各々を取り出すことで，異なる温度域での太陽大気の様子を知ることが可能である．極端紫外線域での太陽全面パトロール観測で代表的なものは，1995年に打ち上げられたSOHO（Solar and Heliospheric Observatory）衛星搭載のEIT（Extreme ultraviolet Imaging Telescope）である．比較的高い空間分解能の画像が実現されたことで，活動領域の磁気ループの構造を詳細に調べることが可能となった．また，「ようこう」衛星に比べ高い時間分解能データが得られることや衛星軌道が太陽－地球系での力学的平衡点（ラグランジュ点L1）を周回するため日陰（夜）がないという利点から，太陽フレアの際に現れる波動現象（もしくはそれに付随する現象）やコロナ減光（ディミング）など，数々の新しい見地を我々に与えて続けている．2006年に打ち上げられたSTEREO衛星にも同様の望遠鏡SECCHI/EUVIが搭載され，また2010年2月に打ち上げられ

holes, ejections, etc. using the SXT has made great contributions to space weather studies. In addition, long-term observation through the SXT has shown that the solar coronal activity varies according to the solar cycle, thereby promoting the importance of solar full-disk observations in X-rays. The Japanese satellite *Hinode*, which is the successor to *Yohkoh*, carries an on-board X-Ray Telescope (XRT) that enables us to obtain X-ray images that have a high spatial resolution and are highly sensitive to low-temperature plasma. On the other hand, the Solar X-ray Imager (SXI) on board the GOES is also monitoring the Sun in X-rays.

Solar full-disk observations in EUVs are also important to examine the active phenomena in the TR and corona. In EUVs, there are many emission lines of ions. Since each line corresponds to a particular temperature, observations in each line using narrow-band filters give us information about the temperatures in different regions of the solar atmosphere. The Extreme-ultraviolet Imaging Telescope (EIT) on board the Solar and Heliospheric Observatory (SOHO), which was launched in 1995, is representative of the telescope used for solar full-disk observations in EUVs. The spatial resolution of the full-disk data is higher than of the *Yohkoh*/SXT data, and this enables us to examine in detail the magnetic loop structures in active regions and/or in other regions. Moreover, the SOHO/EIT is continuously making high-cadence observations of the Sun without nighttime (i.e., shade by the Earth itself), and it has revealed many new aspects of solar active phenomena such as the occurrence of wave (or wave-like) phenomena and coronal dimming. The Sun-Earth Connection Coronal and Heliospheric Investigation

た SDO（Solar Dynamic Observatory）衛星搭載の AIA（Atmospheric Imaging Assembly）では，高い空間分解能に加え，約 10 秒という高時間分解能での極端紫外線全面太陽像観測が行われている．

太陽表面の活動現象をパトロールする別の観測手段として，電波での太陽全面観測がある．国立天文台野辺山太陽電波観測所にある野辺山電波ヘリオグラフでは，1992 年の観測開始以降，高い稼働率で安定して太陽全面観測を続けている．野辺山電波ヘリオグラフでは，17 GHz および 34 GHz という 2 つの観測周波数帯をもち，特に太陽フレアに伴って 1 MeV 程度という高エネルギーにまで加速された相対論的電子が磁場内を旋回する際に放射する，「非熱的なジャイロシンクロトロン放射」を効率よく観測できるよう設計されている．他方で，この周波数帯はフィラメント（プロミネンス）からの熱的な放射にも感度がある．このことから，天候に左右されない電波域の特性を生かし，地上観測ながらフレアやプロミネンス噴出現象などをパトロールすることができる．また，長期間の安定した稼動から，1 太陽活動周期にわたる活動現象の活動度の変動などを調べている．

(SECCHI) developed for the Solar Terrestrial Relation Observatory (STEREO) mission, which was launched in 2006, uses a telescope — EUVI (Extreme Ultraviolet Imager) — to observe the Sun in EUVs. The SECCHI-EUVI has observed the full-disk sun stereoscopically by using two widely spaced satellites. The Atmospheric Imaging Assembly (AIA) on board the Solar Dynamic Observatory (SDO), which was launched in February 2010, is also equipped with EUV telescopes, and it is observing the full-disk sun with a high spatial and temporal resolution.

Solar full-disk observation in microwaves is another procedure to monitor active phenomena on the solar surface. The Nobeyama Radioheliograph (NoRH) at the Nobeyama Solar Radio Observatory, National Astronomical Observatory of Japan, has been continuously obtaining solar full-disk images with a high operational availability since its initiation in 1992. In particular, the NoRH can effectively observe the nonthermal gyrosynchrotron radiation at frequencies of 17 and 34 GHz; such radiation is generated by relativistic electrons that are accelerated to high energy levels of up to $\sim$ 1 MeV during a solar flare and make spiral motions around a magnetic field. On the other hand, these frequencies are also sensitive to the thermal emissions from filaments/prominences on the solar surface. Moreover, in microwaves, the solar radiation can come to the ground, unaffected by the weather conditions. Therefore, although the NoRH is a ground-based telescope, we can use it to constantly monitor active phenomena such as flares and prominence eruptions and also examine the variability of active phenomena over each solar cycle.

第 4 章
11 年間の観測のまとめ

Chapter 4
Summary of the Observations Carried Out over a Period of 11 Years

表 4.1 は 1992 年 5 月から 2003 年 4 月までの期間におけるフレア監視望遠鏡 (FMT) のおよその観測時間を月別に示したものである.

1992 年 5 月から 1996 年 9 月までについてはビデオレコードから実観測時間を集計し,また,1996 年 10 月から 2003 年 4 月までについてはデジタル画像を元にして実観測時間を求めた.これを元に,飛騨天文台における昼間の各月のおよその晴天率を計算してみると,11 年間の平均晴天率は約 25 % で,最大値は約 56 % (1994 年 8 月),最小値は約 7 % (2001 年 1 月) であった.晴天率が高い傾向にあるのは 4 月,5 月,8 月および 10 月で,いずれも 11 年間の平均値が 30 % を超えている.逆に,梅雨に当たる 6 月と冬期の 12 月〜2 月は平均晴天率が 25 % 未満と低くなっている.

イベントの検出と分類は,ビデオ再生画像を目視することで行った.Hα−0.8

Table 4.1 shows the approximate FMT observation time for each month during the period from May 1992 through April 2003.

The actual observation times from May 1992 through September 1996 were obtained by visually reviewing the video records, and the observation times from October 1996 through April 2003 were obtained by analyzing the digitized images. From the data in Table 4.1, we have calculated the approximate observation rates, i.e., the ratio of the observation time to the daytime, at the Hida Observatory for each month. The average observation rate over 11 years was approximately 25%, with the highest and lowest rates being 56% (August 1994) and 7% (January 2001), respectively. The observation rates in April, May, August, and October were relatively high, with the average rates in these months over the 11-year period exceeding 30%. On the other hand, the average rates in June (when the rainy season in Japan sets in), December,

表4.1 FMTの観測時間（単位：時間）.
Table 4.1 FMT observation time (unit: hours).

Year	Month												Total
	Jan	Feb	Mar	Apr	May	Jun	Jul	Aug	Sep	Oct	Nov	Dev	
1992					91	99	122	104	93	82	112	45	748
1993	29	40	127	111	178	111	46	86	82	115	96	62	1083
1994	68	50	102	164	162	133	154	212	116	115	102	50	1428
1995	46	64	91	107	120	65	92	161	90	133	61	64	1094
1996	24	59	73	109	105	62	139	122	91	107	34	66	991
1997	41	70	107	122	97	92	77	123	79	127	94	67	1096
1998	23	86	93	82	114	77	68	56	75	64	74	66	878
1999	36	50	84	106	137	67	73	122	57	81	76	57	946
2000	45	27	68	50	108	91	119	129	107	92	75	68	979
2001	17	57	76	152	133	51	145	118	90	98	93	33	1063
2002	20	65	91	108	109	116	96	144	97	119	55	30	1050
2003	42	73	75	103									293
Avg.	36	58	90	110	123	88	103	125	89	103	79	55	11649

Åの画像から，差しわたし約1万km以上の大きさで，静穏領域の明るさに比して増光／減光が認められるものを取り上げ，その面積，形態を基準にして分類した．大気擾乱の程度にもよるが，最大輝度が静穏領域の明るさの±10％以上の増光／減光を示した現象で寿命が2分以上の現象は，ほぼ網羅している．イベントの分類の概略は表4.2のとおりである．（詳細は第5章参照）．

January, and February (when Japan is in the winter season) are below 25%.

The detection and classification of the events were made by replaying the videos and visually analyzing the recorded images. Specifically, the images in Hα−0.8Å were analyzed for the presence of any events that had a length of 10,000 km or more and whose brightness or darkness was in high contrast with the brightness of the quiescent region. The events thus found were then classified according to their area and form. Although it depends on the degree of atmospheric disturbance, we have found almost all the phenomena that had a brightness/darkness of ±10% or more when compared to the brightness of the quiescent region and lasted for two minutes or longer. An outline of the classification of events is shown in Table 4.2.

We detected more than 24,000 events

表 4.2 イベントの分類（概略）.

ディスクイベント				
		形態／変化	分類略号	補足説明
暗い模様のイベント	新たに出現したもの	点状	IA	短命活動領域や視線方向に噴出したサージ等
		線状	IB	ジェット状サージ
	既存のものの活動	部分的な形態変化	IIC	フィラメントの部分消失や内部運動の発生
		構造の大きな動的変化	IID	大規模フィラメント噴出
明るく輝くイベント		点状	E	輝点フレア
		リボン状	F	ツーリボンフレア
リムイベント				
		形態	分類略号	補足説明
プロミネンス活動		サージ型	A	ジェット状伸縮
		噴出型	B	加速を受けて噴出
		崩壊型	C	徐々に崩壊消滅
		部分的変化	D	部分的増光あるいは崩壊
		ループ型	E	ポストフレアループ等

上記の 11 年間で 2 万 4,000 件を超えるイベントが検出されている．本書に添付された DVD にはその全イベントの情報（観測日時，イベントの種別，規模など）を独自の書式で収録している．

以下では，イベントの種別ごとに，適宜代表例を紹介しながら，FMT による 11 年の観測成果を概説する．

4.1 フレア

フレアは太陽面爆発とも呼ばれる急激な増光現象である．フレアは黒点付近で発生し，そのエネルギー源は黒点近傍の太陽大気中に蓄えられた磁気エネルギーである．大きなフレアになると，10 万 km 平方の巨大な空間で発生し，数時間

within 11 years. The observational data of these events (such as the observation date, type, and size of each event) have been compiled in original formats (see Chapter 5), which are included in the DVD provided along with this book.

The following sections summarize the results of the observations that were made using the FMT over the 11-year period, showing representative examples for each class of events.

4.1 Flare

The solar flare, or simply, a flare, is a phenomenon in which there is a sudden brightening in a specific region on the Sun's surface. Flares occur near sunspots and are activated by the magnetic energy stored in the solar atmosphere near the sunspots. A flare can grow to be as large as $100,000 \times 100,000$ km^2 and can last for

Table 4.2 Event classification (outline).

Disk events					
			Form/Change	Code	Supplementary description
Dark pattern events	New emergence		Point	IA	Short-lived active region, line-of-sight eruption of surge, etc.
			Line	IB	Jet-like surge
	Activation of existing feature		Partial change in form	IIC	Partial disappearance or internal movement of the filament
			Large, dynamic change in structure	IID	Large-scale filament eruption
Brightening event			Point	E	Bright-point flare
			Ribbon	F	Two-ribbon flare
Limb events					
			Form	Code	Supplementary description
Prominence activities			Surge	A	Jet-like expansion and contraction
			Eruption	B	Accelerated eruption
			Disruption	C	Gradual disruption and disappearance
			Partial change	D	Partial brightening or disruption
			Loop	E	Post-flare loop etc.

から10時間も続く．巨大フレアは，しばしば2つのペアになった細長いリボン状を呈する．このようなフレアはツーリボンフレア（two ribbon flare）と呼ばれる．ペアになった発光領域は磁極のN極とS極に対応している．リボンが3本以上出現することもある．

代表例として，FMTで2001年4月10日に観測されたフレアを図4.1に示す．GOESクラスはX2.3である．これはツーリボンフレアの時間的発展を示す典型的な例であり，これまでいくつも論文で研究対象に取り上げられている（詳しくは第6章の"Apr 10, 2001 03:30 – 06:21 UT Flare, Filament Eruption"のコメントを参照）．

up to 1-10 h. Large-scale flares often look like a pair of long, thin ribbons. Such flares are called two-ribbon flares. The paired light-emitting regions respectively correspond to the north and south magnetic polarities. The occurrence of flares having more than two ribbons is also possible.

As a representative example, a flare that was observed with the FMT on April 10, 2001, is shown in Fig. 4.1. Its GOES class was X2.3. This is a typical example that demonstrates the temporal development of a two-ribbon flare, and it has been the subject of study in many scientific papers. For more information, refer to the comment on "Apr 10, 2001, 03:30- 06:21 UT Flare, Filament Eruption" in Chapter 6.

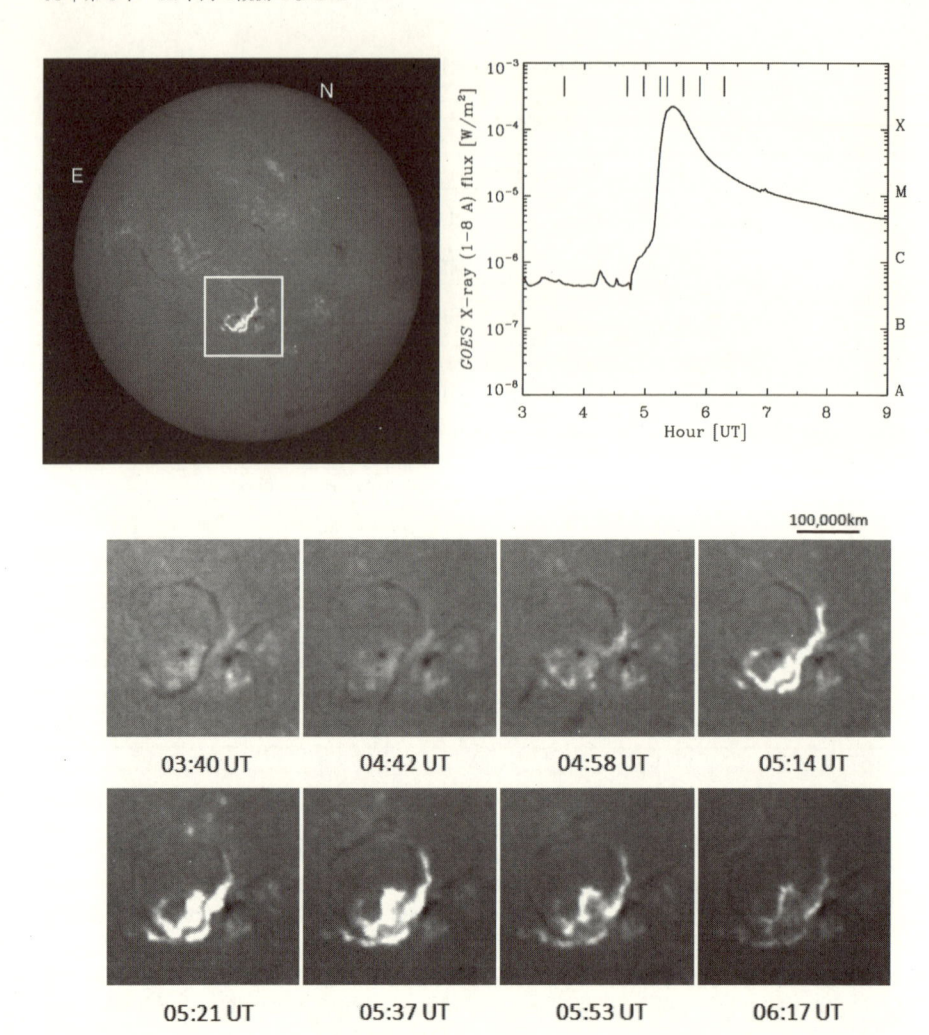

54 | 第4章 11年間の観測のまとめ

03:40 UT **04:42 UT** **04:58 UT** **05:14 UT**

05:21 UT **05:37 UT** **05:53 UT** **06:17 UT**

図4.1 2001年4月10日に観測されたツーリボンフレア．上段左：太陽全面Hα線中心像，上段右：GOES X-ray flux，下段：フレアの時間発展を示す時系列画像（Hα線中心像）．口絵参照．

Fig. 4.1 A two-ribbon flare observed on April 10, 2001. Upper-left panel: solar full-disk image in Hα center; upper right panel: GOES X-ray flux; lower panels: sequential images showing the temporal development of the flare in Hα center. Refer to the frontspiece.

　FMT画像でのフレアの検出は以下のようにした．Hα−0.8Åで増光が認められたもので，差しわたしの大きさが1万km以上で，寿命が2分以上であり，最

In identifying a flare from the FMT images, we classified a flare as that phenomenon whose brightness is distinguishable in Hα−0.8Å and exceeds that of the quiescent region by 10%,

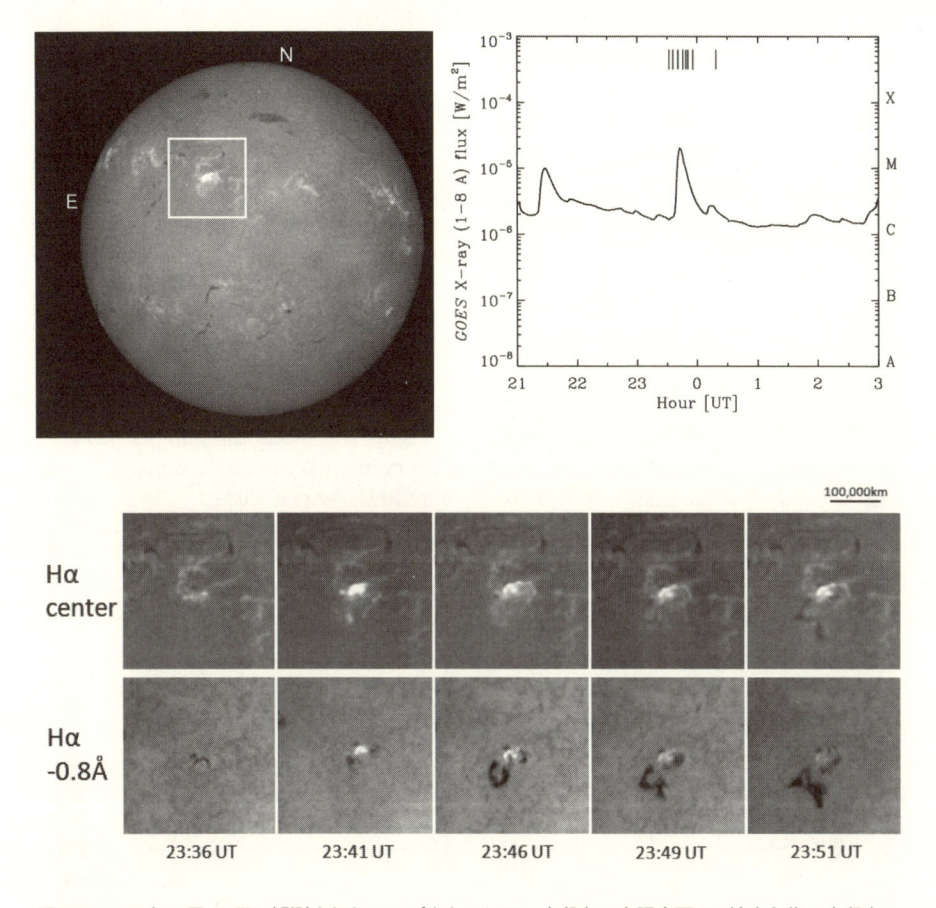

図 4.2　2000 年 6 月 15 日に観測されたコンパクトフレア．上段左：太陽全面 Hα 線中心像，上段右：GOES X-ray flux，下段：フレアの時間発展を示す時系列画像（Hα 線中心像と Hα − 0.8 Å 像）．

Fig. 4.2 A compact flare observed on June 15, 2000. Upper-left panel: Ssolar full-disk image in Hα center; upper-right panel: GOES X-ray flux; lower panels: sequential images showing a temporal change of the flare in Hα center and Hα −0.8Å.

大輝度が静穏領域の 10％を超えるものをフレアとした．このうち，明るい領域の差しわたし長さが約 2 万 km 以下のものは輝点であると分類した（表 4.2 の分類略号 E）．また，明るい領域がそれ以上のものは，フレアであるとした（分類略号 F）．

whose maximum size is greater than 10,000 km, and which lasts for more than 2 min. Furthermore, any flare having an area of brightness of approximately 20,000 km or less was classified as a bright point (classification code: E). A flare having an area of brightness greater than 20,000 km was classified as a normal flare (classification code: F).

もう1つ，コンパクトフレアと呼ばれる比較的小規模なフレアの例を図4.2に示す．このフレアは2000年6月15日に観測されたもので，GOESクラスはM2.0である．このフレアにはモートン波（4.5節）が伴っていたという報告もある（詳しくは第6章の"Jun 15, 2000 23:33-00:20 UT Flare, Filament Eruption"のコメントを参照）．FMTで観測されたモートン波は，このように，面積は大きくないが非常にインパルシブなフレアに伴って生じたものが多い．

図4.3のグラフは，1992年5月から2003年4月までの各年にFMTで観測されたフレアの数を示したものである．

As another example, a relatively small type of flare (called a compact flare) is shown in Fig. 4.2. This flare was observed on June 15, 2000, with a GOES class of M2.0. Some reports pointed out that this flare was accompanied by Moreton waves (to be described later). For more information, refer to the comment of "Jun 15, 2000, 23:33-00:20 UT Flare, Filament Eruption" in Chapter 6. The Moreton waves that have been detected using the FMT are often associated with such small, yet very impulsive, flares.

Figure 4.3 shows the number of flares observed with the FMT each year from May 1992 to April 2003.

This graph demonstrates that the number of flares decreased to the minimum level in 1996 and then

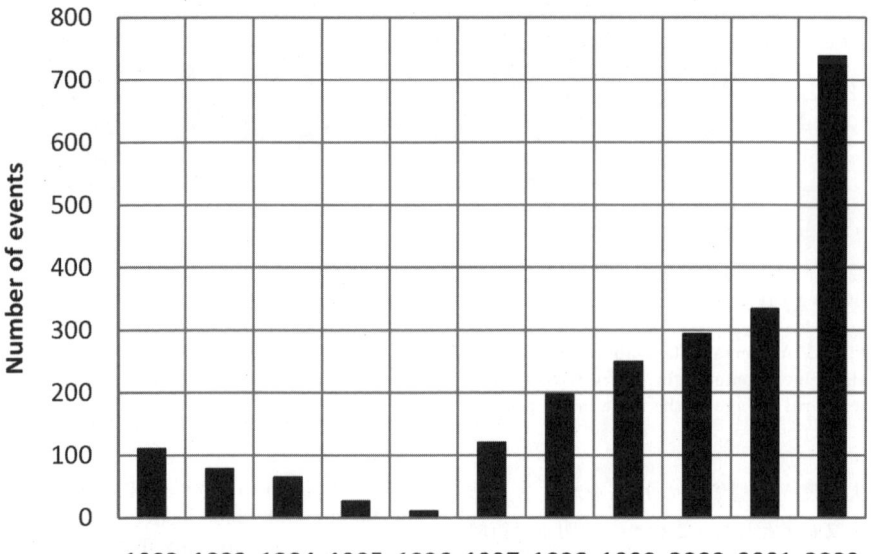

図4.3 FMTで観測されたフレアの数．各年は5月から翌年4月までを1年として表記してある．例えば「1992」は1992年5月から1993年4月までの1年間を示す．

Fig. 4.3 Number of flares observed with the FMT. It should be noted that each year in this context begins with May and ends with April, the following year. For example, 1992 corresponds to the period from May 1992 through April 1993.

このグラフを見ると，フレアの数は1996年に最も少なくなった後，2002年まで年を追うごとに増加している．フレアの発生数は太陽活動とともに増減する．すでに述べたように，太陽活動は約11年周期で変動するといわれており，1995〜1996年は太陽活動の極小期に当たっていた．

4.2 サージ

サージは，ガスが細長く絞られた形でジェット状に噴出する現象である．その速度は数10 km/sから200 km/s，長さは数万kmから10万km程度である．噴出したガスは上昇時と同じ経路をたどって噴出点に落下することが多い．

代表例として，FMTで1999年8月25日に観測されたサージの時間発展を図4.4に示す．GOESクラスはM3.6である．この例は寿命の短いコンパクトフレアにともなって複数のサージが噴出したものである．

この図では特にHα+0.8Åの観測像とHα-0.8Åの観測像の違いに注目したい．まず，Hα-0.8Å像を見ると，画像の中央付近に黒い構造物（サージ）が出現し（01:35 UT），見かけの長さ約10万kmにまで発達している（01:45 UT）．その後，この構造物は急速に衰退し，02:13 UTには完全に消滅している．これらの画像からサージが01:45 UT頃まで上昇段階にあったことがわかる．一方，Hα+0.8Å像にはサージの下降段階の様子が見て取れる．詳しく見ると，01:45 UTまで構造物は観測されていないが，その後，2〜3本の筋状の黒い構造物が出現し（01:49-01:55 UT），その一部は

continued to increase until 2002. The number of flares varies with the activity of the Sun. As already stated, the Sun is said to have an activity cycle of approximately 11 years. The solar activity was at the minimum level during the period from 1995 to 1996.

4.2 Surge

The surge is a jet of gas erupted in a long, collimated form. The erupted gas travels at a velocity of 10-200 km/s over a distance ranging from a few tens of thousands to one hundred thousand kilometers. In many cases, the erupted gas eventually falls back to the point of eruption along the same trajectory in the descending phase as in the ascending phase.

As a representative example, the temporal development of a surge observed with the FMT on August 25, 1999, is shown in Fig. 4.4. Its GOES class was M3.6. This event involved the ejection of a few surges from a short-lived, compact flare.

One important point to note in this figure is the difference between the images observed in Hα-0.8Å and those observed in Hα+0.8Å. The images in Hα-0.8Å show the emergence of a dark structure near the center of the image (01:35 UT); these images show that the dark structure evolved to an apparent length of approximately 100,000 km (01:45 UT) and then rapidly disappeared (02:13 UT). This suggests that the ascending phase of the surge from the Sun's surface lasted until approximately 01:45 UT. On the other hand, the images in Hα+0.8Å reflect the descending phase of the surge. No dark structure was clearly observed until 01:45 UT, after

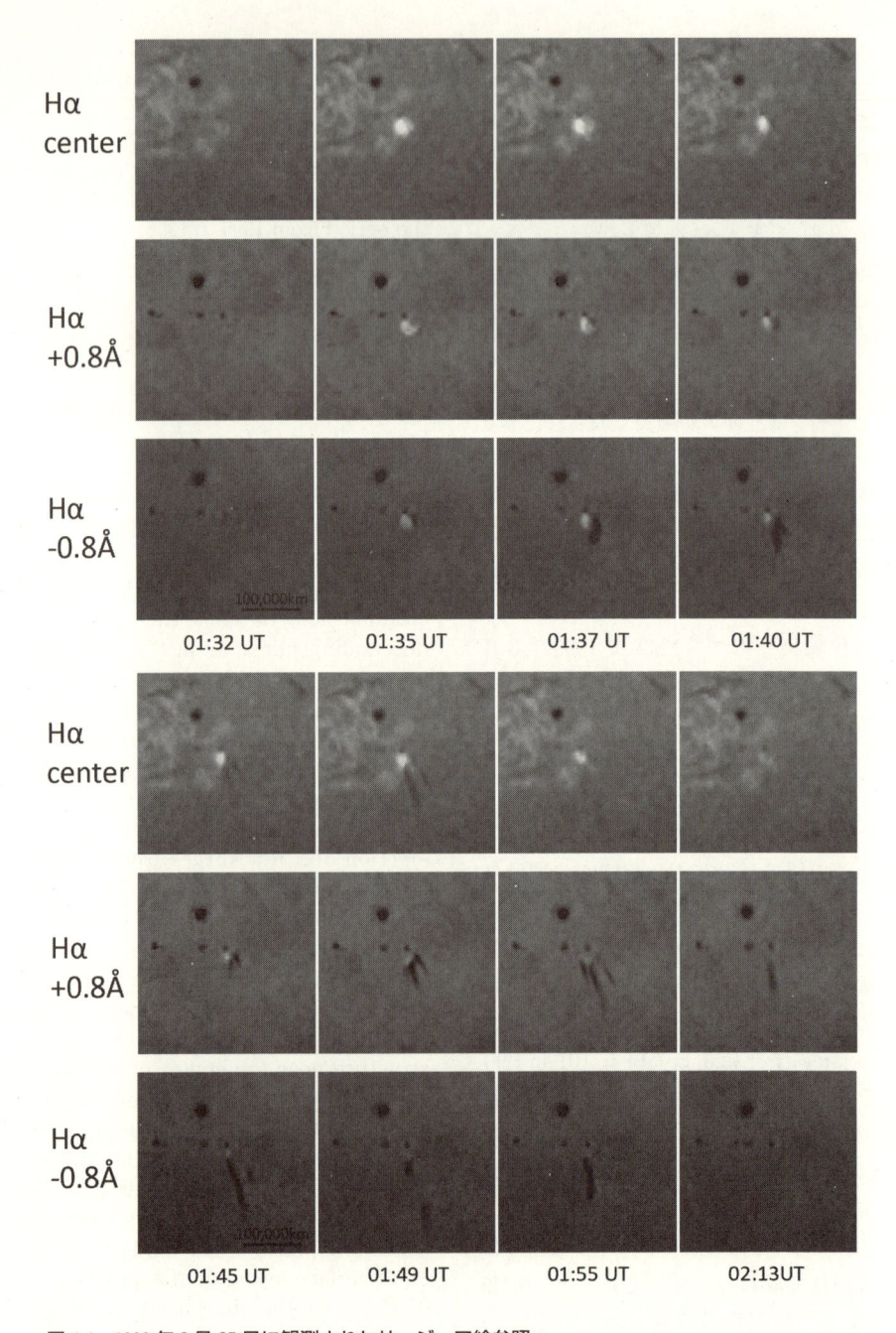

58 | 第 4 章　11 年間の観測のまとめ

図 4.4　1999 年 8 月 25 日に観測されたサージ．口絵参照．

Fig. 4.4　A surge observed through the FMT on August 25, 1999. Refer to the frontspiece.

02:13 UT の時点でもまだ観測されている.

FMT の Hα-0.8Å 画像では，サージは線状に伸縮する暗い構造として検出される（分類略号 IB）. サージが丁度観測者の向きに伸縮した場合は，暗い点状のものが現れて消えてゆくように見える. この場合は，分類略号 IA のグループに分類している.

図 4.5 のグラフは，1992 年 5 月から 2003 年 4 月までの各年に FMT で観測されたサージ（分類略号 IB のみ）の数を示したものである. このグラフと前出のフレアのグラフを比較すると，サージはフレアよりも発生件数がはるかに多いことがわかる. また，サージ発生頻度の変遷の太陽活動周期との相関は，フレアの

which a few spike-like features rapidly emerged (01:49–01:55 UT), and a portion of these features persisted even at 02:13UT.

In the FMT images in Hα-0.8Å, a surge can be detected as a dark structure that linearly expands and contracts (classification code: IB). A surge that expands or contracts in the direction of the line of sight of the observer appears to be the sudden emergence or disappearance, respectively, of a dark spot. Such a surge is classified into group IA in our classification system.

Figure 4.5 shows the number of surges (type: IB) observed with the FMT in each year from May 1992 to April 2003. Comparing this graph with the previous graph (Fig. 4.3) reveals that the occurrence of surges is far more frequent

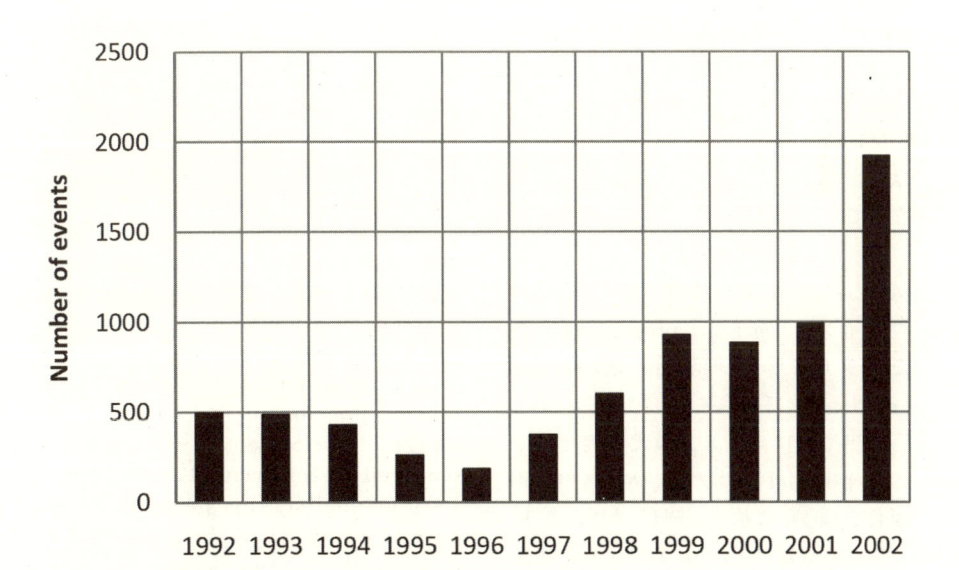

図 4.5　FMT で観測されたサージの数. 各年は 5 月から翌年 4 月までを 1 年として表記してある. 例えば「1992」は 1992 年 5 月から 1993 年 4 月までの 1 年間を示す.

Fig. 4.5 Number of surges observed with the FMT. It should be noted that each year in this context begins with May and ends with April, the following year. For example, 1992 corresponds to the period from May 1992 through April 1993.

それに比べて低くなっている．

4.3 プロミネンス

FMTの画像で太陽の縁（リム）から少し浮かんだ（あるいは突き出た）明るく見える構造がプロミネンス（紅炎）である．その正体は，200万度のコロナに浮かんだ数千から1万度程度の冷たいプラズマである．プロミネンスは磁場の力で浮いていると考えられているが，その磁場構造には謎が多い．プロミネンスはリム上では明るく見えるが，太陽面（ディスク）上にくると黒い筋模様に見える．これはダークフィラメント（dark filament）または暗条と呼ばれる．

プロミネンスは静穏型プロミネンスと活動型プロミネンスに大別できる．静穏型プロミネンス（quiescent prominence）は安定で，中には数ヶ月も太陽面上に存在しているものもある．一方，静穏型プロミネンスが突然不安定化して上昇・消滅することもある．この現象は噴出型プロミネンス（eruptive prominence）またはプロミネンス噴出（prominence eruption）と呼ばれる．

噴出型プロミネンスの代表例として，2000年11月4日のプロミネンスの時間変化を図4.6に示す．FMTイベントリストでの分類はB1（ガスが徐々に加速されていくプロミネンス噴出）である．

噴出型プロミネンスは，ヘリカルな（螺旋状の）チューブ形状（またはループ形状）を保ちつつ，膨張・上昇する．巨大なプロミネンスの場合，噴出される質量

than that of flares. The correlation of the variation in the occurrence frequency of surges with the solar-activity cycle is lower than that of the variation in the occurrence frequency of flares.

4.3 Prominence

A prominence can be identified in the FMT solar images as a structure that is slightly afloat from the edge (limb) of the Sun. A prominence is actually a cold plasma, whose temperature ranges from several thousand to ten thousand Kelvin, floating in the corona, whose temperature is two million Kelvin. The prominence is said to be supported by magnetic forces, but its magnetic structure remains a mystery. Prominences look bright when on the solar limb, but appear as dark structures when seen on the solar disk. The latter type is called a dark filament, or simply, a filament.

Prominences are categorized into quiescent prominences and active prominences. A quiescent prominence is a stable structure that is found on the Sun's surface and can last for over a few months. A quiescent prominence may suddenly become unstable, and then ascend and disappear. This phenomenon is called an eruptive prominence or a prominence eruption.

As a representative example of an eruptive prominence, Fig. 4.6 shows the temporal development of a prominence observed on November 4, 2000. In the FMT event list, this prominence is classified as B1 (i.e., a gradually accelerated eruptive prominence).

Eruptive prominences expand and ascend while maintaining the form of helical (spiral) tubes or loops. An extremely large-scale prominence can eject a huge amount of mass of the order of 100 million to one billion tons (10^{14}-

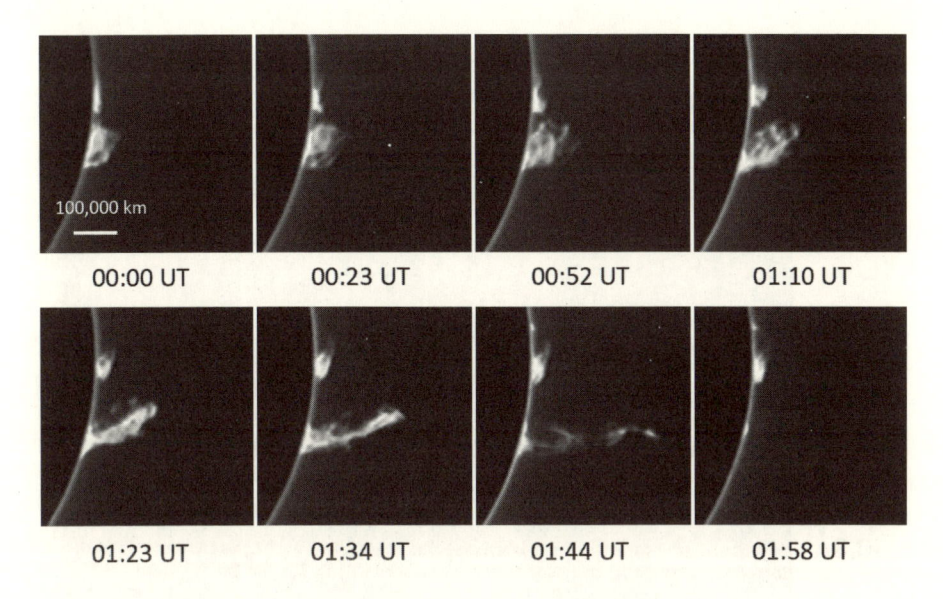

00:00 UT	00:23 UT	00:52 UT	01:10 UT
01:23 UT	01:34 UT	01:44 UT	01:58 UT

図 4.6　FMT で 2000 年 11 月 4 日に観測されたプロミネンス．口絵参照．

Fig. 4.6 Temporal development of a prominence observed with the FMT on November 4, 2000. Refer to the frontpiece.

が 1 億〜10 億 t（10^{14}〜10^{15} g）に達することがある．噴出速度は数 100 km/s である．

　噴出型プロミネンスは，太陽面上ではダークフィラメントの突然消失という現象となって現れる．これをフィラメント消失という．あるいは，「突然消失」を意味するフランス語 disparitionbrusque を略して DB と呼ぶこともある（英語では sudden disappearance という）．フィラメント消失はフレアを伴うことも多い．

4.4 フィラメント活動

　本書では，フィラメントになんらかの変化があった現象のうち，フィラメント消失以外のものを「フィラメント活動（filament activity または filament

10^{15} g). The eruption speed of a prominence is normally of the order of a few to several hundred kilometers per second.

An eruptive prominence that occurs on the solar disk appears as a sudden disappearance of a dark filament. This is called filament disappearance, and is also sometimes represented by the notation DB, an abbreviation for "disparition brusque", a French word that means "sudden disappearance". Filament disappearance is often accompanied by a flare.

4.4　Filament Activity

In this book, any changes in a filament that cannot be categorized as filament disappearance are classified as filament activity (or filament activation). This type of event constitutes a substantial

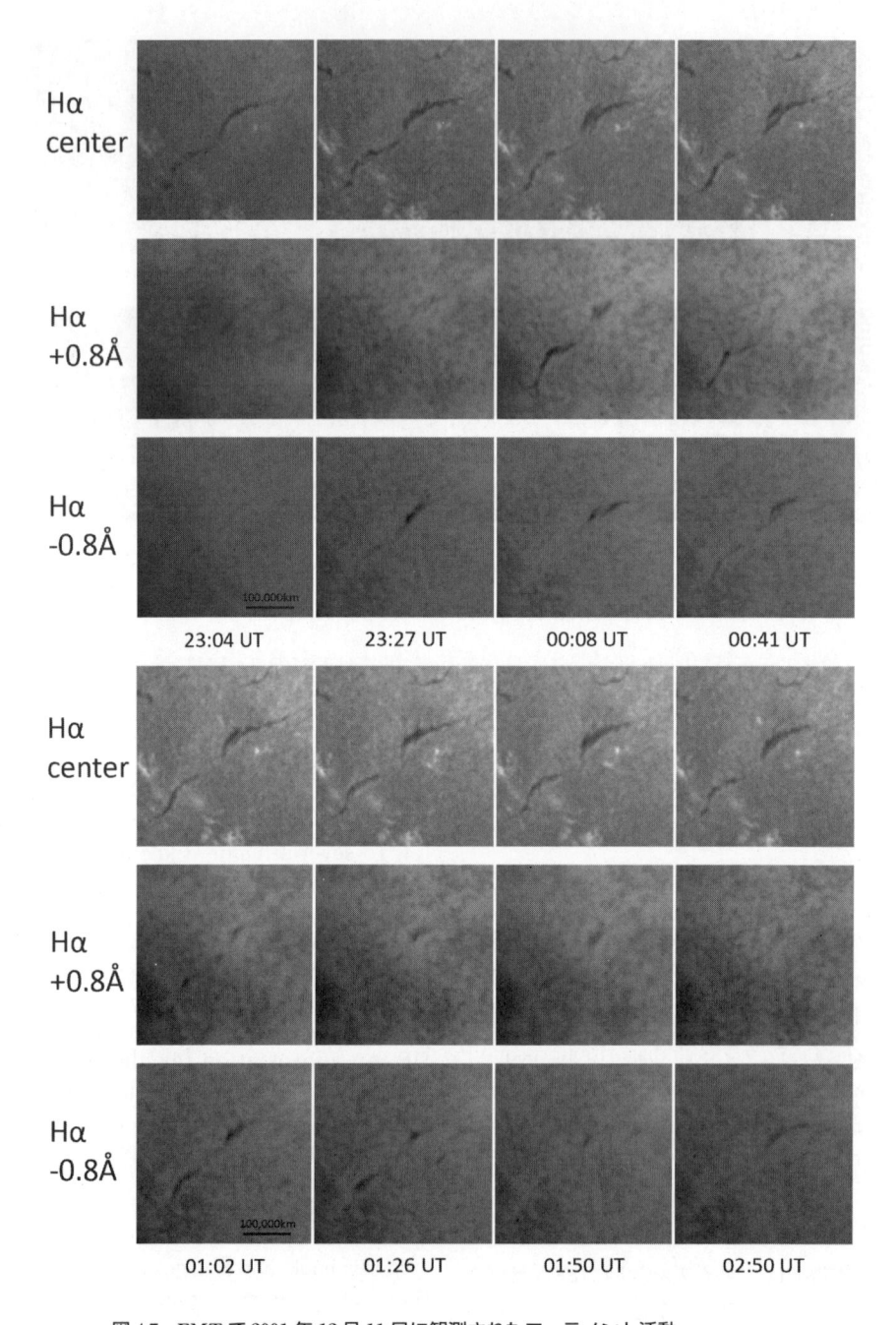

図 4.7　FMT で 2001 年 12 月 11 日に観測されたフィラメント活動.

Fig. 4.7　A filament activity observed with the FMT on December 11, 2001.

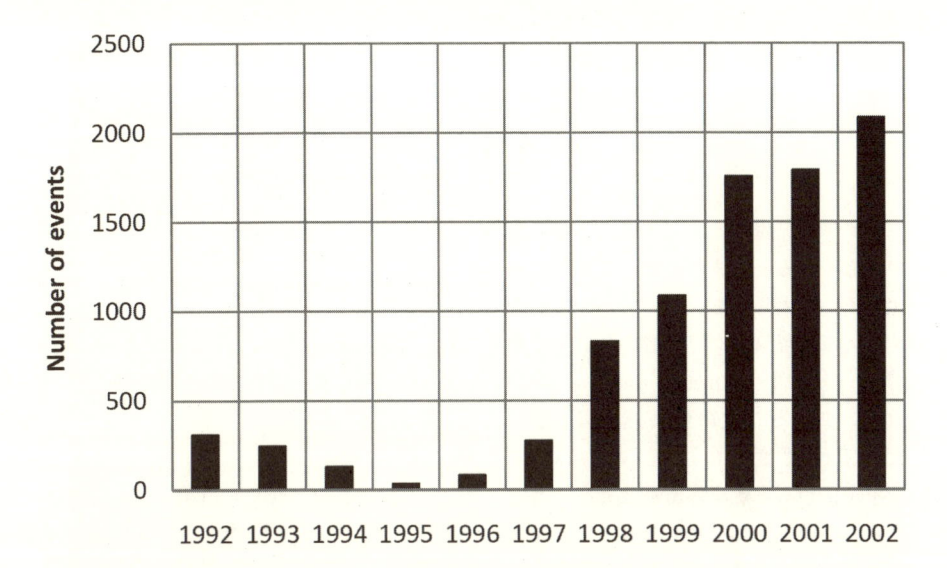

図4.8 FMTで観測されたフィラメント活動の数. 各年は5月から翌年4月までを1年として表記してある. 例えば「1992」は1992年5月から1993年4月までの1年間を示す.

Fig. 4.8 Number of filament activities observed with the FMT. It should be noted that each year in this context begins with May and ends with April, the following year. For example, 1992 corresponds to the period from May 1992 through April 1993.

activation)」という名称で分類している. この現象はイベントリスト中かなりの割合を占める.

上記のような定義上, フィラメント活動には様々な形態の現象が含まれる. 1つの典型例として, フィラメント全体の形状はあまり変化しないが, その内部でガスが激しく運動するというものがあげられる. このようなフィラメント活動の例を図4.7に示す. この現象は2001年12月11日から12日にかけて観測されたもので, FMTイベントリストではIICと分類されている. この現象にはX線の顕著な増光が伴っていなかったため, GOESクラスは付与されていない.

図4.8のグラフは, 1992年5月から

fraction of the event list.

Since the definition of filament activity is quite broad, various forms of events fall into this category. One typical type involves a filament that contains a violent flow of gas, but undergoes a minor change in its overall shape. An example of this type of filament activity is shown in Fig. 4.7. This phenomenon was observed from December 11 to 12, 2001, and categorized as IIC in the FMT event list. No GOES class is assigned to this event since it was not accompanied by any significant increase in the soft X-ray intensity.

Figure 4.8 shows the number of filament activities observed with the FMT in each year from May 1992 to April 2003.

2003年4月までの各年にFMTで観測されたフィラメント活動の数を示したものである．

4.5 モートン波

フレアの発生初期をHα線による動画で詳しく見ると，まれに波がフレアの発生場所から太陽面を伝わっていくのが見える．この現象は，発見者の名にちなんでモートン波（Moreton wave）と呼ばれる．モートン波の伝幡速度は500〜1,500 km/s 程度であり，その正体はコロナ中を伝わる電磁流体衝撃波であると考えられている．ただし，モートン波発生の原因については，フレアに関係するという以外，まだよくわかっていない．

モートン波が発生すると，その波面の伝播範囲に存在するフィラメントが揺れ動くことがある．この現象はウィンキング・フィラメント（winking filament）またはフィラメント振動（filament oscillation）と呼ばれる．ウィンキング・フィラメントはモートン波の発生を捕らえる上で重要な現象である．例えば，モートン波の波面を直接観測することが難しい場合でも，ウィンキング・フィラメントが観測されれば，間接的にモートン波の発生を知ることができる．

当天文台では，FMTで撮影された太陽画像の解析により数多くのモートン波を発見した．その数は，1992年5月から2003年4月までの間に世界で報告されたすべてのモートン波の約3分の1にのぼる．FMTで観測されたモートン波およびウィンキング・フィラメントのリストを表4.3に示す．なお，この表にはFMTで2003年5月以降に観測された

4.5 Moreton Waves

A close examination of videos showing the initial phase of a flare observed in the Hα line occasionally reveals the occurrence of waves propagating from the location of the flare site along the Sun's surface. These waves are called Moreton waves, named after the American astronomer Gail Ernest Moreton. Moreton waves propagate at speeds of approximately 500-1,500 km/s and are considered to be magnetohydrodynamic shock waves that propagate through the corona. However, the generating mechanism of Moreton waves is not yet clearly known.

When a Moreton wave occurs, its wave front sometimes drives an oscillation of a filament located within its propagation zone. This phenomenon is called filament oscillation, and the oscillating filament is referred to as a winking filament. The presence of a winking filament is an important clue for detecting Moreton waves. For example, the occurrence of a Moreton wave can be indirectly confirmed from the presence of a winking filament, even if the wave front of the wave cannot be directly observed.

We have discovered a considerable number of Moreton waves by analyzing the solar images taken using the FMT. The number of Moreton waves discovered by us during the period from May 1992 through April 2003 approximates to one-third of all the Moreton waves reported in the world during the same period. A list of Moreton waves and winking filaments observed with the FMT is shown in Table 4.3. It should be noted that this table also

表 4.3 FMT で観測されたモートン波.
Table 4.3 List of Moreton waves observed through the FMT.

Date	Flare				Moreton wave and/or winking filament(*)
	Peak time	Position	NOAA AR	Class	
1997/11/03	04:38	S20 W13	8100	C8.6	MW
1997/11/04	06:02	S14 W33	8100	X2.1	MW & WF
1998/08/08	03:17	N17 E74	8299	M3.0	MW
1999/02/16	03:12	S23 W14	8458	M3.2	MW & WF
2000/03/03	02:14	S15 W60	8882	M3.8	MW
2000/06/04	22:10	N21 E37	9026	M3.2	WF
2000/06/15	23:43	N19 E19	9040	M2.0	MW
2000/07/16	06:14	S08 W25	9082	C3.8	MW
2001/04/10	05:26	S23 W09	9415	X2.3	WF
2001/05/12	23:35	S17 E00	9455	M3.0	MW
2001/05/13	03:04	S18 W01	9455	M3.6	MW
2001/05/13	23:09	S15 W13	9455	C1.1	WF
2001/05/21	03:20	N22 E08	9461	C9.0	WF
2001/12/19	02:32	N09 E37	9742	C4.9	MW
2002/07/18	07:44	N19 W30	0030	X1.8	WF
2002/08/22	01:57	S07 W62	0069	M5.4	MW
2002/10/04	22:43	N13 E43	0139	M2.7	MW
2003/05/27	23:07	S07 W17	0365	X1.3	MW
2004/11/03	03:35	N09 E45	0696	M1.6	MW
2005/08/03	05:06	S13 E45	0794	M3.4	MW

(*) MW=Moreton wave, WF=winking filament

モートン波3件も含まれている.

また，代表例として，FMT で 1997 年 11 月 4 日に観測されたモートン波の時間発展を図 4.9 に示す．この図の画像は，Hα+0.8Å 像に波面を強調するための画像処理（差分処理）を行ったものである（差分処理については第 6 章の 6.1.3 を参照）．

includes three cases of Moreton waves observed with the FMT in or after May 2003.

As a representative example, the temporal development of a Moreton wave observed with the FMT on November 4, 1997, is shown in Fig. 4.9. The images in the lower panels of this figure are the running difference images produced by processing the images taken in Hα+0.8Å to highlight the wave fronts (see Chapter 6, Section 6.1.3 for more details on the image processing method).

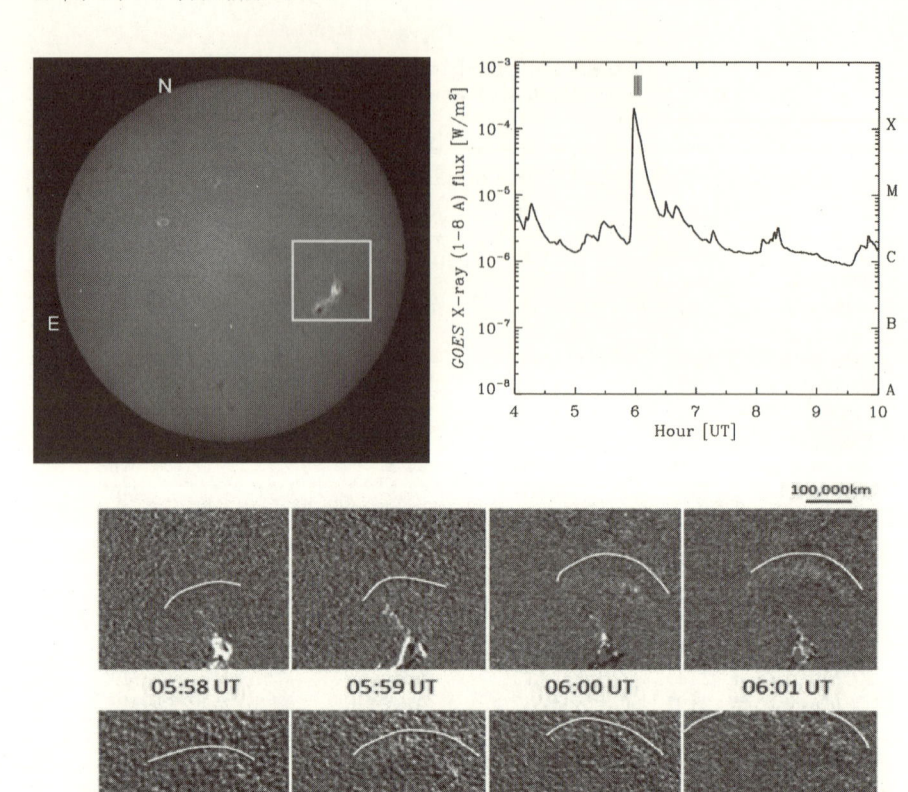

図 4.9 FMT で 1997 年 11 月 4 日に観測されたモートン波. 上段左：太陽全面 Hα 線中心像, 上段右：GOES X-ray flux, 下段：フレアの時間発展を示す時系列画像（差分画像）.

Fig. 4.9 Moreton wave observed with the FMT on November 4, 1997. Upper-left panel: solar full-disk image in Hα center; upper-right panel: GOES X-ray flux; lower panels: sequential images showing the temporal development of the flare (running difference images).

第5章
イベントリスト

Chapter 5

Event Lists

　第2章でも述べたように，当天文台では，フレア監視望遠鏡（FMT）で検出された太陽円盤上の全活動現象（2003年4月まで）と太陽の縁外の全活動現象（2004年8月まで）の特徴や分類を示したイベントリストを作成し，下記のWebサイトを通じて全世界に公開している．
http://www.kwasan.kyoto-u.ac.jp/observation/event/fmt/index.html

　FMTで検出されたイベントの数は膨大であり，そのすべてを本書に掲載することは困難である．そこで本書では，このイベントリスト集をDVDに収録して提供することにした．

　イベントリストには，FMTのHα−0.8Å画像において検出された太陽円盤上の現象を収録したリストと，Hαプロミネンス画像において検出された太陽外縁の現象を収録したリストの2種類がある．

　As already stated in Chapter 2, we have created a collection of event lists showing the physical characteristics and classification of each of the disk events (until April 2003) and limb events (until August 2004). These lists are made available to the public through our website.
http://www.kwasan.kyoto-u.ac.jp/observation/event/fmt/index.html

　Although an enormous number of events have been detected using the FMT, the scope of this book does not allow all of them to be included in it. Accordingly, we have included these event lists in the DVD attached to this book.

　There are two sets of event lists. The first set shows the solar-disk events detected through the FMT images in Hα−0.8Å, and the second set shows the solar limb events detected in the images in H-alpha prominence. The following sections explain the meaning of the codes and symbols used in these lists.

以下，これらのリストで使われている記号の意味を説明する．

5.1 Hα−0.8Å イベントリスト（ディスクイベント）

このリストは，FMT の青色側ウィング像の上で確認された表 4.2 のすべてのディスクイベントを，各月ごとに収録している（宇宙天気環境には，太陽表面からの噴出現象，即ち青方偏移した現象が強く関係していると考えられる）．

リスト中には，各イベントの開始・終了時刻（START TIME, END TIME），太陽表面上の位置（POSITION：太陽面上緯度と太陽円盤中心を通る子午線からの経度），活動性の種類（CLASS），規模（SIZE），所属する活動領域の NOAA 番号（NOAA（米国立海洋大気庁）により活動領域として認定された領域に属するイベントにのみ記載），GOES 軟 X 線データでのピーク強度（リボン型フレアとして分類されたイベントに対して，その発生時間帯に確認された軟 X 線フレア，もしくは発生時刻直前（60 分以内）の軟 X 線フレアの軟 X 線ピーク強度を記載．A, B, C, M, X の順に，より強くなる）を記載している．

注 1：時刻の不確定性について（時刻に付けられる場合がある添字の定義）
B … 現象の発生時刻が不明瞭であることを示す．もしくは，当該日の観測開始時刻に，既に現象が始まっていたことを示す（英語の before の頭文字：B）.
A … 現象の終了時刻が不明瞭であることを示す．もしくは，当該日の観測終了時刻に，現象が継続中であったことを示

5.1 Active Phenomena in Hα −0.8Å (Disk Events)

All transient disk phenomena detected in the blue-shifted images taken using the FMT from May 1992 through April 2003, which are described in Table 4.2, are listed for each month (the blue-shifted images are especially useful to identify the eruptive solar active phenomena that affect the solar-terrestrial environment).

In these event lists, the start and end times, position on the solar disk (solar latitude & longitude with reference to the meridian that passes through the disk center), class of activity, size, NOAA active region number (only for the events that occurred in the regions that were designated as active regions by the NOAA), and peak intensity of the soft X-ray flux measured by the GOES (peak intensities of the soft X-ray flares that occurred simultaneously or just before—less than 60 min—the ribbon-type flares) of each event are described.

NOTE 1: Uncertainty of time (Definition of the characters appended to time)
B … the start-time is not clear, or the event started before the start of the observation period ("B" stands for "Before")
A … the end-time is not clear, or the event continued even after the end of the observation period ("A" stands for "After")

す（英語の after の頭文字：A）．

注2：活動性の種類（CLASS）欄中の分類記号の定義

Ⅰ … 既存のHαフィラメントがない場所に，一時的な輝度低下が見られる現象
Ⅱ … 既存のHαフィラメントの何らかの活動，消滅に伴い，一時的な輝度低下が見られる現象
A；点状の暗い構造物．主に視線方向に沿って発生したサージなどが考えられる
B；線状に伸びる暗い構造物，もしくはサージ
C；既存の暗い構造物が視線垂直方向の運動を見せずに，その場でさらに暗くなる現象
D；既存の構造物がその位置や形状を変化させる，動的な現象
E；点状のフレア（一時的な輝点の発生）
　E1：輝点が1カ所のみ確認されるもの
　E2：輝点が2カ所確認されるもの
　Em：輝点が3カ所以上確認されるもの
F；リボン構造を伴うフレア
　F1：明るいリボン状構造が1本のみ確認されるもの
　F2：明るいリボン状構造が2本確認されるもの
　Fm：明るいリボン状構造が3本以上確認されるもの
！；同じ場所から繰返し間欠的に現象が発生したことを示す
X；上記の分類には当てはめられない現象
/ と，；分類記号が複数付けられる場合，各々以下の例のような意味で用いる．
　[IB/IC]：いずれのタイプか明確には

NOTE 2: Definition of the classification characters used in the CLASS field of the event list

Ⅰ … a transiently darkening feature that is not related to any pre-existing Hα filaments
Ⅱ … a transiently darkening feature that is related to the activation or disappearance of a pre-existing Hα filament
A : a point-like dark feature that, in many cases, is identified with a surge along the line of sight
B : a linearly growing dark feature or surge
C : an in situ darkening feature that shows no apparent transverse motion
D : a dynamical feature that changes its position and shape
E : a small flare (transiently brightening points)
　E1 : one bright point
　E2 : two bright points
　Em : multiple bright points
F : a large flare (transiently brightening ribbons)
　F1 : one bright ribbon
　F2 : two bright ribbons
　Fm : multiple bright ribbons
！: intermittently spouting or recurrent features observed at the same location
X : not definitely classified
/ and , ; Separation symbols between classes
　[IB/IC] : the class cannot be clearly determined (= or)
　[IB, IC] : simultaneous occurrence of two types of events in the same region (= and)

判別し難いことを示す（= or）
[IB,IC]：両方のタイプの現象が同じ領域で同時に発生していることを示す（= and）

注3：規模（SIZE）の定義
S … 太陽面上で5°×5°よりも小さな領域に納まる大きさの現象
M … 太陽面上で5°×5°～10°×10°の領域に納まる大きさの現象
L … 太陽面上で10°×10°の領域には納まらない大きさの現象

注4：NOAA番号に付けられる場合がある添字の定義
［例］
1234C + 1：領域が減衰しNOAA番号が活動領域リストから外された翌日に，当該領域で現象が起きたことを示す．
1234C + 3：領域が減衰しNOAA番号が活動領域リストから外された3日後に，当該領域で現象がおきたことを示す．

5.2 Hαプロミネンスイベントリスト（リムイベント）

このリストは，FMTのHαプロミネンス像の中で，何らかの活動性を示す現象を各月ごとに集録している（地球に直接飛来する物質の起源となる現象は少ないと思われる．太陽活動現象の高さ方向の構造を解析する上で重要な情報を含んでいる）．

リスト中には，各イベントの開始・最大・終了時刻（START TIME, MAXIMUM TIME, END TIME），縁上での位置（POSITION：太陽面上での緯度と，東西いずれの縁か），活動性の種類

NOTE 3：Definition of Size
S … a small feature that is smaller than 5 × 5 deg on the solar surface.
M … a medium feature whose size is between 5 × 5 deg and 10 × 10 deg on the solar surface.
L … a large phenomenon whose size is larger than 10 × 10 deg on the solar surface.

NOTE 4：Definition of the characters appended to the NOAA number
[e.g.]
1234C + 1：An event that occurred in the region which had decayed and whose NOAA number 1234 had been removed from the NOAA list the day before.
1234C + 3：An event that occurred in the region which had decayed and whose NOAA number 1234 had been removed from the NOAA list 3 days before.

5.2 Active Phenomena in Prominence Images (Limb Events)

All the phenomena whose activities are apparent in the limb-prominence images taken using the FMT from May 1992 to August 2004 are listed for each month (these limb events are expected to be useful in studying the radial or vertical structures of solar active phenomena).

In these event lists, the start, maximum, and end times; position on the solar limb (solar latitude and east/west limb); type of activity; and height and length of each phenomenon are described.

（CLASS），高さ・長さ（HEIGHT, LENGTH）を記載している．

注１：時刻の不確定性について（CLASS に付けられる場合がある記号の定義）
＊：現象の発生時刻が不明瞭であることを示す．もしくは，当該日の観測開始時刻に，既に現象が始まっていたことを示す
＃：現象の終了時刻が不明瞭であることを示す．もしくは，当該日の観測終了時刻に，現象が継続中であったことを示す

注２：活動性の種類（CLASS）欄中の分類記号の定義
A．サージ型；ジェット状の噴出現象，かつ，ほとんどのガスが太陽面上に下降して戻るもの
　A1：噴出時と同じ経路に沿ってガスが下降するもの
　A2：ガスがループを描いて別の地点に向かって下降するもの
B．プロミネンス噴出：既存のプロミネンスが高速で噴出・飛翔する現象
　B1：ガスが脱出速度に向かって徐々に加速されていくもの
　B2：スプレーのようにガスが最初から爆発的に加速されているもの
C．プロミネンス崩壊：既存のプロミネンス全体が壊れて消滅する現象
　C1：ゆっくりとした上昇運動を見せた後に崩落するもの
　C2：上昇運動なしに，薄れ消えたり下降したりするもの
D．プロミネンスの部分的活動：部分的に輝度が増す現象，もしくは一部のみが崩壊する現象

NOTE 1：Uncertainty of time (Definition of the characters appended to the CLASS code)
＊：the start time is not clear, or the event started before the start of the observation period
＃：the end time is not clear, or the event continued after the end of the observation period

NOTE 2：Definition of the classification characters used in the CLASS field of the event list
A. a surge；a jet-like ejection, most of whose plasma returns back to the solar surface along a magnetic line of force)
　A1：flow back of the ejected plasma along the ejection path
　A2：flow back of the ejected plasma to another foot point over the apex of a loop
B. eruptive prominence; high-speed eruption of a pre-existing prominence
　B1：gradually accelerated to the escaping velocity
　B2：explosively accelerated to the escaping velocity (spray type)
C. disruptive prominence; disintegration of a pre-existing prominence
　C1：break and slip down after slow rise
　C2：fade out or fall down with no rising motion
D. partially activated prominence; partial brightening or breaking down of a pre-existing prominence

E. ループ型プロミネンス
　E1：単一ループ構造を見せるもので，ループ頂上付近で輝度の上昇が見られる現象
　E2：ポストフレアループ（明るい複数のループ構造が成長していく様子を見せるもの）
　E3：明るい複数の小さなループの塊と思われる山型構造が成長していく現象
F. それ以外の現象・構造
　F1：安定しているが，並みはずれた大きさ，あるいは形状を持つプロミネンス
　F2：上述の分類には当てはまらない，大きな運動を示すプロミネンス

注3：高さ（HEIGHT）と長さ（LENGTH）の定義
高さ：当該現象発生時間中の，太陽の縁からの最大到達高度をMmの単位で記載（1 Mm = 1,000 km）
長さ：当該現象発生時間中の，太陽の縁に沿った最大長さをMmの単位で記載

E. loop prominence
　E1：loop-top brightening followed by a single-loop formation
　E2：post-flare loops (growing multiple-loop system)
　E3：a growing bright mound or a compact-loop cluster
F. others
　F1：a stable but exceptionally large prominence or in extraordinary shape
　F2：unclassifiable prominence with large motion

NOTE 3：Definition of height and length
Height：maximum height measured from the solar limb in Mm(= 1,000 km)
Length：maximum length along the solar limb in Mm(= 1,000 km)

第 6 章 代表的イベント

Chapter 6 Representative Events

この章では，フレア監視望遠鏡で観測された膨大な数のイベントの中から代表的なイベントを選び，ディスクイベントとリムイベント（プロミネンス）に分けて紹介する．

6.1 イベント図版に関する全体的な説明

6.1.1 イベントの選択基準

本章で紹介するイベントは以下のようなものである．
・顕著な視覚的変化（急激な増光，フィラメントやプロミネンスの激しい運動，長大な構造の出現など）が見られたもの．
・強いX線増光を伴っていたもの（GOES衛星による軟X線観測で「Xクラス」に分類されたもの）．
・モートン波（あるいはそれに伴うフィラメント振動）が観測されたもの．
・学術論文で解析の対象となったもの．
これらの基準の少なくとも1つを満た

This chapter presents a number of representative events selected from the enormous number of events observed through the FMT. The events are categorized into disk events and limb events (prominences).

6.1 General Descriptions of the Illustrations

6.1.1 Event Selection Criteria

The events presented in this chapter are classified as follows:
・Events that showed a remarkable visual change (such as a sudden brightening, violent motion of a filament or prominence, or emergence of an extremely long or large structure).
・Events that were associated with a strong X-ray brightening (specifically, events that were classified into the "X" class on the basis of the results of the soft X-ray observation by the GOES).
・Events that were accompanied by the occurrence of Moreton waves.
・Events that were selected as the targets of analysis in academic papers.

したイベント（ディスクイベントおよびプロミネンス）を発生日時の順に並べたものが本章の図版セクションである。

6.1.2 ディスクイベントの図版

ディスクイベントの図版はイベント毎に見開き2ページで構成されている．図版には基本情報（左ページ上段），コメント（右ページ上段）および時系列画像（両ページ下段）という3つのエリアがある．このうち，基本情報エリアについては後で説明する．

コメントエリアには現象の説明が日本語と英語で簡潔に記載されている（コメントの中で触れられている論文の書誌情報については本章の末尾の参考文献リストを参照）．

時系列画像エリアにはイベントの時間発展が拡大画像で示されている．上段がHα線中心での観測画像，中段がHα+0.8Å，下段がHα-0.8Åでの観測画像である．

図6.1は基本情報エリアの例である．このエリアの一番上には図版のタイトルが表示されており，その下に太陽全面像とX線強度のグラフ（GOESプロット）が配置されている．太陽全面像の下にはイベント情報が記載されている．以下，このエリアの各項目について説明する．

タイトル

各図版にはイベントの分類，発生日，開始時刻および終了時刻を示すタイトルが付けられている．時刻は世界標準時（UT）である．2種類以上のイベントが

The events selected according to these criteria are presented in the illustrated sections in the order of date and time of occurrence.

6.1.2 Disk Event

Each disk event is illustrated in a two-page format, which contains the following: a basic information field (on the upper half of the left page), comment field (on the upper half of the right page), and time-series image field (on the lower half of both the pages). The contents of the basic information field will be described later.

The comment field contains a brief description of the event in both English and Japanese. For bibliographic information on the papers referenced in the comment, please refer to the reference list at the end of this chapter.

The time-series image field shows the temporal development of the events by means of a series of enlarged images. The images in the upper, middle, and lower rows are the images observed in Hα center, Hα+0.8Å, and Hα-0.8Å, respectively.

Figure 6.1 shows an example illustrating the contents of the basic information field. The title of the event is mentioned at the top of the field, below which the solar full-disk image and X-ray intensity graph (GOES plot) of the event are shown; additional event information is also provided below the solar full-disk image. A brief description of the different sections that make up the illustration is given below.

Title

The title shows the type, occurrence date, start time, and end time of the event. The times are given in universal time (UT). If two or more different types of events are found simultaneously, the

Chapter 6　Representative Events | 75

Feb 20, 1994 00:17－02:47 UT　　Flare, Filament Eruption ◄── タイトル / Title

太陽全面像 / Solar Full-disk

GOES プロット / GOES plot

GOES Class = M4.0　NOAA 7671
N01,W00　P = -19.0[deg]　B0= -7.0[deg] ◄── イベント情報 / Event information

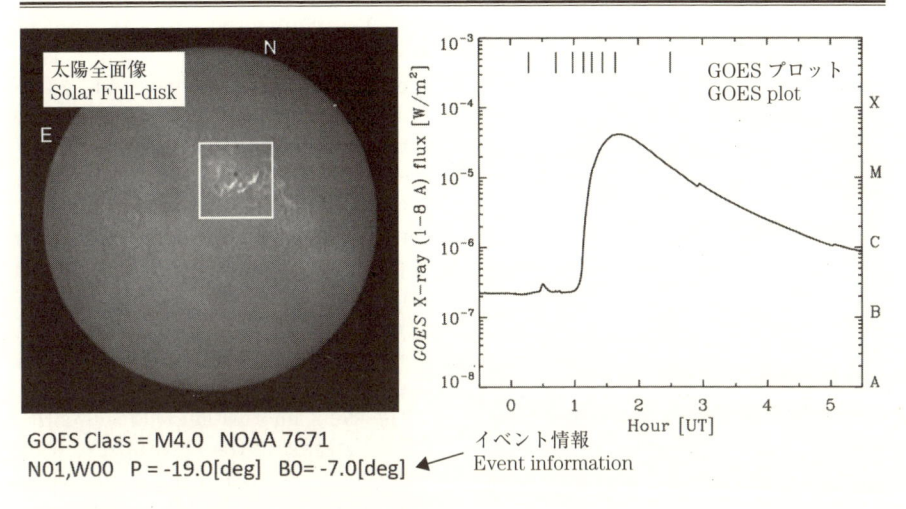

図 6.1　ディスクイベント図版の基本情報エリア．

Fig. 6.1 Contents of the basic information field in the illustration of disk events.

発生している場合は，特に注目すべきイベントの分類を並べて表示してある．現象の開始時刻が不明瞭である場合は時刻の末尾に「B」を付し，現象の終了時刻が不明瞭である場合は時刻の末尾に「A」を付してある（5.1 節の注 1 を参照）．

太陽全面像

イベント発生中に撮影された多数の太陽全面像の中から 1 枚を代表画像として示したものである．画像の中にある文字「N」は太陽の自転軸の北を示している．また，図の左右方向が東になる（図中の文字「E」を参照）．画像中の白い枠は時系列画像エリアに示された各画像の視野に対応している．

most notable two events are chosen. The letter "B" is appended to the start time of an event whose start time is unclear, and the letter "A" is appended to the end time of an event whose end time is indeterminable (see Note 1 in Section 5.1 of Chapter 5).

Solar full-disk image

This is a representative image selected from a number of solar full-disk images taken during the development of the event. The letter "N" in this image indicates the north of the axis of solar rotation. The left side of the image is the east (see the letter "E" in Fig. 6.1). The box shown in the image corresponds to the field of view of the enlarged images displayed in the time-series image field.

GOES プロット

GOES プロットは，米国の GOES 衛星により観測された太陽の軟X線の強度を示すグラフである．本章の GOES プロットはイベントの継続時間を含む6時間をカバーしている．グラフの上部にある短い線の列は，時系列画像エリアに示された各画像の撮影時刻を示している．

イベント情報

イベント情報の欄には GOES Class，NOAA，イベント発生位置および太陽の自転軸の傾き（P, B0）が示されている．各項目の説明は以下の通りである．

GOES Class：イベントの発生に伴って生じた軟X線増光のピークの高さを示す分類．強度の大きい順に X，M，C，B および A の5クラスに大別される．これらのクラス記号と，それに付した数値によってピークの高さが示される（例えば図6.1ではM4.0）．同じクラス内では数値が大きいほどX線が強い．二重ハイフン（--）はイベントにX線増光が伴っていなかったことを示す．

NOAA：米国立海洋大気庁（National Oceanic and Atmospheric Administration = NOAA）が太陽面上の活動領域に付与した識別番号．二重ハイフン（--）はイベントが NOAA の指定した活動領域とは無関係な場所で起きたことを示す．

イベント発生位置：太陽面上におけるイベントの発生位置．記号 N および S は太陽面緯度（北緯，南緯）を示し，E および W は経度（東経，西経）を示す．

GOES plot

The GOES plot is a graph showing the soft X-ray intensity of the Sun as observed by the GOES, which is a geostationary satellite operated by the United States. The GOES plots in this chapter show the changes in the X-ray intensities observed over a six-hour period that includes the event duration period. The short vertical lines at the top of the graph indicate the points in time of the images displayed in the time-series image field.

Event information

The event information field shows the GOES class, NOAA number, and occurrence position of the event, along with the inclination (P, B0) of the axis of rotation of the Sun. A brief description of these parameters is given below.

The GOES class is a classification that indicates the peak height of a soft X-ray brightening associated with the occurrence of an event. There are five major classifications: X, M, C, B, and A, in the decreasing order of the X-ray intensity. The peak height is indicated by one of these letters, followed by a numerical value (e.g., M4.0 in Fig. 6.1). A larger numerical value implies a stronger X-ray emission. Two hyphens (--) are used for an event that is not accompanied by any significant X-ray brightening.

The NOAA number is an identification number assigned by the National Oceanic and Atmospheric Administration (NOAA) to each active region on the solar disk. Two hyphens (--) are used for an event that occurs at a location that is not associated with any NOAA-designated active region.

The occurrence position shows the location of occurrence of the event on the solar disk. The letters N and S and

PとB0：太陽の自転軸の傾きを示す角度（単位は度）．Pは天の北極に対する東西方向の傾きを表し，東に傾いているときにプラス（+）の値となる．B0は黄道面に対する傾きを表し，北極（N）が手前（観測者側）に見えるときにプラス（+）の値となる．

6.1.3 モートン波の図版

モートン波は独立した現象ではなく，フレアなど他のイベントに付随して生じる現象である．そこで，モートン波を伴うイベントについては，まず現象の大元となったイベント（ほとんどの場合，フレア）の図版を示し，その次にモートン波の通常の図版と差分画像の図版を示すという特別な形式を取ることにした．差分画像（running difference image）とは，ある時点での撮影画像とその直前の撮影画像（例えば1分前の画像）との差分をピクセルごとに計算することによって得られる画像である．

図6.2はモートン波の差分画像の図版の例である．差分画像の図版は見開き2ページで構成されており，図6.2はその1ページ分だけを示している．各ページには同じ拡大画像を示す2つの時系列画像エリアが上段と下段にそれぞれ配置されている．

差分画像では，2つの時点の間で値（光の強さ）が大きく変化したピクセルが特に強調される（周囲より明るく，あるい

the letters E and W, each of which is followed by a two-digit number, show the latitude (north/south) and longitude (east/west), respectively, on the solar surface.

P and B0 are the inclination angles of the axis of rotation of the Sun (in degrees). P is the angle of inclination in the east-west direction with respect to the celestial north and takes positive values when the axis is inclined eastward. B0 is the inclination of the axis with respect to the ecliptic plane and takes positive values when the north pole of the Sun is visible from the Earth.

6.1.3 Moreton Wave

A Moreton wave is not an isolated phenomenon, but always accompanies some other event such as a flare. Taking this into account, we have prepared a special set of illustrations for the events accompanied by Moreton waves. This set starts with an illustration of the source event (in most cases, a flare), followed by two illustrations showing the Moreton wave, one having the same layout as that of the normal H-alpha images and the other having the layout of the running difference images. The running difference images are the images produced by calculating the pixel-by-pixel difference between two images, one taken at a given point in time and the other taken immediately before the first image (e.g., one minute earlier).

Figure 6.2 shows an example of the illustration of running difference images of a Moreton wave. The running difference images are illustrated in a two-page format in which two time-series image fields, upper and lower, are provided. It should be noted that Fig. 6.2 shows only one of the two pages.

In the running difference images, a pixel that has significantly changed its

78 | 第6章　代表的イベント

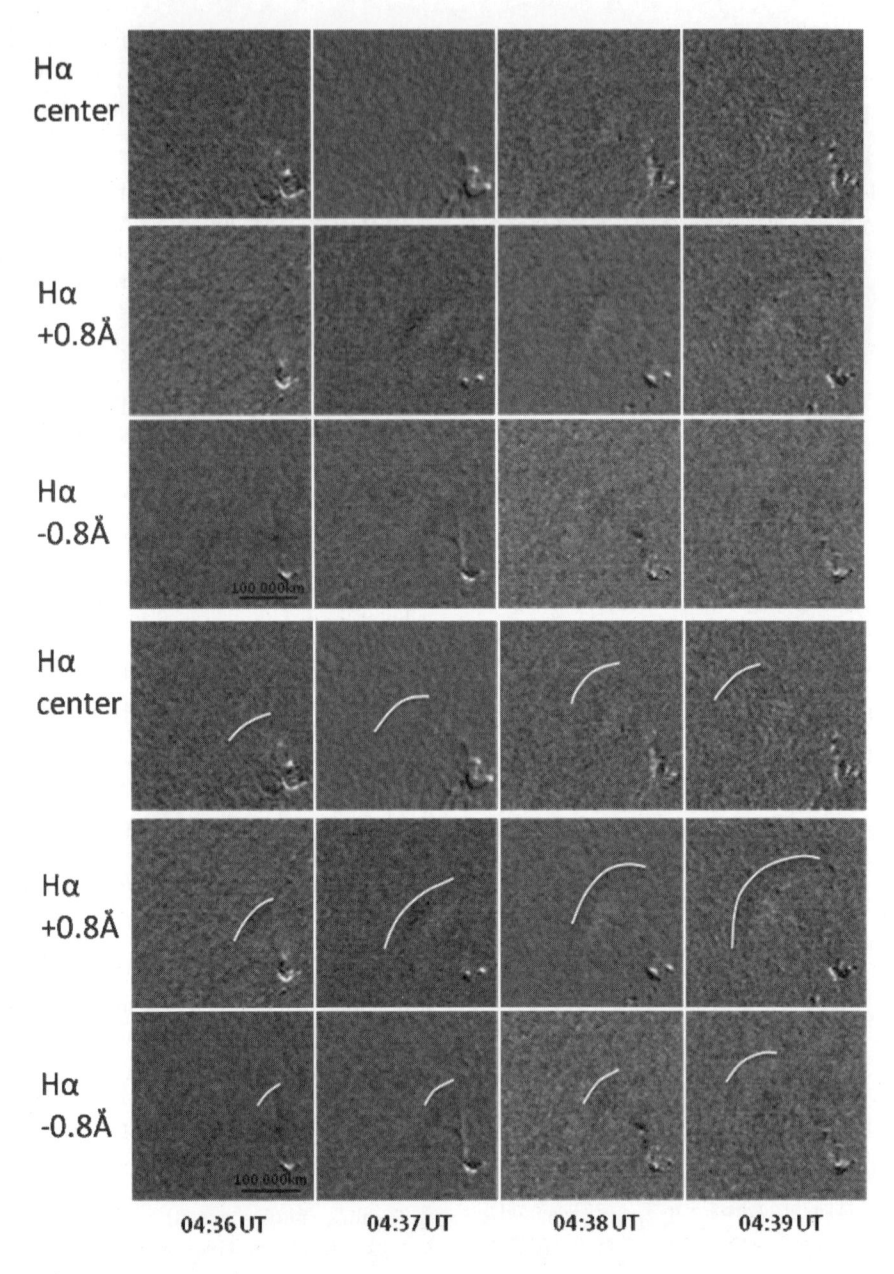

図 6.2　モートン波の差分画像の図版．下段の画像に書き込まれた曲線は各時点での波面を示している．

Fig. 6.2 Running difference images of a Moreton wave. The curved lines in the lower section show the wave front at each point in time.

Dec 28, 2001 23:35−06:17 UT Post Flare Loop ← タイトル / Title

太陽全面像 / Solar Full-disk

00:29 UT

GOES プロット / GOES plot

ES20-50
P = 4.1[deg] B0= -2.5[deg]   ← イベント情報 / Event information

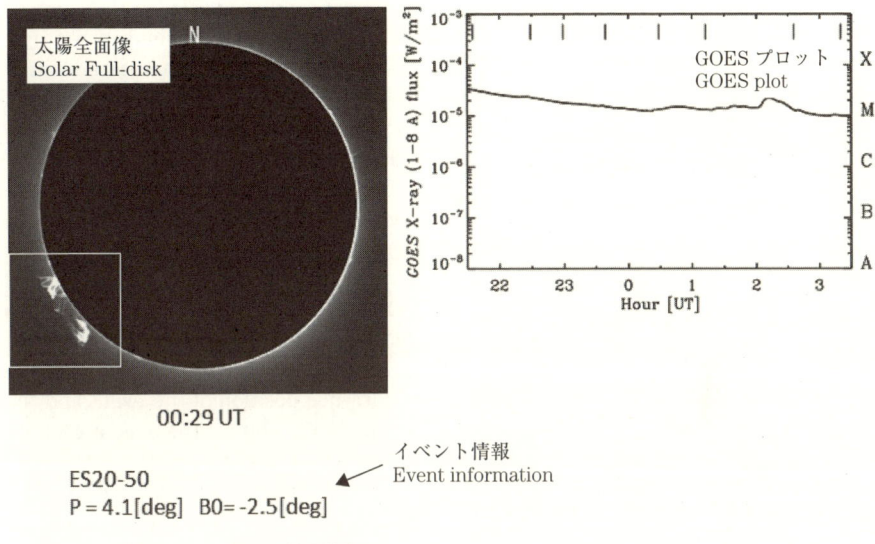

図 6.3 プロミネンス図版の基本情報エリア．

Fig. 6.3 Contents of the basic information field in the illustration of prominence events.

は暗くなる)．このような処理を行うと画像中のモートン波の波面が見やすくなる．さらに下段の差分画像では，円弧状の線（ガイドライン）を書き込むことにより波面の位置や伝播方向をよりわかりやすく示している．

6.1.4 リムイベントの図版

リムイベントの図版はイベント毎に1ページで構成されている．ディスクイベント図版と同様に，リムイベント図版にも基本情報，コメントおよび時系列画像という3つのエリアがある．基本情報エ

value (intensity of light) between the two points in time appears to be particularly highlighted (that is to say, it becomes brighter or darker than the surrounding pixels). Such an image processing method makes the wave front of the Moreton wave more apparent. The lower section of the illustration shows the same set of running difference images as that in the upper section, but the former shows curved lines (guide lines) to indicate the position of the wave front and the propagating direction.

6.1.4 Limb Event (Prominence)

Limb events are illustrated in a one-page format. The illustration of limb events is similar to that of the disk events in that it also carries the following: basic information field, comment field, and time-series image field. The contents of

リアについては後で説明する．コメントエリアの内容はディスクイベント図版のものと基本的に同じである．時系列画像エリアでは，FMTのプロミネンスモードで撮影された8時点のHα線中心画像によりイベントの時間発展を示している．

図6.3はリムイベント図版の基本情報エリアの例である．このエリアに含まれる情報のうち，タイトル，太陽全面像およびGOESプロットはディスクイベント図版で説明したものと基本的に同じである．

イベント情報としては，太陽の自転軸の傾き（P, B0）とイベント発生位置だけが示されている．PとB0の意味は先にディスクイベントの例で説明したとおりである．イベントの発生位置は，太陽面上の東西（EW）および南北（SN）を表すアルファベット2文字と緯度を表す2桁の数字で表されている．例えば図6.3の「WS41」は，「太陽の西側（W）リムの南緯41度（S41）」を意味している．

6.2 代表的ディスクイベント

この節ではFMTで観測された代表的なディスクイベントを紹介する．表6.1に代表的ディスクイベントの一覧を示す（開始時刻，終了時刻，X線強度，NOAA番号，位置，種類および規模については第5章の5.1節および第4章の表4.2を，またPおよびB0については本章の6.1.2をそれぞれ参照）．

the basic information field will be described later. The contents of the comment field are basically the same as those of the comment field in the disk event illustration. The time-series image field shows the temporal development of the event by means of eight images taken using the FMT in the prominence mode.

Figure 6.3 is an example showing the contents of the basic information field in the limb event illustration. The contents of the title, solar full-disk image, and GOES plot are the same as those for the disk events.

The event information field shows the occurrence position of the event and the inclination (P, B0) of the axis of rotation of the Sun. P and B0 are the same as in the illustration of disk events. The occurrence position of an event is denoted by two letters indicating the east/west (EW) and north/south (SN) on the solar disk, followed by two digits showing the latitude. For example, "WS41" in Fig. 6.3 means that the event occurred on the western limb at latitude 41 south.

6.2 Representative Disk Events

This section shows the representative disk events observed through the FMT. Table 6.1 is a summary list of the representative disk events. Please refer to Section 5.1 of Chapter 5 and Table 4.2 of Chapter 4 for the definitions of start time, end time, X-ray intensity, NOAA number, position, class, and size. P and B0 have been defined in Subsection 6.1.2 of this chapter.

表 6.1 代表的ディスクイベント一覧.
Table 6.1 Representative Disk Events.

イベント種別 Event Type	日付 Date	開始時刻 − 終了時刻 StartTime-End Time	X 線強度 X − Ray Intensity	NOAA 番号 NOAA No.	位置 Position	P, B0	種類 Class	規模 Size
Filament Eruption	1992/11/5	00:45 − 02:15	− −	7325,7332	S20 W17	23.7, 3.9	IID	L
Filament Activity	1992/12/21	01:38 − 04:37	− −	− −	S35 E04	7.3, −1.7	IIC	L
Flare, Filament Eruption	1993/3/20	00:16 − 01:14	M1.8	7448	N16 W38	−25.0, −7.1	ID,F2	L
Flare, Filament Eruption	1993/5/14	21:57 − 00:09	M4.4	7500	N15 W52	−21.3, −2.8	ID,F2	L
Filament Eruption	1993/5/15	22:29 − 00:27	− −	− −	S28 E30	−21.0, −2.6	IID	L
Flare, Surge	1993/6/6	07:07 − 07:46	C1.8	7518	S10 W22	−13.6, 0.0	IB,Fm	L
Flare, Surge	1993/6/7	05:43 − 06:44	C5.1	7518	S08 W33	−13.1, 0.0	Fm, IB!	L
Filament, Eruption	1993/9/25	23:03 − 23:51	− −	7585	S06 W09	25.5, 7.0	ID/IID	L
Filament Eruption	1993/10/21	02:51 − 05:19	− −	7605	S15 W08	25.9, 5.4	IID	L
Flare, Filament Eruption	1994/1/5	06:04 − 07:05	M1.0	7646,7647	S10 W22	0.2, −3.5	IID,Fm	L
Flare, Filament Eruption	1994/2/20	00:17 − 02:47	M4.0	7671	N01 W00	−19.0, −7.0	F2,IID	L
Surge	1994/7/12	22:46 − 23:50	− −	7746	N12 W65	2.2, 4.0	IB,E1	L
Filament Eruption	1994/9/5	06:16 − 08:00	− −	7773	S05 W03	22.0, 7.2	IID	L
Surge	1995/2/11	23:54 − 00:45	− −	7838	N11 E35	−15.9, −6.6	IB	L
Flare, Filament Eruption	1997/5/12	04:27 − 05:26	C1.3	8038	N20 W09	−21.8, −3.0	IID,F2	L
Flare, Filament Eruption	1997/8/29	22:31 − 00:47	M1.4	8076	N30 E16	20.3, 7.1	ID,Fm	L
Surge	1997/11/3	01:51 − 02:45	C1.0	8100	S20 W16	24.2, 4.2	IB,F1	L
Flare	1997/11/3	04:34 − 05:05	C8.6	8100	S19 W22	24.2, 4.2	IB,F1	M
Moreton wave	1997/11/3	04:35 − 04:47	C8.6	8100	S19 W22	24.2, 4.2	− −	

イベント種別 Event Type	日付 Date	開始時刻-終了時刻 StartTime-EndTime	X線強度 X-Ray Intensity	NOAA番号 NOAA No.	位置 Position	P, B0	種類 Class	規模 Size
Flare, Filament Eruption	1997/11/4	06:00-06:55	X2.1	8100	S18 W39	24.0, 4.1	ID,F2	L
Moreton wave	1997/11/4	05:57-06:06	X2.1	8100	S18 W39	24.0, 4.1		
Winking Filament	1997/11/4	06:06-06:35	X2.1	8100	S18 W39	24.0, 4.1	--	
Flare, Filament Eruption	1998/4/11	04:15-05:41	C2.1, C3.0	8198	S37 E35	-26.2, -5.9	IID,Fm	L
Flare, Surge	1998/4/16	01:05-01:26	C9.5	8203	N31 W26	-26.0, -5.5	IB,F1	L
Flare, Filament Eruption	1998/4/29	05:11-05:40	B7.6, B7.7	8210	S10 E32	-24.5, -4.4	IID,F1	L
Flare, Surge	1998/8/8	03:00-03:30	M3.0	8299	N16 E80	13.4, 6.2	F1,IB	M
Moreton wave	1998/8/8	03:14-03:26	M3.0	8299	N16 E80	13.4, 6.2	--	
Flare, Filament Eruption	1998/9/20	02:00-05:28	M1.8	8340	N20 E70	24.8, 7.1	IID,Fm	L
Filament Activity	1998/12/24	22:44B-06:35A	--	--	N08 E60	6.1, -2.0	IIC	L
Two ribbon flare in spotless region	1999/2/9	03:07-05:22	C2.3	8453	S27 W39	-15.1, -6.5	IIC,F2	L
Flare, Filament Eruption	1999/2/16	01:42-04:05	M3.2	8458	S27 W18	-17.6, -6.9	IID,F2	L
Moreton wave	1999/2/16	02:53-03:02	M3.2	8458	S27 W18	-17.6, -6.9	--	
Filament Eruption	1999/2/16	02:41-05:24	--	8457	N33 W20	-17.6, -6.9	IID	L
Flare, Filament Eruption	1999/6/1	06:24-07:08	C6.2	8562,8557	S23 E17	-15.6, -0.7	F1,IID	L
Flare	1999/8/2	21:22-22:00	X1.4	8647,8645	S19 W50	11.0, 5.8	IIC/IID,Fm	M
Flare, Surge	1999/8/25	01:34-02:03	M3.6	8673,8674	S27 E12	19.0, 7.0	F1,IB	L
Flare, Filament Eruption	1999/12/28	00:41-01:27	M4.5	8806	N19 W51	4.3, -2.5	F2,IID,IB!	L
Flare, Filament Eruption	2000/3/3	02:11-03:14	M3.8	8882	S20 W70	-22.1, -7.2	F1,IID,IB	L

イベント種別 Event Type	日付 Date	開始時刻- 終了時刻 StartTime- End Time	X線強度 X-Ray Intensity	NOAA 番号 NOAA No.	位置 Position	P, B0	種類 Class	規模 Size
Moreton wave	2000/3/3	02:11- 02:21	M3.8	8882	S20 W70	-22.1, -7.2	--	
Flare, Filament Eruption	2000/6/4	22:05- 23:22	M3.2	9026	N20 E40	-14.3, -0.3	F1,IID	L
Filament Eruption	2000/6/6	01:02- 08:36A	C2.4	9026	N35 E30	-13.4, 0.0	IID	L
Flare, Filament Eruption	2000/6/15	23:33- 00:20	M2.0	9040	N16 E17	-9.7, 1.1	F1,IID	L
Moreton wave	2000/6/15	23:37- 23:43	M2.0	9040	N16 E17	-9.7, 1.1	--	
Flare	2000/7/16	06:09- 06:37	C3.8	9082	S10 W25	4.2, 4.5	F1,IID	M
Moreton wave	2000/7/16	06:06- 06:16	C3.8	9082	S10 W25	4.2, 4.5	--	
Flare, Filament Eruption	2000/7/19	06:42- 08:03	M6.4	9087	S10 E12	5.5, 4.7	IIC,IID, Fm	L
Surge	2000/7/23	04:27- 04:57	--	9097	N05 E16	7.2, 5.1	IB	L
Surge	2000/10/1	01:54- 02:56	C2.1	9176	S08 E25	26.0, 6.7	E1,IB	L
Filament Activity	2000/10/24	03:18- 05:40	--	--	S01 E52	25.6, 5.1	IIC	L
Filament Activity	2000/11/3	21:59B- 02:46	--	9212,9213, 9218	N12 W03	24.1, 4.1	IIC/IID	L
Flare, Filament Eruption	2000/11/23	05:32- 06:19	C5.4	9238,9231	S27 W41	18.9, 1.8	IID,F2	L
Flare, Filament Eruption	2000/11/24	04:47- 05:30	X2.0	9236	N20 W08	18.5, 1.7	IID,F2	L
Filament Eruption	2000/11/29	00:20- 01:45	--	--	S46 W58	16.8, 1.1	IID	L
Flare, Filament Eruption	2001/3/21	02:31- 03:22	M1.8	9373	S06 W70	-25.3, -7.0	F1, IB!	L
Flare, Filament Eruption	2001/4/10	03:30- 06:21	X2.3	9415	S20 W16	-26.3, -6.0	F2,IID	L
Winking Filament	2001/4/10	05:37- 05:51	X2.3	9415	S20 W16	-26.3, -6.0	--	
Filament Eruption	2001/4/22	23:43- 00:24	--	--	N33 W33	-25.4, -6.9	IID	L
Flare, Surge	2001/5/12	23:27- 23:43	M3.0	9455	S17 E02	-21.8, -3.0	F1,IID, IB	M

イベント種別 Event Type	日付 Date	開始時刻－ 終了時刻 StartTime- End Time	X線強度 X-Ray Intensity	NOAA 番号 NOAA No.	位置 Position	P, B0	種類 Class	規模 Size
Moreton wave	2001/5/12	23:27－23:34	M3.0	9455	S17 E02	-21.8, -3.0	--	
Flare	2001/5/13	03:01－03:18	M3.6	9455	S18 W01	-21.5, -2.9	F1,IID	L
Moreton wave	2001/5/13	03:01－03:09	M3.6	9455	S18 W01	-21.5, -2.9	--	
Flare, Filament Eruption	2001/5/21	03:08－03:36A	C9.0	9461	N24 E09	-19.2, -2.0	IID,F1	L
Winking Filament	2001/5/21	03:18－03:33	C9.0	9461	N37 E10	-19.2, -2.0	--	
Flare, Filament Eruption	2001/9/25	04:27－06:09	M7.6	9628	S22 E01	25.5, 7.0	IID, Fm,E2	L
Flare	2001/10/19	00:51－01:38	X1.6	9661	N17 W21	26.0, 5.6	F2,IIC	L
Filament Eruption	2001/11/10	21:54－02:40	C5.3	9692,9685	N02 W59	26.3, 6.1	IID,F2	L
Flare, Filament Eruption	2001/11/20	01:29－02:46	C3.4	9704	S15 E06	19.9, 2.2	IID,F1	L
Filament Activity	2001/12/11	23:04－03:23	--	9734	S09 E14	11.5, -0.5	IIC	L
Flare	2001/12/19	02:30－03:21A	C4.9	9742	N13 E32	8.3, -1.4	F1,IID, IB	L
Moreton wave	2001/12/19	02:30－02:40	C4.9	9742	N13 E32	8.3, -1.4	--	
Flare, Filament Eruption	2002/2/4	05:45－07:20A	M1.5, M2.3	9809	S03 E33	-13.3, -6.2	F2,IID	L
Two ribbon flare in spotless region	2002/3/17	23:56－01:54A	--	9870	S08 W18	-24.9, -7.1	IIC	L
Two ribbon flare in spotless region	2002/4/4	01:53－05:32	C9.8	--	S26 E53	-26.3, -6.4	IID,F2	L
Flare, Filament Eruption	2002/4/4	06:46－07:44	C3.0	9888	S08 W00	-26.3, -6.4	IID,F1	L
Flare	2002/7/18	07:41－08:18A	X1.8	10030	N18 W33	4.9, 4.6	F2,IID, IB	L
Winking Filament	2002/7/18	07:53－07:55A	X1.8	10030	N45 W45	4.9, 4.6	--	
Filament Activity	2002/8/16	05:54－07:00	--	10069	S07 E13	16.3, 6.7	IIC	L

イベント種別 Event Type	日付 Date	開始時刻- 終了時刻 StartTime- End Time	X線強度 X-Ray Intensity	NOAA 番号 NOAA No.	位置 Position	P, B0	種類 Class	規模 Size
Flare, Surge	2002/8/21	05:31- 06:20	X1.0	10069	S13 W48	17.9, 6.9	Fm,IIC, IB	L
Flare	2002/8/22	01:13- 02:57A	M5.4	10069	S16 W53	18.2, 6.9	IID,E1, Fm	L
Moreton wave	2002/8/22	01:50- 01:58	M5.4	10069	S16 W53	18.2, 6.9	--	
Surge	2002/8/30	04:52- 05:18	C5.7	10087	S10 W28	11.6, 7.2	E1!, IB!	L
Flare, Filament Eruption	2002/9/18	22:30- 23:52	C1.7	10119	S22 W00	24.5, 7.2	IID,F2	L
Flare	2002/10/4	22:36- 23:14	M2.7	10139	N10 E43	26.1, 6.6	F2	-
Moreton wave	2002/10/4	22:38- 22:44	M2.7	10139	N10 E43	26.1, 6.6	--	
Two ribbon flare in spotless region	2003/2/12	23:21- 00:32	B4.5	10283	N08 W17	-16.2, -6.7	IIC,F2	
Flare, Filament Eruption	2003/3/13	01:53- 02:34	C1.3	10311	S13 E08	-24.0, -7.2	IID,F2	L
Two ribbon flare in spotless region	2003/4/26	23:49B- 03:07	C6.5	10346	N15 E51	-24.9, -4.6	IID,F1, F2	L
Flare, Filament Eruption	2003/4/27	00:16- 02:28	C9.3	10342, 10338	N15 W65	-24.9, -4.6	ID/ IID,Fm	L
Flare	2003/5/27	22:56- 23:26	X1.3	10365	S07 W17	-17.4, -1.3	F2,IID, F1	L
Moreton wave	2003/5/27	23:01- 23:10	X1.3	10365	S07 W17	-17.4, -1.3	--	
Flare	2003/7/17	08:19- 08:39	C9.8	10412	N14 E12	4.3, 4.5	F2	
Moreton Wave, Winking Filament	2003/7/17	08:19- 08:25	C9.8	10412	N14 E12	4.3, 4.5		
Winking Filament	2003/7/17	08:21- 08:33	C9.8	10412	N07 E12	4.3, 4.5		
Flare, Filament Eruption	2005/8/3	04:54- 05:32	M3.4	10794	S13 E42	11.6, 5.9	F2,IID	L
Moreton wave	2005/8/3	05:01- 05:08	M3.4	10794	S13 E42	11.6, 5.9	--	

86 | 第6章　代表的イベント

Nov 05, 1992 00:45−02:15 UT   Filament Eruption

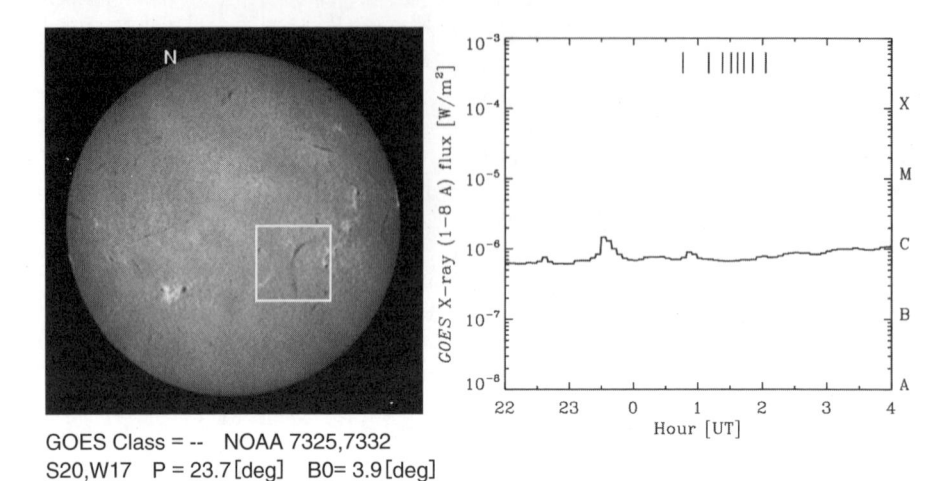

GOES Class = --   NOAA 7325,7332
S20,W17   P = 23.7[deg]   B0= 3.9[deg]

Hα center

Hα +0.8Å

Hα −0.8Å

100,000km

00:46 UT　　01:10 UT　　01:23 UT　　01:31 UT

A long dark filament erupted and disappeared in a time span of about 1 hour. At the first phase, the filament showed slow rise with a speed of 5 km/s. Just at the time of soft-X-ray brightening, the filament was abruptly accelerated and started their rapid rise motion. Comparative studies of Yohkoh soft-X ray images and Hα images of the event were done by Kurokawa et al. (1995), McAllister et al. (1996) and Morimoto et al. (2003,a,b). According to them, a bright soft X-ray arcade appeared striding over the filament channel after the filament eruption. The foot points of the arcade loops were located at the bright two-ribbon patches observed in Hα. In this event, heating up of the filament gas to 10,000 K was observed by the 17GHz radio observation by Hanaoka and Shinkawa (1999).

長大なフィラメントが上昇しはじめ，1時間程度でフィラメントが消滅した．上昇初期は5km/sのゆっくりとしたものであったが，軟X線の増光時に急激に加速され140km/sもの速度になった．この現象の「ようこう」衛星の軟X線画像とHα画像の比較研究が，Kurokawa et al.（1995），McAllister et al.（1996）とMorimoto et al.（2003,a,b）によってなされた．その結果によると，Hαフィラメントが消滅した後，元のフィラメント位置を跨ぐような形で明るい軟X線アーケード構造が現れた．そのアーケードの両足元は，元のフィラメント位置の両脇に現れた2つのツーリボン状の明るい帯状領域に繋がっていた．また，Hanaoka and Shinkawa (1999) によって，フィラメント上昇時，その温度が1万度まで加熱されていたことが17GHzの観測から報告されている．

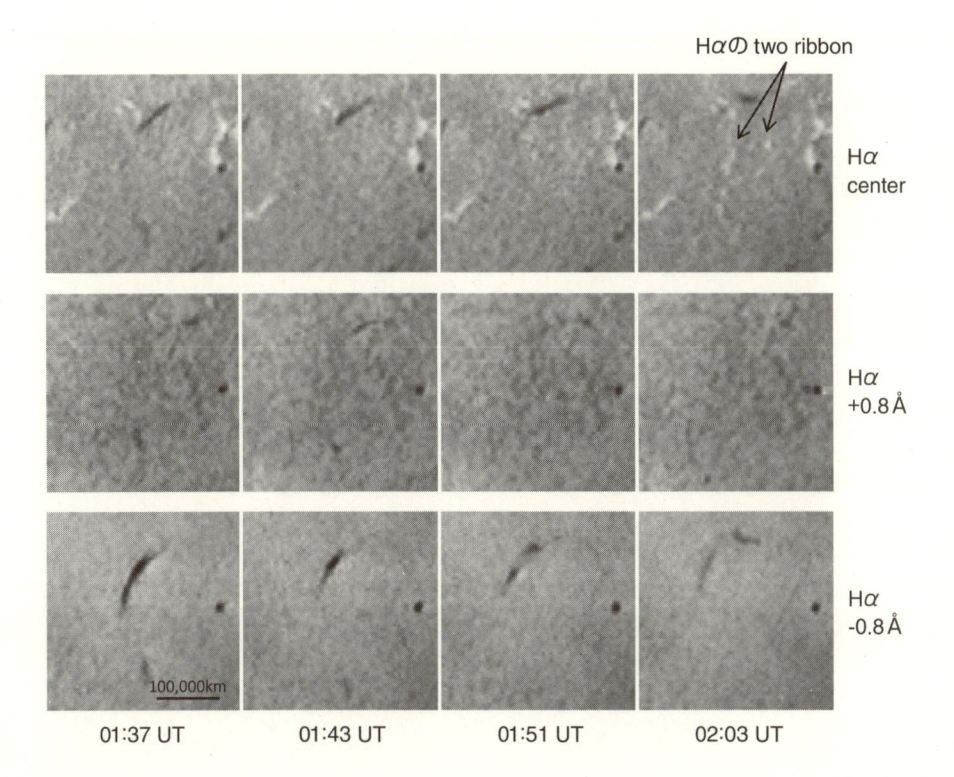

88 | 第6章 代表的イベント

Dec 21, 1992 01:38−04:37 UT Filament Activity

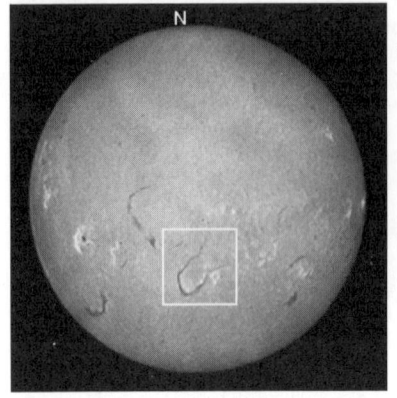

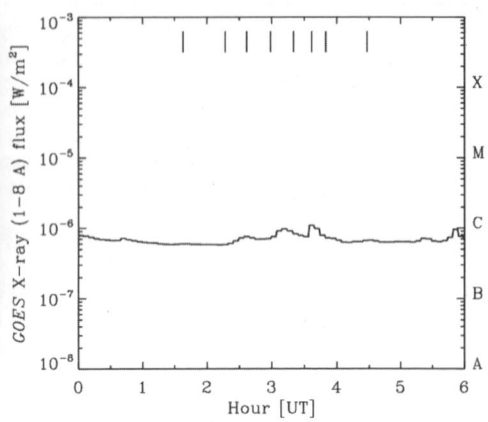

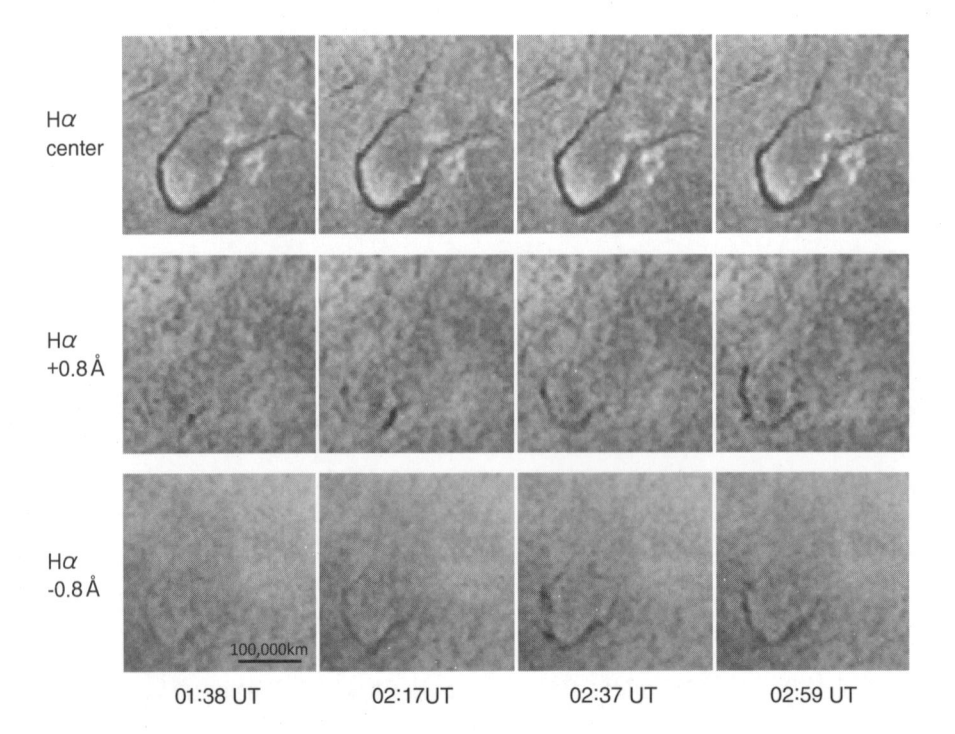

GOES Class = --　NOAA --
S35,E04　P = 7.3[deg]　B0= -1.7[deg]

Hα center

Hα +0.8Å

Hα -0.8Å

100,000km

01:38 UT　　02:17UT　　02:37 UT　　02:59 UT

Chapter 6 Representative Events | 89

A curved long filament of 500 Mm length was activated for 3 hours. Internal turbulent motions were excited intermittently without any large-scale morphological changes. Small Hα bight points were observed at both sides of the filament.

50万kmもの長さをもつ長大で湾曲した形のフィラメントが3時間の間活発化した．フィラメント全体の形はほとんど変えずに，内部乱流運動が引きこされていた．フィラメント両脇に時々Hα輝点が現われていた．

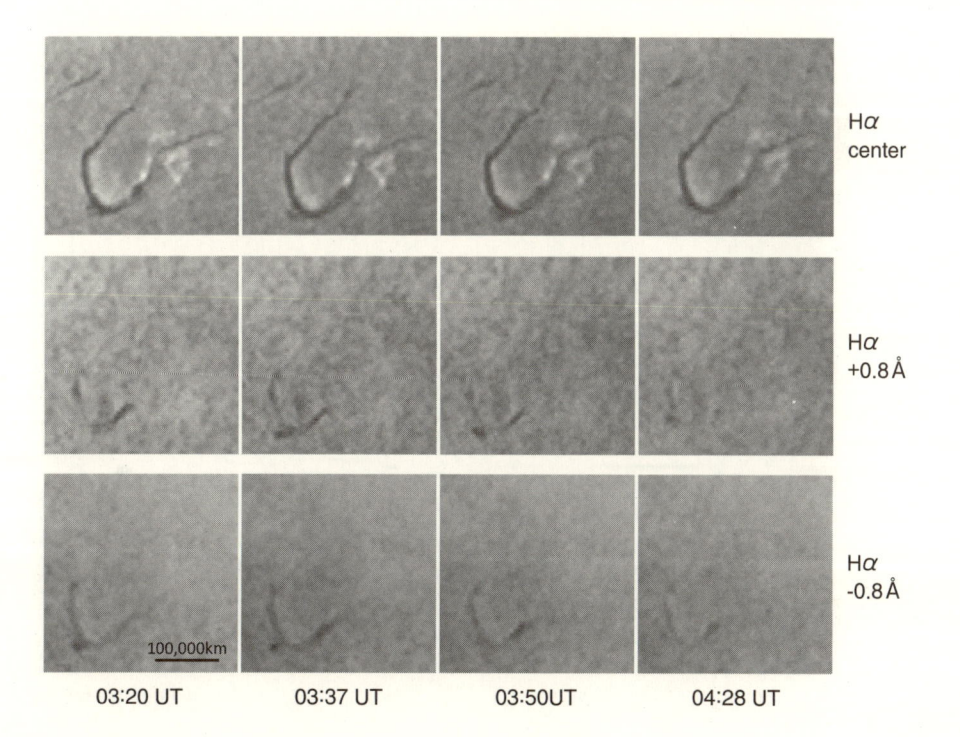

				Hα center
				Hα +0.8Å
				Hα -0.8Å
03:20 UT	03:37 UT	03:50 UT	04:28 UT	

Mar 20, 1993 00:16−01:14 UT   Flare, Filament Eruption

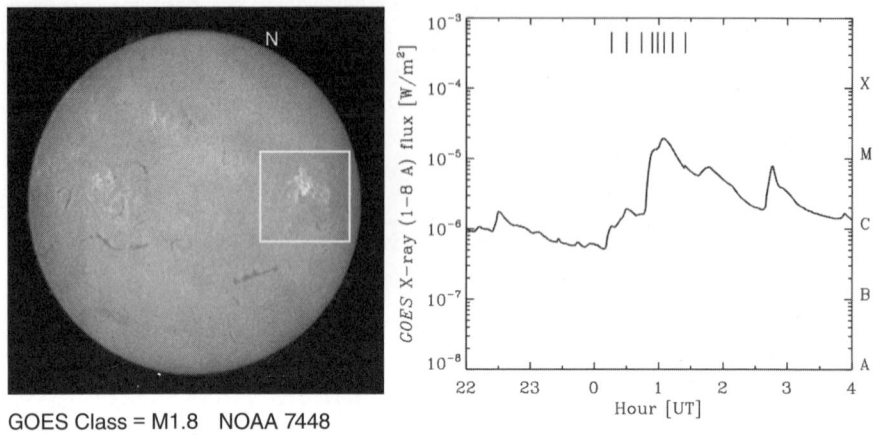

GOES Class = M1.8   NOAA 7448
N16,W38   P = -25.0 [deg]   B0= -7.1 [deg]

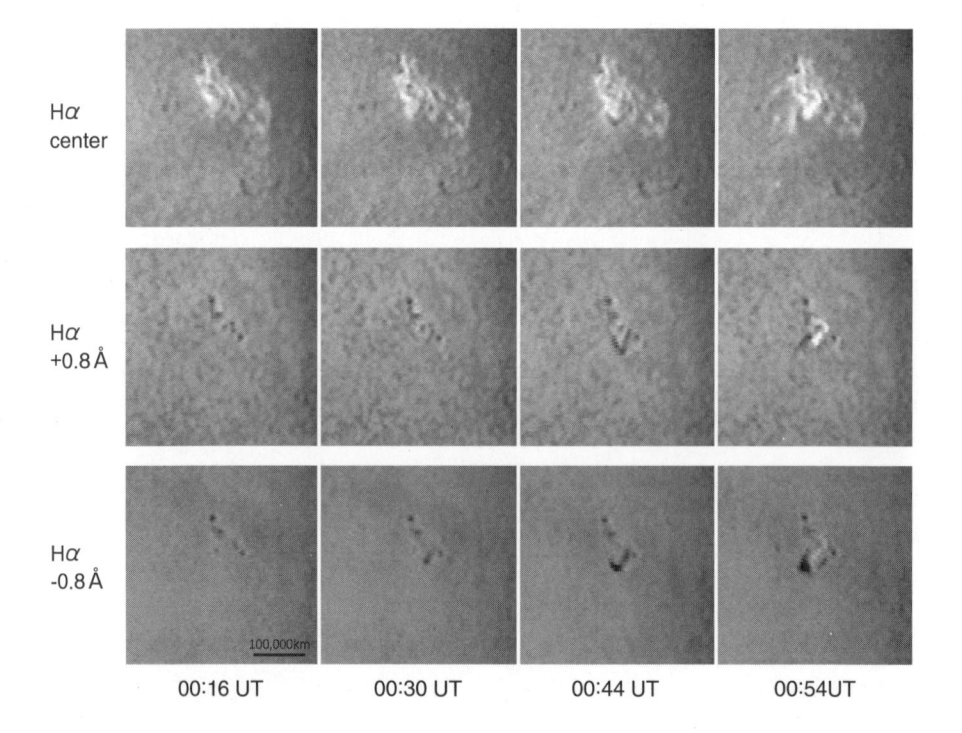

Hα center

Hα +0.8Å

Hα −0.8Å

100,000 km

00:16 UT   00:30 UT   00:44 UT   00:54 UT

Following the eruption of an active region filament, an M1.8 flare occurred in the active region NOAA7448. The two-ribbon patches seen in Hα were extremely bright and had wide Hα emission profiles in kernel parts, as the brightness in both Hα wings was high. The quiescent filament located south of the region seems to be activated by the flare explosion since 01:05 UT.

NOAA7448 で，活動領域フィラメントが爆発的上昇を始めた後 M1.8 のフレアが発生した．Hα 線で見えるツーリボン領域は極めて明るく，そのカーネル部は Hα±0.8Å 像でも明るく見えることから幅の広い Hα 輝線プロファイルを持っていたものと思われる．活動領域の南にあった静穏フィラメントが 01:05 UT 以降，このフレア爆発により活発化したように見受けられる．

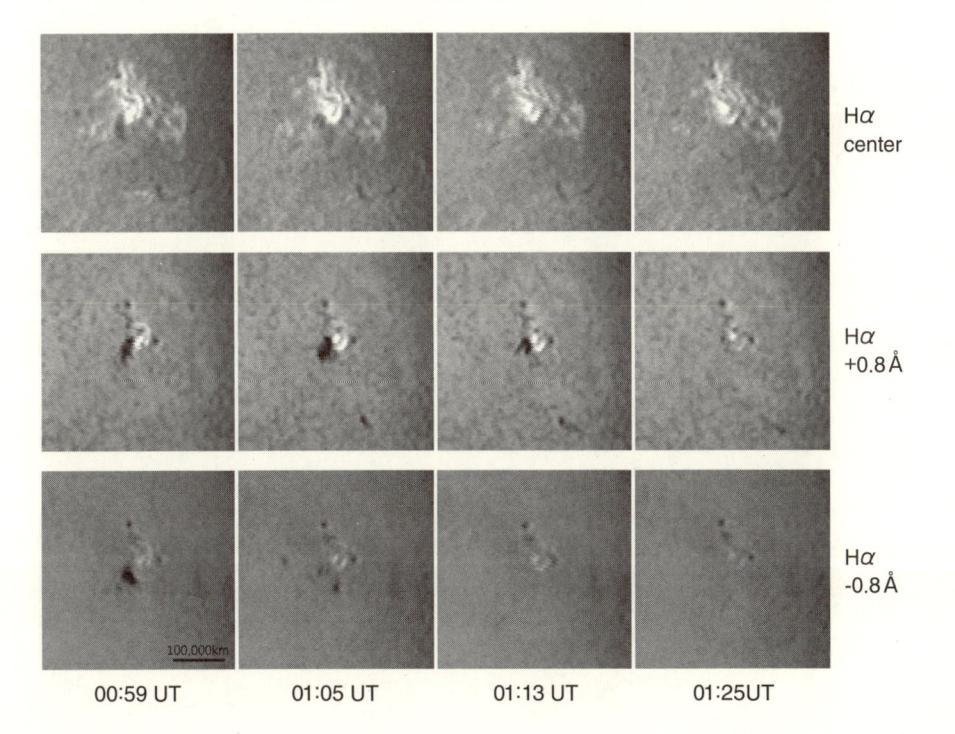

May 14, 1993 21:57−00:09 UT   Flare, Filament Eruption

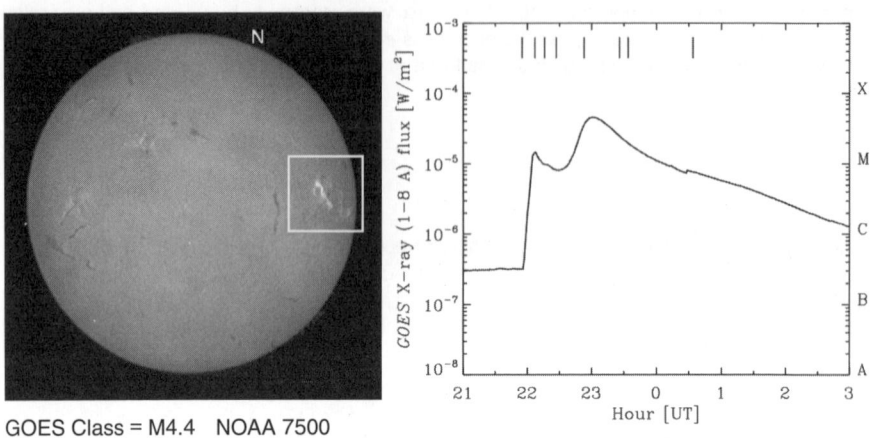

GOES Class = M4.4   NOAA 7500
N15,W52   P = -21.3 [deg]   B0 = -2.8 [deg]

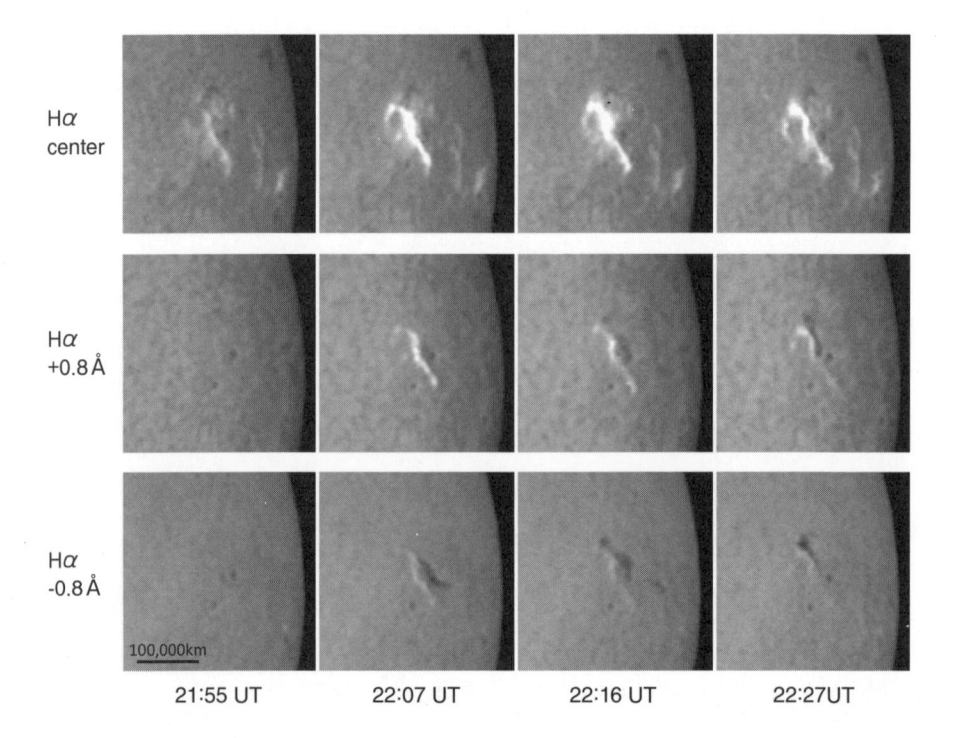

Hα center

Hα +0.8Å

Hα -0.8Å

100,000km

21:55 UT   22:07 UT   22:16 UT   22:27 UT

Following the eruption of an active region filament, an M4.4 flare occurred in the active region NOAA7500. At the first stage of the flare, the two-ribbon patches were located near the magnetic neutral line. The bright patch in positive magnetic polarity was displaced from that in the negative polarity in the direction of the magnetic neutral line. At the main explosive phase of the flare, the two ribbon patches in both polarities extended their lengths parallel to the direction of the neutral line, and started to move and separate to each other. From the development of the Hα ribbons, it seems that the coronal magnetic field above the neutral line changed from the sigmoidal one to the arcade-type during the course of the flare.

活動領域フィラメントの上昇後，NOAA7500 で M4.4 クラスのフレアが発生した．このフレアの初期には，磁気中性線の近傍に Hα ツーリボンが現われた．このときは 2 つのリボン状の明るい領域の位置は，中性線方向に互いに大きくずれていた．ところが，引き続いて起きたフレアの主爆発時には，リボン状の明るい領域が磁気中性線の方向に延伸し，お互い正面に相対する形となった後，相互に分離するような動きを示した．このような振る舞いから判断すると，コロナ中の磁場は，フレアの発展に応じてシグモイド型からアーケード型に変化したものと思われる．

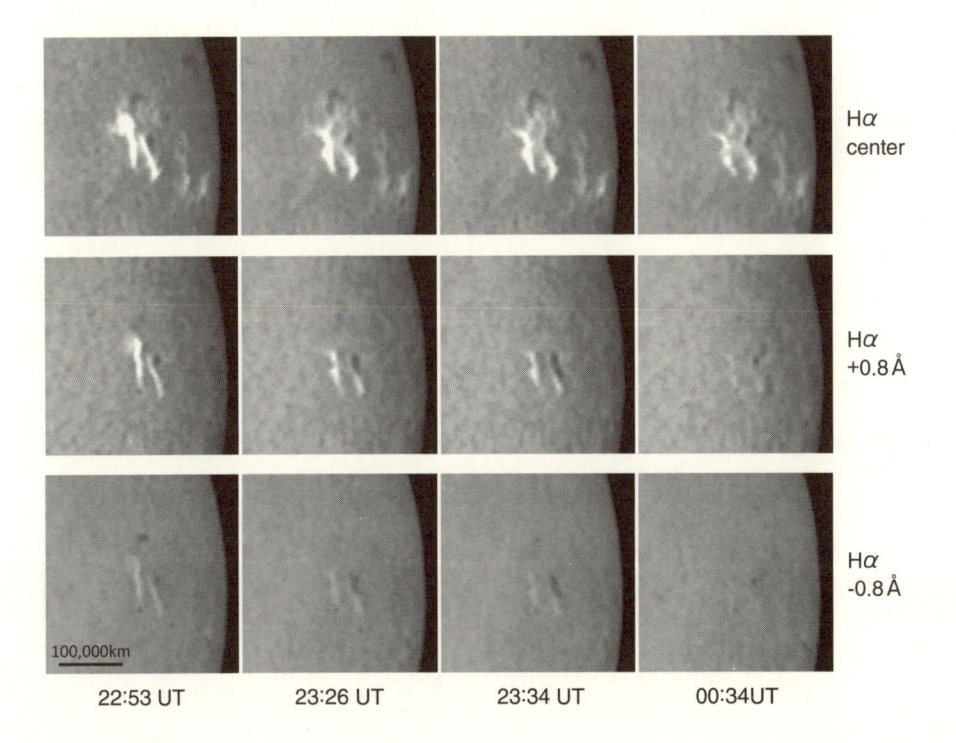

94 | 第6章 代表的イベント

May 15, 1993　22:29－00:27 UT　Filament Eruption

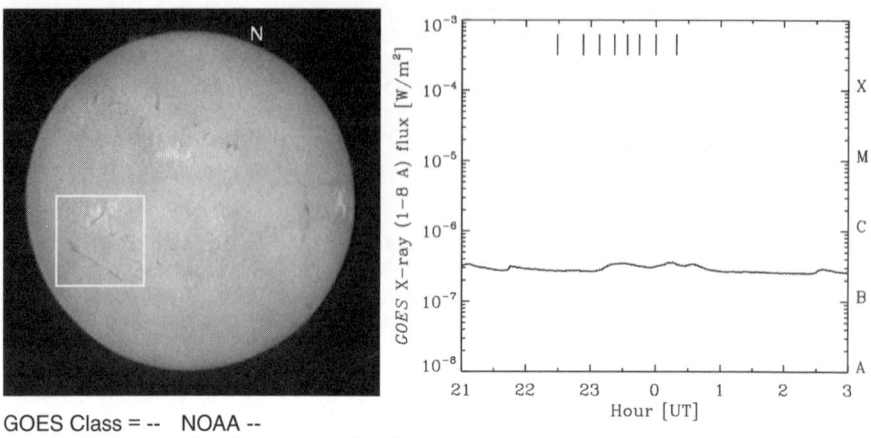

GOES Class = --　NOAA --
S28,E30　P = -21.0 [deg]　B0= -2.6 [deg]

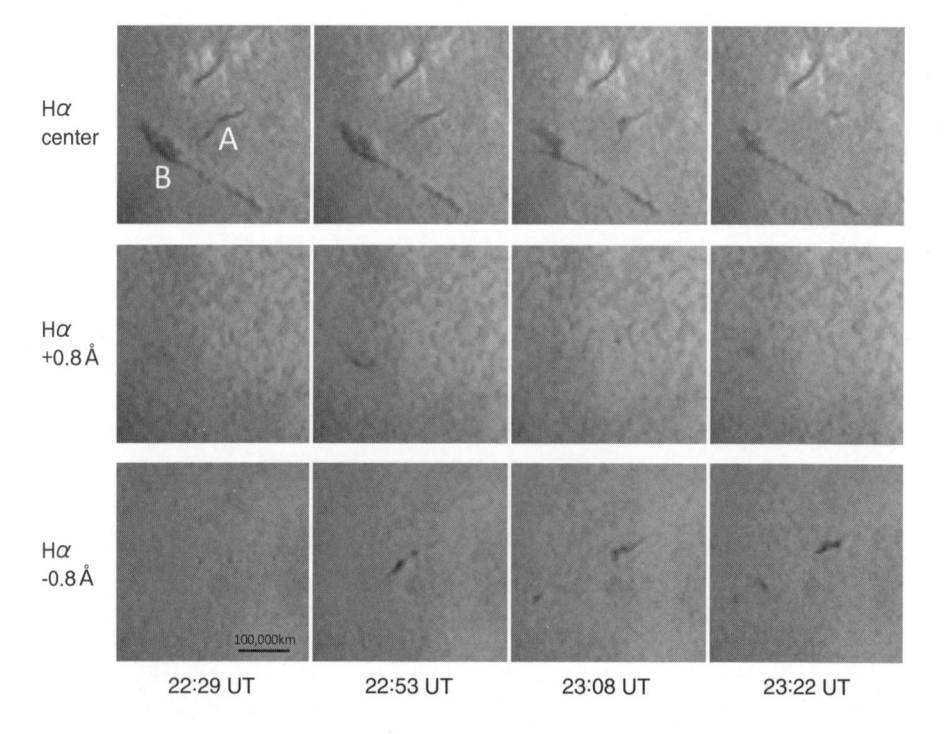

Hα center

Hα +0.8 Å

Hα -0.8 Å

100,000 km

22:29 UT　　22:53 UT　　23:08 UT　　23:22 UT

In the two quiescent filaments located at the center and the south in the sub-panel field, erupting motions occurred for 2 hours. According to Morimoto et al. (2003b), about half of the filament gas in both filaments were erupted to the interplanetary space due to the activities. In Yohkoh/SXT, bright arcades were formed in the corona over the filament locations.

サブパネル視野の中央および南側にある２つの静穏フィラメント（AとB）で，２時間にわたってフィラメントの膨張上昇運動が起こった．Morimoto et al. (2003b) によると，それぞれのフィラメント内のガスの約半分が惑星間空間に噴出した．また，この現象によって，２つのフィラメント上空のコロナに，明るいアーケード状の構造が形成されたことが Yohkoh/SXT により観測されている．

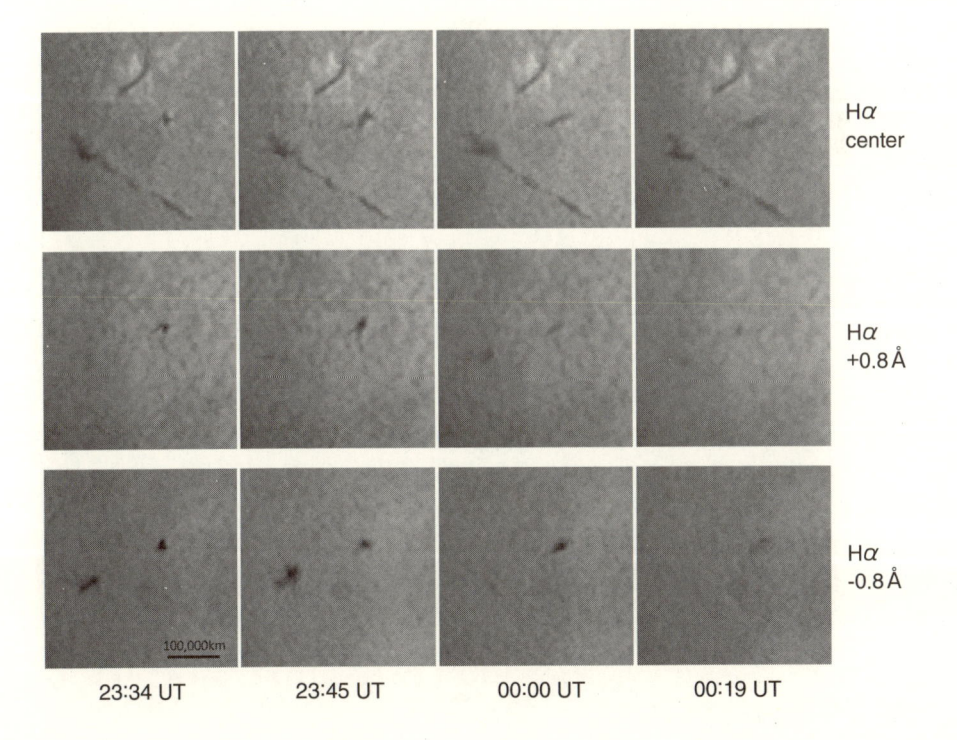

96 | 第6章 代表的イベント

Jun 06, 1993 07:07−07:46 UT Flare, Surge

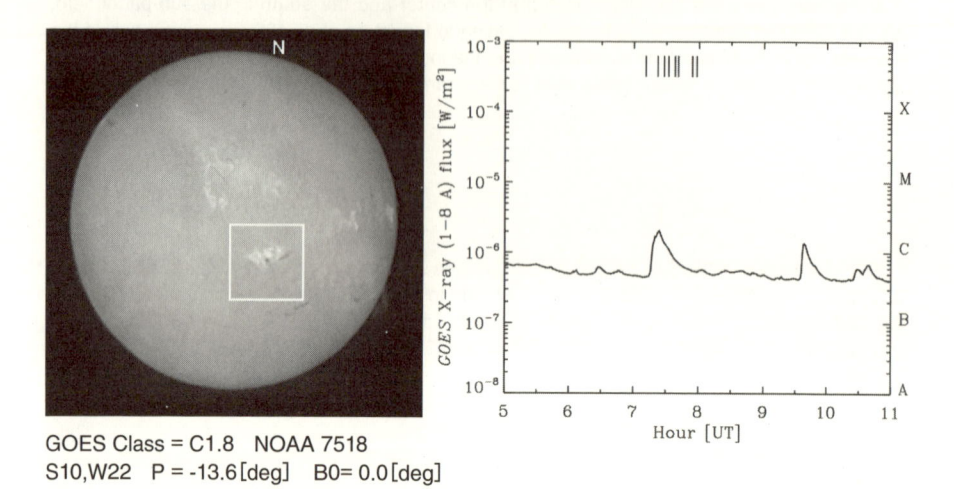

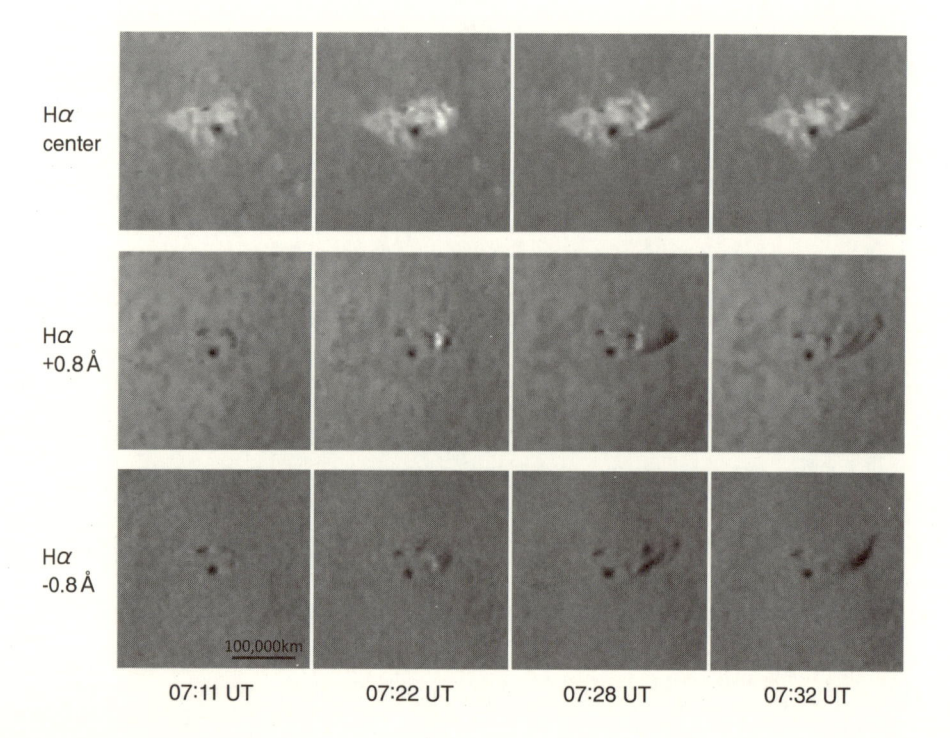

GOES Class = C1.8   NOAA 7518
S10,W22   P = -13.6 [deg]   B0= 0.0 [deg]

Hα center

Hα +0.8 Å

Hα -0.8 Å

100,000 km

07:11 UT   07:22 UT   07:28 UT   07:32 UT

Chapter 6 Representative Events | 97

A compact C1.8 flare occurred in NOAA7518. Just simultaneously, a surge of gas was ejected from the flare kernel with a speed of about 160 km/s. In the region, similar activities were observed to occur on the next day.

活動領域 NOAA7518 において，C1.8 のコンパクトフレアが発生した．フレア爆発と同時に，フレアのカーネルから約 160 km/s の速さでサージが噴出した．この領域では，次の日も同様な現象が観測されている．

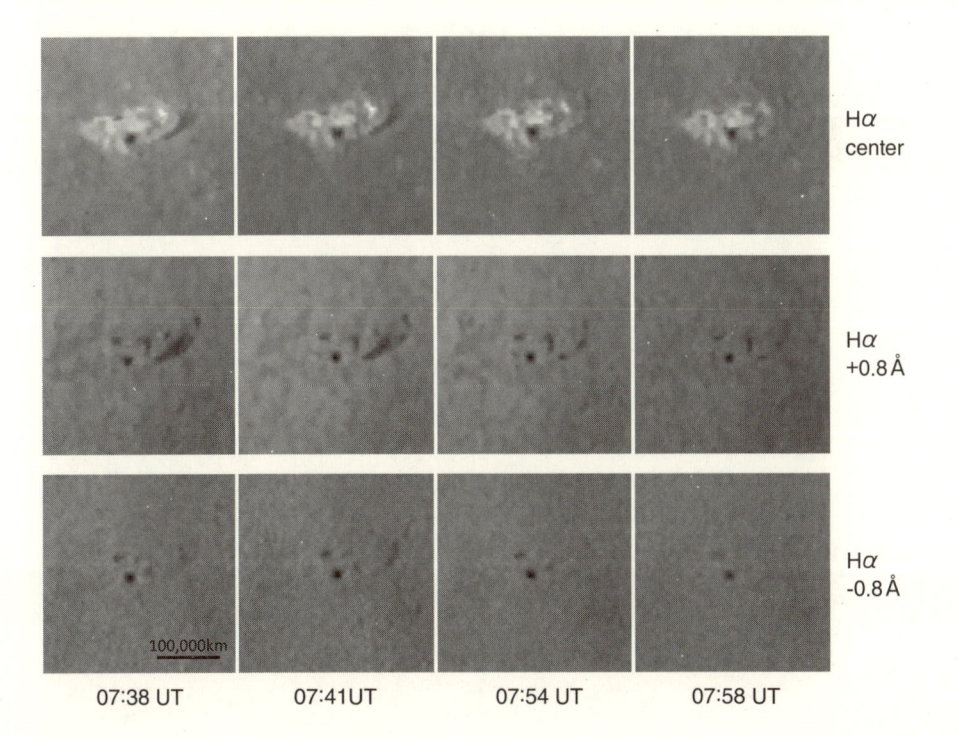

Hα center

Hα +0.8Å

Hα -0.8Å

100,000km

07:38 UT   07:41UT   07:54 UT   07:58 UT

Jun 07, 1993 05:43−06:44 UT Flare, Surge

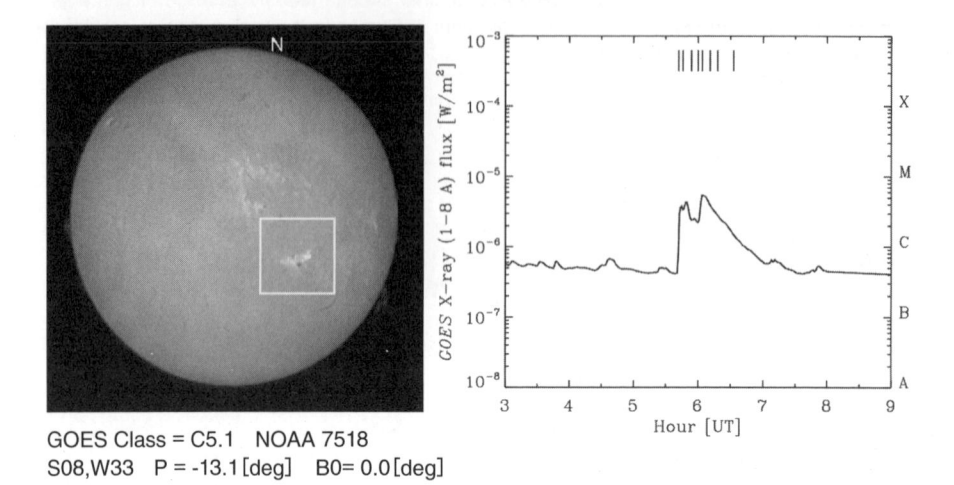

GOES Class = C5.1   NOAA 7518
S08,W33   P = −13.1 [deg]   B0= 0.0 [deg]

Hα center

Hα +0.8Å

Hα −0.8Å

100,000km

05:42 UT   05:46 UT   05:54 UT   06:00 UT

Two compact C5 flares occurred consecutively in NOAA7518. Several strands of surge gases were ejected from the flare kernels with speeds of around 50 km/s. The region repeatedly produced the same type of activity since the previous day.

C5 クラスの 2 つのコンパクトフレアが NOAA7518 で引き続いて発生した．それぞれのフレアカーネルから約 50 km/s の速さでサージが何回も噴出した．この領域では，前日より同じような活動が繰り返し発生した．

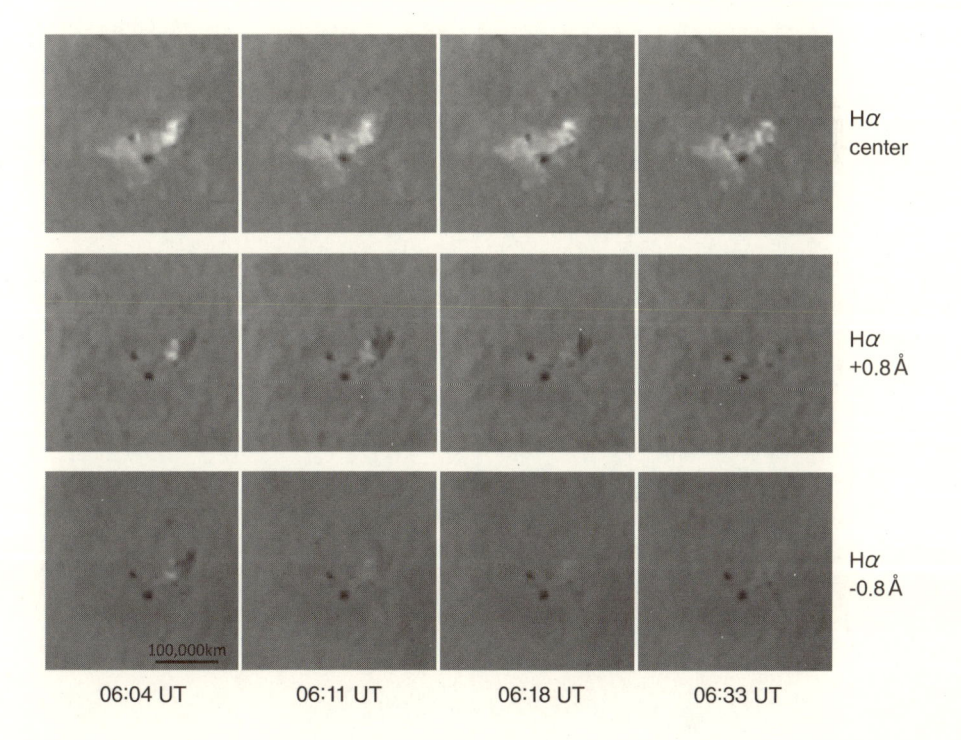

				Hα center
				Hα +0.8Å
				Hα -0.8Å
06:04 UT	06:11 UT	06:18 UT	06:33 UT	

100,000km

Sep 25, 1993 23:03−23:51 UT Filament Eruption

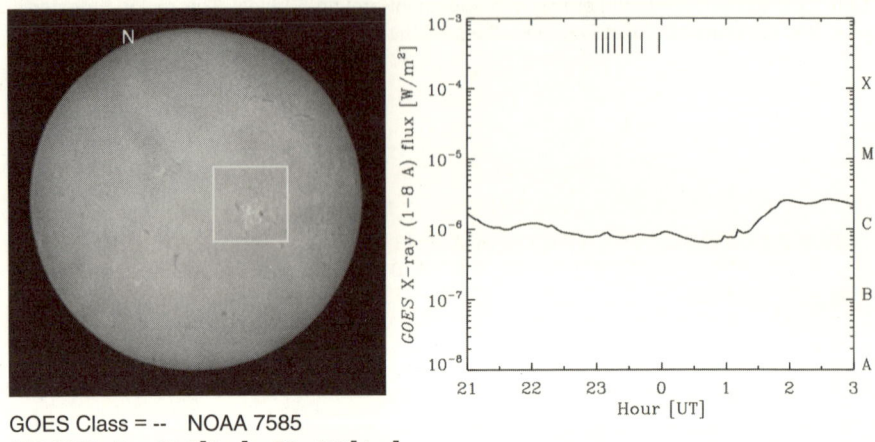

GOES Class = --   NOAA 7585
S06, W09   P = 25.5 [deg]   B0 = 7.0 [deg]

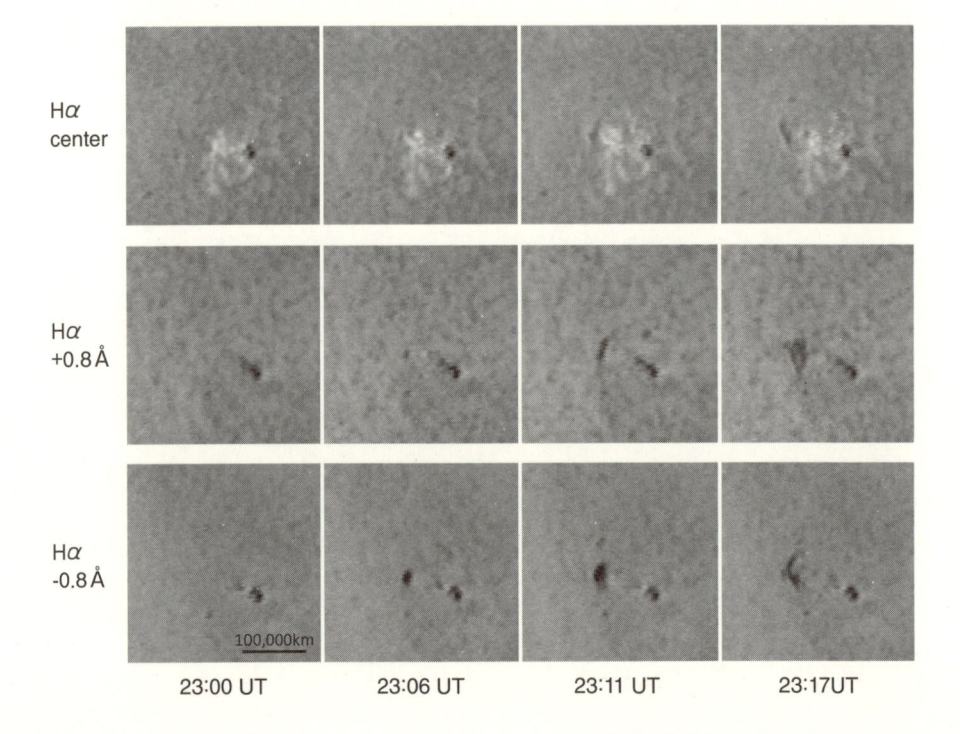

Hα center

Hα +0.8Å

Hα −0.8Å

100,000 km

23:00 UT   23:06 UT   23:11 UT   23:17 UT

Chapter 6 Representative Events | 101

In the northern periphery of the NOAA7585, transient filament activity took place for about 1 hour. Loop-shaped filaments appeared and disappeared intermittently showing the complex internal motions, such as ejections, streaming and etc.

活動領域 NOAA7585 の北の周辺部で，1 時間程度フィラメント活動が活発になった．ループの形をしたフィラメントが生まれては消えるという現象であった．ガスが放出されたり，ループの軸方向に流れ出したりいう複雑な動きを示した．

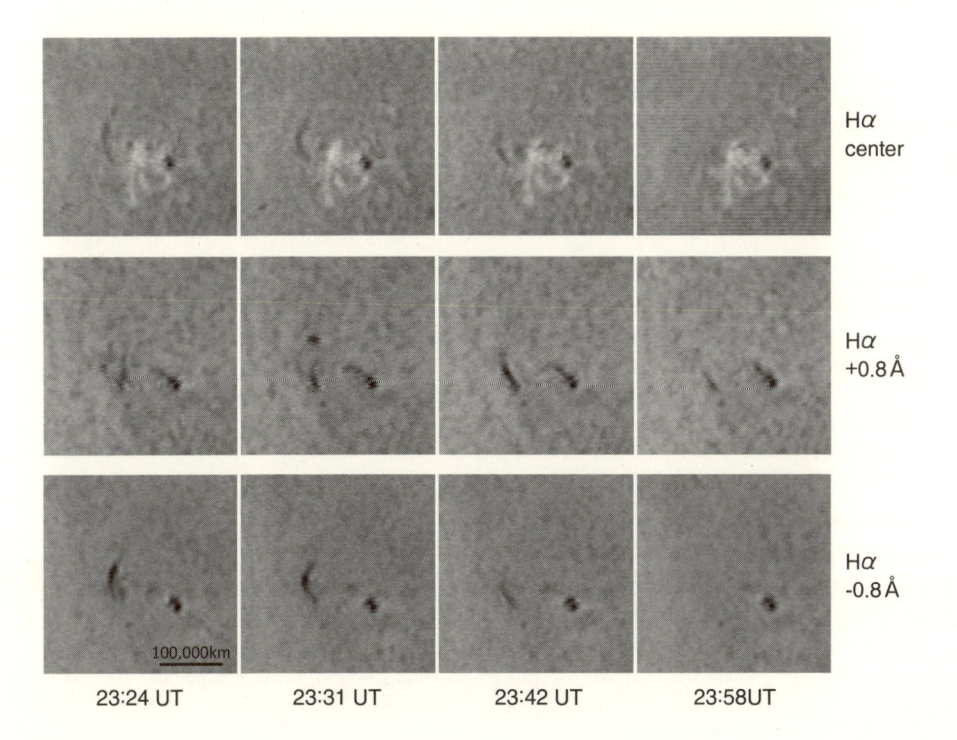

102 | 第6章 代表的イベント

Oct 21, 1993 02:51−05:19 UT Filament Eruption

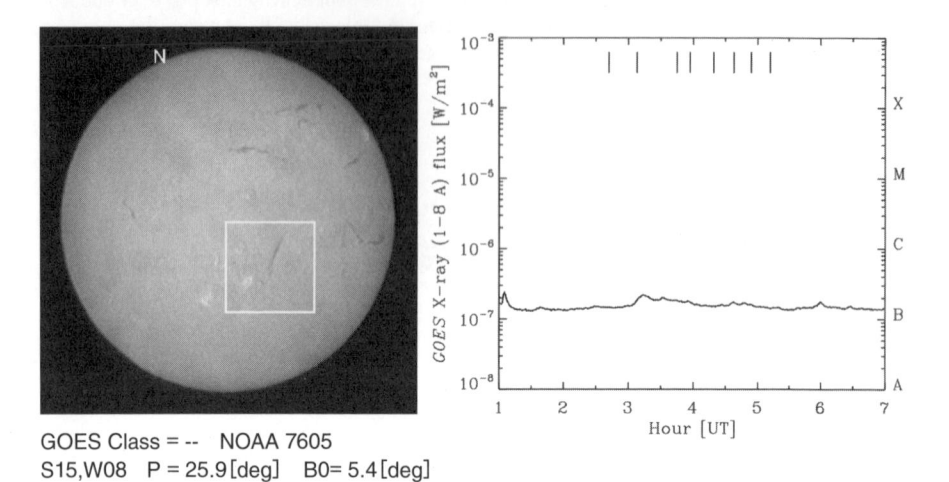

GOES Class = --   NOAA 7605
S15, W08   P = 25.9 [deg]   B0 = 5.4 [deg]

Hα center				
Hα +0.8Å				
Hα -0.8Å				
	02:42 UT	03:08 UT	03:45 UT	03:57 UT

The filament erupted and disappeared in a time span of about 100 min. No bright emissions in Hα were observed after the eruption. In soft X-ray observation, no arcade structure over the filament channel was formed, but several strand of X-ray loops (or X-ray filaments) along the filament channel were observed to form in a long time span of 10 hours according to Kurokawa et al. (1995).

このフィラメントは約 100 分で消失した．フィラメント消失時，消失フィラメントの元の位置の両脇に Hα で明るい模様がときに見られるが，このイベントでは現れなかった．Kurokawa et al. (1995) によると，軟 X 線画像では，アーケード構造は観測されなかった．代わりに，フィラメントの軸に沿う方向のループ状の X 線フィラメントが 10 時間ほどの時間をかけてゆっくりと形成された．

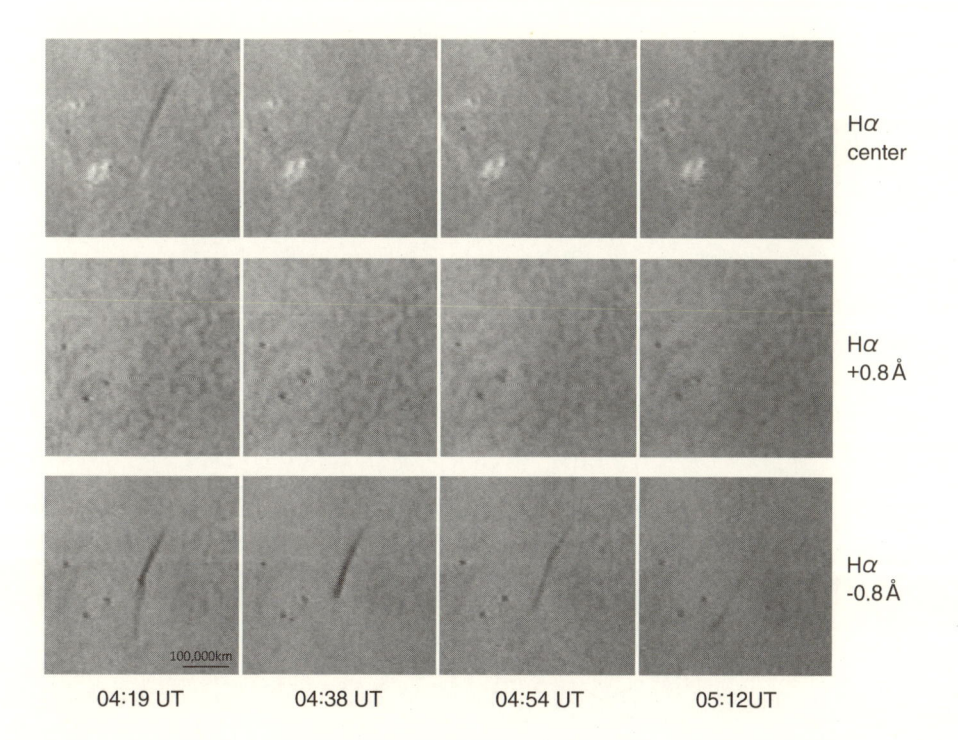

				Hα center
				Hα +0.8Å
				Hα -0.8Å
04:19 UT	04:38 UT	04:54 UT	05:12UT	

Jan 05, 1994 06:04−07:05 UT Flare, Filament Eruption

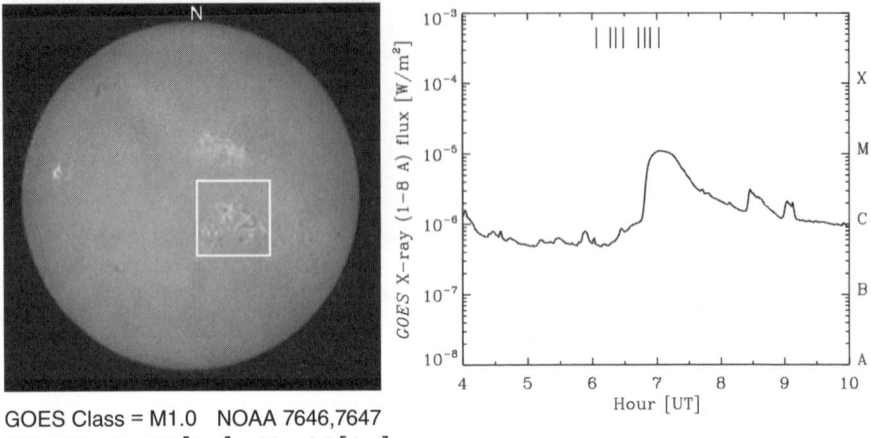

GOES Class = M1.0   NOAA 7646, 7647
S10, W22   P = 0.2 [deg]   B0 = −3.5 [deg]

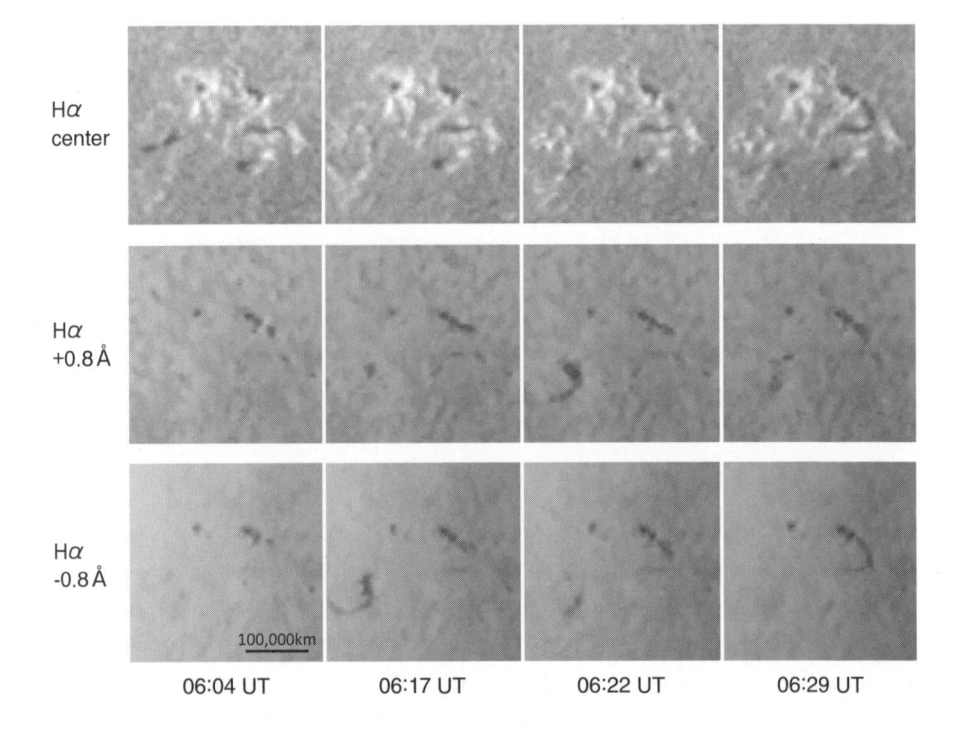

Hα center

Hα +0.8Å

Hα −0.8Å

100,000 km

06:04 UT   06:17 UT   06:22 UT   06:29 UT

Chapter 6 Representative Events | 105

NOAA7646 was a $\beta\gamma$-type active region of complex magnetic field distribution. In the pre-flare stage, two separate active region filaments were activated to be merged into one long filament. After the eruption of one part of this merged filament, an M1.0 flare took place and there appeared several bright patches in Hα.

NOAA7646 は，$\beta\gamma$ 型の複雑な磁場分布をもつ領域であった．フレアが発生する前に，2 つの活動領域フィラメントが融合して 1 つの長いフィラメントとなった．この長いフィラメントの一部が爆発上昇して，M1.0 のフレアを発生させた．Hα では数カ所の明るい領域が現われた．

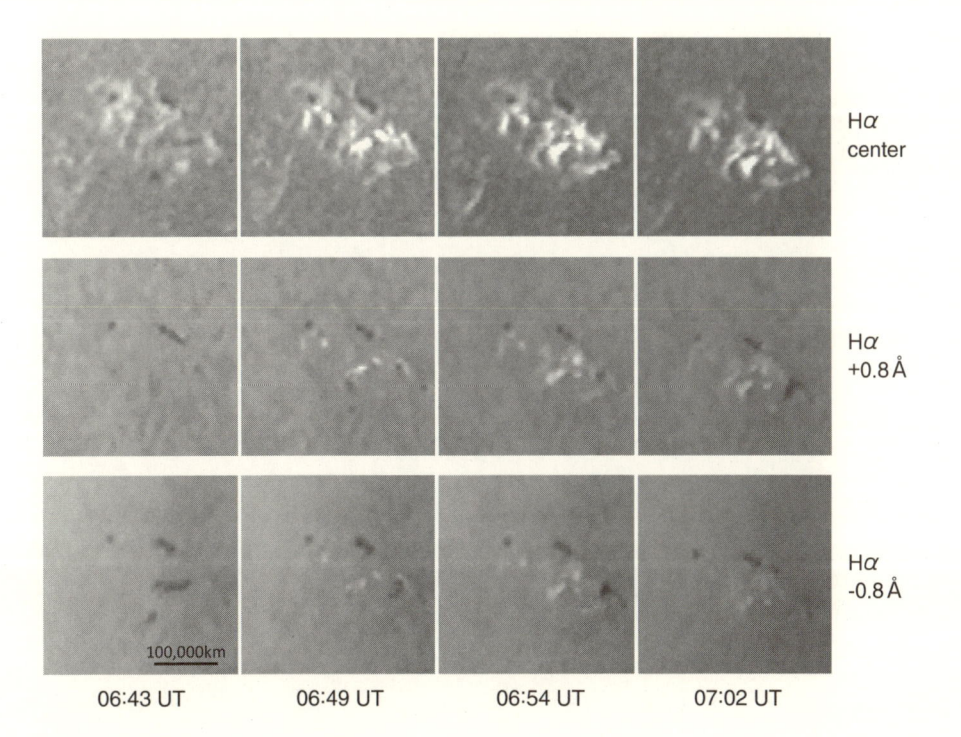

Hα center

Hα +0.8Å

Hα -0.8Å

100,000km

06:43 UT   06:49 UT   06:54 UT   07:02 UT

106 | 第6章 代表的イベント

Feb 20, 1994 00:17−02:47 UT   Flare, Filament Eruption

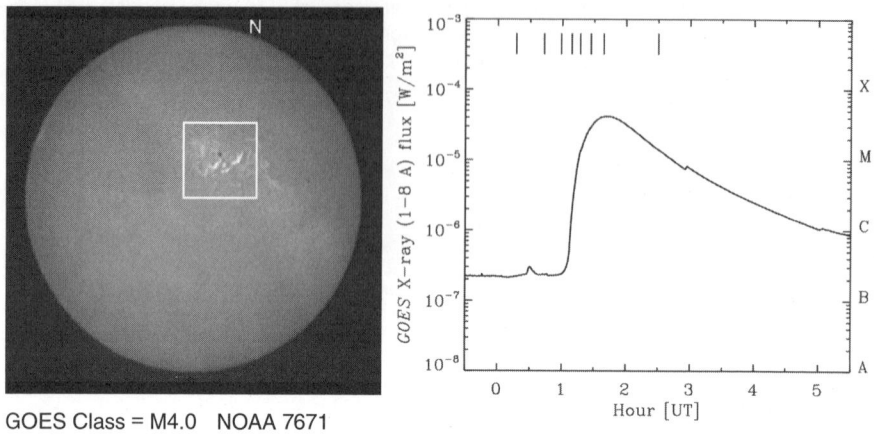

GOES Class = M4.0   NOAA 7671
N01,W00   P = −19.0[deg]   B0= −7.0[deg]

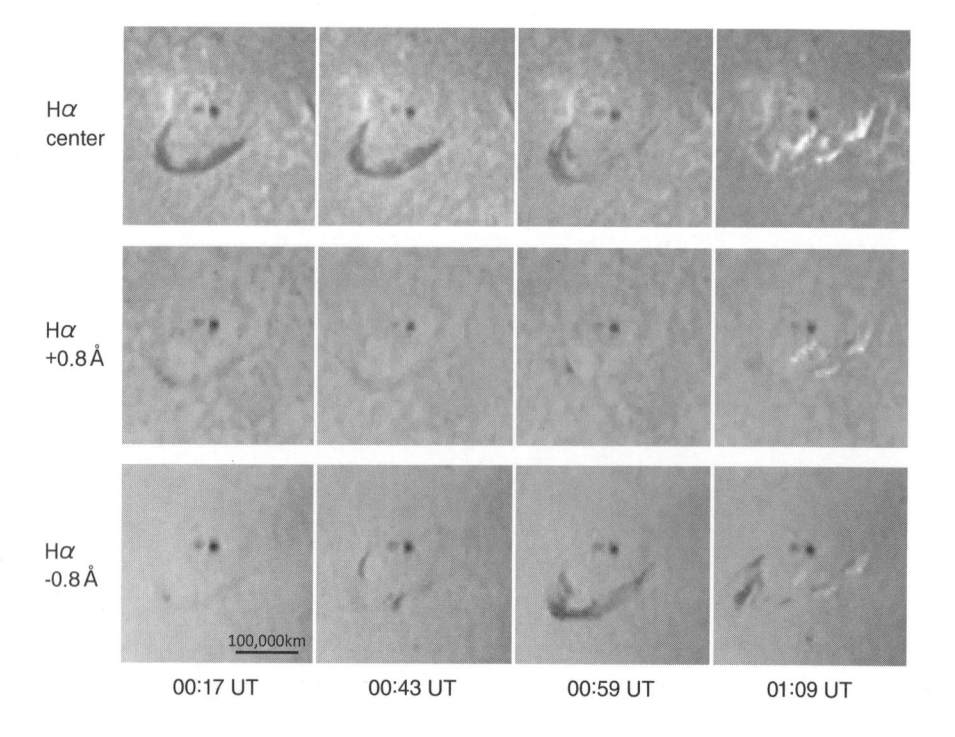

Hα center

Hα +0.8Å

Hα −0.8Å

100,000km

00:17 UT   00:43 UT   00:59 UT   01:09 UT

M4.0 flare occurred after a filament eruption. According to Hanaoka and Shinkawa (1999), soft X-ray loops ran along the filament in highly sheared form at the preflare stage. After the eruption, soft X-ray arcade and two-ribbons in Hα appeared. The event was a typical one of the filament eruption followed by a long duration flare. From the 17 GHz radio observation, it was found that the filament gas was heated up to 10,000 K. In association with the event, an interplanetary shock was observed in situ to propagate by the Geotail satellite and was identified as the source of low energy (< 100 keV) ions acceleration (Koi et al. 2005, Terasawa et al. 2006).

フィラメント爆発の後，M4.0のフレアが発生した．Hanaoka and Shinkawa (1999) によると，フレアの前駆段階では，大きくねじれた軟X線ループがフィラメントの軸に沿う方向に見られた．フィラメント爆発後，軟X線アーケード構造およびHαツーリボンが出現した．この現象は，フィラメント爆発後に長寿命フレアが発生する典型例である．17 GHzでの観測から，上昇するフィラメントガスは1万度まで加熱されていたことが見つけられている．このイベントに伴って，惑星間空間衝撃波がジオテイル衛星によって観測された．この観測から，衝撃波によって100 keV以下の低エネルギー粒子が加速されていることが見出された（Koi et al. 2005, Terasawa et al. 2006）．

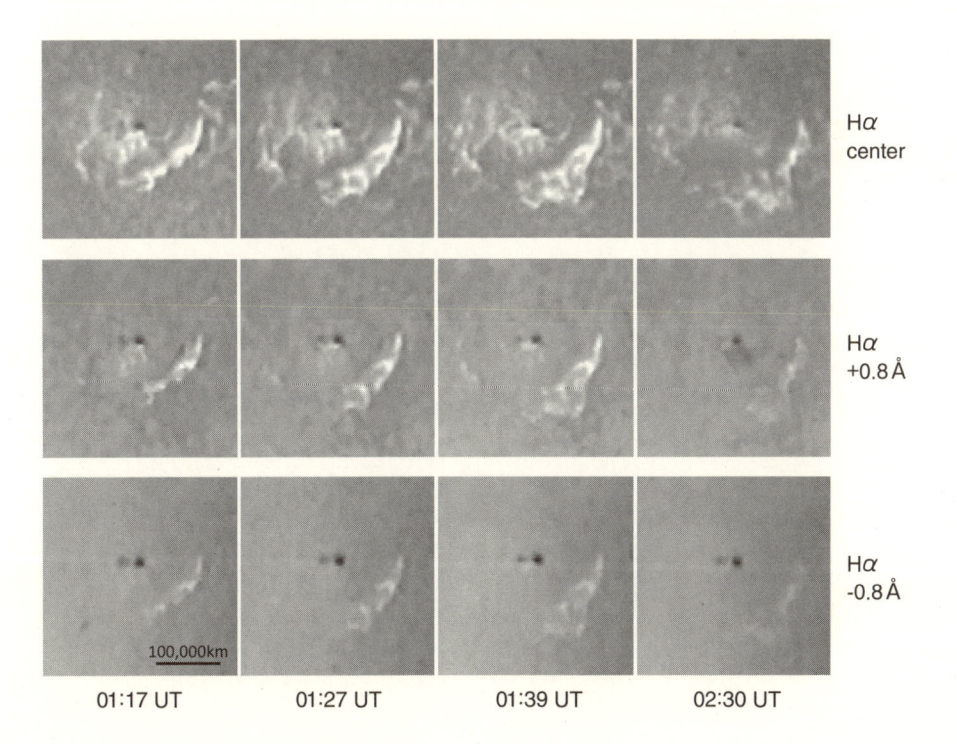

Jul 12, 1994 22:46－23:50 UT Surge

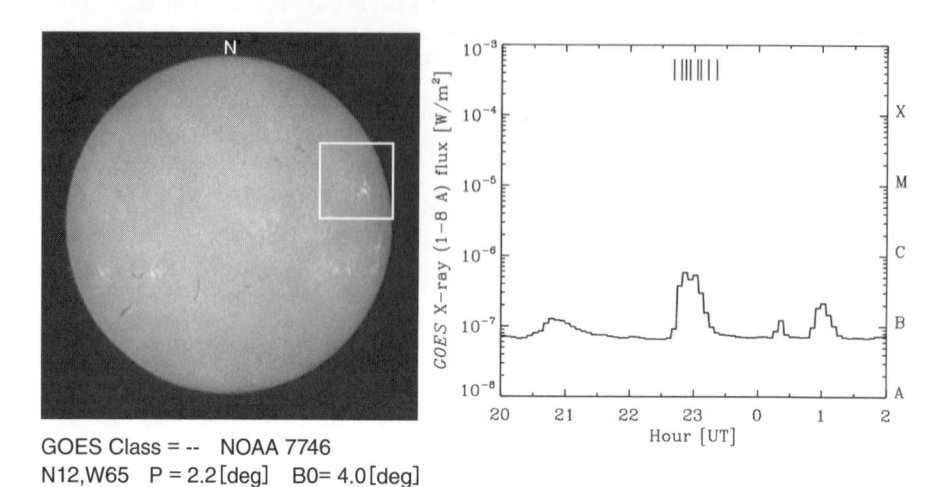

GOES Class = --   NOAA 7746
N12, W65   P = 2.2 [deg]   B0 = 4.0 [deg]

Hα center

Hα +0.8Å

Hα -0.8Å

100,000 km

22:41 UT   22:48 UT   22:52 UT   22:56 UT

From a bright point appeared in NOAA7746, a surge was ejected with an apparent speed of 130 km/s. A soft X-ray enhancement of B6 class was associated with the event.

NOAA7746 に Hα 輝点が現われ，そこからサージが見かけの速度 130 km/s で噴出した．この現象には B6 クラスの X 線増光が伴っていた．

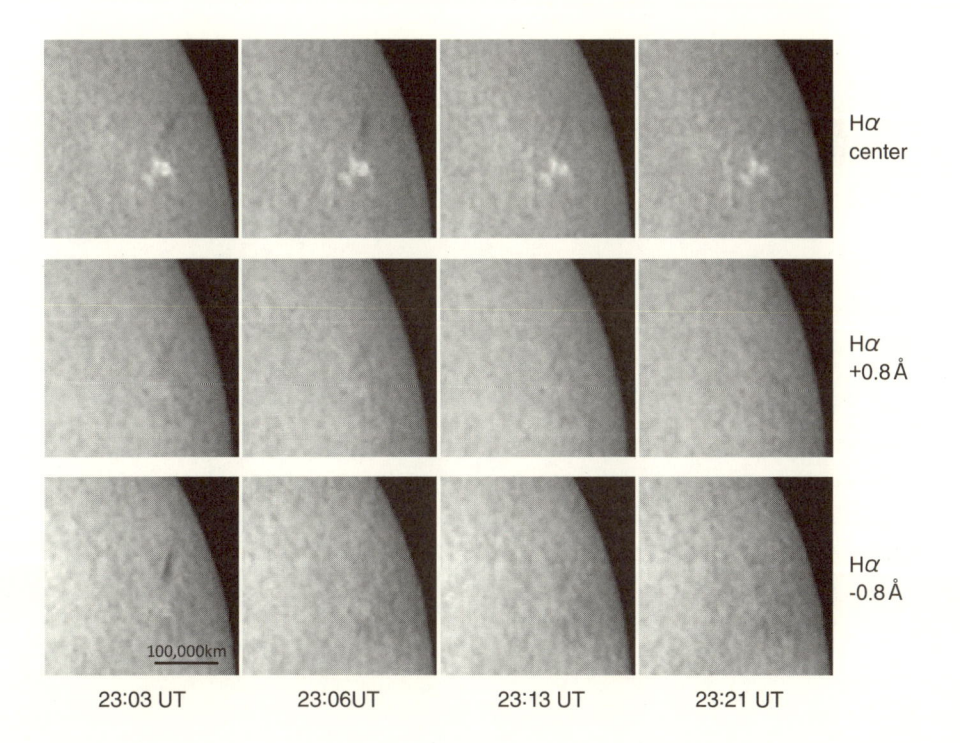

Sep 05, 1994 06:16−08:00 UT   Filament Eruption

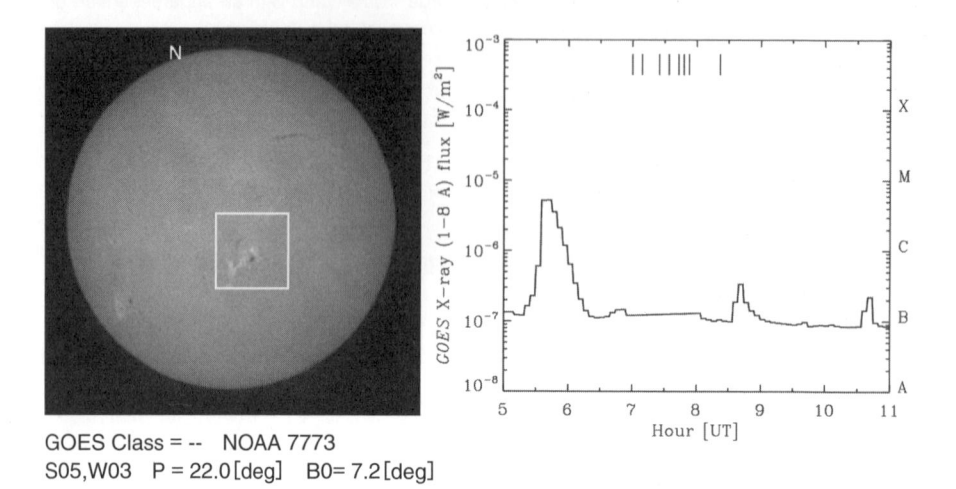

GOES Class = --   NOAA 7773
S05,W03   P = 22.0 [deg]   B0= 7.2 [deg]

Hα center

Hα +0.8Å

Hα −0.8Å

100,000km

07:00 UT   07:09 UT   07:25 UT   07:34 UT

Chapter 6 Representative Events | 111

Associated with the appearance of Hα bright patch in NOAA7773, an active region filament was destabilized for 1 hour. After the complex motions of uplifting and streaming-out, the filament was stabilized again to its original state. No prominent X-ray enhancement was associated with the event.

NOAA7773 に明るい Hα 領域が現われ，近傍の活動領域フィラメントが約 1 時間不安定化した．フィラメントは，上昇や軸方向の流れ出しなどの複雑な運動を示した後，当初の安定状態に落ち着いた．この現象には顕著な X 線増光は伴わなかった．

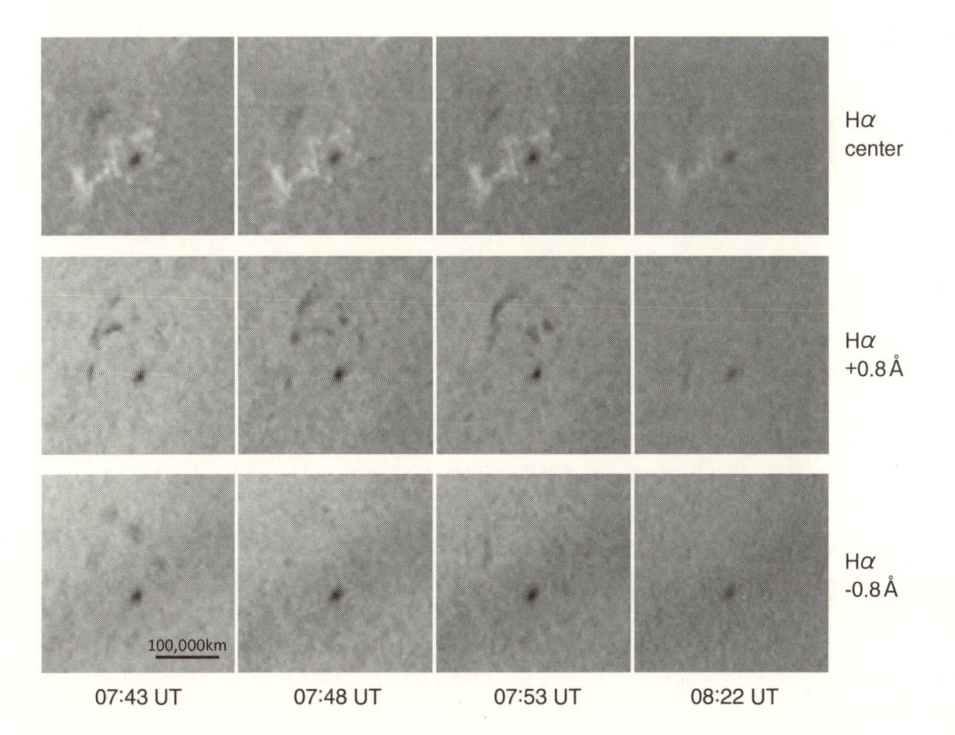

				Hα center
				Hα +0.8Å
				Hα -0.8Å
07:43 UT	07:48 UT	07:53 UT	08:22 UT	

112 | 第6章 代表的イベント

Feb 11, 1995 23:54−00:45 UT　Surge

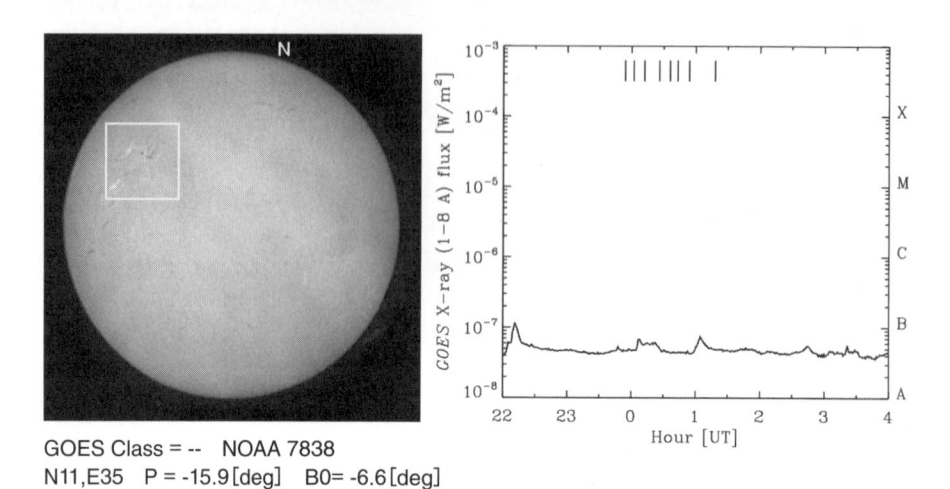

GOES Class = --　NOAA 7838
N11,E35　P = -15.9[deg]　B0= -6.6[deg]

| | 23:54 UT | 00:02 UT | 00:12 UT | 00:26 UT |

Hα center

Hα +0.8Å

Hα -0.8Å

100,000km

Chapter 6 Representative Events | 113

At the periphery of the preceding spot of NOAA7838, a surge activity occurred. The chromospheric gas was ejected from an Hα bright point and returned back to the original location along the same path of ejection.

NOAA7838の先行黒点の脇でサージが発生した．彩層ガスがHα輝点から放出され，放出時と同じ道筋をたどって帰還した．

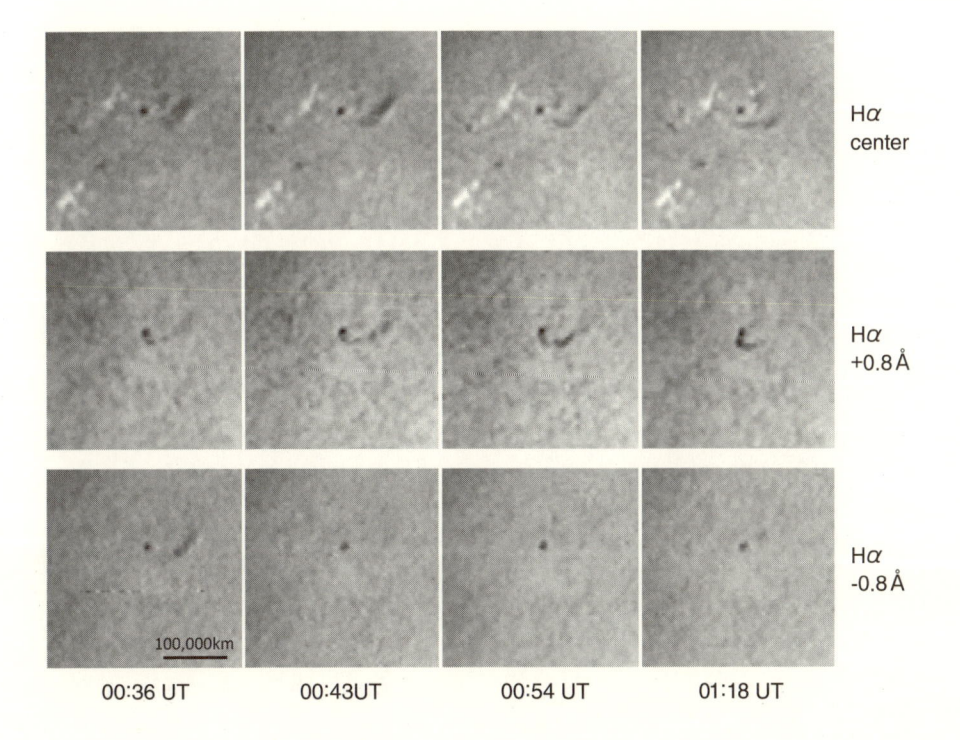

				Hα center
				Hα +0.8Å
				Hα -0.8Å
00:36 UT	00:43 UT	00:54 UT	01:18 UT	

100,000km

May 12, 1997 04:27−05:26 UT   Flare, Filament Eruption

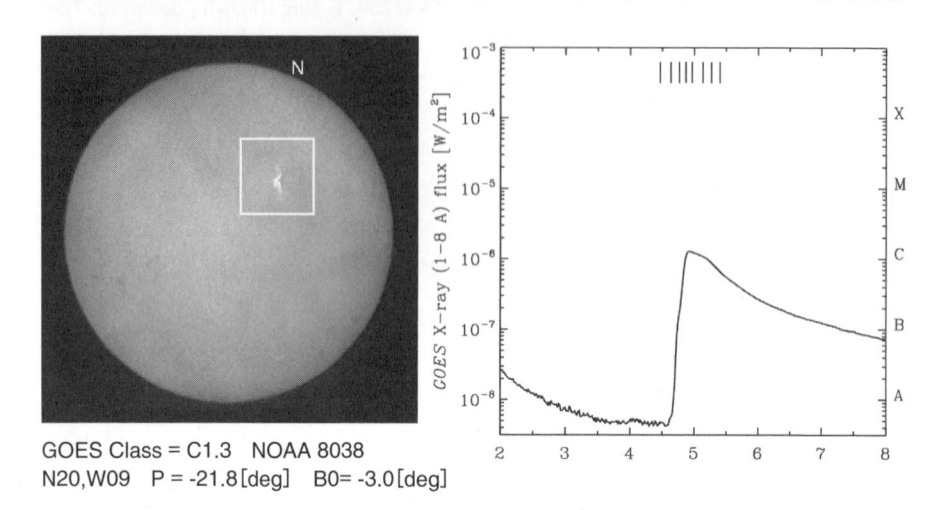

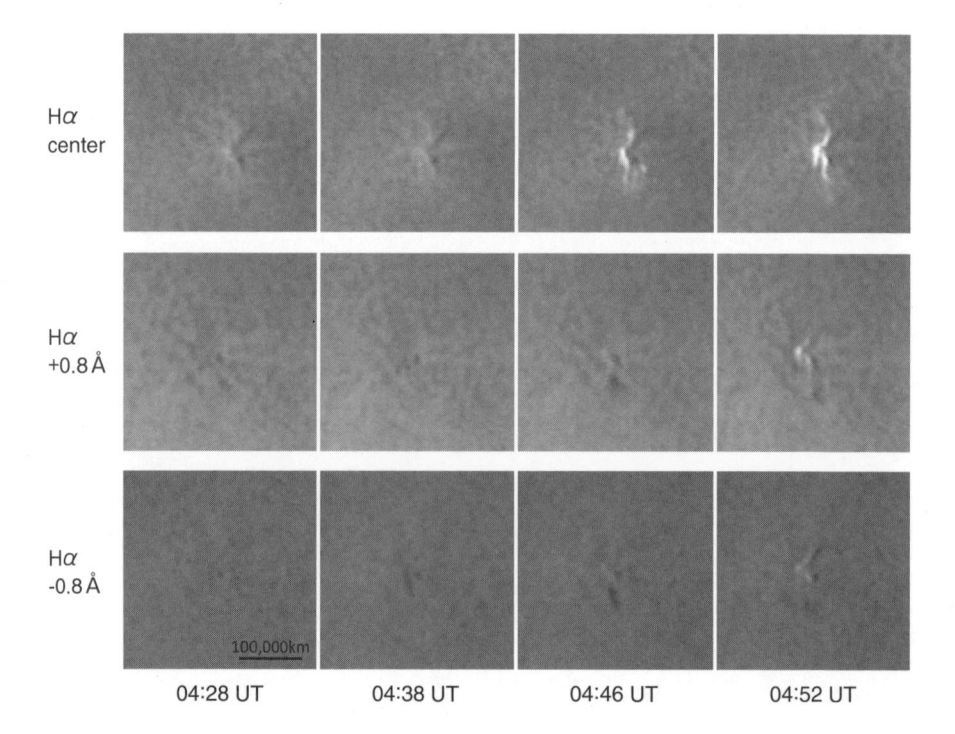

GOES Class = C1.3   NOAA 8038
N20,W09   P = -21.8 [deg]   B0= -3.0 [deg]

Hα center

Hα +0.8 Å

Hα -0.8 Å

100,000 km

04:28 UT   04:38 UT   04:46 UT   04:52 UT

After an Hα filament eruption to the south, a two-ribbon C1.3 flare occurred in NOAA8038. The SXT telescope of Japanese Yohkoh satellite clearly detected the formation and evolution of several cusp-shaped loops of the X-ray arcade in the long lasting LDE phase. From the photospheric magnetic field data and the motion of Hα ribbons, Isobe et al. (2002) succeeded in quantitatively estimating the energy release rate and the magnetic reconnection rate for the first time, and found that the flare model based on the magnetic reconnection in the corona described in Shibata (1999) can nicely explain the evolution of the flare.

　活動領域 NOAA8038 で，Hαフィラメント爆発が起こり C1.3 の X 線クラスのツーリボンフレアが発生した．日本の「ようこう」衛星搭載の SXT 望遠鏡（Yohkoh/SXT）は，このフレア後長時間継続した LDE 現象の間，カスプ形ループでできた X 線アーケードが形成され成長してゆくのを明瞭に捉えた．このときの光球磁場および Hαリボンの動きを解析して，Isobe et al. (2002) はフレア時のエネルギー解放率と磁気リコネクション率を世界で初めて求めることに成功した．また，Shibata（1999）で示されているコロナ中の磁気リコネクションによるフレアモデルが，このフレアの時間発展をよく説明することを見出した．

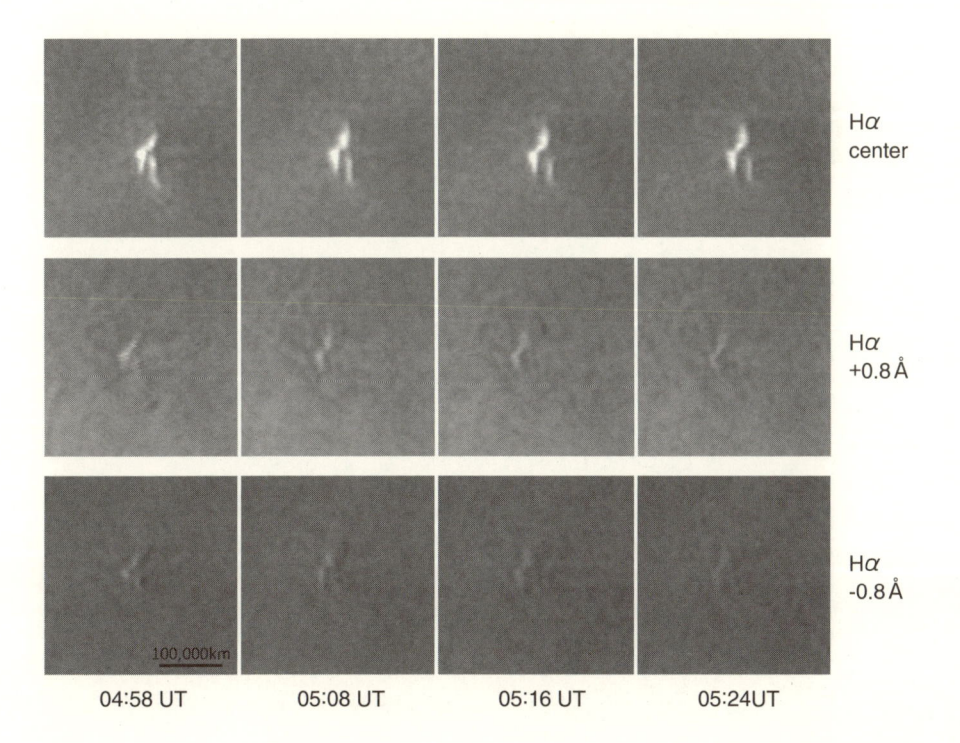

Aug 29, 1997 22:31−00:47 UT   Flare, Filament Eruption

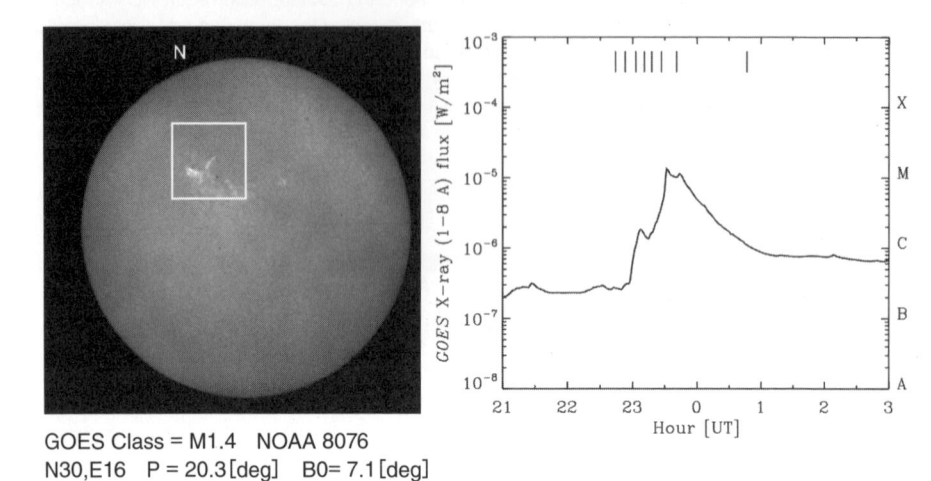

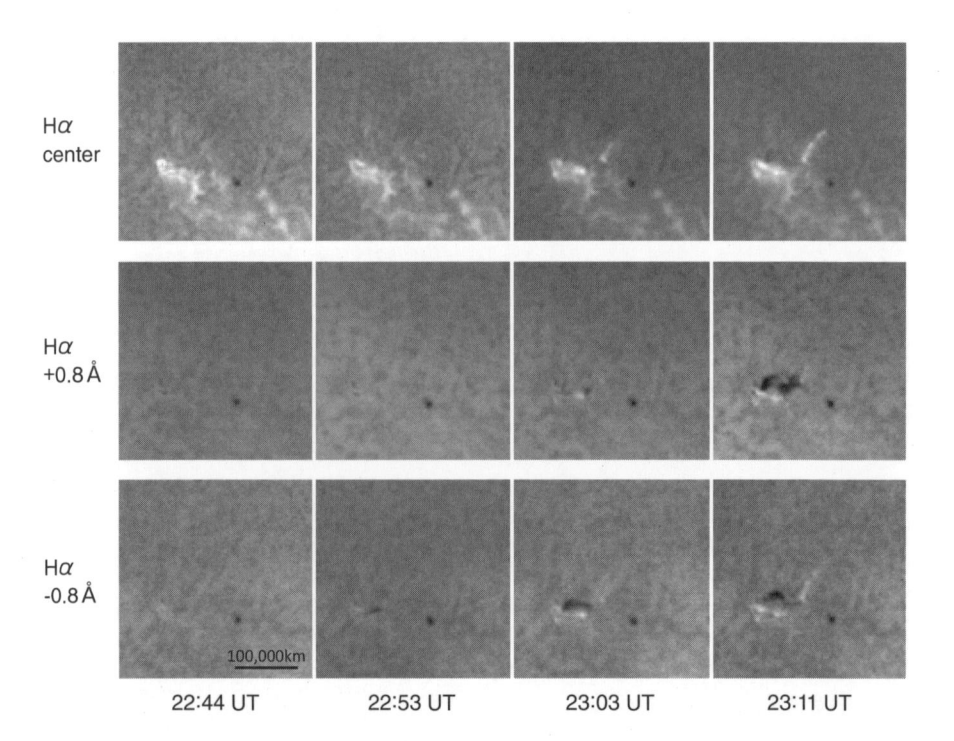

GOES Class = M1.4   NOAA 8076
N30,E16   P = 20.3 [deg]   B0= 7.1 [deg]

Hα center

Hα +0.8 Å

Hα −0.8 Å

100,000 km

22:44 UT   22:53 UT   23:03 UT   23:11 UT

In the following part of the active region, there were compact and bright plage areas. A bright compact flare exploded in the plage area. Simultaneously an active region filament was abruptly activated and erupted. The ejected filament travelled to the preceding part of the region. The flare was observed to explode in two stages. Or it may be consecutive flare explosions in the same area. In the first explosion, a few of flare kernels were observed with the filament eruption. About half an hour later, second explosion occurred with very bright flare kernels. Their Hα emissions, probably, had extremely wide profiles.

活動領域の東側に位置する後続黒点域には，明るいプラージュ域があった．この領域でコンパクトフレアが発生し，同時にこのプラージュ域にあった短いフィラメントが爆発的に噴出した．噴出したフィラメントは先行黒点まで到達して消滅した．このフレアは二段階のエネルギー解放があった．あるいは同じ場所で引き続いて2回発生していた可能性もある．当初はフィラメント噴出と同時に点状の明るいフレアカーネルが数カ所観測された．その約半時間後，極めて明るいフレアが同じ場所で発生した．このフレアカーネルは極めて幅の広いHα輝線であったと思われる．

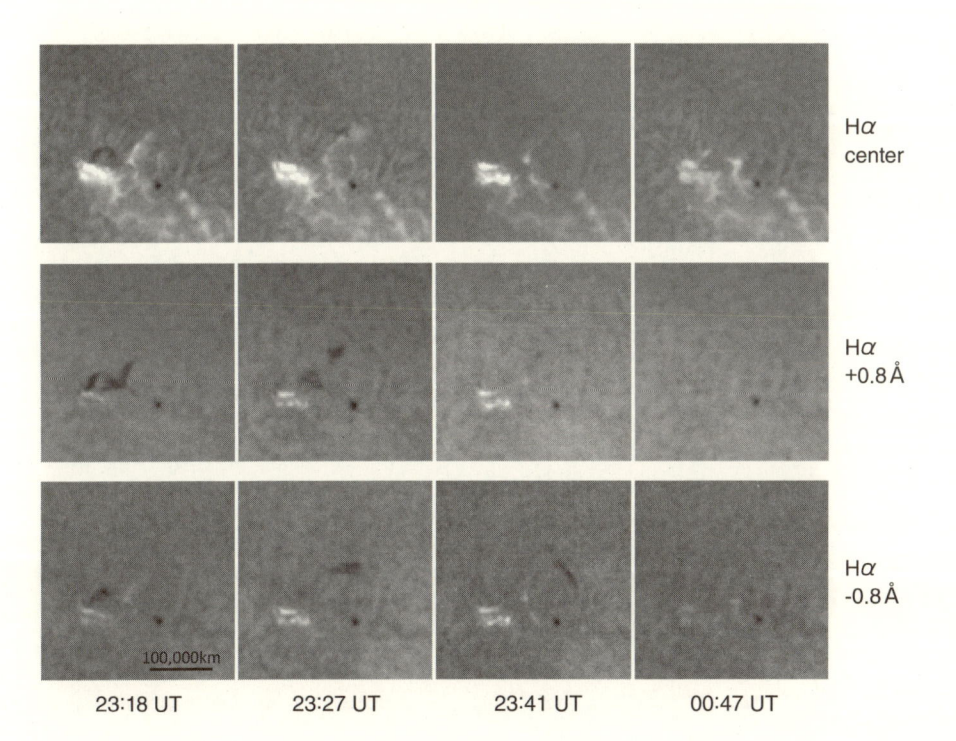

Nov 03, 1997 01:51−02:45 UT Surge

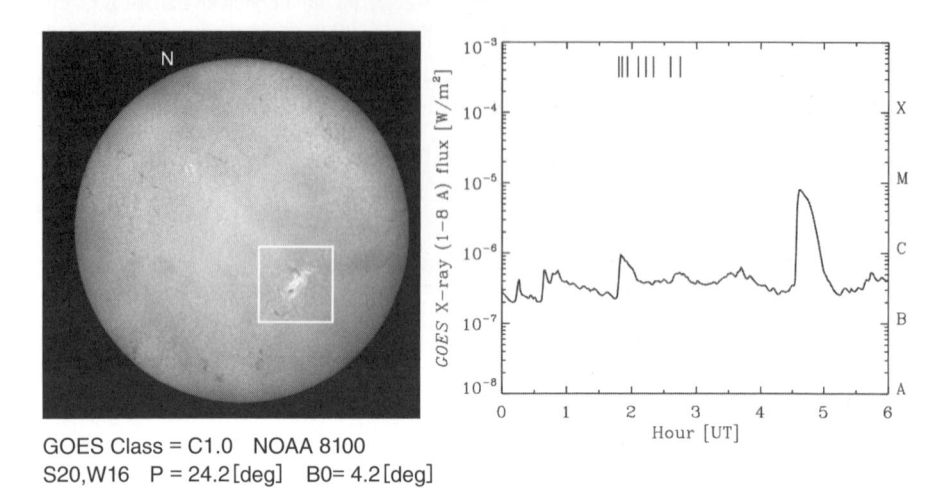

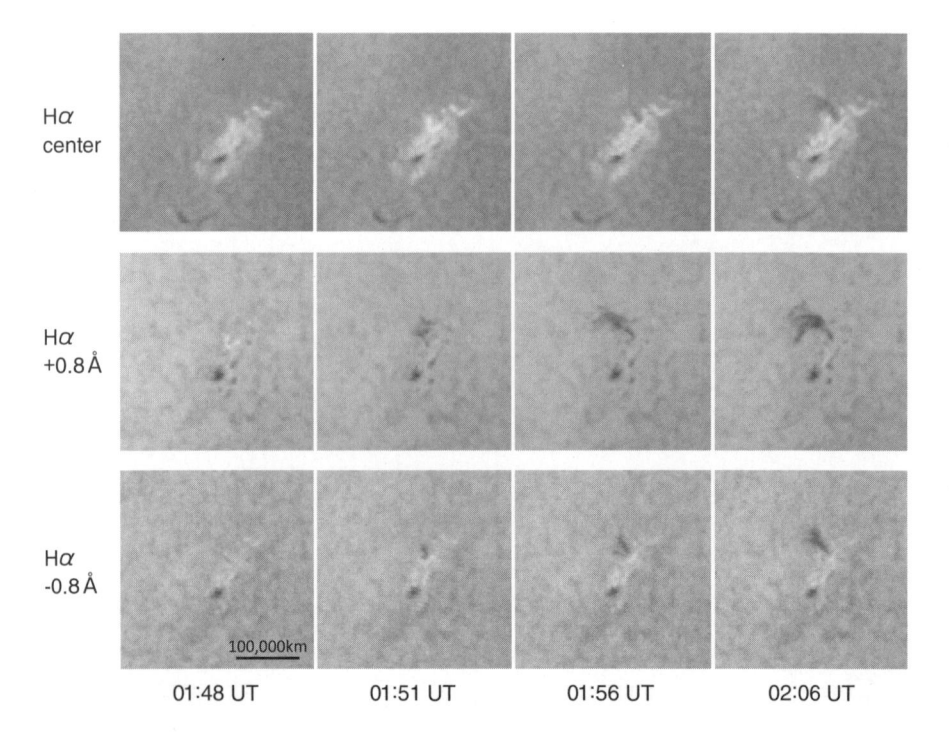

GOES Class = C1.0   NOAA 8100
S20,W16   P = 24.2 [deg]   B0= 4.2 [deg]

Hα center

Hα +0.8 Å

Hα −0.8 Å

100,000 km

01:48 UT   01:51 UT   01:56 UT   02:06 UT

The active region NOAA8100 was activated on Nov 1 by the emergence of satellite polarity region around the preceding spot, and produced several M-class flares in the following days. These flares were proton flares first observed in the solar activity cycle 23 (Mosalam Shaltout, M. A.: 2000). Associated with the flares, Moreton waves were detected by the FMT observations (Narukage et al. 2002, Eto et al. 2002). The surge event occurred in the satellite polarity region, where the M-class flares were produced in the following days. The gas was ejected with a speed of 70 km/s and attained the length of 60-70 Mm.

この活動領域 NOAA8100 は，11 月 1 日に先行黒点近傍にサテライト極性の磁場が現われて活発化し，M クラスのフレアを連発したものである．活動周期サイクル 23 の最初にプロトンフレアを発生した領域であった (Mosalam Shaltout, M. A.:2000)．これらのフレアに伴って，モートン波が発生していることが FMT 観測によって見出されている (Narukage et al. 2002, Eto et al. 2002)．このサージ現象は，後に発生する強大フレアに先駆けてサテライト黒点近傍で発生したもので，70 km/s の速度で噴出し，最大 7 万 km まで長くなった．

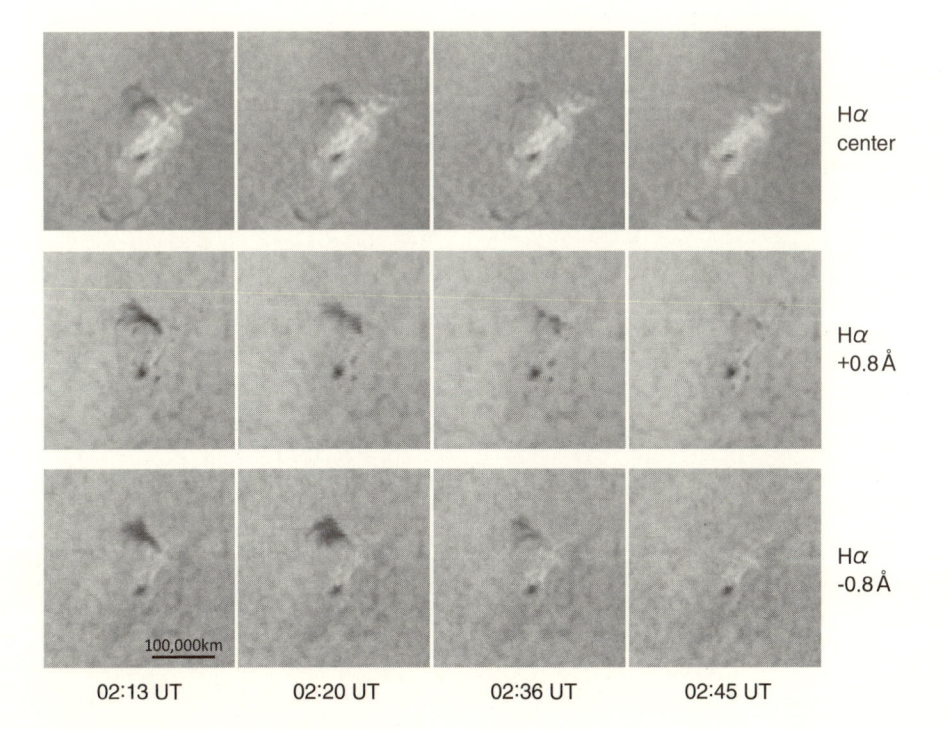

Nov 03, 1997 04:34−05:05 UT   Flare

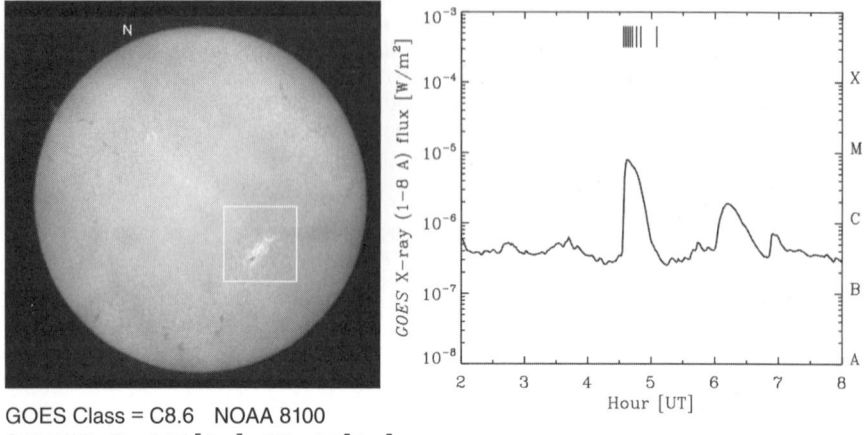

GOES Class = C8.6   NOAA 8100
S19,W22   P = 24.2 [deg]   B0 = 4.2 [deg]

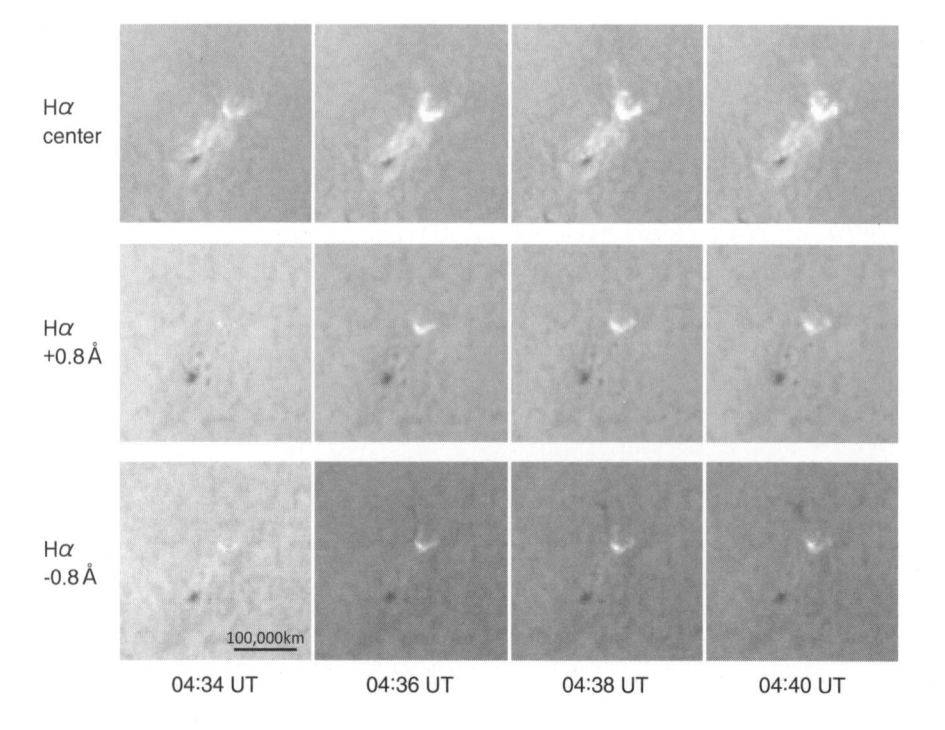

Hα center

Hα +0.8 Å

Hα -0.8 Å

100,000 km

04:34 UT   04:36 UT   04:38 UT   04:40 UT

Chapter 6 Representative Events | 121

NOAA8100 was a magnetic β-type active region on Nov 1, 1997, increased its complexities since Nov 2 and developed to a $\beta\gamma\delta$-type region on Nov 4. In accordance with the development, the region produced many strong M and X class flares. The flare shown here was a C8.6-class compact flare on Nov 3 of half an hour duration, which was associated with gas ejection from the flare kernel. Prominent feature of the event was the association of the emission of Moreton wave from the flare site. The detailed propagation of the Moreton wave is shown in the following pages.

NOAA8100 は，1997 年 11 月 1 日には β 型の活動領域であった．11 月 2 日より磁場分布が複雑になりはじめ，11 月 4 日には $\beta\gamma\delta$ 型に発展した．この発展に応じて，多数の M クラス，X クラスのフレアを頻発した．ここで示されているのは，11 月 3 日に発生した寿命約半時間の C8.6 クラスフレアである．フレアカーネル部からのガス噴出が見られた．このイベントの顕著な特徴は，フレア発生地点からモートン波が放射されたことである．モートン波の伝播の様子は次ページ以降に示されている．

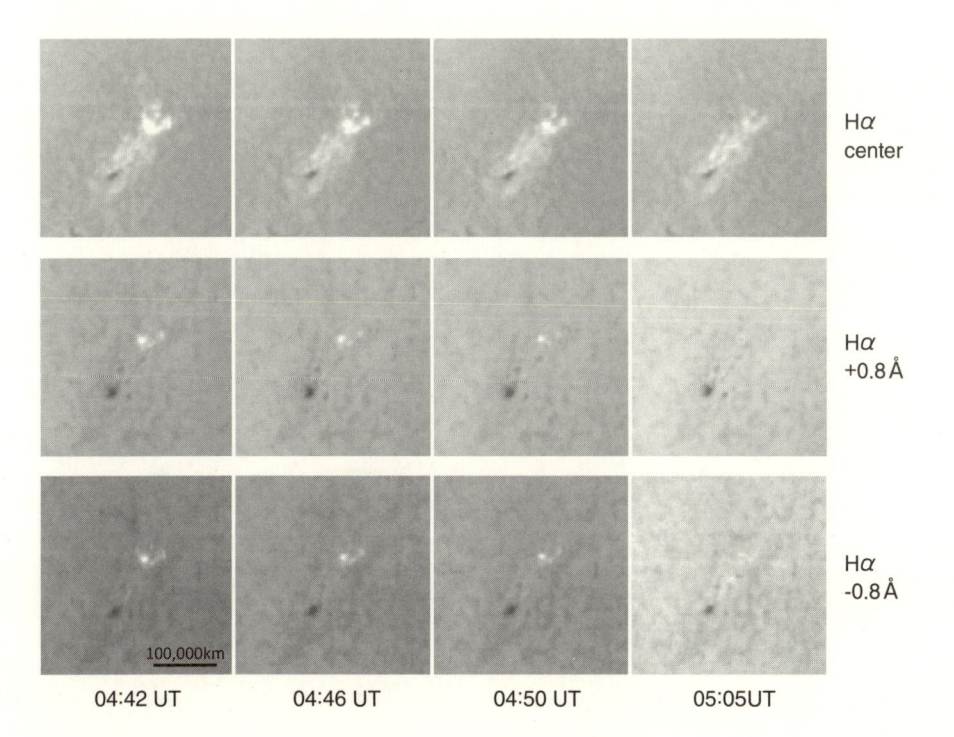

Hα center

Hα +0.8Å

Hα -0.8Å

100,000km

04:42 UT   04:46 UT   04:50 UT   05:05 UT

122 | 第6章 代表的イベント

Nov 03, 1997 04:35−04:47 UT Moreton Wave

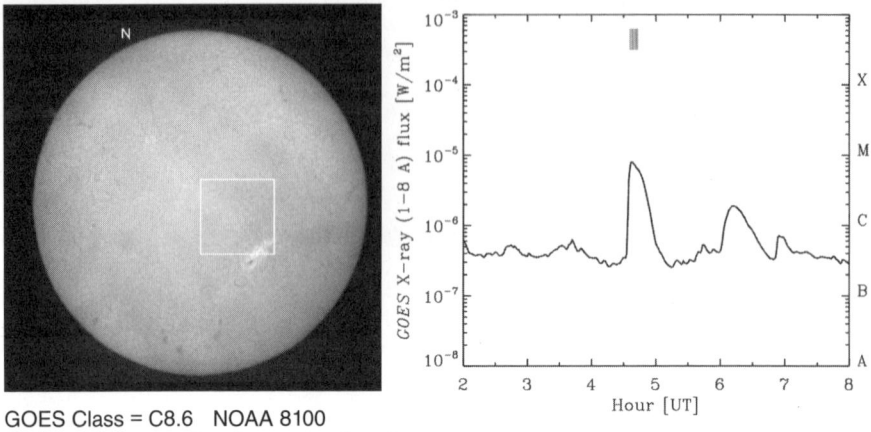

GOES Class = C8.6   NOAA 8100
S19,W22   P = 24.2[deg]   B0= 4.2[deg]

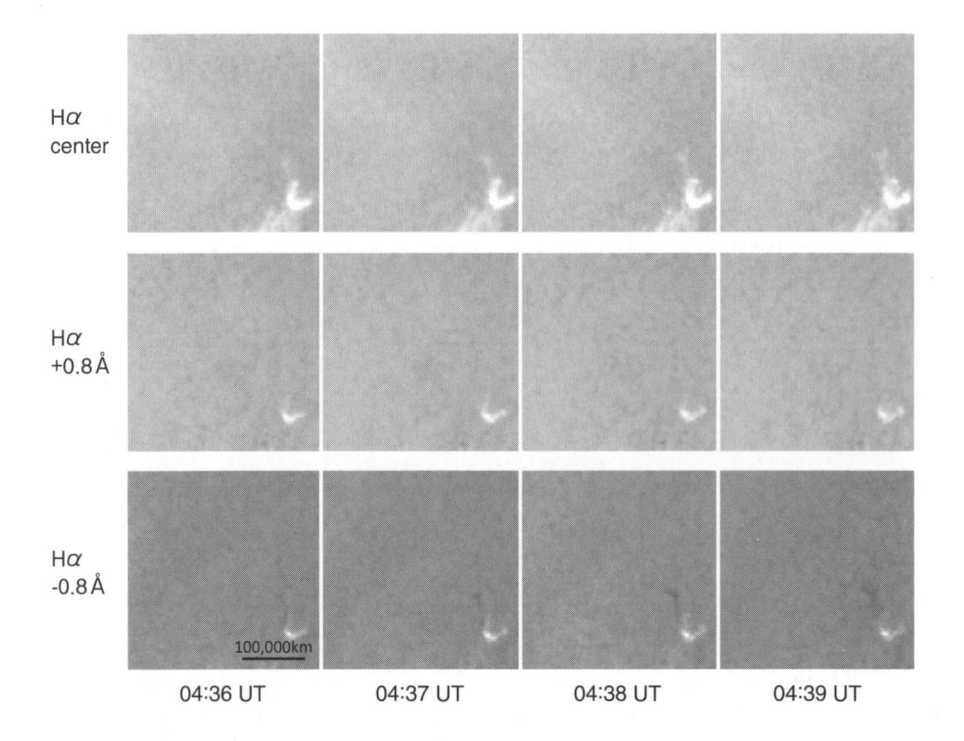

Hα center

Hα +0.8Å

Hα −0.8Å

100,000km

04:36 UT   04:37 UT   04:38 UT   04:39 UT

Chapter 6 Representative Events | 123

In the active region NOAA8100, there occurred a strong C8.6-class flare of half an hour duration, associated with the simultaneous gas ejection from the flare kernel. In the event, a Moreton wave was detected to propagate and expand in a sectorial zone to the North direction from the flare site, with a propagation speed of around 540 km/s (Narukage & Shibata 2004). Associated coronal X-ray wave with a similar speed of propagation as the Moreton wave was also observed by Yohkoh/SXT (Narukage et al. 2002). SOHO/EIT195 detected the associated EIT wave with a propagation speed of around 300 km/s (Warmuth et al. 2004a,b). The Moreton wave in this event was the first one that was observed by the FMT since the start of its observation. (In the next two pages, time-difference images to enhance the visibility of wave fronts are shown with the manually drawn sketch of the wave fronts.)

活動領域 NOAA8100 において，C8.6 クラスで継続時間半時間の強いフレアが発生した．このときフレア発生と同時に，カーネル部からガスが噴出した．さらに，このフレア部から北方に向かって，伝搬速度約 540 km/s で扇状にモートン波が伝搬しているのが観測された（Narukage& Shibata 2004）．Yohkoh/SXT によって，コロナ中をモートン波と同程度の速度で X 線波が伝搬しているのが観測されている（Narukage et al. 2002）．SOHO/EIT195 では，約 300 km/s で伝搬する EIT 波が確認されている（Warmuth et al. 2004a,b）．なお，このイベントは，FMT が観測開始以来，明瞭に観測された最初のモートン波現象である．（次の 2 ページでは，波面を見やすくするように，画像の明るさの時間変化像を示している．また波面位置のスケッチも参考のために付してある）.

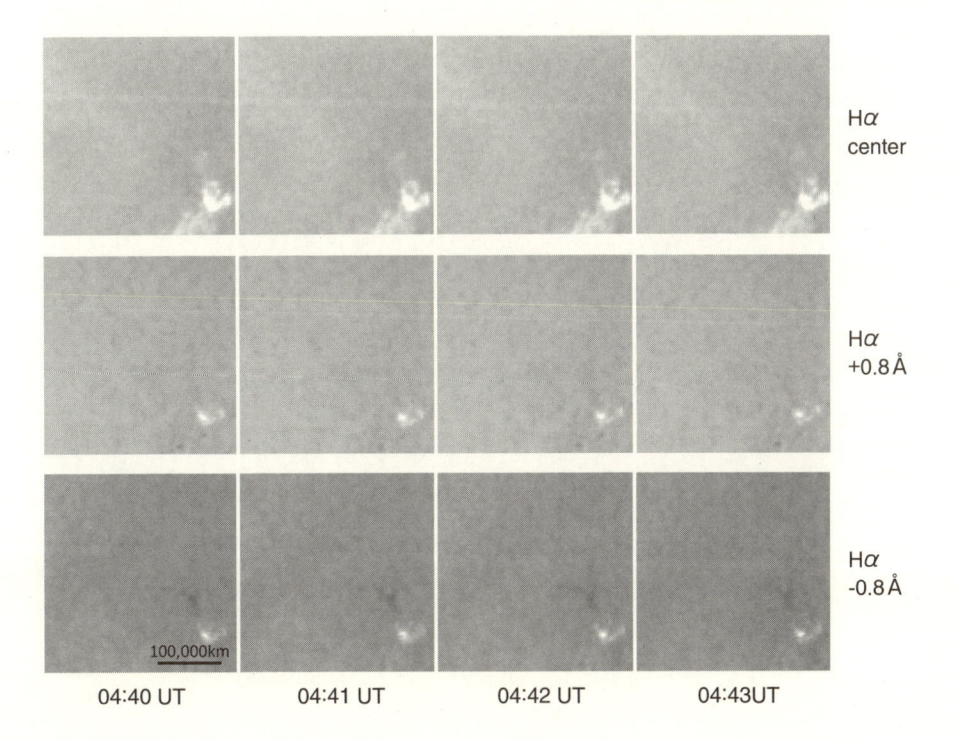

Hα center

Hα +0.8 Å

Hα −0.8 Å

100,000km

04:40 UT   04:41 UT   04:42 UT   04:43 UT

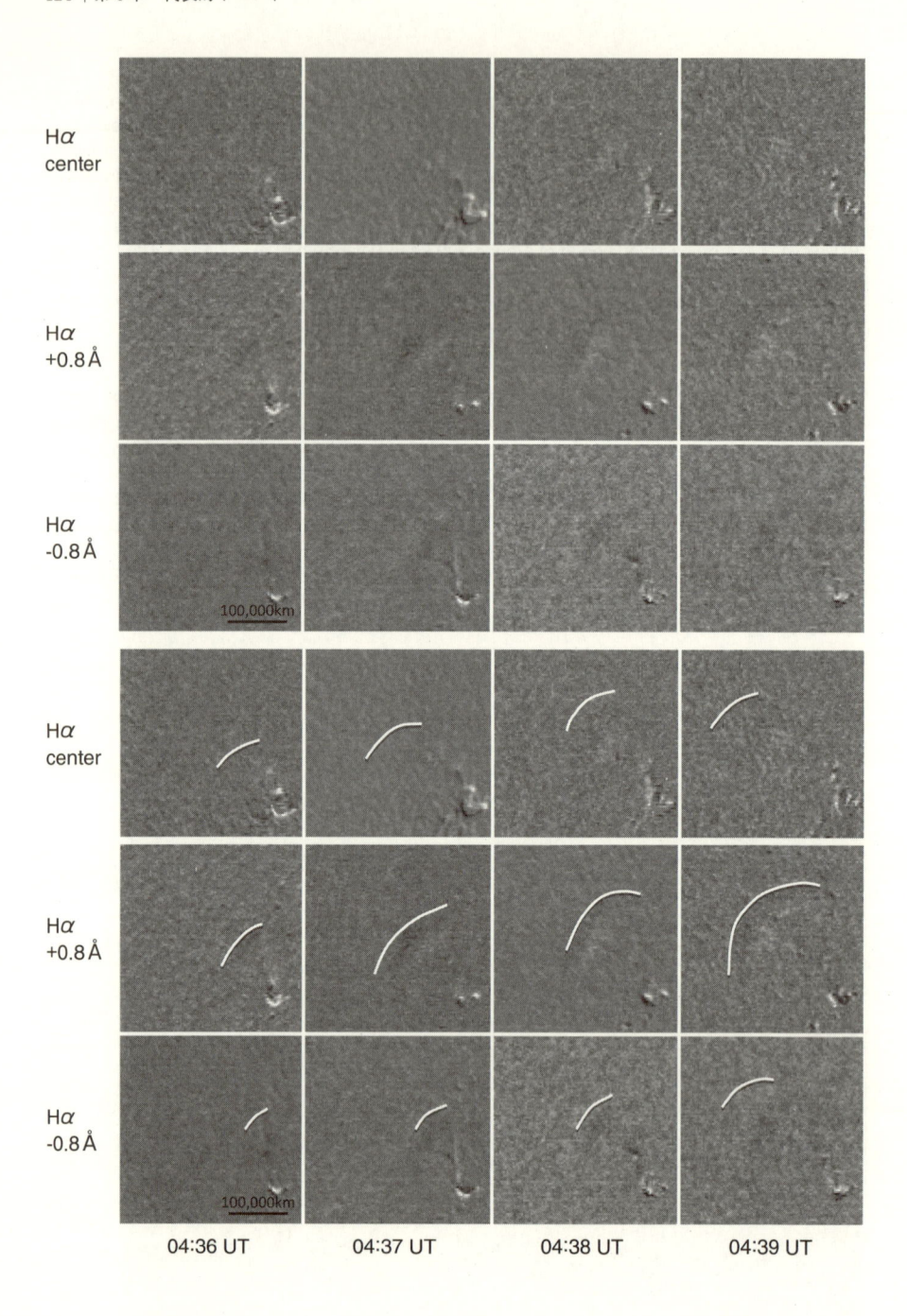

	04:36 UT	04:37 UT	04:38 UT	04:39 UT
Hα center				
Hα +0.8Å				
Hα -0.8Å				
Hα center				
Hα +0.8Å				
Hα -0.8Å				

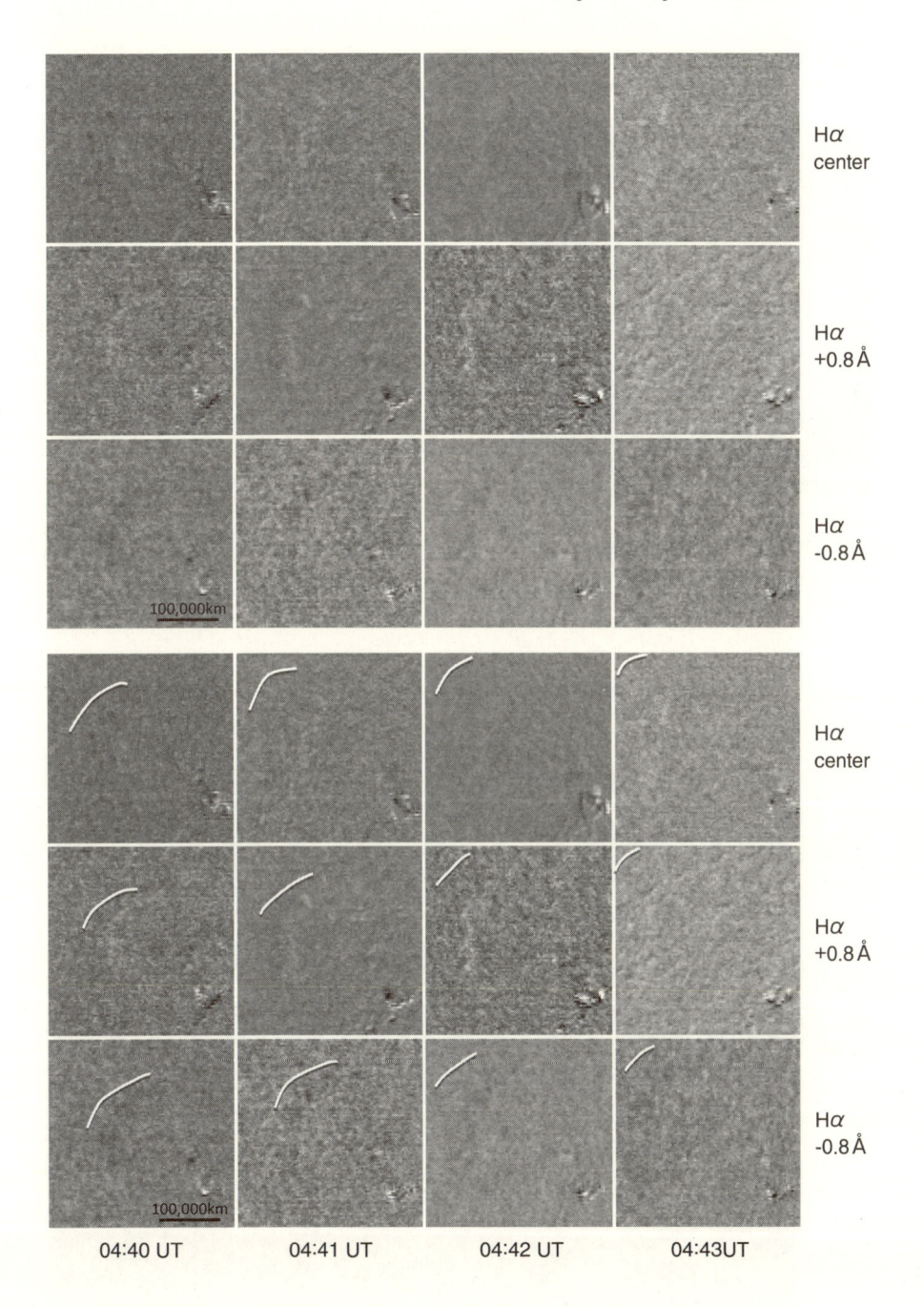

				Hα center
				Hα +0.8Å
				Hα -0.8Å
				Hα center
				Hα +0.8Å
				Hα -0.8Å
04:40 UT	04:41 UT	04:42 UT	04:43 UT	

Nov 04, 1997 06:00−06:55 UT   Flare, Filament Eruption

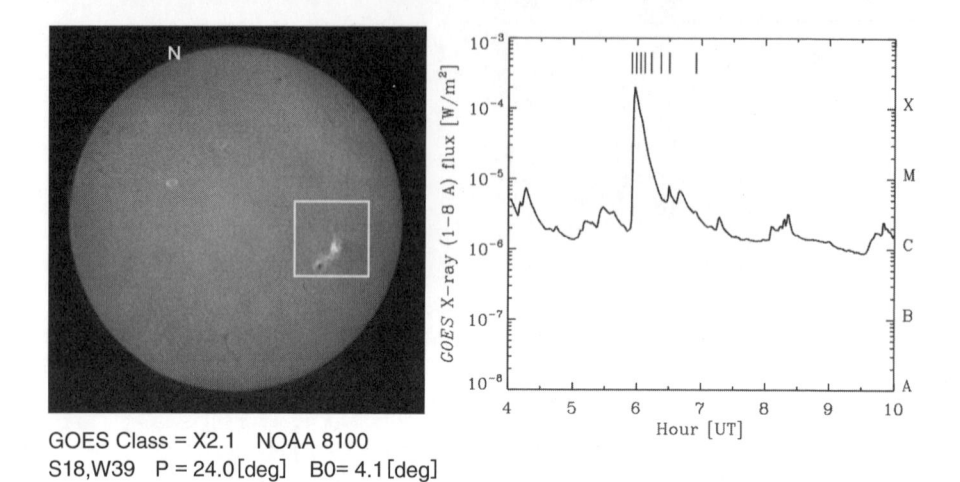

GOES Class = X2.1　NOAA 8100
S18,W39　P = 24.0 [deg]　B0= 4.1 [deg]

Hα center			
Hα +0.8 Å			
Hα −0.8 Å			
05:55 UT	05:59 UT	06:03 UT	06:07 UT

The event is a very strong X2.1 class flare that occurred in a flare productive active region NOAA8100. The flare was a compact one of half an hour duration, which was associated with the filament gas ejection. In this event, a Moreton wave was emitted from the flare site and it excited a filament located far from the active region to periodically oscillate (Winking filament). The propagation of Moreton wave in 1 min cadence is displayed in the following pages.

このイベントは，フレアを頻発した活動領域 NOAA8100 で発生した X2.1 クラスの強烈なフレアである．寿命約半時間のコンパクトフレアで，フィラメントガス噴出を伴っていた．また，モートン波も放出されており，それが伝播して活動領域から遠く離れたところのフィラメントを周期的に振動させるウインキング・フィラメントという現象を引き起こした．モートン波の伝播の様子を 1 分間隔で表示したものを以降のページに示している．

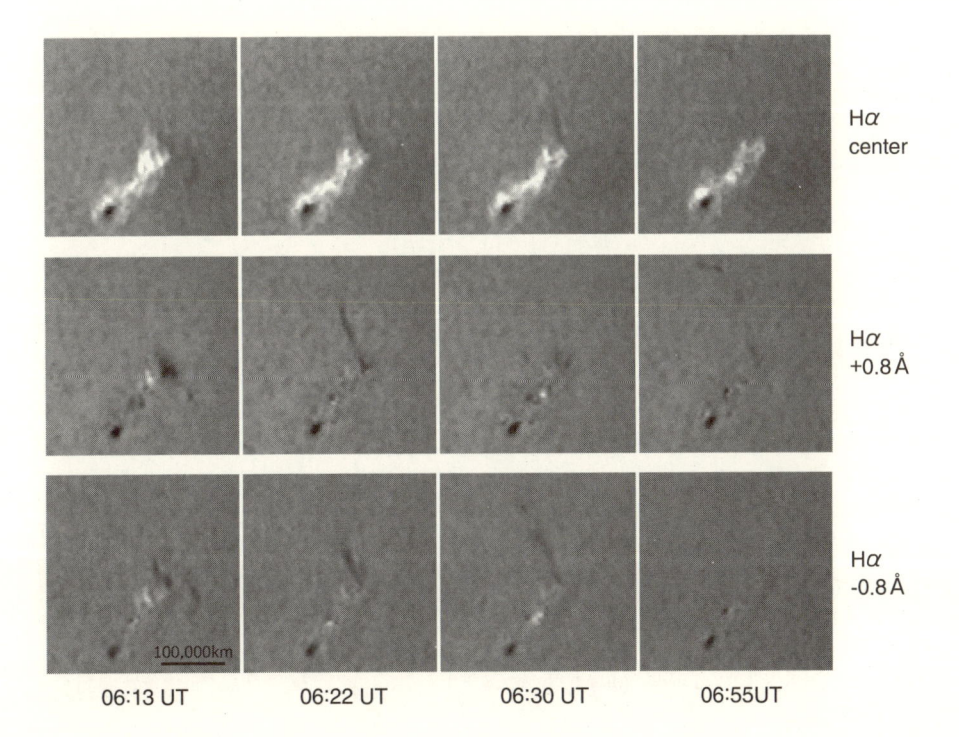

				Hα center
				Hα +0.8Å
				Hα -0.8Å
06:13 UT	06:22 UT	06:30 UT	06:55 UT	

100,000km

Nov 04, 1997 05:57－06:06 UT Moreton Wave

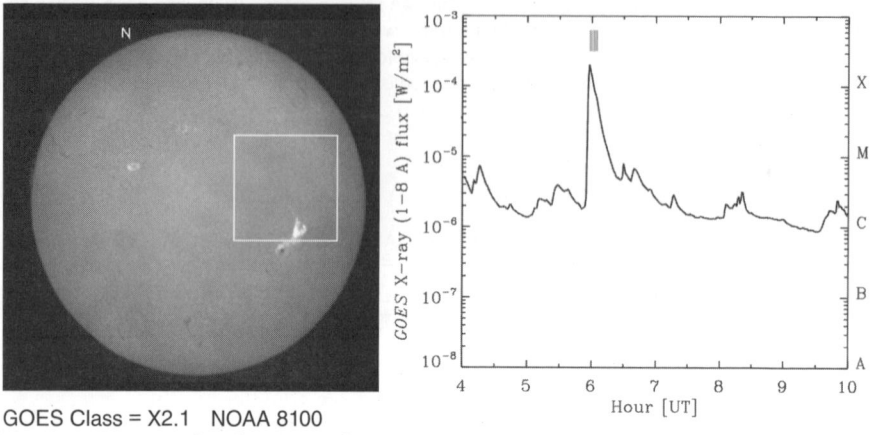

GOES Class = X2.1 NOAA 8100
S18,W39 P = 24.0[deg] B0= 4.1[deg]

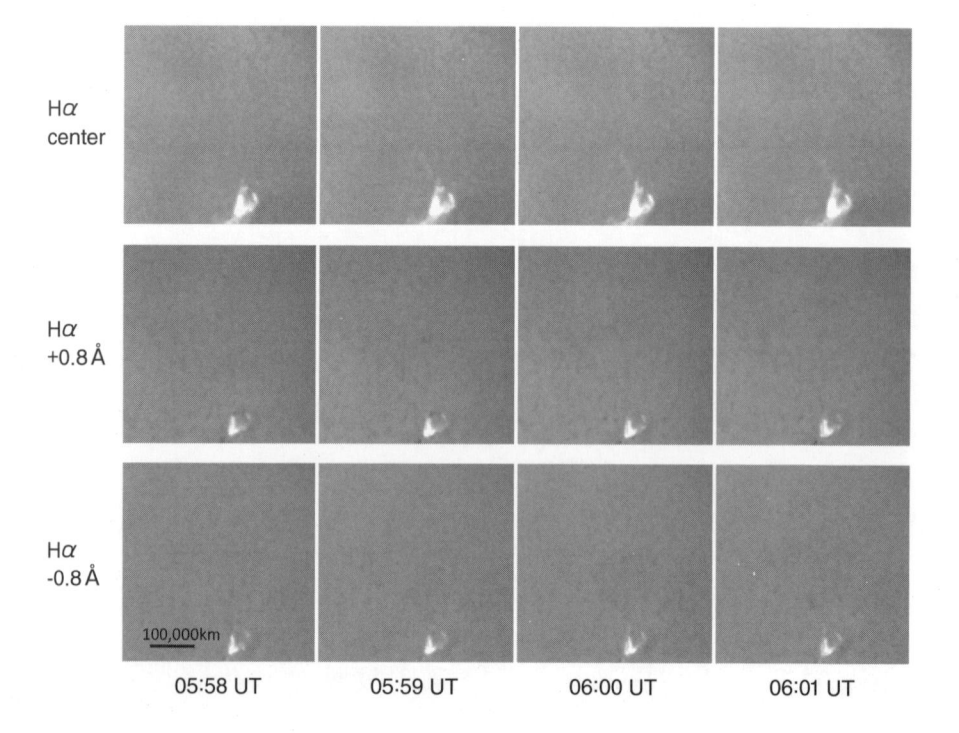

Hα center

Hα +0.8Å

Hα −0.8Å

100,000km

05:58 UT   05:59 UT   06:00 UT   06:01 UT

Chapter 6 Representative Events | 129

Associated with the explosion of an X2.1 class flare, the emission of a Moreton wave was detected. According to the analysis by Eto et al. (2002), the speed of propagation was 715 km/s, while a simultaneously detected EIT wave propagated with a speed of 202 km/s. The difference of propagation speed suggests us the difference of the mode of the waves. The Moreton wave activated a filament oscillation of 17min period at the distant location from the flare site. The event was also discussed by Shibata et al. (2002). The event was included in the statistical study of inter-relations among a varieties of flare waves such as the Moreton waves, EIT waves, Type II radio bursts and HeI 10830 waves by Warmuth et al. (2003, 2004a and 2004b). Their conclusion was that the waves detected in various wavelengths are different manifestation of propagating MHD fast mode shocks. The difference of the propagating speed between the Moreton wave and the EIT wave was interpreted by the deceleration of wave propagation. (In the next two pages, time-difference images to enhance the visibility of wave fronts are shown with the manually drawn sketch of the wave fronts.)

X2.1 クラスのフレアの発生に伴ってモートン波が放出された．Eto et al. (2002) の解析によると，モートン波の伝播速度は 715 km/s であるのに比して，同時に観測された EIT 波の伝播速度は 202 km/s であった．このことから，2 つの波は異なるものであることが示唆されている．また，モートン波は発生位置から遥か離れたところにあった静穏フィラメントに到着したとき，フィラメント振動現象（周期約 17 分）を励起した．このイベントは, Shibata et al. (2002) でも議論されている．一方，Warmuth et al. (2003, 2004a and 2004b) は，この現象も含めて，モートン波，EIT 波，TypeII 電波バースト，HeI波といったフレアに伴う波動現象の相互関係を統計的に調査した．そして，これらは同じ電磁流体力学的ファーストモード衝撃波であろうという結論を得ている．モートン波と EIT 波の伝播速度の相違は，波の伝播が時間的に減速を受けるからであるとしている．（次の 2 ページでは，波面を見やすくするように，画像の明るさの時間変化像を示している．また波面位置のスケッチも参考のために付してある．）

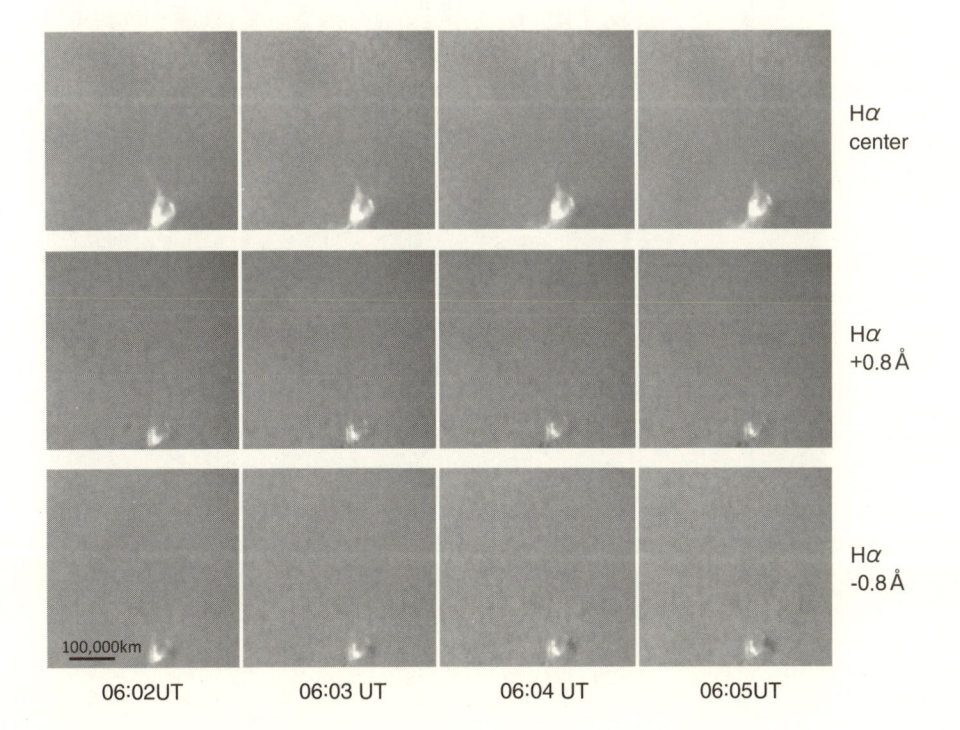

Hα center

Hα +0.8Å

Hα -0.8Å

100,000km

06:02UT　　06:03 UT　　06:04 UT　　06:05UT

130 | 第6章　代表的イベント

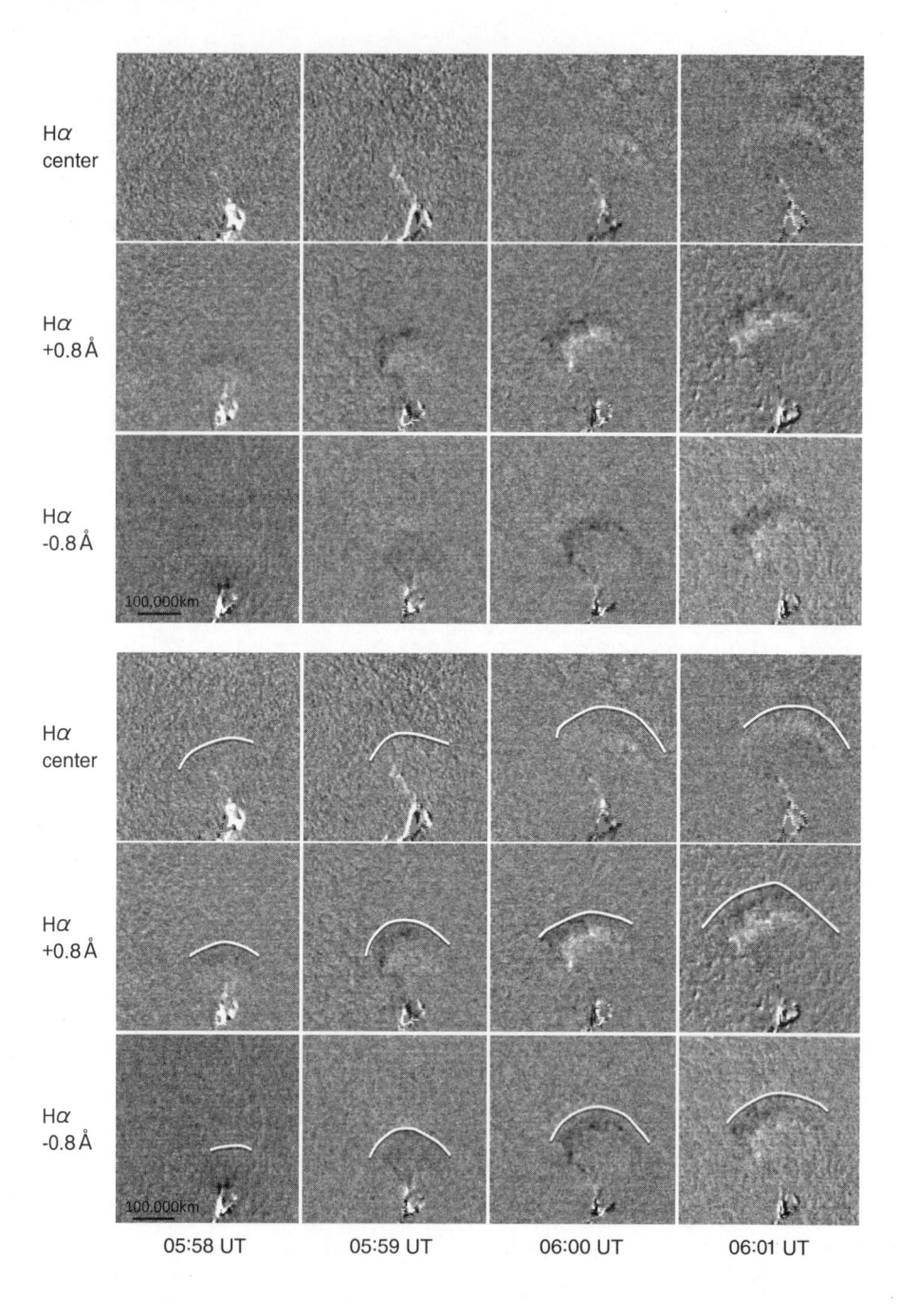

Hα center				
Hα +0.8Å				
Hα −0.8Å	100,000km			
Hα center				
Hα +0.8Å				
Hα −0.8Å	100,000km			
	05:58 UT	05:59 UT	06:00 UT	06:01 UT

Hα center

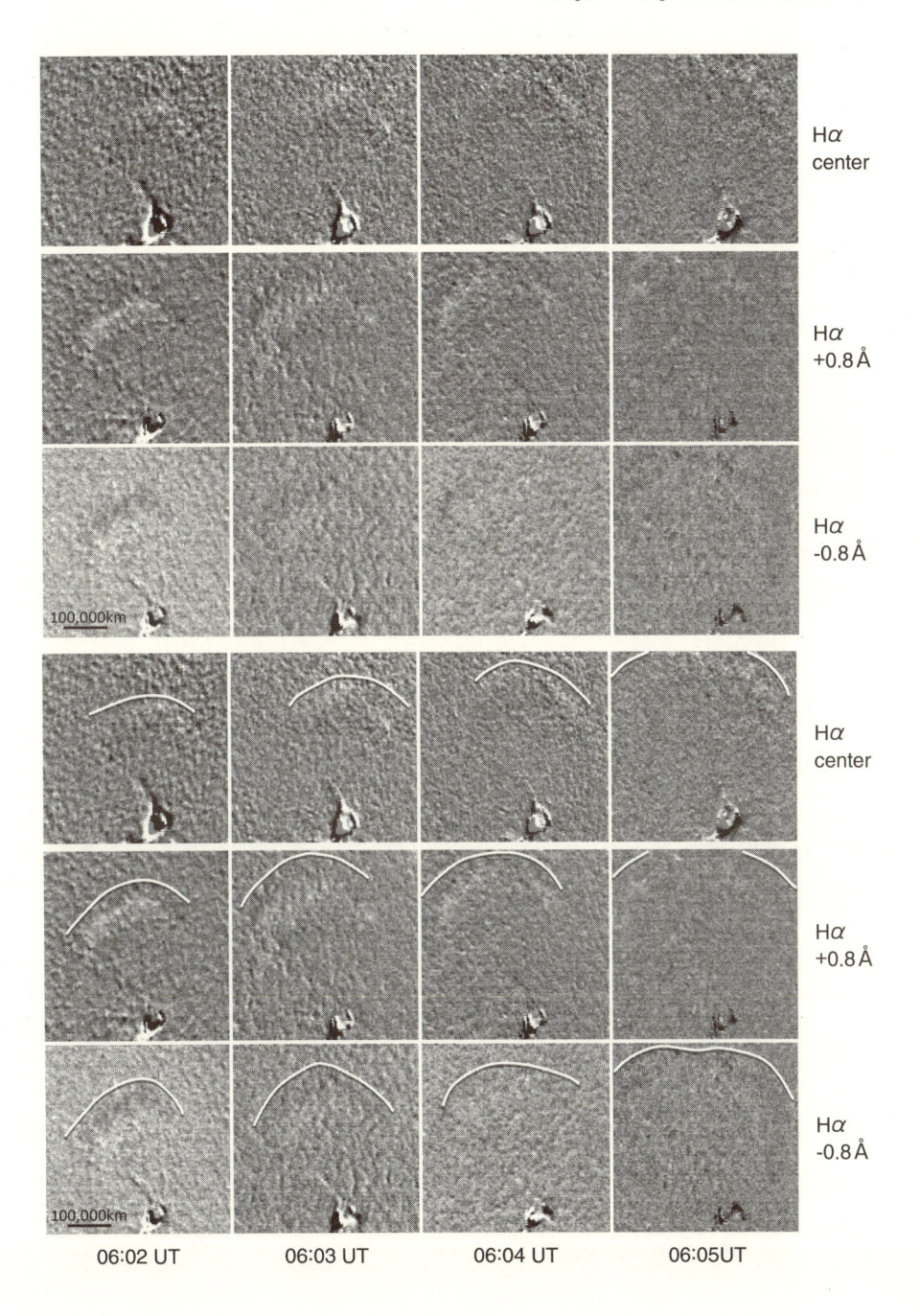

Hα +0.8Å

Hα −0.8Å

100,000km

Hα center

Hα +0.8Å

Hα −0.8Å

100,000km

06:02 UT　　06:03 UT　　06:04 UT　　06:05 UT

Nov 04, 1997 06:06−06:35 UT Winking Filament

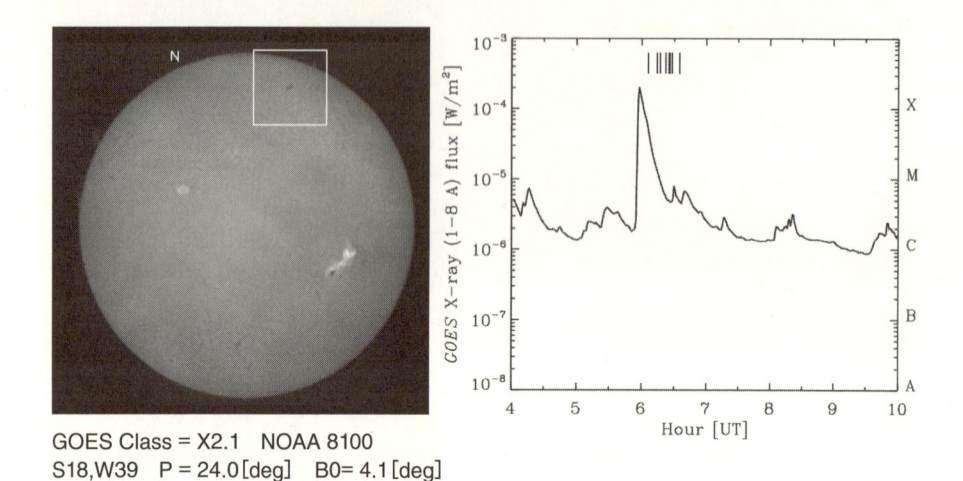

GOES Class = X2.1   NOAA 8100
S18,W39   P = 24.0 [deg]   B0= 4.1 [deg]

Hα center

Hα +0.8 Å

Hα −0.8 Å

100,000 km

06:06 UT   06:14 UT   06:17 UT   06:22 UT

Associated with the explosion of an X class flare, the emission of a Moreton wave was detected in Hα on-band and off-band images. The event was associated with a filament eruption at the flare site. The Moreton wave activated a filament oscillation of 17min period at the distant location from the flare site.

Xクラスのフレアの発生に伴ってモートン波が放出された．モートン波は，Hα線の中心でもウイング波長でも観測された．フレアが発生した活動領域では，フィラメント爆発が観測されていた．また，モートン波は発生位置から遥か離れたところにあった静穏フィラメントに到着したとき，フィラメント振動現象（周期約 17 分）を励起した．

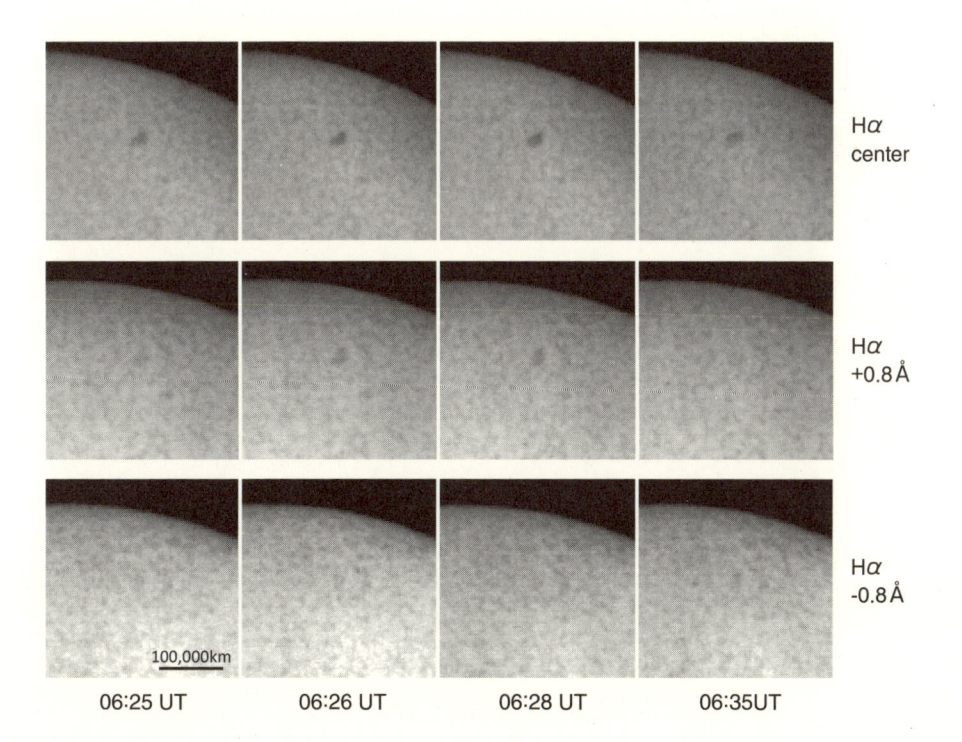

				Hα center
				Hα +0.8Å
				Hα -0.8Å
06:25 UT	06:26 UT	06:28 UT	06:35 UT	

100,000km

134 | 第6章 代表的イベント

Apr 11, 1998　04:15－05:41 UT　　Flare, Filament Eruption

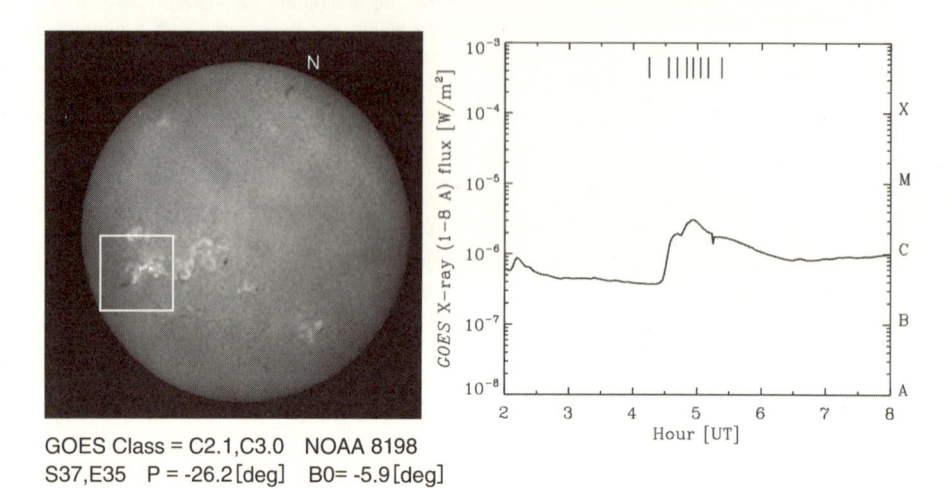

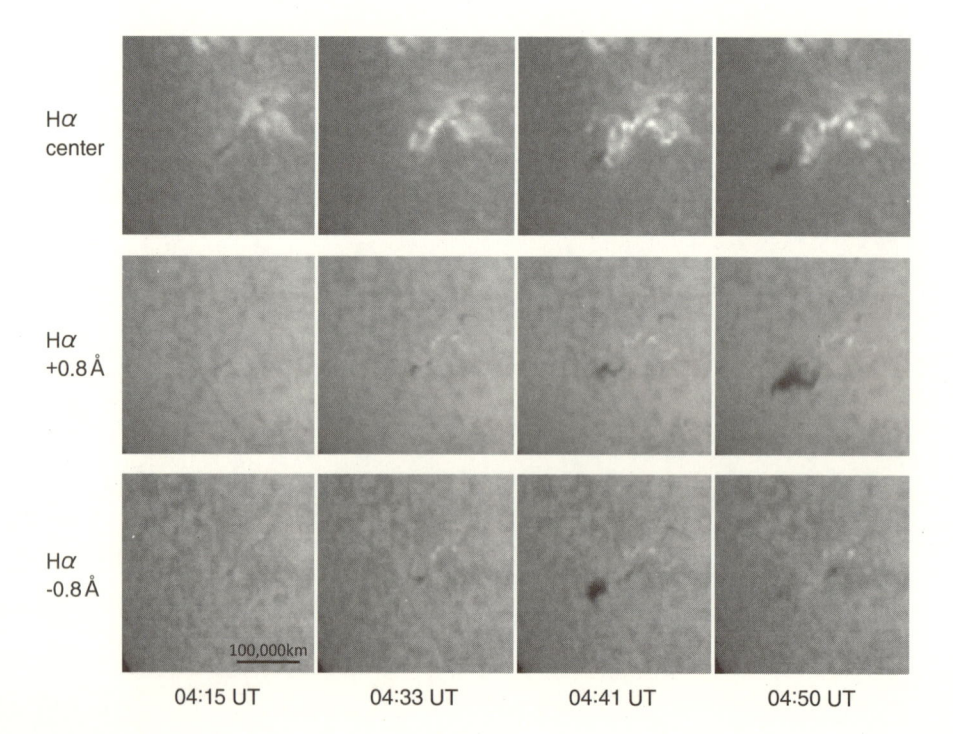

GOES Class = C2.1, C3.0　　NOAA 8198
S37, E35　　P = -26.2 [deg]　　B0 = -5.9 [deg]

Hα center

Hα +0.8Å

Hα -0.8Å

100,000 km

04:15 UT　　04:33 UT　　04:41 UT　　04:50 UT

An active region filament, located in the SE periphery of the active region NOAA8198 was activated and ejected with a speed of about 50 km/s. Simultaneously a C-class flare started to explode at the plage areas in the central part of the region. According to the analysis by Morimoto et al. (2003b), half of the filament was ejected to the interplanetary space and induced a CME event.

活動領域 NOAA8198 の南東にあった活動領域フィラメントが活発化し，約 50 km/s の速度で噴出を始め，それと同時に活動領域中央部のプラージュが散在している場所でフレアが発生した．Morimoto et al. (2003b) によると，このフィラメントの約半分が惑星間空間に放出されて，それによって CME 現象を引き起こした．

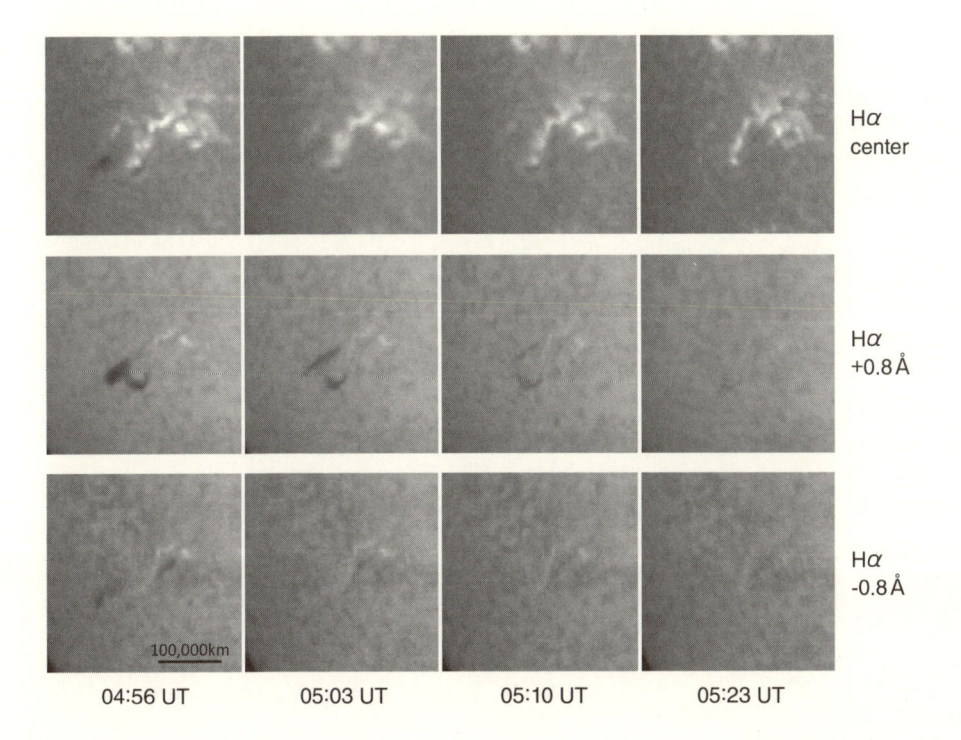

Apr 16, 1998 01:05−01:26 UT   Flare, Surge

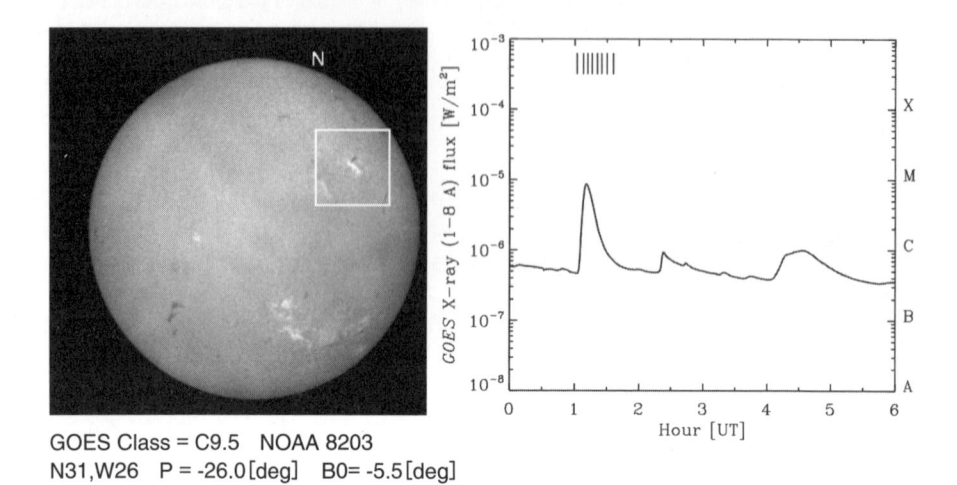

GOES Class = C9.5   NOAA 8203
N31,W26   P = −26.0 [deg]   B0= −5.5 [deg]

Hα center

Hα +0.8Å

Hα −0.8Å

100,000km

01:02 UT   01:08 UT   01:12 UT   01:16 UT

A compact C9.3 flare exploded in an evolving emerging flux region. Nearly simultaneously, a big surge was ejected with a maximum velocity of 260 km/s. A CME event was observed by SOHO/LASCO associated with the surge and their mutual relation was studied by Liu et al. (2005). They proposed that the onset of the CME resulted from the significant restructuring of the large-scale coronal magnetic field as a result of flux emergence in the active region. This surge-CME event strongly suggests that emerging flux may not only trigger a surge but also simultaneously trigger a CME by means of small-scale magnetic reconnection in the lower atmosphere.

活発な磁場浮上域でC9.3クラスのコンパクトフレアが発生した．ほぼ同時に最大260 km/sにもなる速度で大きなサージが噴出した．このサージの発生に伴ってSOHO衛星のLASCO観測によりCME現象が発生していることがLiu et al. (2005) によって確かめられた．彼らは，この現象時に磁気浮上によって上空のコロナの大きなスケールの磁場構造が変化していることを確認した．このことから，彼らは，浮上磁場は磁気リコネクション機構によってサージを発生させたばかりではなく，同時にCMEも引き起こしていたというモデルを提案した．

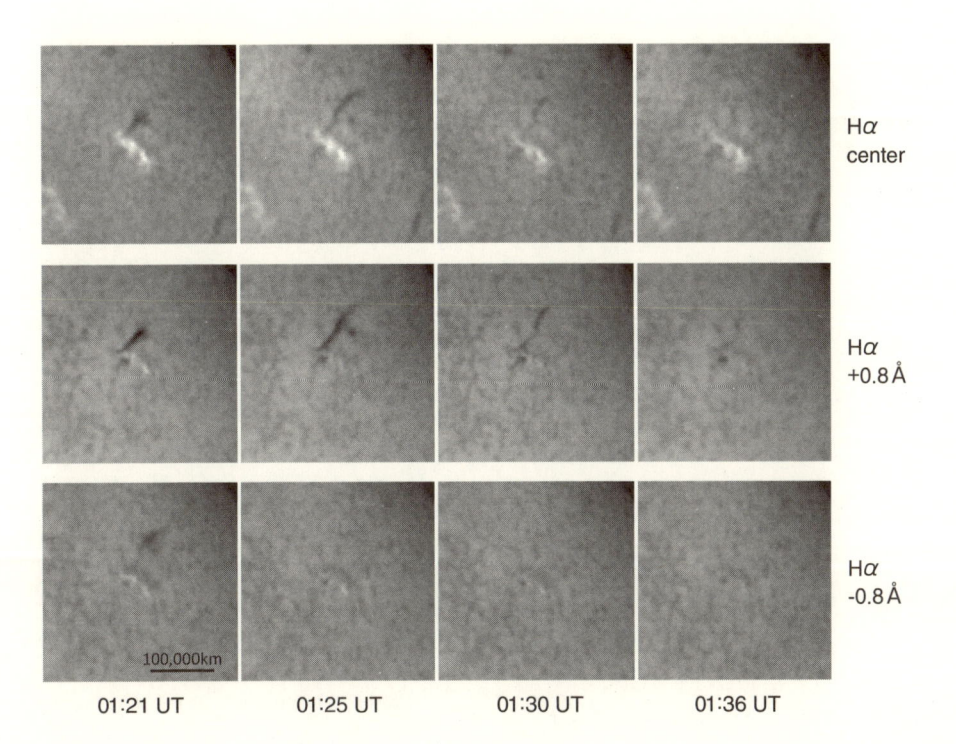

138 | 第6章　代表的イベント

Apr 29, 1998　05:11−05:40 UT　Flare, Filament Eruption

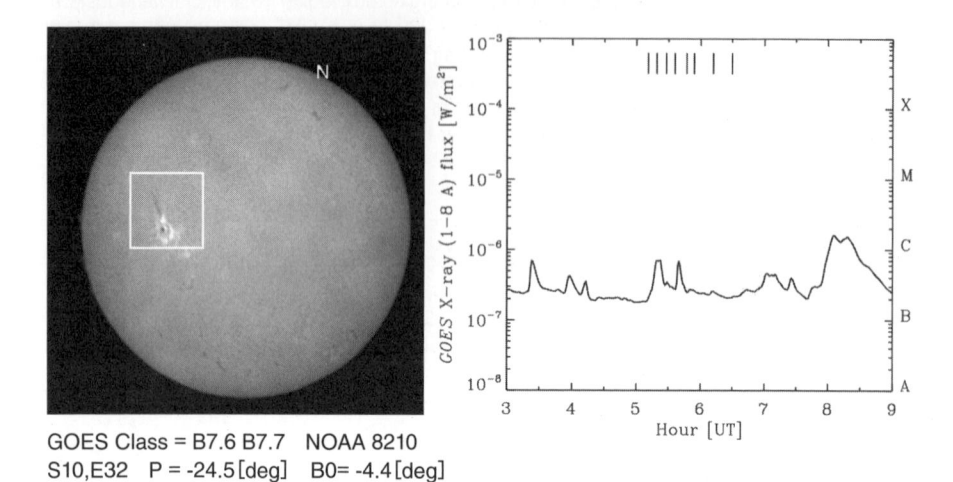

GOES Class = B7.6 B7.7　NOAA 8210
S10,E32　P = −24.5[deg]　B0= −4.4[deg]

Hα center

Hα +0.8Å

Hα −0.8Å

100,000km

05:11 UT　　05:19 UT　　05:28 UT　　05:36 UT

Although the X-ray burst (B7.7) of the flare was not so intense, a rapid Hα filament ejection with a speed of 100 km/s occurred when the flare exploded. According to Morimoto et al. (2003b), about 1/3 of the filament was ejected to the interplanetary space. SOHO/EIT observed the bright arcade formation, dark filamentary ejections and a coronal dimming in association with the event.

このフレアのX線強度はB7.7という弱いものであったが，100 km/s 近い速さの激しいHαフィラメント噴出が伴った．Morimoto et al. (2003b) によると，フィラメントの約1/3が惑星間空間に放出された．また，SOHO/EIT では，この現象に伴って明るいアーケード構造の形成，暗い筋状のガス放出，更にはコロナのディミング（EUV コロナ強度の低下）が発生したことが観測されている．

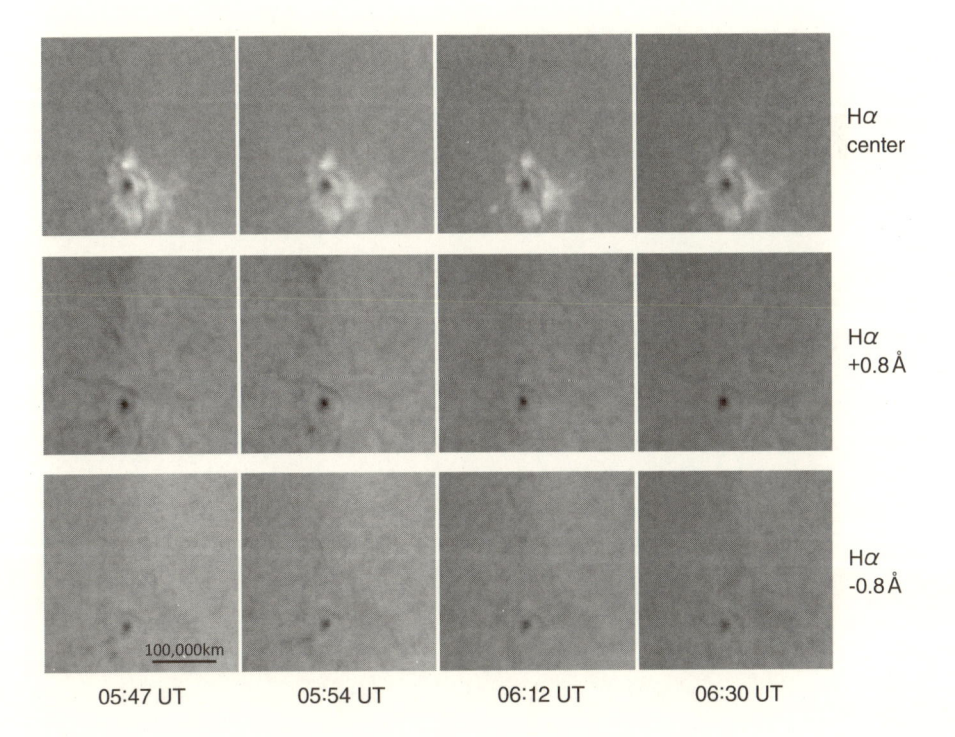

140 | 第6章 代表的イベント

Aug 08, 1998 03:00−03:30 UT Flare, Surge

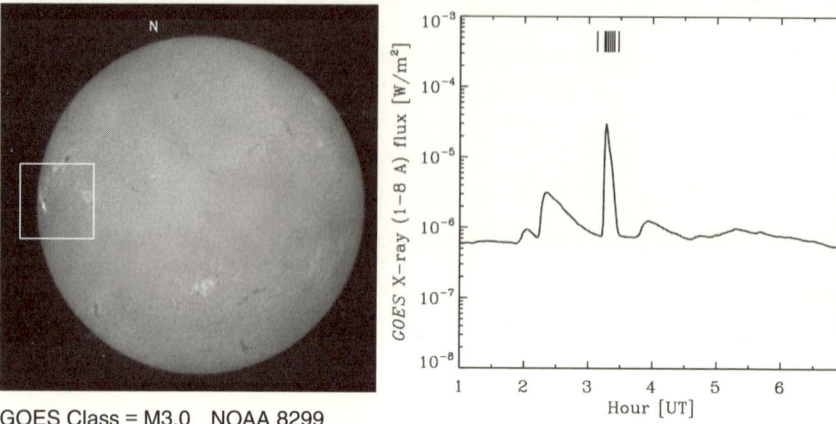

GOES Class = M3.0   NOAA 8299
N16,E80   P = 13.4 [deg]   B0= 6.2 [deg]

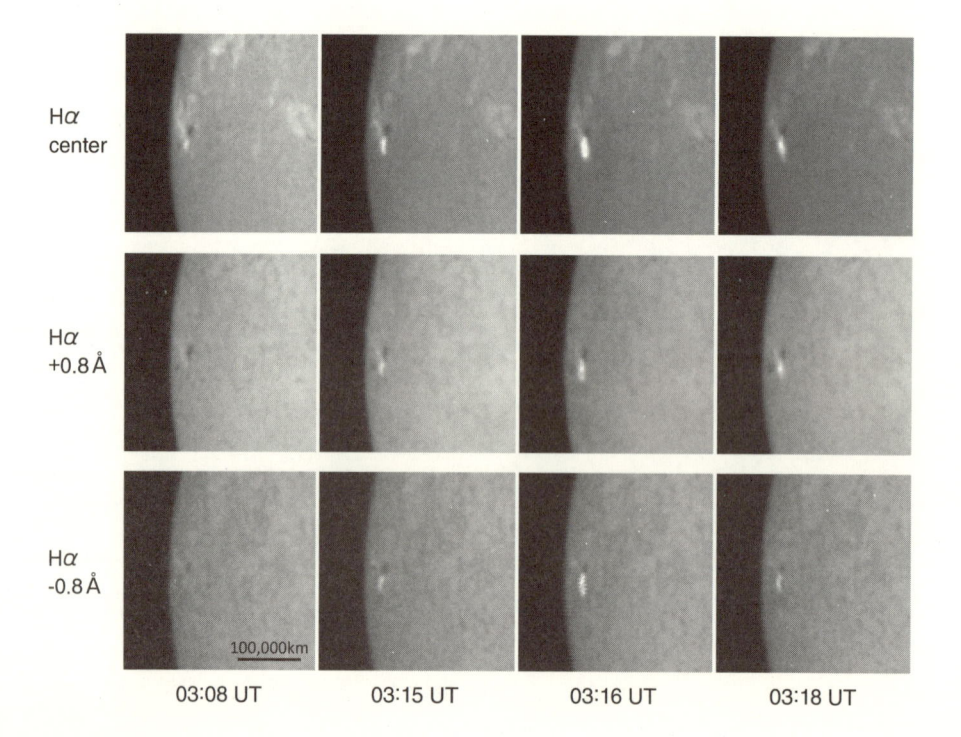

Hα center

Hα +0.8 Å

Hα −0.8 Å

100,000 km

03:08 UT   03:15 UT   03:16 UT   03:18 UT

In the active region NOAA8299, there occurred a compact M3.0 flare of 15min duration. In association with the flare, a surge of chromospheric gas was ejected toward the south with an apparent speed of 60-70 km/s. In the same direction as the surge, a Moreton wave was observed to propagate from the flare kernel. The propagation of Moreton wave in 1 min cadence is displayed in the following pages.

活動領域 NOAA8299 において，M3.0 クラスの寿命 15 分のコンパクトフレアが発生した．フレアに伴って彩層ガスが，60-70 km/s の見かけの速さで南へサージ状に放出された．また，ちょうどガスが放出された方向に，モートン波が伝播してゆくのが観測された．モートン波の伝播の様子は次ページ以降に示されている．

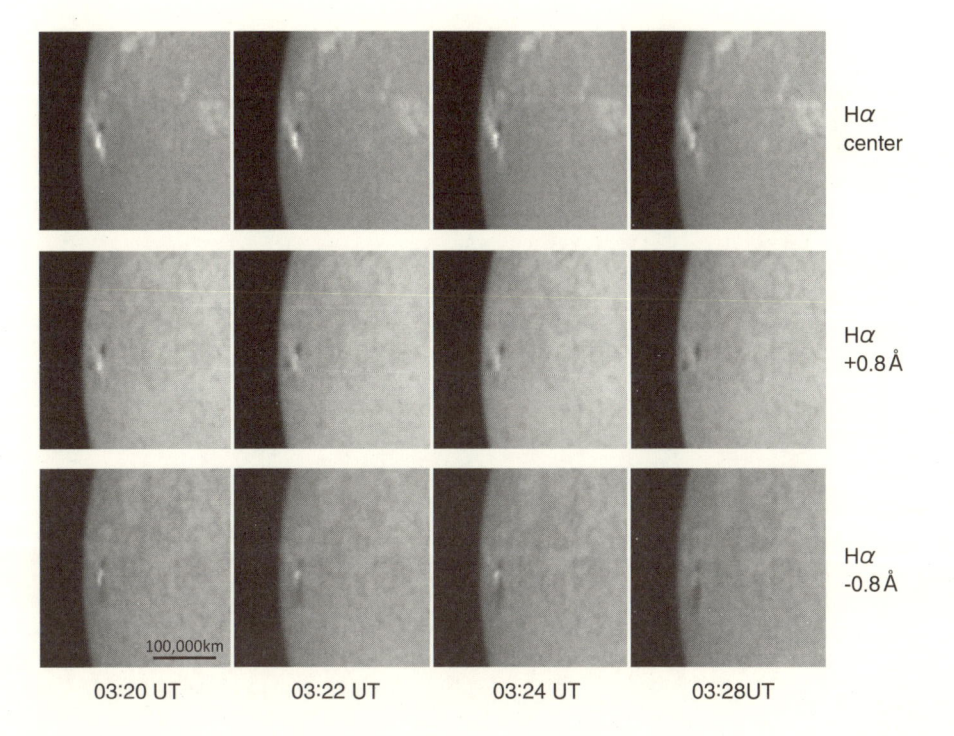

				Hα center
				Hα +0.8Å
				Hα -0.8Å
03:20 UT	03:22 UT	03:24 UT	03:28UT	

100,000km

Aug 08, 1998 03:14−03:26 UT   Moreton Wave

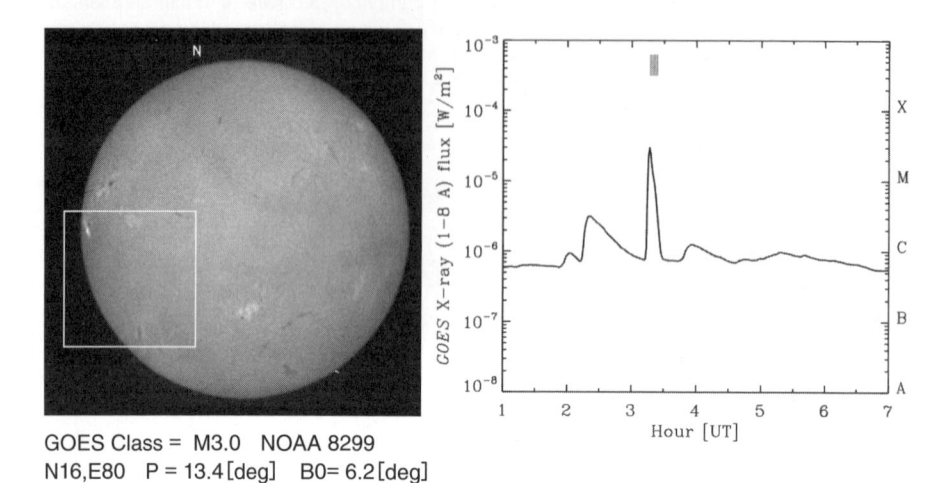

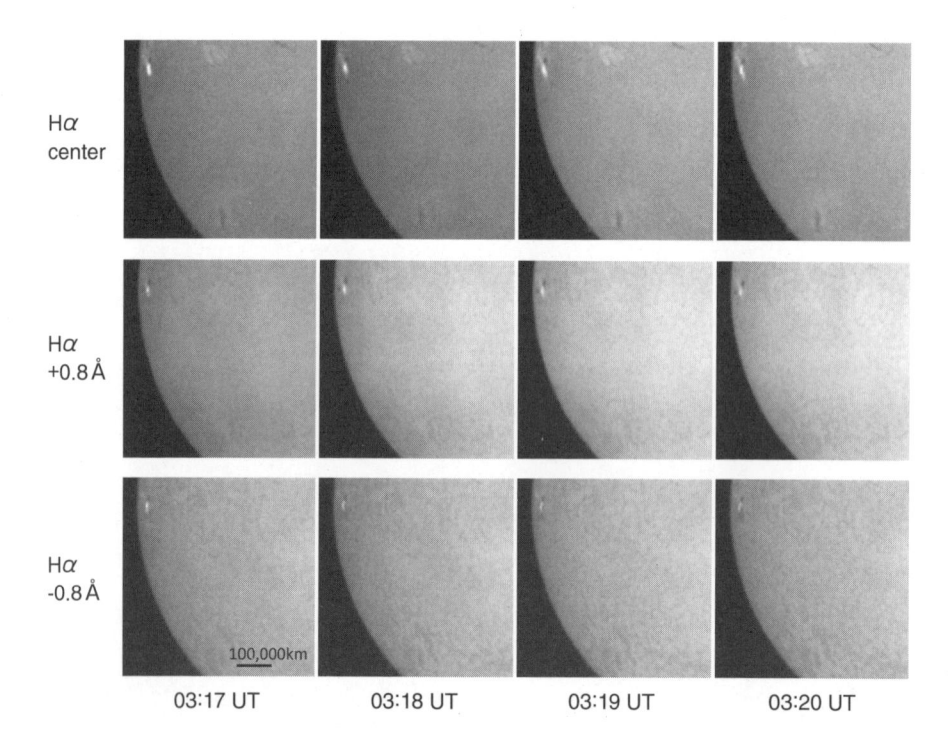

GOES Class = M3.0   NOAA 8299
N16,E80   P = 13.4 [deg]   B0= 6.2 [deg]

Hα center

Hα +0.8 Å

Hα −0.8 Å

100,000 km

03:17 UT   03:18 UT   03:19 UT   03:20 UT

Here is shown the temporal evolution of the Moreton wave emitted in association with the M3.0 flare in NOAA8299. The flare was associated with the ejection of gas. The Moreton wave propagated along the direction of gas ejection with a speed of 800 km/s and reached to the far distance of 350 Mm. According to Warmuth et al. (2004a), Nobeyama Radio Heliograph also detected the propagation of bright wavefront in 17 GHz. On the other hand, no EIT wave was reported in this event. (In the next two pages, time-difference images to enhance the visibility of wave fronts are shown with the manually drawn sketch of the wave fronts.)

活動領域 NOAA8299 で発生した M3.0 クラスのフレアに伴って発生したモートン波の伝播の様子が示されている．このフレアは，サージ状のガス噴出を伴っていた．このモートン波は約 800 km/s の速度でガス噴出の方向に伝搬し，活動領域から 35 万 km の遠方まで到達した．Warmuth et al. (2004a) の解析によると，17 GHz 野辺山太陽電波ヘリオグラフでもモートン波が明るく観測されていた．一方，このイベントでは EIT 波は検出されていないと報告されている．（次の 2 ページでは，波面を見やすくするように，画像の明るさの時間変化像を示している．また波面位置のスケッチも参考のために付してある．）

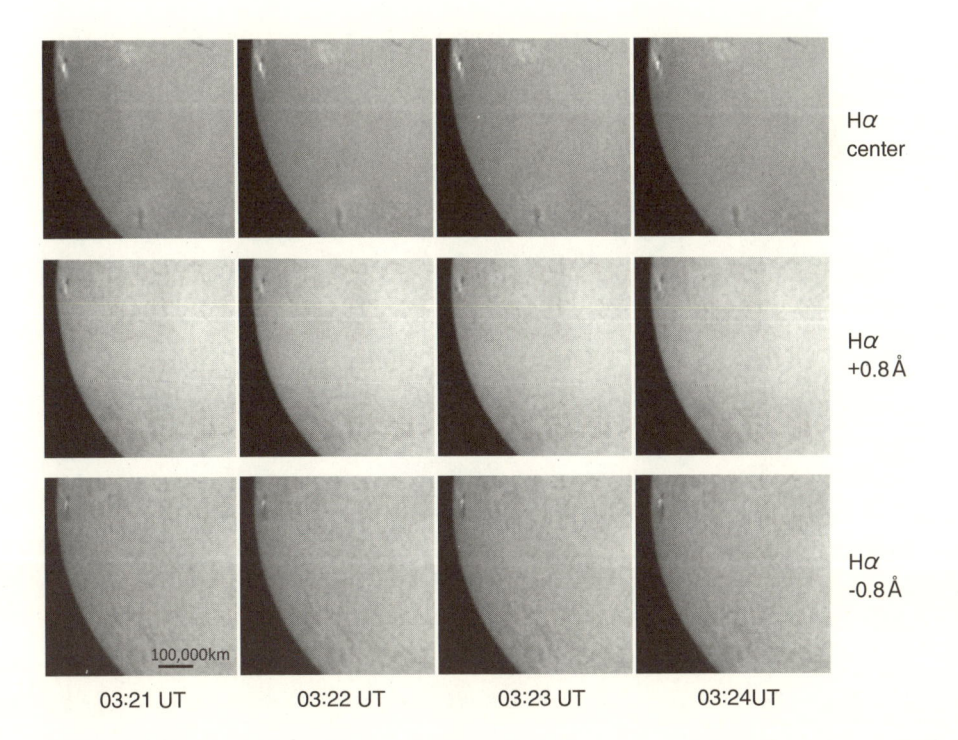

144 | 第6章 代表的イベント

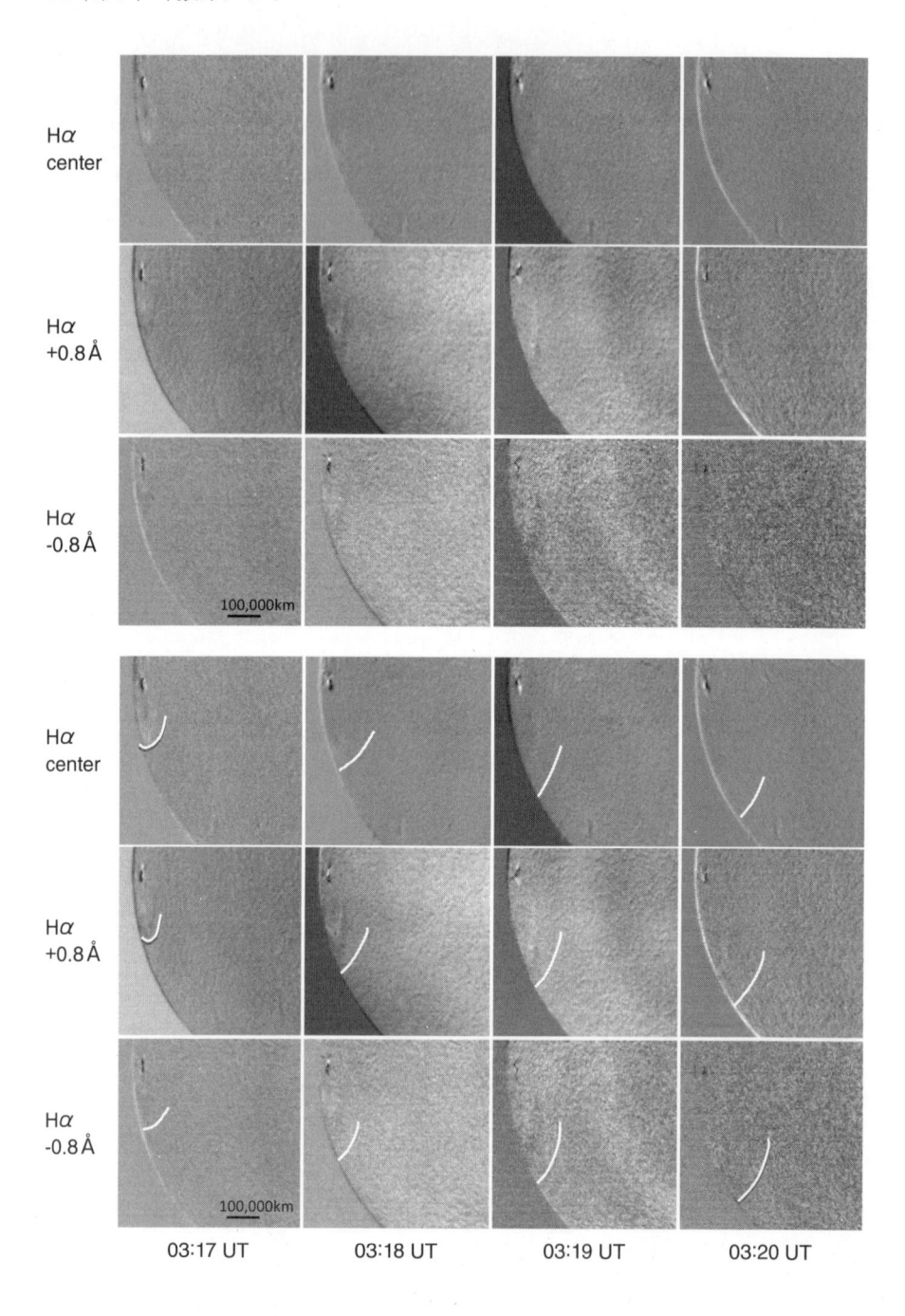

03:17 UT   03:18 UT   03:19 UT   03:20 UT

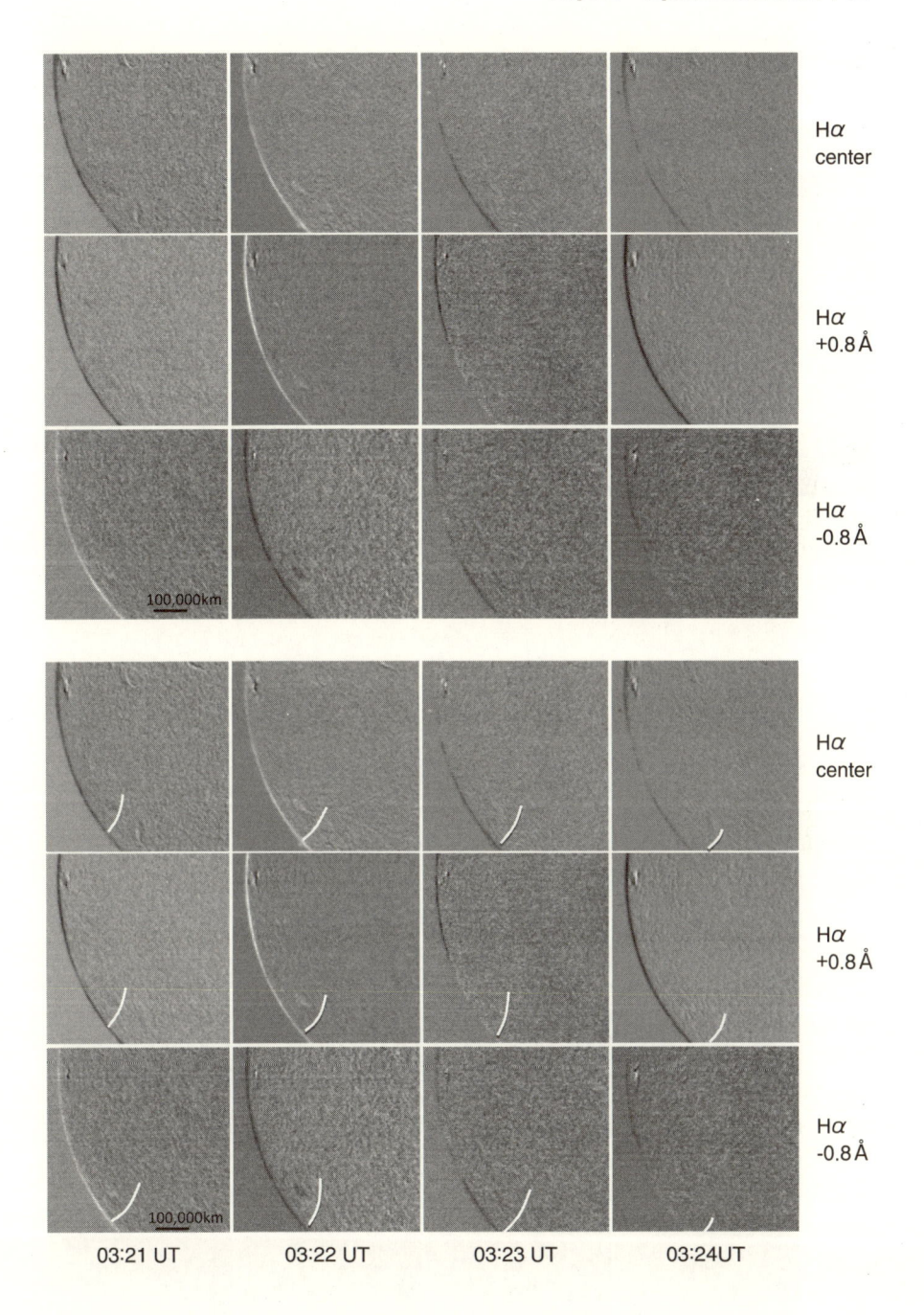

Chapter 6 Representative Events | 145

Hα center

Hα +0.8Å

Hα −0.8Å

100,000km

Hα center

Hα +0.8Å

Hα −0.8Å

100,000km

03:21 UT   03:22 UT   03:23 UT   03:24 UT

Sep 20, 1998 02:00−05:28 UT   Flare, Filament Eruption

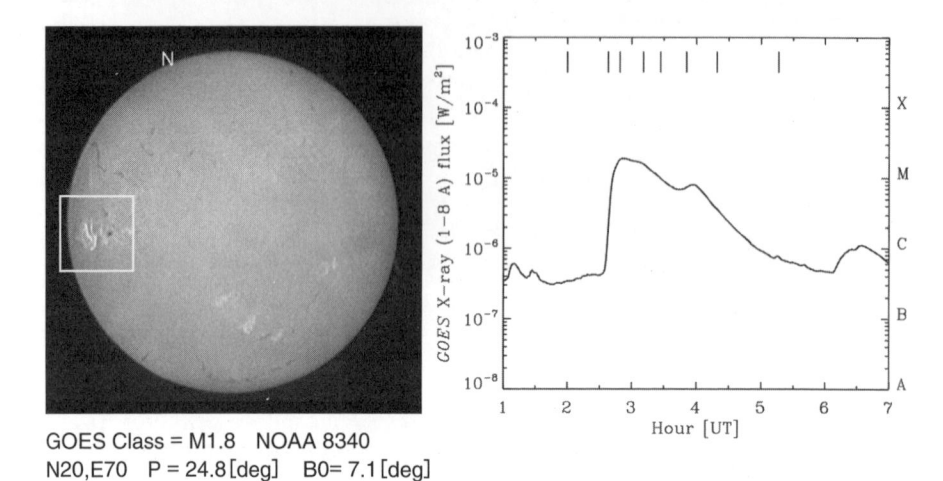

GOES Class = M1.8   NOAA 8340
N20,E70   P = 24.8 [deg]   B0= 7.1 [deg]

Hα center

Hα +0.8 Å

Hα −0.8 Å

100,000 km

02:00 UT   02:38 UT   02:49 UT   03:11 UT

In the east part of the NOAA8340, the magnetic field distribution was a peculiar one where the one polarity region was surrounded by the opposite polarity region, forming a circular magnetic neutral line of 150 Mm diameter. The M1.3 two-ribbon flare occurred in the west part of the circular neutral line. During the development of the flare, Hα filament located in the east half of the circular neutral line was activated and exploded to trigger the second C7 flare explosion. Post flare loops can be seen in Hα wing images over the flaring neutral line. According to Morimoto et al. (2003b), about 2/3 of the Hα filament was expelled to the interplanetary space. A bright arcade formation associated with the event was observed by Yohkoh/SXT.

NOAA8340の東では，磁場分布が特異な形をしていた．1つの磁気極性領域が，反対極性磁場に取り巻かれており，両者の間の磁気中性線が直径15万kmの円形となっていた．M1.3のツーリボンフレアがこの円形の中性線の西側で発生した．このフレアが発達するに伴い，東側の中性線上にあったHαフィラメントが活発化されて噴出して，2番目のC7のフレアを引き起こした．Hαのウイング波長では，フレアを起こした磁気中性線上にポストフレアループが見られた．Morimoto et al. (2003b)によると，元のHαフィラメントの約2/3が惑星間空間に放出された．また，Yohkoh/SXTによって，明るいアーケード状構造がコロナに形成されたことが観測されている．

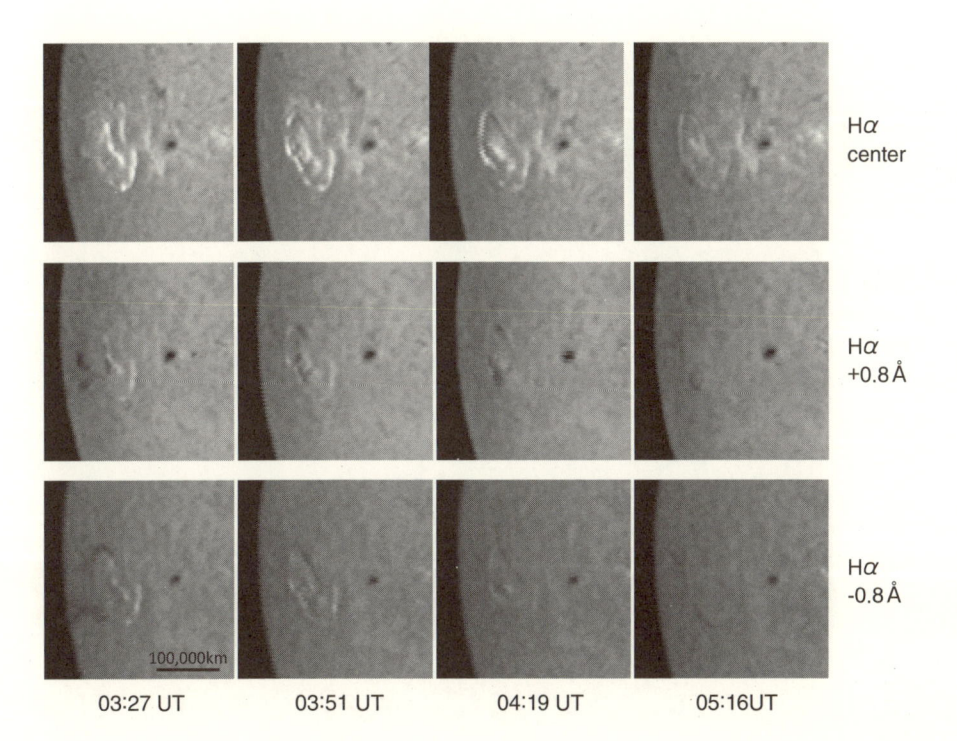

Dec 24, 1998 22:44B－06:35A UT   Filament Activity

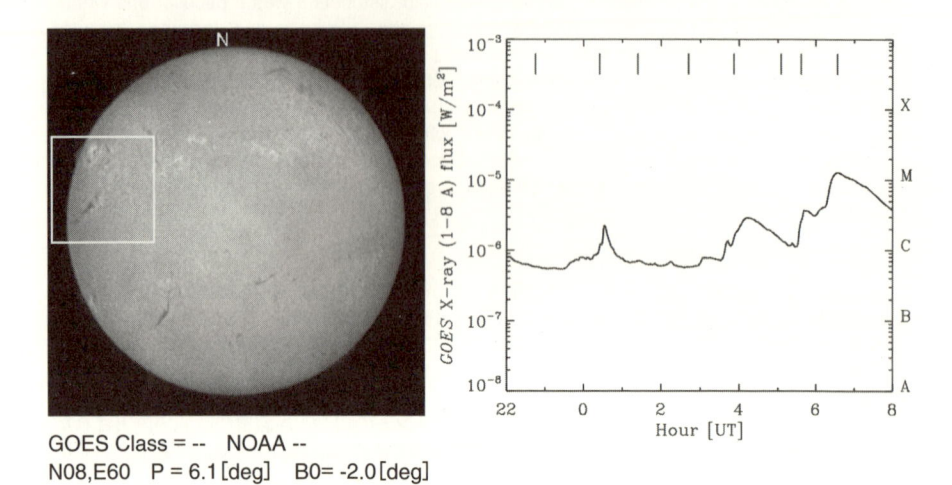

GOES Class = --   NOAA --
N08,E60   P = 6.1 [deg]   B0= -2.0 [deg]

Hα center

Hα +0.8Å

Hα -0.8Å

100,000km

22:45 UT   00:25 UT   01:24 UT   02:42 UT

A dark filament in a long filament channel of 500 Mm length extending from the NOAA8416 was activated in this event for nearly half a day. Although the location of the filament channel did not change so much, the spatial distribution of the Hα cool gas was intermittently changed due to the internal streaming flow along the axis of the filament.

NOAA8416 から延びる 50 万 km の長大なフィラメントチャンネルに存在した Hα ダークフィラメントが，ほぼ半日の間活発化したイベントである．フィラメントチャンネルの位置はほとんど変わらなかったものの，チャンネル内での Hα ガスの空間分布がフィラメントの軸方向の流れによって時々大きく変化した現象である．

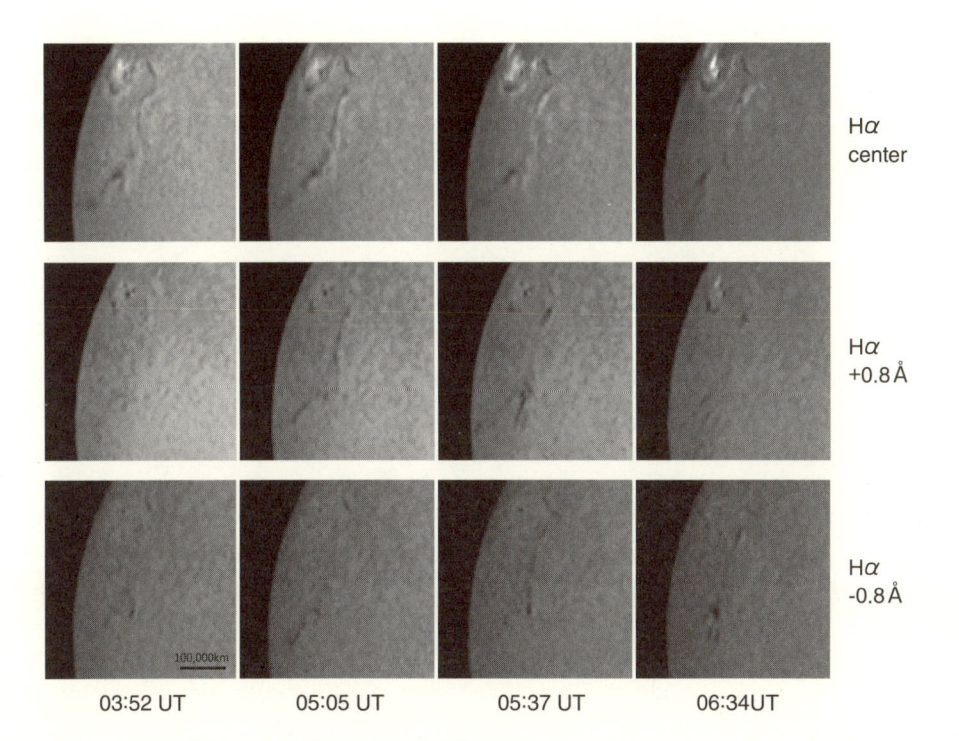

				Hα center
				Hα +0.8 Å
				Hα -0.8 Å
03:52 UT	05:05 UT	05:37 UT	06:34 UT	

Feb 09, 1999 03:07−05:22 UT   Two Ribbon Flare in Spotless Region

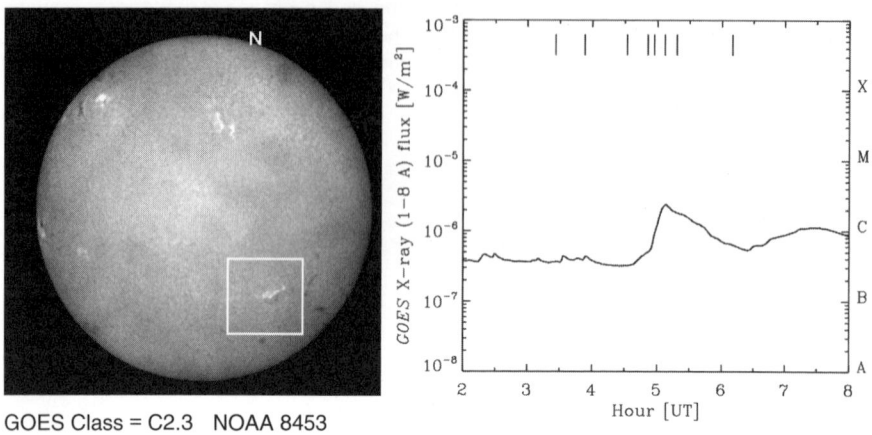

GOES Class = C2.3   NOAA 8453
S27,W39   P = -15.1 [deg]   B0= -6.5 [deg]

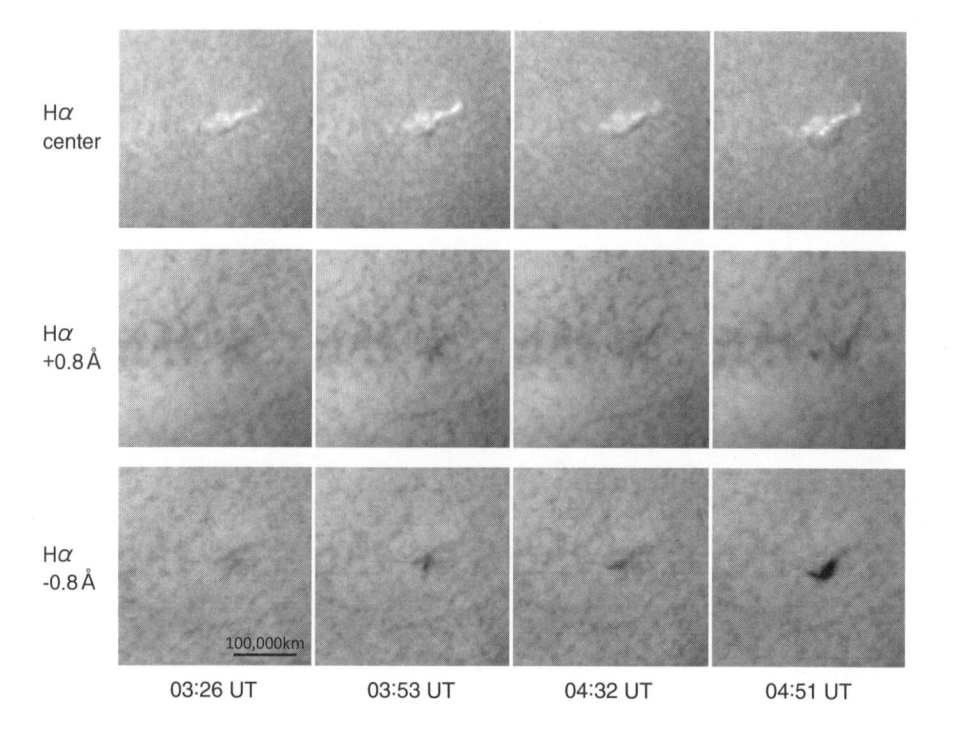

Hα center

Hα +0.8 Å

Hα -0.8 Å

100,000 km

03:26 UT   03:53 UT   04:32 UT   04:51 UT

A two-ribbon-flare occurred after the eruption of the active region filament in a remnant of an old active region NOAA8453. The filament, as a precursor of the flare, showed a complex internal motion during 1.5 hours before the flare explosion.

黒点の消滅した活動領域（NOAA8453）に残された活動領域フィラメントが消滅してツーリボンフレアが発生した．フレア発生前の 1 時間半ほどの間，先駆的なフィラメントの活動が観測された．

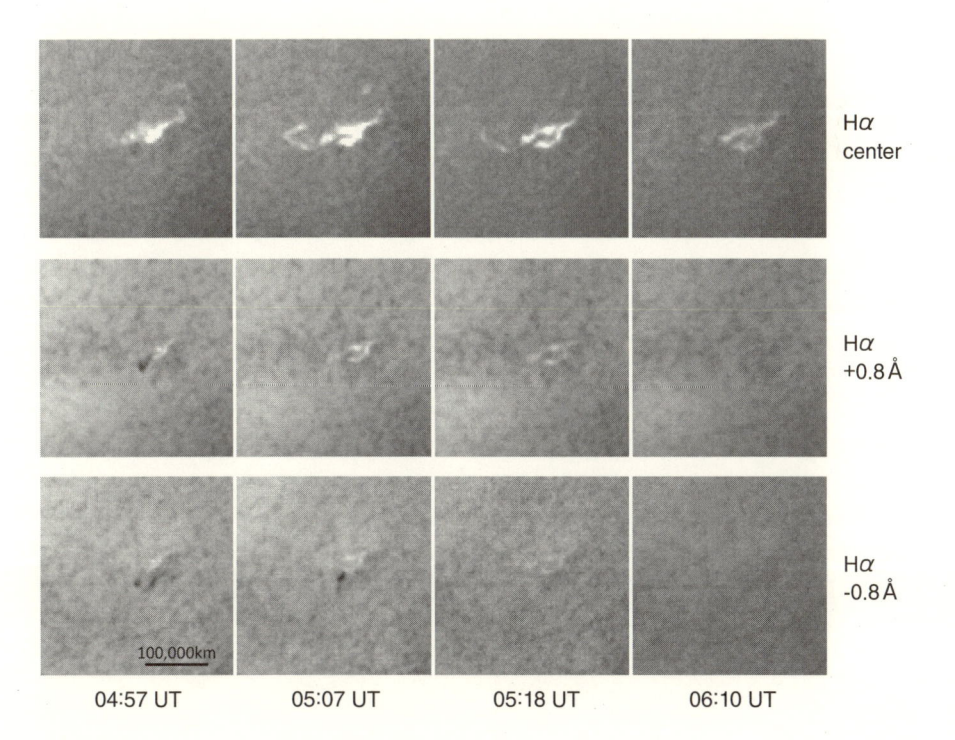

				Hα center
				Hα +0.8Å
				Hα -0.8Å
04:57 UT	05:07 UT	05:18 UT	06:10 UT	

Feb 16, 1999 01:42−04:05 UT   Flare, Filament Eruption

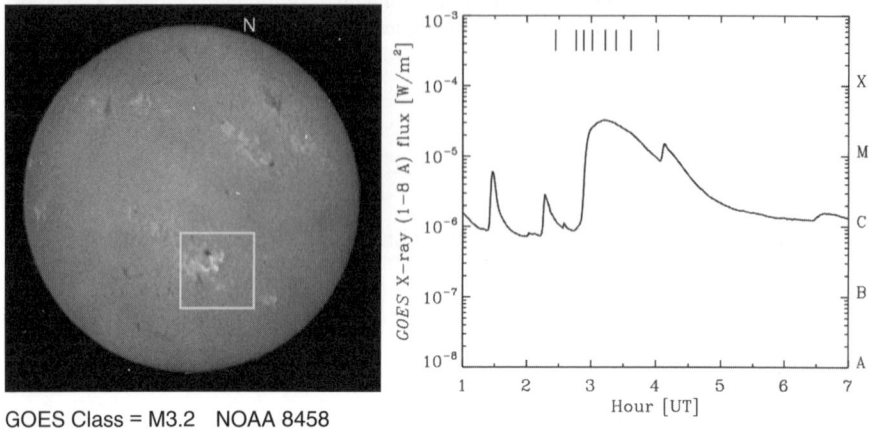

GOES Class = M3.2   NOAA 8458
S27,W18   P = -17.6 [deg]   B0= -6.9 [deg]

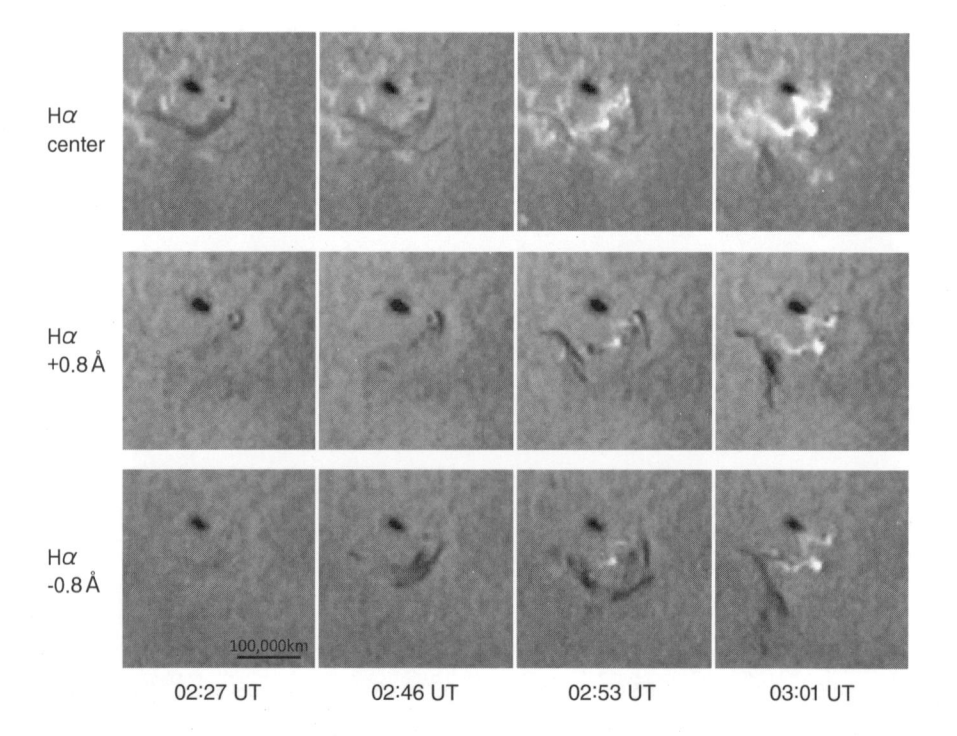

Hα center

Hα +0.8Å

Hα -0.8Å

100,000 km

02:27 UT   02:46 UT   02:53 UT   03:01 UT

The filament eruption started with an M3.2 flare occurrence. Applying the cloud model analysis, Morimoto et al. (2003a) estimated the line-of-sight speed of the filament to be around 57 km/s. Combined with the tangential velocities measured from the displacements of filament blobs, the magnitude of total velocity of filament expansion was around 200 km/s at maximum. The event was really an eruptive one and was associated with the formation of coronal arcade seen in soft X-ray (Morimoto et al. 2003b). A Moreton wave was detected in association with the event. The propagation of Moreton wave in 1 min cadence is displayed in the following pages.

M3.2 クラスのフレア発生に伴ってフィラメントが爆発した．Morimoto et al. (2003a) は，クラウドモデル解析から，その視線方向上昇速度を 57 km/s と求めた．フィラメントの断片の見かけの動きから視線と垂直な方向の速度を求めた結果，爆発の三次元的な速度の大きさは最大 200 km/s に達していた．この現象はその上昇速度が大きく爆発的であり，上昇の後にはコロナ中に明るい軟 X 線アーケード構造が形成されたことが確認されている(Morimoto et al. 2003b)．この現象に伴ってモートン波が発生した．モートン波の伝播の様子は次ページ以降に示されている．

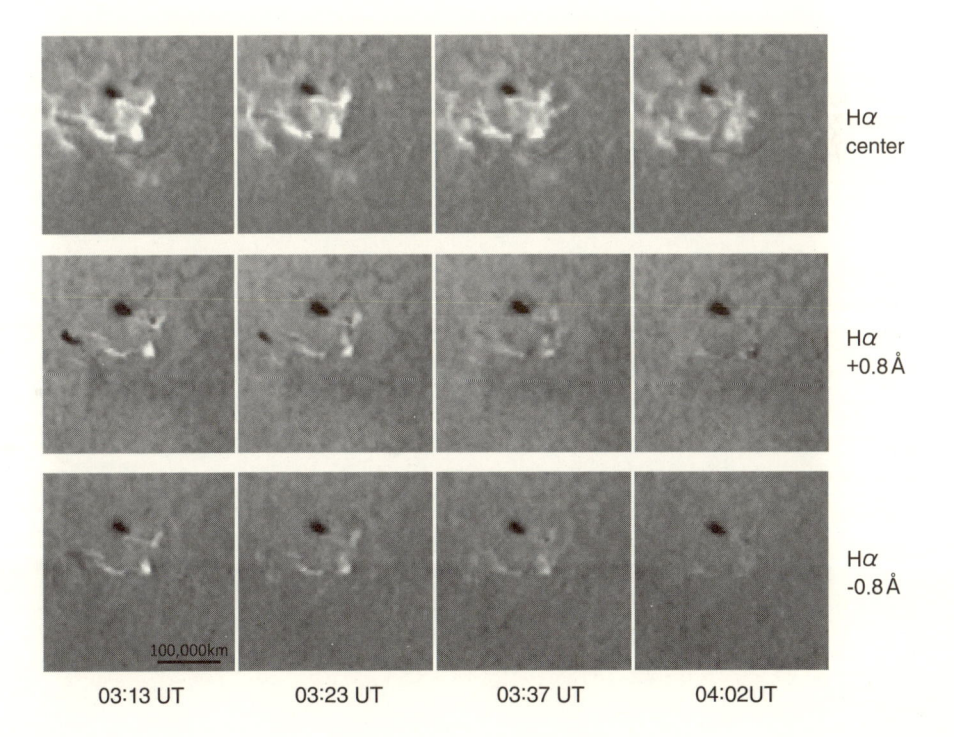

154 | 第6章 代表的イベント

Feb 16, 1999 02:53−03:02 UT　Moreton Wave

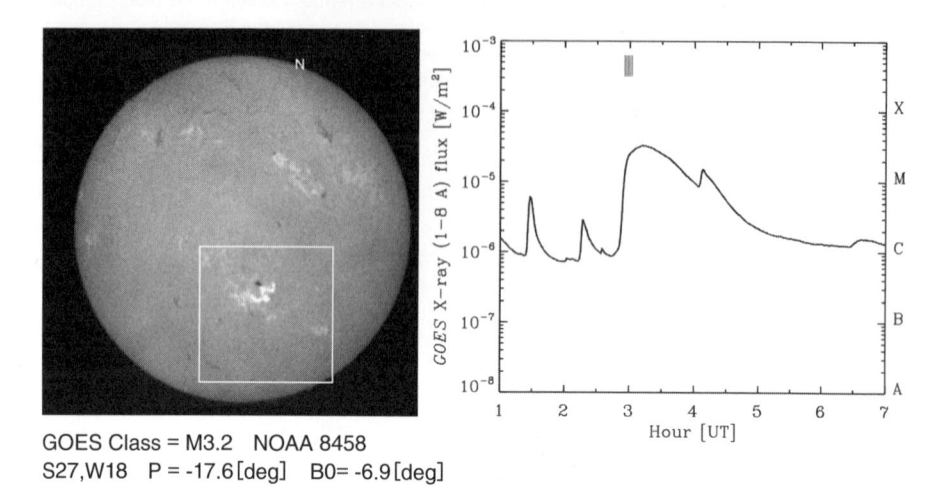

GOES Class = M3.2　NOAA 8458
S27,W18　P = -17.6[deg]　B0= -6.9[deg]

Hα center

Hα +0.8Å

Hα -0.8Å

100,000km

02:55 UT　　02:56 UT　　02:57 UT　　02:58 UT

Here is shown the temporal evolution of the Moreton wave emitted in the M3.2 class flare in NOAA8458. The wave came into appearance at 02:58 UT and propagated westwards with a speed of about 700 km/s. (In the next two pages, time-difference images to enhance the visibility of wave fronts are shown with the manually drawn sketch of the wave fronts.)

ここでは，NOAA8458 で起きた M3.2 クラスのフレアに伴って放射されたモートン波の伝播の様子を示している．モートン波は，02:58 UT に現れて西方向に約 700 km/s の速さで伝搬した．（次の 2 ページでは，波面を見やすくするように，画像の明るさの時間変化像を示している．また波面位置のスケッチも参考のために付してある．)

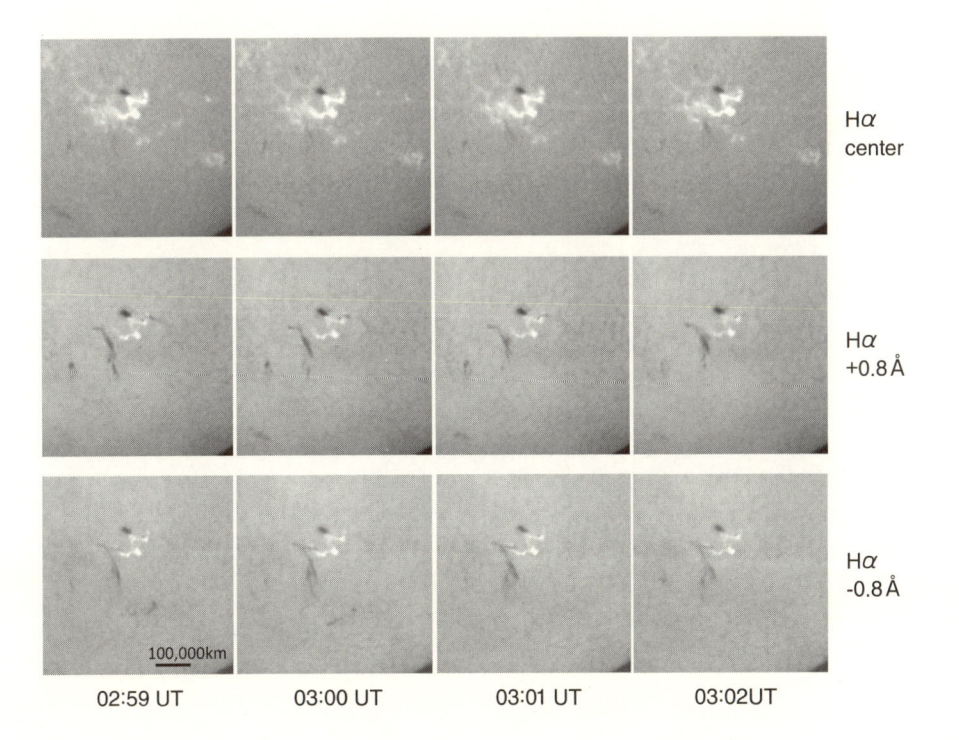

| | 02:55 UT | 02:56 UT | 02:57 UT | 02:58 UT |

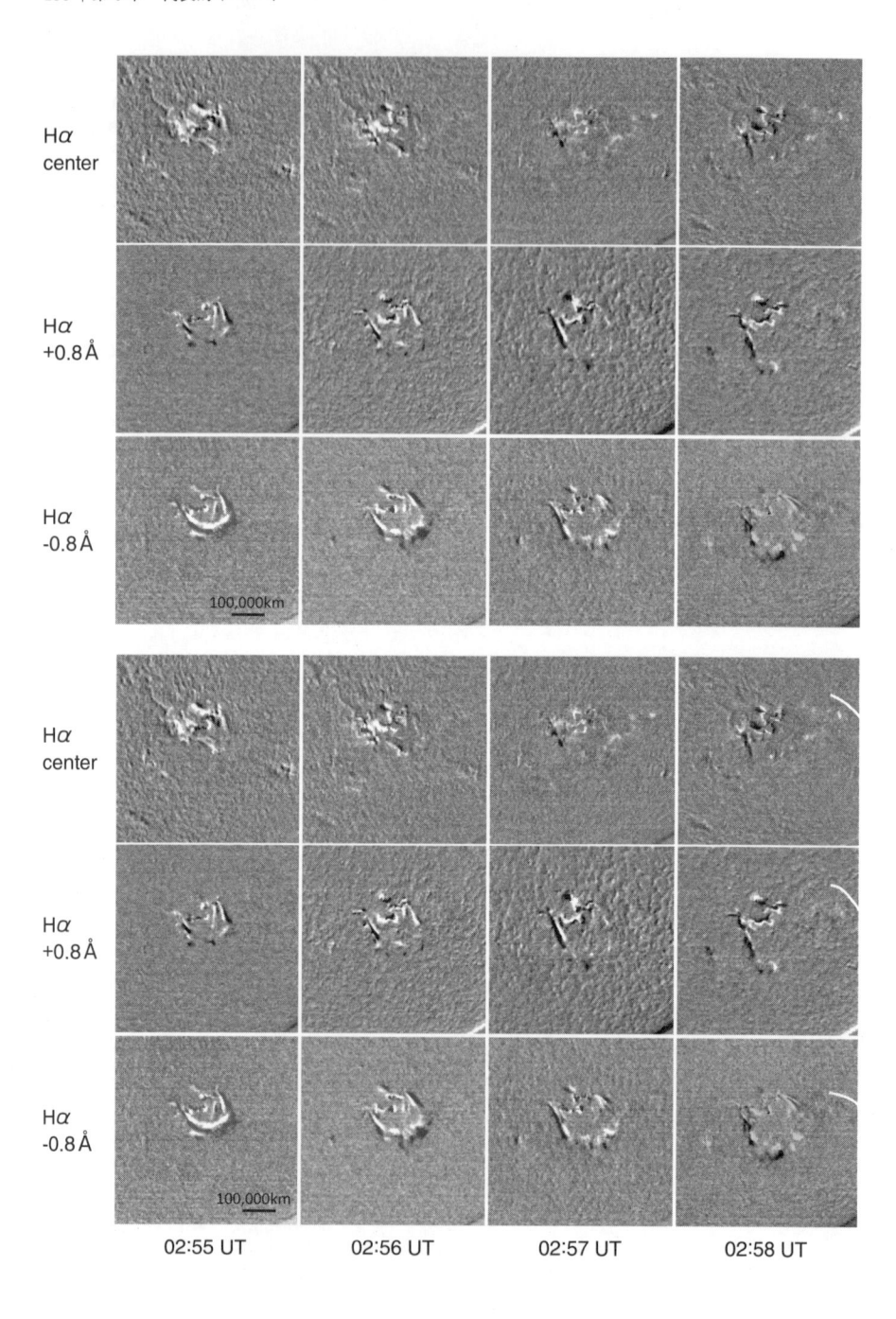

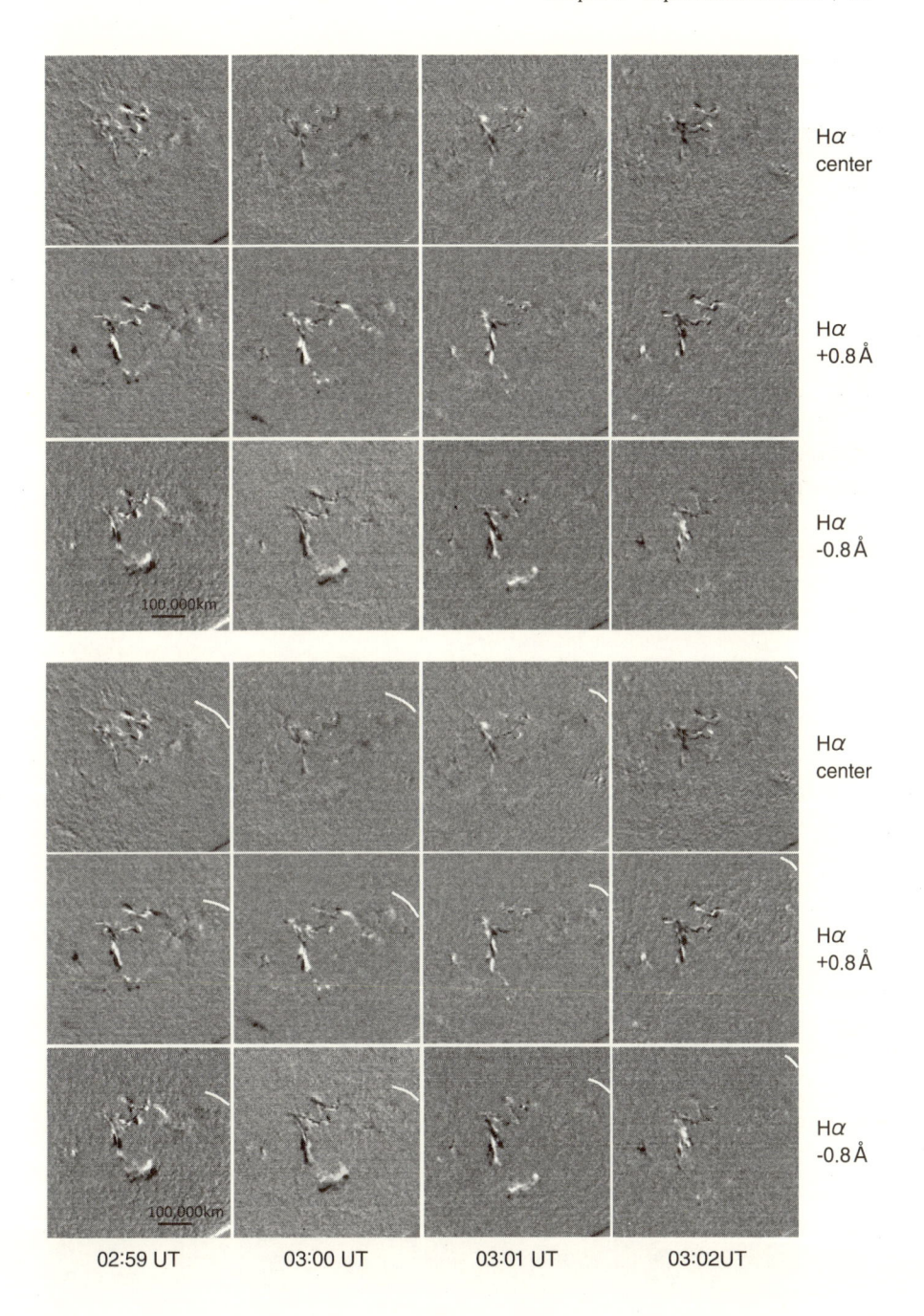

Chapter 6 Representative Events | 157

				Hα center
				Hα +0.8Å
				Hα −0.8Å
				Hα center
				Hα +0.8Å
				Hα −0.8Å
02:59 UT	03:00 UT	03:01 UT	03:02 UT	

Feb 16, 1999 02:41−05:24 UT Filament Eruption

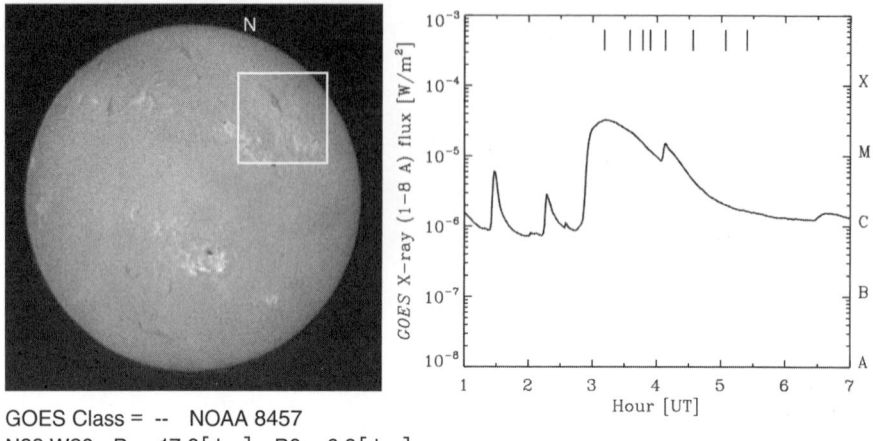

GOES Class = -- NOAA 8457
N33,W20 P = -17.6[deg] B0= -6.9[deg]

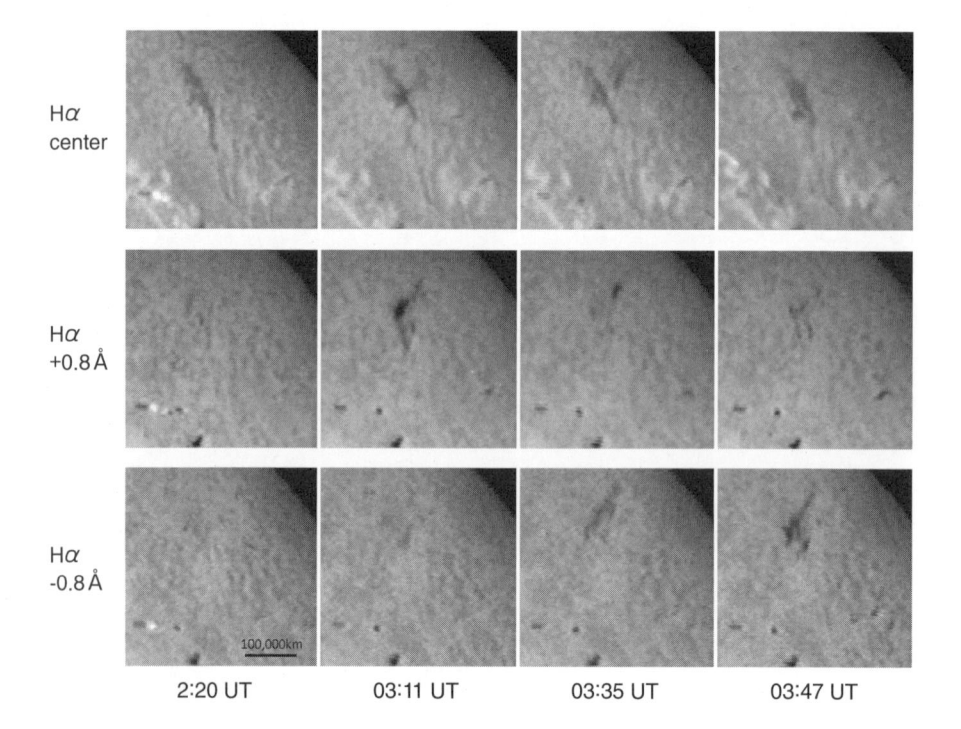

Hα center

Hα +0.8Å

Hα -0.8Å

2:20 UT   03:11 UT   03:35 UT   03:47 UT

In the event, a part of a quiescent filament expanded and exploded. The ejected gases returned back to the original locations. After the event ceased, the quiescent filament was connected to a neighboring active region filament, resulting in a very long active region filament of 500 Mm. The X-ray enhancement at around 02:49 UT was not related to the event but due to a flare in another active region.

静穏フィラメントの一部が膨張噴出したイベントである．噴出したガスは太陽表面に帰還して，最終的には静穏フィラメントは，近くにあった活動領域フィラメントと一体化して長さ 50 万 km ものフィラメントになった．なお，02:49 UT の X 線増光は，この現象には関係がなく他の活動領域でのフレアによるものである．

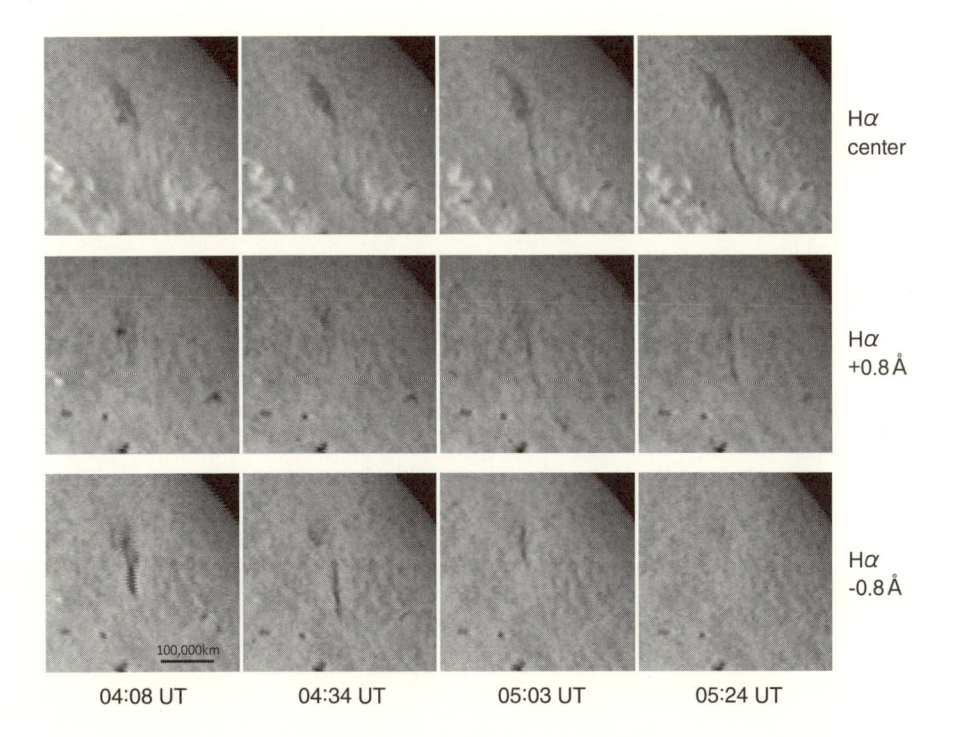

Jun 01, 1999 06:24−07:08 UT Flare, Filament Eruption

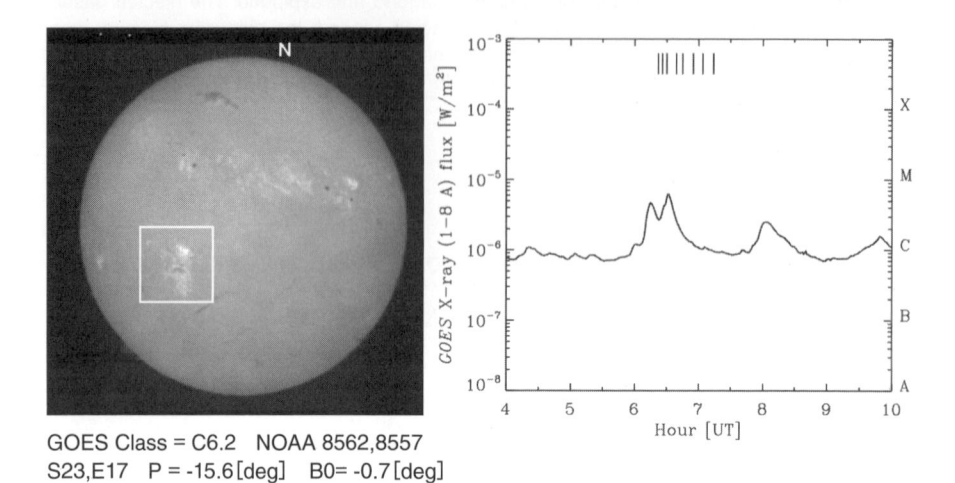

GOES Class = C6.2 NOAA 8562,8557
S23,E17 P = -15.6 [deg] B0= -0.7 [deg]

Hα center

Hα +0.8Å

Hα -0.8Å

100,000km

06:22 UT   06:26 UT   06:30 UT   06:39 UT

The event was a case where two different types of the compact flares occurred contiguously in separate locations of an active region. The first one lasted long for about 50 min. On the other hand, the second one brightened nearly 4 min later and dimmed in 10 min. The latter was associated with a rapid gas ejection and resulted in the formation of a new active region filament at the flare site.

この現象は1つの活動領域内で異なるタイプのコンパクトフレアが2つ引き続いて発生したものである．まず最初に活動領域の北側で比較的長寿命（継続時間約50分）のものが発生した．その約4分後，南側で短寿命（10分弱）のコンパクトフレアが発生した．後者は，激しいフィラメント噴出を起こし現象終了後活動領域フィラメントが新たに形成された．

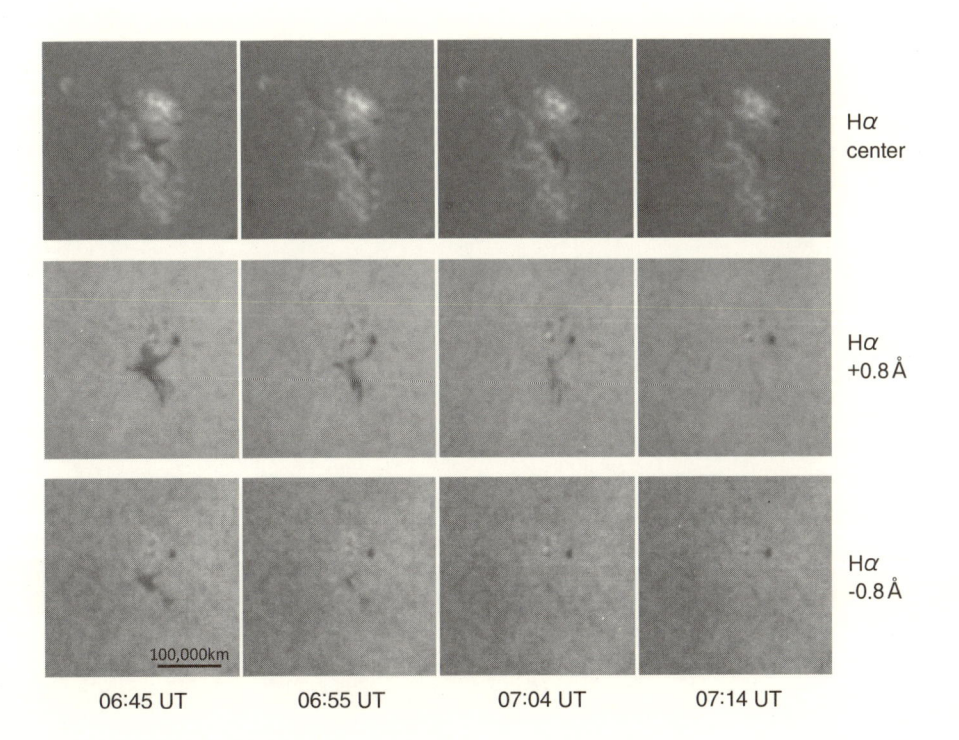

				Hα center
				Hα +0.8Å
				Hα -0.8Å
06:45 UT	06:55 UT	07:04 UT	07:14 UT	

100,000km

Aug 02, 1999 21:22−22:00 UT Flare

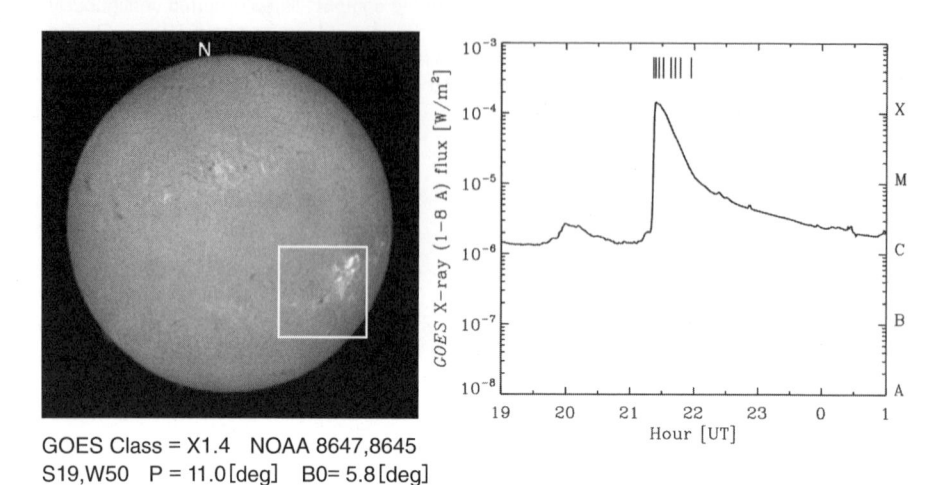

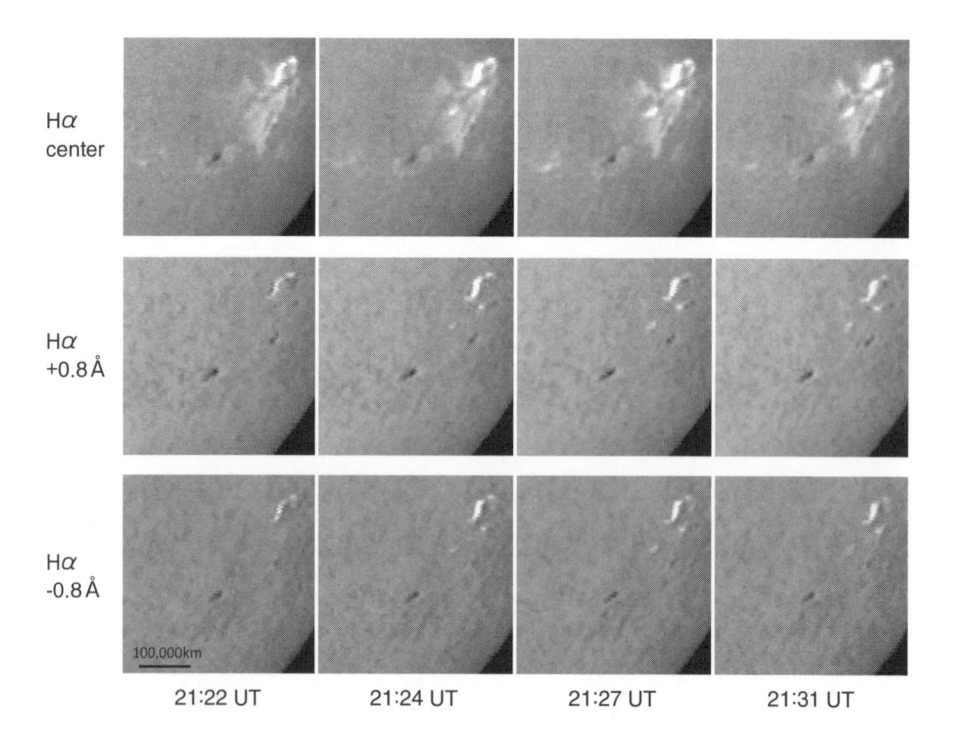

GOES Class = X1.4   NOAA 8647,8645
S19,W50   P = 11.0 [deg]   B0= 5.8 [deg]

Hα center

Hα +0.8Å

Hα −0.8Å

100,000km

21:22 UT   21:24 UT   21:27 UT   21:31 UT

An energetic X1.4 flare occurred in the complex of the β type NOAA8647 and the $\beta\gamma$ type NOAA8645. Although the flare was powerful, no precursoring Hα filament eruption was observed. No prominent CME was observed in association with the event.

GOES クラスが X1.4 の強いフレアが，β 型の NOAA8647 と $\beta\gamma$ 型の NOAA8615 複合活動領域で発生した．フレアそのものは強烈であったものの，先行 Hα フィラメント爆発や CME 現象は観測されなかった．

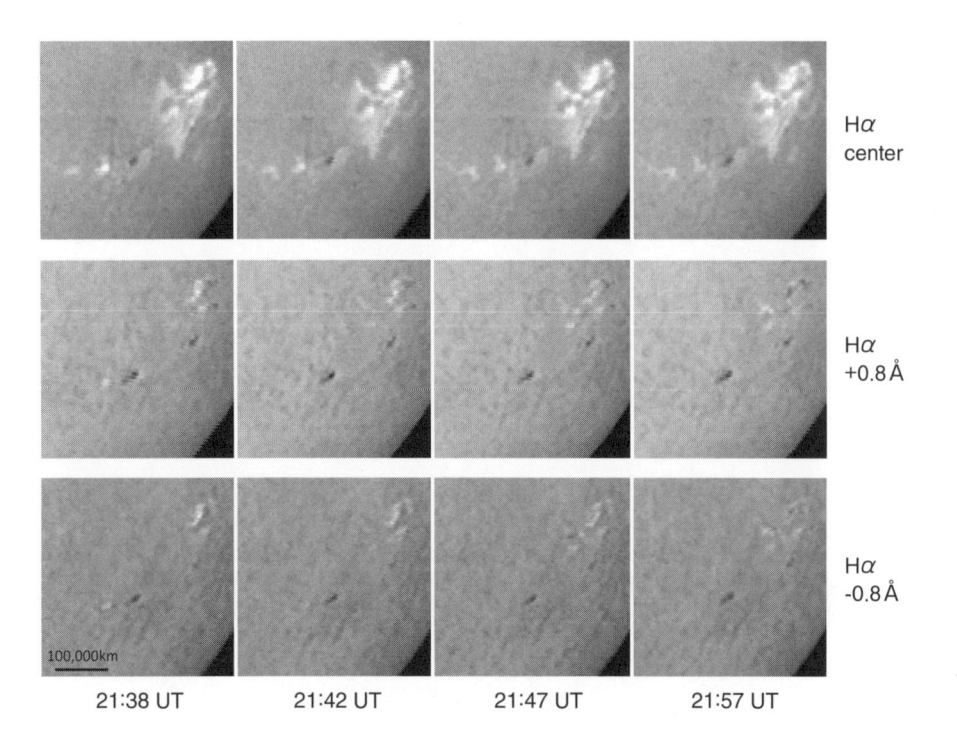

Aug 25, 1999 01:34−02:03 UT   Flare, Surge

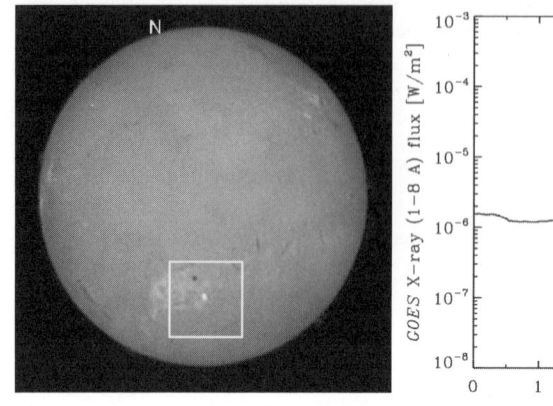

GOES Class = M3.6   NOAA 8673,8674
S27,E12   P = 19.0 [deg]   B0= 7.0 [deg]

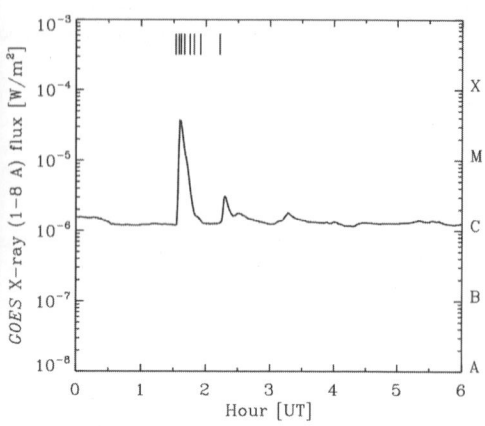

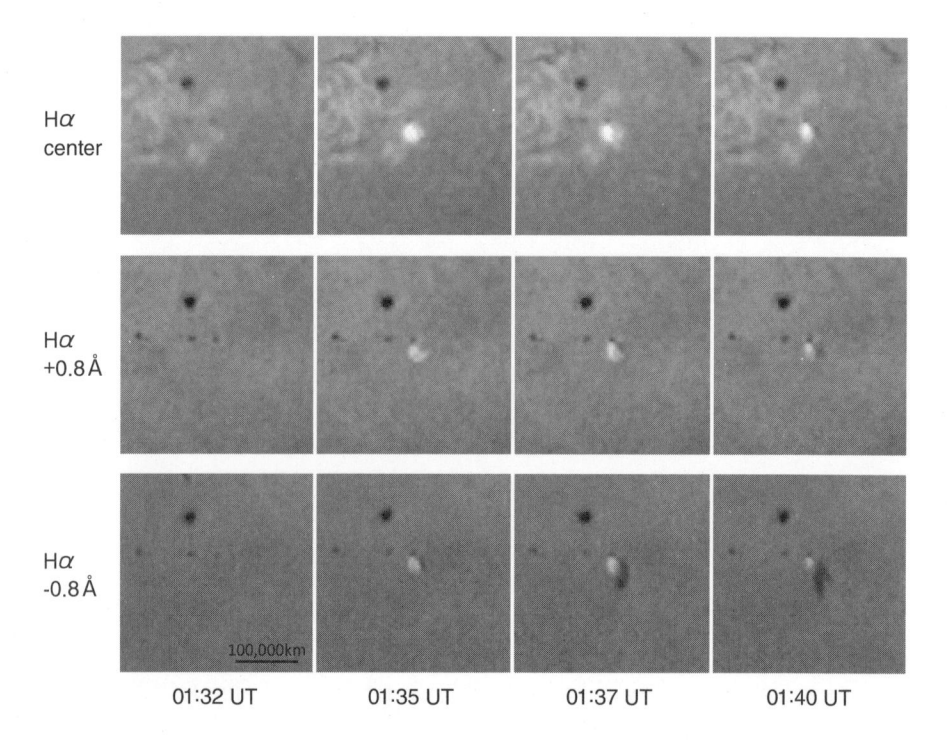

Hα center

Hα +0.8Å

Hα −0.8Å

100,000 km

01:32 UT   01:35 UT   01:37 UT   01:40 UT

Chapter 6　Representative Events | 165

A powerful M3.6 flare occurred in closely located two active regions NOAA8673 and 8674. The X-ray flare was of short duration of 15 min and showed spiky temporal variation without any LDE feature. In Hα images, the flare appeared as a compact one and ejected a few strands of surges. They were ejected with a speed of about 230 km/s and extended to the length of 100 Mm. The event showed a typical behavior of a compact flare explosion associated with mass ejection. The region was a very active and strong-flare productive one which produced an M9.8 flare on Aug 20 (Lee et al. 2001), and another M5.5 flare on Aug 27 (Belkina et al. 2002).

近接した活動領域 NOAA8673，8674 で，M3.6 の X 線強度の高いフレアが発生した．このフレアは寿命 15 分程度の短命なもので X 線での LDE 現象は伴っていなかった． Hα像では，コンパクトなフレアであり，そこからサージが 2-3 本放出された．サージの見かけの延伸速度は 230 km/s，その最大長は 10 万 km に達するものであった．コンパクトフレアからの質量放出現象の典型的な例である．この領域は，東のリム近傍で 8 月 20 日に M9.8 フレアを発生し（Lee et al. 2001），さらに 8 月 27 日にも M5.5 フレアを発生した（Belkina et al. 2002）極めて活発な活動領域であった．

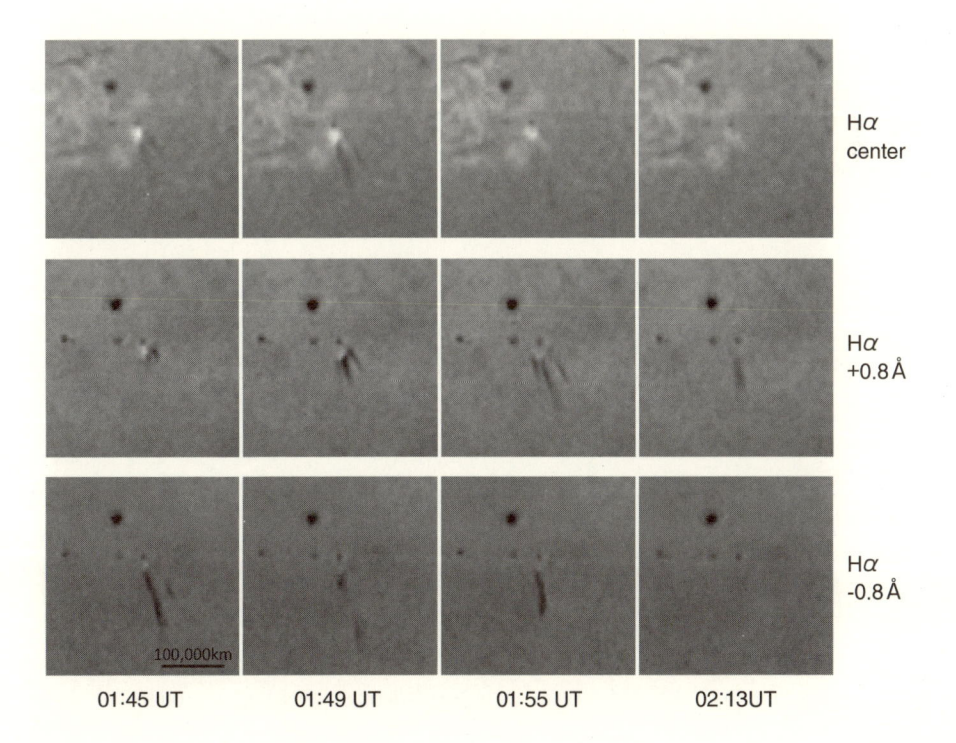

166 | 第6章　代表的イベント

Dec 28, 1999 00:41－01:27 UT   Flare, Filament Eruption

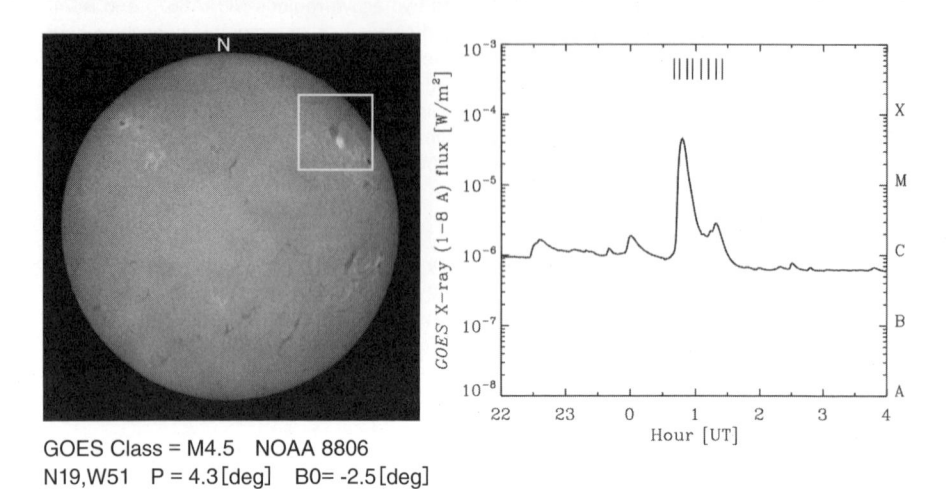

GOES Class = M4.5　NOAA 8806
N19,W51　P = 4.3 [deg]　B0= -2.5 [deg]

Hα center

Hα +0.8Å

Hα -0.8Å

100,000 km

00:40 UT　　00:45 UT　　00:52 UT　　00:57 UT

NOAA8802 was a magnetically complex $\beta\gamma\delta$ type active region, which produced many M-class flares. The compact M4.5 flare shown here ejected several strands of mass from the bright kernels of the flare with speeds of around 100 km/s, nearly simultaneously with the flare explosion. In association with the event, a CME was observed to expand radially with a speed of 500 km/s by SOHO/LASCO/C2.

NOAA8802は$\beta\gamma\delta$型の複雑な磁場を持った領域で多数のMクラスフレアを頻発した．ここで示されているM4.5のコンパクトフレアでは，フレア爆発時に明るいフレアカーネルから約100 km/sの速さでガスが数回放出された．これに伴って，CMEが500 km/sの速さで膨張しているのがSOHO/LASCO/C2で観測されている．

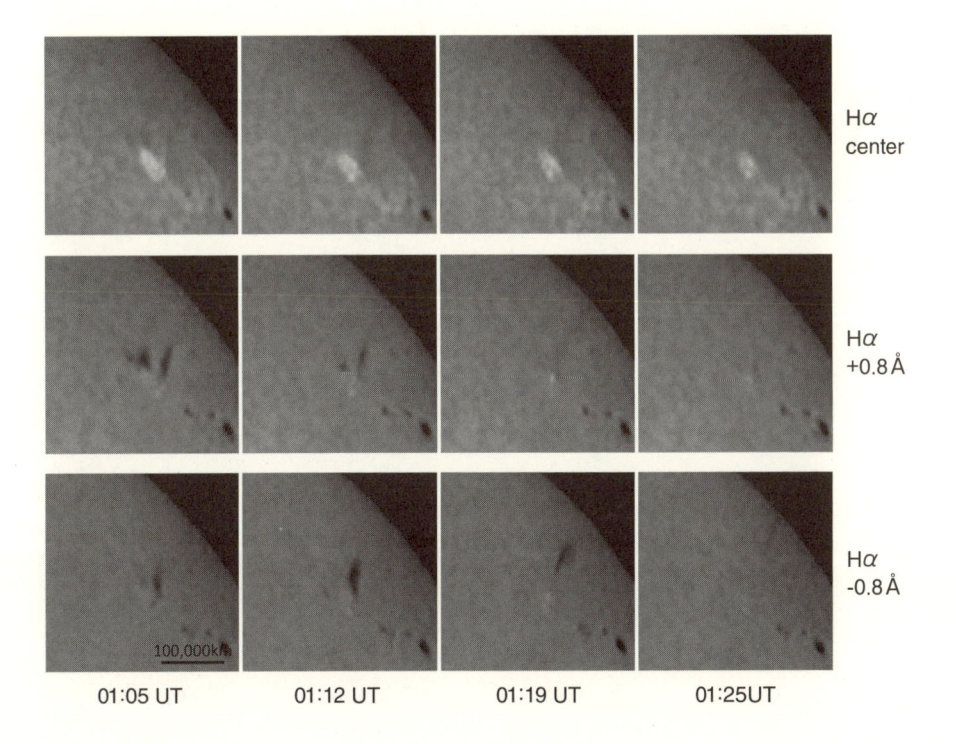

168 | 第6章 代表的イベント

Mar 03, 2000 02:11−03:14 UT   Flare, Filament Eruption

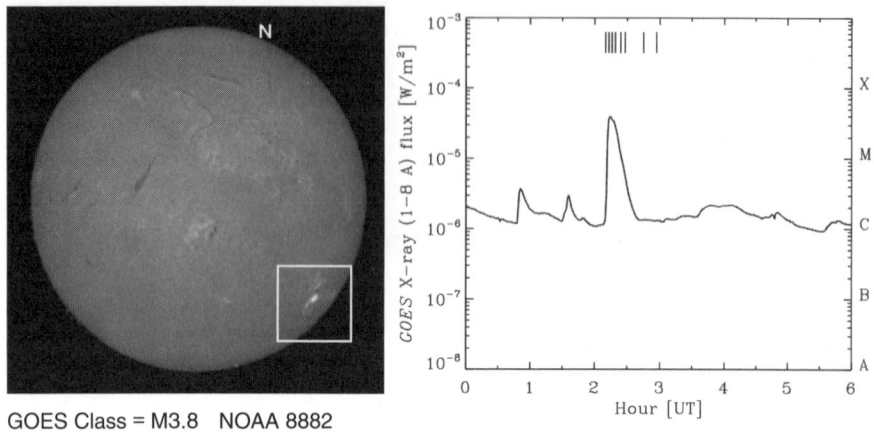

GOES Class = M3.8   NOAA 8882
S20,W70   P = −22.1 [deg]   B0= −7.2 [deg]

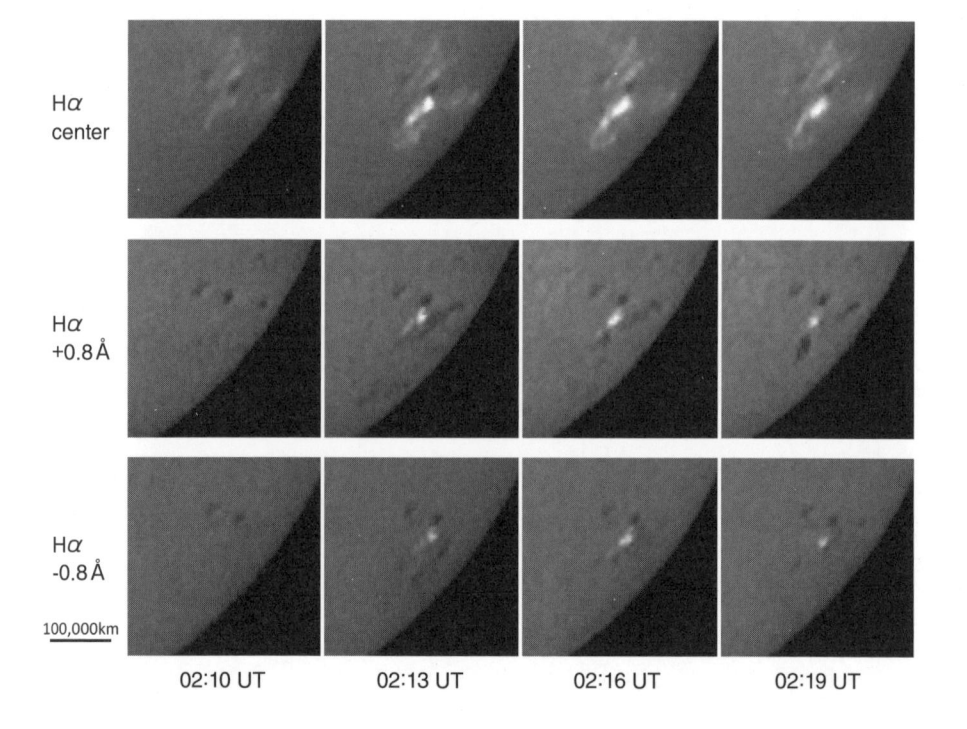

Hα center

Hα +0.8Å

Hα −0.8Å

100,000km

02:10 UT   02:13 UT   02:16 UT   02:19 UT

NOAA8882 was a complex magnetic $\beta\gamma\delta$ type active region and produced many M-class flares. In the compact M3.8 flare shown here, chromospheric Hα gas was ejected with a speed of 200 km/s from the bright kernel of the flare. Emissions of both an EIT wave and a Moreton wave were detected. Moreover, a CME and a type-II radio burst were also observed simultaneously. The propagation of Moreton wave in 1 min cadence is displayed in the following pages.

NOAA8882は，βγδタイプの複雑な磁場をもった領域であって多数のMクラスフレアを頻発した．ここで示されているM3.8のコンパクトフレアでは，明るいフレアカーネルから彩層ガスが200 km/sの速度で放出された．EIT波，およびモートン波が，このイベントに伴って発生した．さらに，CMEおよびII型の電波バーストも同時に観測された．モートン波の伝播の様子は次ページ以降に示されている．

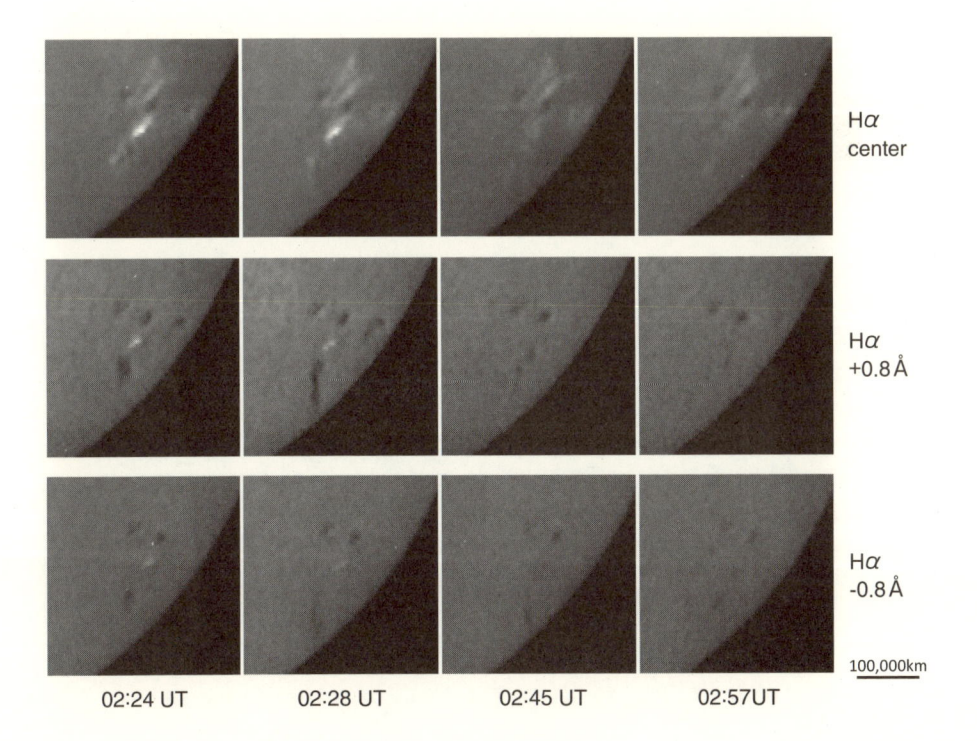

Mar 03, 2000 02:11−02:21 UT Moreton Wave

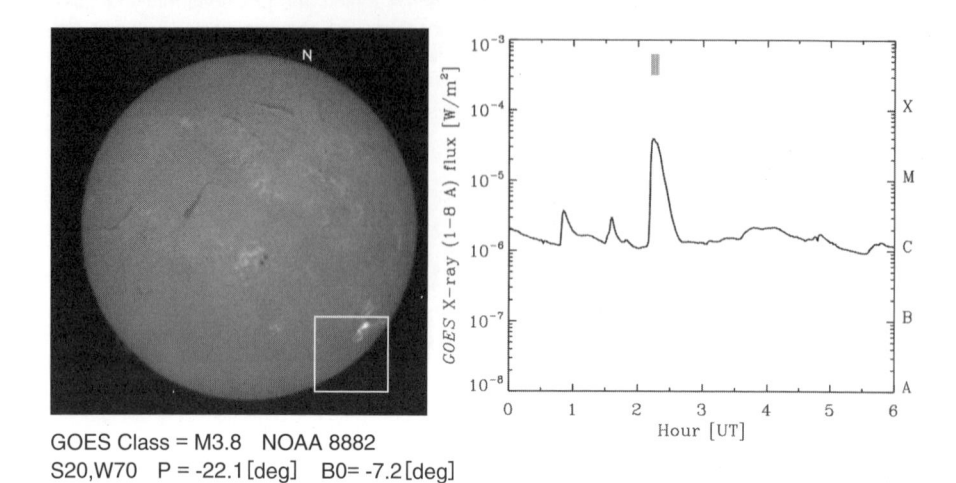

GOES Class = M3.8   NOAA 8882
S20,W70   P = -22.1 [deg]   B0= -7.2 [deg]

Hα center

Hα +0.8Å

Hα -0.8Å

100,000km

02:12 UT   02:13 UT   02:14 UT   02:15 UT

In the compact M3.8 flare that occurred in NOAA8882 shown here, chromospheric Hα gas was ejected with a speed of 200 km/s from the bright kernel of the flare. Both an EIT wave and a Moreton wave were detected to propagate with propagation speeds of 200 km/s and 1,000 km/s, respectively. According to Narukage et al. (2002, 2004), X-ray wave was also detected by Yohkoh/SXT with a propagation speed of 1,400 km/s. A CME was also observed to expand with a speed of 730 km/s in SOHO/LASCO/C2 images. Moreover, a type-II radio burst was observed simultaneously. The shock speed derived from the type-II burst is 1,150 km/s. (In the next two pages, time-difference images to enhance the visibility of wave fronts are shown with the manually drawn sketch of the wave fronts.)

ここで示されている NOAA8882 で発生した M3.2 のコンパクトフレアでは，明るいフレアカーネルから彩層ガスが 200 km/s の速度で放出された．EIT 波，およびモートン波が，この現象に伴って発生した．波の伝播速度はそれぞれ 200 km/s と 1,000 km/s であった．Narukage et al. (2002, 2004) によると，X 線波も 1,400 km/s の速度で伝播しているのが「ようこう」衛星/SXT で観測された．また，SOHO/LASCO/C2 では，見かけの速さ 730 km/s の CME が観測された．さらに，II 型の電波バーストが同時に観測されていた．バーストを起こしている衝撃波の速度は 1,150 km/s であった．（次の 2 ページでは，波面を見やすくするように，画像の明るさの時間変化像を示している．また波面位置のスケッチも参考のために付してある．）

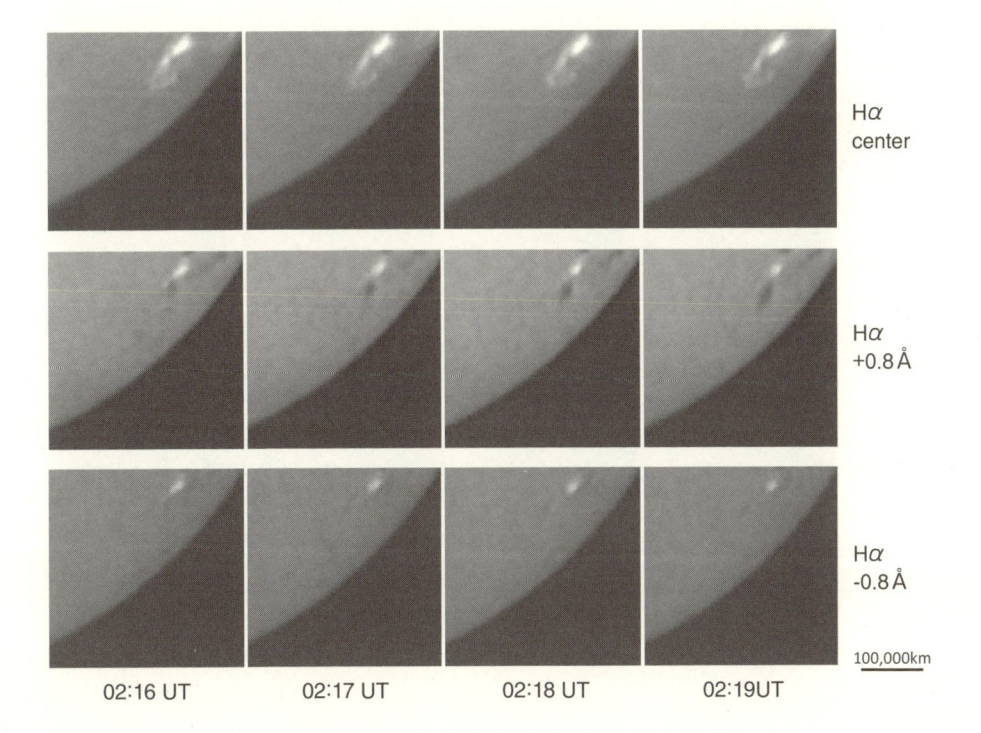

172 | 第6章　代表的イベント

Hα center

Hα +0.8Å

Hα -0.8Å

100,000km

Hα center

Hα +0.8Å

Hα -0.8Å

100,000km

02:12 UT　　02:13 UT　　02:14 UT　　02:15 UT

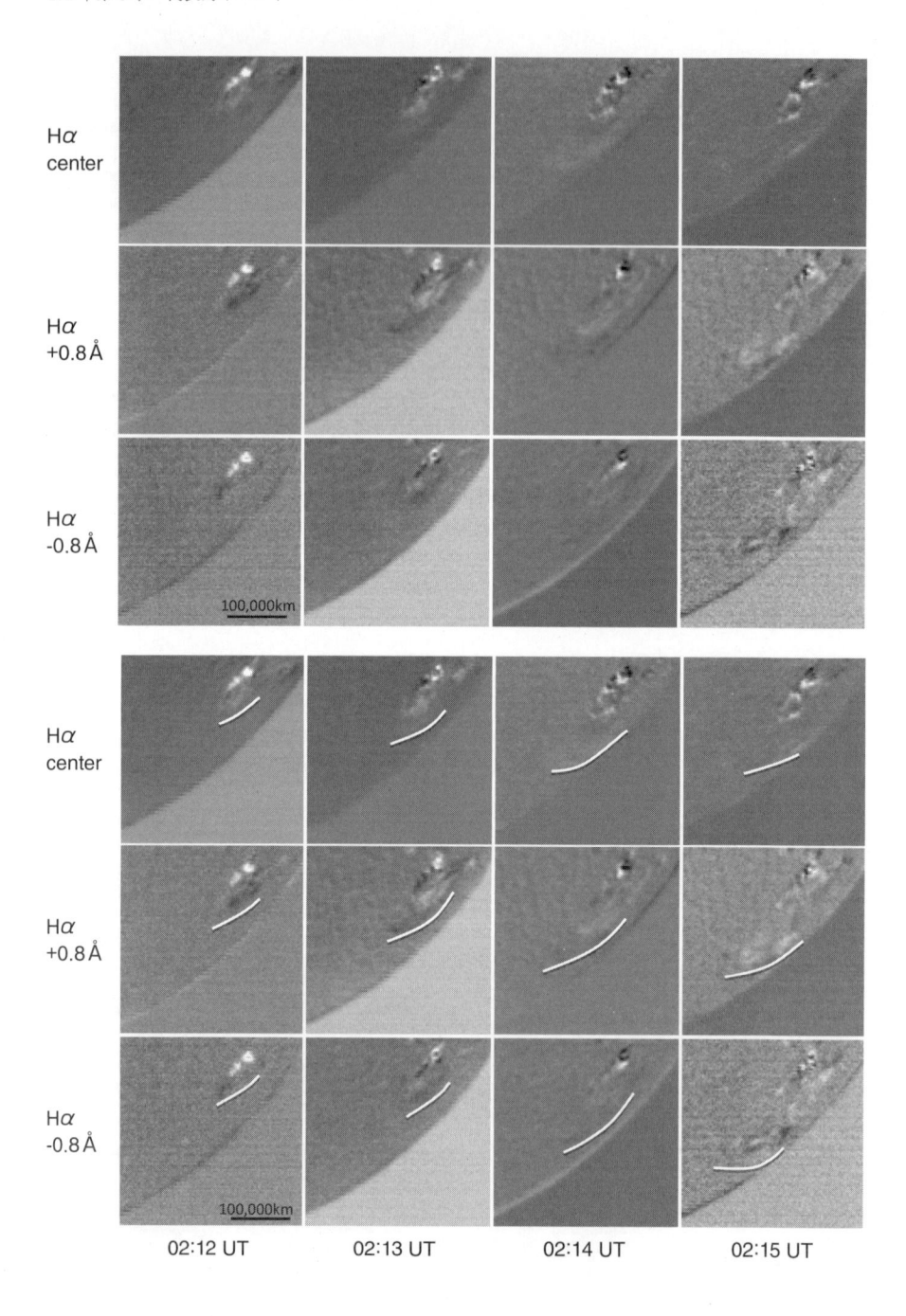

Hα center

Hα +0.8Å

Hα -0.8Å

100,000km

Hα center

Hα +0.8Å

Hα -0.8Å

100,000km

02:16 UT   02:17 UT   02:18 UT   02:19 UT

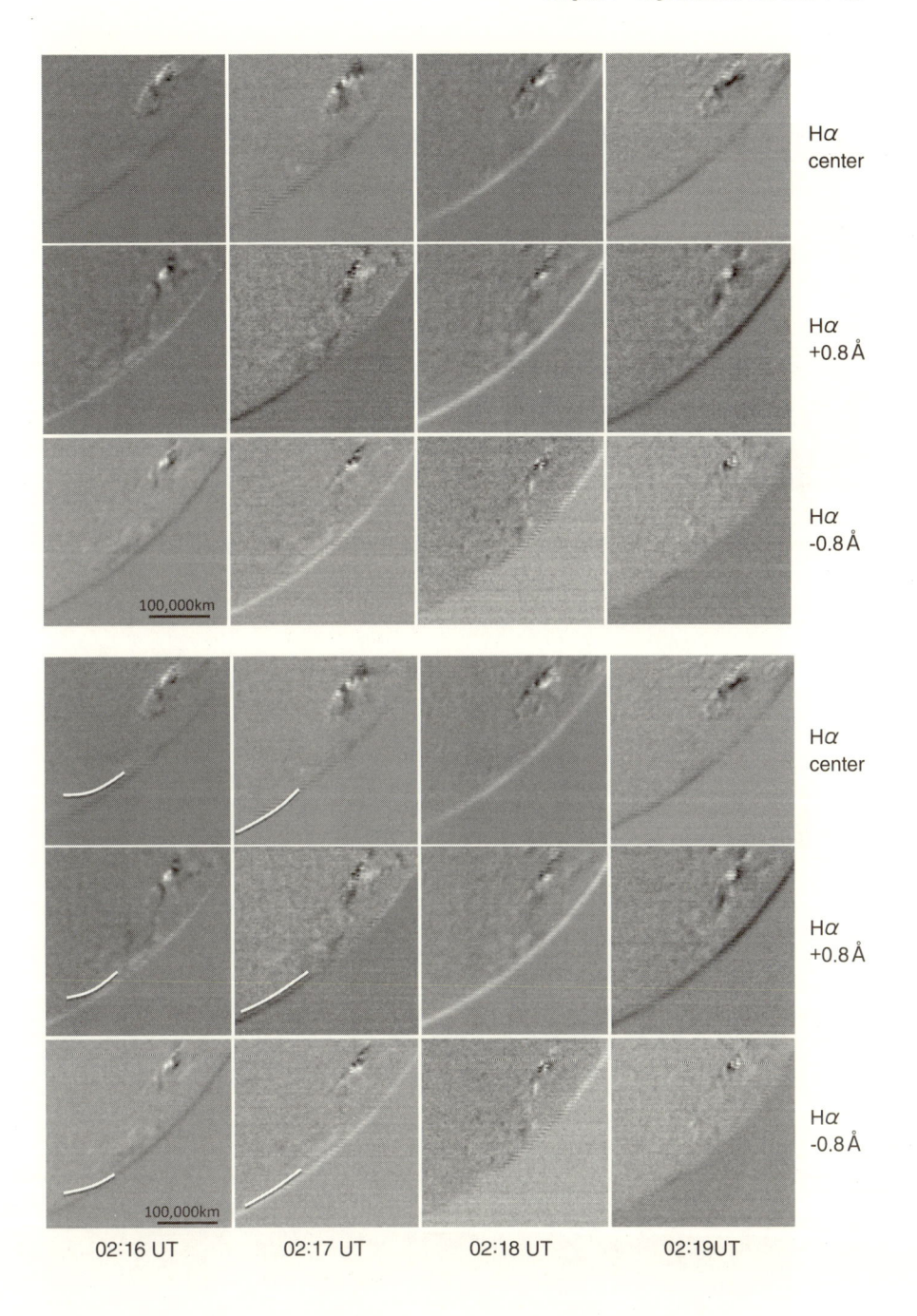

Jun 04, 2000 22:05−23:22 UT   Flare, Filament Eruption

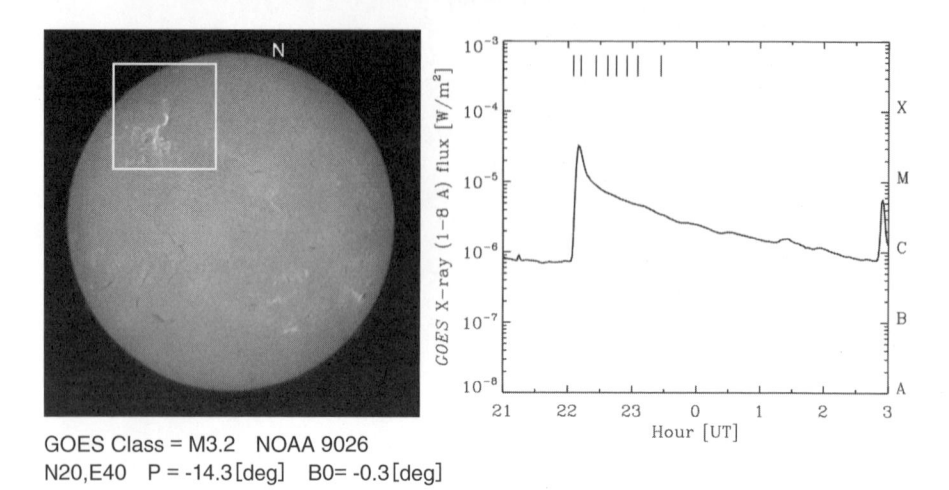

GOES Class = M3.2   NOAA 9026
N20,E40   P = -14.3[deg]   B0= -0.3[deg]

Hα center

Hα +0.8Å

Hα -0.8Å

100,000 km

22:05 UT   22:12 UT   22:26 UT   22:37 UT

NOAA9026 was a very flare-productive $\beta\gamma$ type region, which produced many M-class flares. In the flare of M3.2 shown here, bright SOHO/EIT195 flux ropes were observed to expand with a speed of about 300 km/s, associated with the Hα filament expansion. SOHO/LASCO/C2 also showed an associated CME expansion with a speed of 500 km/s. According to the detailed study of the region development by Kurokawa et al. (2002), the strong activities of the region was ascribed to the emergence of twisted magnetic flux from beneath the deep layers.

NOAA9026 は，極めて活発な領域（β γ 型）であって多数の M クラスフレアを頻発した．ここで示されている M3.2 フレアでは，Hαフィラメント爆発に伴って，SOHO/EIT195 の明るい磁束管が 300 km/s の速度で膨張していることが観測された．また，LASCO/C2 では 500 km/s の速度で CME が膨張していることが見られた．Kurokawa et al. (2002) のこの領域の時間発展の研究によると，この領域の特別に強い活動性はねじれた磁束管が浮上してきたことによるものである．

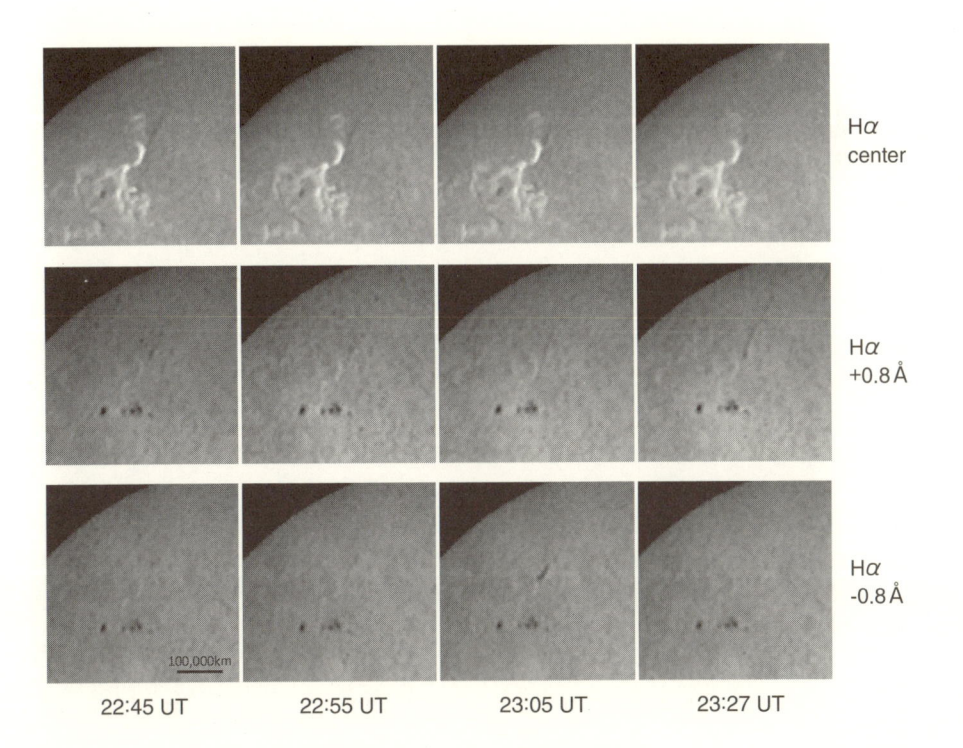

				Hα center
				Hα +0.8 Å
				Hα -0.8 Å
22:45 UT	22:55 UT	23:05 UT	23:27 UT	

Jun 06, 2000 01:02－08:36A UT   Filament Eruption

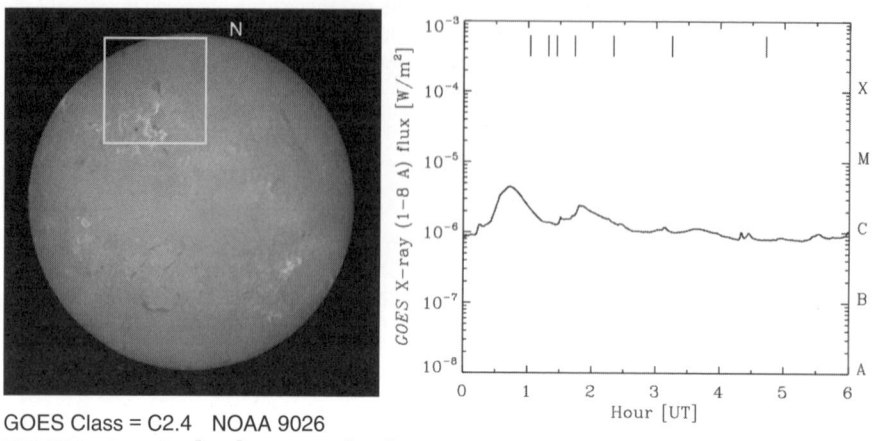

GOES Class = C2.4   NOAA 9026
N35,E30   P = -13.4[deg]   B0= 0.0[deg]

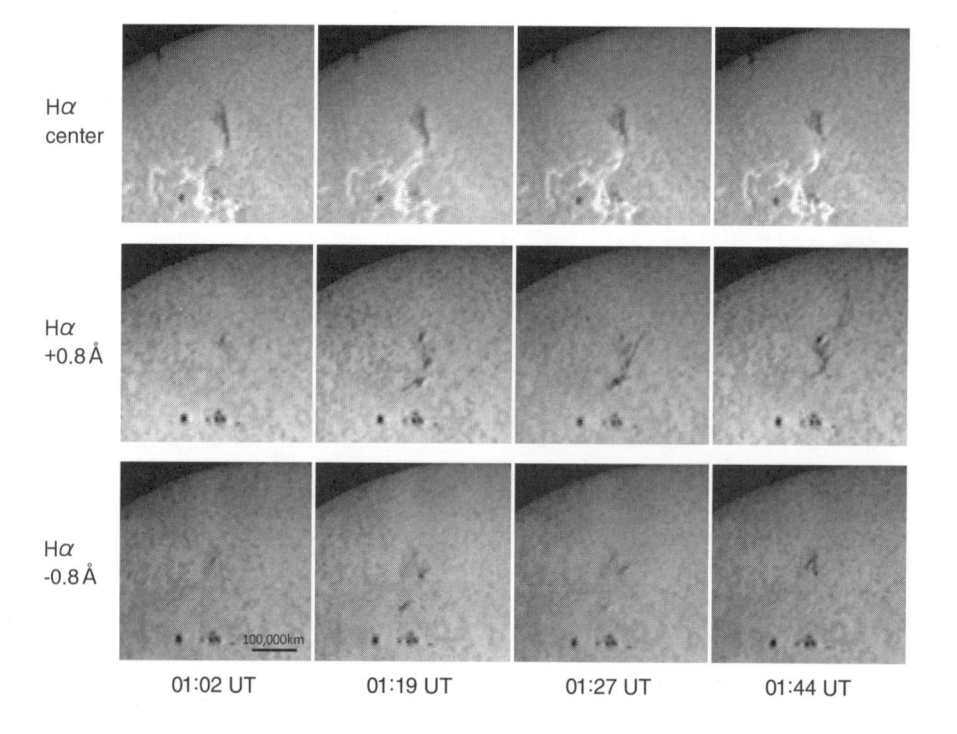

Hα center

Hα +0.8Å

Hα -0.8Å

01:02 UT   01:19 UT   01:27 UT   01:44 UT

NOAA9026 of $\beta\gamma\delta$ magnetic class showed a very high activity since its appearance from the east limb. The event shown here is the filament eruption of 150 km/s apparent speed, located in the northern part of the region. SOHO/EIT195 detected a bright ejection of gas, while SOHO/LASCO/C2 observed a CME with a expansion speed of 300 km/s. In later times, during one hour around 13:00 UT, three strong flares (M2.7, X1.1 and M7.1) exploded consecutively and produced another CME of 800 km/s speed from the region. Akiyama et al. (2002) reported that the second fast CME caught up the first slow one and produced a strong interaction between the two.

$\beta\gamma\delta$型の活動領域 NOAA9026 は，東のリムから現れたときから活発な領域であった．ここで表示されているイベントは，この領域の北縁にあった活動領域フィラメントが 150 km/s の見かけの速さで噴出したものである．これに伴って，SOHO/EIT195 でも明るいガスが噴出しているのが観測された．また，SOHO/LASCO/C2 でも，300 km/s で膨張する CME が観測された．なお，この領域では，後刻 13:00 UT から始まる 1 時間内に，M2.7, X1.1, M7.1 の強烈なフレアが連発し，800 km/s もの高速 CME を発生させた．Akiyama et al. (2002) によると，後発の CME が先発のものに追いつき，両者の間に強い相互作用を生み出した．

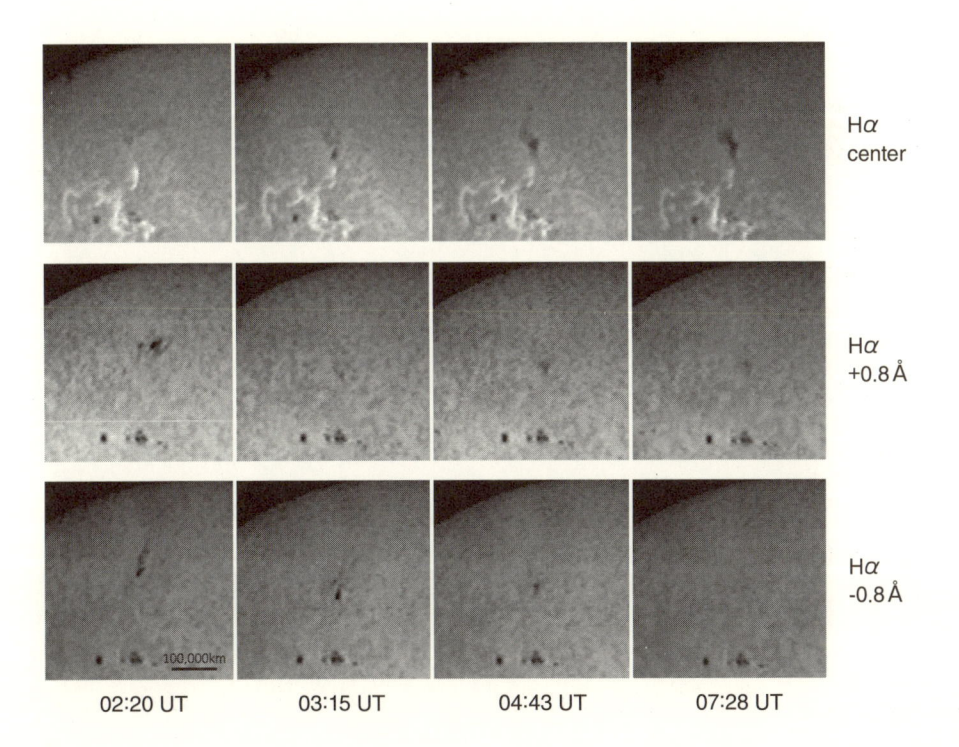

				Hα center
				Hα +0.8Å
				Hα −0.8Å
02:20 UT	03:15 UT	04:43 UT	07:28 UT	

Jun 15, 2000 23:33−00:20 UT   Flare, Filament Eruption

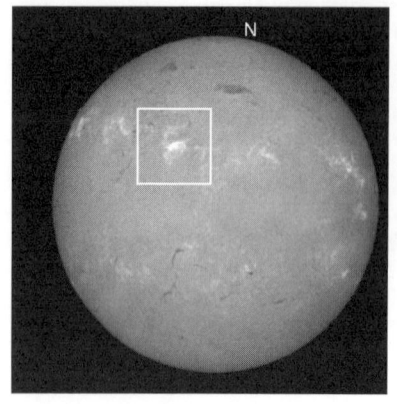

GOES Class = M2.0   NOAA 9040
N16,E17   P = -9.7[deg]   B0= 1.1[deg]

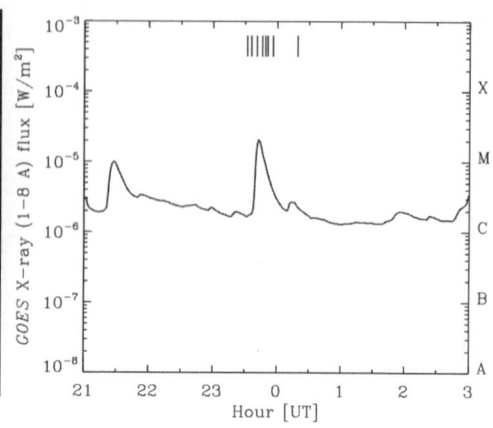

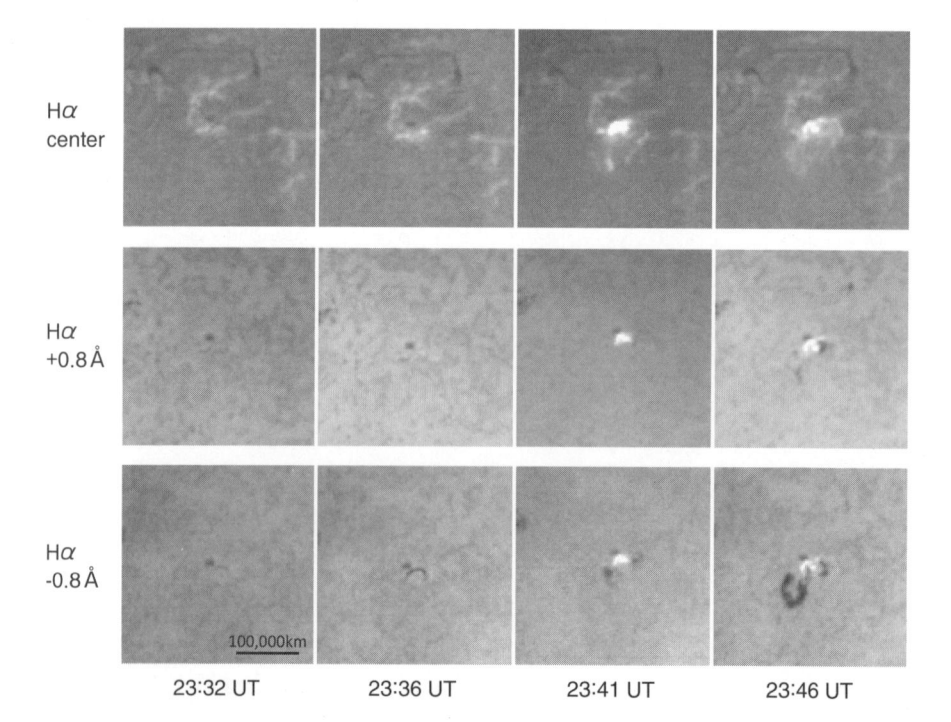

Hα center

Hα +0.8Å

Hα -0.8Å

100,000km

23:32 UT   23:36 UT   23:41 UT   23:46 UT

Chapter 6 Representative Events | 179

The event was a compact M2.0 class flare of half an hour duration. At the time of flare break-up, a loop-like filament came into appearance and exploded with a speed of 140 km/s. In the event, both a Moreton wave and an EIT wave were observed simultaneously. The propagation of Moreton wave in 1 min cadence is displayed in the following pages.

このフレアは，活動領域のプラージュ域で発生した寿命約半時間の M2.0 クラスコンパクトフレアである．ループ状のフィラメント膨張の後，140 km/s の速さでフィラメント爆発を起こした．フレアに伴って，モートン波と EIT 波が放出されたのが観測されている．モートン波の伝播の様子は次ページ以降に示されている．

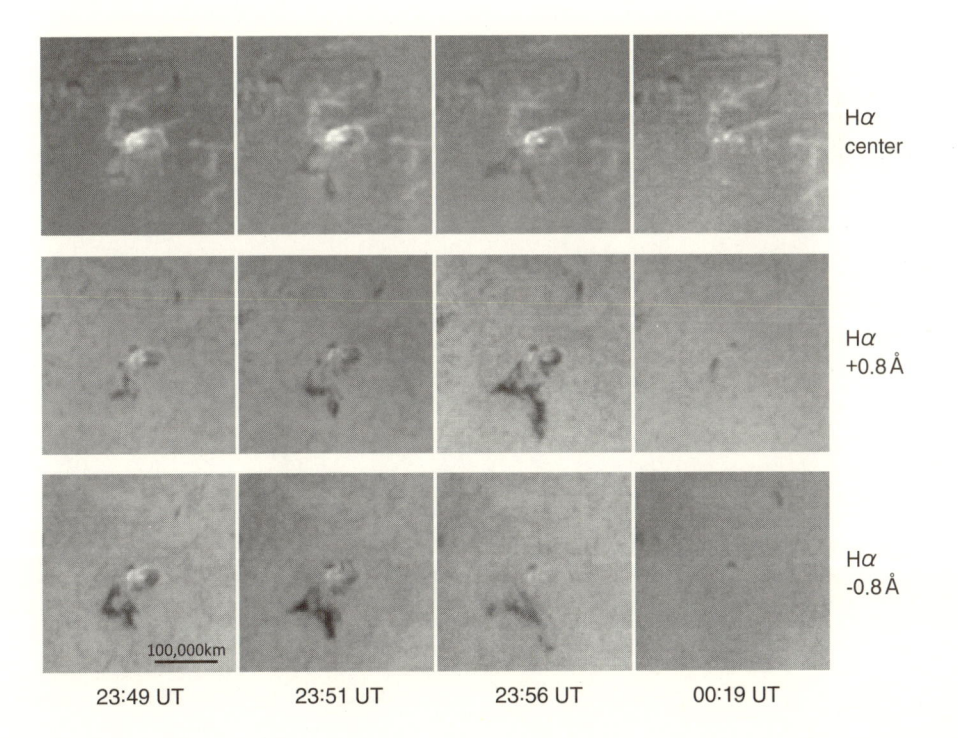

23:49 UT	23:51 UT	23:56 UT	00:19 UT	Hα center
				Hα +0.8Å
				Hα −0.8Å

100,000km

Jun 15, 2000 23:37−23:43 UT   Moreton Wave

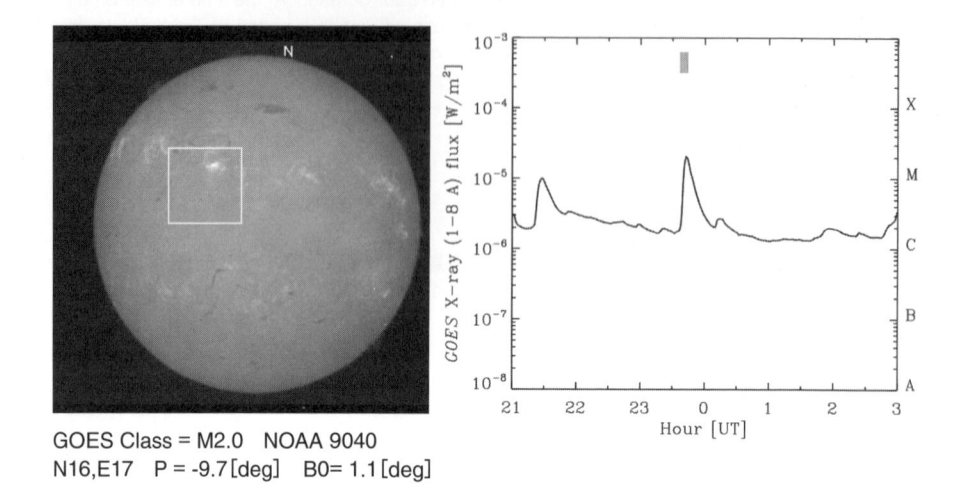

GOES Class = M2.0   NOAA 9040
N16,E17   P = -9.7 [deg]   B0= 1.1 [deg]

Hα center

Hα +0.8 Å

Hα -0.8 Å

100,000 km

23:37 UT   23:38 UT   23:39 UT   23:40 UT

Okamoto et al. (2004) observed that the M2.0 flare in NOAA9040 emitted both a Moreton wave and an EIT wave simultaneously. As both of the waves showed the similar propagating speed of about 350 km/s, it was suggested that the both waves will be different manifestation of a common propagating disturbance, probably a MHD fast-mode shock. (In the next two pages, time-difference images to enhance the visibility of wave fronts are shown with the manually drawn sketch of the wave fronts.)

Okamoto et al. (2004) は，NOAA9040 で発生した M2.0 クラスのフレアによりモートン波と EIT 波が同時発生したことを観測した．両者の伝播速度が約 350 km/s でほぼ等しいことから，両者は同じ 1 つの擾乱であろう，そしておそらくは電磁流体学的ファストモード衝撃波であろうと判断された．(次の 2 ページでは，波面を見やすくするように，画像の明るさの時間変化像を示している．また波面位置のスケッチも参考のために付してある．)

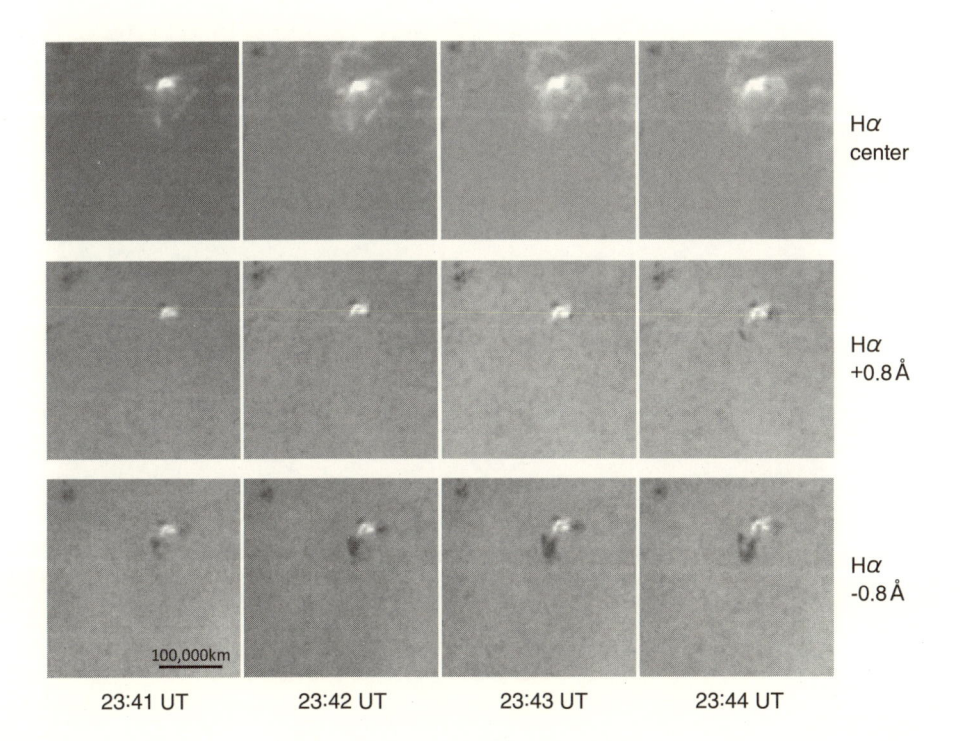

Hα center

Hα +0.8Å

Hα −0.8Å

100,000km

23:41 UT 23:42 UT 23:43 UT 23:44 UT

182 | 第6章　代表的イベント

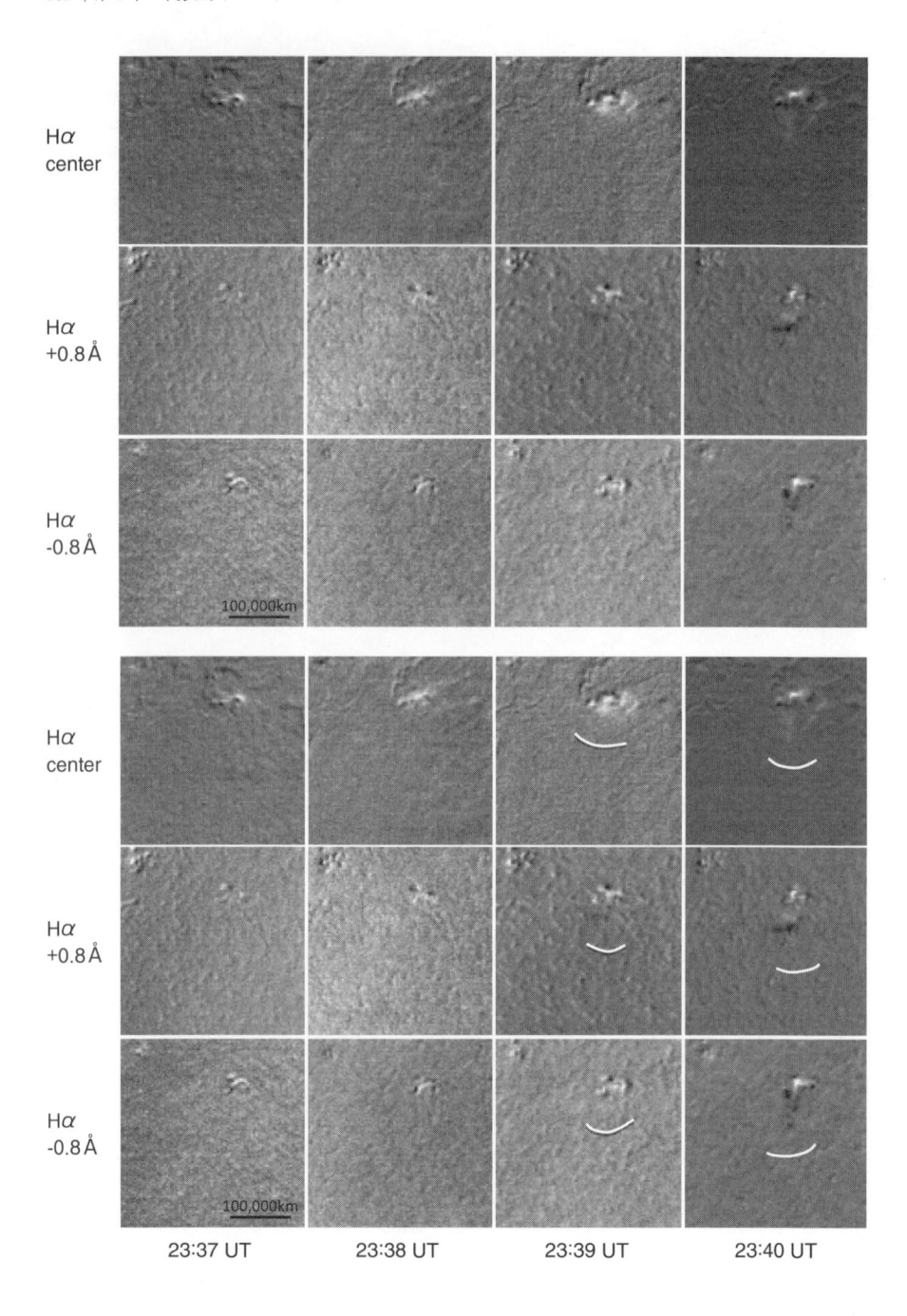

Chapter 6 Representative Events | 183

Hα center

Hα +0.8Å

Hα -0.8Å

100,000km

Hα center

Hα +0.8Å

Hα -0.8Å

100,000km

23:41 UT   23:42 UT   23:43 UT   23:44 UT

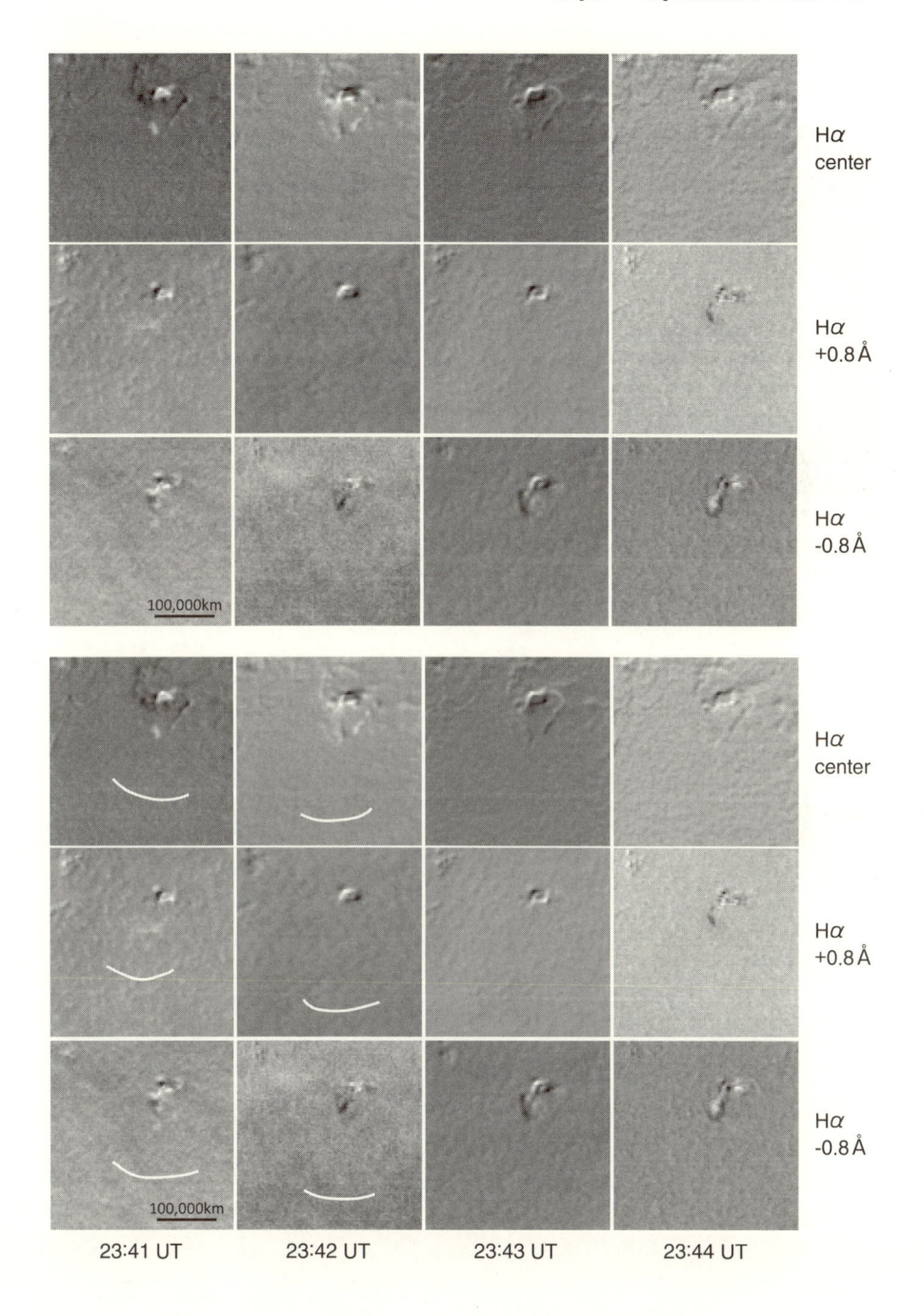

184 | 第6章　代表的イベント

Jul 16, 2000 06:09−06:37 UT　Flare

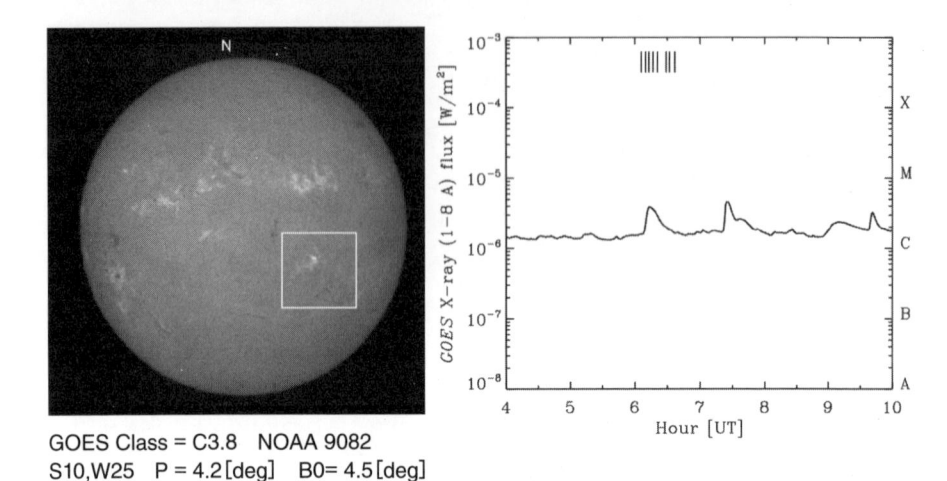

GOES Class = C3.8　NOAA 9082
S10, W25　P = 4.2 [deg]　B0 = 4.5 [deg]

Hα center

Hα +0.8Å

Hα −0.8Å

100,000 km

06:06 UT　　06:10 UT　　06:13 UT　　06:17 UT

Chapter 6 Representative Events | 185

In a β-type active region NOAA9082, there occurred a C3.8 flare. In the explosive phase of the flare, a Moreton wave was detected in FMT video movies to propagate from the flare site to the North. The propagation of Moreton wave in 1 min cadence is displayed in the following pages.

β型の活動領域 NOAA9082 で，C3.8 クラスのフレアが発生した．FMT ビデオムービーでは，フレアの爆発時にモートン波が現れて北の方向に伝播してゆくのが観測された．モートン波伝播の 1 分間隔の様子は次ページ以降に示されている．

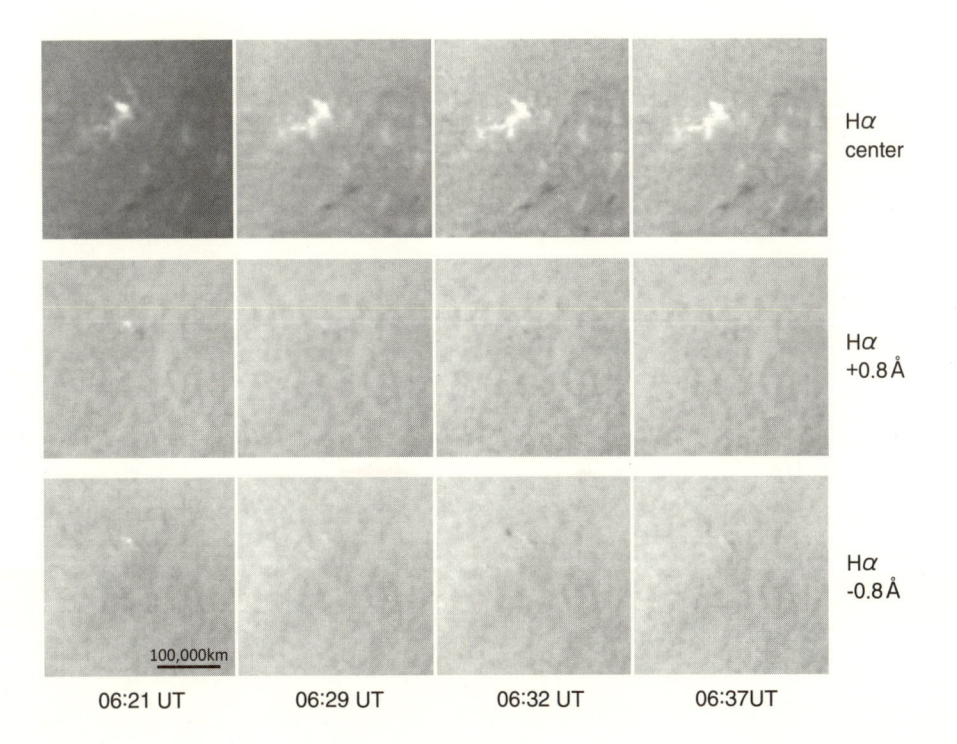

				Hα center
				Hα +0.8Å
				Hα -0.8Å
06:21 UT	06:29 UT	06:32 UT	06:37 UT	

100,000km

186 | 第6章　代表的イベント

Jul 16, 2000　06:06−06:16 UT　Moreton Wave

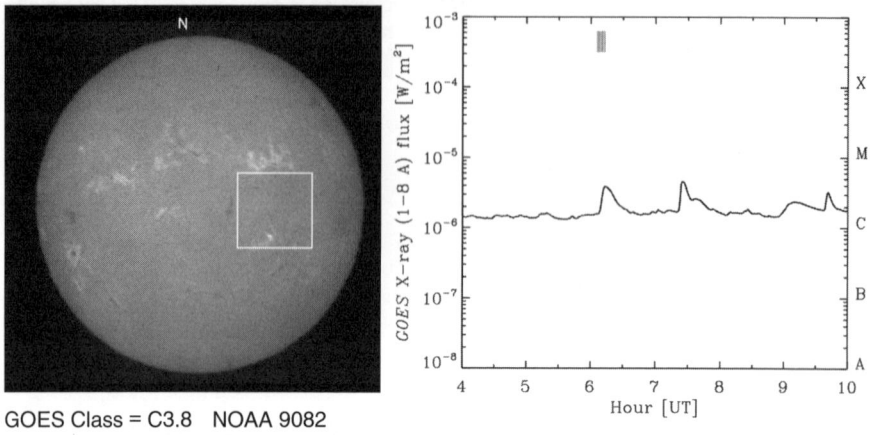

GOES Class = C3.8　NOAA 9082
S10,W25　P = 4.2 [deg]　B0 = 4.5 [deg]

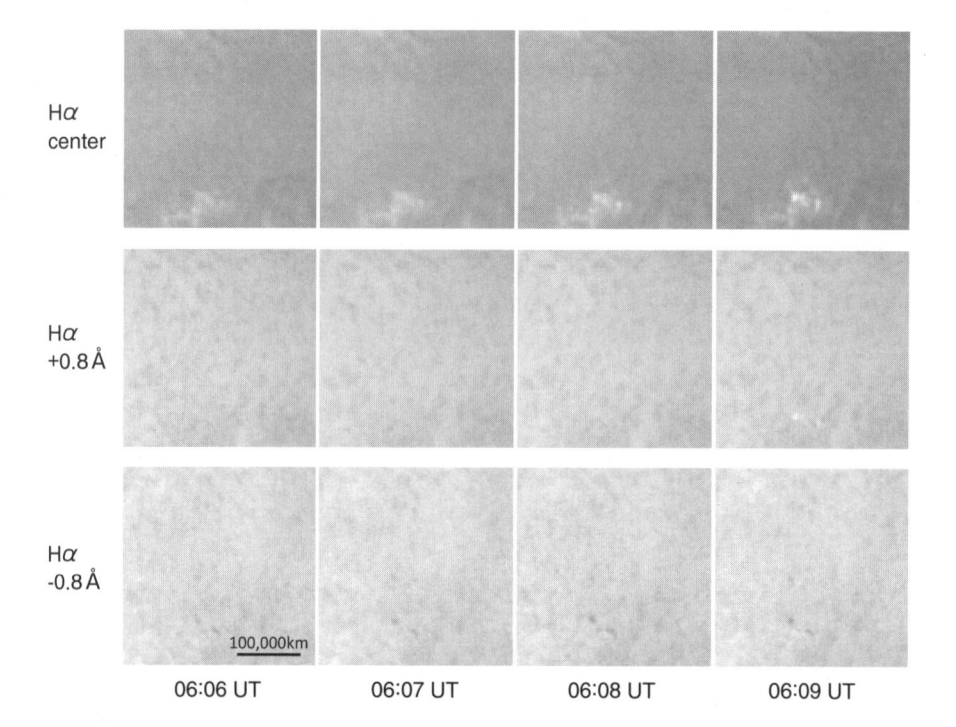

Hα center

Hα +0.8 Å

Hα −0.8 Å

100,000 km

06:06 UT　06:07 UT　06:08 UT　06:09 UT

Here is shown the temporal evolution in 1 min cadence of the Moreton wave emitted from the C3.8 flare in the NOAA9082. It propagated to the North with a speed of 360 km/s. No EIT wave was observed with the SOHO satellite in this event. (In the next two pages, time-difference images to enhance the visibility of wave fronts are shown with the manually drawn sketch of the wave fronts.)

NOAA9082 で発生した C3.8 クラスのフレアから放射されたモートン波の 1 分間隔での伝播の様子を示している．その波は，速さ 360 km/s で北の方角に伝播した．このイベントに伴う可能性のある EIT 波は SOHO では観測されていなかった．（次の 2 ページでは，波面を見やすくするように，画像の明るさの時間変化像を示している．また波面位置のスケッチも参考のために付してある．）

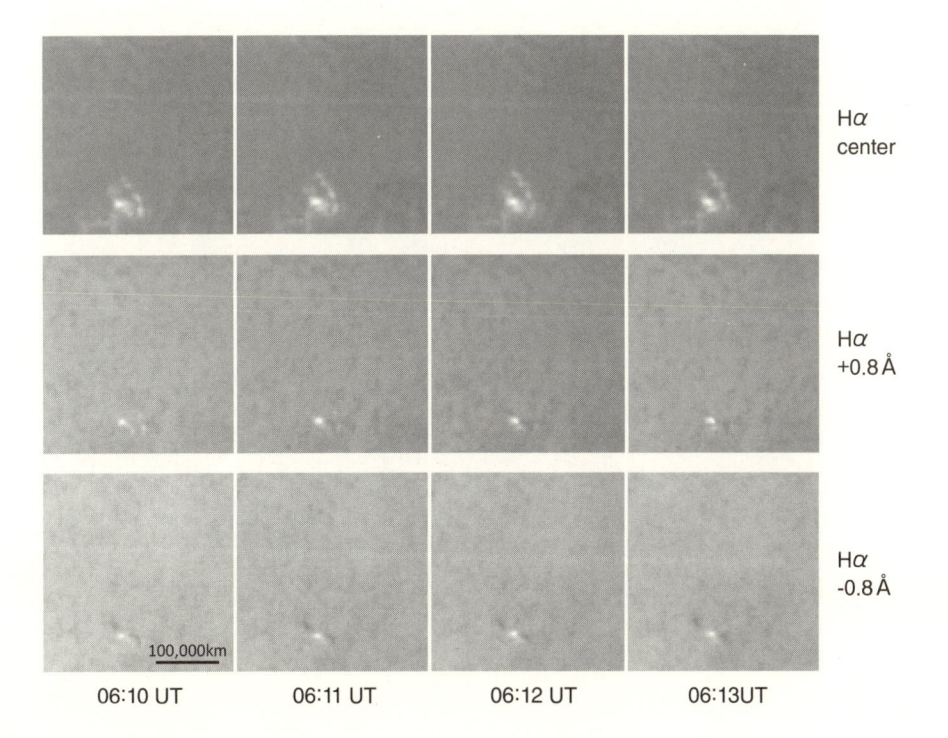

				Hα center
				Hα +0.8 Å
				Hα −0.8 Å
06:10 UT	06:11 UT	06:12 UT	06:13 UT	

100,000 km

188 | 第6章　代表的イベント

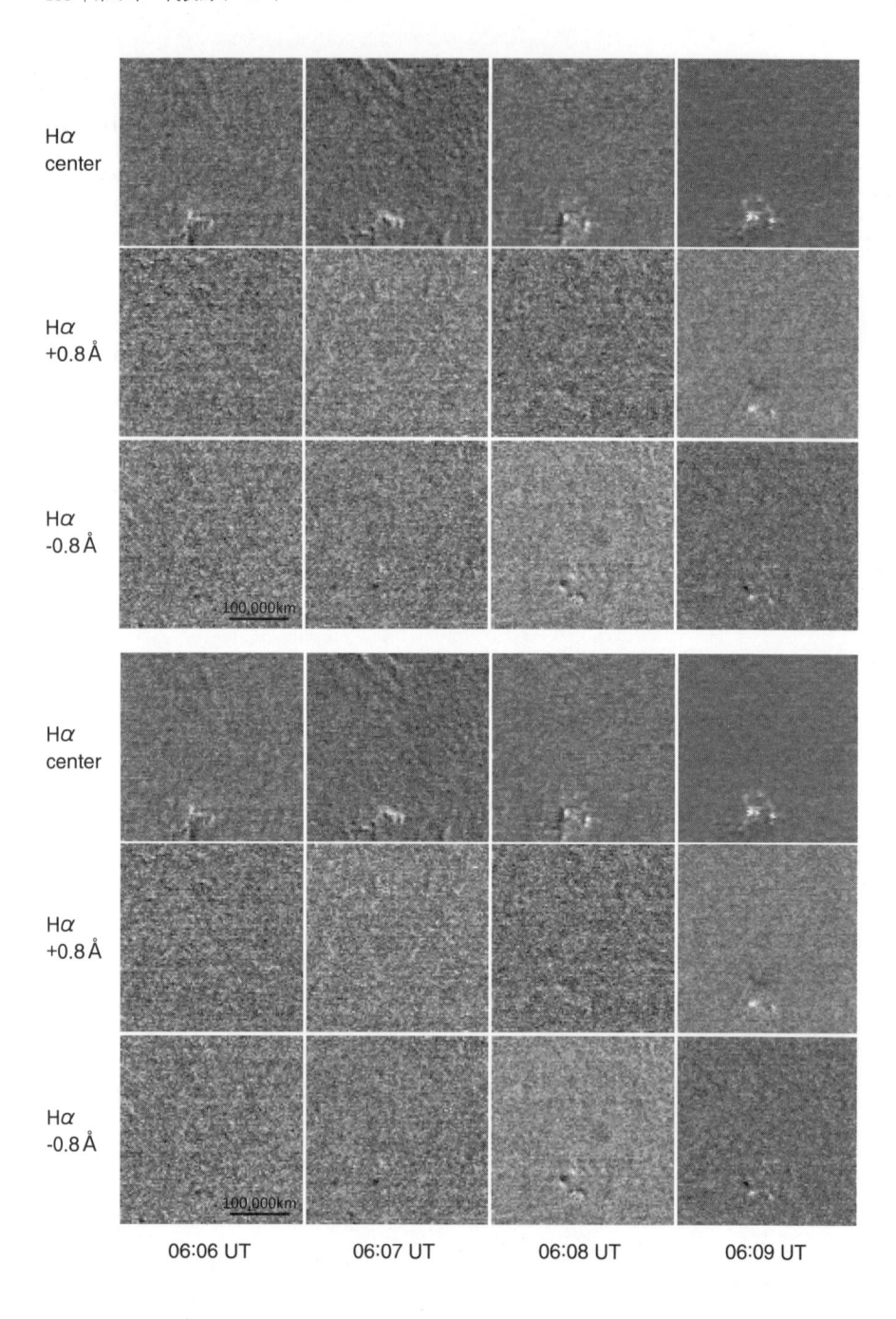

Hα center

Hα +0.8Å

Hα -0.8Å

100,000km

Hα center

Hα +0.8Å

Hα -0.8Å

100,000km

06:06 UT　　06:07 UT　　06:08 UT　　06:09 UT

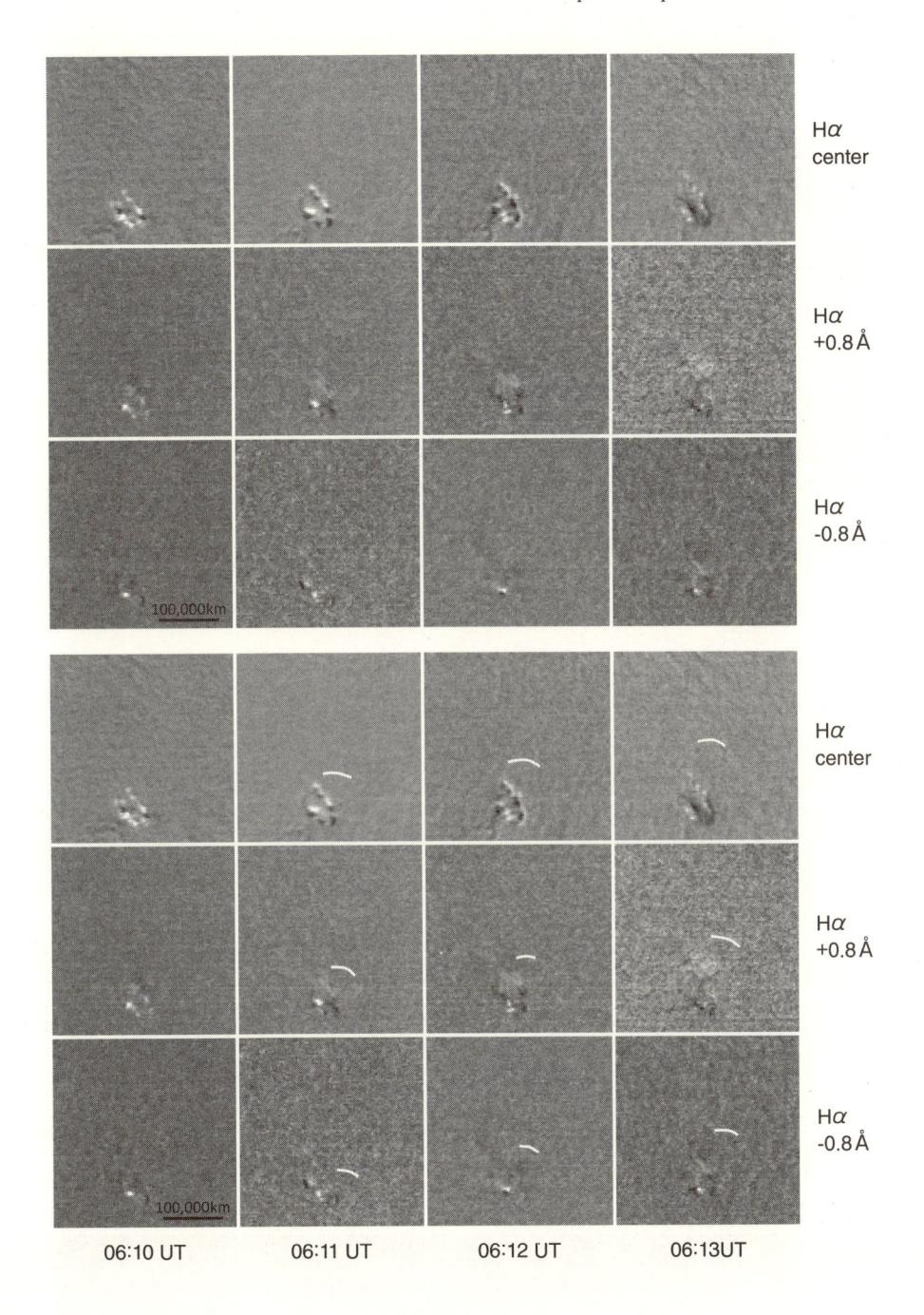

Chapter 6 Representative Events | 189

Hα center

Hα +0.8Å

Hα -0.8Å

100,000km

Hα center

Hα +0.8Å

Hα -0.8Å

100,000km

06:10 UT   06:11 UT   06:12 UT   06:13 UT

Jul 19, 2000 06:42−08:03 UT   Flare, Filament Eruption

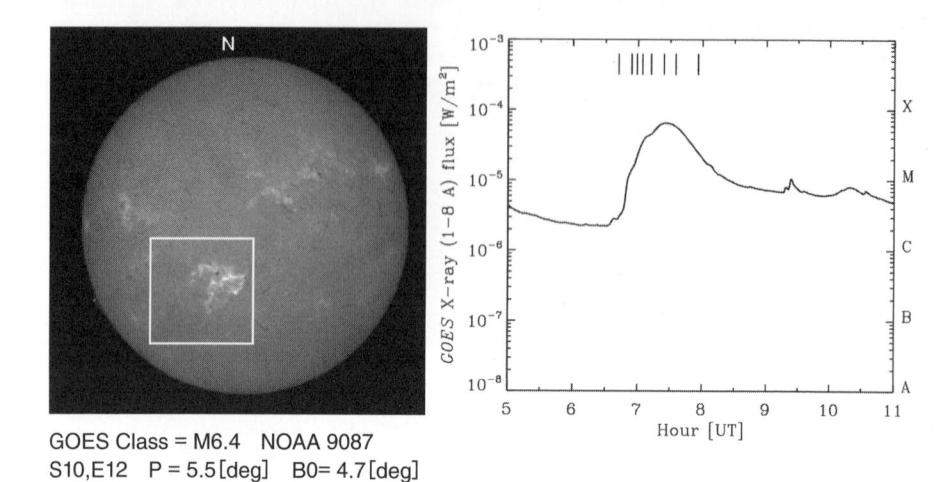

GOES Class = M6.4   NOAA 9087
S10,E12   P = 5.5[deg]   B0= 4.7[deg]

Hα center

Hα +0.8Å

Hα −0.8Å

100,000km

06:42 UT   06:54 UT   06:59 UT   07:04 UT

The event was a complex one where two different types of flares occurred simultaneously side by side in an active region. An active region filament located far from the center of the region got activated, and then a short-lived (5 min) compact flare occurred associated with an explosive mass ejection with a speed of 170 km/s near the main spot. Just simultaneously on the other side of the spot, another flare started to brighten. The latter flare was a typical two-ribbon flare, which decayed gradually in a long time-span.

　この現象は1つの活動領域内の隣り合う場所で異なるタイプのフレアが同時発生した複雑なものである．まず活動領域の外延部にあるフィラメントが活動をし始め，引き続いて黒点近傍で170 km/sのフィラメント噴出と共に継続時間5分程度の短命なコンパクトフレアが発生した．そして，同時に別の場所（黒点南西部）でもフレアが発生した．後者のフレアは典型的なツーリボンフレアでゆっくりと減衰するものであった．

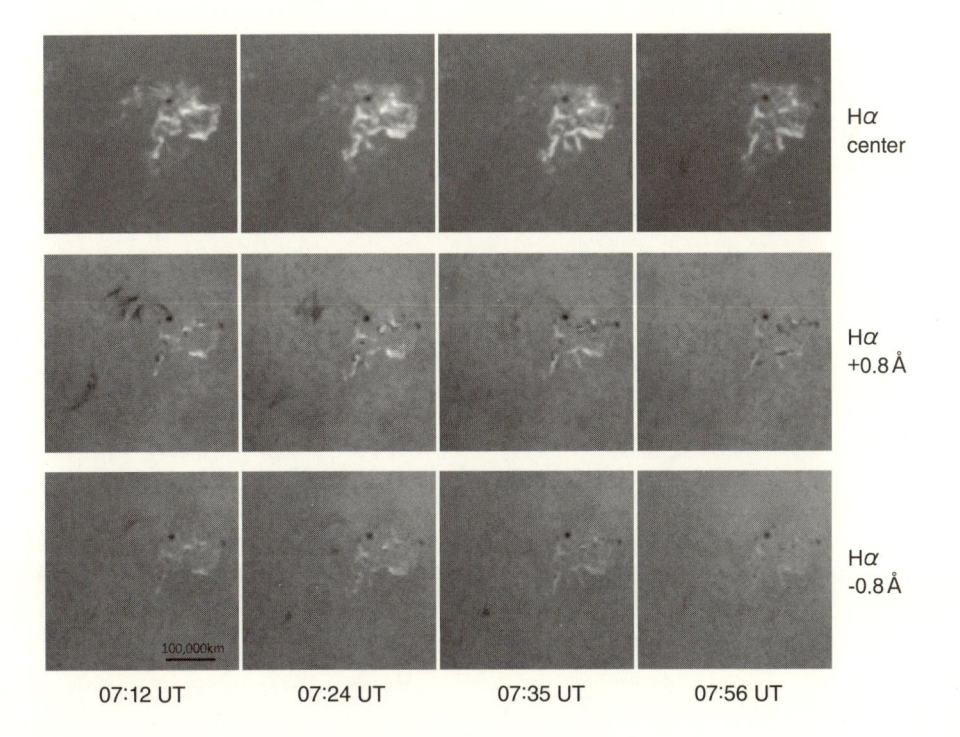

192 | 第6章　代表的イベント

Jul 23, 2000 04:27−04:57 UT　Surge

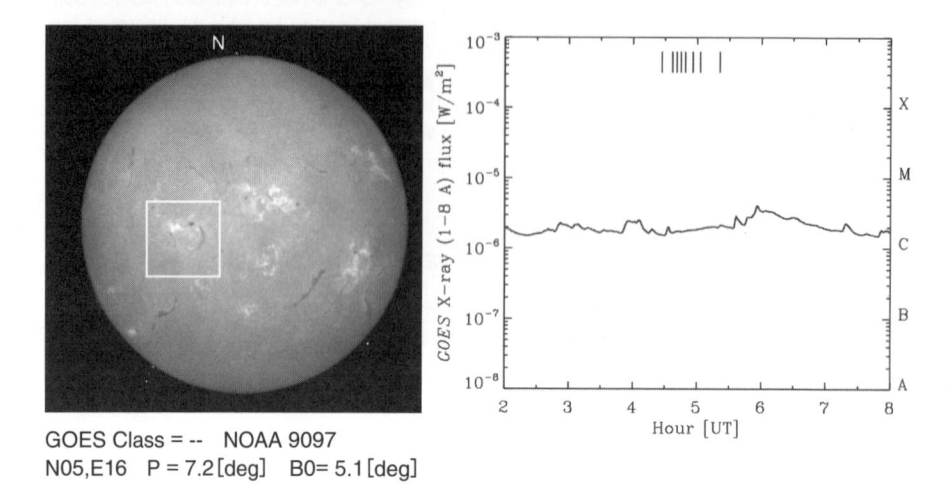

GOES Class = --　NOAA 9097
N05,E16　P = 7.2[deg]　B0= 5.1[deg]

Hα center

Hα +0.8Å

Hα −0.8Å

100,000km

04:27 UT　　04:37 UT　　04:41 UT　　04:45 UT

The event was a surge from a bright point in the chromosphere. As the bright point was observed in all the monochromatic channels of the Hα, the spectral profile of the Hα emission should be very wide. The gas was ejected along a curved line in a speed of larger than 150 km/s. According to Liu et al. (2005), the ejected surge gas was trapped in the "empty" filament channel and suddenly formed a new filament. The filament stayed as an active region filament for about 1 day. This type of a new filament formation was first found and reported with the FMT imaging data.

彩層に輝点が現われその点からガスが噴出するサージ現象である．この輝点はHα線の両ウイング像でも見られたため，波長幅の大きな輝線スペクトルを示していると思われる．ガスは150 km/s以上の速度で曲線状の軌跡をたどって噴出した．Liu et al. (2005) によると，噴出したガスは「空の」フィラメントチャンネルに捉えられて新しいフィラメントを形成した．このフィラメントはほぼ1日間活動領域内に滞在した．このようなフィラメント形成プロセスは，この現象のFMT観測で初めて見出された．

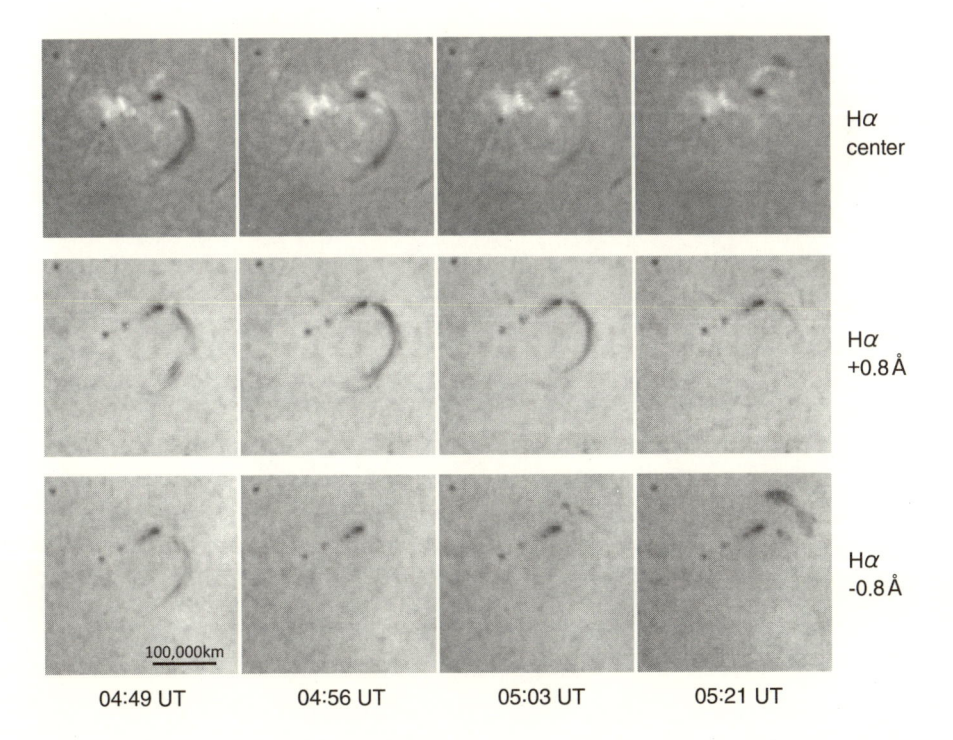

194 | 第6章 代表的イベント

Oct 01, 2000 01:54−02:56 UT Surge

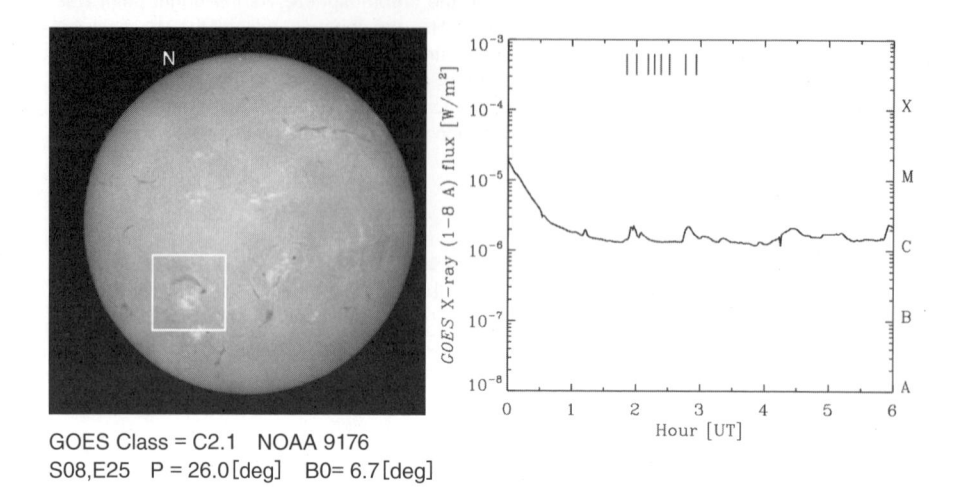

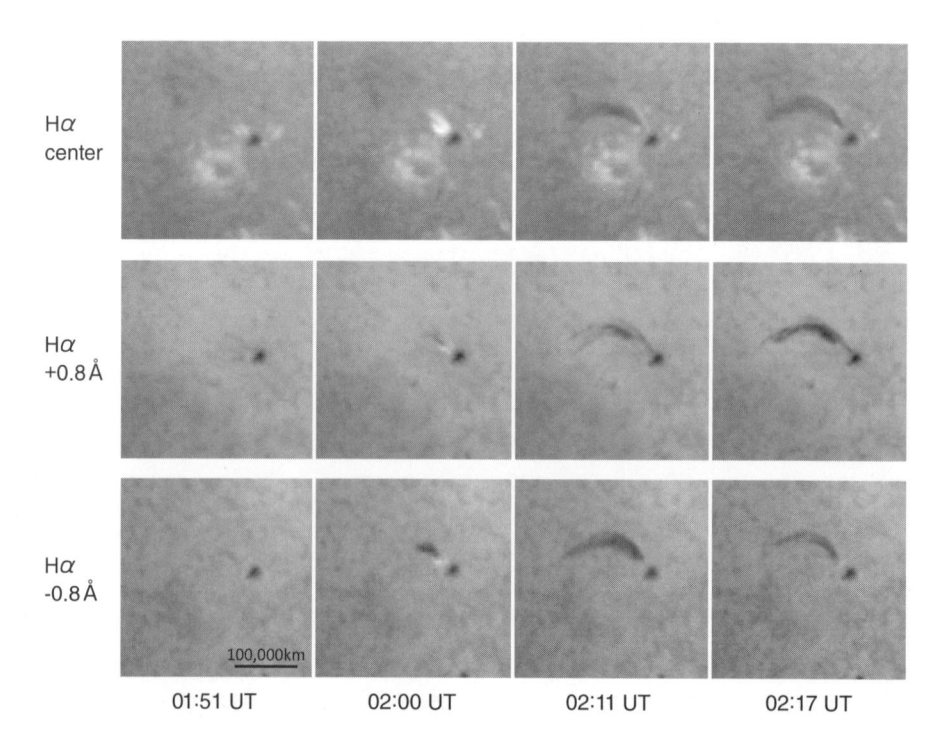

GOES Class = C2.1   NOAA 9176
S08,E25   P = 26.0[deg]   B0= 6.7[deg]

Hα center

Hα +0.8Å

Hα −0.8Å

100,000km

01:51 UT   02:00 UT   02:11 UT   02:17 UT

There were a few of satellite magnetic regions around the preceding spot of the active region NOAA9176. There occurred a compact C2.1 flare in one of the satellite polarity regions located north of the spot. In the chromosphere, a surge was observed to be ejected with a speed of 150 km/s from the bright flaring region. The Hα emission profile at the root region of the surge was very wide. The ejected gas moved along a curved trajectory, attained the length of 150 Mm, and then receded to the original location along the same trajectory.

活動領域 NOAA9176 の先行黒点の周辺にサテライト極性の磁場領域があり，その近傍で X 線クラス C2.1 のコンパクトなフレアが発生した．このとき，彩層の明るいフレア領域が現われ，そこからサージ現象が発生してガスが 150 km/s の速度で噴出した．サージの根元の Hα 線は幅の広い輝線スペクトルであった．ガスは湾曲した曲線に沿って噴出し，約 15 万 km まで延伸した後，同じ軌跡をたどって帰還した．

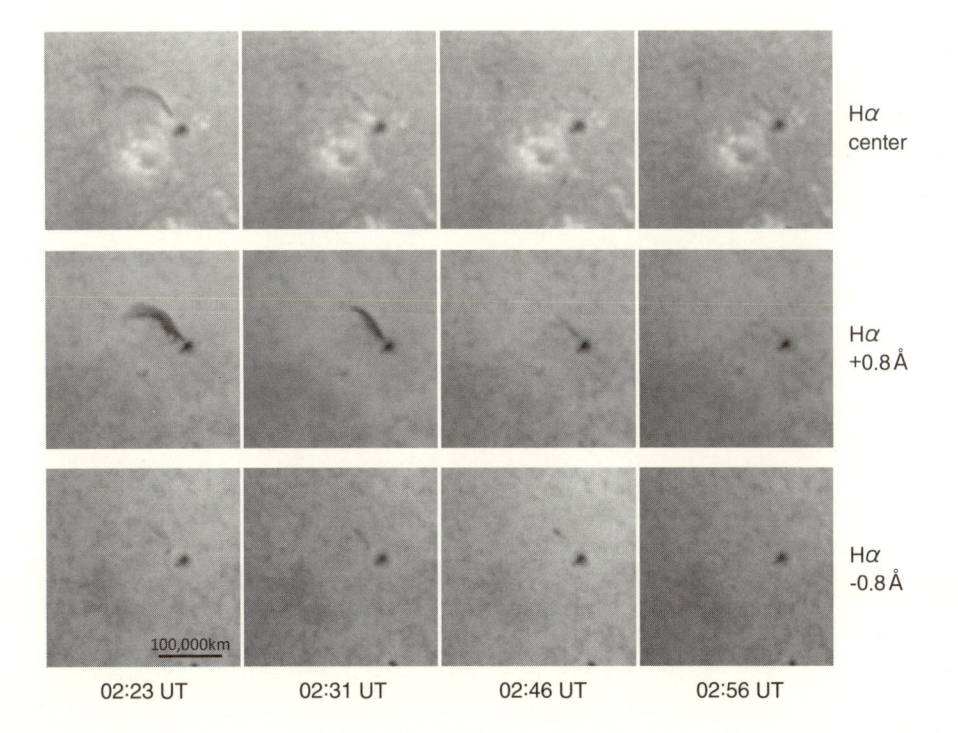

Oct 24, 2000 03:18−05:40 UT   Filament Activity

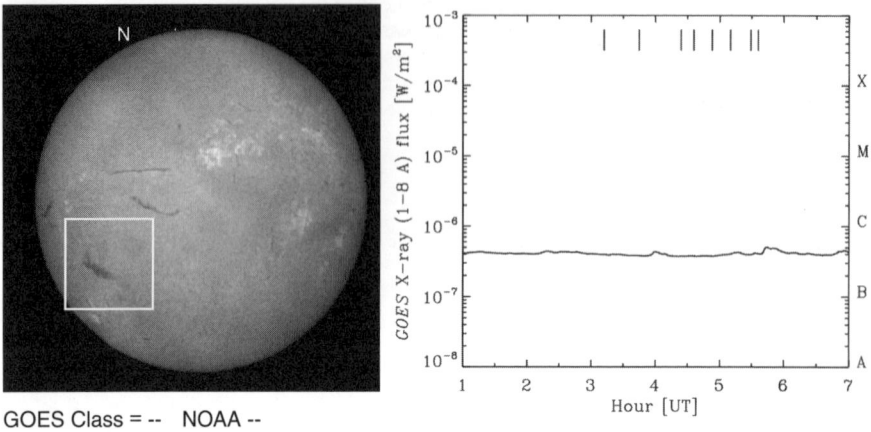

GOES Class = --   NOAA --
S01,E52   P = 25.6 [deg]   B0= 5.1 [deg]

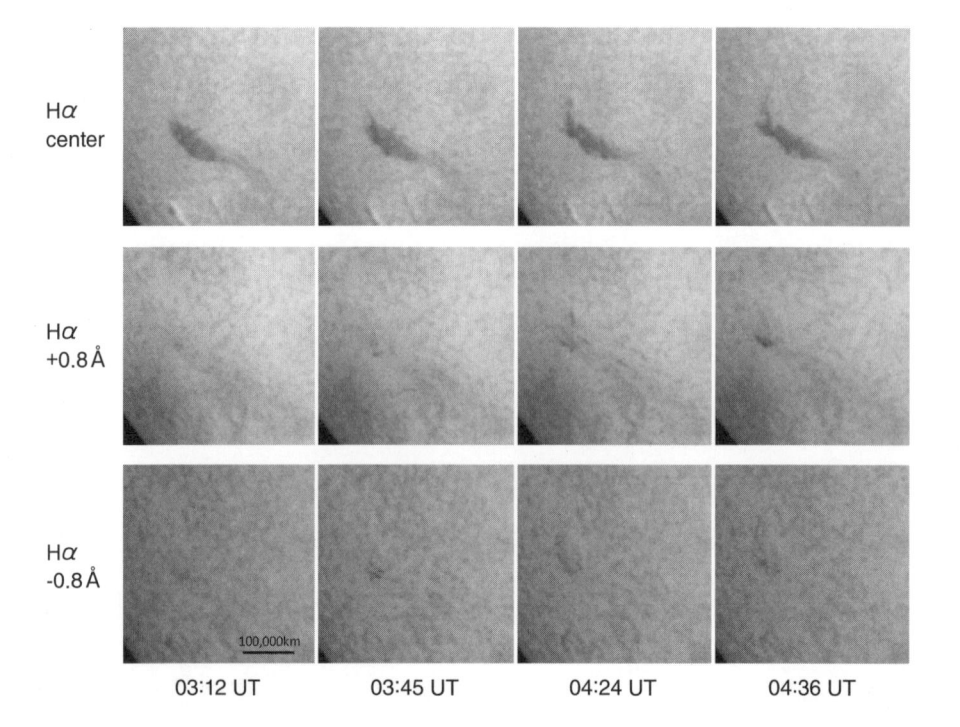

Hα center

Hα +0.8Å

Hα −0.8Å

100,000km

03:12 UT   03:45 UT   04:24 UT   04:36 UT

The eastern part of the quiescent filament was activated and the filament gas continued to stream out of the main body during the period of 2.5 hours. Neither EIT195 brightening nor prominent coronal X-ray enhancement was observed in association with the event.

静穏フィラメントの東端が活発化し，そこから 2 時間半にわたってフィラメントのガスが流れ出す現象が継続した．この現象に付随する EIT195 増光現象，X 線増光現象は見られなかった．

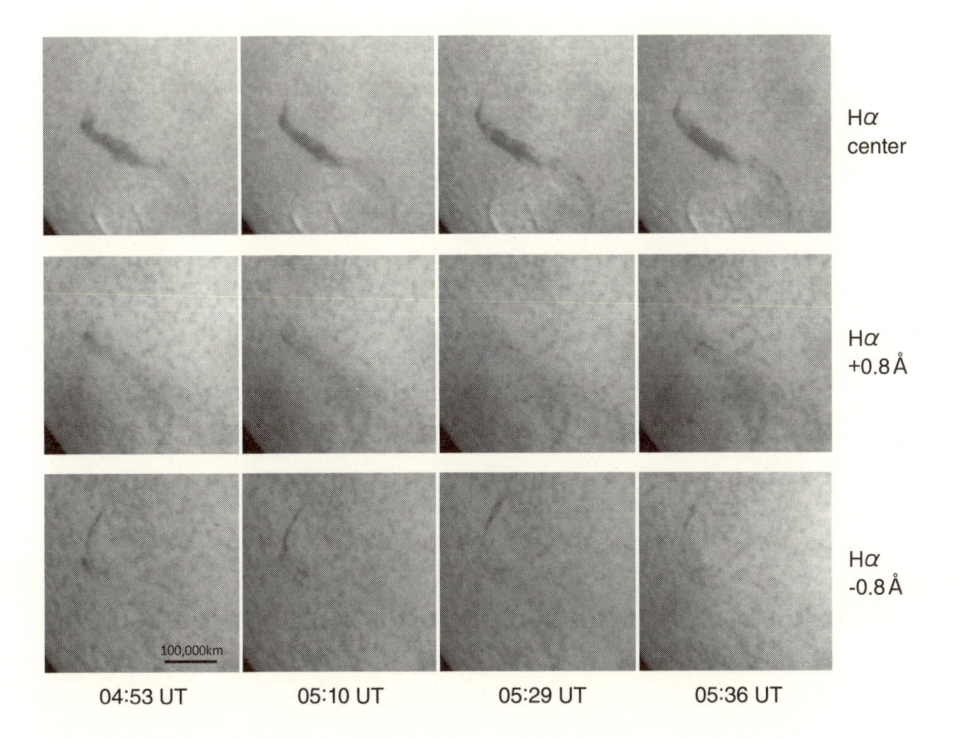

Nov 03, 2000 21:59B−02:46 UT   Filament Activity

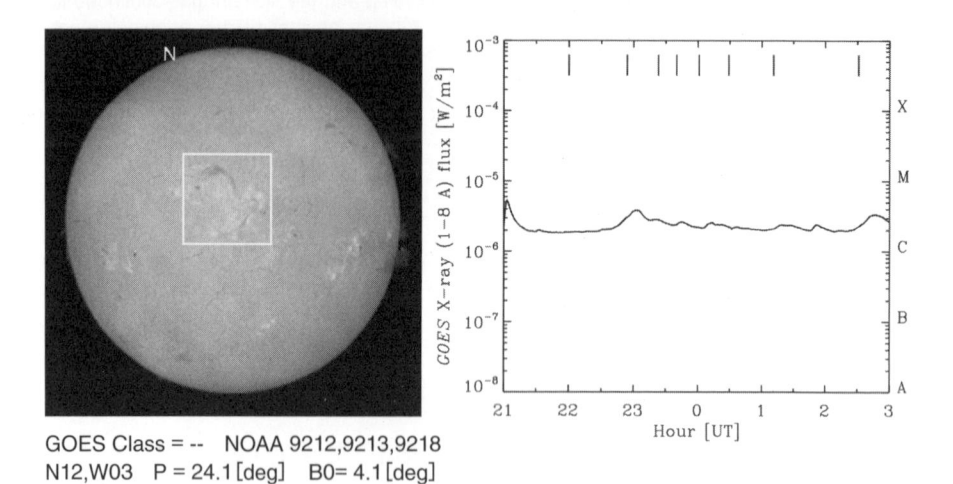

GOES Class = --   NOAA 9212,9213,9218
N12,W03   P = 24.1 [deg]   B0= 4.1 [deg]

Hα center

Hα +0.8Å

Hα −0.8Å

100,000km

22:00 UT   22:54 UT   23:23 UT   23:40 UT

An active region filament located between NOAA9218 and a neighboring old active region was activated for around 5 hours. Although the basic structure of the filament did not changed so much, internal turbulent motions and streaming flows along the axis of the filament were observed to occur. During the activation, coronal loops seen in SOHO/EIT195 overlying the filament channel were observed to expand intermittently. The activation, probably, was excited by the large-scale magnetic reconfiguration in the overlying corona. The C3.5 X-ray burst occurred at 23:00 UT is due to the flare erupted near the preceding spots of the region.

NOAA9218と近傍の古い活動領域との間にあった活動領域フィラメントが約5時間にわたって活発化した．フィラメントの全体の形はあまり変わらなかったが，その内部の乱流運動や軸に沿う方向の流れが引き起こされた．この活動の間，SOHO/EIT195ではフィラメント上空のコロナループが時々膨張するのが観測された．この活動現象は，上空コロナでの磁場が再構成される過程で発生したものと思われる．なお，23:00 UTのC3.5クラスのX線バーストは，この領域の先行黒点近傍で発生したフレアによるものである．

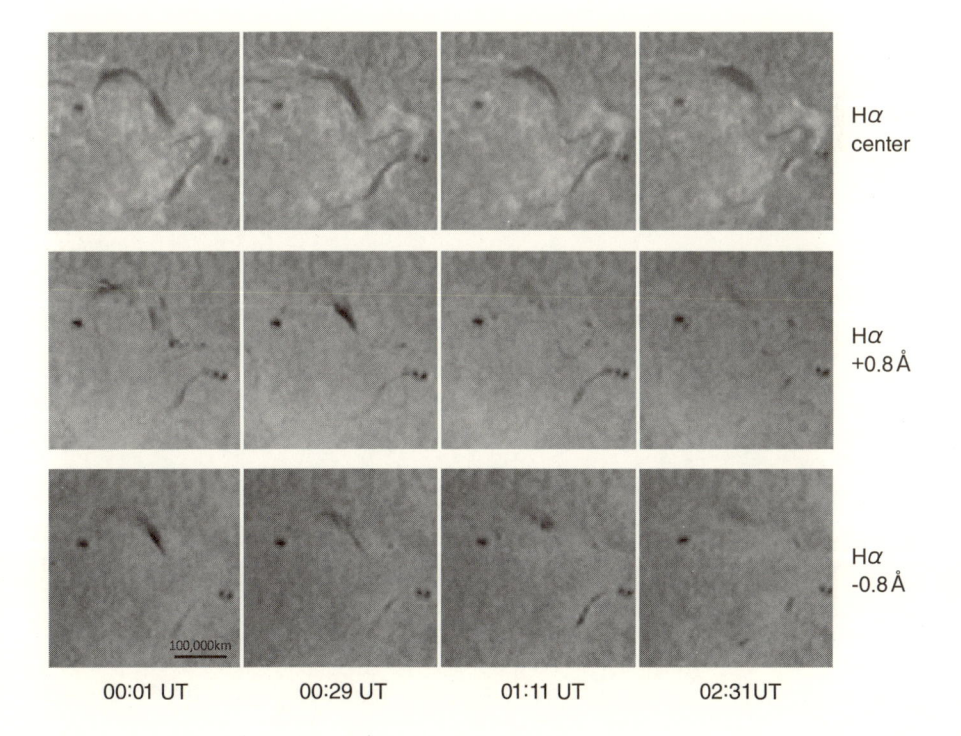

Nov 23, 2000 05:32−06:19 UT   Flare, Filament Eruption

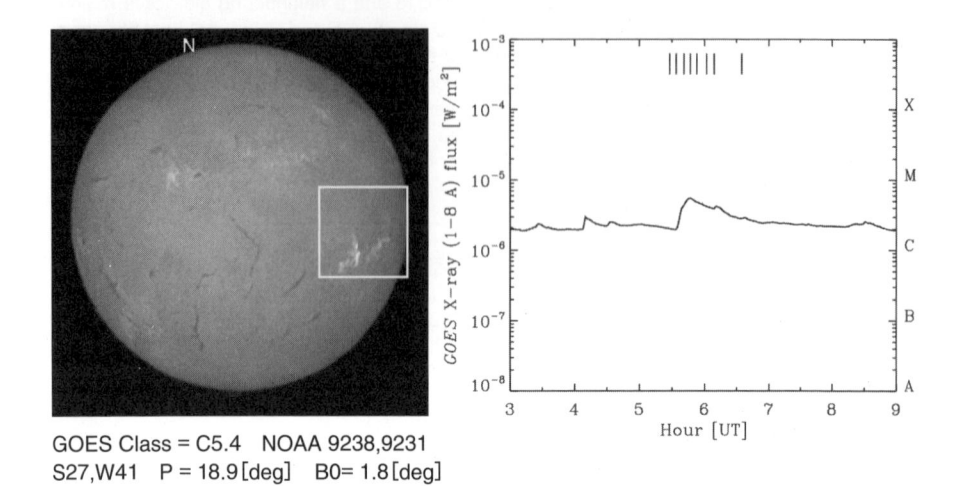

GOES Class = C5.4   NOAA 9238,9231
S27,W41   P = 18.9 [deg]   B0= 1.8 [deg]

Hα center

Hα +0.8 Å

Hα −0.8 Å

100,000 km

05:28 UT   05:34 UT   05:41 UT   05:47 UT

In an α-type active region NOAA9238, a part of an active region filament started to erupt and trigger a two-ribbon flare. The flare explosion activated the remaining part of the filament and ejected out the filament gas along the filament channel. The expelled gas was observed brightly in SOHO/EIT195 images to move along the same direction as the Hα gas. In SOHO/LASCO/C2, an associated halo-type CME was observed to expand with a speed of 280 km/s.

α型の活動領域 NOAA9238 で，活動領域フィラメントの一部が爆発的に上昇し，ツーリボンフレアを引き起こした．フレア爆発に伴って活動領域フィラメントの残りの部分が活発化し，フィラメントチャンネルの軸方向に Hαガスを放出した．SOHO/EIT195 像では，放出されたガスが明るく励起されていた．SOHO/LASCO/C2 では，この現象に伴うハロー型の CME が 280 km/s の速度で膨張していることが観測された．

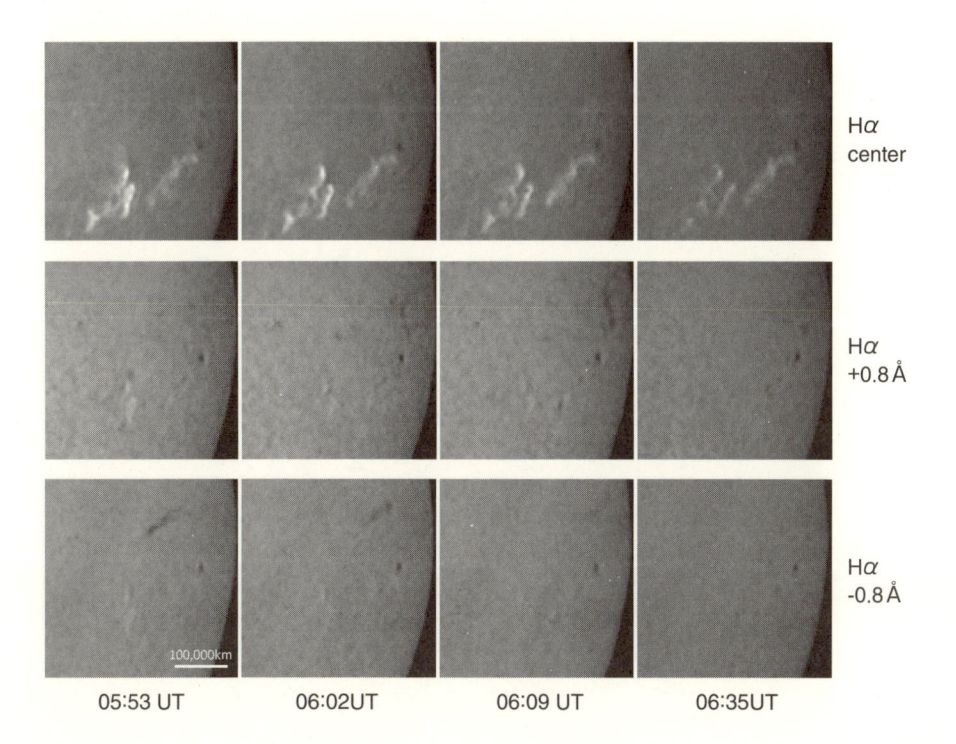

Nov 24, 2000 04:47−05:30 UT   Flare, Filament Eruption

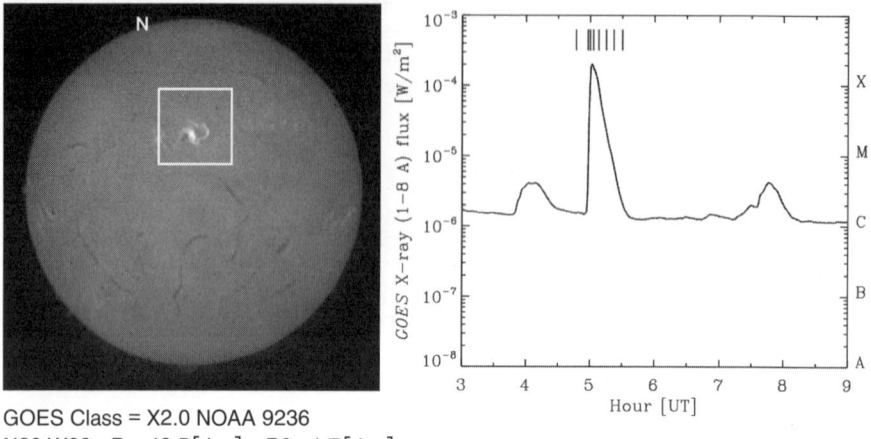

GOES Class = X2.0 NOAA 9236
N20,W08   P = 18.5 [deg]   B0 = 1.7 [deg]

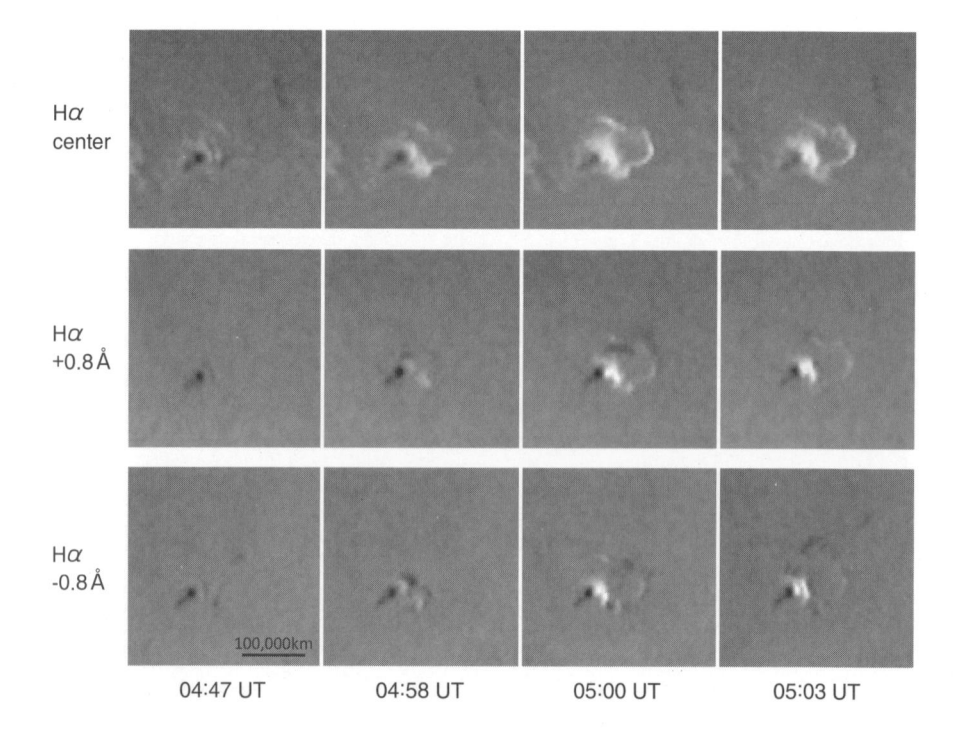

Hα center

Hα +0.8Å

Hα −0.8Å

100,000 km

04:47 UT   04:58 UT   05:00 UT   05:03 UT

NOAA9236 produced three homologous flares : X2.0 at 04:55 UT, X2.3 at 14:51 UT and X1.8 at 21:43 UT on Nov. 24, 2000. The development of the first one observed with the FMT is shown here. The three flares were comprehensively studied with the multi-wavelength data by Takasaki et al. (2004). They confirmed that all the parameters derived from the observation, such as the separation speeds of two-ribbon, filament rise speed, soft-X ray intensity, and hard-X ray intensity were consistent with the predicted behavior of the coronal magnetic reconnection model of the flare.

活動領域 NOAA9236 は，2000 年 11 月 24 日に 3 個の同形フレア（ホモロガスフレア）を発生した．04:55 UT に X2.0 クラス，14:51 UT には X2.3 クラス，そして 21:43 UT に X1.8 クラスという強いフレアであった．ここでは，FMT で観測された 04:55 UT のものを例示している．この 3 個のフレアは，多波長データを用いて Takasaki et al. (2004) によって詳しく研究された．その結果，ツーリボンの分離速度，フィラメント上昇速度，軟 X 線強度および硬 X 線強度変化という観測された諸量が，フレアのコロナ磁気リコネクションモデルによって矛盾無く説明できることが確認された．

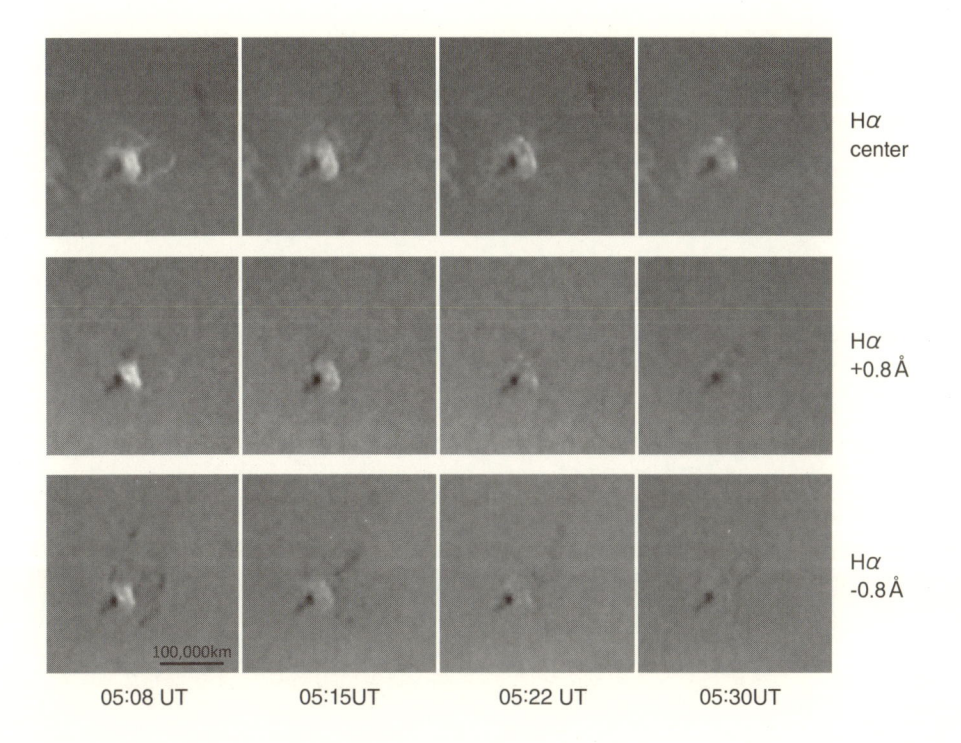

Nov 29, 2000 00:20−01:45 UT   Filament Eruption

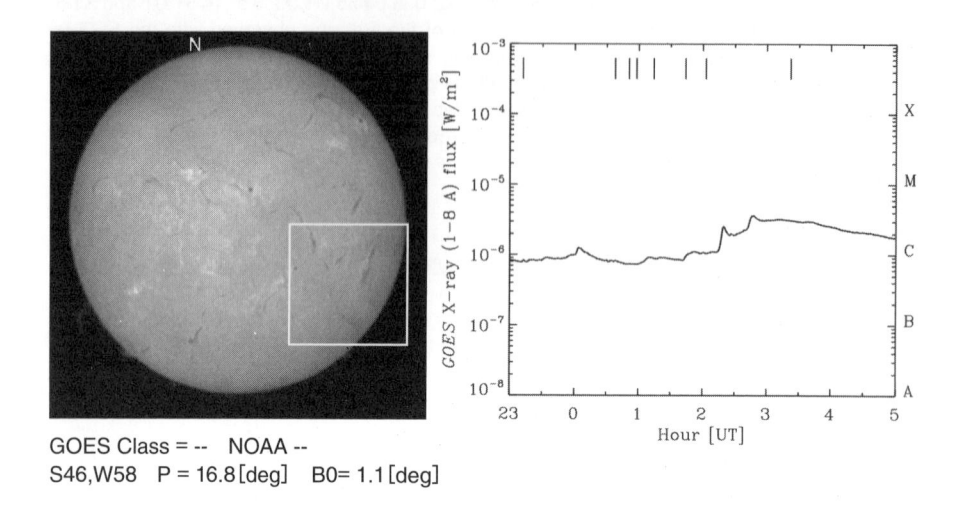

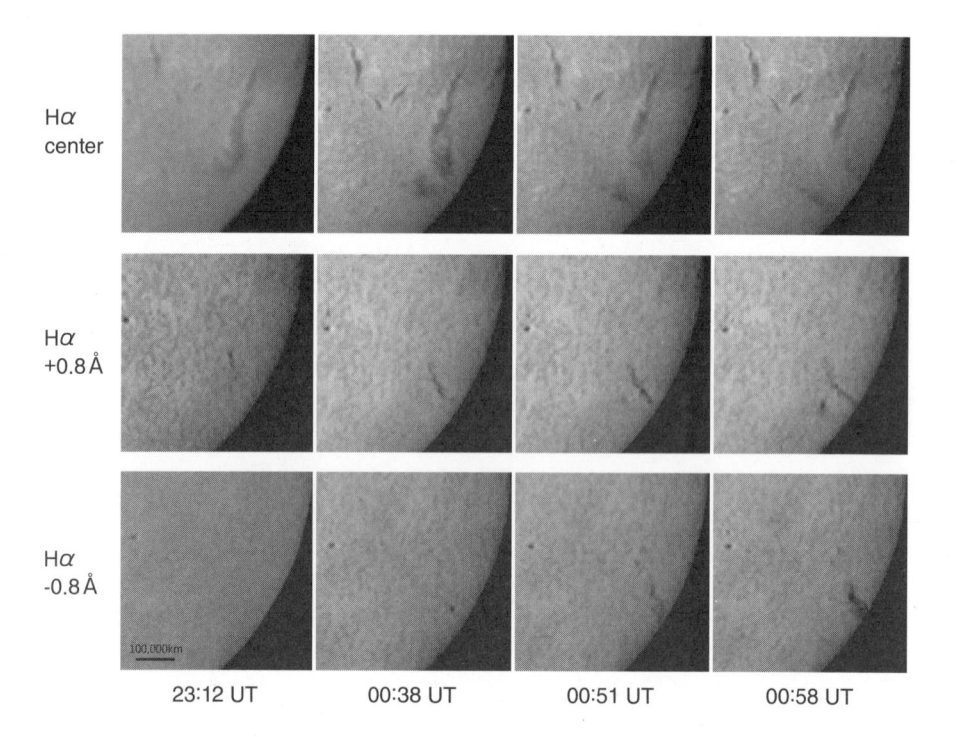

GOES Class = --   NOAA --
S46,W58   P = 16.8 [deg]   B0= 1.1 [deg]

	23:12 UT	00:38 UT	00:51 UT	00:58 UT
Hα center				
Hα +0.8 Å				
Hα −0.8 Å				

A long quiescent filament of 300 Mm length was activated and completely disappeared in 1.5 hours. After the eruptive disappearance of the southern half of the filament, the filament gas in the remaining northern half flowed out along the axial direction of the filament. In the event, no prominent chromospheric brightening was observed in both sides of the filament channel.

30万kmにわたる長大な静穏フィラメントが活発化して上昇し，1時間半の間に消失したものである．フィラメントの南側半分が上昇を始めて消失した後，北半分のフィラメントガスがその軸方向に流れ出して完全に消失した．この現象では，フィラメントの両脇の彩層加熱現象はほとんど見られなかった．

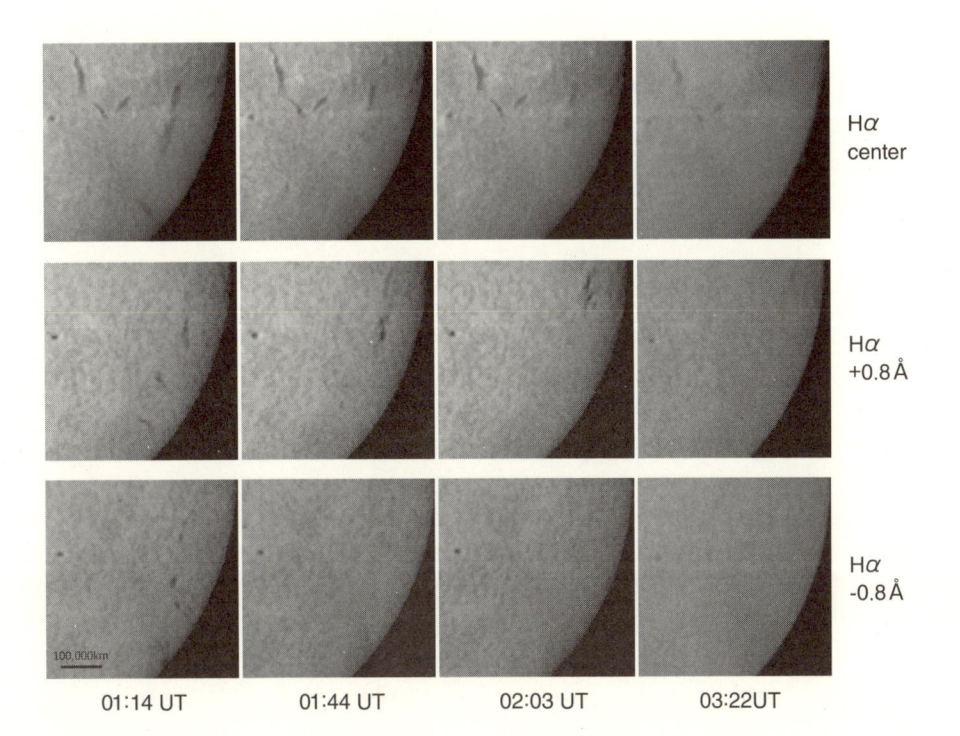

Mar 21, 2001 02:31−03:22 UT   Flare, Filament Eruption

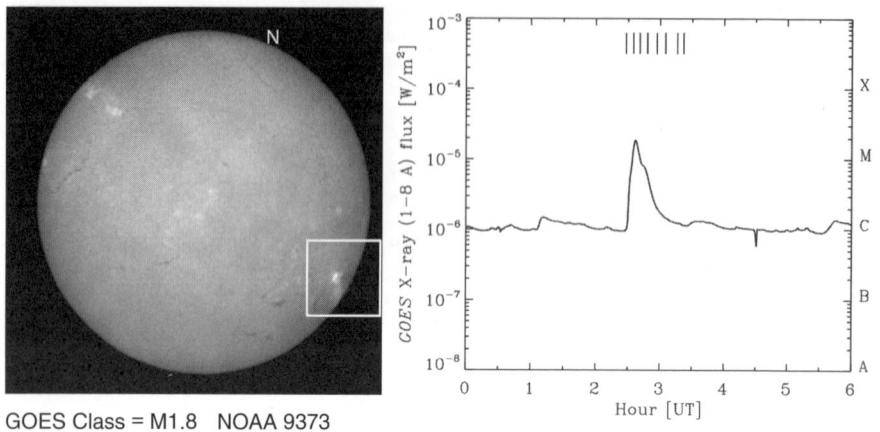

GOES Class = M1.8   NOAA 9373
S06,W70   P = -25.3[deg]   B0= -7.0[deg]

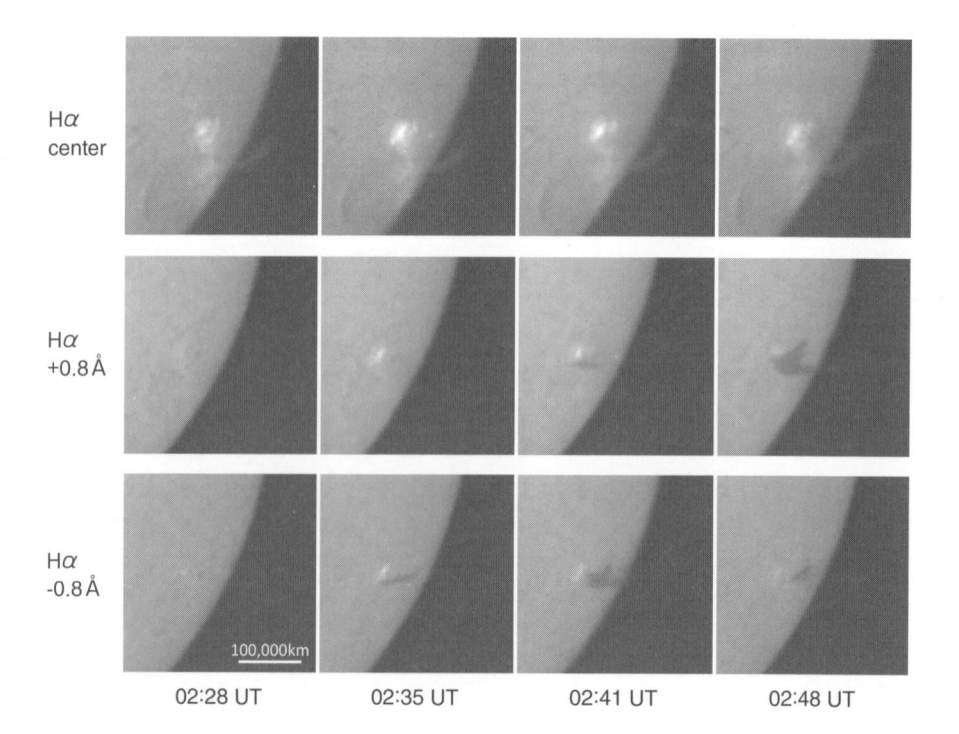

Hα center

Hα +0.8Å

Hα -0.8Å

100,000km

02:28 UT   02:35 UT   02:41 UT   02:48 UT

NOAA9373 was activated by newly emerging magnetic flux on Mar 18 and the magnetic type changed from the β to the γ type. A compact M2.2 flare occurred on Mar 21. From the flaring region, 2-3 filaments erupted with a speed of 80 km/s. One foot of the filament was anchored on the surface, while the other foot was cut off and ascended to coronal heights. The tether cut filament was observed to rise to the height of 250 Mm by SOHO/EIT195. A CME event also was observed in association with the event by LASCO/C2. The flare was of short duration and not followed by any LDE event.

β型の安定したNOAA9373領域の後行磁場領域に，3月18日に新たな磁気浮上領域が現われγ型領域となってこの領域が活発化した．3月21日にM2.2クラスのコンパクトフレアが発生した．このフレア領域から，フィラメントが80 km/sの速度で噴出した．フィラメントの片足が太陽面に固定され，他の足が切り離されて上昇する形であった．SOHO/EIT195では上空コロナに向かって25万kmの高さまで伸びたたことが観測されている．また，LASCO/C2ではこれに伴うCMEが観測されている．なお，このフレアは寿命が短くLDE現象を伴っていなかった．

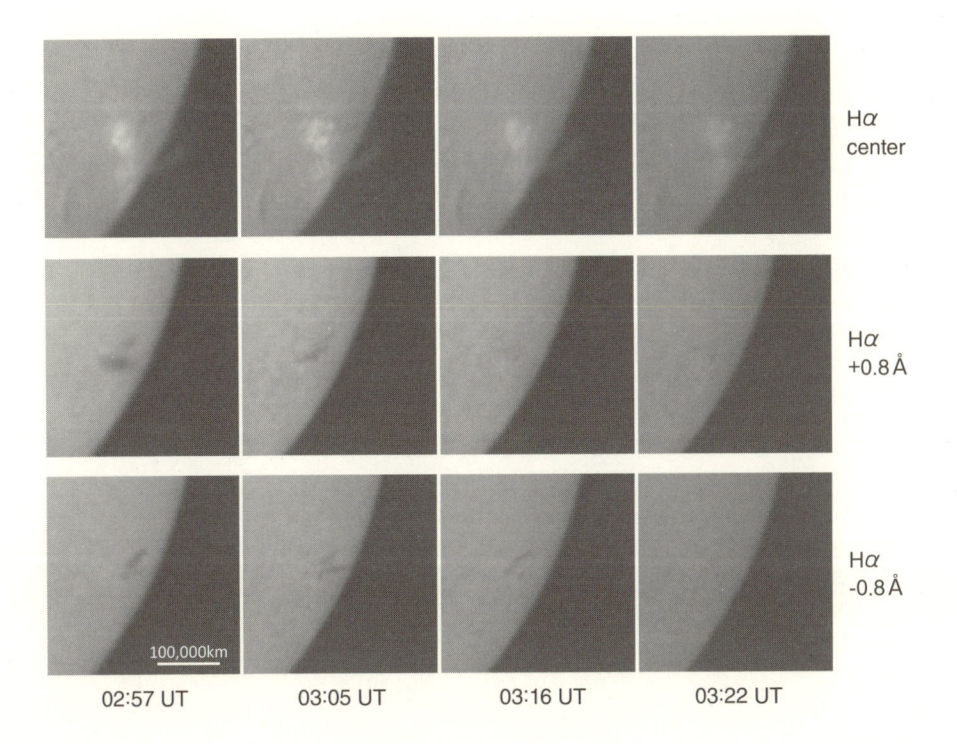

Apr 10, 2001 03:30−06:21 UT   Flare, Filament Eruption

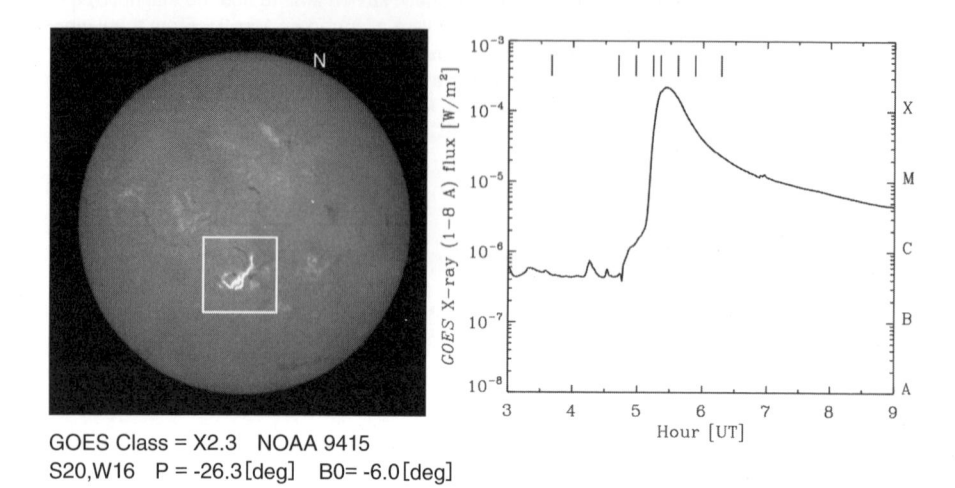

GOES Class = X2.3   NOAA 9415
S20,W16   P = −26.3 [deg]   B0 = −6.0 [deg]

Hα center

Hα +0.8 Å

Hα −0.8 Å

100,000 km

03:40 UT   04:42 UT   04:58 UT   05:14 UT

The event was one of the strongest flare explosion (X2.3 class) and showed the typical development of the filament-eruption triggered two-ribbon flare. After the filament eruption, two ribbons of chromospheric flare brightened brilliantly and separated to each other along the time. In the later phase, down-flows along the numerous post-flare loops could be seen in red wing filter images of Hα. Analysis of Kwasan observation by Asai et al. (2003, 2004) showed that conjugate Hα kernels were connected by EUV loops as predicted by the magnetic reconnection model and that the energy release rates estimated by the observed data were consistent with the model. In this event, an EIT wave was detected clearly and was observed to drive filament oscillations of 27 min period in distant locations on the solar surface, although the amplitude of possibly associated Moreton wave was very small (Okamoto et al. 2004). It is interesting that all the filaments located in the propagating path of the EIT wave were not always excited to oscillate.

この現象はもっとも強烈なフレアに匹敵する強度（X2.3）を示した．また，フィラメント消失によって起動されたツーリボンフレアの時間発展を示す典型的な例である．フィラメント上昇後，フレアが始まると彩層部で極めて明るい2つのリボンが現われ，時間の経過と共にリボンの間隔が広がっていった．その後，ポストフレアループが現われそのループに沿って上空のコロナガスが冷却されて流れ落ちていることが，Hαの長波長側のフィルター像で見られた．Asai et al. (2003, 2004) は，花山天文台での観測からHαカーネルのペアーがEUVで観測されるコロナループの両足に対応していることからフレアの磁気リコネクションモデルが正しいこと，また，観測データから導いたエネルギー解放率が上記モデルで説明できることを示した．また，このイベントでは，EIT波が発生しており，活動領域から遠く離れた静穏フィラメントの振動（周期約27分）を励起したことが確認されている（Okamoto et al. 2004）．モートン波も随伴していた可能性があるがその振幅は小さく確認はできなかった．なお，興味深いことに，EIT波の伝播経路にあったフィラメントがすべて振動を示したわけではなかった．

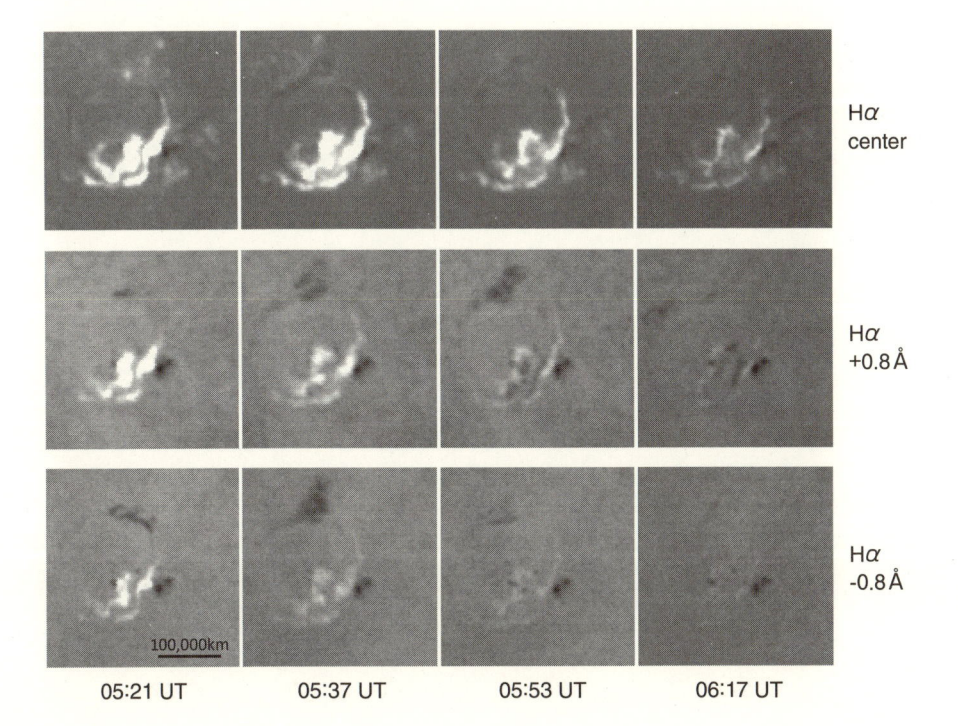

210 | 第6章 代表的イベント

Apr 10, 2001 05:37−05:51 UT Winking Filament

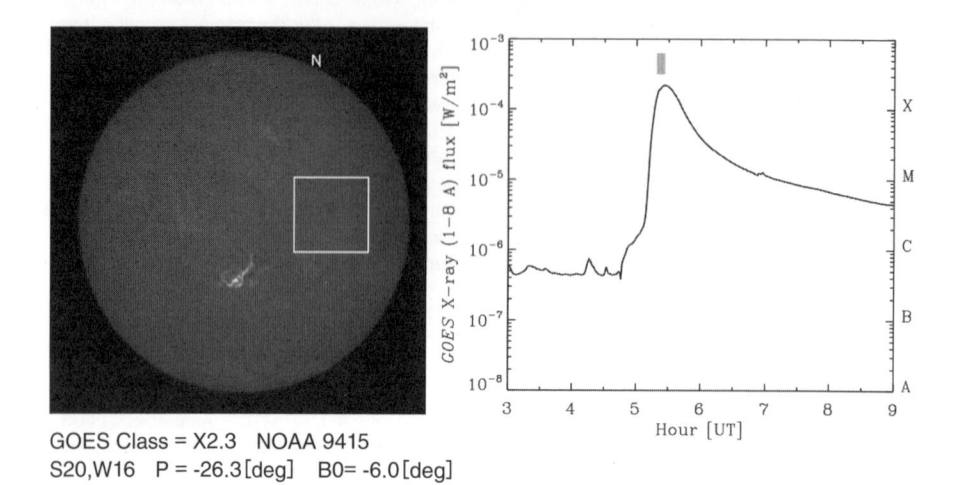

GOES Class = X2.3　NOAA 9415
S20,W16　P = −26.3[deg]　B0= −6.0[deg]

Hα center

Hα +0.8Å

Hα −0.8Å

100,000 km

05:37 UT　　05:41 UT　　05:45 UT　　05:49 UT

The event was one of the strongest flare explosion (X2.3 class) and showed the typical development of the filament-eruption triggered two-ribbon flare. In this event, an EIT wave was detected clearly and was observed to drive filament oscillations of 27min period in distant locations on the solar surface, although the amplitude of possibly associated Moreton wave was very small (Okamoto et al. 2004). It is interesting that all the filaments located in the propagating path of the EIT wave were not always excited to oscillate.

この現象はもっとも強烈なフレアに匹敵する強度 (X2.3) を示した．また，フィラメント消失によって起動されたツーリボンフレアの時間発展を示す典型的な例である．このイベントでは，EIT 波が発生しており，活動領域から遠く離れた静穏フィラメントの振動（周期約 27 分）を励起したことが確認されている (Okamoto et al. 2004). モートン波も随伴していた可能性があるがその振幅は小さく確認はできなかった．なお，興味深いことに，EIT 波の伝播経路にあったフィラメントがすべて振動を示したわけではなかった．

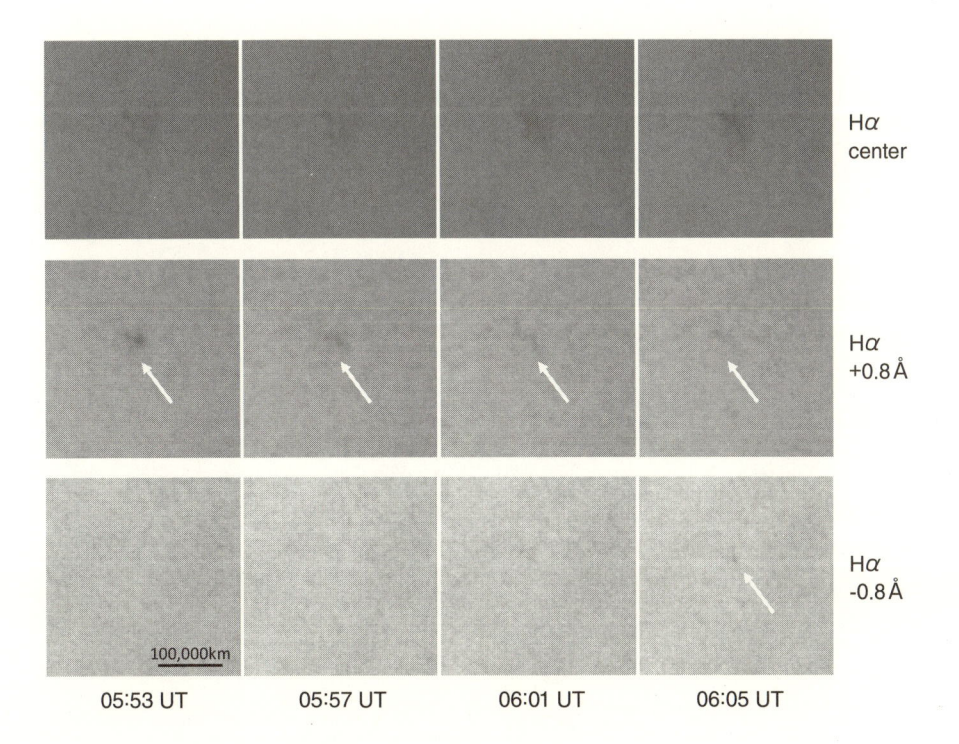

Apr 22, 2001 23:43−00:24 UT   Filament Eruption

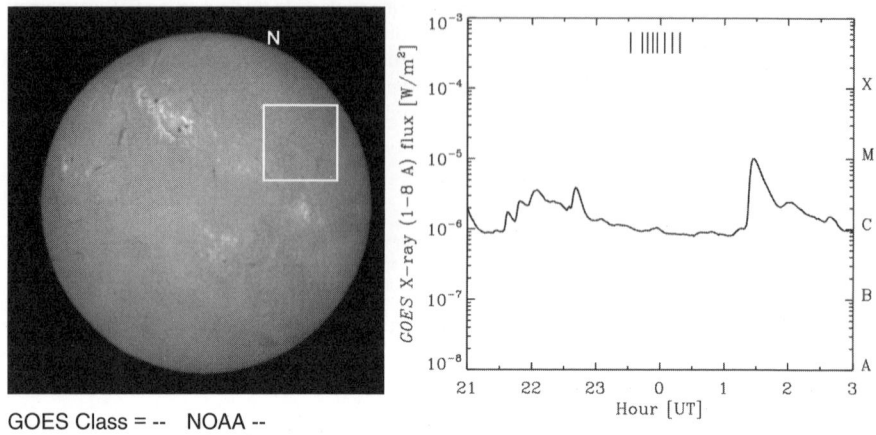

GOES Class = --   NOAA --
N33,W33   P = -25.4[deg]   B0= -6.9[deg]

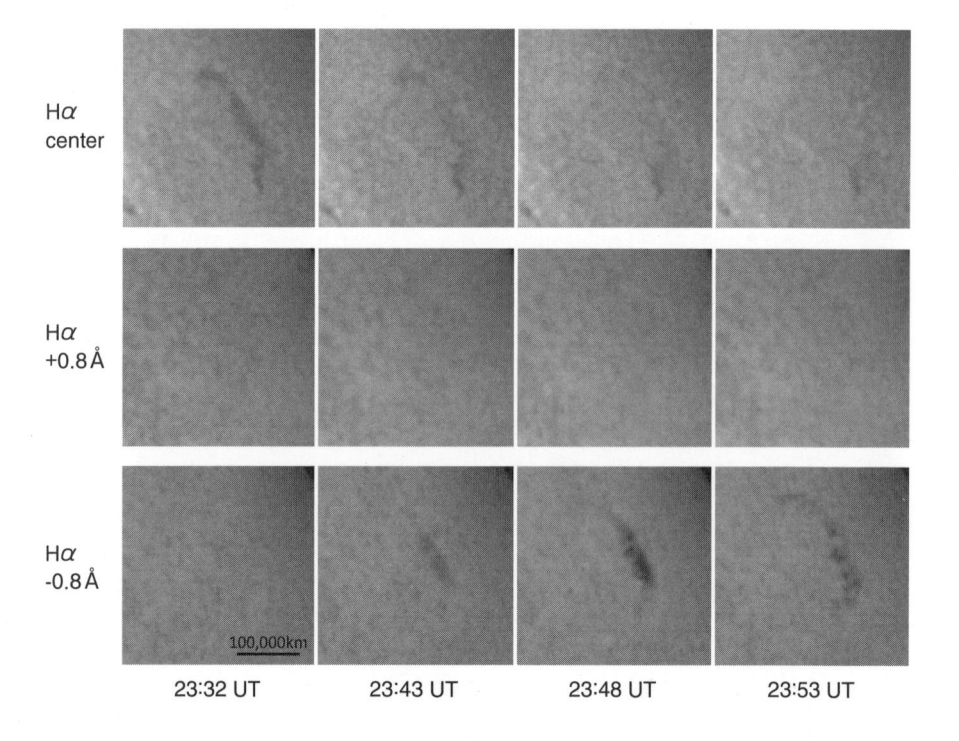

Hα center

Hα +0.8Å

Hα -0.8Å

100,000km

23:32 UT   23:43 UT   23:48 UT   23:53 UT

The event is a typical example of the slow eruption of quiescent filament. While both ends of the filament kept their locations, its main body rose up, expanded towards the corona and then completely disappeared in 40 min. SOHO/LASCO detected an associated CME event which expanded with a speed of around 500 km/s.

この現象は静穏フィラメントのゆっくりとした上昇消滅の典型例である．フィラメントの両端位置は固定されているものの，その本体が上昇しつつ大きくなり終には完全に消滅した．この間 40 分であった．SOHO/LASCO では，この現象に伴う CME 現象が観測されている．その速度は約 500 km/s であった．

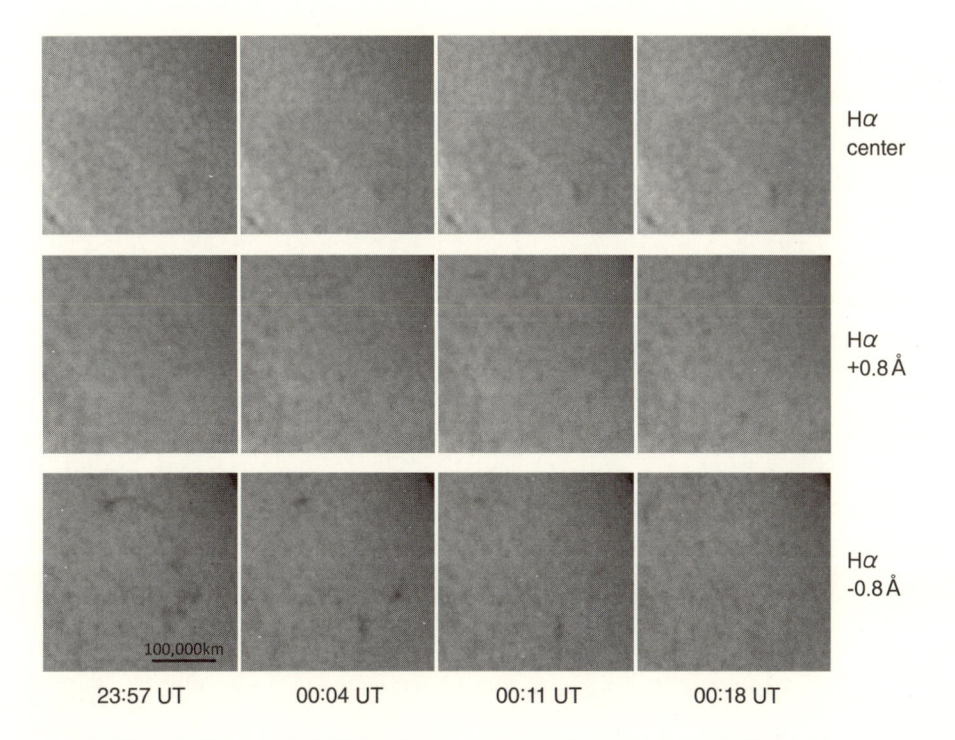

				Hα center
				Hα +0.8Å
100,000km				Hα −0.8Å
23:57 UT	00:04 UT	00:11 UT	00:18 UT	

May 12, 2001　23:27−23:43 UT　Flare, Surge

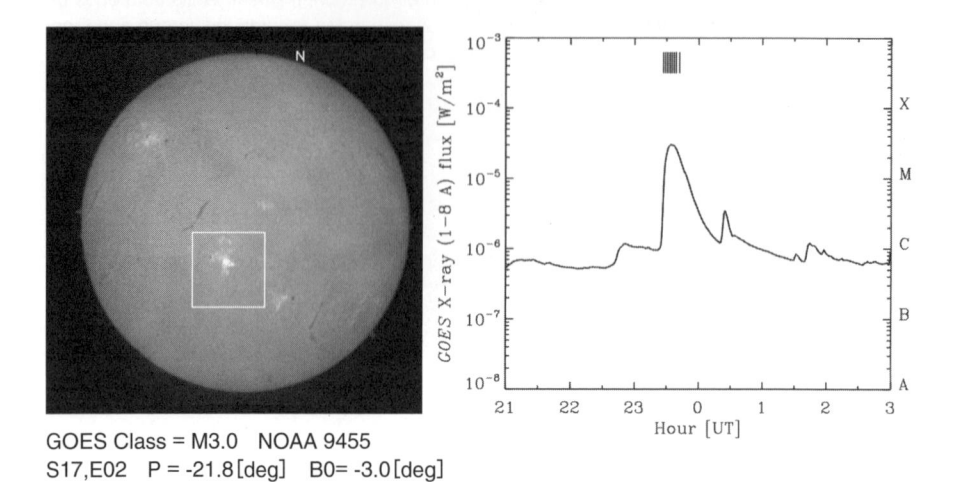

GOES Class = M3.0　NOAA 9455
S17,E02　P = −21.8[deg]　B0= −3.0[deg]

	23:27 UT	23:29 UT	23:31 UT	23:33 UT
Hα center				
Hα +0.8Å				
Hα −0.8Å				

100,000km

In the β-type active region NOAA9455, there occurred a compact M3.0 flare of 1hour duration. From the flare kernel, a surge was ejected with an apparent speed of 130 km/s. The direction of ejection was nearly parallel to the solar surface. Associated with the flare and the surge ejection, a Moreton wave also was emitted along the direction of the surge. The propagation of Moreton wave in 1 min cadence is displayed in the following pages.

β型の活動領域 NOAA9455 で，寿命約 1 時間の M3.0 クラスのコンパクトフレアが発生した．このフレアのカーネル部から見かけの速さ 130 km/s のサージが噴出した．その噴出方向は，ほぼ太陽表面に沿った方向であった．このフレア，サージ噴出にともなって，モートン波が放射され，サージ噴出とほぼ同じ方向に伝播した．モートン波の伝播の様子は次ページ以降に示されている．

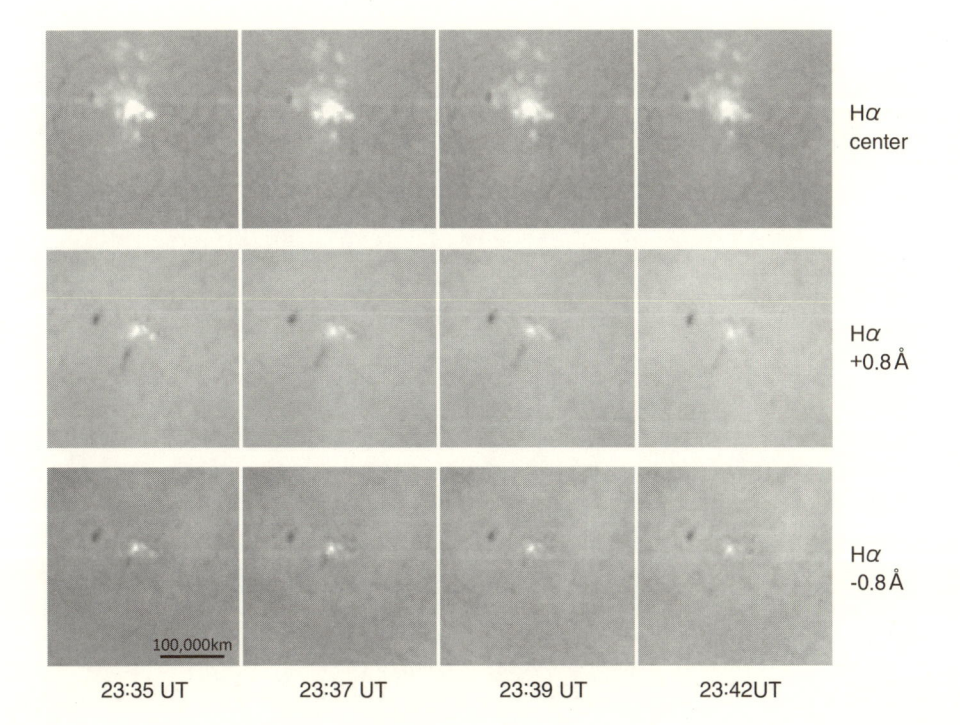

				Hα center
				Hα +0.8Å
				Hα −0.8Å
23:35 UT	23:37 UT	23:39 UT	23:42 UT	

100,000km

May 12, 2001 23:27−23:34 UT Moreton Wave

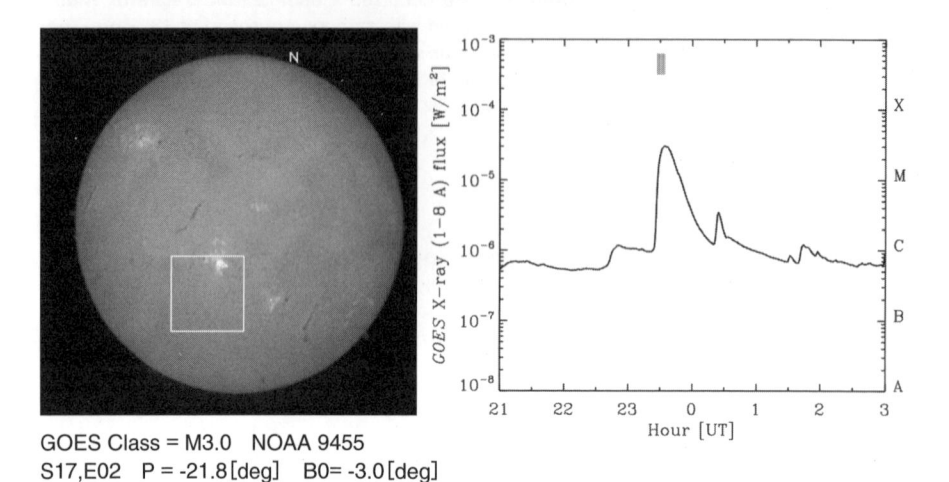

GOES Class = M3.0   NOAA 9455
S17,E02   P = -21.8[deg]   B0= -3.0[deg]

Hα center

Hα +0.8Å

Hα -0.8Å

100,000km

23:27 UT   23:28 UT   23:29 UT   23:30 UT

Associated with an M3.0 flare and the surge ejection in NOAA9455, a Moreton wave was emitted along the direction of the surge. The propagation speed was around 650 km/s. Unfortunately, the association of an EIT wave or a CME was uncertain, as the EUV observation with SOHO was not done during the day. (In the next two pages, time-difference images to enhance the visibility of wave fronts are shown with the manually drawn sketch of the wave fronts.)

NOAA9455 で発生した M3.0 クラスフレア，サージ噴出にともなって，モートン波が放射された．その方向はサージ噴出方向であって，伝搬速度は約 650 km/s であった．SOHO による極端紫外線での観測はこの期間行われていなかったため，EIT 波あるいは CME がこのイベントに伴っていたかどうかは不明である．（次の 2 ページでは，波面を見やすくするように，画像の明るさの時間変化像を示している．また波面位置のスケッチも参考のために付してある．）

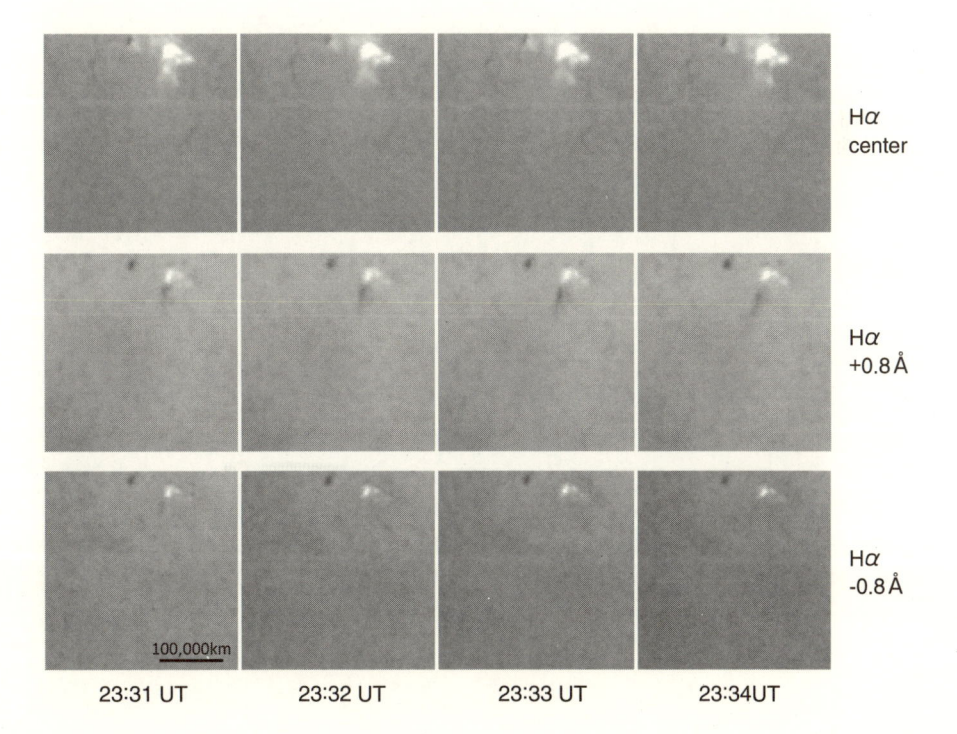

				Hα center
				Hα +0.8Å
				Hα -0.8Å
23:31 UT	23:32 UT	23:33 UT	23:34 UT	

100,000km

218 | 第6章　代表的イベント

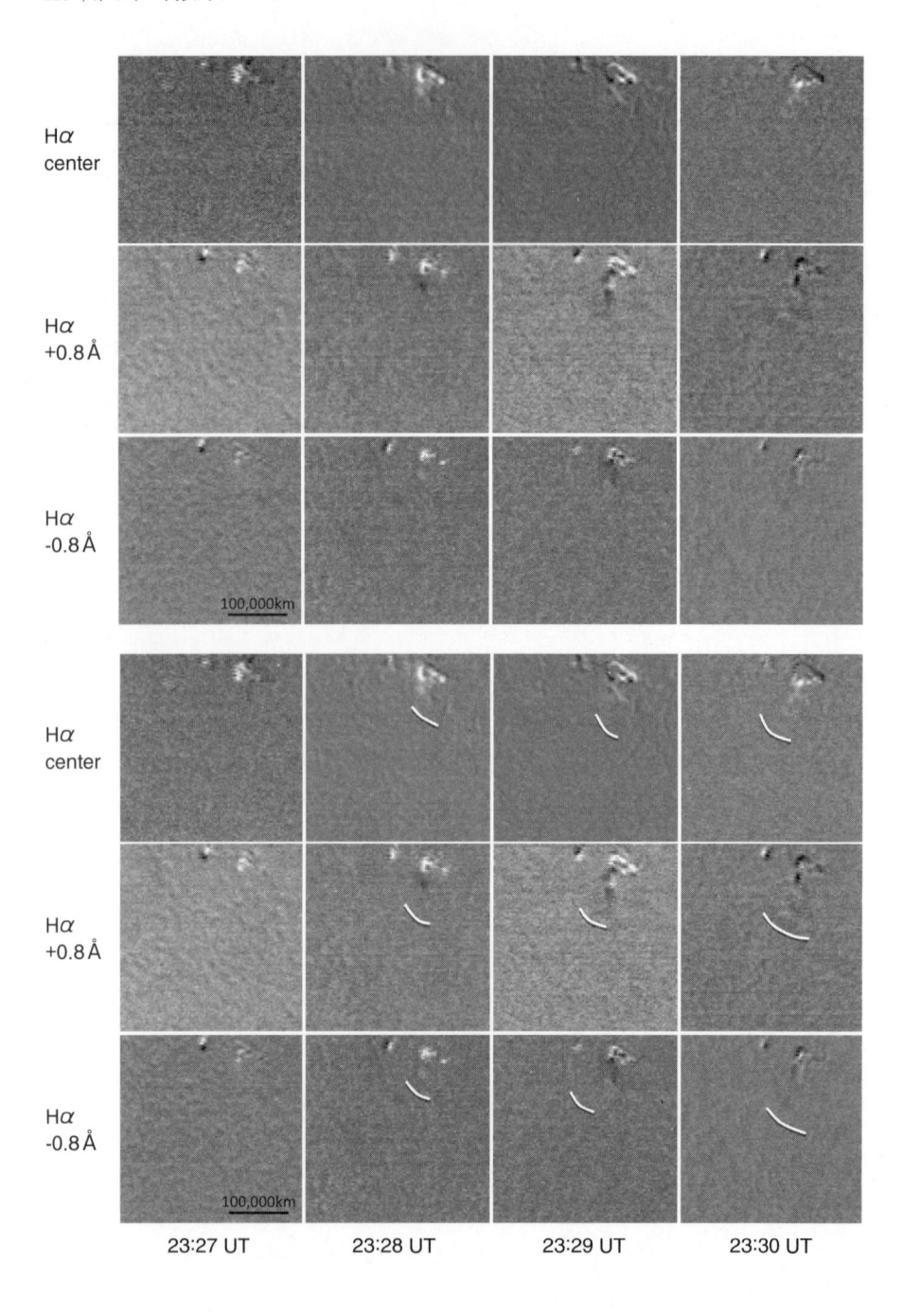

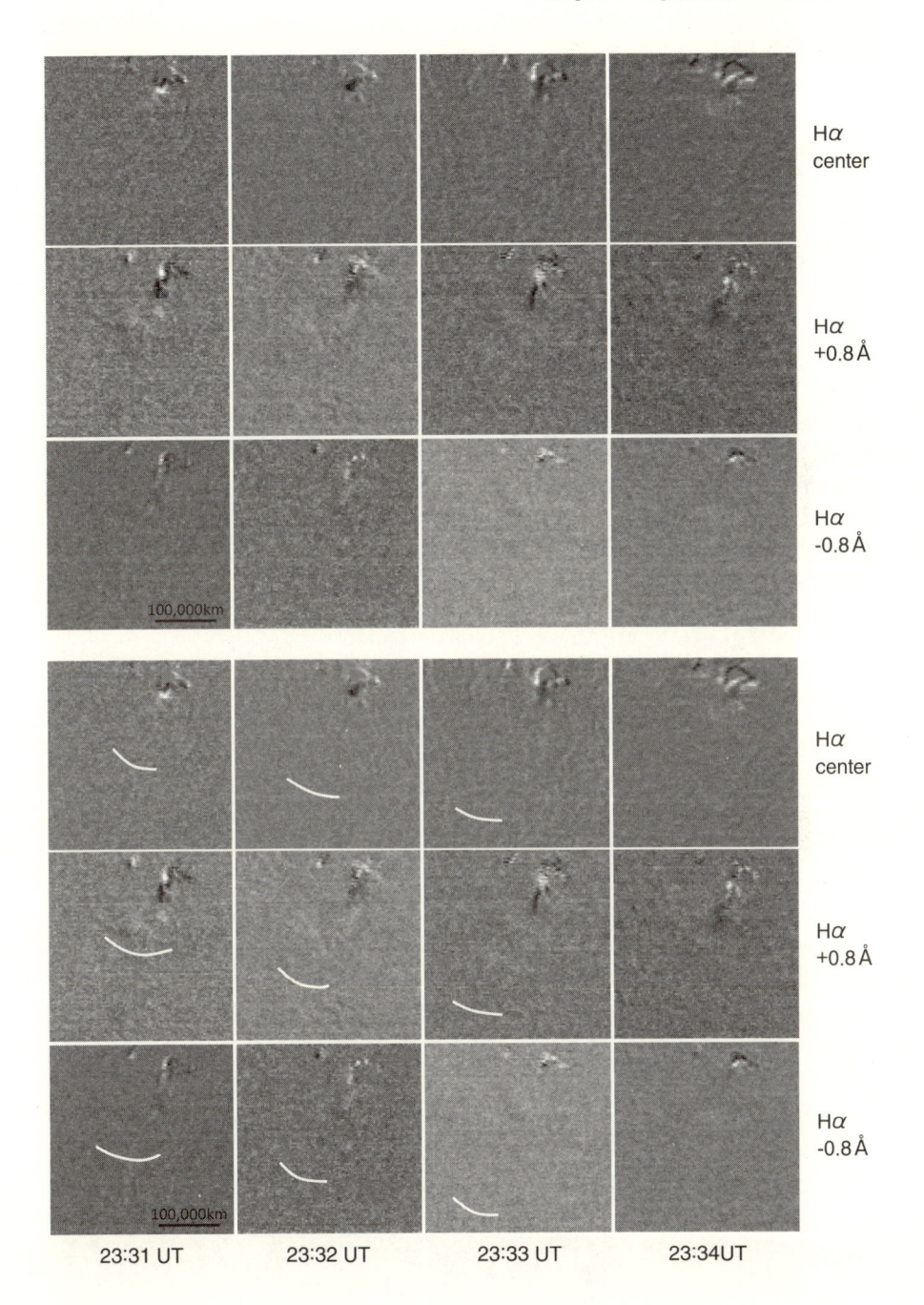

Chapter 6 Representative Events | 219

Hα center

Hα +0.8Å

Hα −0.8Å

100,000km

Hα center

Hα +0.8Å

Hα −0.8Å

100,000km

23:31 UT   23:32 UT   23:33 UT   23:34 UT

May 13, 2001 03:01−03:18 UT Flare

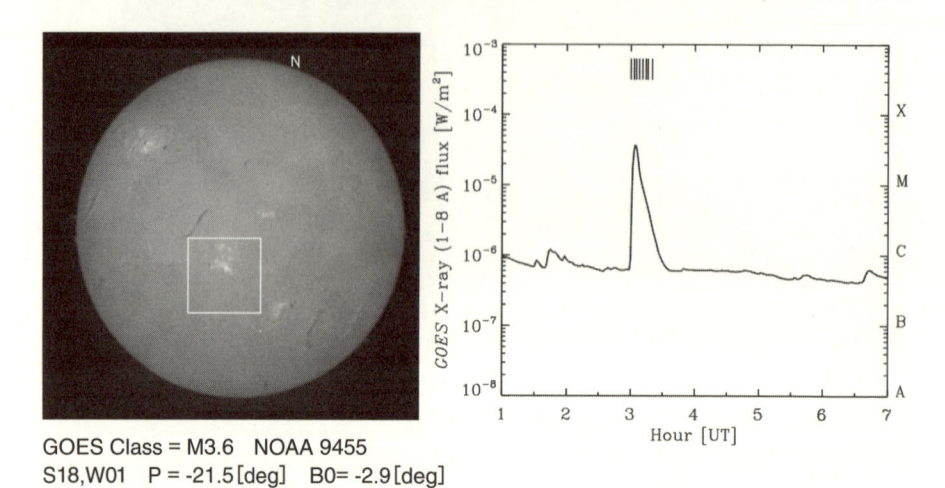

GOES Class = M3.6   NOAA 9455
S18,W01   P = -21.5[deg]   B0= -2.9[deg]

Hα center

Hα +0.8Å

Hα −0.8Å

100,000km

03:00 UT   03:03 UT   03:05 UT   03:08 UT

In 3.5 hours after the explosion of M3.0 flare in the NOAA9455 on May 12, 2001, another M3.6 flare exploded in the same active region. Since the second flare showed the very similar behavior (compactness of bright area, mass ejection and Moreton wave emission) as the first one, the pair of these two flares was a nice case of homologous flares. The propagation of Moreton wave in 1 min cadence is displayed in the following pages.

2001年5月12日にNOAA9455でM3.0クラスのフレアが発生した3.5時間後,同じ領域内のほぼ同じ場所でもう1つのM3.6クラスのフレアが発生した.明るい領域がコンパクトであること,ガスの放出があること及びモートン波が放射されたことという共通の振る舞いを示したことから,この2つのフレアはホモロガスフレア(同形フレア)の格好の例である.モートン波の伝播の様子は次ページ以降に示されている.

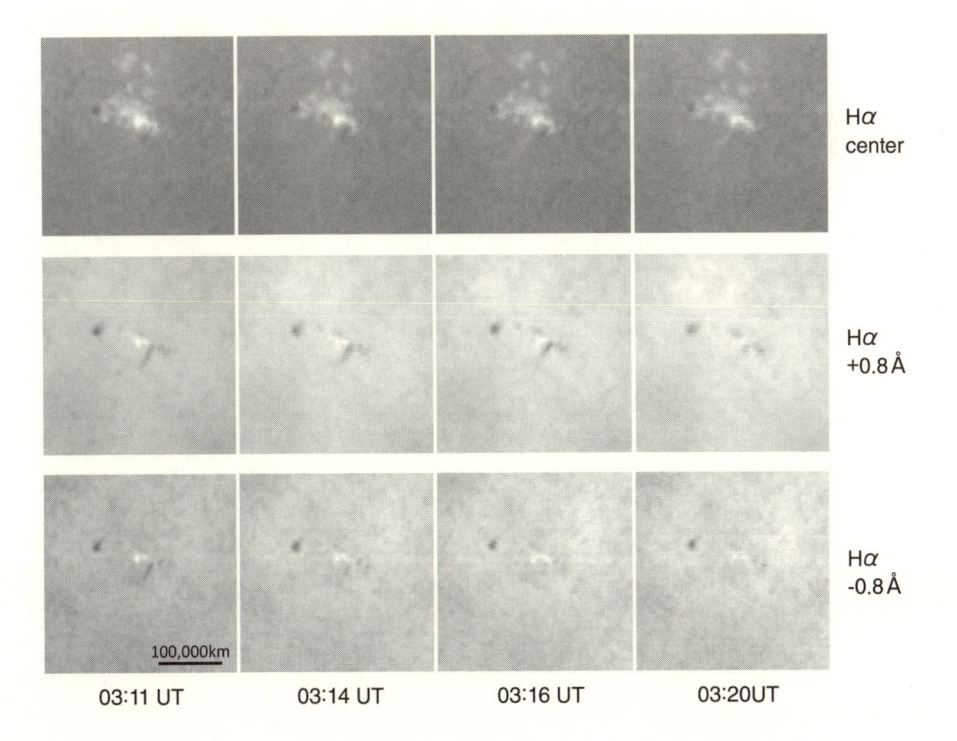

May 13, 2001 03:01−03:09 UT   Moreton Wave

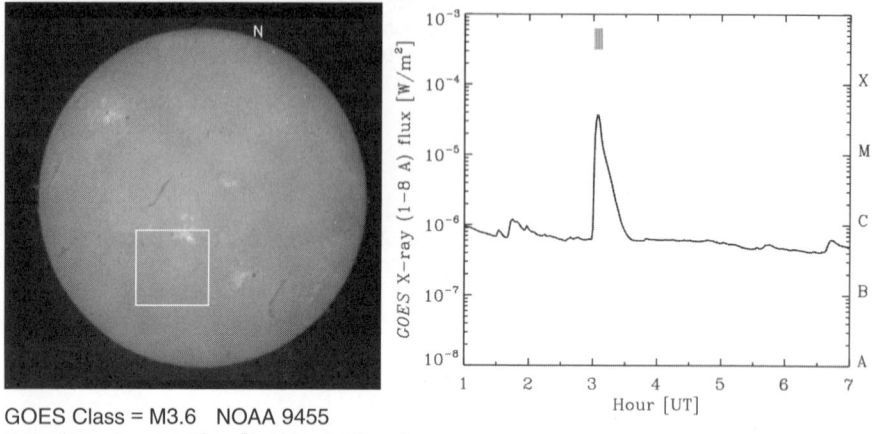

GOES Class = M3.6　NOAA 9455
S18,W01　P = -21.5[deg]　B0= -2.9[deg]

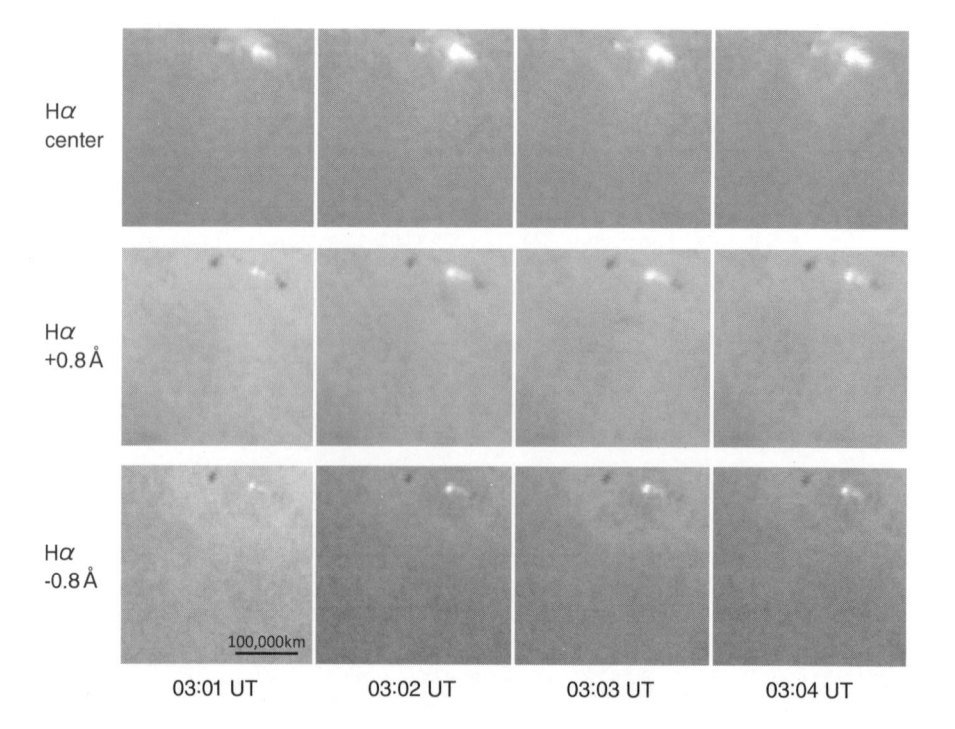

Hα center

Hα +0.8Å

Hα -0.8Å

100,000km

03:01 UT　　03:02 UT　　03:03 UT　　03:04 UT

The temporal evolution of the Moreton wave emitted by the second flare of the homologous flares in the NOAA9455 is shown in 1 min cadence. In this event, the gas ejection speed was around 160 km/s, while the propagation speed of Moreton wave was around 350 km/s. (In the next two pages, time-difference images to enhance the visibility of wave fronts are shown with the manually drawn sketch of the wave fronts.)

NOAA9455 で起こったホモロガスフレア（同形フレア）で2回目に発生したモートン波の1分ごとの時間変化を示している．このイベントでは，ガス噴出の速さは約 160 km/s であり，モートン波の伝搬速度は 350 km/s であった．（次の2ページでは，波面を見やすくするように，画像の明るさの時間変化像を示している．また波面位置のスケッチも参考のために付してある．）

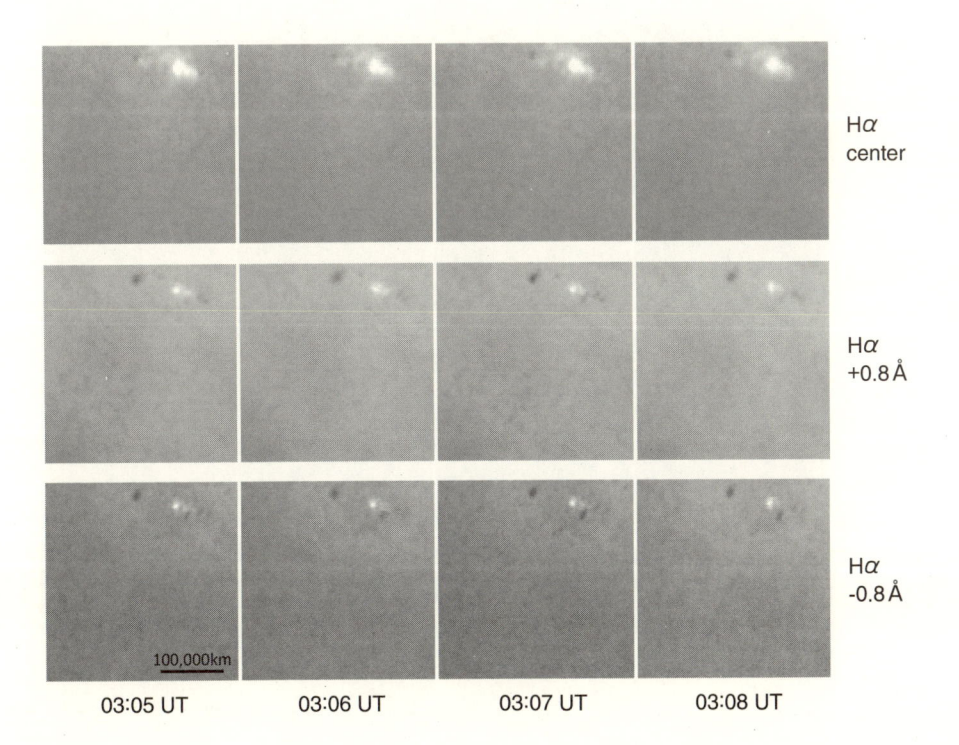

224 | 第6章 代表的イベント

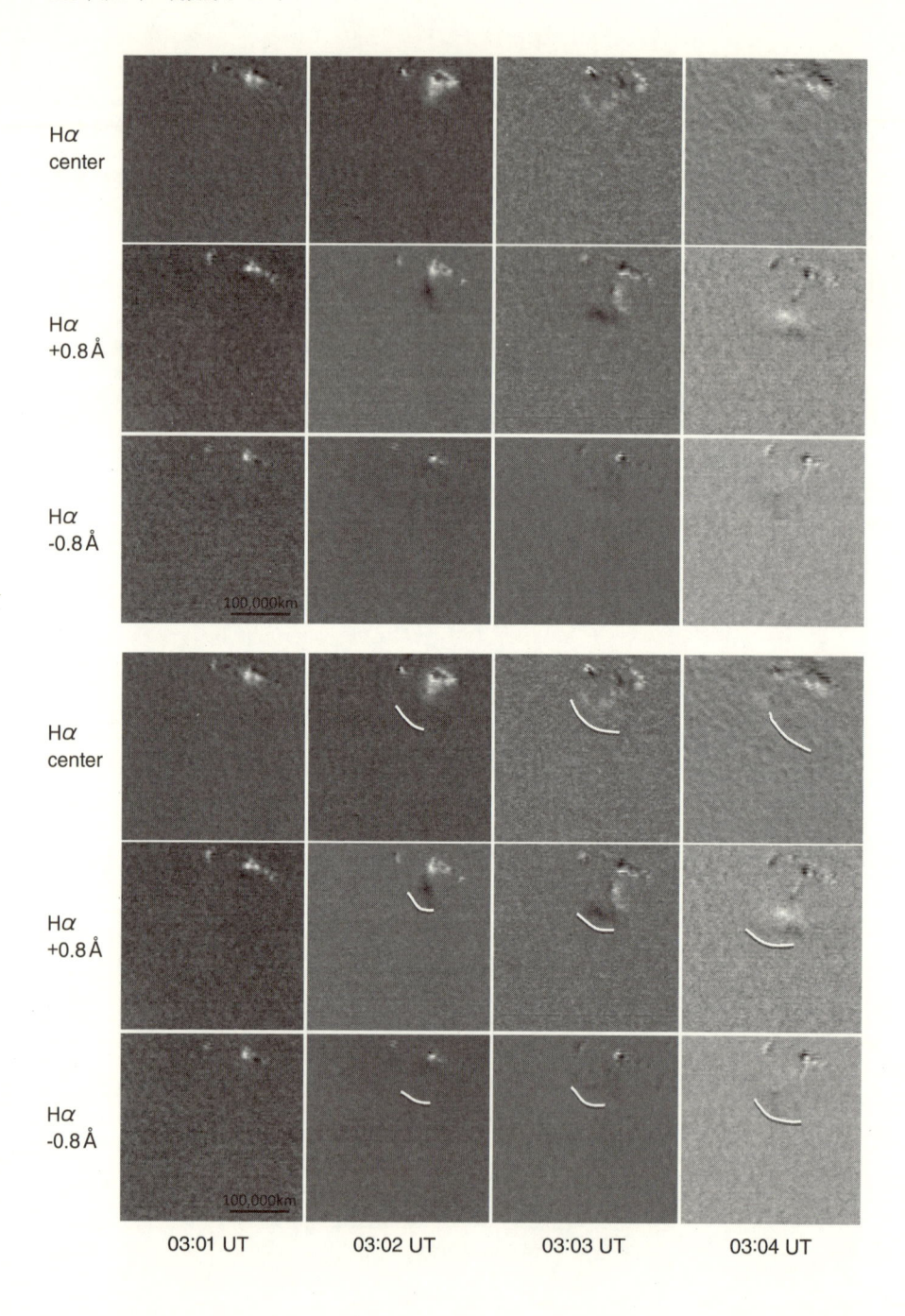

	03:01 UT	03:02 UT	03:03 UT	03:04 UT
Hα center				
Hα +0.8Å				
Hα -0.8Å				
Hα center				
Hα +0.8Å				
Hα -0.8Å				

100,000km

Hα
center

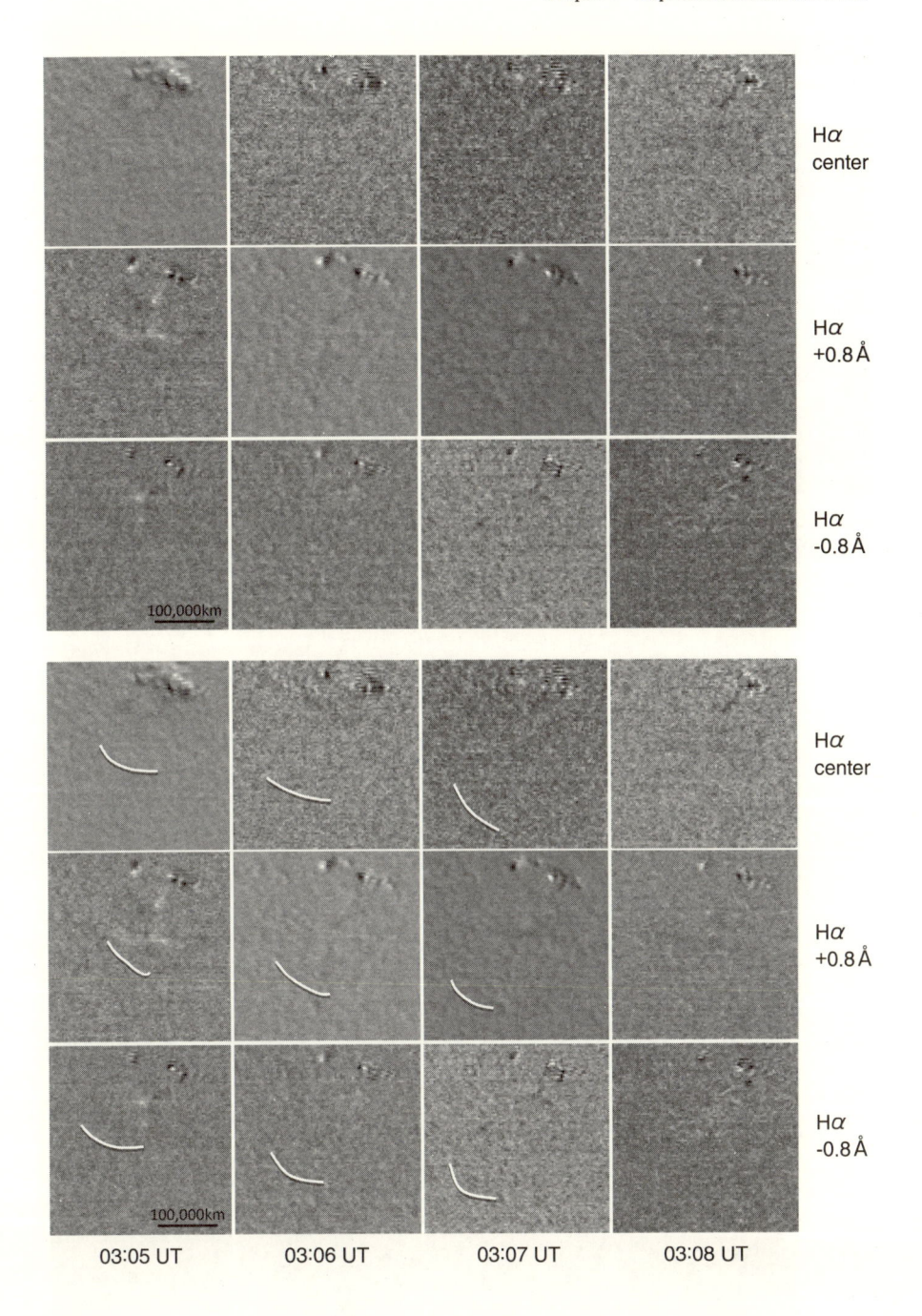

Hα
+0.8Å

Hα
−0.8Å

100,000km

Hα
center

Hα
+0.8Å

Hα
−0.8Å

100,000km

03:05 UT   03:06 UT   03:07 UT   03:08 UT

May 21, 2001 03:08−03:36A UT   Flare, Filament Eruption

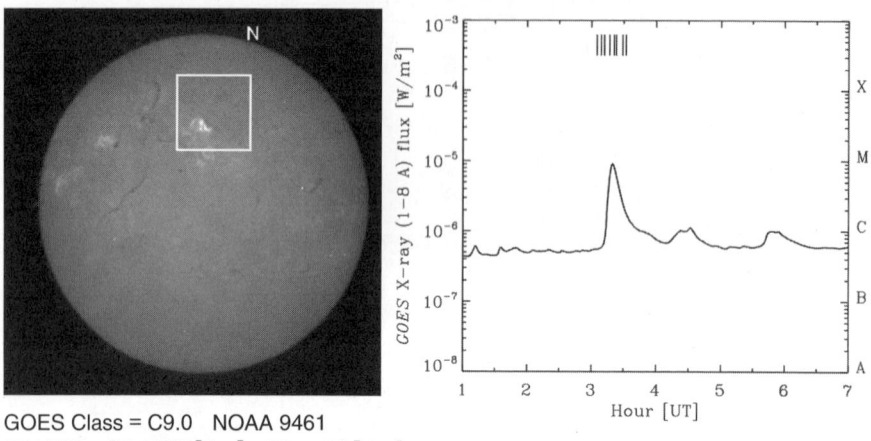

GOES Class = C9.0   NOAA 9461
N24,E09   P = -19.2 [deg]   B0= -2.0 [deg]

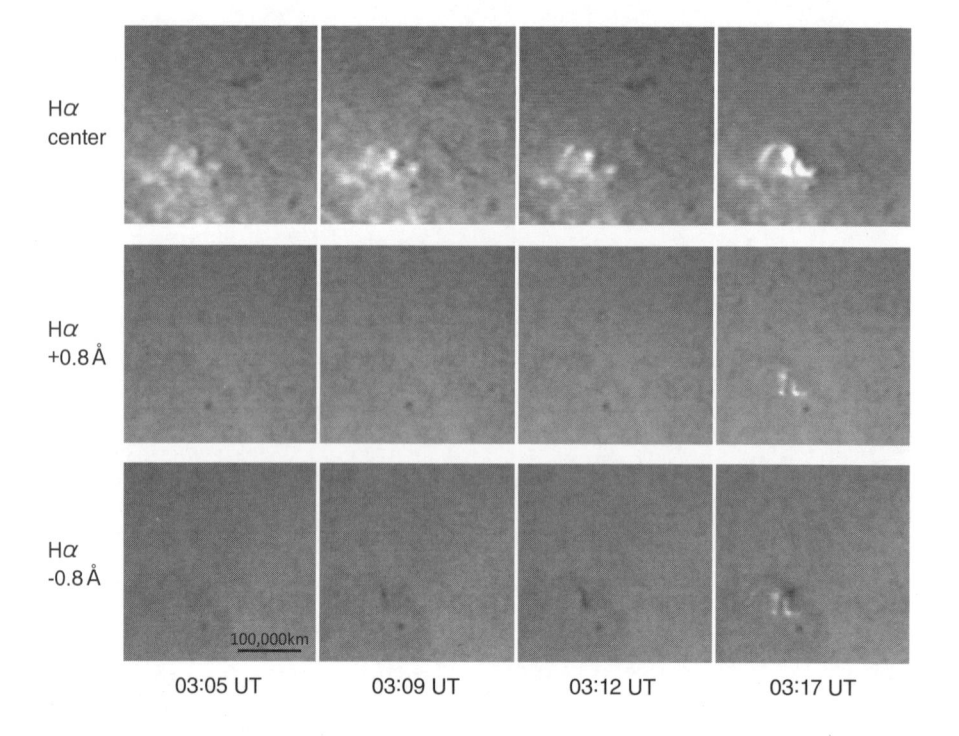

Hα center

Hα +0.8Å

Hα -0.8Å

100,000km

03:05 UT   03:09 UT   03:12 UT   03:17 UT

After several minutes of pre-heating, a compact flare gave rise to brighten in the active region NOAA9461. A nearby active region filament was ejected out from the active region with an extension speed of 150 km/s. Then the gas continued to move with the same speed, travelled a long distance of 400 Mm and finally disappeared in Hα. In SOHO/EIT195 and SOHO/LASCO/C2 images, the cool gas was detected to move farther and leave the sun with the speed of around 400 km/s. A quiescent filament located 150 Mm apart from the active region in the direction of ejection seems to be driven to oscillate at 03:19 UT.

活動領域 NOAA9461 において前兆的な加熱の後，コンパクトフレアが発生した．同時に近傍にあった既存の活動領域フィラメントの一部が 150 km/s の速度で放出された．その後放出ガスは，ほぼ同じ速度を保って 40 万 km の遠方まで到達し Hα 線観測では消滅した．一方，SOHO の EIT195 および LASCO/C2 では，ガスが更に加速されて約 400 km/s の速さで太陽から放出されていたことが観測されている．なお，フィラメント噴出方向の約 15 万 km 離れた位置に元々存在した静穏フィラメントが 03:19 UT に振動を開始した模様である．

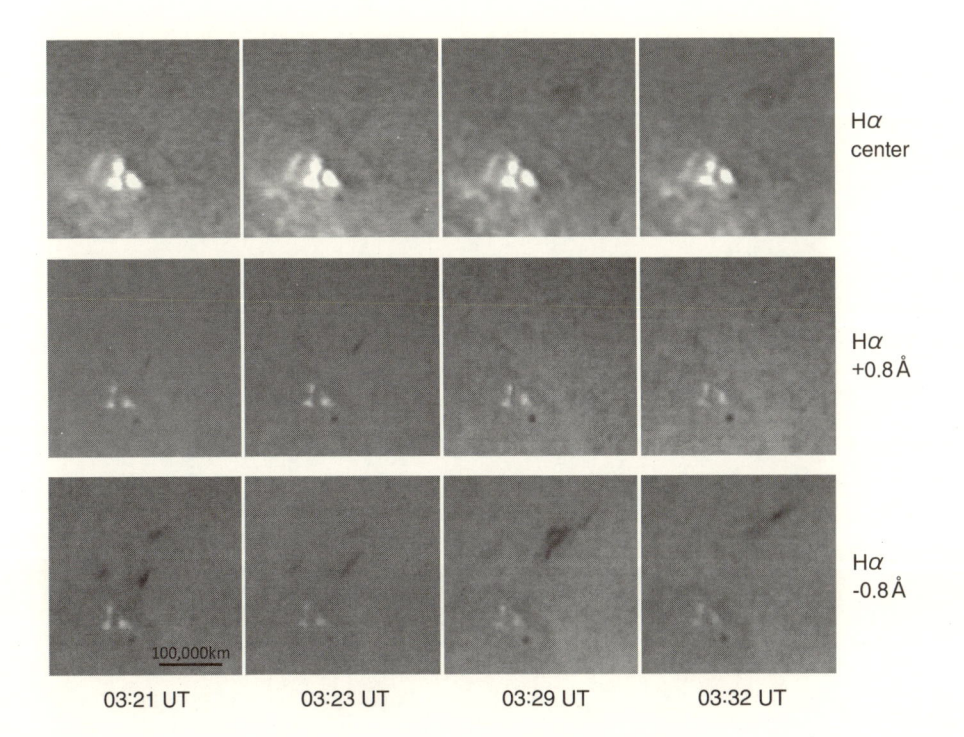

May 21, 2001 03:18−03:33 UT Winking Filament

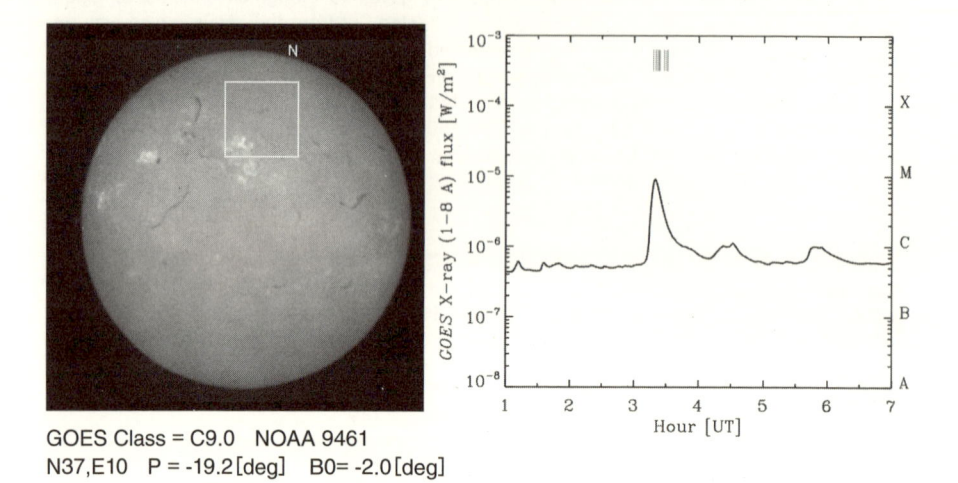

GOES Class = C9.0   NOAA 9461
N37,E10   P = -19.2 [deg]   B0= -2.0 [deg]

Hα center

Hα +0.8Å

Hα -0.8Å

100,000km

03:18 UT   03:20 UT   03:22 UT   03:24 UT

After several minutes of pre-heating, a compact flare gave rise to brighten in the active region NOAA9461. A nearby active region filament was ejected out from the active region with an extension speed of 150 km/s. A quiescent filament located 150 Mm apart from the active region in the direction of ejection seems to be driven to oscillate at 03:19 UT.

活動領域 NOAA9461 において前兆的な加熱の後，コンパクトフレアが発生した．同時に近傍にあった既存の活動領域フィラメントの一部が 150 km/s の速度で放出された．フィラメント噴出方向の約 15 万 km 離れた位置に元々存在した静穏フィラメントが 03:19 UT に振動を開始した模様である．

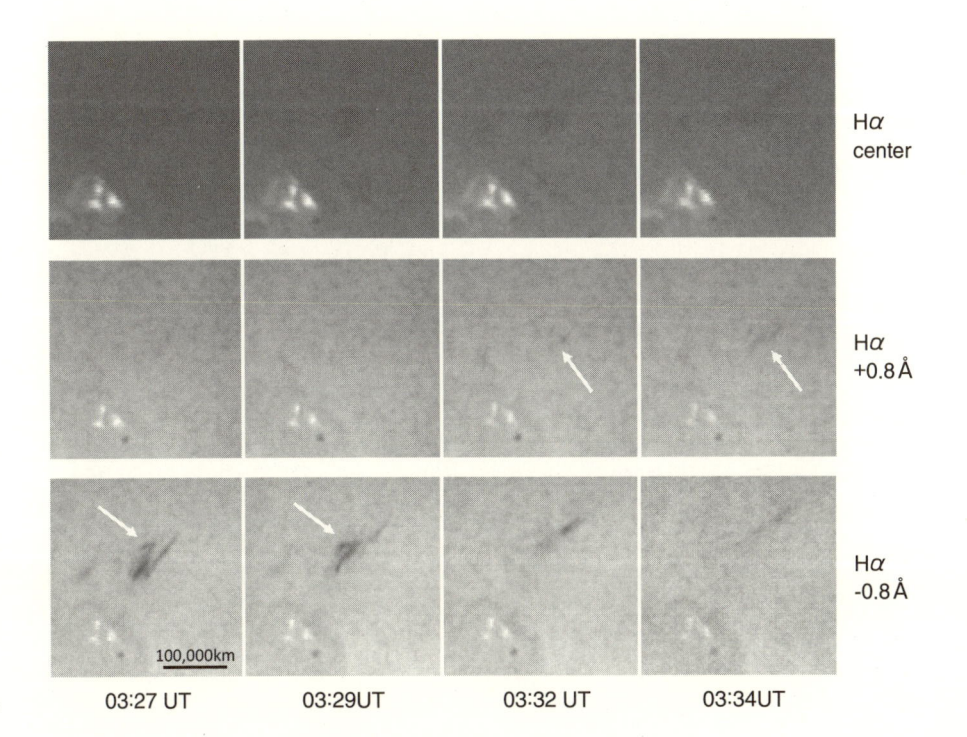

Sep 25, 2001 04:27−06:09 UT Flare, Filament Eruption

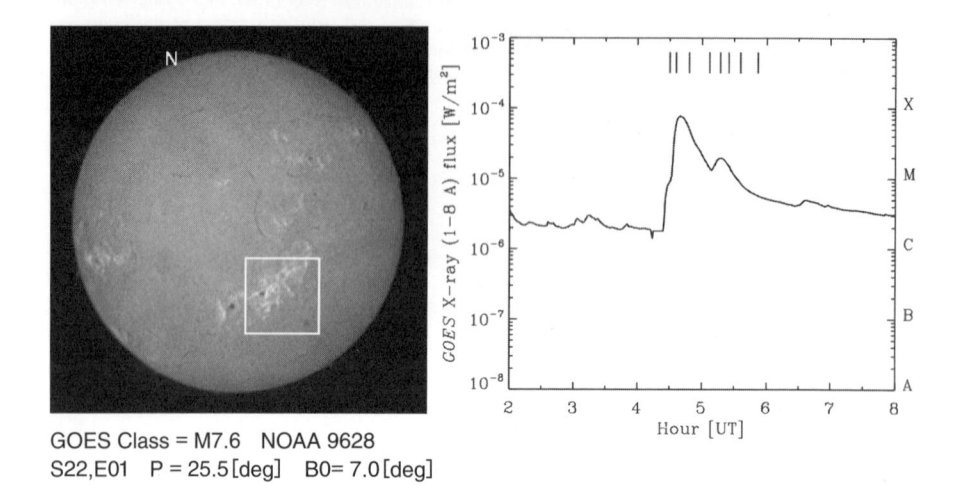

GOES Class = M7.6 NOAA 9628
S22,E01 P = 25.5 [deg] B0= 7.0 [deg]

	04:30 UT	04:36 UT	04:48 UT	05:07 UT
Hα center				
Hα +0.8Å				
Hα −0.8Å				

100,000 km

Two flares occurred consecutively in the $\beta\gamma\delta$ type active region NOAA9628. During the decay phase of the first flare (M7.6), an active region filament, displaced from the first flare, started to expand with a speed of 40 km/s and lead to the explosion of the second flare (M2.0). The filament expanded with a speed of 60 km/s after the second explosion and finally disintegrated completely. An expanding loop, in association with the filament expansion, was detected by EIT195 Å imager, while no prominent CME was found by LASCO/C2.

$\beta\gamma\delta$型の活動領域 NOAA9628 で2つのフレアが連続して発生した．最初の M7.6 フレアの減衰期に別の場所にあった活動領域フィラメントが 40 km/s の速度で膨張を始め，そのフィラメントの両脇に2番目の M2.0 フレアが発生した．フィラメントは2番目のフレア爆発後，加速して 60 km/s の速度で膨張を続け，終にはすべて消滅した．このフィラメント膨張に伴って，EIT195Å 像で膨張するループが見られた．LASCO/C2 では，目立った CME は発生していなかった．

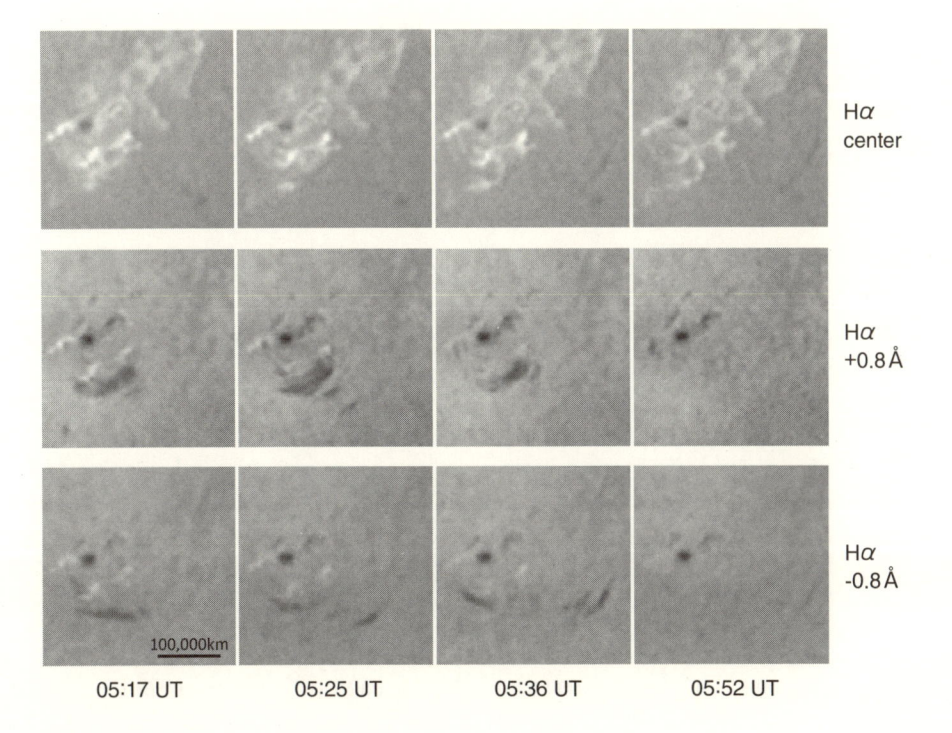

Oct 19, 2001 00:51−01:38 UT Flare

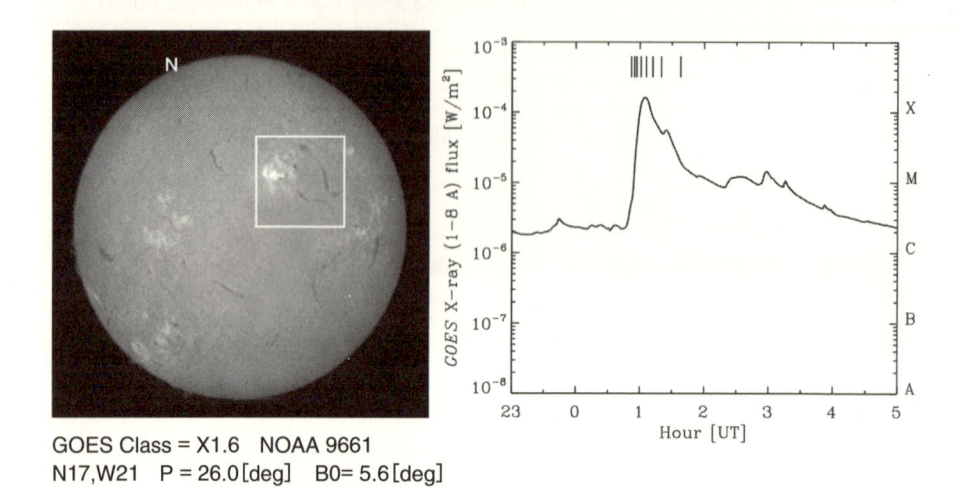

GOES Class = X1.6   NOAA 9661
N17,W21   P = 26.0[deg]   B0= 5.6[deg]

Hα center

Hα +0.8Å

Hα −0.8Å

100,000km

00:52 UT   00:55 UT   00:57 UT   01:01 UT

An energetic X1.6 flare occurred in the complex β γ δ type active region NOAA9661. Although no prominent Hα filament eruption was observed, a Halo-type CME was observed to expand with a speed of 550 km/s, by the SOHO/LASCO. Ishii et al (2005) found that the magnetic neutral line showed a continuous rotation during the high activity of the region and suggested that the emergence of twisted magnetic flux bundles were the energy source for strong flares.

β γ δ 型の活動領域 NOAA9661 で X1.6 の強烈なフレアが発生した．先行する Hα フィラメント爆発は伴わなかったものの，ハロー型の CME が約 550 km/s の速度で膨張するのが SOHO/LASCO で観測された．Ishii et al. (2005) は，この活動領域が高い活動性を示していた間，この領域の磁気中性線が常に回転するような動きを示したことを見出した．そして，ねじれた磁力線が浮上してきていることがフレア活動の源であろうと示唆している．

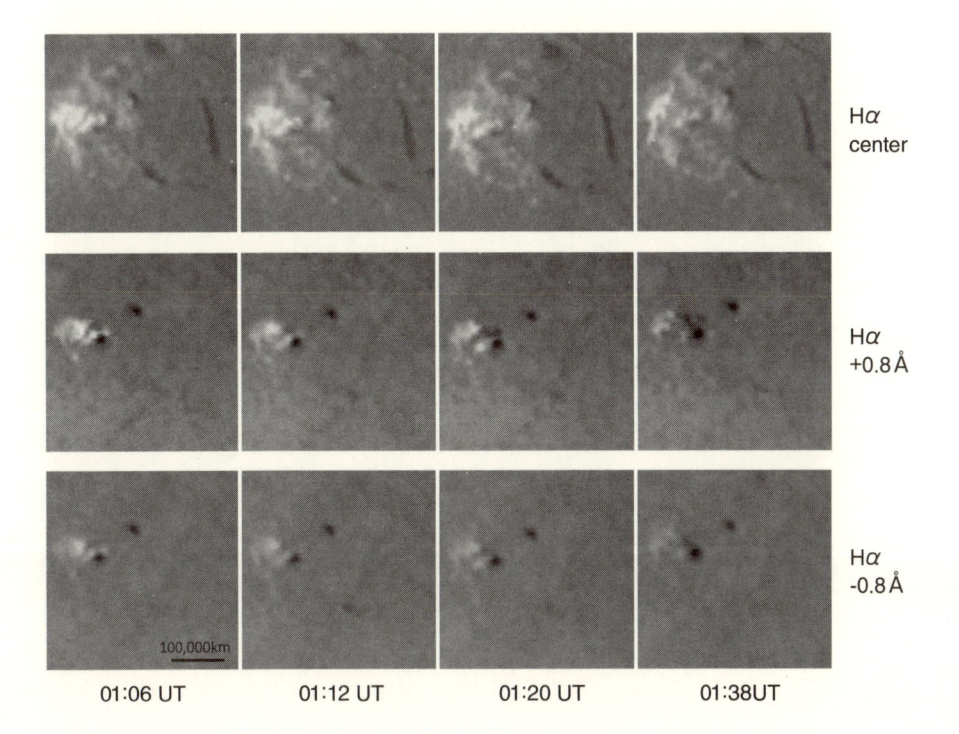

234 | 第6章 代表的イベント

Nov 10, 2001 21:54−02:40 UT   Filament Eruption

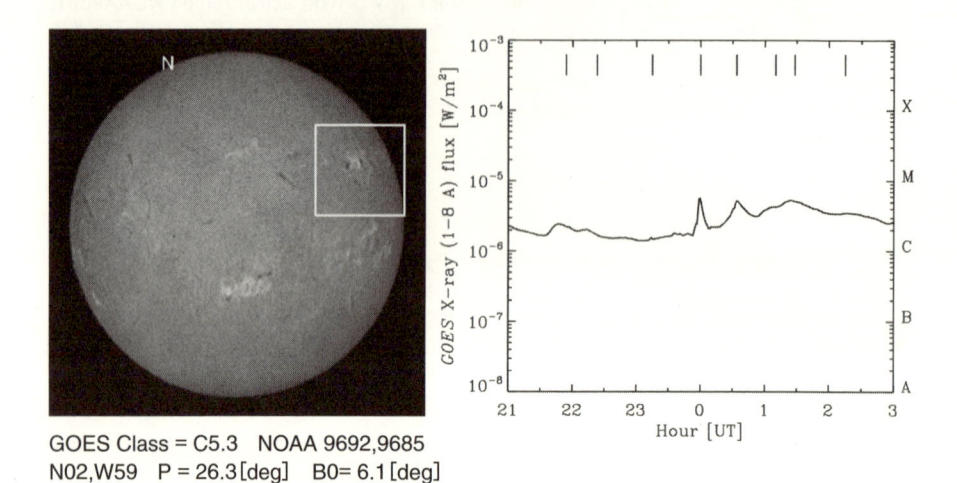

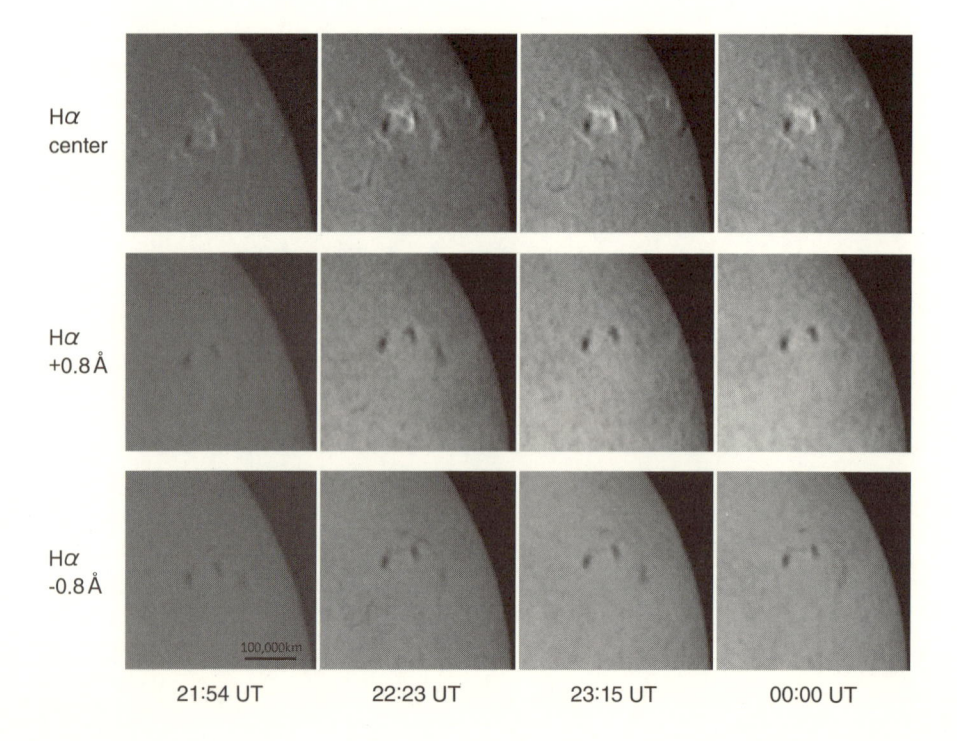

GOES Class = C5.3   NOAA 9692,9685
N02,W59   P = 26.3[deg]   B0= 6.1[deg]

	21:54 UT	22:23 UT	23:15 UT	00:00 UT
Hα center				
Hα +0.8Å				
Hα −0.8Å				

100,000km

A part of an active region filament located west of the region erupted and triggered a two-ribbon C5.3 flare. Remaining part of the filament was activated to erupt sequentially and finally disappeared. In accordance with the spatial advancement of the filament eruption, faint brightening in Hα was observed to extend parallel to the axis of the filament.

活動領域の西側に位置していた活動領域フィラメントの一部（黒点近傍）が爆発を起こし，ほぼ同時にその両脇に C5.3 クラスのツーリボンフレアを発生した．このフィラメントの黒点から離れた残りの部分も順次上昇して消滅した．そして，このフィラメントの上昇活動の移動に応じて，フィラメント近傍の彩層が淡く光り延伸してゆくのが見られた．

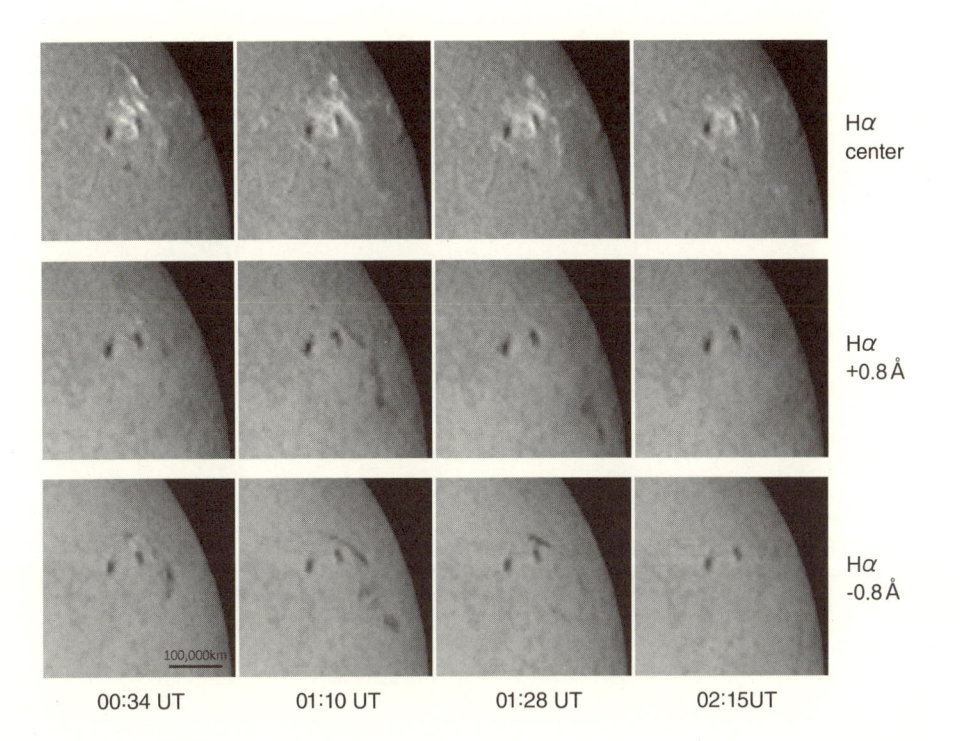

Nov 20, 2001 01:29−02:46 UT   Flare, Filament Eruption

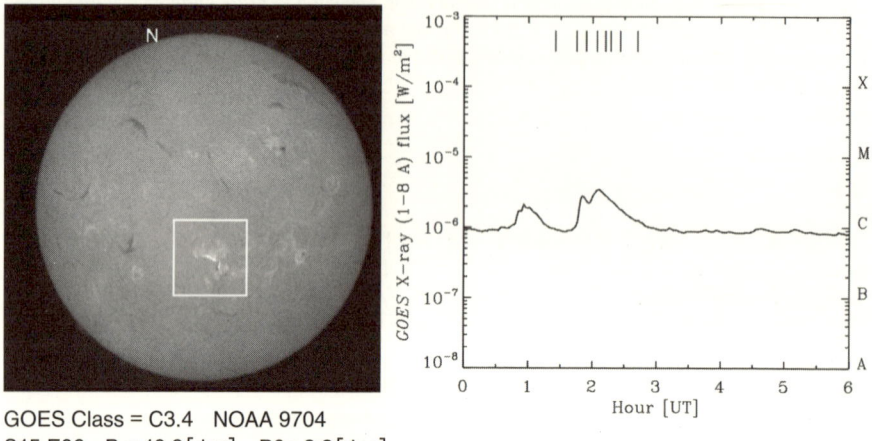

GOES Class = C3.4   NOAA 9704
S15,E06   P = 19.9 [deg]   B0= 2.2 [deg]

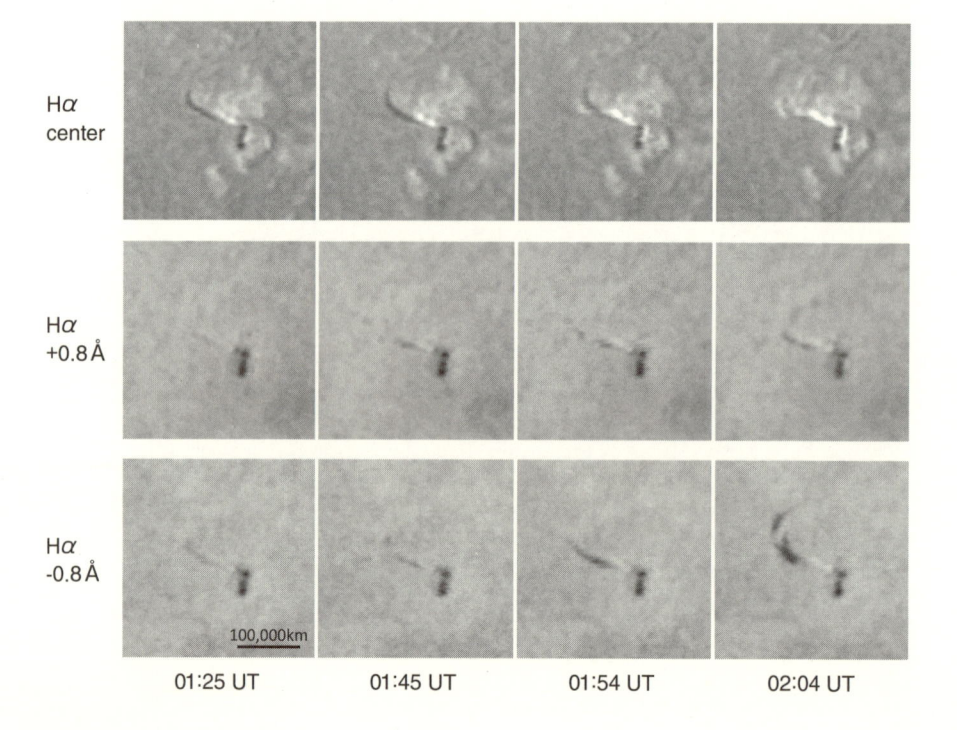

Hα center

Hα +0.8Å

Hα −0.8Å

100,000 km

01:25 UT   01:45 UT   01:54 UT   02:04 UT

The active region NOAA9704 was a long-lived one, which kept the complex magnetic structure of βγδ type for three solar rotations and produced many M-class flares. Tian et al. (2005) detected the continuous clockwise rotation of the region and argued that the continuous clockwise rotation of the long-lived active region may be a manifestation that a highly right-hand twisted and kinked flux tube was emerging through the photosphere and chromosphere into the corona. The event shown here was a C3.4 flare of sigmoid configuration which occurred near the magnetic neutral line in the region. The active region filament was driven to eject along the neutral line with a speed of 40-50 km/s and formed a long Hα filament. The elongation of the filament excited the brightening in EIT195 along the filament channel. From the radio burst observation of the region on the other day, Kundu et al. (2003) speculated that some kind of reconnection in the multiple flux systems that exist between the flaring active region and the erupting filament may be the trigger of the flare.

NOAA9704 は，太陽自転の3周期もの長い期間 βγδ 型の複雑な磁場構造を維持して，多数の M クラスのフレアを発生した領域である．Tian et al. (2005) は，この領域が長い期間にわたって時計回りに回転していることを観測し，そのことから右手系で捩じられてキンクができた磁束管が光球，彩層，コロナへと浮上してきているとの考えを提出している．このページで示されているイベントは，それらの一連の活動の一部であって，シグモイドとよばれる S 字型をした C3.4 クラスのフレアである．このとき，領域内の磁気中性線に沿って存在したフィラメントが，その中性線に沿う方向に 40-50 km/s の速度で放出され，長い Hα フィラメントとなった．これに応じて，フィラメントチャンネルにそって，EIT195 ではループ状の明るい構造が現れた．電波観測から，Kundu et al. (2003) は，この領域内の他の日時の同様なフレアの観測から，複数の磁気ループ相互間のリコネクションがフレア，フィラメント放出を引き起こしたとの考えを提出している．

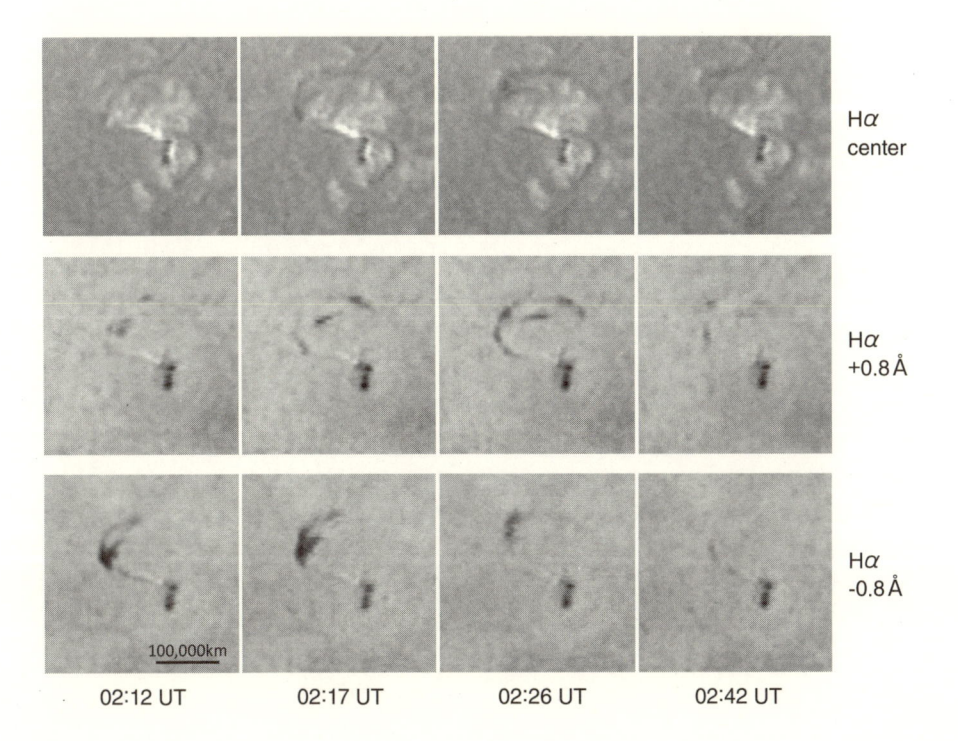

				Hα center
				Hα +0.8Å
				Hα −0.8Å
02:12 UT	02:17 UT	02:26 UT	02:42 UT	

238 | 第6章 代表的イベント

Dec 11, 2001 23:04−03:23 UT   Filament Activity

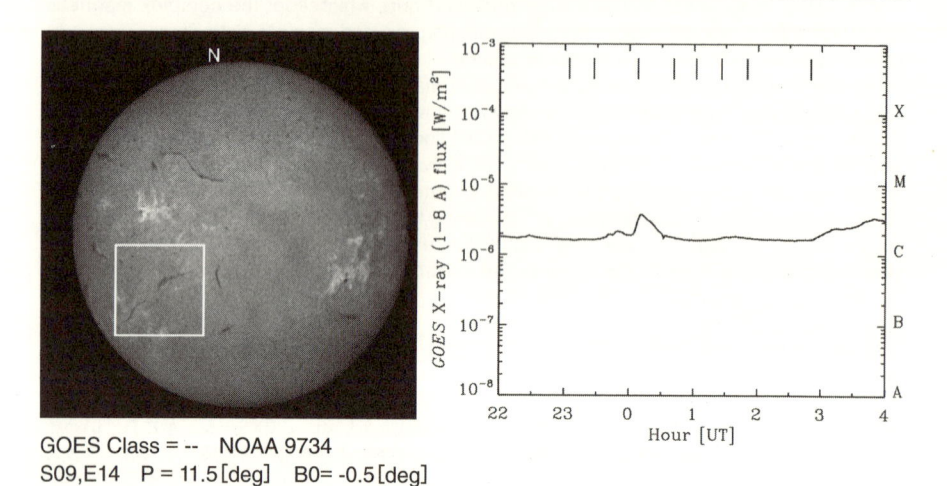

GOES Class = --   NOAA 9734
S09, E14   P = 11.5 [deg]   B0= -0.5 [deg]

Hα center

Hα +0.8Å

Hα -0.8Å

100,000km

23:04 UT   23:27 UT   00:08 UT   00:41 UT

A long Hα filament of 300 Mm length was activated for a period of 2.5 hours. While the overall morphology of the filament did not changed so much, internal gas motions along the axis of the filament were excited in the event. In EIT195 images, the filament was located in the central part of a large scale filament channel. A faint EIT195 loop was observed to appear along the filament in association with the activation.

30万 km もの長いフィラメントが2時間半にわたって活発化した現象である．フィラメント全体の形は変わらずに，フィラメント内部で軸方向に沿う形でガスの流れが観測された．EIT195 像では，このフィラメントは大規模なフィラメントチャンネルの中に位置しており，この活動の発生時に，淡い EIT ループがフィラメントに沿う形で現れた．

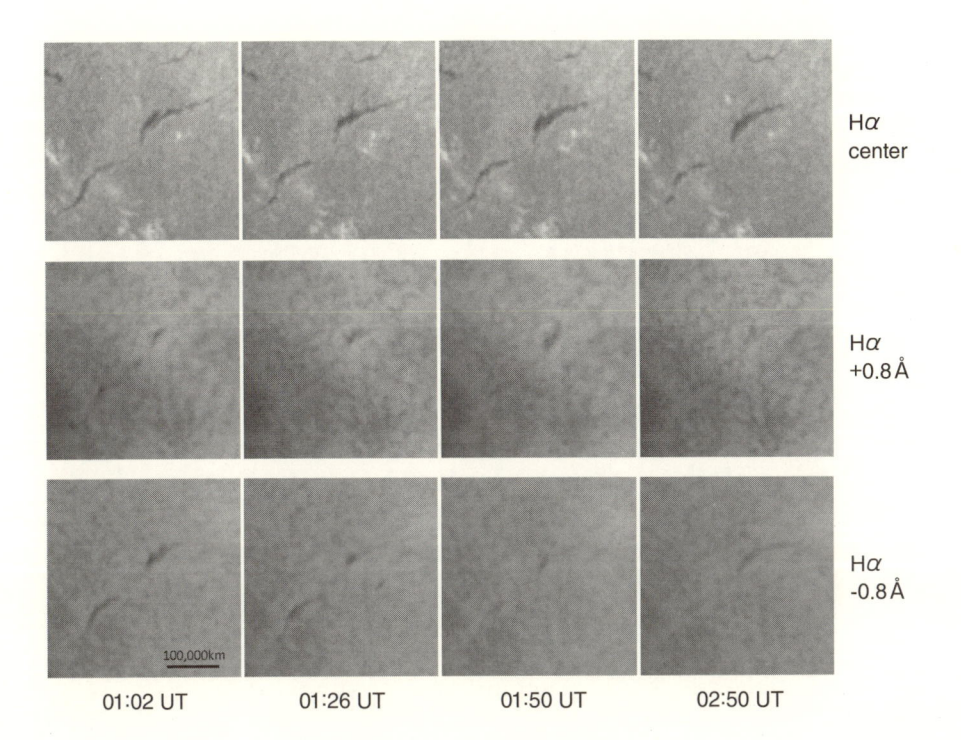

Dec 19, 2001　02:30－03:21A UT　Flare

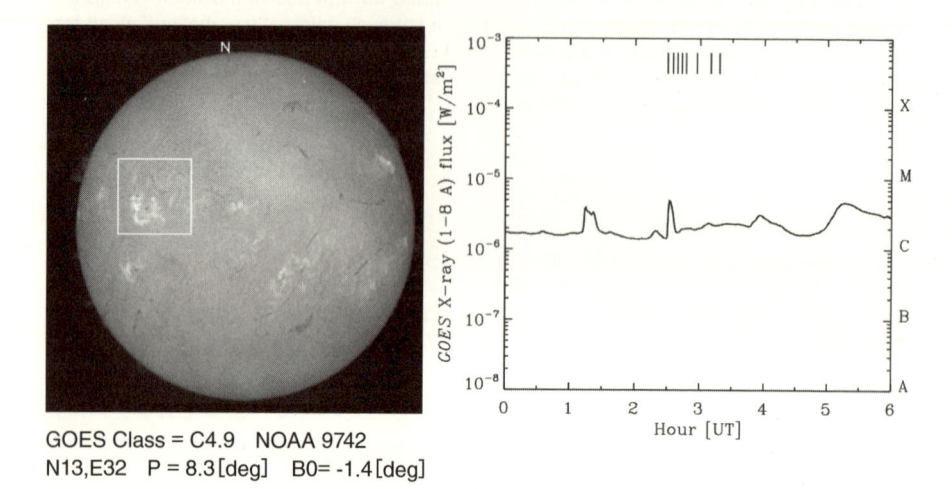

GOES Class = C4.9　NOAA 9742
N13,E32　P = 8.3[deg]　B0= -1.4[deg]

| | 02:30 UT | 02:35 UT | 02:39 UT | 02:43 UT |

Hα center

Hα +0.8Å

Hα -0.8Å

100,000km

Chapter 6 Representative Events | 241

In the complex βγ-type active region NOAA9742, there occurred a C4.9 flare of 10min duration. The flare was a compact one, triggered the mass ejection from the kernel and emitted a Moreton wave. The propagation of Moreton wave in 1 min cadence is displayed in the following pages.

βγ型の複雑な磁場構造を持つ活動領域 NOAA9742 で寿命 10 分の C4.9 フレアが発生した．このフレアはコンパクトなタイプであり，カーネル部からガスが噴出し，モートン波を放出した．モートン波の伝播の様子は次ページ以降に示されている．

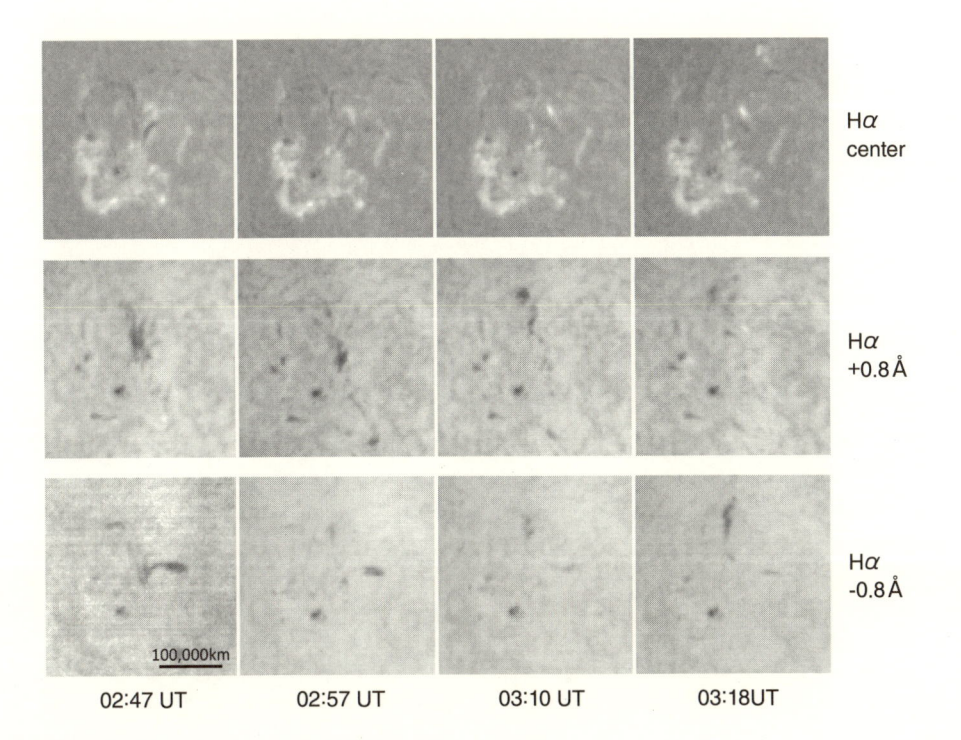

Hα center

Hα +0.8Å

Hα -0.8Å

100,000km

02:47 UT 02:57 UT 03:10 UT 03:18UT

242 | 第6章　代表的イベント

Dec 19, 2001　02:30−02:40 UT　Moreton Wave

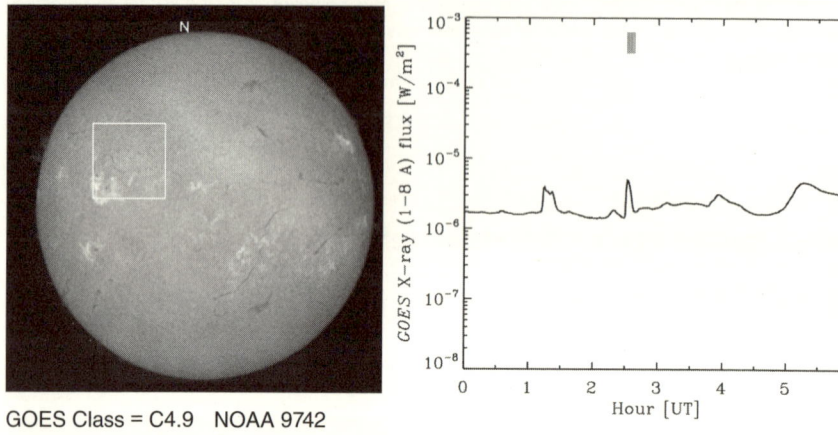

GOES Class = C4.9　　NOAA 9742
N13,E32　　P = 8.3[deg]　　B0= -1.4[deg]

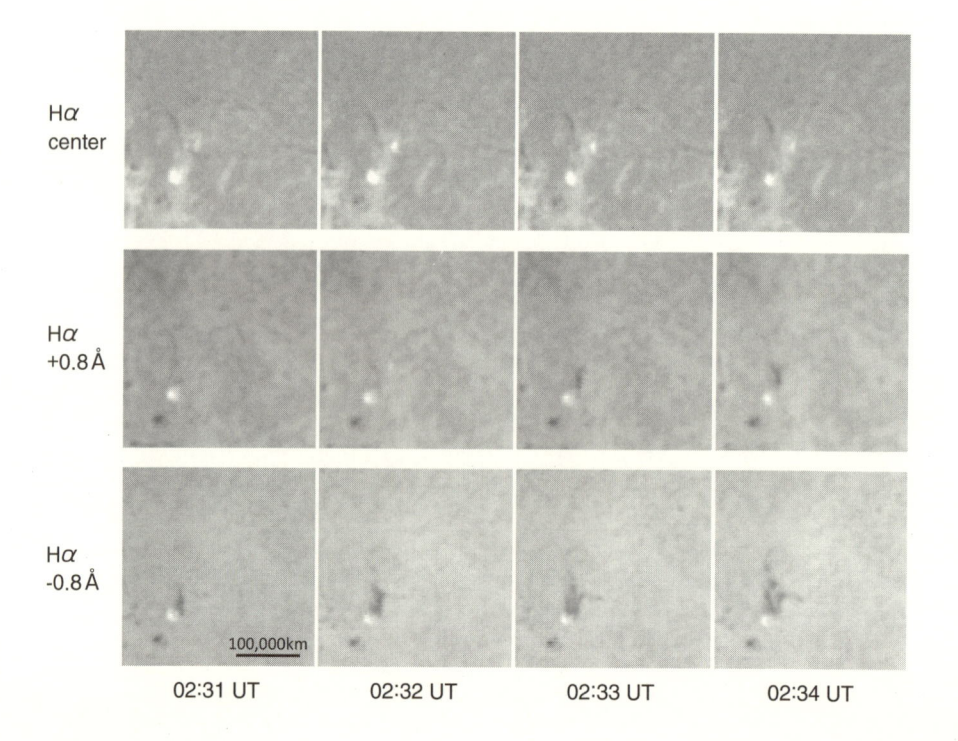

Hα center

Hα +0.8Å

Hα -0.8Å

100,000km

02:31 UT　　02:32 UT　　02:33 UT　　02:34 UT

Chapter 6 Representative Events | 243

Here is shown the temporal evolution of the Moreton wave emitted by the C4.9 class compact flare in NOAA9742. The wave propagated with a speed of 650 km/s. (In the next two pages, time-difference images to enhance the visibility of wave fronts are shown with the manually drawn sketch of the wave fronts.)

NOAA9742 で発生した C4.9 クラスのフレアに伴うモートン波の伝播の様子が示されている。その伝搬速度は約 650 km/s であった。（次の 2 ページでは，波面を見やすくするように，画像の明るさの時間変化像を示している．また波面位置のスケッチも参考のために付してある．）

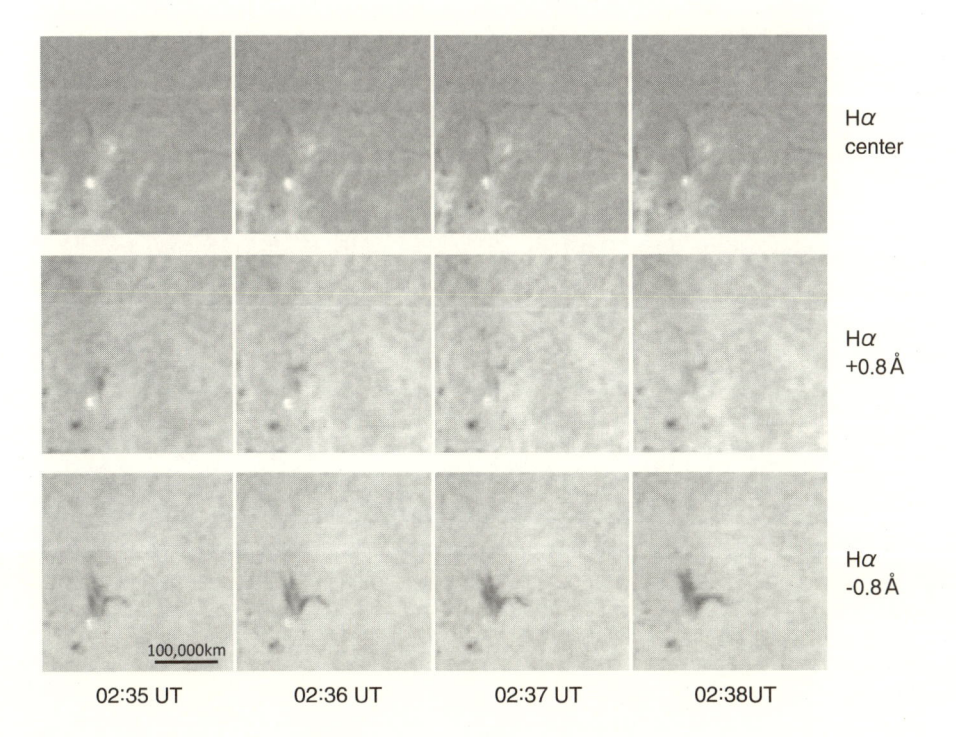

Hα center

Hα +0.8Å

Hα −0.8Å

100,000 km

02:35 UT　　02:36 UT　　02:37 UT　　02:38 UT

244 | 第6章　代表的イベント

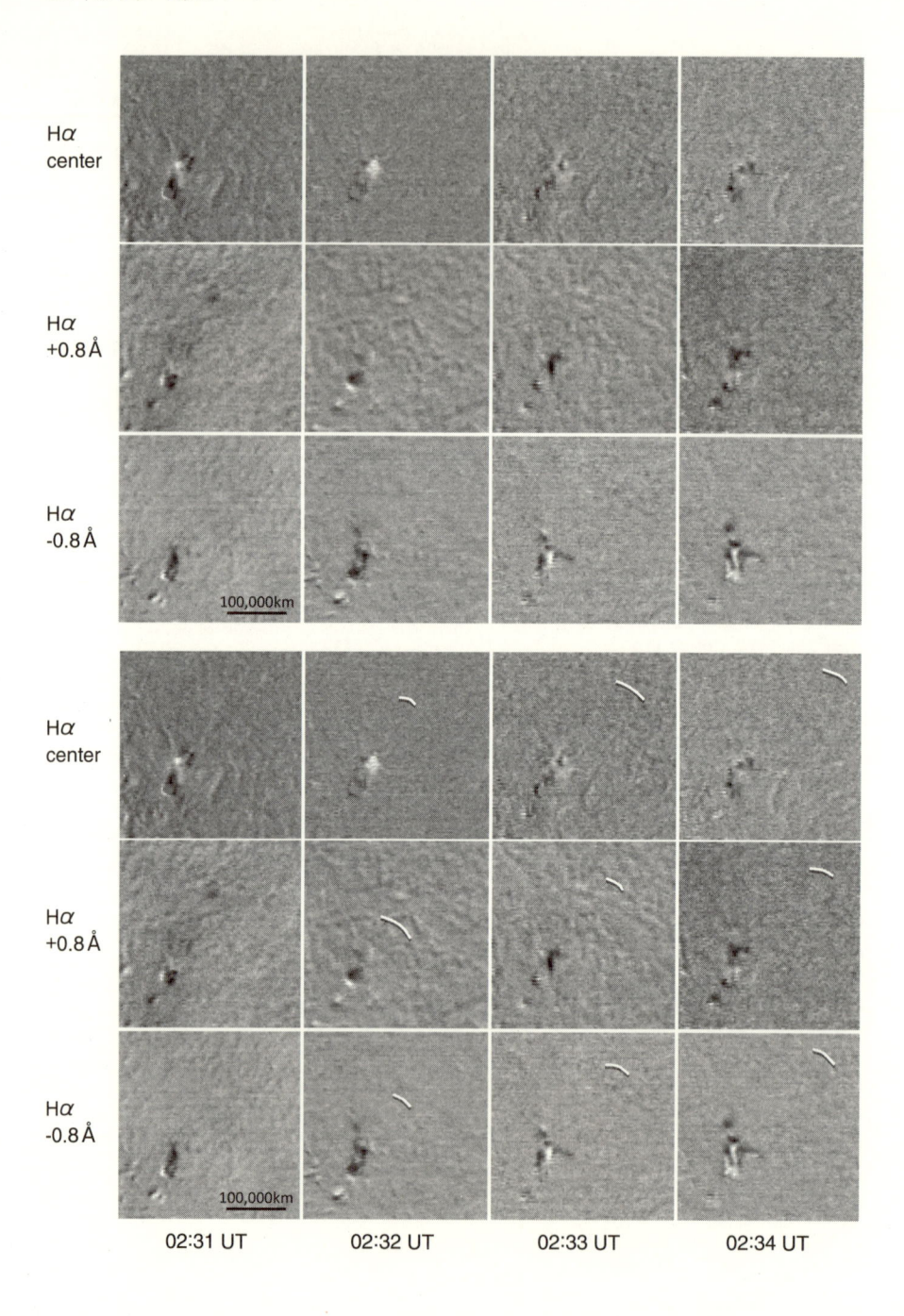

Hα center

Hα +0.8Å

Hα −0.8Å

100,000km

Hα center

Hα +0.8Å

Hα −0.8Å

100,000km

02:31 UT　　02:32 UT　　02:33 UT　　02:34 UT

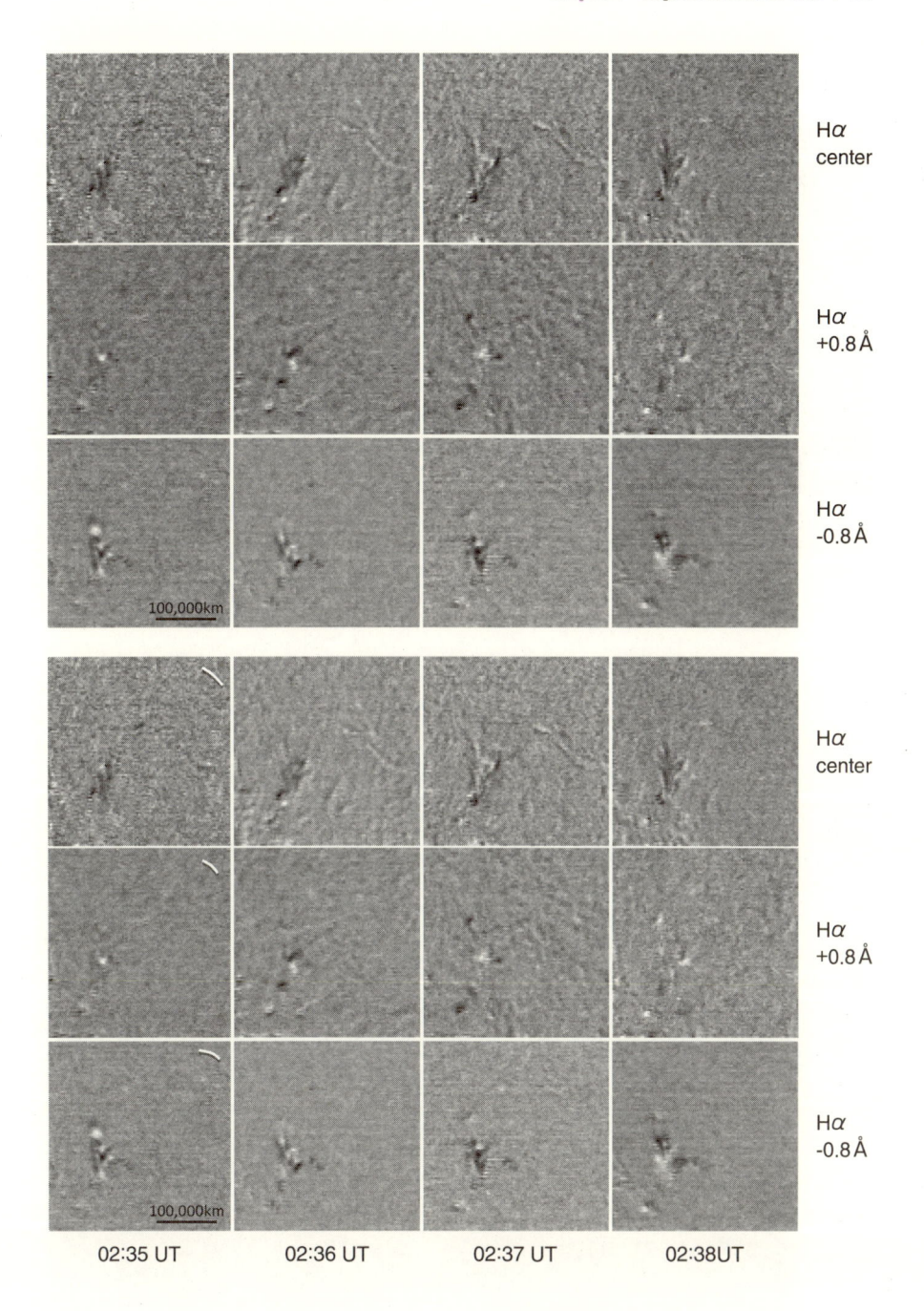

Chapter 6 Representative Events | 245

Hα center

Hα +0.8Å

Hα -0.8Å

100,000km

Hα center

Hα +0.8Å

Hα -0.8Å

100,000km

02:35 UT   02:36 UT   02:37 UT   02:38 UT

246 | 第6章 代表的イベント

Feb 04, 2002 05:45−07:20A UT   Flare, Filament Eruption

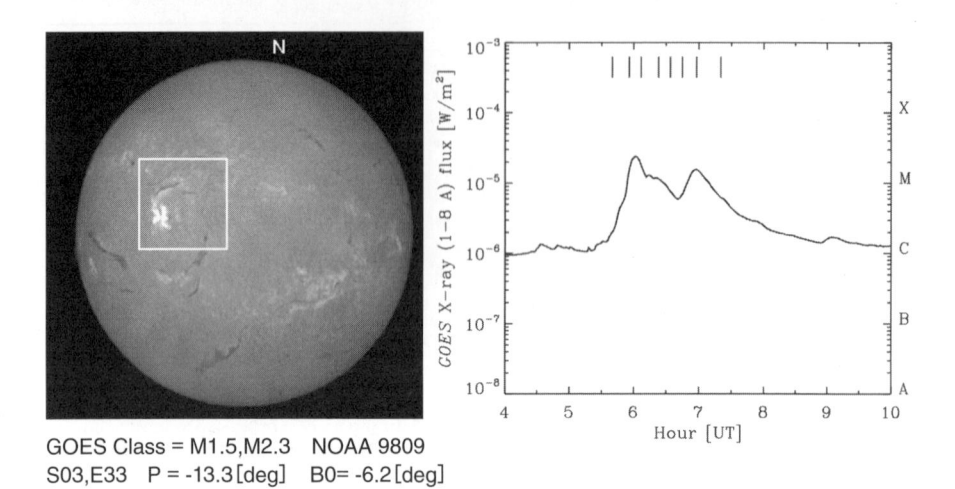

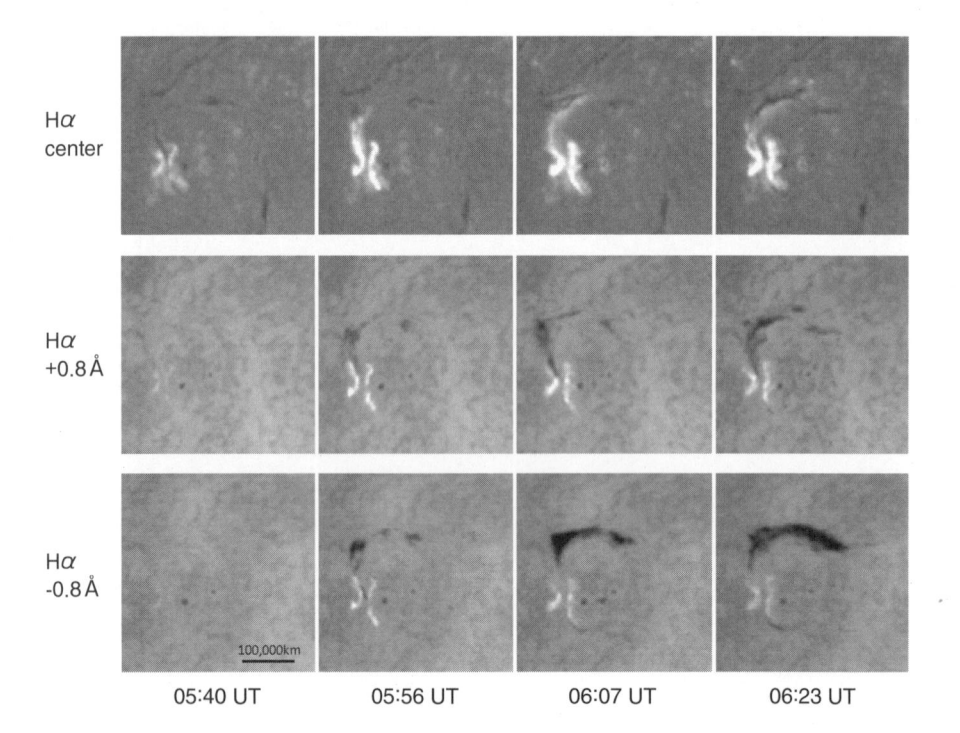

GOES Class = M1.5, M2.3   NOAA 9809
S03, E33   P = -13.3 [deg]   B0= -6.2 [deg]

Hα center

Hα +0.8Å

Hα -0.8Å

100,000km

05:40 UT　　05:56 UT　　06:07 UT　　06:23 UT

Two large-scale and bright two-ribbon flares occurred consecutively in a β-type active region NOAA9809. Associated with the first flare, an explosive mass ejection with an initial speed of 150 km/s was driven from the end part of a pre-existent central filament along the direction of the filament axis. Bright loops in EIT195 appeared in parallel to the Hα filament. The ejected gas extended nearly parallel to the solar surface, excited another filament channel and formed a very long filament of 300 Mm with active internal motions. It showed a signature of rotation around the axis. The second flare was occurred in a way to re-brighten the ribbons of the first one. In this case, EIT195 images showed the arcade formation of bright loops connecting the two ribbons.

ベータ型の黒点 NOAA9809 で，大規模で明るいツーリボンフレアが 2 回連続して発生した．最初のフレアに伴って，2 つのリボン間に元々あったフィラメントの一部がフィラメントの軸方向に爆発的に放出された．このとき EIT195 像では，フィラメントに沿う形の明るいループ構造が現れていた．放出ガスは，初期速度が約 150 km/s であり，太陽表面に沿うような形で伸びた．この放出ガスは遠方のフィラメントチャンネルを刺激して，30 万 km の長さにわたって活発な内部運動を示す長大なフィラメントを形成した．その際，フィラメント軸の周りの回転運動を示した．2 回目のフレアは，1 回目のリボンを再び輝化するような型であった．ただし，EIT195 像では，2 つのリボンを結ぶアーケード状ループを形成した．

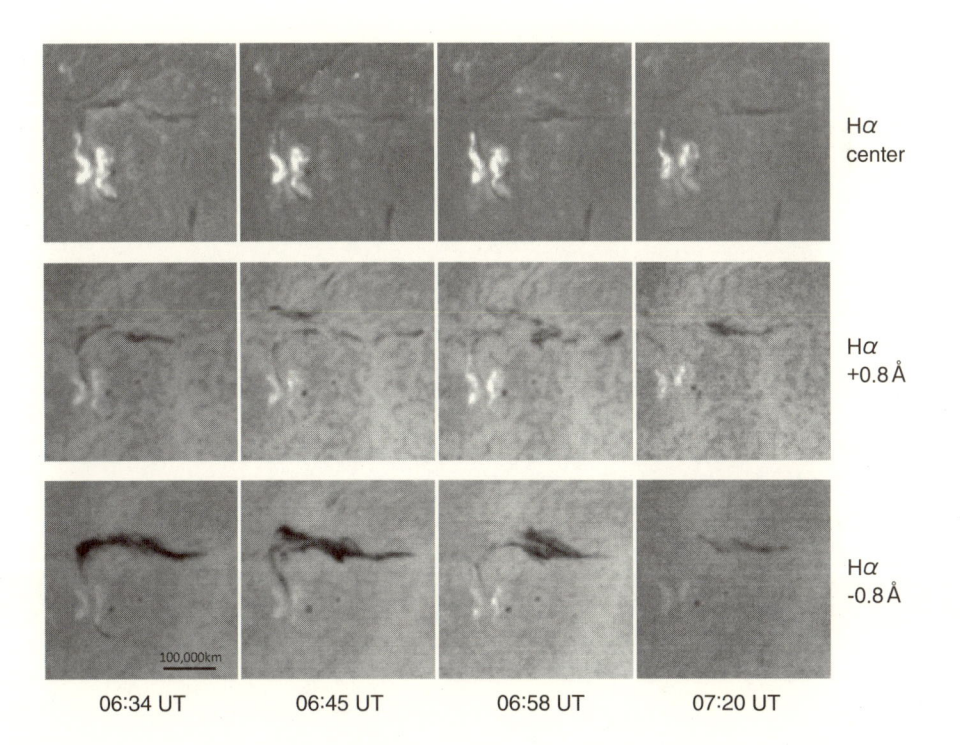

06:34 UT	06:45 UT	06:58 UT	07:20 UT	Hα center / Hα +0.8Å / Hα -0.8Å

100,000km

248 | 第6章 代表的イベント

Mar 17, 2002 23:56−01:54A UT   Two Ribbon Flare in Spotless Region

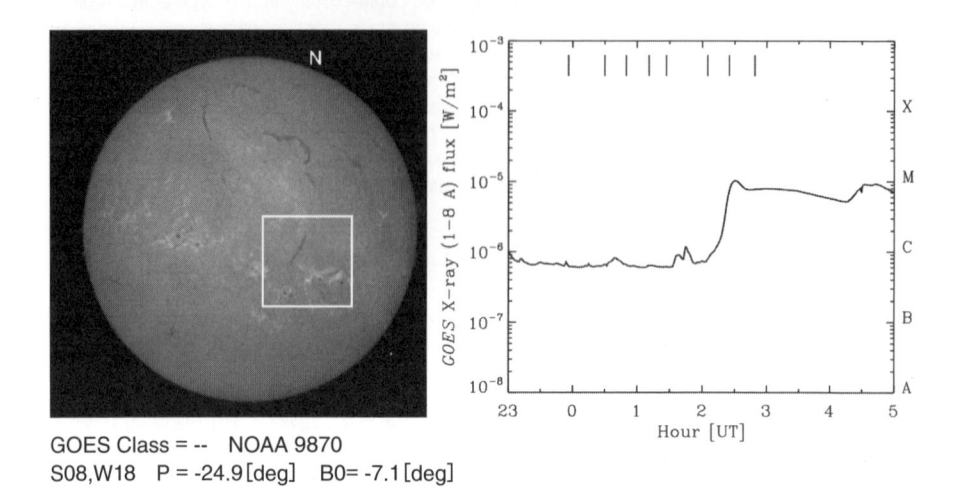

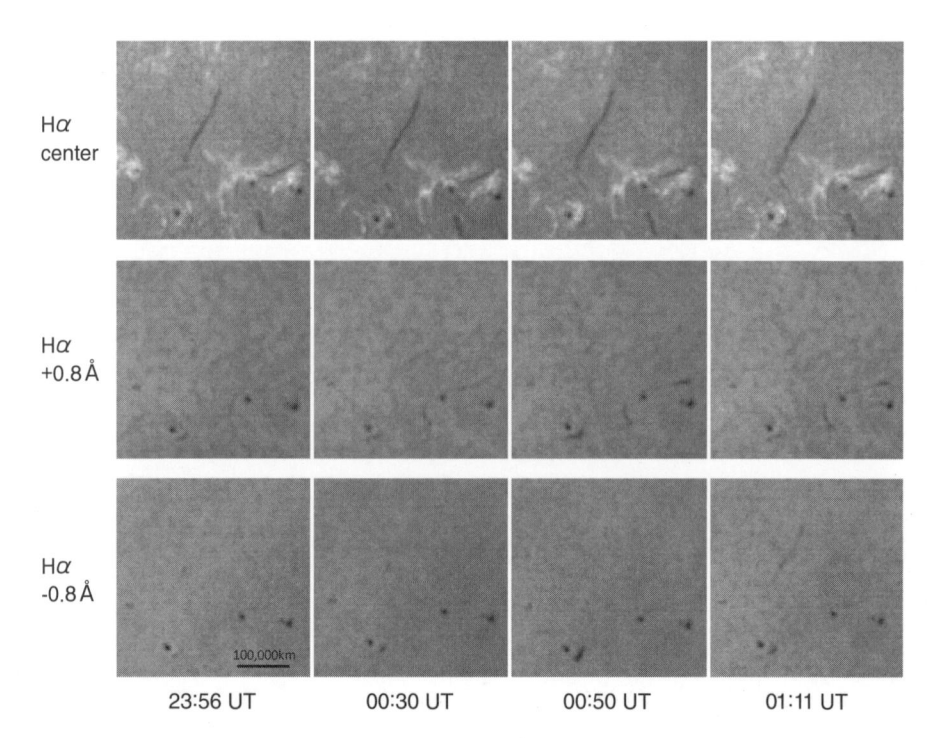

GOES Class = --   NOAA 9870
S08,W18   P = -24.9[deg]   B0= -7.1[deg]

Hα center

Hα +0.8Å

Hα -0.8Å

100,000km

23:56 UT   00:30 UT   00:50 UT   01:11 UT

A filament located among three active regions (NOAA 9866, 9870 and 9873) disappeared in 2 hours. NOAA9866 (west side) was a δ-type active region, while NOAA9873 (east side) was a newly emerged region of 2-days-old. After the disappearance, two ribbon-patches started to gradually brighten at both sides of the disappeared filament. Bright arcade loops in EIT195 images were observed to develop over the location of the filament. Nearly simultaneously with the two-ribbon brightening, another M1.0 flare occurred in NOAA9866. An associated CME event was observed by the LASCO/C2. (Note: Dark shadows in the lower right corner of the images are not real but due to instrumental troubles.)

3つの活動領域（NOAA 9866, 9870 および 9873）に囲まれた領域にあったフィラメントが2時間程度の時間をかけて消滅した．西側のNOAA9866はδ型で活発な領域であり，東側のNOAA9873は2日前に誕生した若い領域である．フィラメント消失後，フィラメントのあった場所の両脇にツーリボン状の領域がゆっくりと輝き出した．明るいアーケードループがフィラメントのあった場所に形成されてゆくのが EIT/195 で観測された．このツーリボン状の領域の発生とほぼ同時に南西の活動領域で別のM1.0フレアが発生した．また，この現象にCMEが伴っていることもLASCO/C2で観測された．（注：画像の右下角にある影はフィルターの一時的なトラブルのためである．）

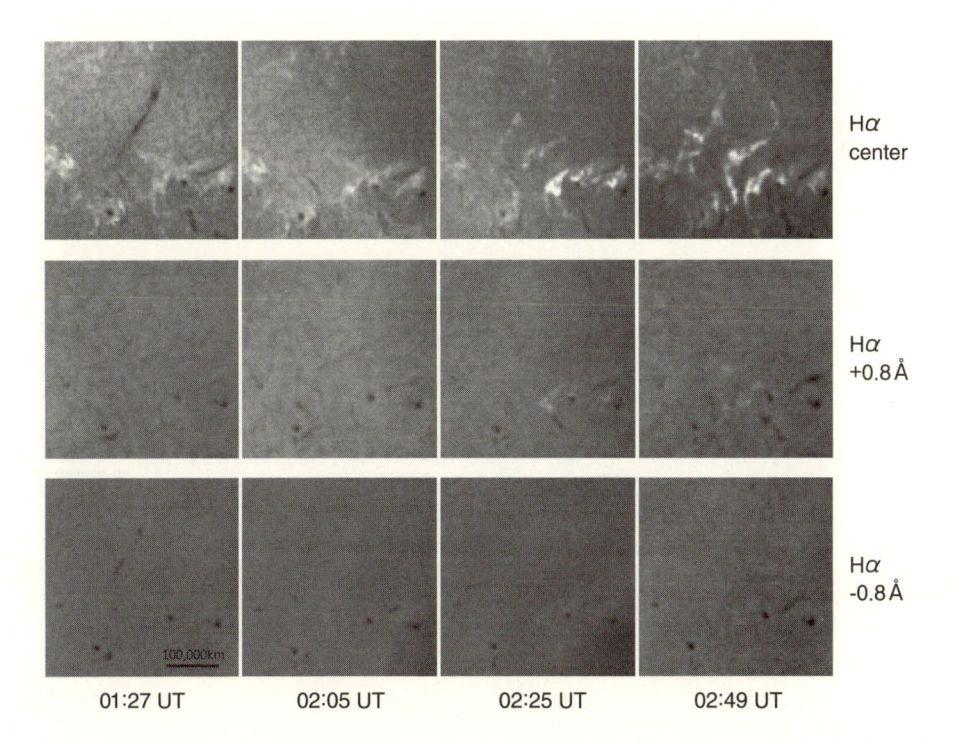

250 | 第 6 章 代表的イベント

Apr 04, 2002 01:53−05:32 UT   Two Ribbon Flare in Spotless Region

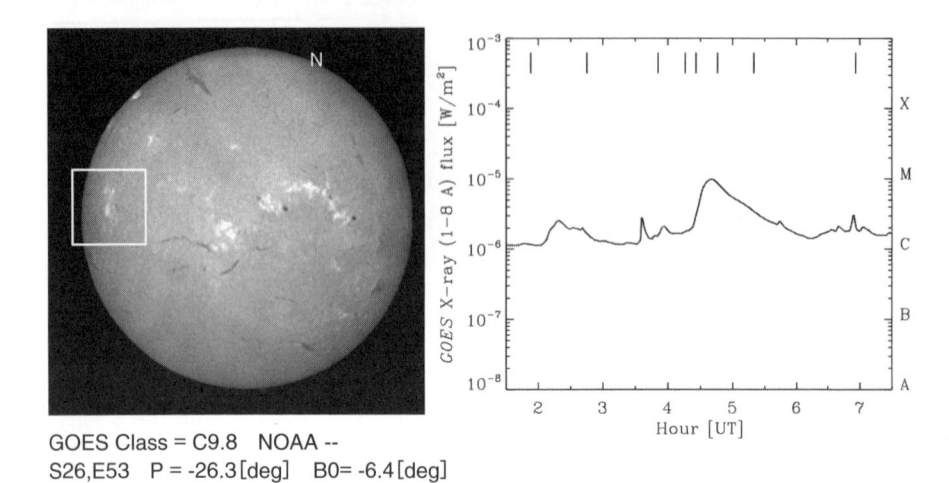

GOES Class = C9.8   NOAA --
S26, E53   P = -26.3 [deg]   B0 = -6.4 [deg]

| Hα center |
| Hα +0.8 Å |
| Hα −0.8 Å |

100,000 km

01:53 UT   02:45 UT   03:51 UT   04:16 UT

A two-ribbon C9.8 flare occurred after the eruption of the active region filament in a remnant of an old active region, where there remained no sunspots. The filament, as a precursor of the flare, showed a complex internal motion during 2 hours before the flare explosion. Bright arch formation in SOHO/EIT195 images was detected and a CME was clearly observed by LASCO/C2 in association with the flare.

黒点の消滅した古い活動領域に残された活動領域フィラメントが爆発的に上昇して消滅した．それに伴って C9.8 クラスのツーリボンフレアが発生した．フレア発生前の 2 時間ほどの間，先駆的なフィラメントガスの複雑な運動が観測された．このフレアに伴って，SOHO/EIT195 で明るいアーチ状の構造が現れたことが観測された．また，明るい CME も発生していることが LASCO/C2 で確認されている．

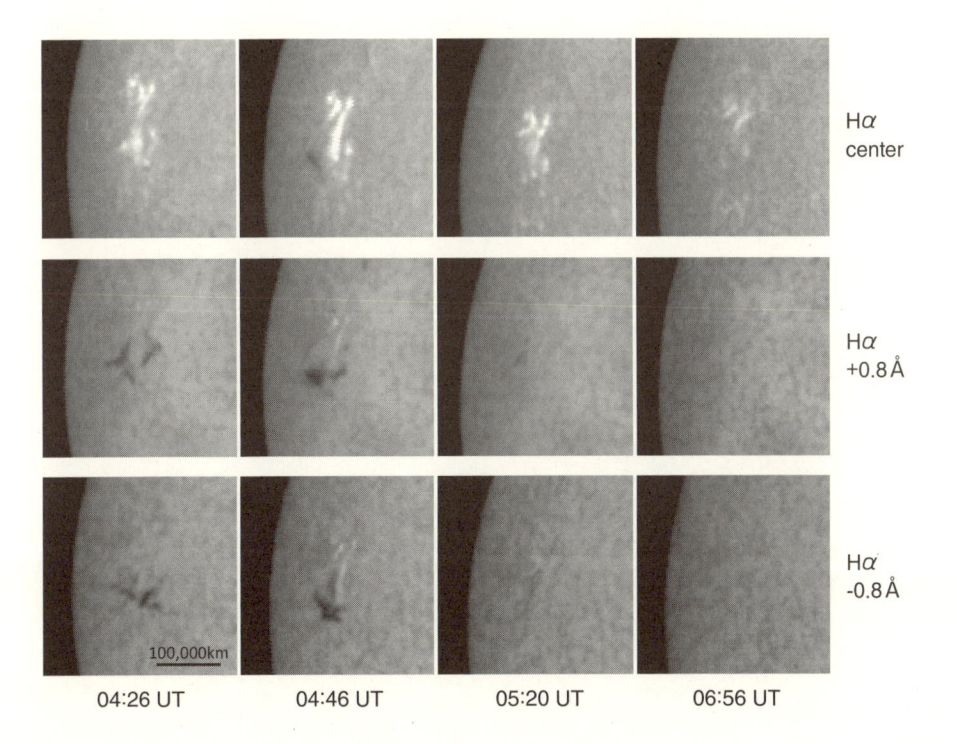

Apr 04, 2002 06:46−07:44 UT   Flare, Filament Eruption

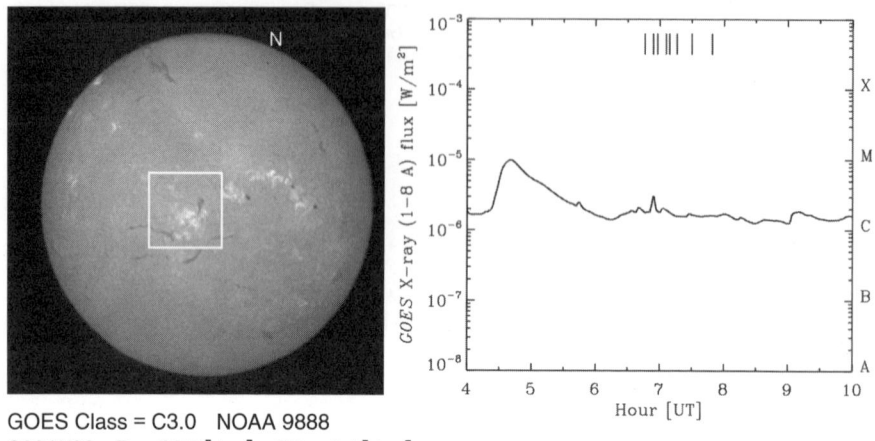

GOES Class = C3.0   NOAA 9888
S08,W00   P = -26.3[deg]   B0= -6.4[deg]

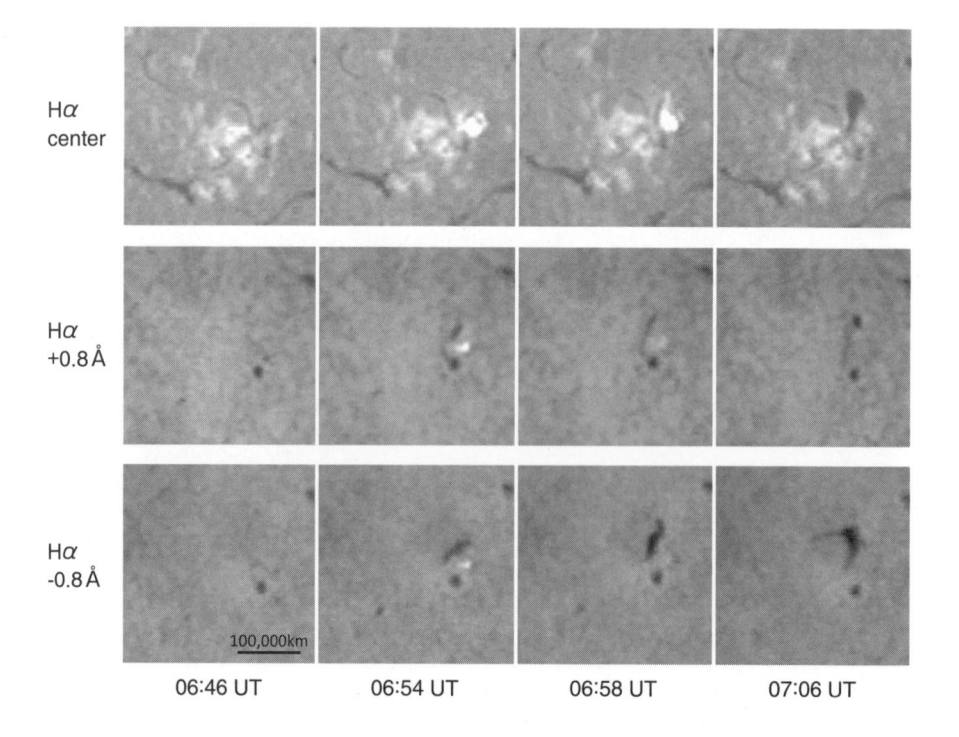

Hα center

Hα +0.8Å

Hα -0.8Å

100,000km

06:46 UT   06:54 UT   06:58 UT   07:06 UT

A short-lived compact flare with a lifetime of 12 min occurred in the β-type active region NOAA9888. The location of the flare was near the satellite polarity area of the preceding spot. Simultaneously a part of the pre-existent active region filament was destabilized and ejected with an extension speed of 80 km/s. After the 15 min of extension phase, the ejected gas returned back along the same trajectory to the solar surface and faded away.

β型の活動領域 NOAA9888 で寿命 12 分弱のコンパクトフレアが発生した．フレアは先行黒点周辺のサテライト極性域で発生した．このとき，近傍にあった既存の活動領域フィラメントの一部が不安定となり，約 80 km/s の速度で放出された．放出されたガスは約 15 分間延伸した後，ほぼ同じ軌道を辿って戻った後消滅した．

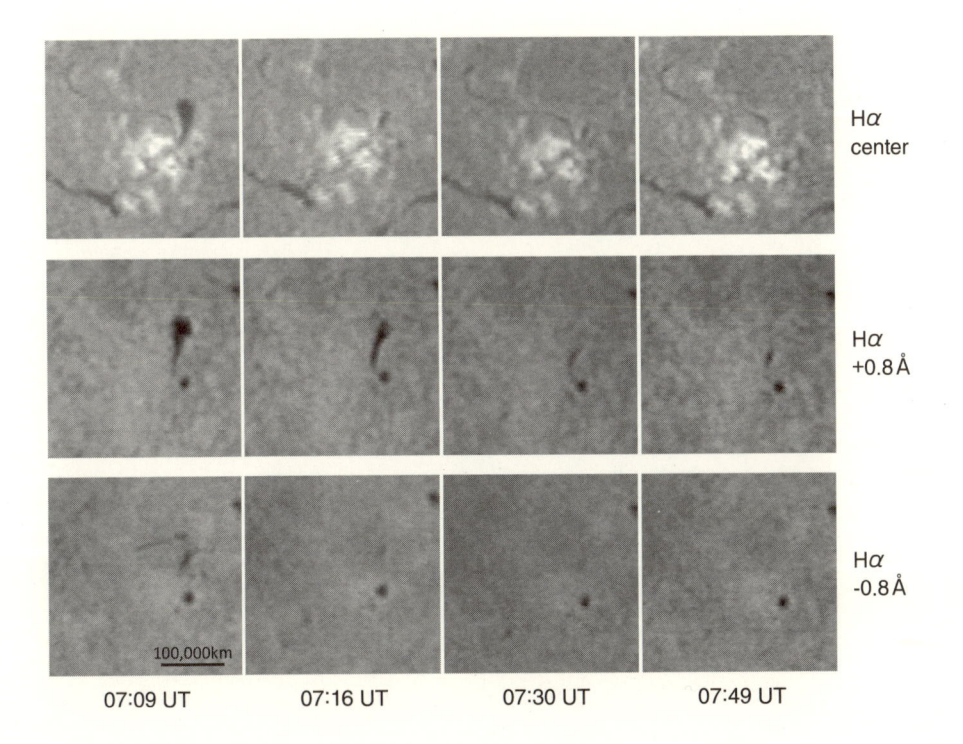

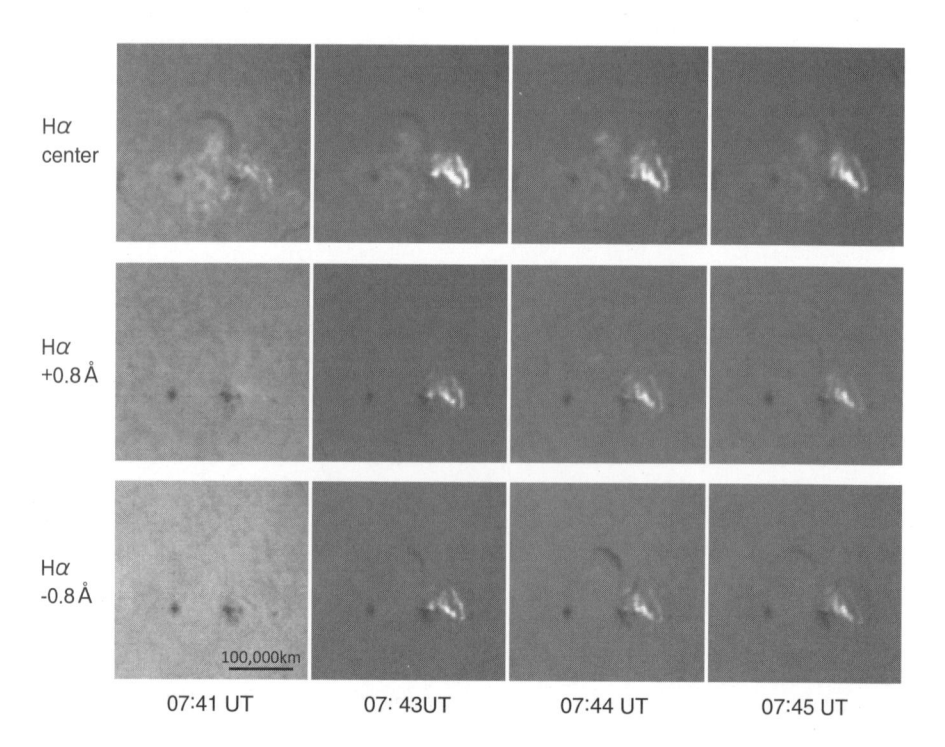

Jul 18, 2002 07:41−08:18A UT　Flare

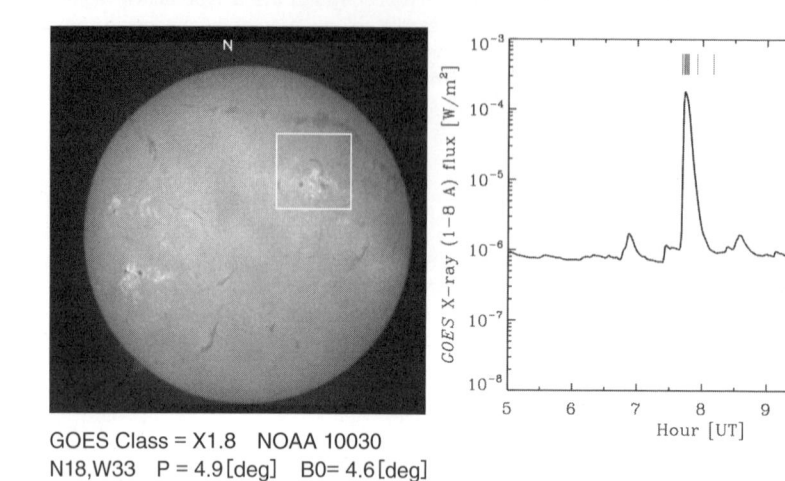

GOES Class = X1.8　NOAA 10030
N18,W33　P = 4.9 [deg]　B0= 4.6 [deg]

Hα center

Hα +0.8Å

Hα −0.8Å

100,000km

07:41 UT　　07:43UT　　07:44 UT　　07:45 UT

The active region NOAA10030 was a flare-productive $\beta\gamma\delta$-type active region. The large preceding spot showed a counter-clockwise motion during the highly active phase of 6 days and helicity accumulation (Park et al. 2008; Tian et al. 2008). In the event shown here, it produced an X1.8 flare of 30min duration. In the later phase of the flare, a big surge of gas was ejected with an apparent speed of 100 km/s. Associated with the flare eruption, it was observed that an active region filament in the northern part of the active region was driven to oscillate with a period of about 9 min.

活動領域 NOAA10030 は，βγδ型の領域でフレアを頻発した．6 日間にわたる活発なフレア活動を示した時期に，大きな先行黒点が反時計回りの回転運動を示し，磁気ヘリシティーが蓄積されていた（Park et al. 2008; Tian et al. 2008）．ここで示されているイベントでは，30 分の寿命のX1.8 クラスフレアの様子が示されている．フレアの後期には，100 km/s の速さを示したサージガス噴出が観測されている．また，フレアに伴って，活動領域北側にあった活動領域フィラメントが周期約 9 分で振動しているのが観測された．

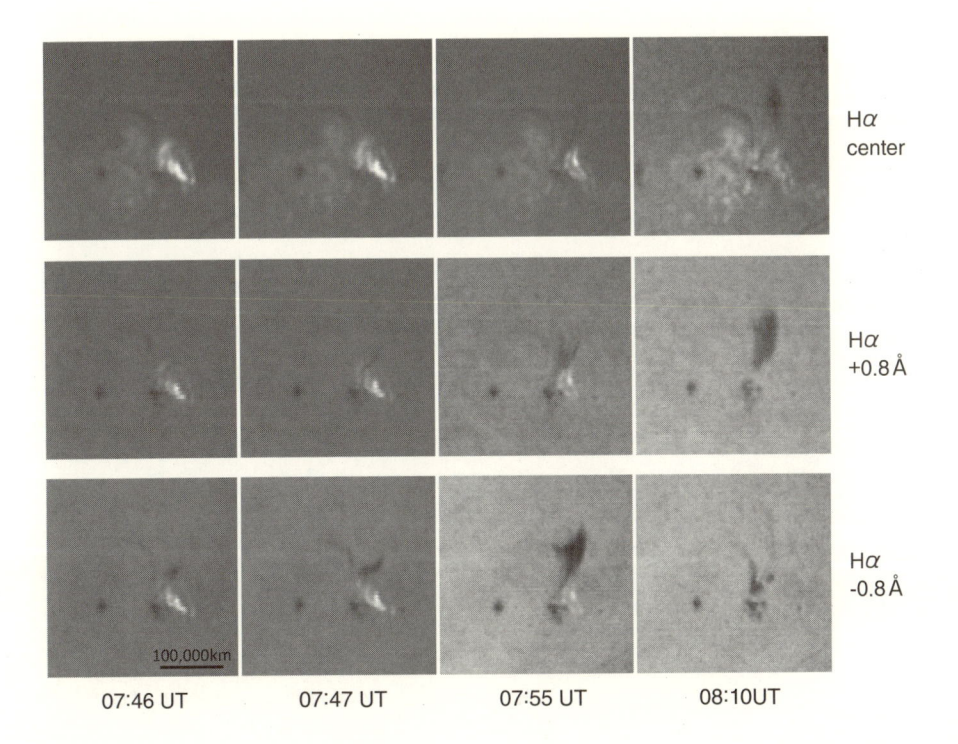

Jul 18, 2002 07:53−07:55A UT　Winking Filament

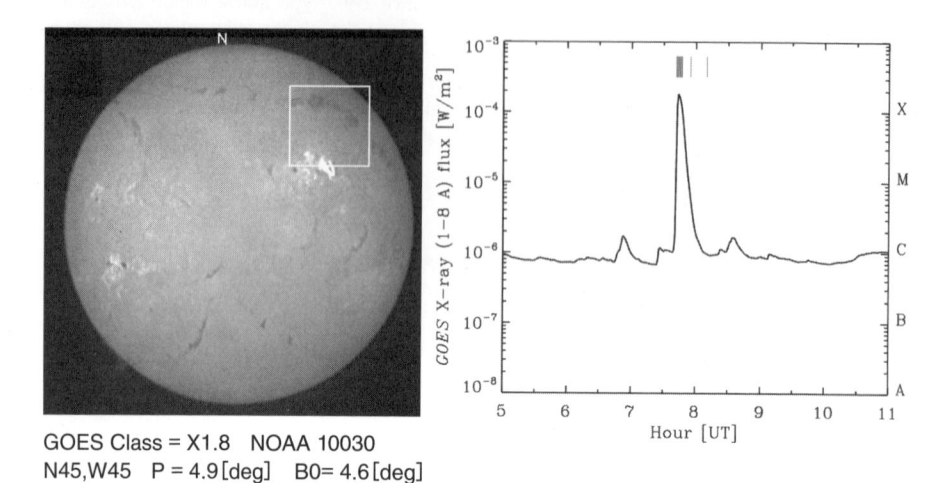

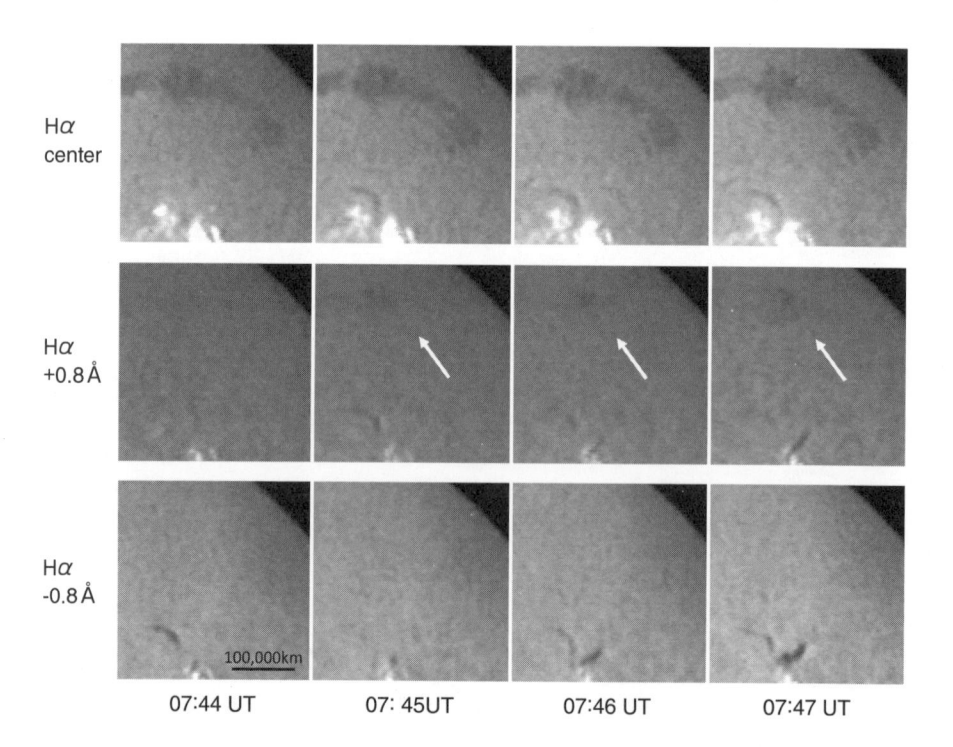

GOES Class = X1.8　NOAA 10030
N45,W45　P = 4.9 [deg]　B0 = 4.6 [deg]

Hα center

Hα +0.8Å

Hα −0.8Å

100,000km

07:44 UT　　07:45 UT　　07:46 UT　　07:47 UT

In the event shown here, the $\beta\gamma\delta$-type active region NOAA10030 produced an X1.8 flare of 30min duration. In the later phase of the flare, a big surge of gas was ejected with an apparent speed of 100 km/s. Associated with the flare eruption, it was observed that an active region filament in the northern part of the active region was driven to oscillate with a period of about 9 min.

ここで示されているイベントは，$\beta\gamma\delta$型の活動領域 NOAA10030 で発生した寿命 30 分の X1.8 クラスフレアである．フレアの後期には，100 km/s の速さを示したサージガス噴出が観測されている．また，フレアに伴って，活動領域北側にあった活動領域フィラメントが周期約 9 分で振動しているのが観測された．

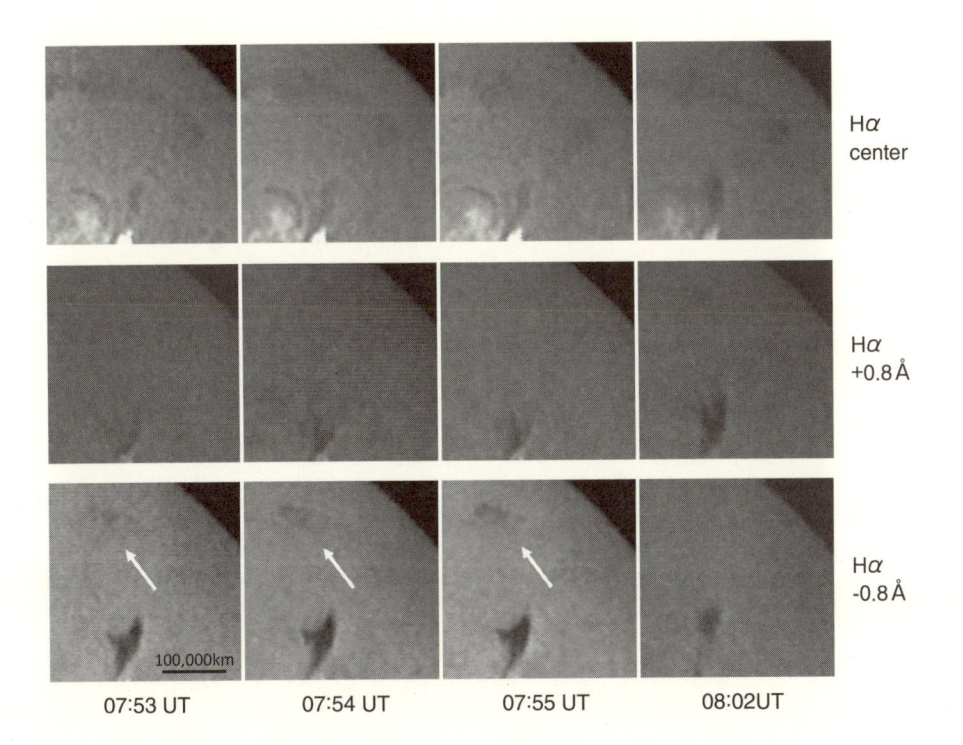

Hα center

Hα +0.8Å

Hα -0.8Å

100,000km

07:53 UT　　07:54 UT　　07:55 UT　　08:02UT

258 | 第6章 代表的イベント

Aug 16, 2002 05:54−07:00 UT   Filament Activity

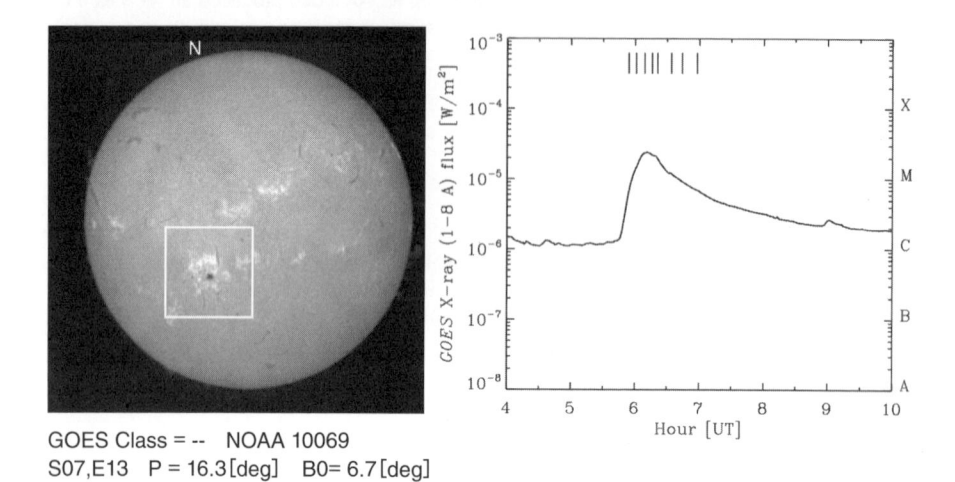

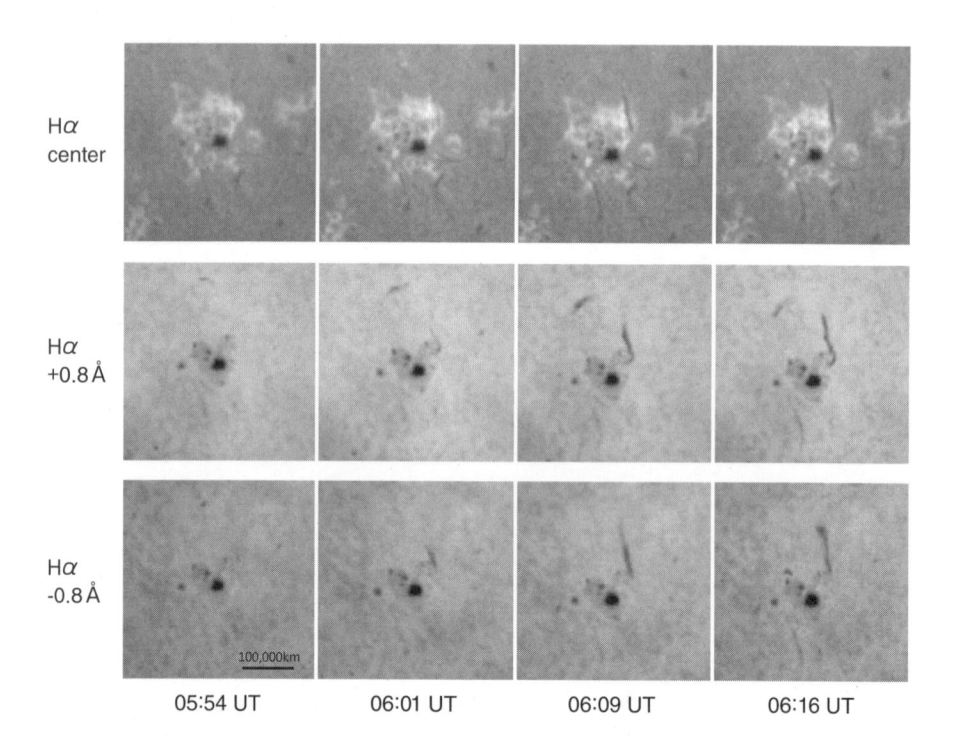

GOES Class = --   NOAA 10069
S07,E13   P = 16.3[deg]   B0= 6.7[deg]

Hα center

Hα +0.8Å

Hα −0.8Å

100,000km

05:54 UT   06:01 UT   06:09 UT   06:16 UT

NOAA10069 developed to the δ-type magnetic configuration on Aug 14, and produced an M6 class flare in the S-E part of the region around 12:00 UT on Aug 16. The event shown here is a filament activity that occurred in a displaced part from the flare site 6 hours before the flare. Along the northern periphery of the active region, there appeared a semi-circular filament. It showed internal motions along the filament for about 20 min and then faded out. In SOHO/EIT190 images, a similar shaped filament was observed to brighten in association with the cool Hα filament. (Note: The M2 flare shown in the GOES plot around 06:00 UT occurred in another displaced region NOAA10061 near the west limb and not related to the present event.)

NOAA10069 は，8月14日からδ型となって複雑な磁場構造となった．そして，8月16日の 12:00 UT 頃，領域の南東部で M6 クラスのフレアを発生した．このページで表示されている現象は，この領域内の別の場所でフレアの6時間ほど前に発生したフィラメント活動である．領域の北側の境界に沿って円弧状のフィラメントが発生し，20分ほどの間内部運動を示した後，消滅した．このとき，同じ円弧状の明るいフィラメントが SOHO/EIT190 像でも観測された．（なお，06:00 UT 頃の M2 クラスのフレアは西リムにあった NOAA10061 で発生したもので，このイベントとは無関係である．）

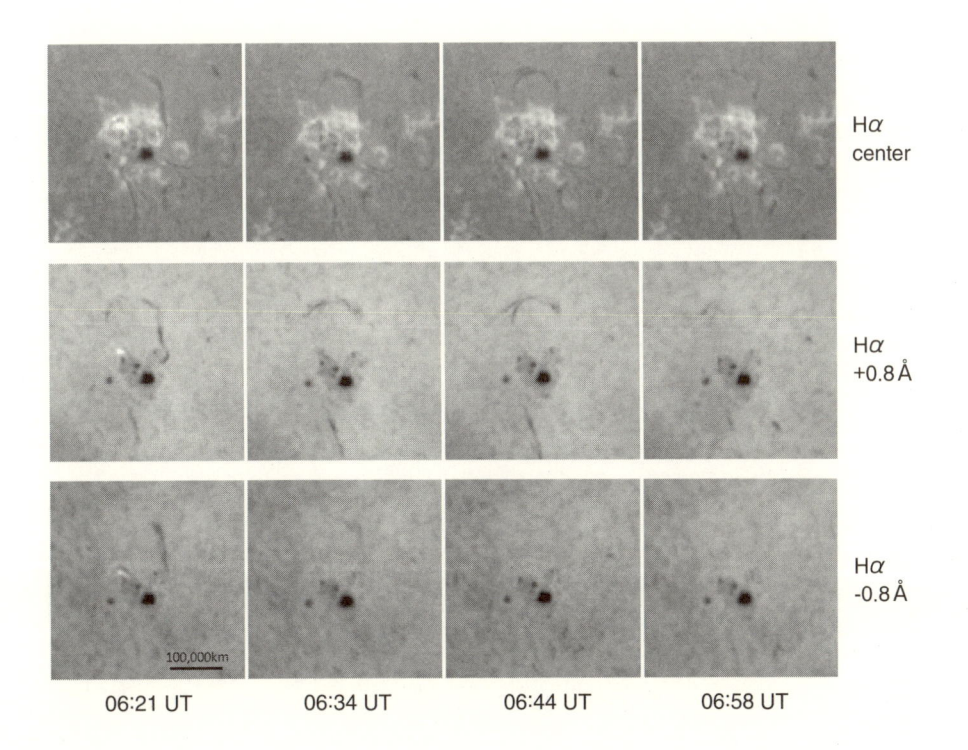

Hα center

Hα +0.8 Å

Hα −0.8 Å

06:21 UT　　06:34 UT　　06:44 UT　　06:58 UT

Aug 21, 2002 05:31−06:20 UT   Flare, Surge

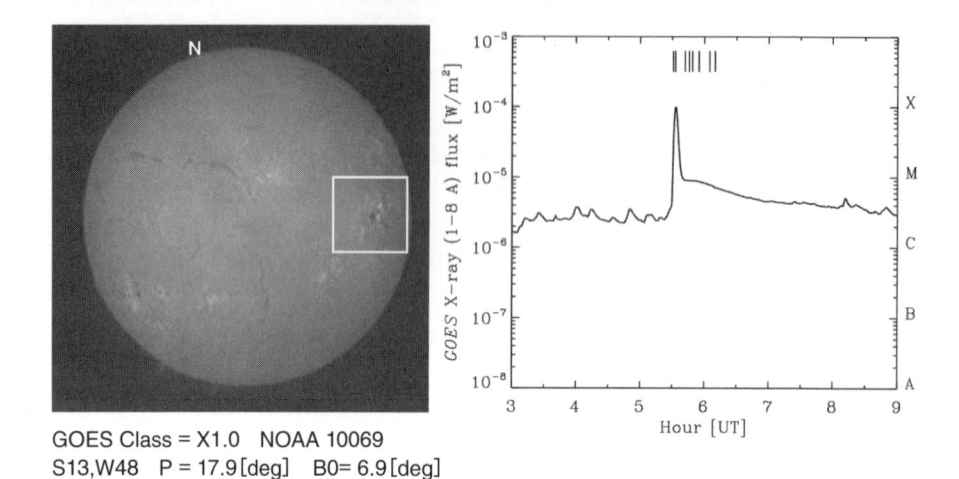

GOES Class = X1.0   NOAA 10069
S13,W48   P = 17.9 [deg]   B0= 6.9 [deg]

Hα center

Hα +0.8 Å

Hα −0.8 Å

100,000 km

05:31 UT   05:33 UT   05:42 UT   05:46 UT

The event was a strong and compact X1.0 flare class in the region NOAA10069 of high activity. After the start of the flare, it triggered the Hα surge mass ejection with a speed of around 80 km/s. SOHO/EIT195 observed the ejection of hot gas nearly along the direction of Hα gas.

このイベントは，非常に活発な NOAA10069 で起こった X1.0 の強いコンパクトフレアである．フレア発生後，80 km/s の速度で Hαガスがサージとして放出された．SOHO/EIT195 でも，高温のガスが Hαガスとほぼ同じ方向に放出されたのが観測されている．

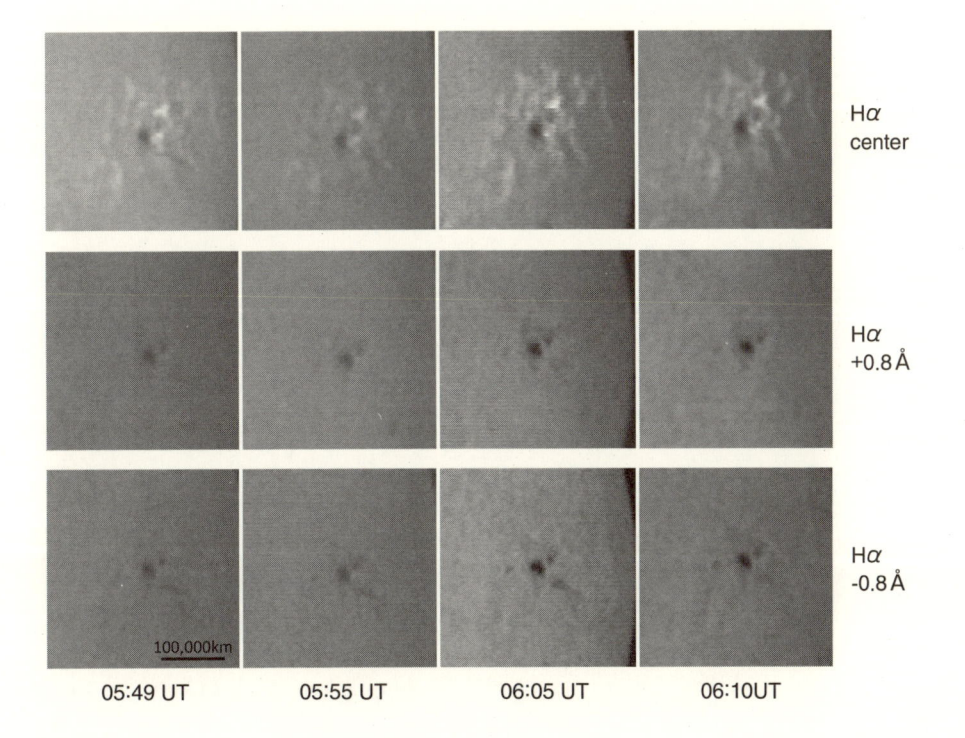

				Hα center
				Hα +0.8Å
				Hα -0.8Å
05:49 UT	05:55 UT	06:05 UT	06:10 UT	

Aug 22, 2002 01:13−02:57A UT Flare

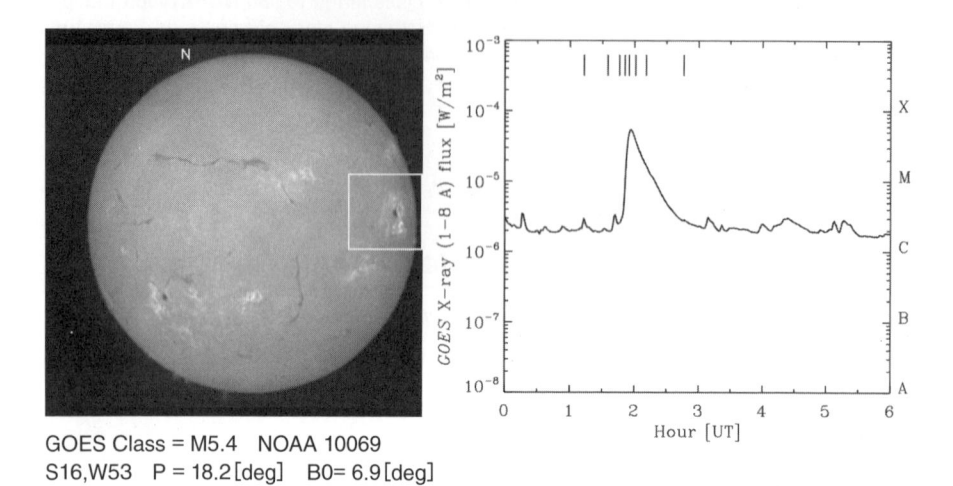

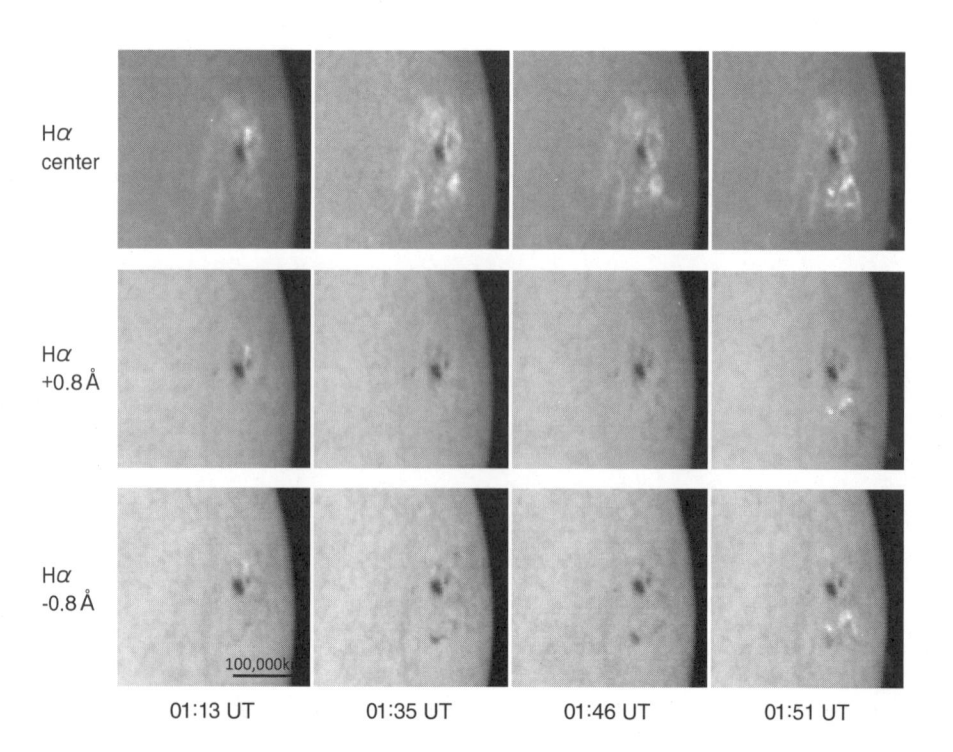

GOES Class = M5.4   NOAA 10069
S16,W53   P = 18.2 [deg]   B0= 6.9 [deg]

Hα center

Hα +0.8 Å

Hα −0.8 Å

01:13 UT   01:35 UT   01:46 UT   01:51 UT

The active region NOAA10069 was a $\beta\gamma\delta$-type region and produced many M-class flares during its passage on the solar disk. The flare shown here was a M5.4-class one and triggered a filament eruption. An EIT wave was observed to propagate with a speed of 130 km/s, while the Moreton wave was detected by FMT. It propagated to the south along the solar limb with a speed of about 1,300 km/s. The propagation of Moreton wave in 1 min cadence is displayed in the following pages.

活動領域 NOAA10069 は $\beta\gamma\delta$ 型の領域で，この領域が観測されている間に多数の M クラスフレアを頻発した．ここで示されているフレアは M5.4 クラスのものであり，フィラメント爆発を引き起こした．同時に EIT 波が放出され，130 km/s で伝搬していることが観測されている．一方，FMT では，モートン波が太陽の縁に沿う方向に南に向かって約 1,300 km/s で伝搬するのが観測された．モートン波の伝搬の様子は次ページ以降に示されている．

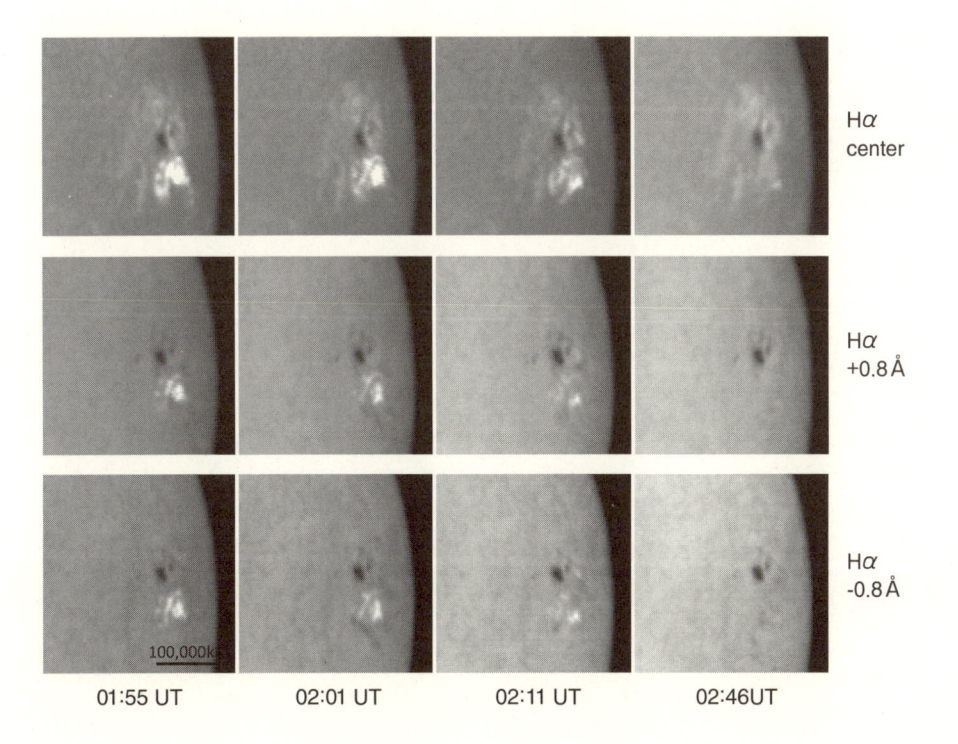

Aug 22, 2002 01:50−01:58 UT   Moreton Wave

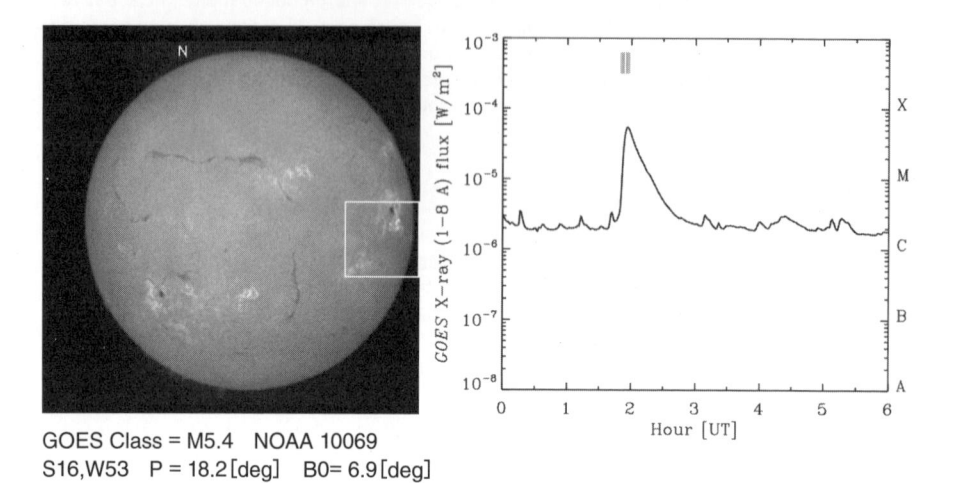

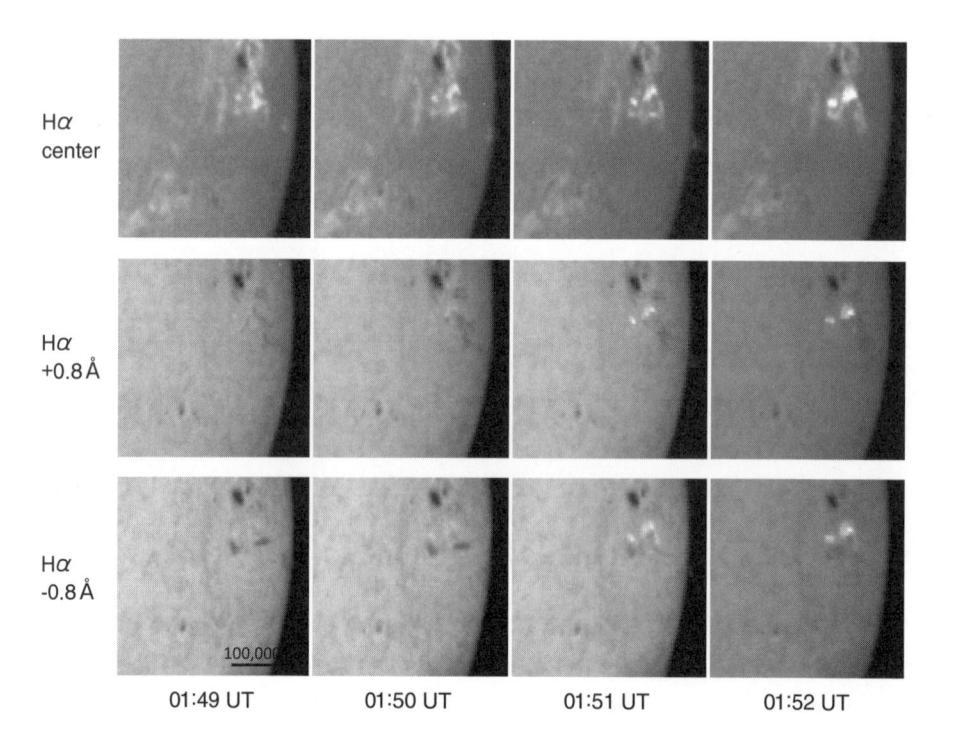

GOES Class = M5.4   NOAA 10069
S16,W53   P = 18.2 [deg]   B0 = 6.9 [deg]

Hα center

Hα +0.8 Å

Hα −0.8 Å

100,000

01:49 UT   01:50 UT   01:51 UT   01:52 UT

Chapter 6 Representative Events | 265

Here is shown the temporal evolution of the Moreton wave emitted by the M5.4-class flare in NOAA10069, with a speed of about 1,300 km/s. Simultaneous type II and V radio bursts were also observed, which indicates the emission of a shock wave to the corona. Associated with the event, a prominent CME was observed by SOHO/LASCO to expand with a speed of 1,800 km/s. (In the next two pages, time-difference images to enhance the visibility of wave fronts are shown with the manually drawn sketch of the wave fronts.)

ここでは，活動領域 NOAA10069 で発生した M5.4 クラスのフレアによって放射されたモートン波が約 1,300 km/s の速さで伝播する様子を示している．この現象には，Type II, V の電波バーストが同時に観測されており，コロナ中を衝撃波が伝搬したものと考えられる．また，このイベントに伴って，SOHO/LASCO で 1,800 km/s もの速度で広がる明るい CME が観測されている．（次の 2 ページでは，波面を見やすくするように，画像の明るさの時間変化像を示している．また波面位置のスケッチも参考のために付してある．）

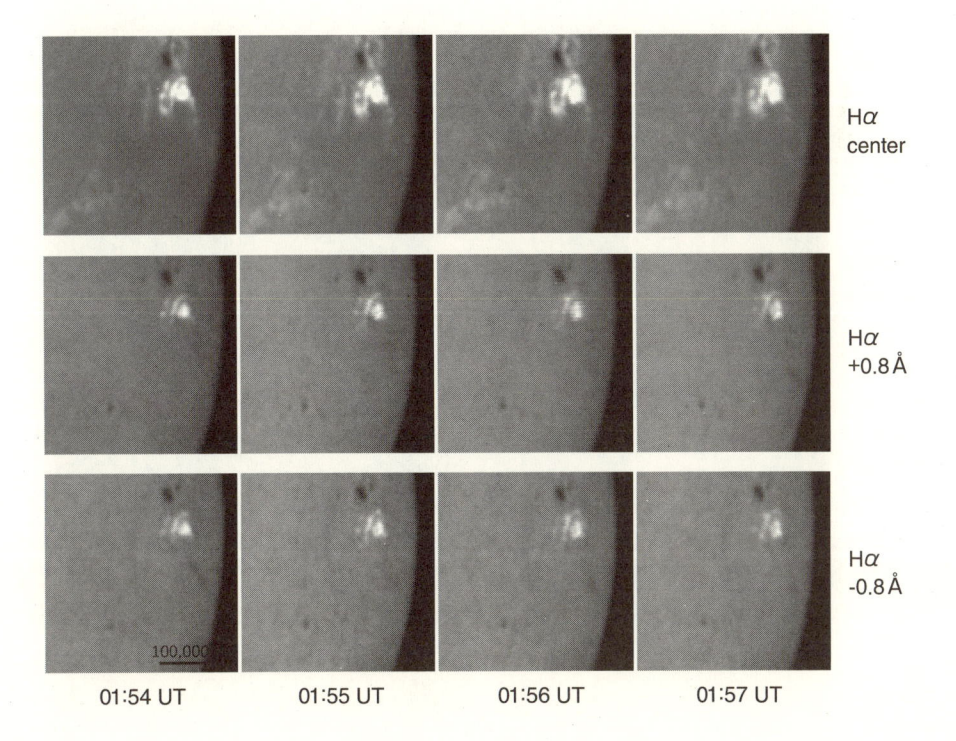

				Hα center
				Hα +0.8Å
				Hα -0.8Å
01:54 UT	01:55 UT	01:56 UT	01:57 UT	

266 | 第 6 章　代表的イベント

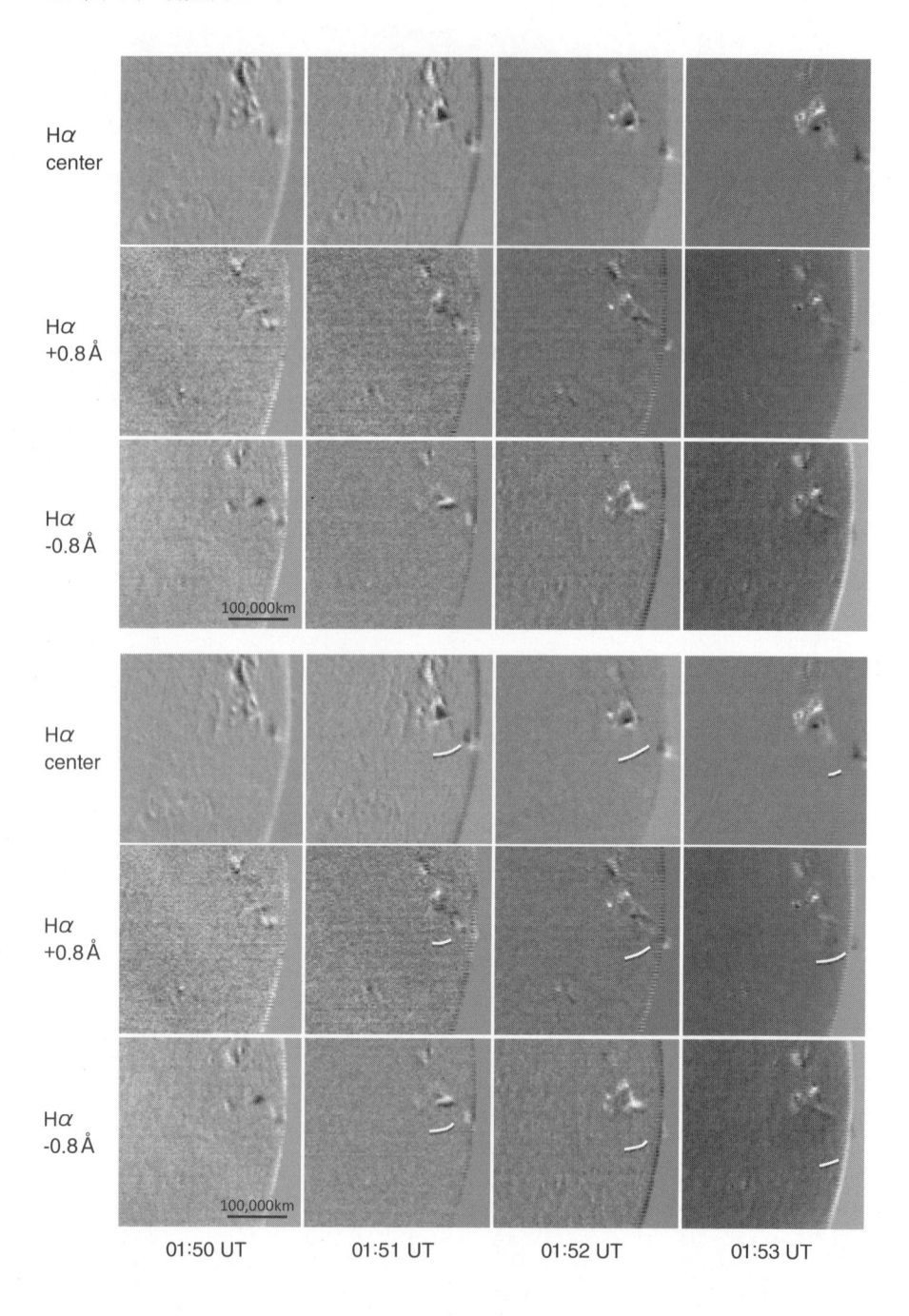

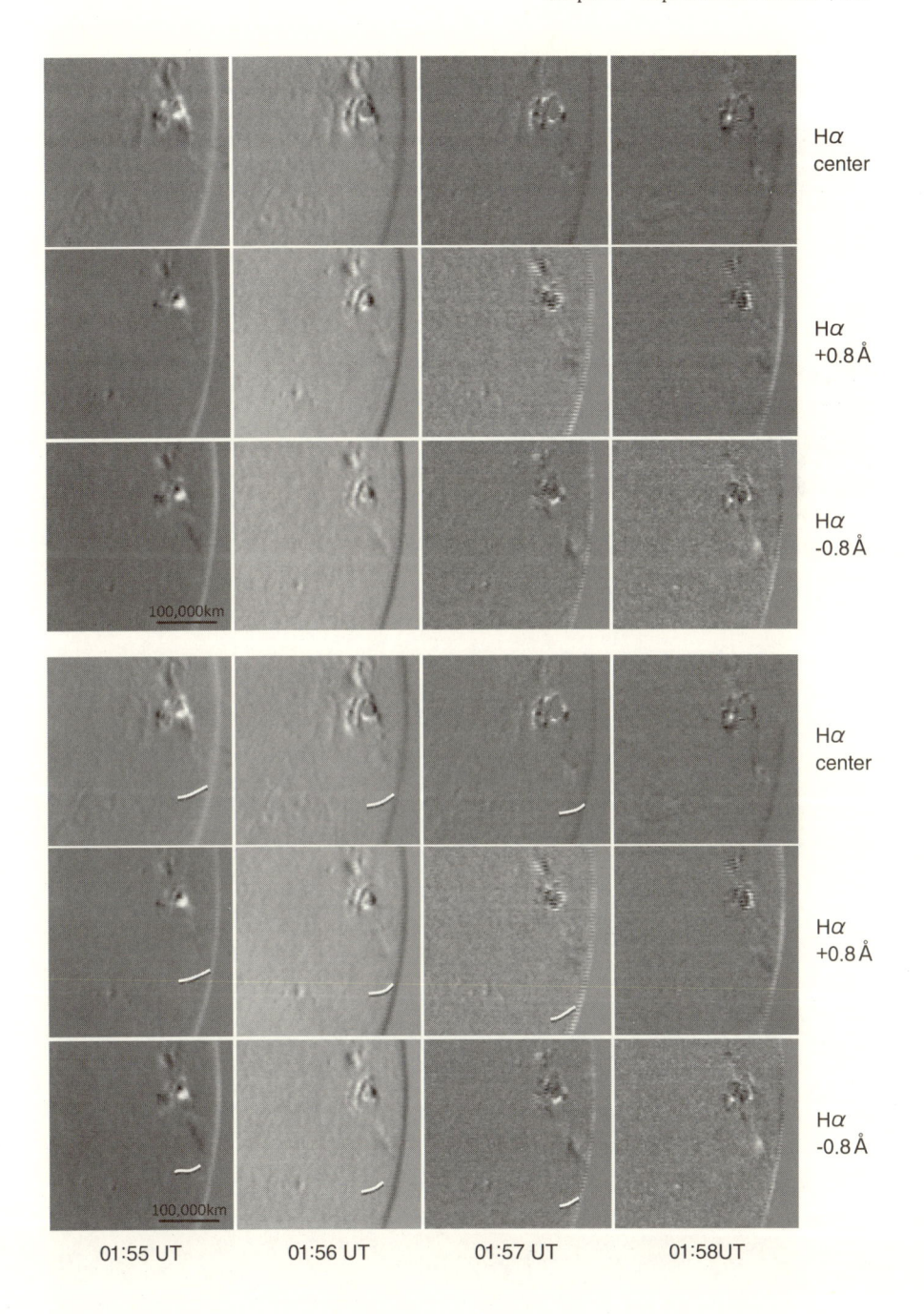

Chapter 6 Representative Events | 267

Hα center

Hα +0.8 Å

Hα −0.8 Å

100,000 km

Hα center

Hα +0.8 Å

Hα −0.8 Å

100,000 km

01:55 UT   01:56 UT   01:57 UT   01:58 UT

Aug 30, 2002 04:52−05:18 UT Surge

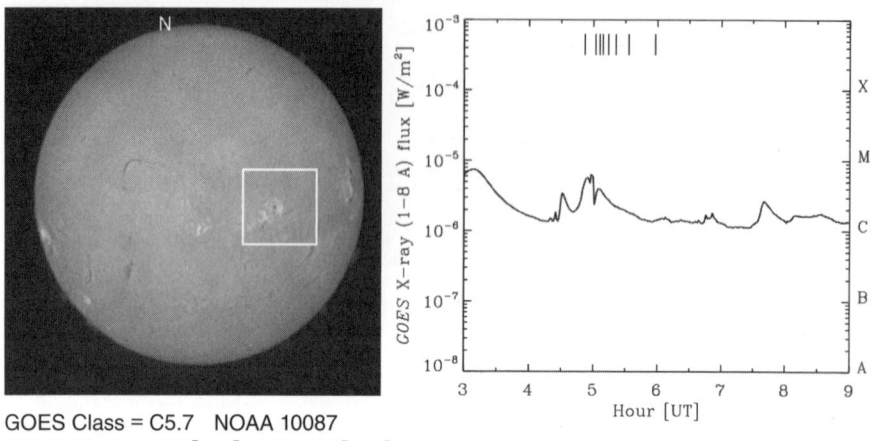

GOES Class = C5.7   NOAA 10087
S10,W28   P = 11.6 [deg]   B0 = 7.2 [deg]

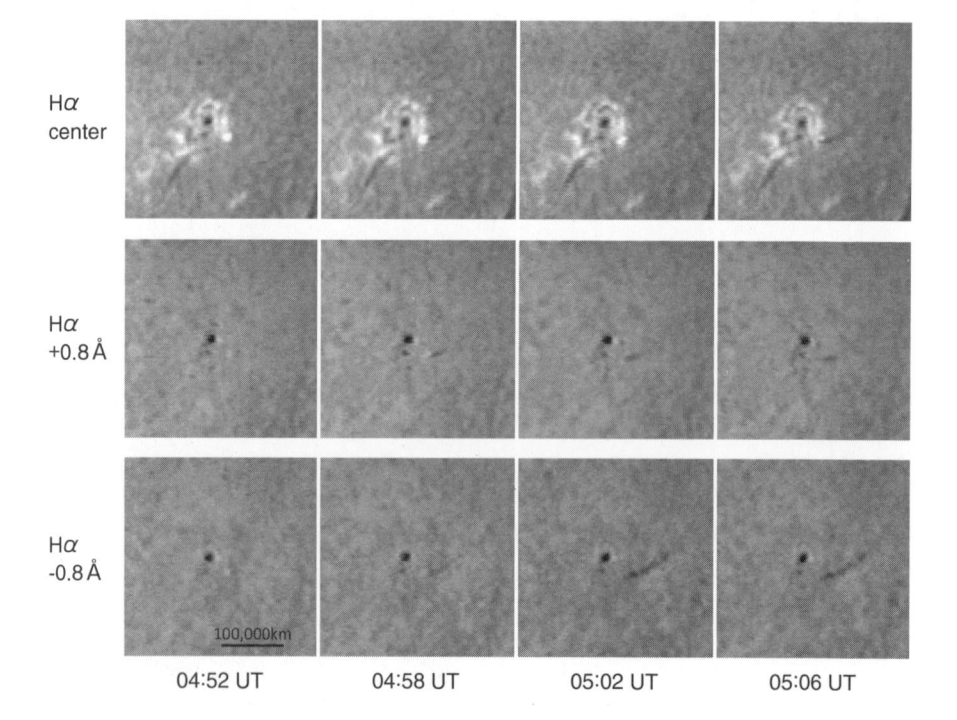

Hα center

Hα +0.8 Å

Hα −0.8 Å

100,000 km

04:52 UT   04:58 UT   05:02 UT   05:06 UT

NOAA10087 was a $\beta\gamma\delta$ type active region, although its activity level was not so high. The event shown here is a surge event, where chromospheric gas was ejected with a extension speed of 200 km/s from a transiently brightened point in Hα, along a curved trajectory. The bright point was located at one of the satellite polarity regions dispatched in the west side of the main spot.

NOAA10087 は $\beta\gamma\delta$ 型の活動領域であったが，その活動性はそれほど高くないものであった．ここで示されているのはサージイベントである．Hα 線像で見ると，輝点が現れそこから約 200 km/s の速さで彩層ガスが放出された．ガスは曲線状の軌道をたどって伸びきった後，同じ道筋で帰還した．活動領域の主黒点の西側には反対の極性を持った領域が散在しており，このサージの根元の輝点はこのうち 1 つの場所に位置していた．

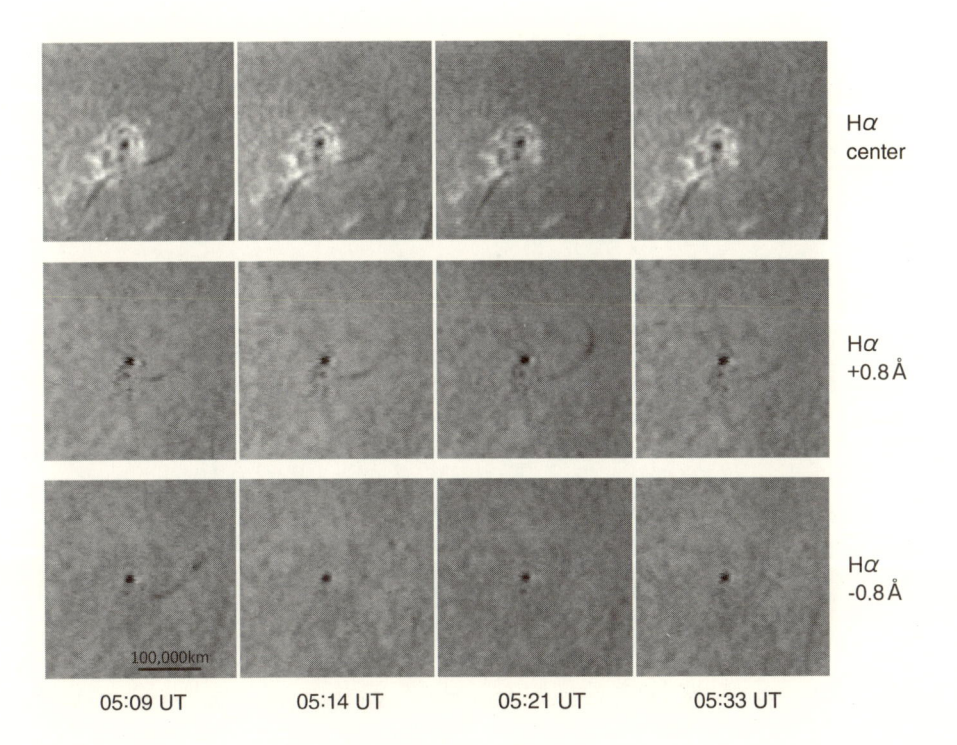

				Hα center
				Hα +0.8Å
				Hα −0.8Å
05:09 UT	05:14 UT	05:21 UT	05:33 UT	

Sep 18, 2002 22:30−23:52 UT   Flare, Filament Eruption

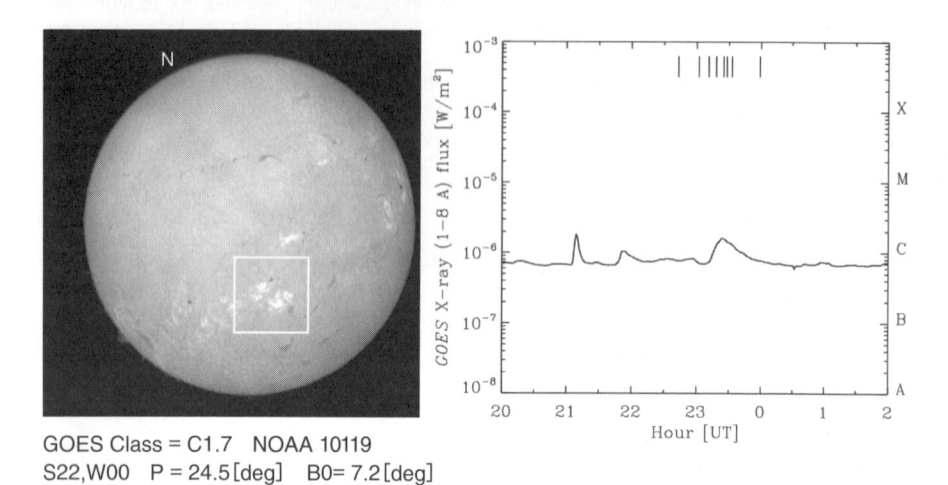

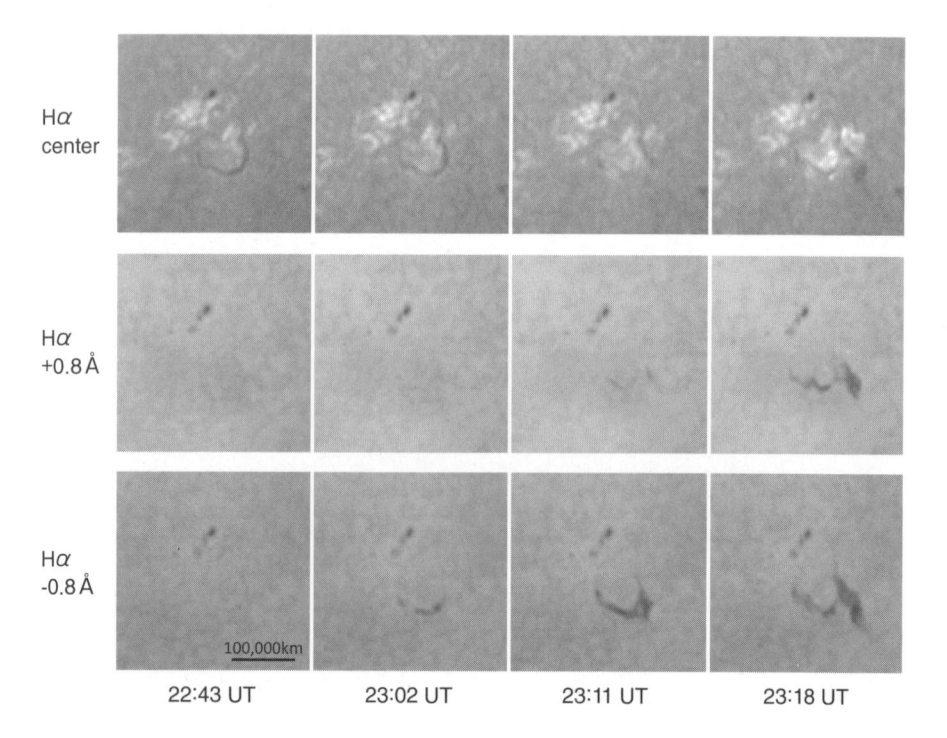

GOES Class = C1.7   NOAA 10119
S22,W00   P = 24.5[deg]   B0= 7.2[deg]

Hα center

Hα +0.8Å

Hα −0.8Å

100,000km

22:43 UT   23:02 UT   23:11 UT   23:18 UT

A two-ribbon flare exploded after the eruption of the central part of a dark filament, which was located in an old active region at the south of the region NOAA10119. Initially the filament was activated and showed a gradual expansion, and then it started to explosively rise up with a speed of around 120 km/s. At the explosive phase of the filament, the flare abruptly brightened and developed two ribbons of bright patches on both sides of the filament location. The associated CME was observed by LASCO/C2 to expand with a speed of around 1,000 km/s.

活動領域 NOAA10119 の南側にあった古い活動領域内のフィラメントの中央部が爆発消滅してツーリボンフレアを引き起こした．当初フィラメントは徐々に上昇していたが，突然爆発的に約 120 km/s の速度で上昇を開始した．この爆発期に合わせて Hα フレアが輝きだし，フィラメントの両脇に 2 つのフレアリボンが現われた．LASCO/C2 により，この現象に伴って CME が約 1,000 km/s の速度で膨張していることが観測されている．

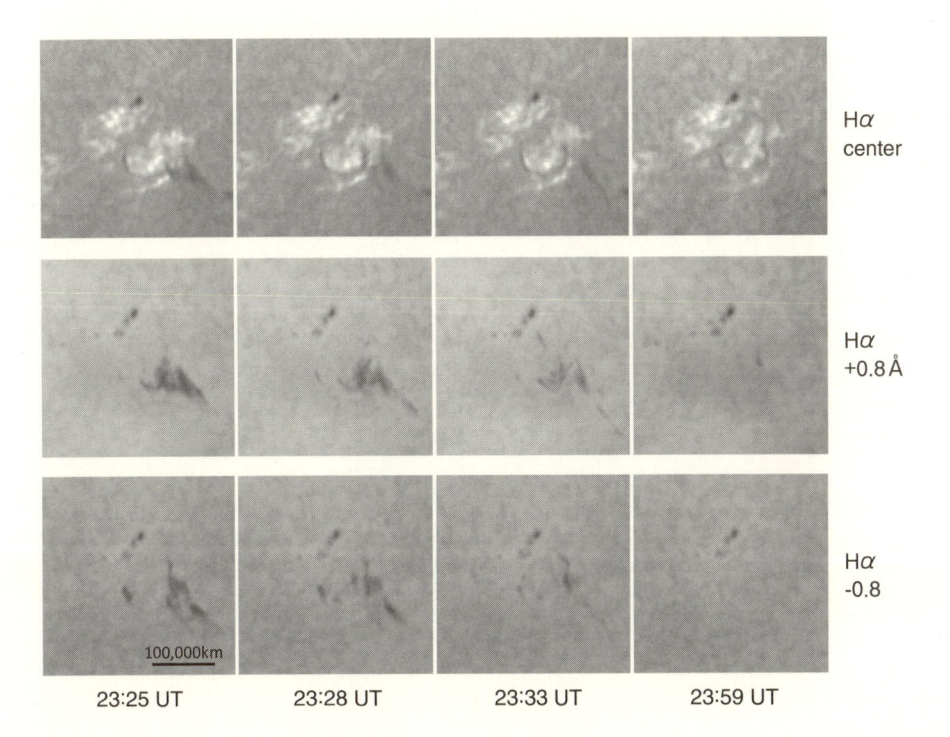

272 | 第6章 代表的イベント

Oct 04, 2002 22:36−23:14 UT　Flare

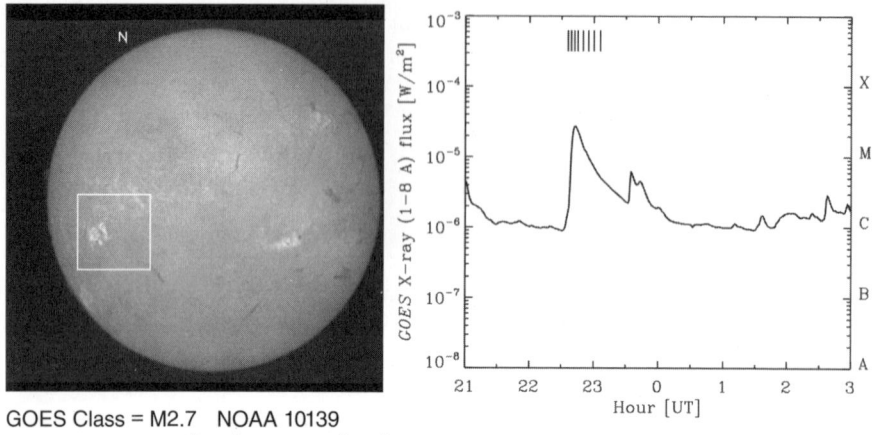

GOES Class = M2.7　NOAA 10139
N10, E43　P = 26.1 [deg]　B0 = 6.6 [deg]

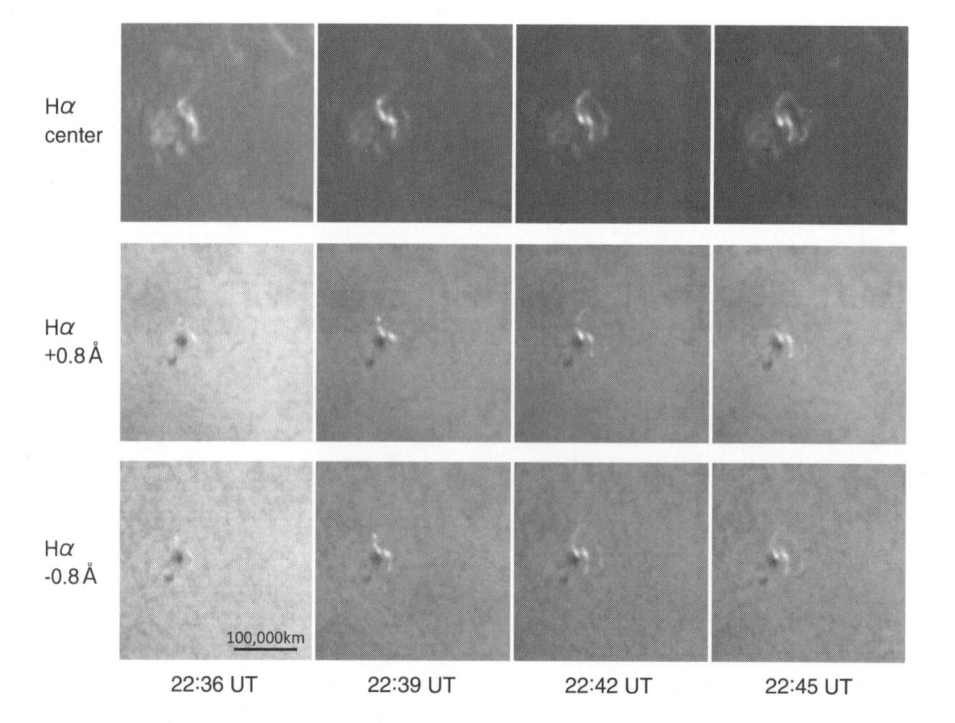

Hα center

Hα +0.8 Å

Hα −0.8 Å

100,000 km

22:36 UT　　22:39 UT　　22:42 UT　　22:45 UT

In the active region NOAA10139, there occurred a strong M2.7 flare. A Moreton wave was observed to propagate from the flare kernel with a propagation speed of 350 km/s. No prominent filament eruption was associated in the event, different from the other Moreton wave events. The propagation of Moreton wave in 1 min cadence is displayed in the following pages.

活動領域 NOAA10139 で M2.7 の強いフレアが発生した．このときモートン波が発生し，フレアカーネルから速さ 350 km/s で伝搬した．他のモートン波の現象とは異なって，このイベントには激しいフィラメント爆発は見られなかった．モートン波の伝播の様子は次ページ以降に示されている．

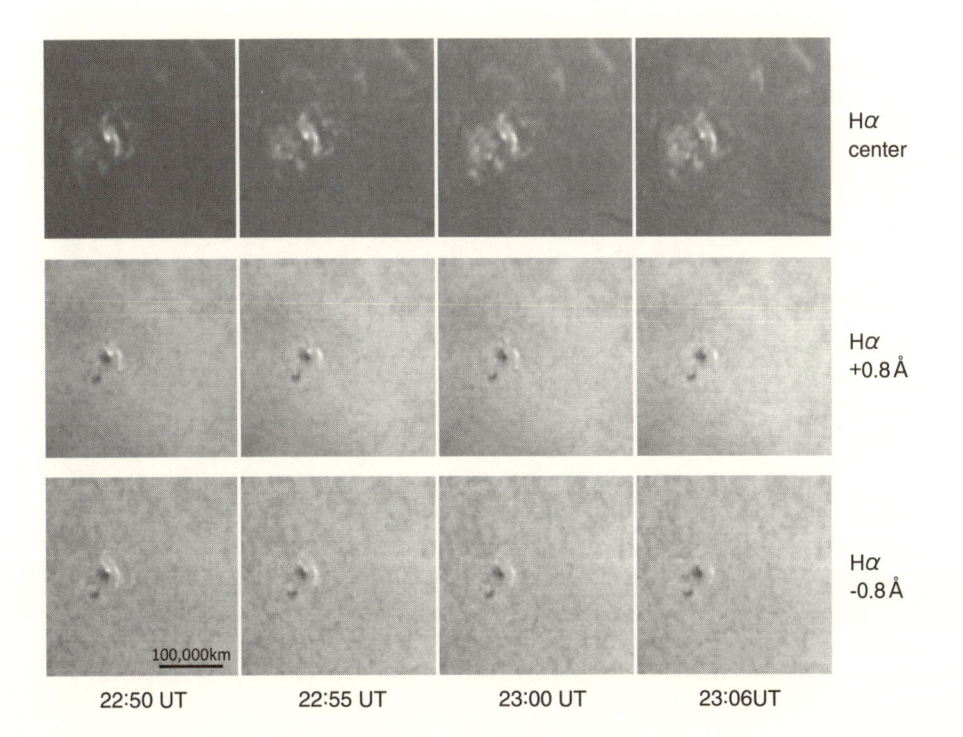

Oct 04, 2002 22:38−22:44 UT   Moreton Wave

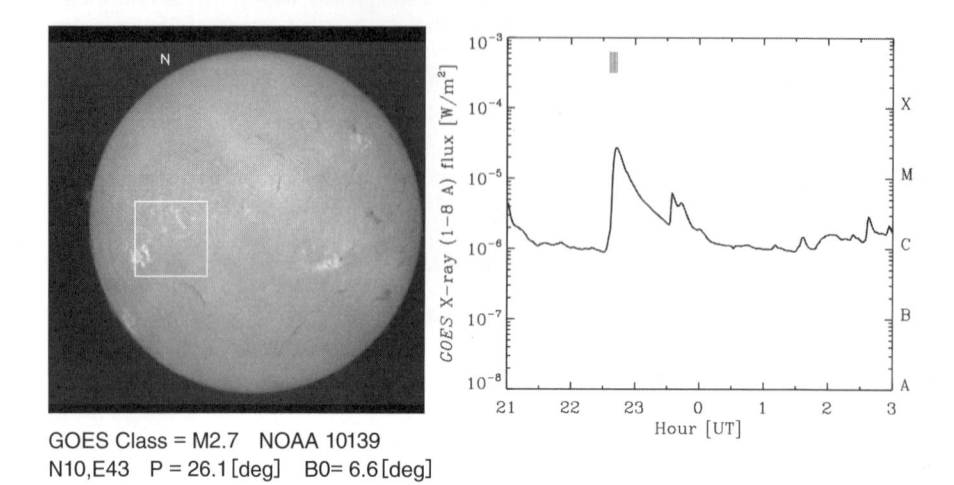

GOES Class = M2.7   NOAA 10139
N10,E43   P = 26.1 [deg]   B0= 6.6 [deg]

Hα center

Hα +0.8 Å

Hα −0.8 Å

100,000 km

22:36 UT   22:37 UT   22:38 UT   22:39 UT

Chapter 6 Representative Events | 275

Here is shown the temporal evolution of the Moreton wave emitted by the M2.7 flare in NOAA10139. The wave propagated in a sectorial region from the flare kernel toward SW direction, with a speed of 350 km/s. The wave propagation can be seen only in the video-movie, as the amplitude of the wave was not so high. (In the next two pages, time-difference images to enhance the visibility of wave fronts are shown with the manually drawn sketch of the wave fronts.)

活動領域 NOAA10139 で発生した M2.7 フレアによって放射されたモートン波の時間変化が示されている．モートン波は，フレアカーネルから南東方面へ扇形上に，速さ 350 km/s で伝搬した．なお，モートン波の振幅は大きくなく，ビデオムービーでやっと検出できる程度のものであった．（次の２ページでは，波面を見やすくするように，画像の明るさの時間変化像を示している．また波面位置のスケッチも参考のために付してある．）

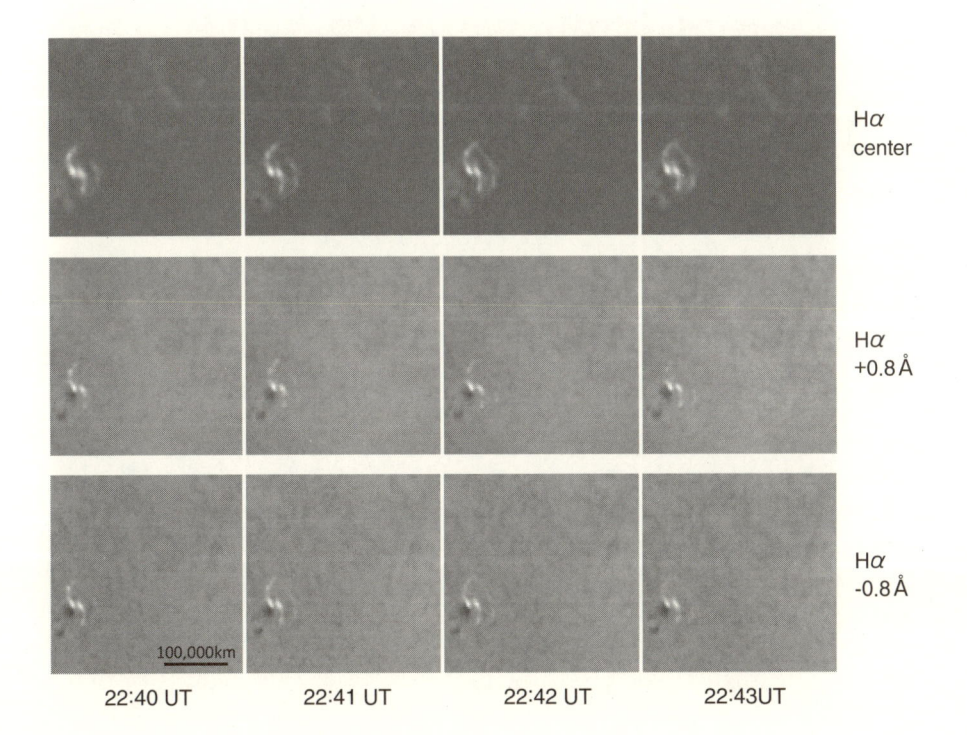

				Hα center
				Hα +0.8Å
				Hα −0.8Å
22:40 UT	22:41 UT	22:42 UT	22:43 UT	

100,000km

276 | 第6章　代表的イベント

	22:36 UT	22:37 UT	22:38 UT	22:39 UT
Hα center				
Hα +0.8Å				
Hα -0.8Å				
Hα center				
Hα +0.8Å				
Hα -0.8Å				

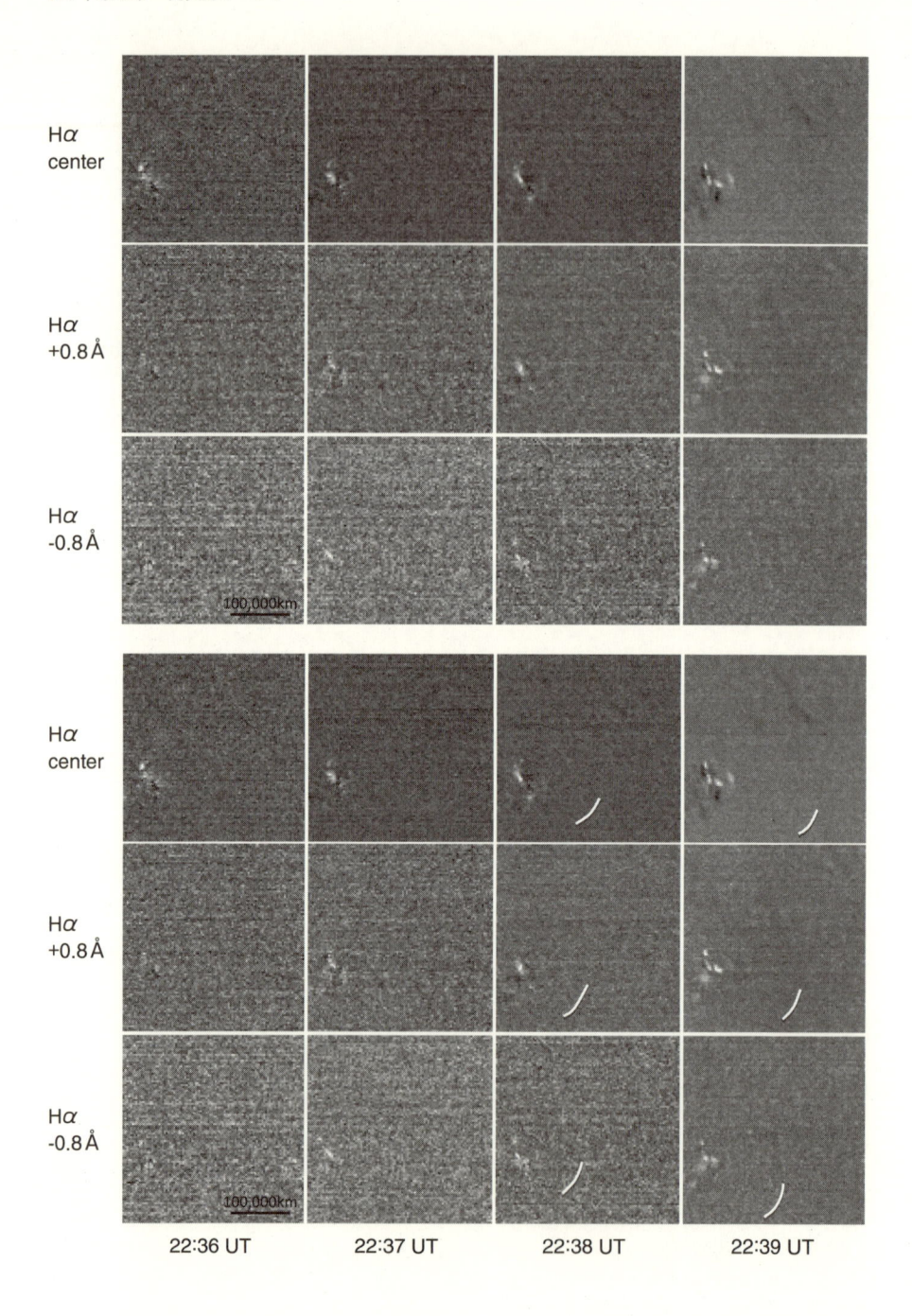

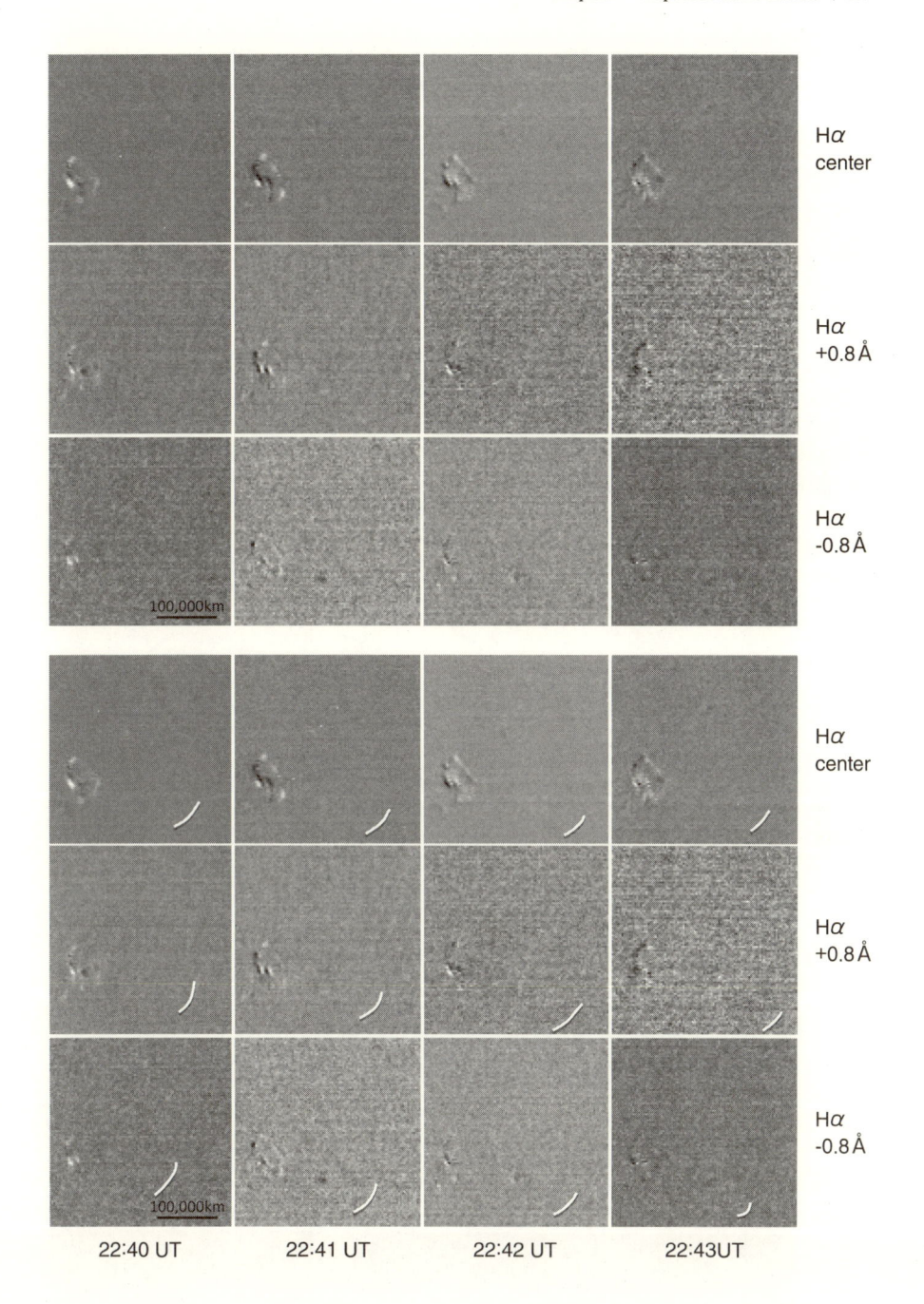

Hα
center

Hα
+0.8Å

Hα
−0.8Å

100,000km

Hα
center

Hα
+0.8Å

Hα
−0.8Å

100,000km

22:40 UT   22:41 UT   22:42 UT   22:43 UT

Feb 12, 2003 23:21B−00:32 UT Two Ribbon Flare in Spotless Region

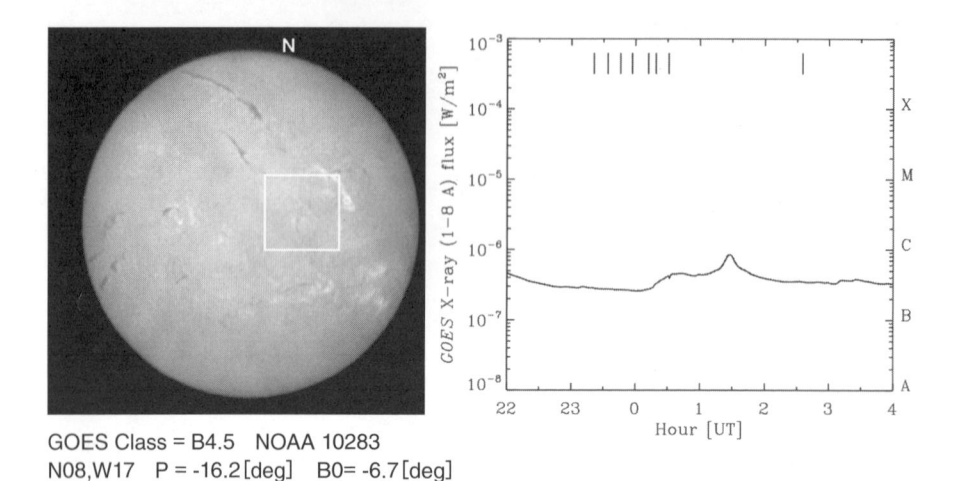

GOES Class = B4.5   NOAA 10283
N08,W17   P = -16.2 [deg]   B0 = -6.7 [deg]

Hα center

Hα +0.8 Å

Hα -0.8 Å

100,000 km

23:21 UT   23:34 UT   23:46 UT   23:57 UT

An active region filament in an old β-type active region NOAA10283 started to slowly erupt and disappeared in 50 min. After the eruption of the filament, a small-scale two-ribbon flare in Hα appeared at both side of the filament. In SOHO/EIT191 images, compact and bright arcade loops were observed to develop connecting the two-ribbon locations.

減衰したβ型の活動領域 NOAA10283 内の活動領域フィラメントがゆっくりと上昇を始め，50分後には消滅した．この後，フィラメントがあった位置の両脇に小規模の Hαツーリボンフレアが発生した．このとき SOHO/EIT191 像では，Hαツーリボンを結ぶアーケード形のループ構造が形成されているのが観測されている．

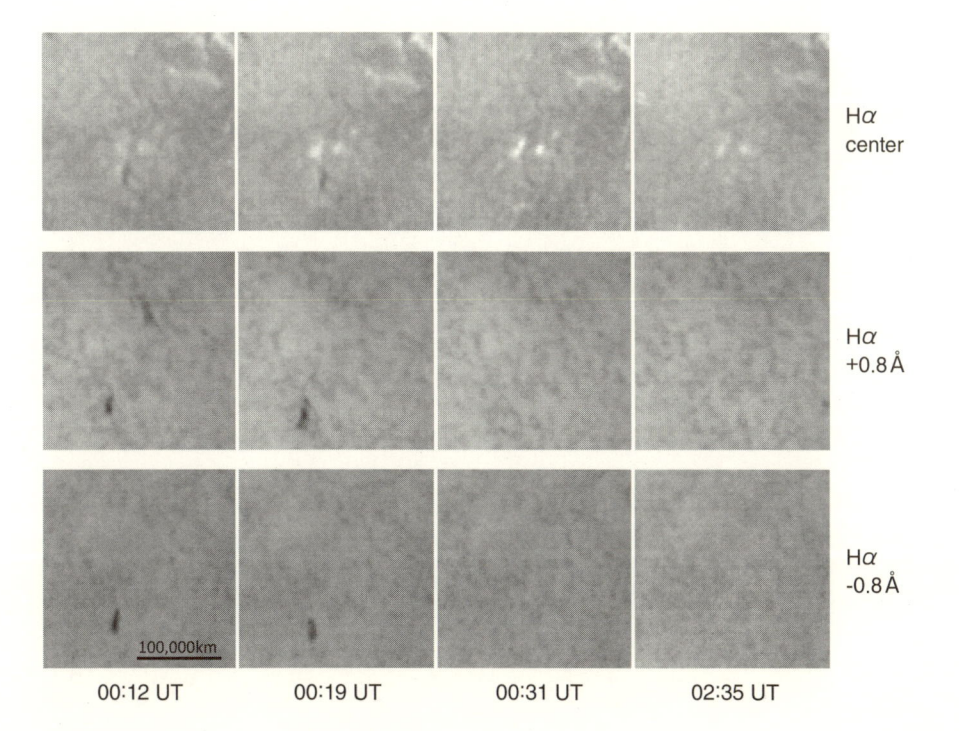

Mar 13, 2003 01:53−02:34 UT   Flare, Filament Eruption

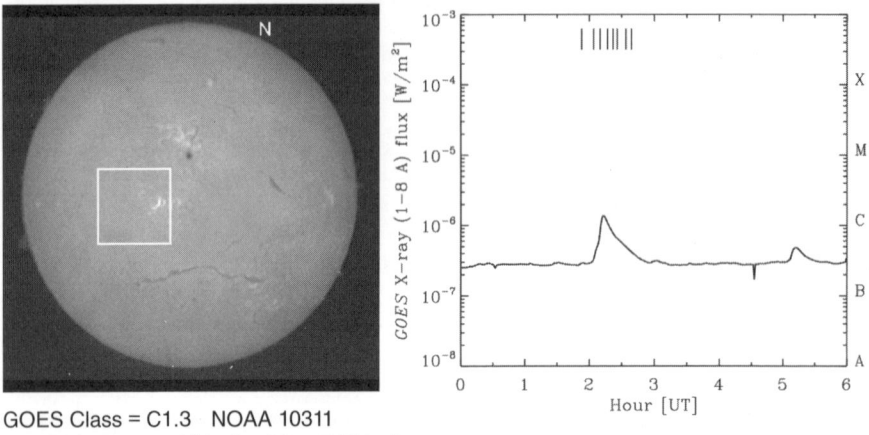

GOES Class = C1.3   NOAA 10311
S13,E08   P = -24.0 [deg]   B0= -7.2 [deg]

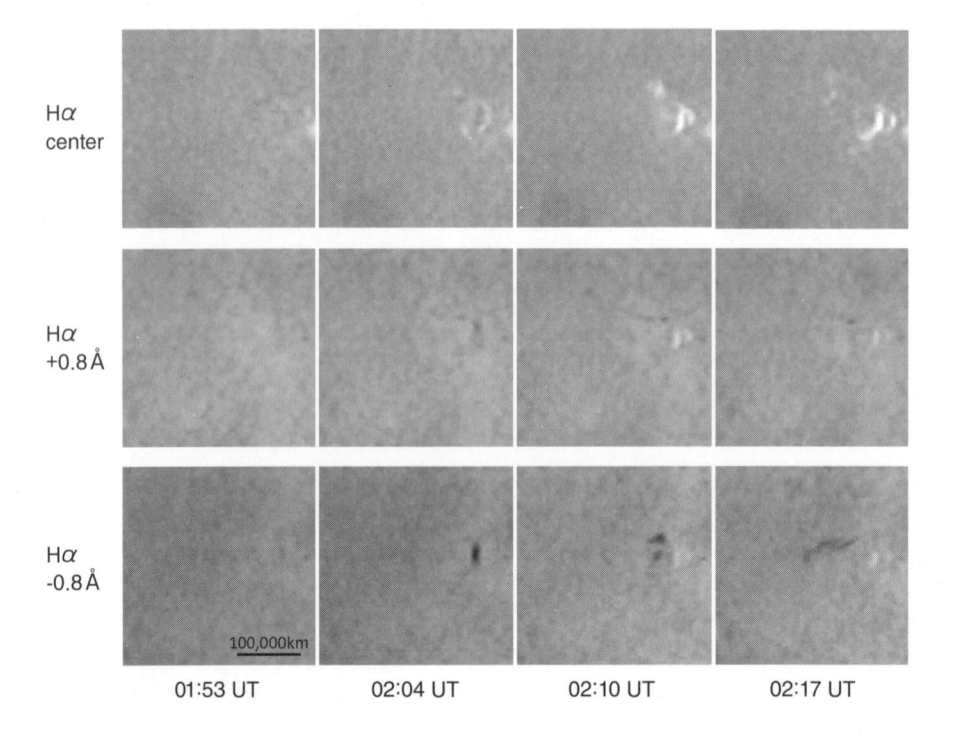

	01:53 UT	02:04 UT	02:10 UT	02:17 UT
Hα center				
Hα +0.8Å				
Hα −0.8Å				

About 5 min after the start of filament expansion, a two-ribbon flare occurred in the active region NOAA10311. The filament expanded with a speed of 200 km/s, nearly parallel to the solar surface. In SOHO/EIT195Å images, hot ejecta were also detected to explode in association with the cool filament expansion. No prominent CMEs were detected by the SOHO/LASCO/C2.

フィラメント膨張の開始後 5 分ほど後に，活動領域 NOAA10311 でツーリボンフレアが発生した．フィラメントは太陽表面とほぼ平行に 200 km/s の速度で膨張した．SOHO/EIT195Å 画像では，低温のフィラメント膨張に伴って熱いガスも同様に放出されていたことが観測された．なお，SOHO/LASCO/C2 の画像では，CME 現象は観測されなかった．

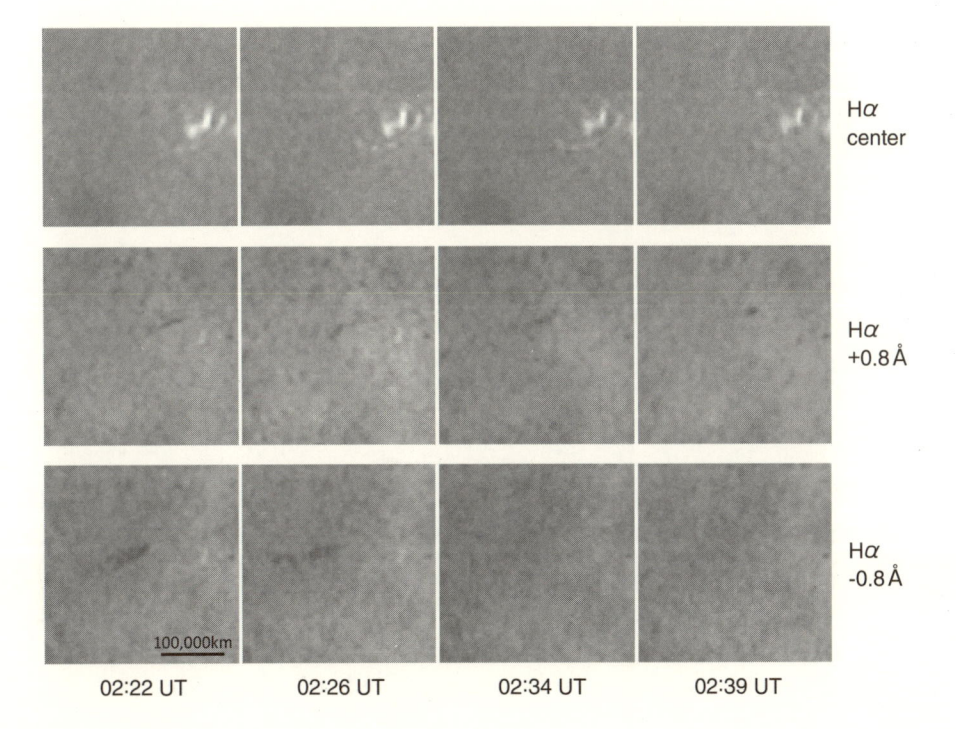

Apr 26, 2003 23:49B−03:07 UT   Two Ribbon Flare in Spotless Region

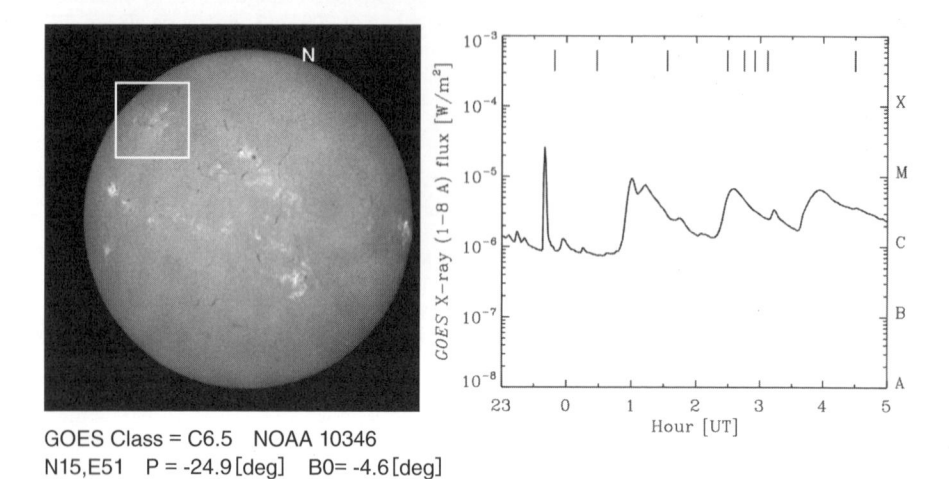

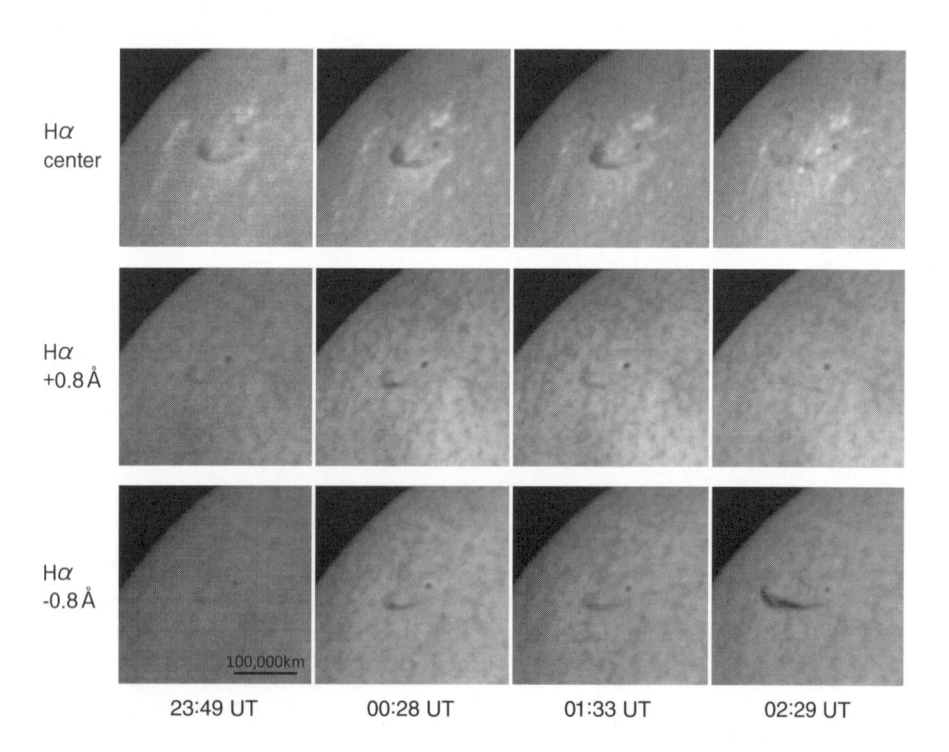

GOES Class = C6.5   NOAA 10346
N15,E51   P = -24.9[deg]   B0= -4.6[deg]

Hα center

Hα +0.8Å

Hα -0.8Å

100,000km

23:49 UT   00:28 UT   01:33 UT   02:29 UT

A two-ribbon flare occurred after the eruption of the active region filament in an old active region NOAA10346. The filament, as a precursor of the flare, showed a complex internal motion during 2 hours before the flare explosion. The filament expanded in an arch shape with its top speed of above 50 km/s. In LASCO/C2 images, a CME of twisted flux rope morphology was detected to rise and expand with a speed of about 400 km/s, in association with the filament eruption. The C-class X-ray bursts started at 00:50 and 03:37 UT occurred in another active region NOAA10338. The X-ray burst of C6.5 around 02:22 UT is related to this event.

安定した活動領域 NOAA10346 内の活動領域フィラメントが上昇し，その後ツーリボンフレアが発生した．フレア発生前の2時間程度の間，先駆現象としてフィラメントは複雑な内部運動を示していた．フィラメントはアーチ状に 50 km/s 以上の速度で膨張上昇した．LASCO/C2 では，ねじれた磁束管の形で，CME が約 400 km/s の速度で上昇膨張していることが観測された．なお，00:50 と 03:37 UT から始まる C クラスの X 線バーストは別の領域 NOAA10338 で発生したものである．この現象に対応するものは 02:22 UT 頃の C6.5 の X 線バーストである．

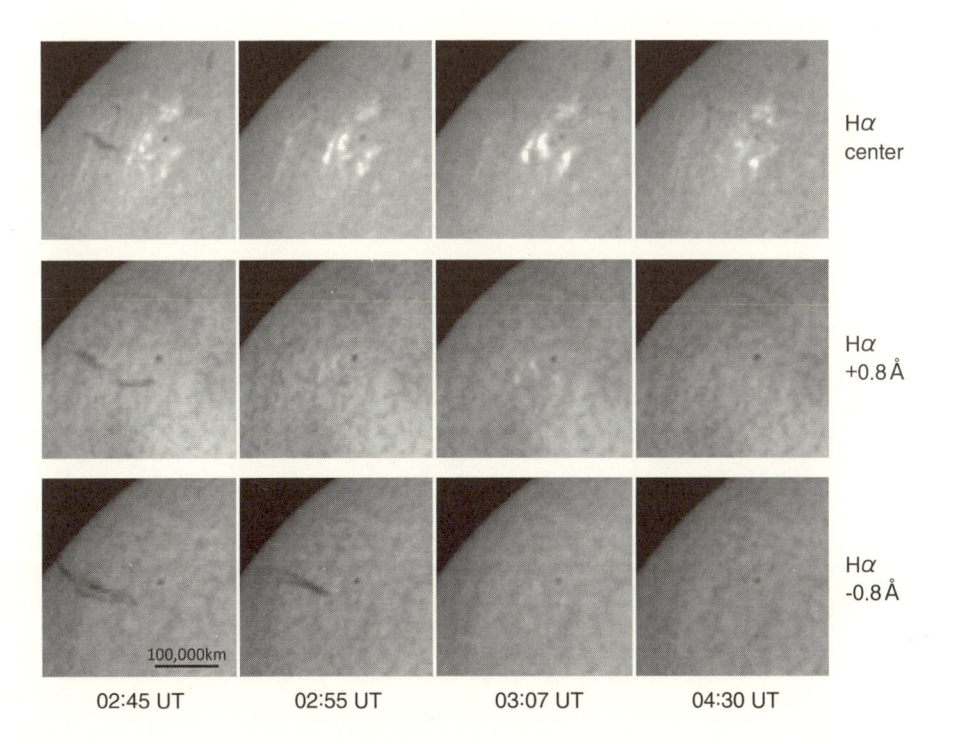

Apr 27, 2003　00:16−02:28 UT　　Flare, Filament Eruption

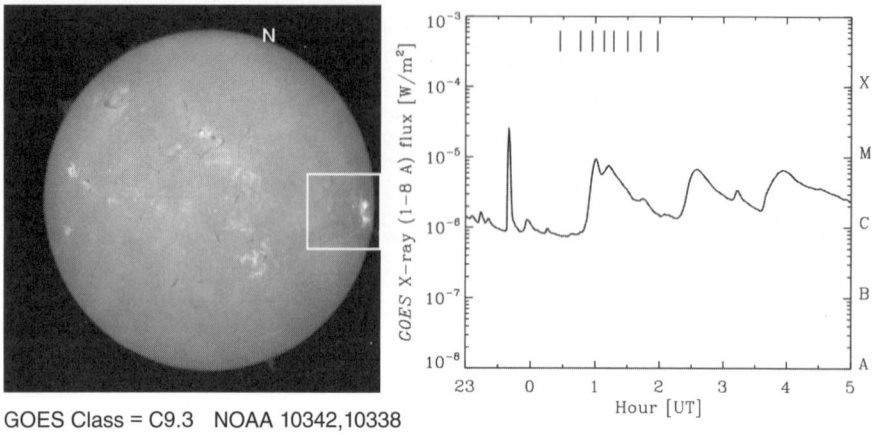

GOES Class = C9.3　NOAA 10342, 10338
N15, W65　P = −24.9 [deg]　B0 = −4.6 [deg]

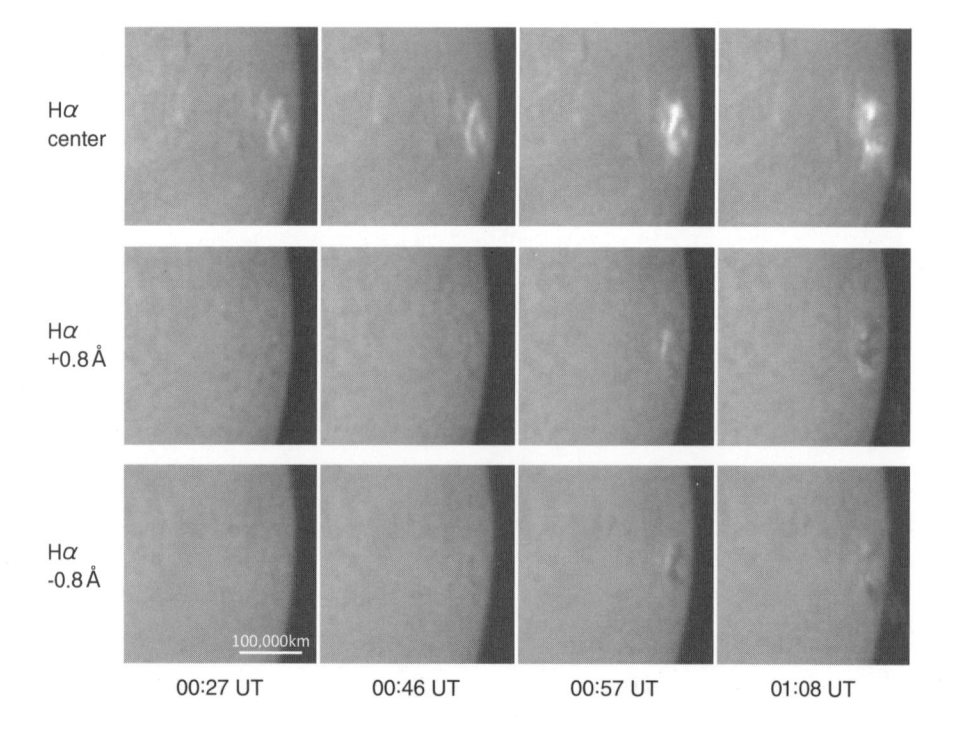

Hα center

Hα +0.8 Å

Hα −0.8 Å

100,000 km

00:27 UT　　00:46 UT　　00:57 UT　　01:08 UT

NOAA10338 was born on Apr 19 on the Sun. At the west part of an old active region, many new magnetic regions floated up consecutively and formed a δ-type sunspot on Apr 26 by fusion to the pre-existent magnetic field. During the development, the active region produced many strong M-class flares. The event shown here is a C9.3 flare on Apr 27. In the active region located near the west limb, a filament eruption occurred and triggered a two-ribbon flare. The filament expanded in a giant arch shape and exploded beyond the limb of the solar disk. Associated with the event, a CME event was observed. Kardapolova et al. (2008) summarized the EUV and radio polarization data of the event and confirmed that the filament eruption can be explained with a model of filament eruption by the instability of magnetic twisted flux rope.

NOAA10338 は，4 月 19 日に太陽表面上に現れた．古い活動領域の西側に磁気浮上域が次々と発生し，古い領域の N 極磁場と新しい領域の S 極磁場が融合し 4 月 26 日には δ 型の黒点となった．この間，この領域では M クラスの強いフレアが頻発した．このページで表示されている現象は，4 月 27 日に発生した C9.3 クラスのフレアである．太陽の西の縁近くに位置していたこの領域でフィラメント爆発が発生しそれに伴ってツーリボンフレアが発生した．フィラメントは大きなアーチの形で膨張爆発し太陽縁外へ四散した．このフィラメント爆発に伴って，CME が観測されている．Kardapolova et al. (2008) は，極端紫外線と電波偏波観測から導出された磁場構造を元にして，ねじれた磁束管の不安定性によりフィラメント爆発が引き起こされるというモデルがこの現象を説明できることを示した．

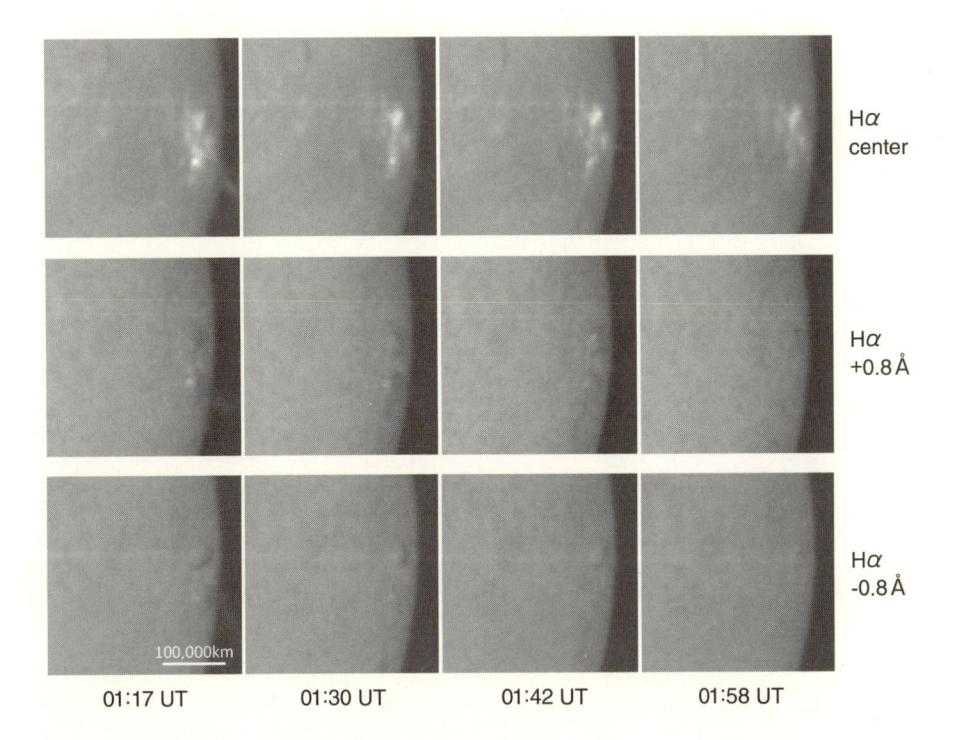

May 27, 2003　22:56－23:26 UT　Flare

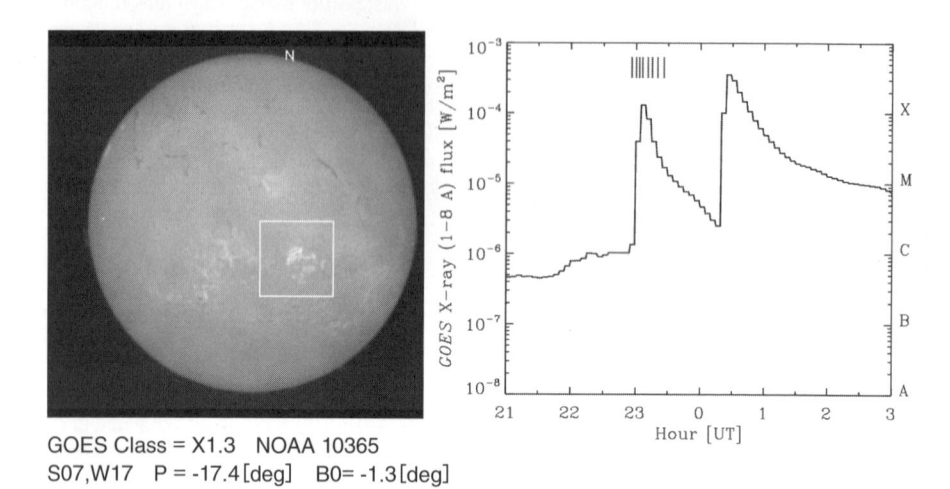

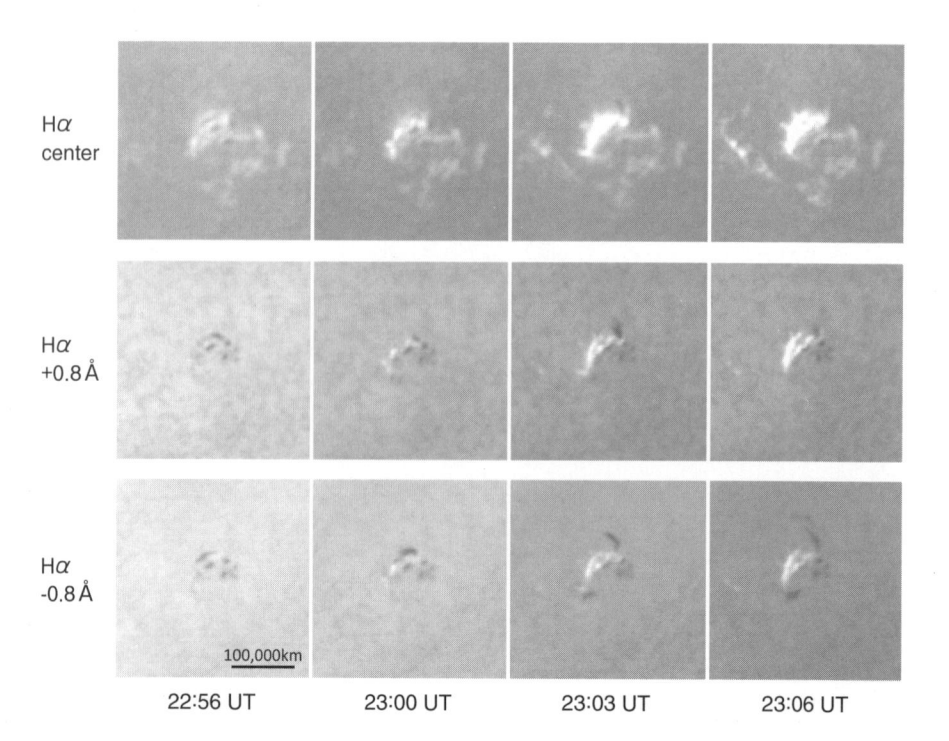

GOES Class = X1.3　NOAA 10365
S07,W17　P = -17.4[deg]　B0= -1.3[deg]

Hα center

Hα +0.8Å

Hα -0.8Å

100,000km

22:56 UT　　23:00 UT　　23:03 UT　　23:06 UT

The active region NOAA10365 got activated to be a $\beta\gamma\delta$-type region and produced many M-class flares since May 25, 2003. The activation and flare production was due to the additional emergence of twisted magnetic flux or injection of helicity (Chae et al. 2004; Chandra et al. 2009). The event shown here was an X1.3 class flare initiated by a filament eruption. Emission of Moreton wave was observed by FMT in the event. The propagation of Moreton wave in 1 min cadence is displayed in the following pages.

活動領域 NOAA10365 は，2003 年 5 月 25 日以降 $\beta\gamma\delta$ 型の領域となって活発化し，M クラスフレアを頻発した．この活発化は，ねじれた磁気ループの浮上あるいは磁気ヘリシティーの追加注入によるものと考えられている（Chae et al. 2004; Chandra et al. 2009）．ここで示されているイベントは，活動領域フィラメントの爆発によって X1.3 クラスフレアが発生したものである．このとき FMT によってモートン波が放射されるのが観測された．モートン波の伝播の様子は次ページ以降に示されている．

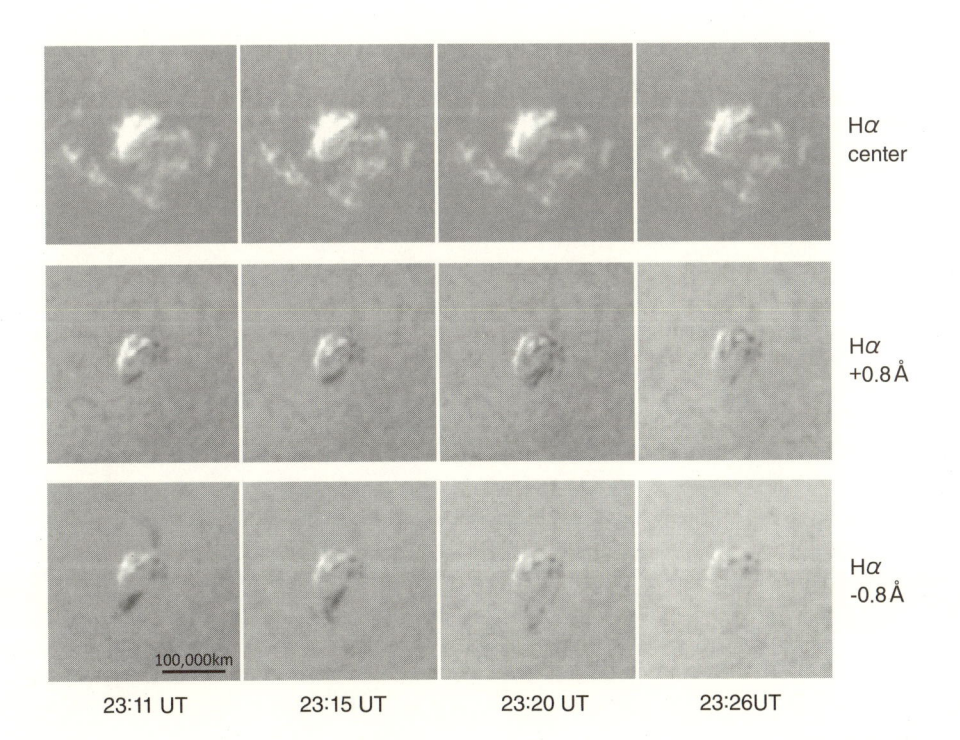

Hα center

Hα +0.8Å

Hα -0.8Å

100,000km

23:11 UT   23:15 UT   23:20 UT   23:26 UT

288 | 第6章 代表的イベント

May 27, 2003 23:01－23:10 UT   Moreton Wave

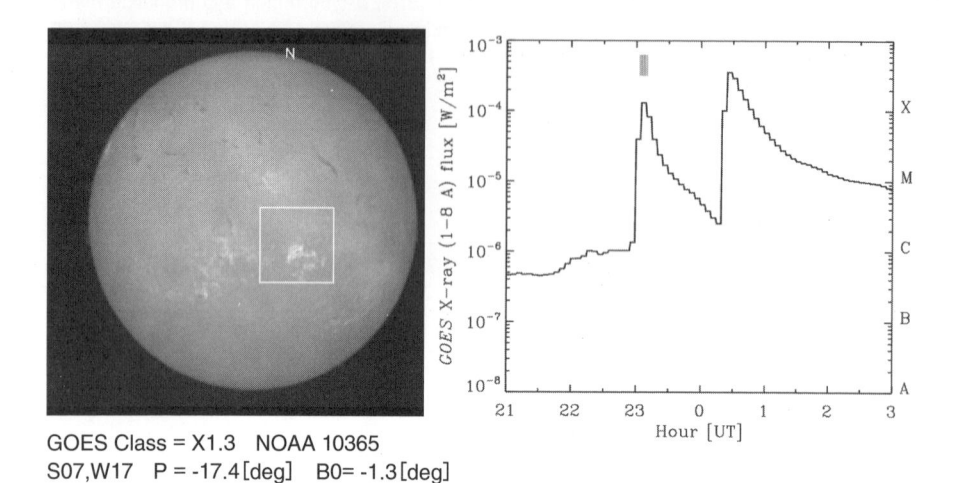

GOES Class = X1.3   NOAA 10365
S07,W17   P = -17.4 [deg]   B0 = -1.3 [deg]

Hα center

Hα +0.8 Å

Hα -0.8 Å

100,000 km

23:03 UT   23:04 UT   23:05 UT   23:06 UT

Here is shown the temporal evolution of the arc-shaped Moreton wave emitted in the X1.3 class flare in NOAA10365. One wave front propagated northwards with a speed of 350 km/s, nearly in the same direction as the filament eruption, while the other one propagated southwards with a speed of 400 km/s. The event is a very rare case of Moreton wave that has two oppositely propagating wave fronts. (In the next two pages, time-difference images to enhance the visibility of wave fronts are shown with the manually drawn sketch of the wave fronts.)

活動領域 NOAA10365 で発生した X1.3 クラスのフレアで放射されたモートン波の時間変化が示されている．モートン波の 1 つの波面はフィラメント爆発と同じ方向である北のほうに向かって 350 km/s の速さで伝播し，もう 1 つの波面は南側に 400 km/s の速さで伝搬した．このイベントはモートン波の波面が，対向する 2 方向に伝搬する珍しい例である．（次の 2 ページでは，波面を見やすくするように，画像の明るさの時間変化像を示している．また波面位置のスケッチも参考のために付してある．）

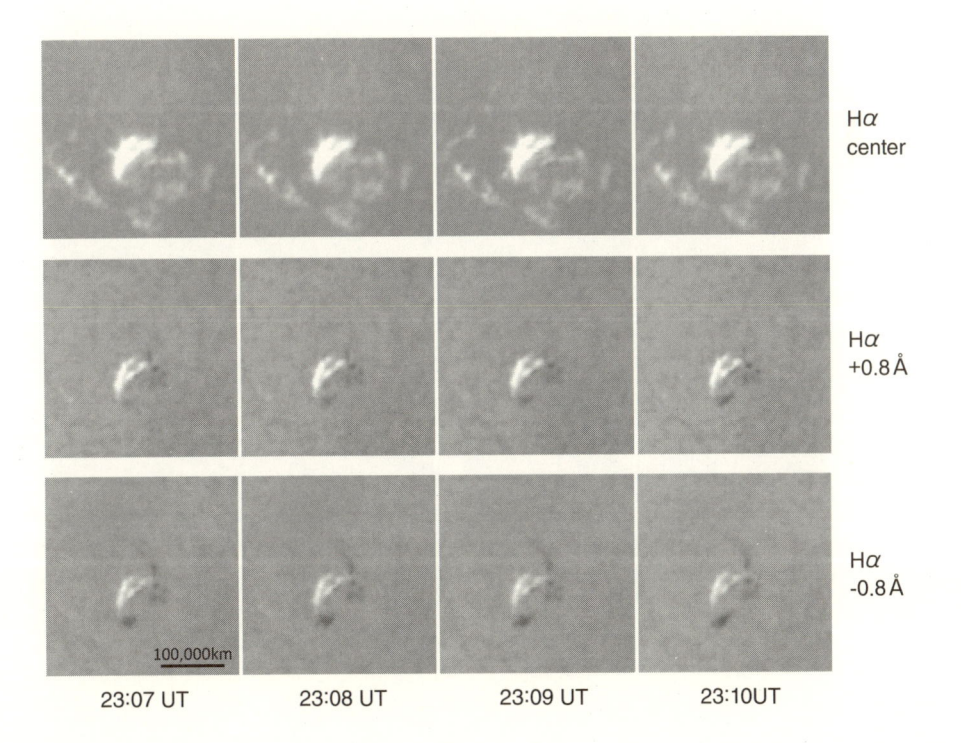

290 | 第6章 代表的イベント

Hα center

Hα +0.8Å

Hα −0.8Å

100,000km

Hα center

Hα +0.8Å

Hα −0.8Å

100,000km

23:03 UT　　23:04 UT　　23:05 UT　　23:06 UT

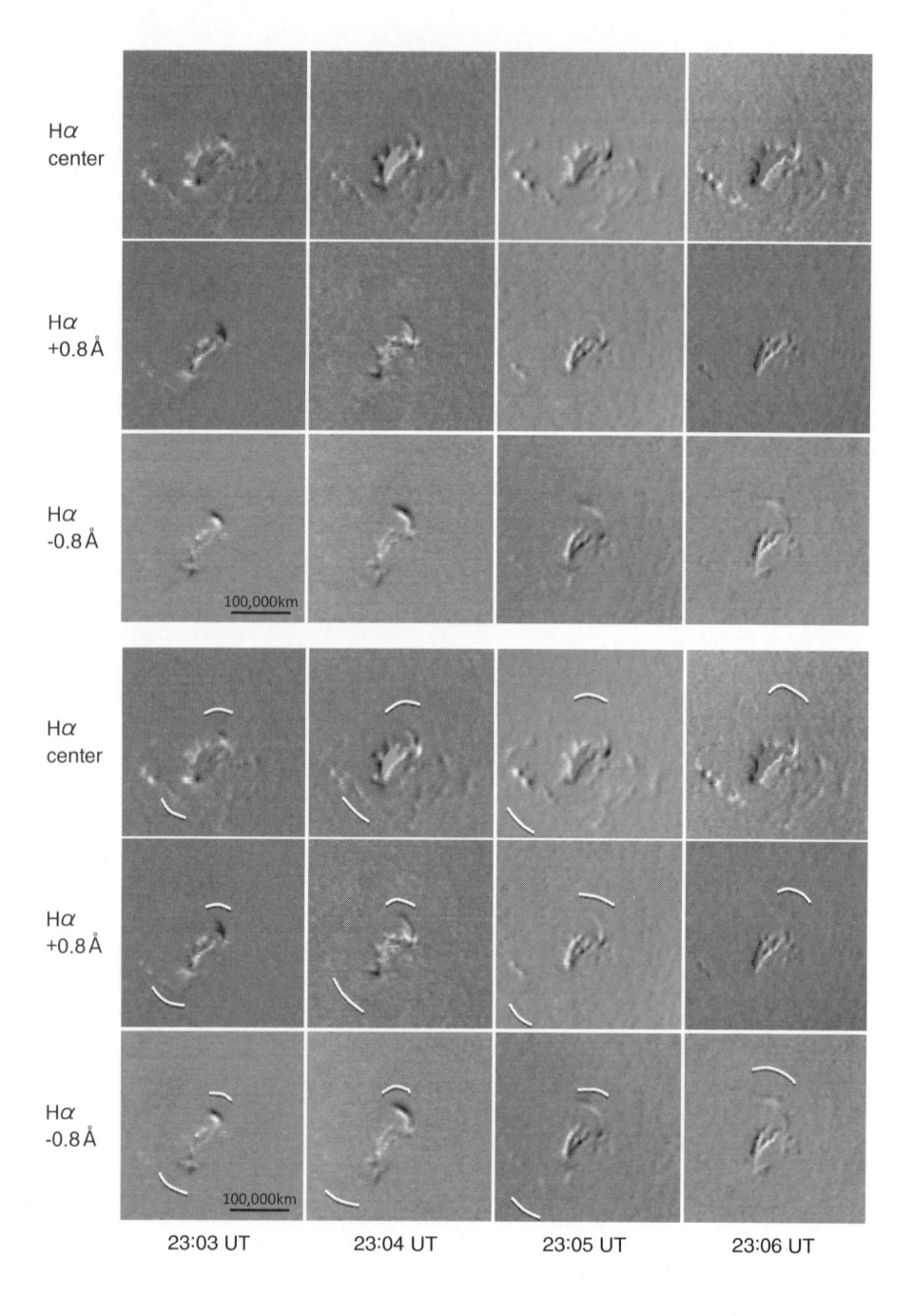

Chapter 6 Representative Events | 291

Hα center

Hα +0.8Å

Hα −0.8Å

Hα center

Hα +0.8Å

Hα −0.8Å

23:07 UT   23:08 UT   23:09 UT   23:10 UT

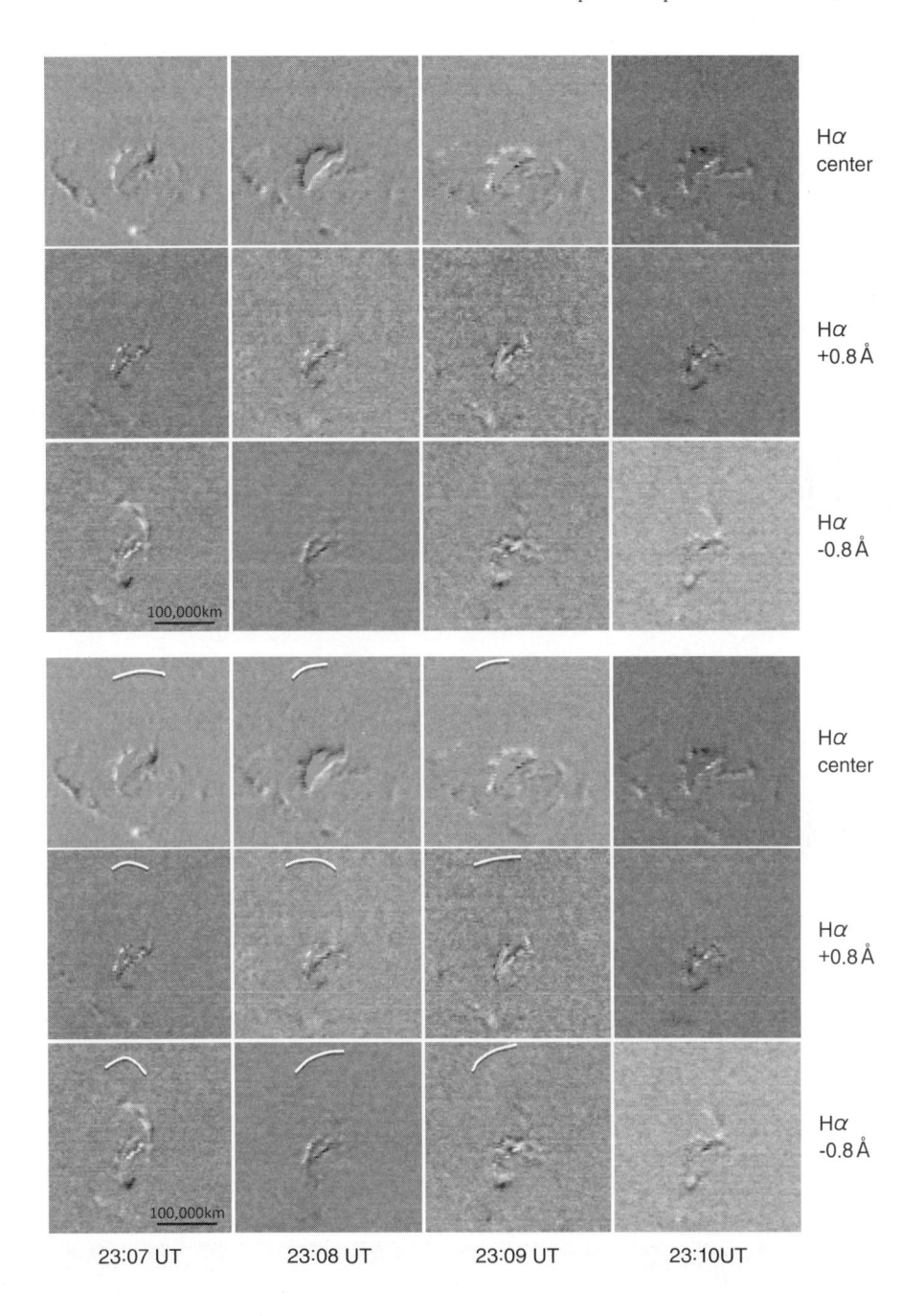

Jul 17, 2003 08:19−08:39 UT Flare

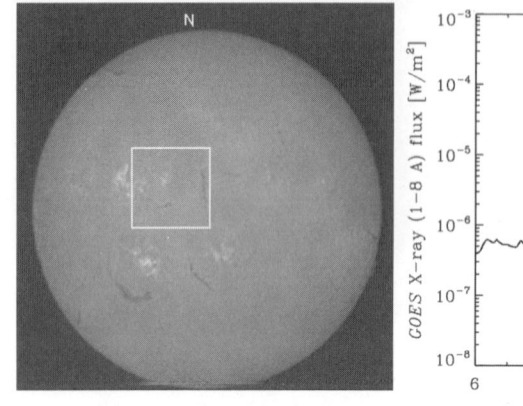

GOES Class = C9.8   NOAA 10412
N14,E12   P = 4.3[deg]   B0= 4.5[deg]

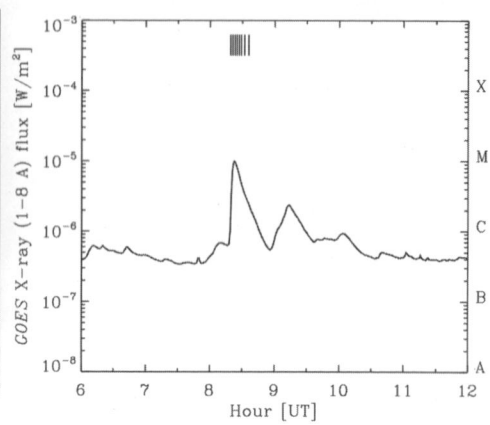

Hα center

Hα +0.8Å

Hα −0.8Å

100,000km

08:19 UT   08:21 UT   08:23 UT   08:25 UT

NOAA10412 was born on July 16, 2003 and produced a C9.8 flare on July 17. Associated with the flare, a Moreton wave was detected to propagate from the flare site to the south with a propagation speed of 940 km/s, on the video movies of FMT data. The propagation of Moreton wave in 1 min cadence is displayed in the following pages.

NOAA10412 は，2003 年 7 月 16 日に誕生し，翌日の 7 月 17 日に C9.8 クラスのフレアを発生させた．このフレアに伴ってモートン波が発生し，南側に 940 km/s の速度で伝搬するのが FMT ビデオムービーで観測された．モートン波の伝播の様子は次ページ以降に示されている．

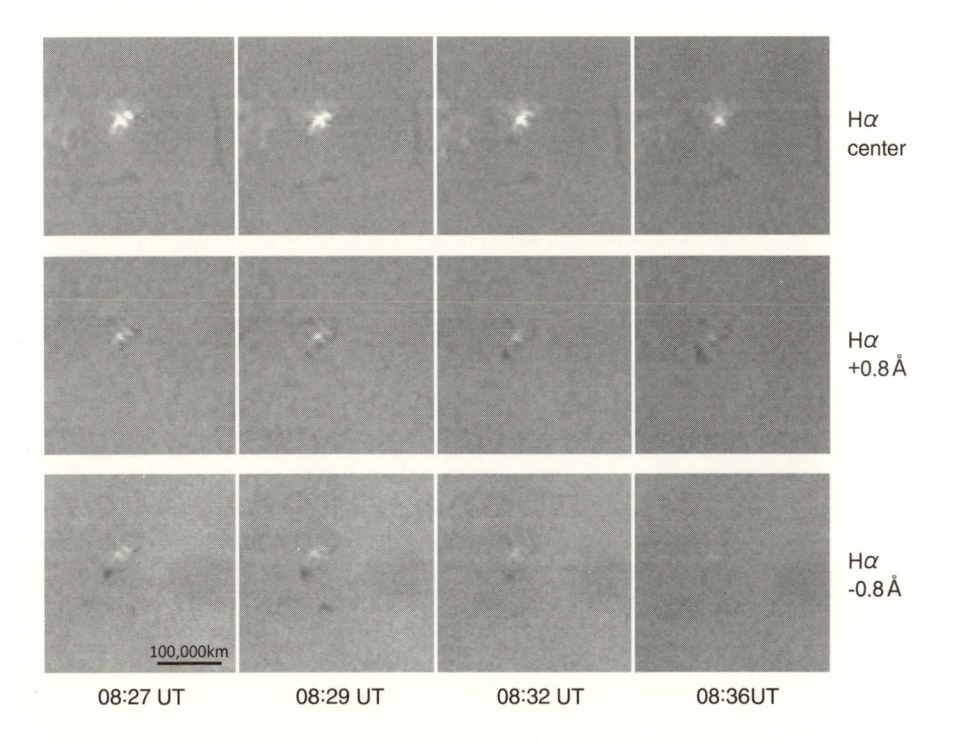

Jul 17, 2003 08:19−08:25 UT   Moreton Wave , Winking Filament

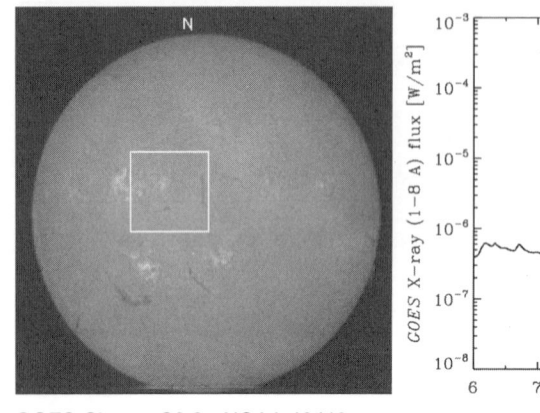

GOES Class = C9.8   NOAA 10412
N14, E12   P = 4.3 [deg]   B0 = 4.5 [deg]

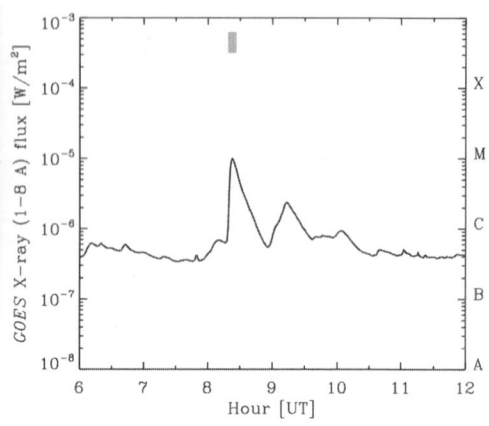

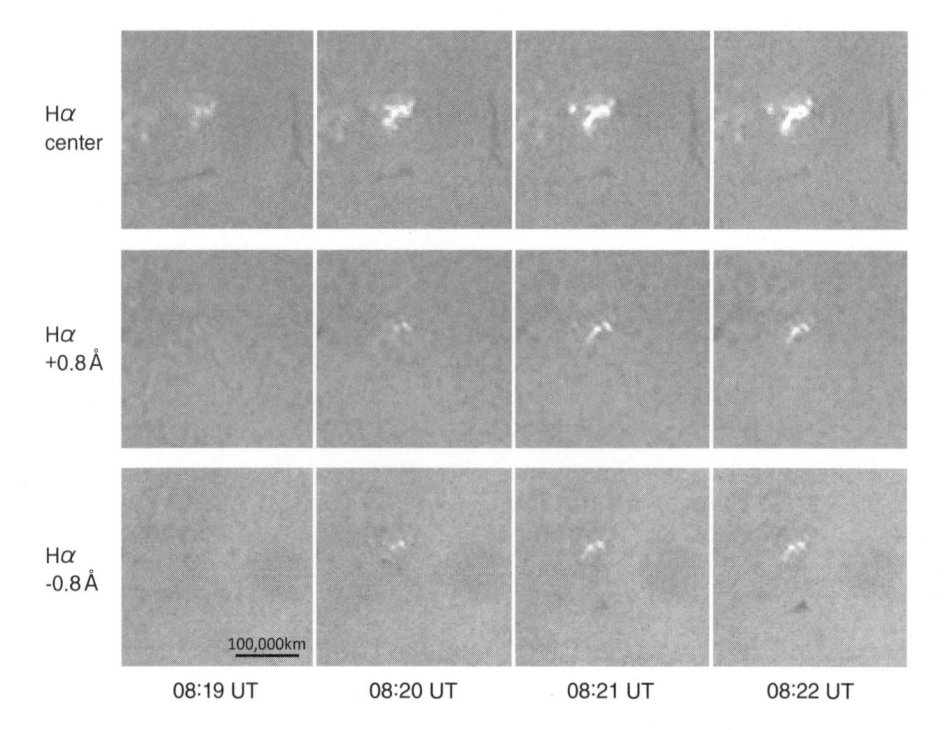

Hα center

Hα +0.8Å

Hα −0.8Å

100,000km

08:19 UT   08:20 UT   08:21 UT   08:22 UT

Here is shown the temporal evolution of the Moreton wave emitted by a C9.8 flare in NOAA10412. It propagated from the flare site to the south with a propagation speed of 940 km/s. As the amplitude of the wave was not so high, the propagation could be identified on the video movies of FMT data. When the wave hit a pre-existent filament located to the south of the region, it triggered an oscillatory filament motion (i.e. Winking filament) of 7min period. (In the next two pages, time-difference images to enhance the visibility of wave fronts are shown with the manually drawn sketch of the wave fronts.)

NOAA10412 の C9.8 クラスのフレアに伴って放射されたモートン波の時間変化を示している。波は，南側に 940 km/s の速度で伝搬していた．波の振幅が大きくないため，FMT ビデオムービーでのみ観測された．この波が活動領域の南方にあったフィラメントを直撃したとき周期約 7 分のフィラメント振動現象（ウインキングフィラメント）を引き起こした．（次の 2 ページでは，波面を見やすくするように，画像の明るさの時間変化像を示している．また波面位置のスケッチも参考のために付してある．）

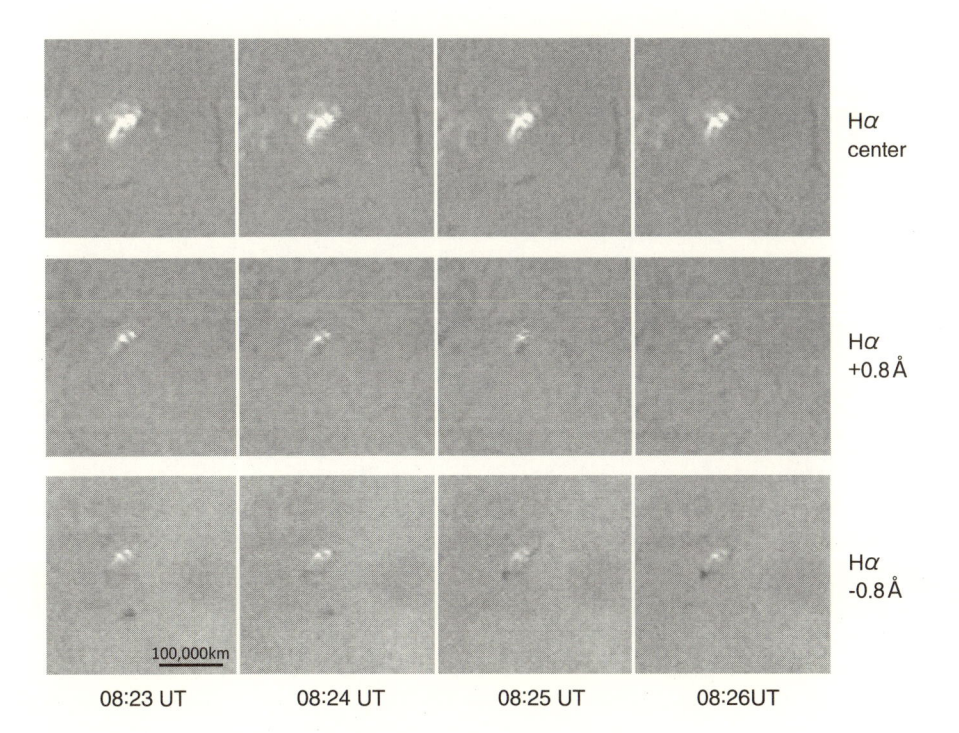

296 | 第6章 代表的イベント

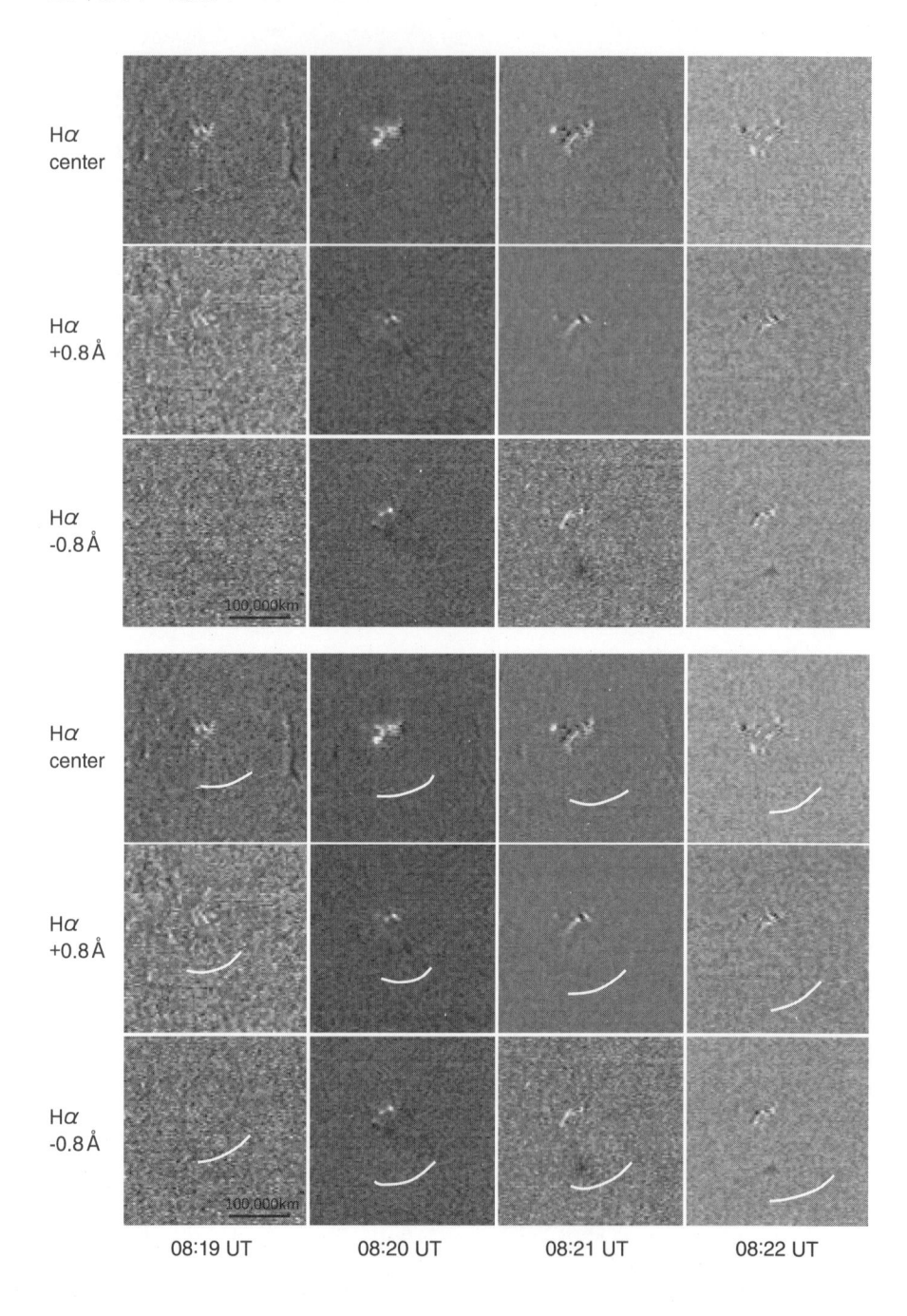

Hα center

Hα +0.8Å

Hα -0.8Å

100,000km

Hα center

Hα +0.8Å

Hα -0.8Å

100,000km

08:19 UT　　08:20 UT　　08:21 UT　　08:22 UT

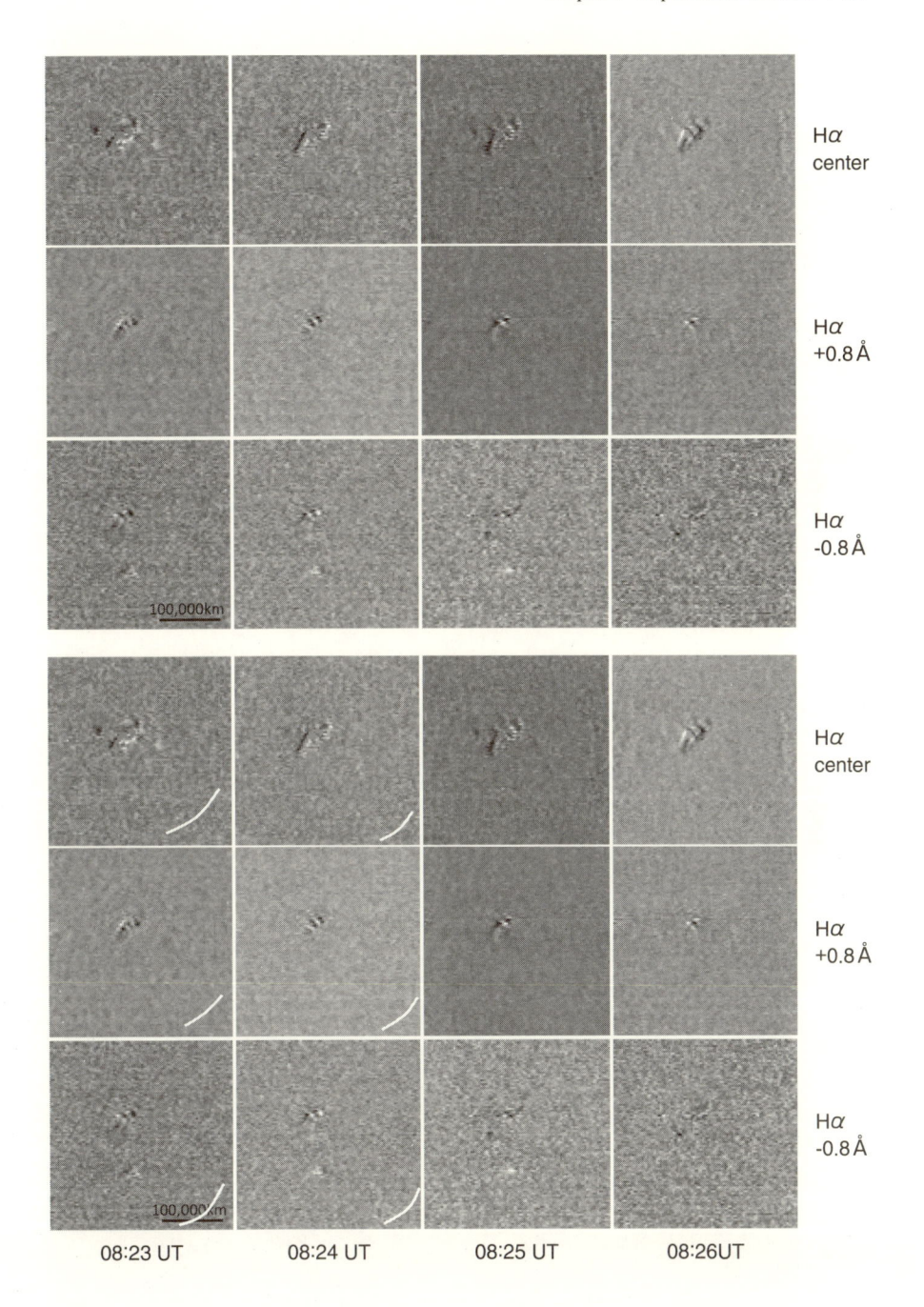

Chapter 6 Representative Events | 297

Hα center

Hα +0.8Å

Hα -0.8Å

100,000 km

Hα center

Hα +0.8Å

Hα -0.8Å

100,000 km

08:23 UT   08:24 UT   08:25 UT   08:26 UT

Jul 17, 2003 08:21−08:33 UT   Winking Filament

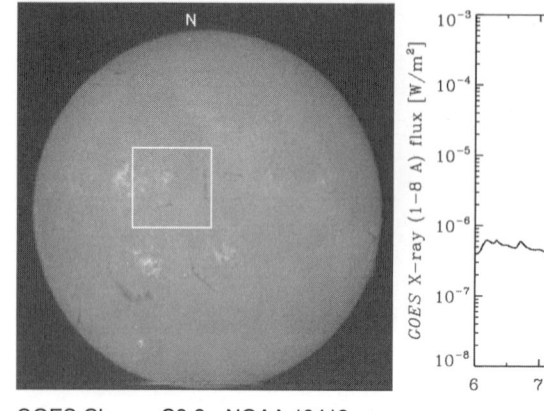

GOES Class = C9.8   NOAA 10412
N07,E12   P = 4.3 [deg]   B0= 4.5 [deg]

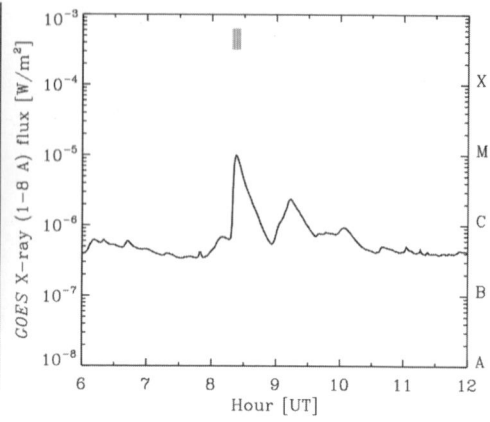

Hα center

Hα +0.8Å

Hα −0.8Å

100,000km

08:21 UT   08:23 UT   08:25 UT   08:27 UT

NOAA10412 was born on July 16, 2003 and produced a C9.8 flare on July 17. Associated with the flare, a Moreton wave was detected to propagate from the flare site to the south with a propagation speed of 940 km/s, on the video movies of FMT data. When the wave hit a pre-existent filament located to the south of the region, it triggered an oscillatory filament motion (i.e. Winking filament) of 7min period.

NOAA10412 は，2003 年 7 月 16 日に誕生し，翌日の 7 月 17 日に C9.8 クラスのフレアを発生させた．このフレアに伴ってモートン波が発生し，南側に 940 km/s の速度で伝搬するのが FMT ビデオムービーで観測された．この波が活動領域の南方にあったフィラメントを直撃したとき周期約 7 分のフィラメント振動現象（ウインキングフィラメント）を引き起こした．

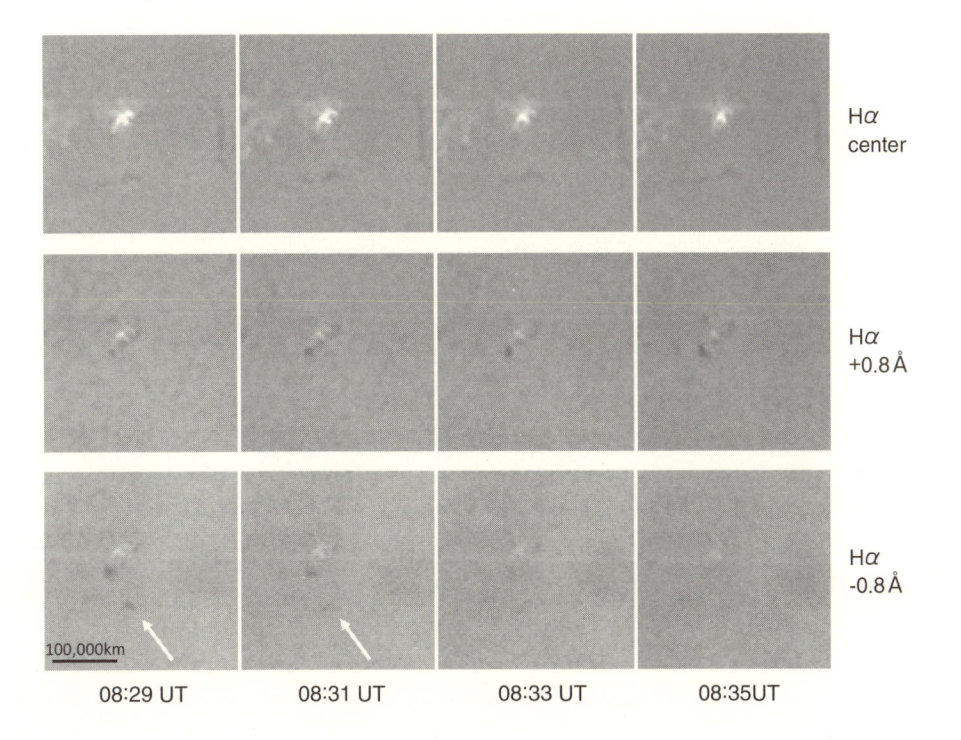

Aug 03, 2005 04:54−05:32 UT   Flare, Filament Eruption

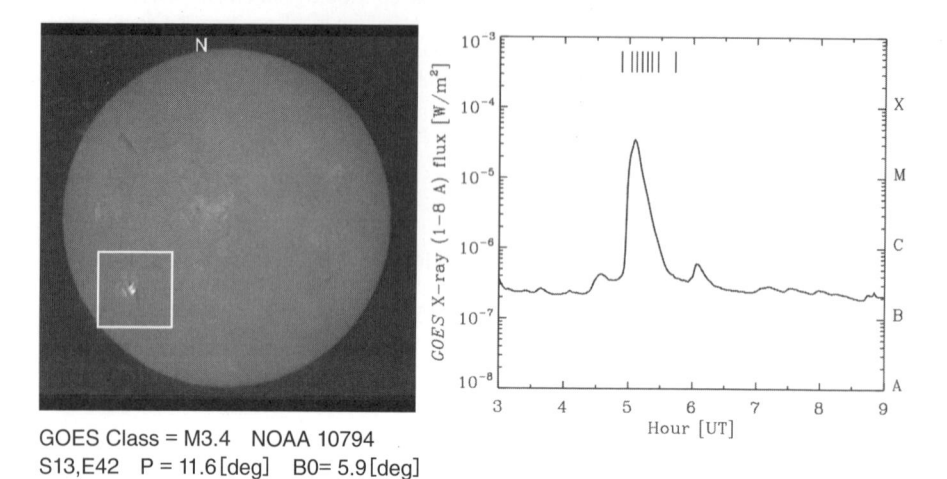

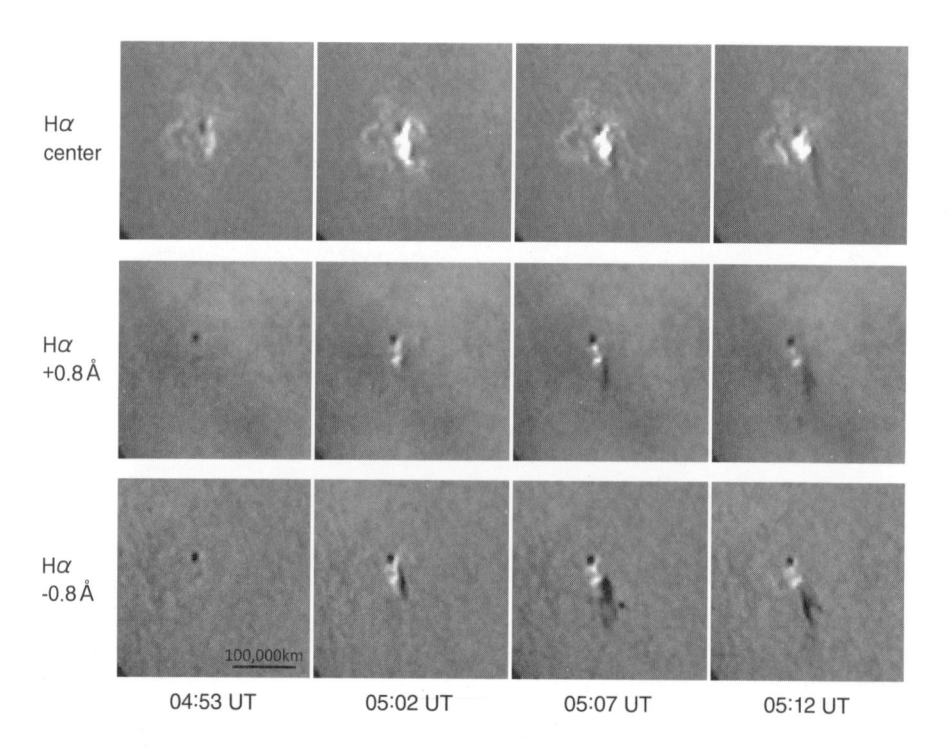

GOES Class = M3.4   NOAA 10794
S13,E42   P = 11.6 [deg]   B0 = 5.9 [deg]

Hα center

Hα +0.8Å

Hα −0.8Å

100,000km

04:53 UT   05:02 UT   05:07 UT   05:12 UT

In the β-type active region NOAA10794, there occurred a compact M3.4 flare. From the bright flaring area in the chromosphere, a few strands of gas were ejected with a speed of around 100 km/s. Narukage et al (2008) closely studied the event with Hida-SMART/Hα images, RHESSI/Hard-X data, and Hiraiso/Radiospectrograph data. According to them, filamentary gas was ejected three times successively. And Moreton waves (MHD shocks) were emitted three times successively in accordance with the gas ejections. The propagation of Moreton wave in 1 min cadence is displayed in the following pages.

β型の活動領域 NOAA10794 で，M3.4 のコンパクトなフレアが発生した．彩層フレア域から，複数本のガスが約 100 km/s の速度で放出された．Narukage et al. (2008) は，このフレアについて，飛騨天文台 SMART データ，RHESSI 衛星の硬 X 線データ，平磯電波分光観測データを用いて詳細な解析を行った．その結果，このフレアから 3 回ガス噴出が連続して発生し，それに応じてモートン波（MHD 衝撃波）が 3 回放射されていることが発見された．モートン波の伝播の様子は次ページ以降に示されている．

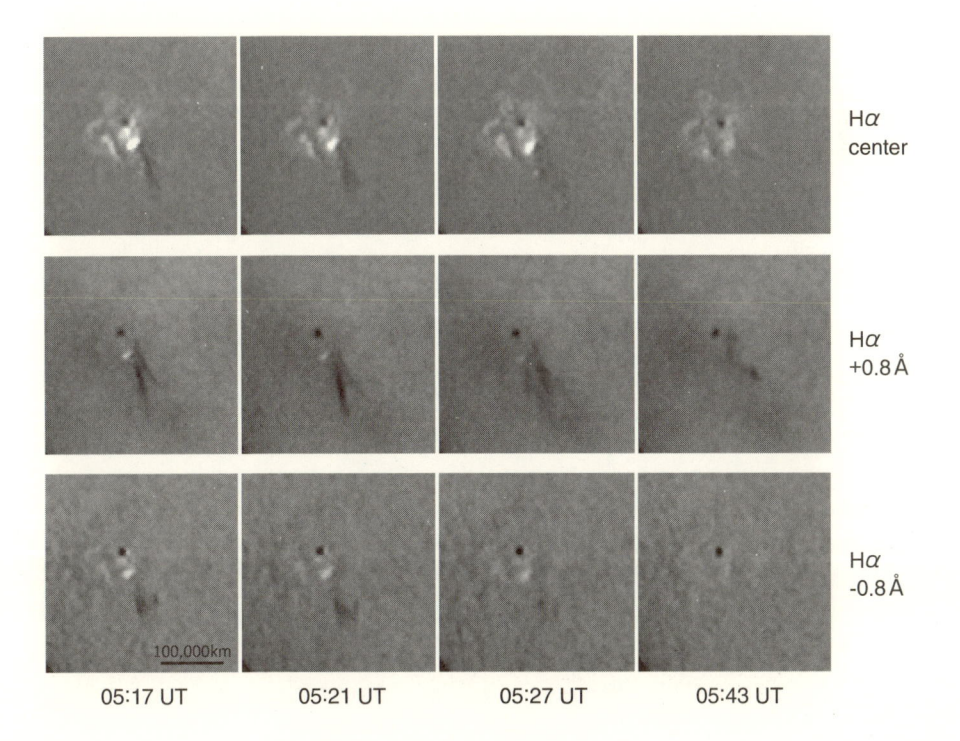

Aug 03, 2005 05:01−05:08 UT Moreton Wave

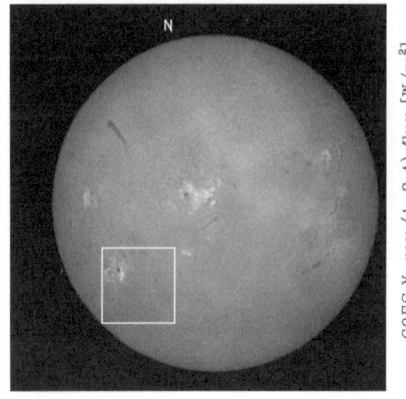

GOES Class = M3.4   NOAA 10794
S13,E42   P = 11.6 [deg]   B0= 5.9 [deg]

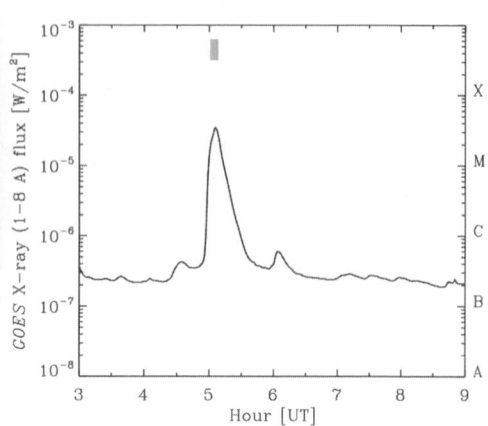

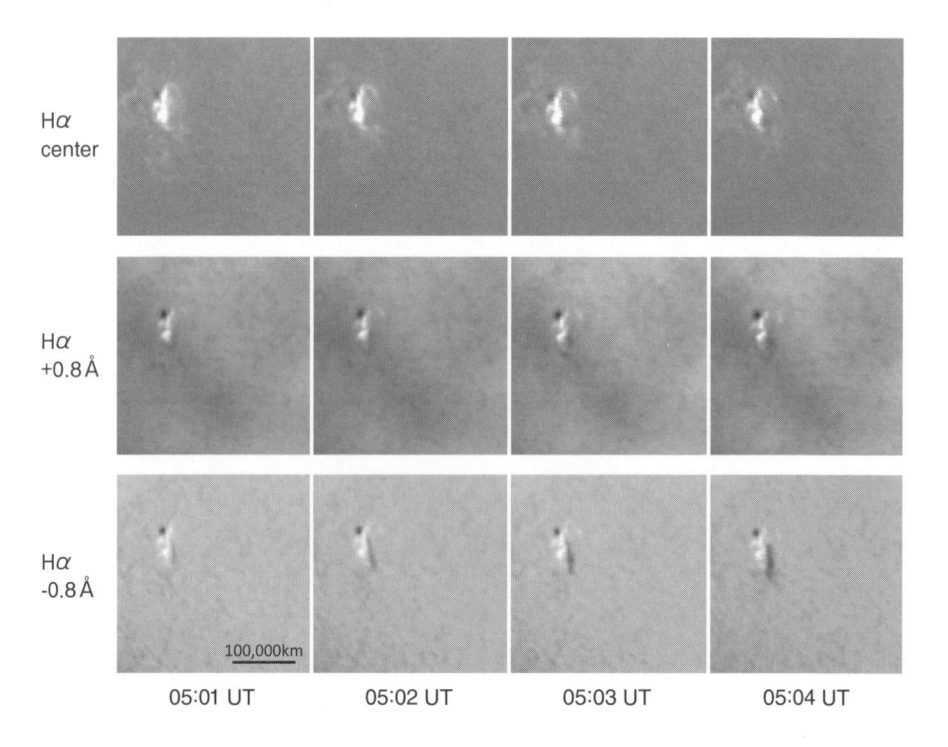

Hα center

Hα +0.8Å

Hα −0.8Å

100,000 km

05:01 UT   05:02 UT   05:03 UT   05:04 UT

The Moreton waves were emitted three times successively in association of the M3.4 flare in NOAA10794. Here is shown the temporal evolution of the thirdly appeared wave. According to Narukage et al (2008), the speeds of Moreton waves (shocks) ranged from 500 to 1,200 km/s. When the first shock was surpassed by the second higher-speed shock, the shock-shock interaction produced strong radio emissions. Moreover, firstly ejected gas was accelerated by the gas-shock interaction when the thirdly emitted shock surpassed the filament gas. These physically important interactions were firstly observed and found in this interesting event. (In the next two pages, time-difference images to enhance the visibility of wave fronts are shown with the manually drawn sketch of the wave fronts.)

活動領域 NOAA10794 で発生した M3.4 フレアに伴って，モートン波が 3 回連続して放出された．ここでは 3 回目に現れたモートン波の時間変化を示している．Narukage et al (2008) によると，モートン波に対応する衝撃波の伝播速度は 500-1,200 km/s というものであった．この現象では，最初の衝撃波に 2 回目の衝撃波が追突し，2 つの衝撃波相互作用の結果強い電波が放射されることが初めて確認された．さらに，最初に放出されたフィラメントガスを 3 回目の衝撃波が追い越してゆくとき，ガスが加速されるというプラズマ－衝撃波相互作用も発見された．この現象は物理的に重要なこれらの相互作用が初めてとらえられた興味あるイベントである．（次の 2 ページでは，波面を見やすくするように，画像の明るさの時間変化像を示している．また波面位置のスケッチも参考のために付してある．）

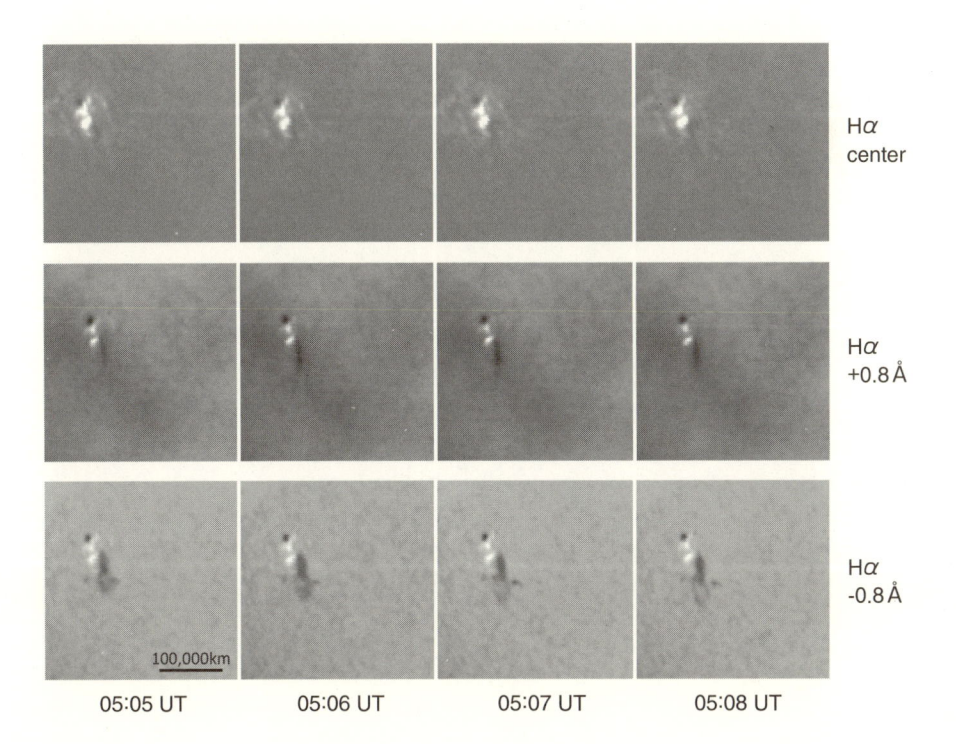

304 | 第6章 代表的イベント

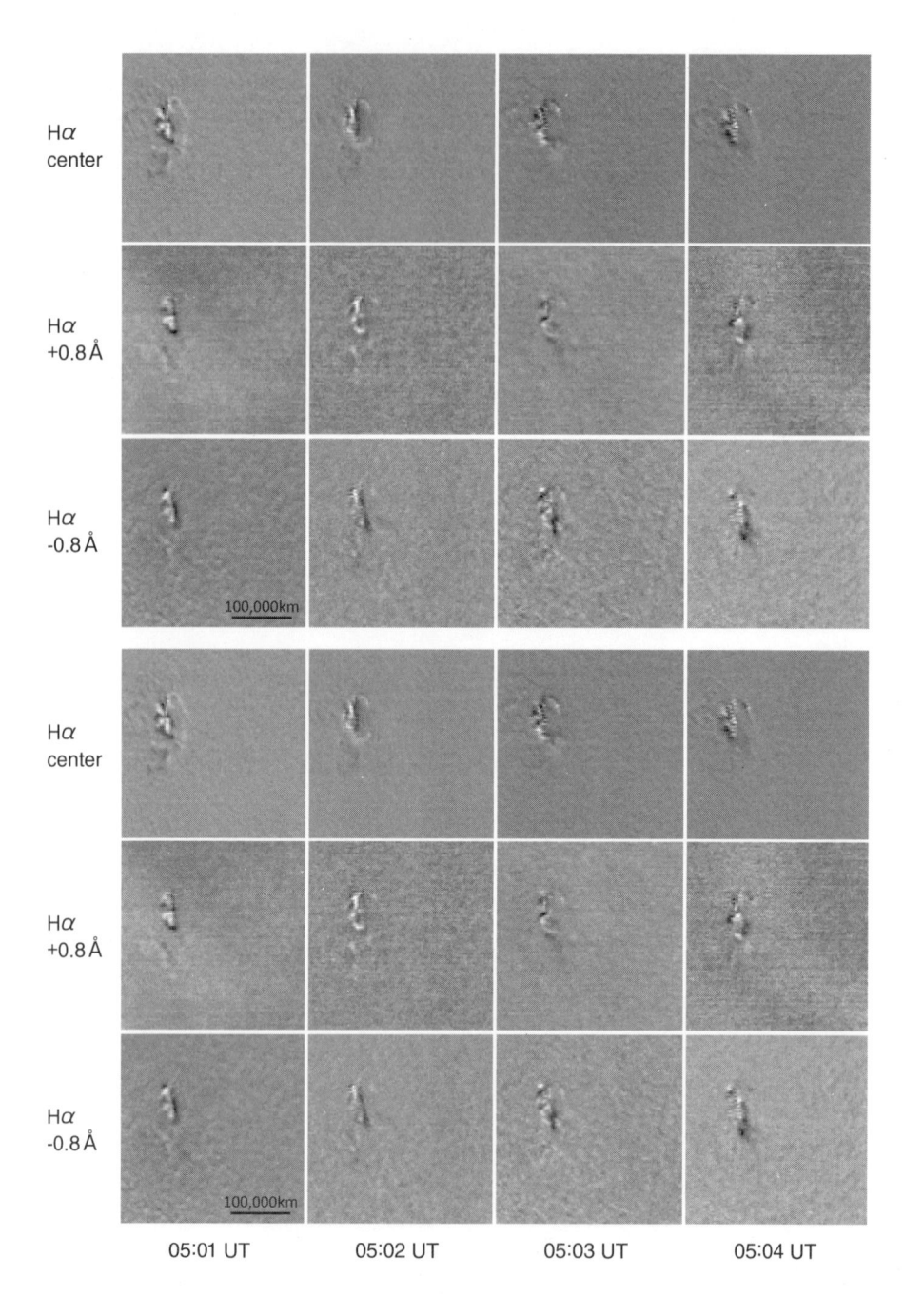

Hα
center

Hα
+0.8Å

Hα
-0.8Å

100,000km

Hα
center

Hα
+0.8Å

Hα
-0.8Å

100,000km

05:01 UT　　　05:02 UT　　　05:03 UT　　　05:04 UT

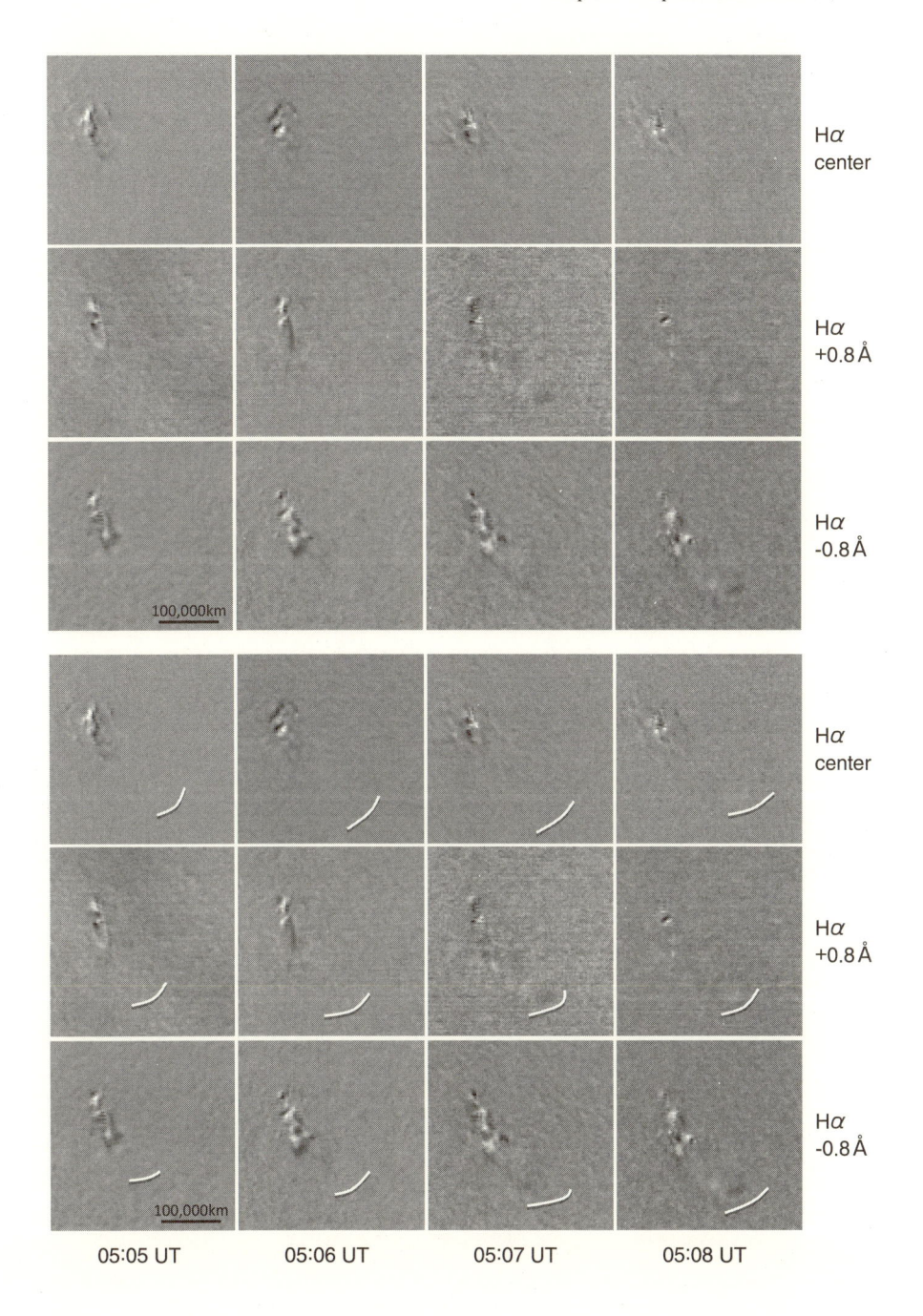

Chapter 6 Representative Events | 305

Hα center

Hα +0.8Å

Hα -0.8Å

100,000km

Hα center

Hα +0.8Å

Hα -0.8Å

100,000km

05:05 UT   05:06 UT   05:07 UT   05:08 UT

6.3 代表的リムイベント（プロミネンス）

この節ではFMTで観測された代表的なリムイベント（プロミネンス）を紹介する．表6.2に代表的リムイベントの一覧を示す（開始時刻，終了時刻，位置，種類，高さおよび長さについては第5章の5.2節および第4章の表4.2を，またPおよびB0については第6章の6.1.2をそれぞれ参照）．

6.3 Representative Limb Events (Prominences)

This section shows the representative limb events (prominences) observed through the FMT. Table 6.2 is a summary list of the representative limb events. Please refer to Section 5.2 of Chapter 5 and Table 4.2 of Chapter 4 for the definitions of start time, end time, position, class, height, and length. P and B0 have been defined in Subsection 6.1.2 of this chapter.

表 6.2 代表的リムイベント一覧.
Table 6.2 Representative Limb Events.

イベント種別 Event Type	日付 Date	開始時刻 − 終了時刻	位置 Position	P, B0	種類 Class	高さ Height (Mm)	長さ Length (Mm)
Prominence Eruption	1992/7/16	07:41 − 08:27	WS 41	4.5, 4.2	B1	243	243
Prominence Eruption	1992/7/30	23:50 − 01:01	EN 17	10.1, 5.7	＊B1	490	730
Prominence Eruption	1995/4/9	23:05 − 00:15	ES 32	−26.3, −6.1	＊B1	243	243
Prominence Eruption	1999/2/8	06:30 − 07:17	ES 40	−14.8, −6.1	#B1	−	105
Prominence Eruption	2000/5/4	04:36 − 05:46	WS 17	−23.5, −3.8	C2	162	162
Prominence Eruption	2000/8/3	07:44 − 08:42	WN 20	11.7, −5.9	B2	405	162
Prominence Eruption	2000/11/4	00:00 − 01:58	WS 30	23.9, 4.0	B1	35	35
Post Flare Loop	2001/12/28	23:35 − 06:17	ES 20 −50	−4.1, −2.5	F1 or F2	122	405
Prominence Eruption	2002/4/15	02:50 − 03:20	NW 15 −20	−26.1, −5.6	B2	122	81
Prominence Eruption	2002/4/22	00:06 − 00:30	WS 20	−25.5, −5.0	B2	122	41
Prominence Eruption	2002/5/12	02:52 − 03:35	NW 10 −20	−21.8, −3.0	F2	203	162
Prominence Eruption	2002/9/3	23:07 − 23:40	NE 0−20	21.5, 7.2	B2 or D1	122	122
Prominence Eruption	2002/11/18	23:07 − 03:38	NW 0 −10	20.6, 2.5	F1	365	324
Surge Prominence	2002/11/23	01:25 − 02:00	NE 10 −15	19.0, 1.9	A1	122	41
Prominence Eruption	2003/4/9	23:28 − 00:19	WS 0−10	−26.3, −6.1	A2 or B2	162	81
Prominence Eruption	2003/4/27	01:02 − 03:29	NW 0−15	−24.9, −4.6	F2		324

Jul 16, 1992 07:41−08:27 UT Prominence Eruption

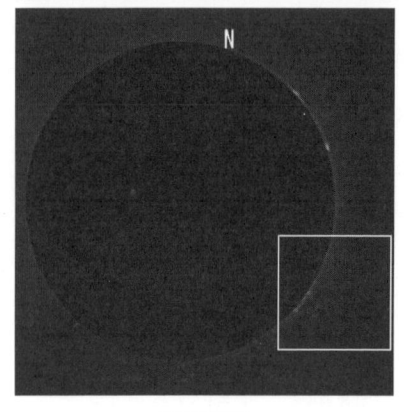

07:59 UT

WS 41
P = 4.2 [deg]   B0 = 4.5 [deg]

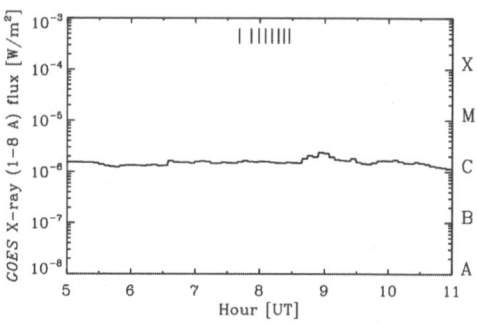

A prominence rose up and exploded away with an average speed of around 100 km/s. As one of the foot point of the prominence was nearly fixed, the original location of the prominence, probably, was near the limb.

プロミネンスがリムから現れ，上昇膨張して消失した．平均速度は約 100 km/s であった．片方の足元が固定位置にあったことから，このプロミネンスは元々リム近くにあったものと思われる．

Temporal Evolution in Hα center

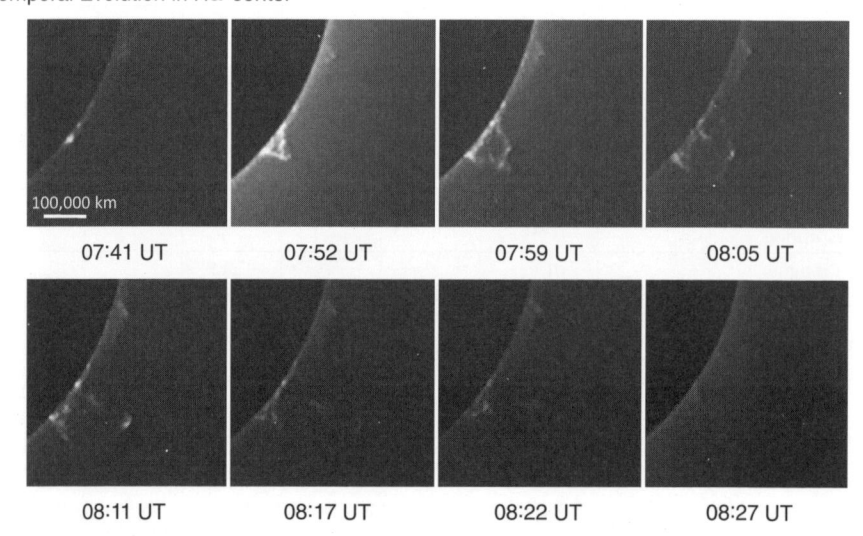

07:41 UT   07:52 UT   07:59 UT   08:05 UT

08:11 UT   08:17 UT   08:22 UT   08:27 UT

Jul 30, 1992 23:50−01:01 UT Prominence Eruption

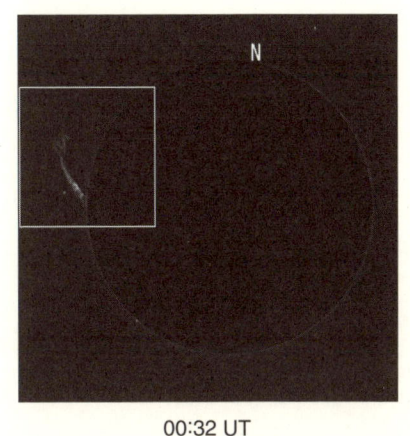

00:32 UT

EN 17
P = 10.1 [deg]   B0= 5.7 [deg]

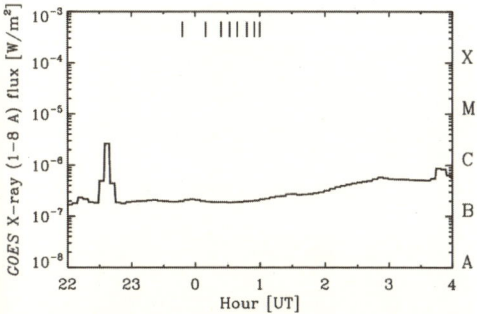

A gigantic disk filament erupted as an entity and expanded with a speed of about 100 km/s crossing over the limb. According to the analysis by Hanaoka et al. (1994) on Yohkoh SXT and Nobeyama 17 GHz radio-heliographic observations, a hot (T=3.5 MK) coronal arcade was formed at the original location of the filament after the eruption and increased its height with a speed of 5 km/s.

巨大なフィラメントが一体として上昇，膨張拡大し100 km/sの速度を示すプロミネンス爆発として太陽縁外に現われた．Hanaoka et al. (1994) によるようこう衛星の軟X線望遠鏡および野辺山での17 GHz電波ヘリオグラフ観測の解析から，フィラメント爆発後，フィラメントの当初位置に350万度の高温のコロナアーケード構造が形成され，5 km/sの速度でゆっくりと上昇していたことがわかった．

Temporal Evolution in Hα center

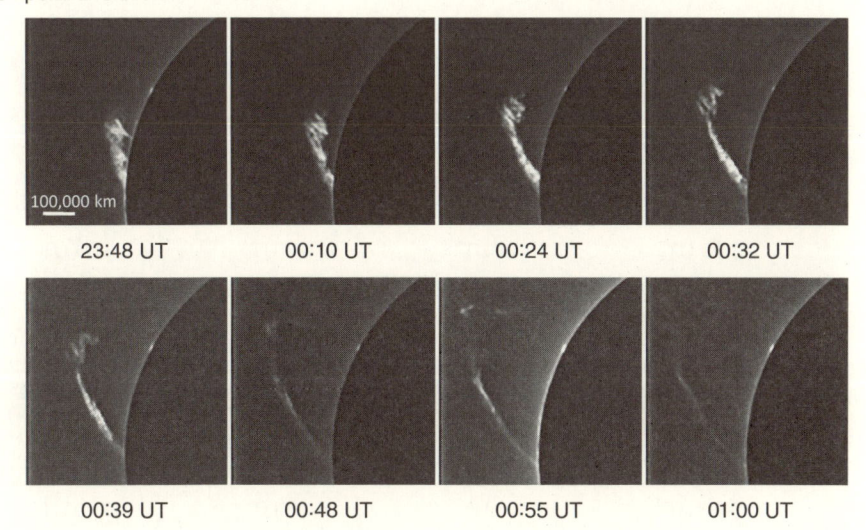

100,000 km

23:48 UT   00:10 UT   00:24 UT   00:32 UT

00:39 UT   00:48 UT   00:55 UT   01:00 UT

Apr 09, 1995 23:05−00:15 UT Prominence Eruption

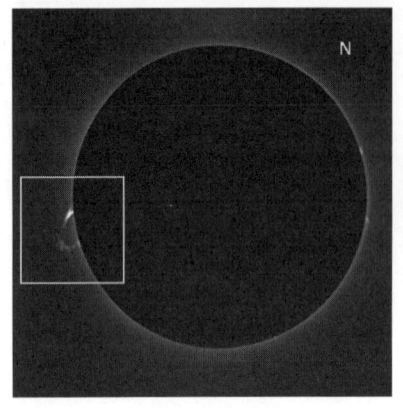

23:27 UT

ES 32
P =-26.3 [deg]   B0= -6.1 [deg]

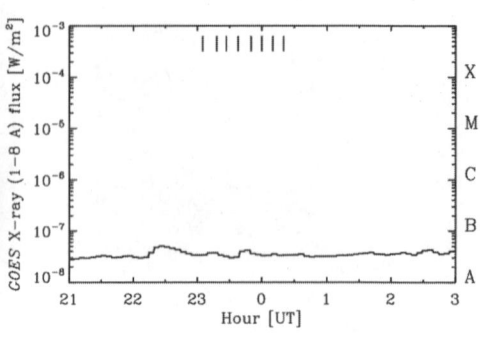

A prominence loop expanded with a speed of 37 km/s and faded away in about 80 min.

ループ状のプロミネンスが 37 km/s の速さで膨張し,その後消滅した.活動時間はほぼ 80 分であった.

Temporal Evolution in Hα center

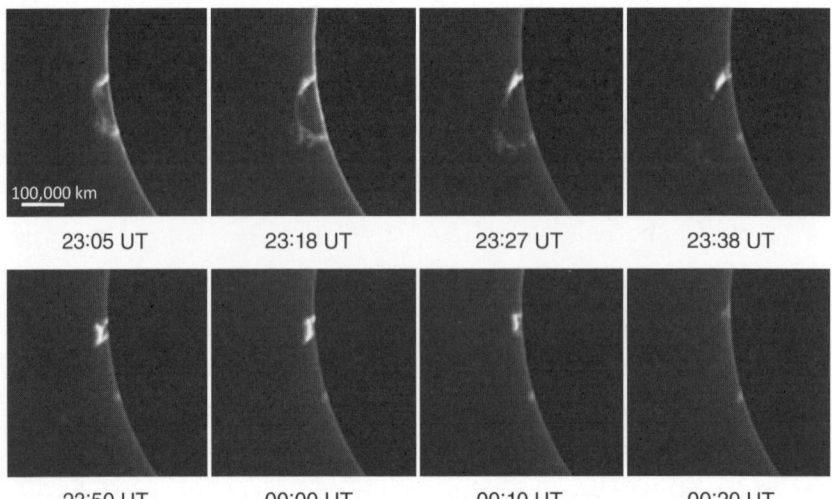

100,000 km

23:05 UT 23:18 UT 23:27 UT 23:38 UT

23:50 UT 00:00 UT 00:10 UT 00:20 UT

Feb 08, 1999 06:30−07:17 UT Prominence Eruption

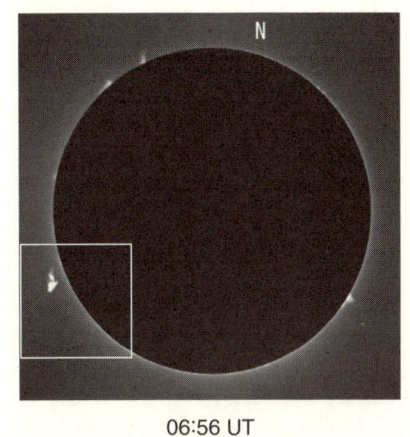

06:56 UT

ES40
P = -14.8 [deg]   B0= -6.1 [deg]

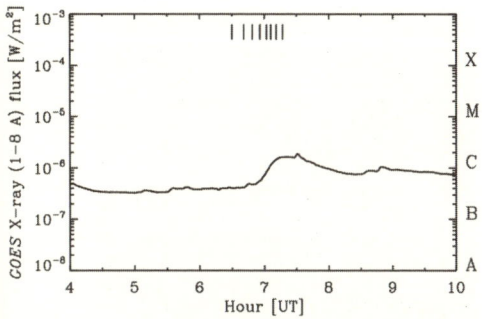

The prominence appeared above the limb slowly (30 km/s) at the initial phase, was accelerated to 170 km/s, and then exploded away in about 50 min. The event was, probably, related to a flare (C1.8 class) in the NOAA8458 just behind the east limb.

当初 30 km/s のゆっくりとした速度でリム外に現れたプロミネンスが，170 km/s まで加速されて膨張後消滅した．この間 50 分であった．この現象は，東のリムの向こう側にいた NOAA8458 での C1.8 フレアに付随したものと考えられる．

Temporal Evolution in Hα center

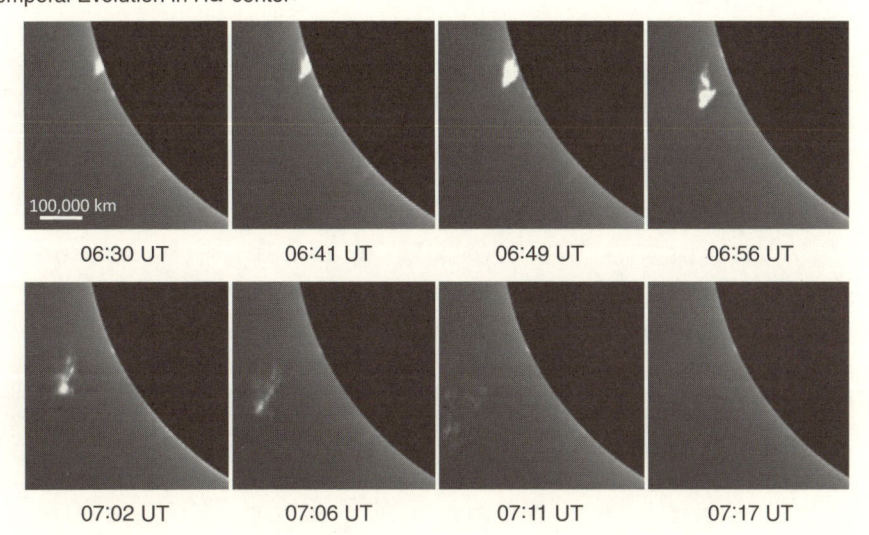

06:30 UT   06:41 UT   06:49 UT   06:56 UT

07:02 UT   07:06 UT   07:11 UT   07:17 UT

May 04, 2000 04:36 − 05:46 UT Prominence Eruption

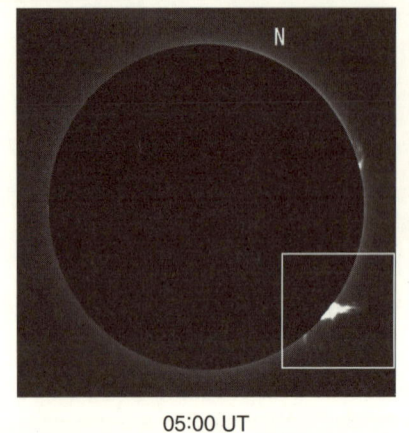

05:00 UT

WS17
P = −23.5 [deg]   B0= −3.8 [deg]

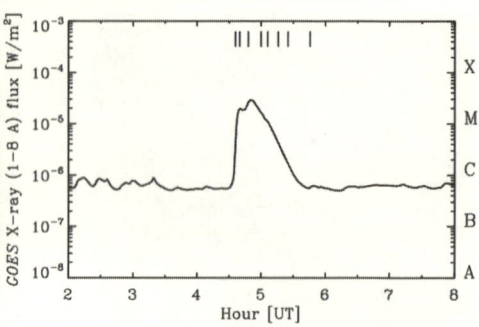

When an M2.8 flare occurred in NOAA8977, a bright EUV loop was observed above the limb by SOHO/EIT. In a few minutes, Hα prominence gas came into appearance and exploded with a speed of 170 km/s. An associated CME of bubble structure was observed to expand with a speed of about 1,500 km/s.

NOAA8977 で M2.8 クラスのフレアが発生したとき，まず SOHO/EIT で明るい極端紫外線ループがリム上に現れた．2-3 分後に Hα プロミネンスが現れ 170 km/s の速度で爆発した．このとき，泡状の CME が 1,500 km/s の速度で膨張してゆくのが観測された．

Temporal Evolution in Hα center

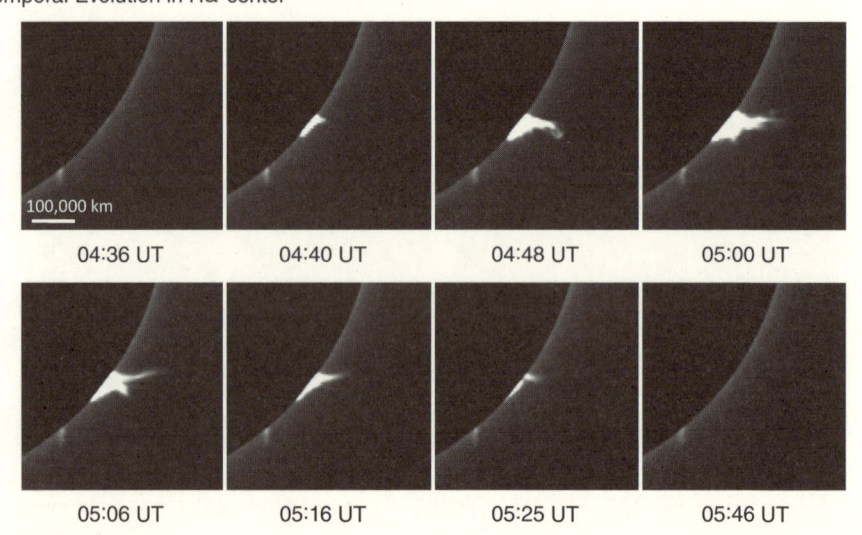

100,000 km

04:36 UT   04:40 UT   04:48 UT   05:00 UT

05:06 UT   05:16 UT   05:25 UT   05:46 UT

Aug 03, 2000 07:44−08:42 UT Prominence Eruption

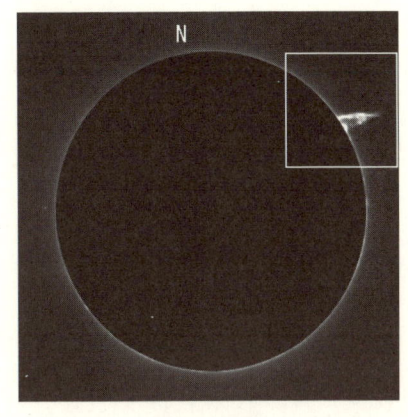

08 : 14UT

WN20
P = 11.7 [deg]   B0 = 5.9 [deg]

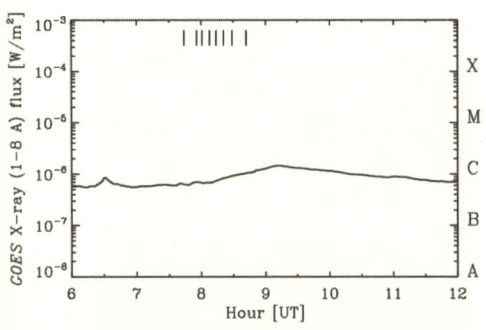

A long quiescent filament near the west limb slowly rose up and appeared as an erupting prominence. The initial rise speed of 40 km/s was accelerated to 140 km/s in later phase. SOHO/EIT observed EUV brightening on the disk, while SOHO/LASCO detected an associated helical CME event, the expansion speed of which was around 780 km/s.

西リム近くのディスク内に位置していた長い静穏型フィラメントが上昇して，リム外にプロミネンスとして現れた．当初 40 km/s の上昇速度が，後には 140 km/s に加速された．SOHO/EIT では，極端紫外線光の明るい斑点がフィラメントの当初位置に見えており，SOHO/LASCO では，螺旋形の CME が速度 780 km/s で膨張するのが確認されている．

Temporal Evolution in Hα center

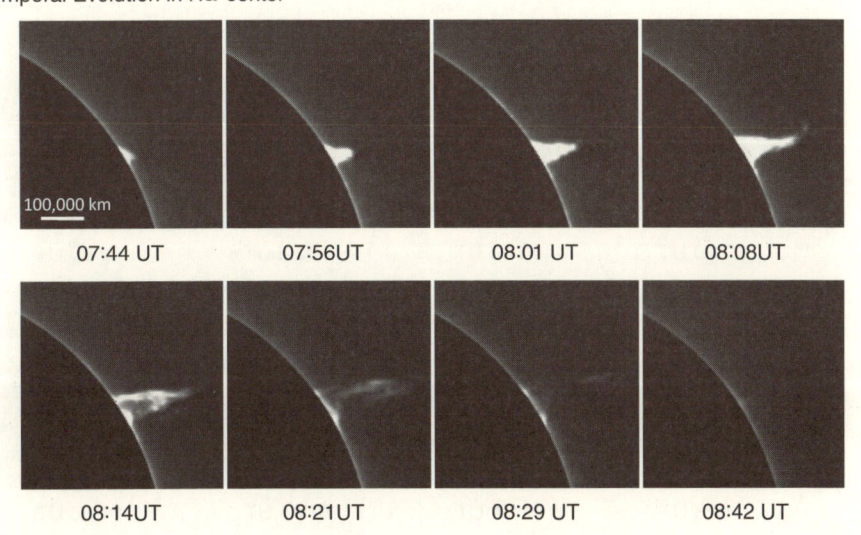

100,000 km

07:44 UT   07:56 UT   08:01 UT   08:08 UT

08:14 UT   08:21 UT   08:29 UT   08:42 UT

Nov 04, 2000 00:00−01:58 UT Prominence Eruption

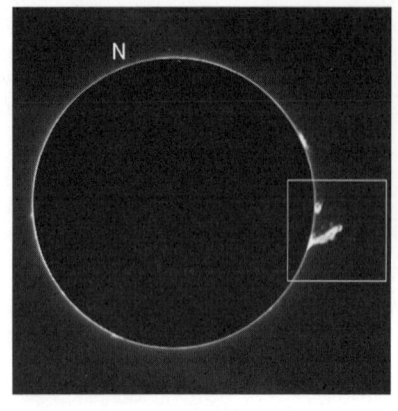

01:23 UT

WS30
P = 23.9 [deg]　B0= 4.0 [deg]

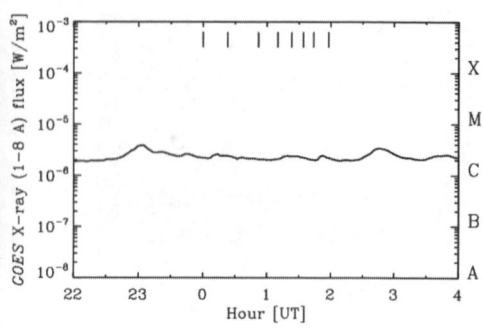

SOHO/EIT observed the initial filament rise 4 hours before the appearance at the limb. After the slow rise of 19 km/s, the prominence exploded when one of the footpoint of the filament tether-cut. SOHO/LASCO observed an associated CME of bubble structure, whose speed of expansion was around 780 km/s.

SOHO/EIT 観測によると，ディスク内のフィラメントがゆっくりと上昇し，4時間後リムでプロミネンスとして現れた．当初は 19 km/s の上昇速度を示したが，プロミネンスの片方の足元で切り離しが起こって爆発的に上昇した．SOHO/LASCO では，780 km/s で膨張する泡構造の CME が伴っていることが捉えられた．

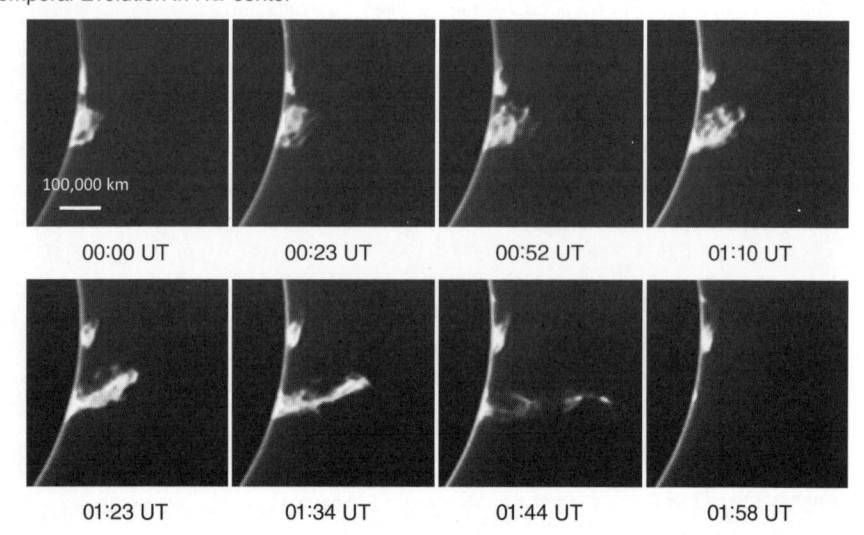

Temporal Evolution in Hα center

00:00 UT　　00:23 UT　　00:52 UT　　01:10 UT

01:23 UT　　01:34 UT　　01:44 UT　　01:58 UT

Dec 28, 2001 23:35−06:17 UT Post Flare Loop

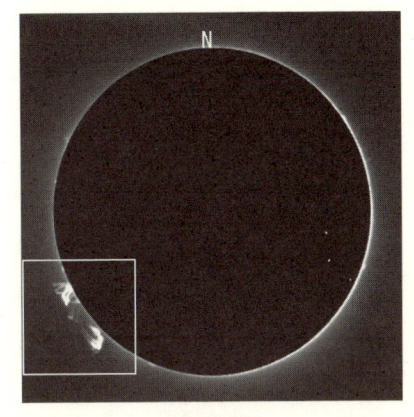

00:29 UT

ES20-50
P = 4.1 [deg]   B0= -2.5 [deg]

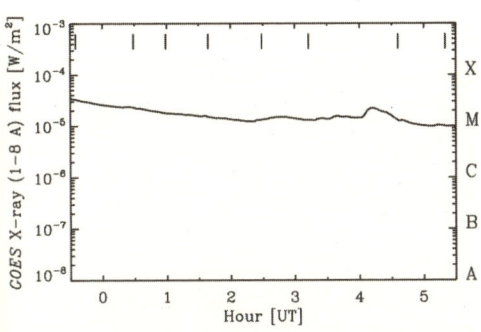

The event was a post flare loop system produced by an X3 class flare on Dec 28, 2001 at 20:00 UT. The source active region NOAA9767 was located behind the east limb at the time of the flare. SOHO/LASCO detected a tremendously large-scale CME on Dec 28 20:30 UT.

この現象は，2001年12月28日の20:00 UTに発生したX3クラスのフレアによって発生したポストフレアループである．フレアを起こした領域NOAA9767は，このときは東のリムの裏側に位置していた．SOHO/LASCOでは，12月28日20:30 UTに極めて大規模なCME現象が観測されている．

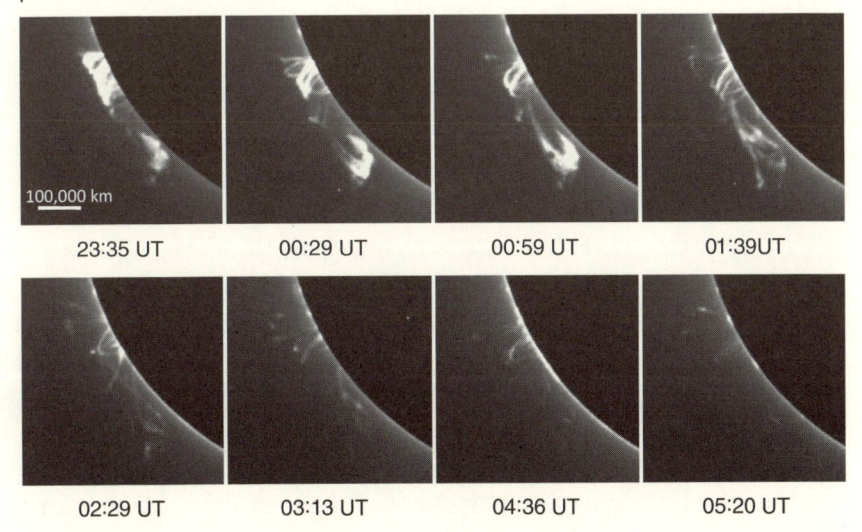

Temporal Evolution in Hα center

100,000 km

23:35 UT   00:29 UT   00:59 UT   01:39 UT

02:29 UT   03:13 UT   04:36 UT   05:20 UT

Apr 15, 2002 02:50−03:20 UT Prominence Eruption

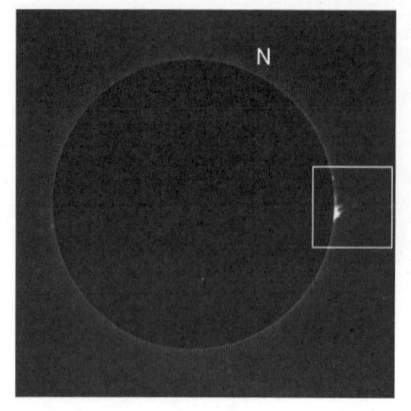

03:04 UT

WN15-20
P = -26.1 [deg]   B0= -5.6 [deg]

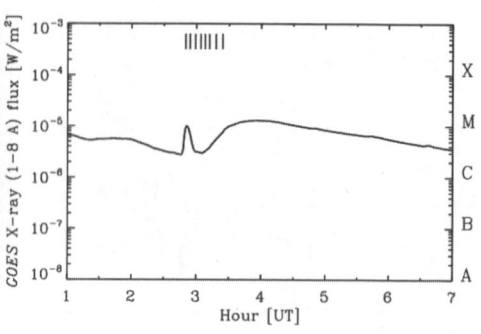

A C9.8 flare occurred in a δ-type active region NOAA9893, located near the west limb. Several ejections of Hα gas occurred over the limb with speeds of 100-150 km/s. SOHO/LASCO observed the expansion of an associated CME with a speed of 760 km/s.

西のリム近くのδ型活動領域 NOAA9893 で C9.8 クラスのフレアが発生した．このとき，100-150 km/s の速度でリム外に Hα ガスが放出された．このとき SOHO/LASCO では，760 km/s で CME が膨張するのが観測された．

Temporal Evolution in Hα center

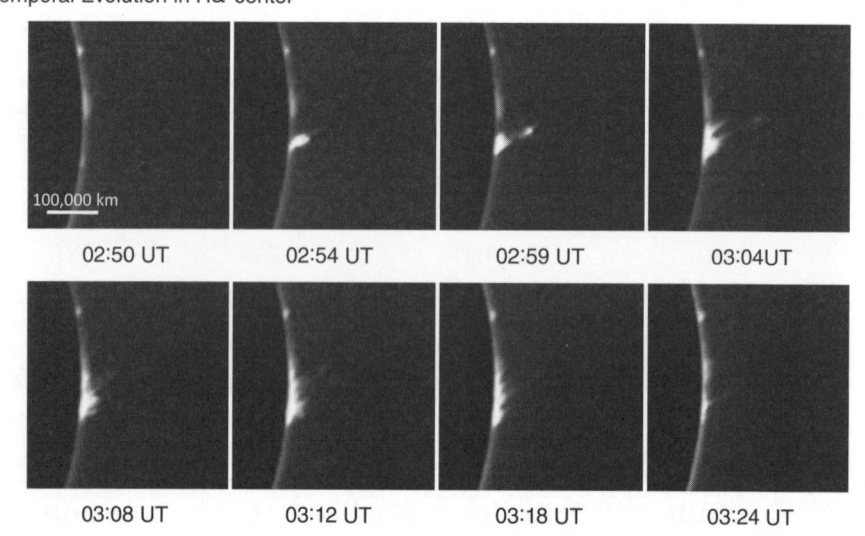

02:50 UT   02:54 UT   02:59 UT   03:04 UT

03:08 UT   03:12 UT   03:18 UT   03:24 UT

Apr 22, 2002 00:06−00:30 UT Prominence Eruption

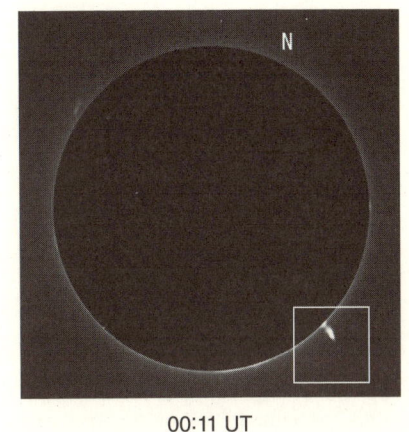

00:11 UT

WS20
P = -25.5 [deg]   B0= -5.0 [deg]

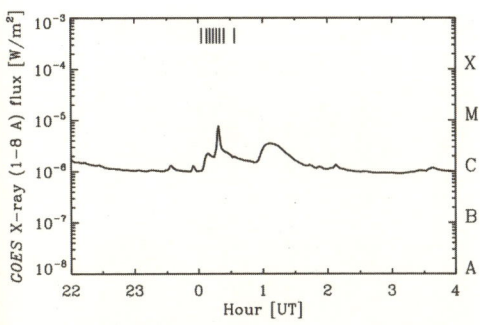

After the west limb passage of a flare-productive region NOAA9906, a C7.7 flare occurred behind the limb, and ejected several blobs of Hα gas with speeds of 100-120 km/s. SOHO/LASCO detected an associated CME with a very high speed of 1,200 km/s.

活発にフレアを引き起こした領域 NOAA9906 が西のリムを越えた後で，C7.7 クラスのフレアを発生させた．そのとき 100-120 km/s の速さで Hα ガスを放出した．また，SOHO/LASCO では，1,200 km/s もの速さで CME が膨張しているのが観測された．

Temporal Evolution in Hα center

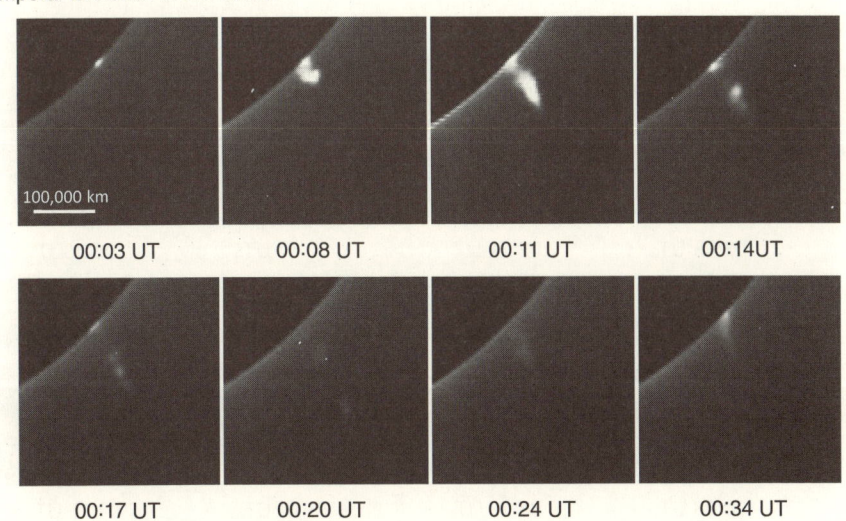

100,000 km

00:03 UT   00:08 UT   00:11 UT   00:14 UT

00:17 UT   00:20 UT   00:24 UT   00:34 UT

May 12, 2002 02:52−03:35 UT Prominence Eruption

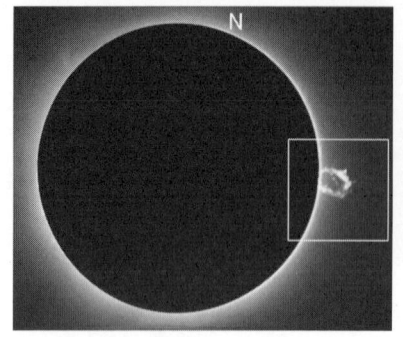

03:03 UT

WN10-20
P = -21.8 [deg]   B0= -3.0 [deg]

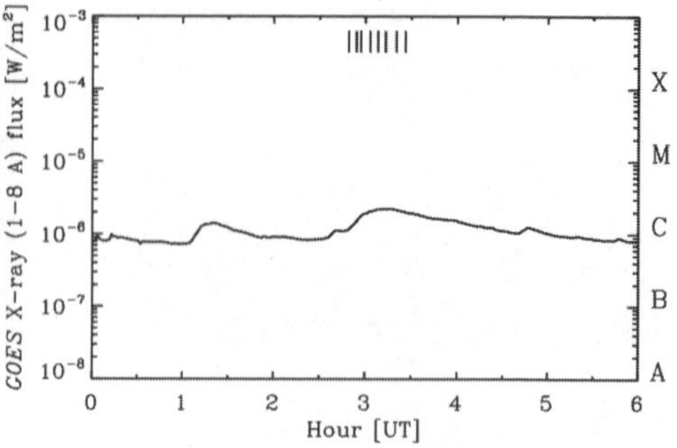

A prominence of helical tube shape erupted with a speed of 200 km/s. SOHO/LASCO detected an associated CME of bubble shape expanding with a speed of around 500 km/s.

ねじじられたチューブのような形のプロミネンスが 200 km/s の速さで膨張上昇した．このとき，SOHO/LASCO では，500 km/s の速さで泡構造をした CME が広がっているのが観測された．

Temporal Evolution in Hα center

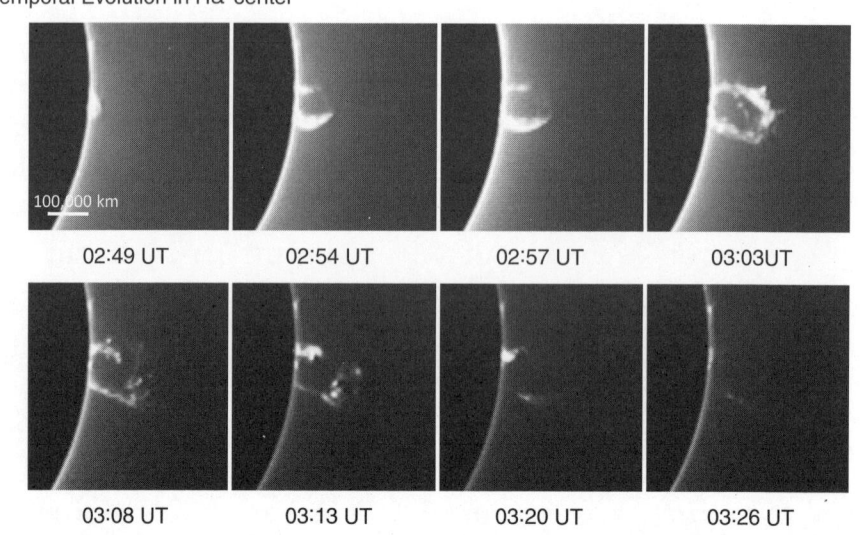

02:49 UT　　02:54 UT　　02:57 UT　　03:03 UT

03:08 UT　　03:13 UT　　03:20 UT　　03:26 UT

Sep 03, 2002 23:07−23:40 UT Prominence Eruption

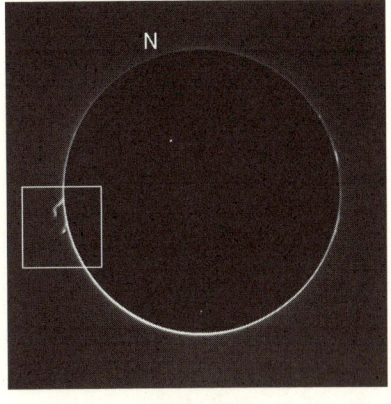

23:20 UT

EN 0-20
P = 21.5 [deg]   B0 = 7.2 [deg]

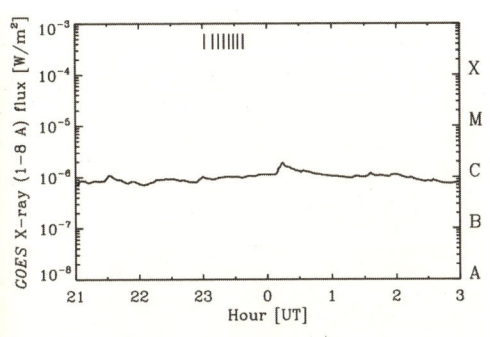

A prominence was lifted up with a speed of 50 km/s to the height of about 50 Mm, and then it faded out, draining the gas along the both legs. No prominent EUV activities, nor the CME association were reported by SOHO/EIT and LASCO.

プロミネンスが 50 km/s の速度で 5 万 km の高さまで持ち上げられた後、その両足からガスを流し落として消滅した。極端紫外線像では大きな変化はなく、CME も伴ってはいなかったことが SOHO/EIT と LASCO の観測から報告されている。

Temporal Evolution in Hα center

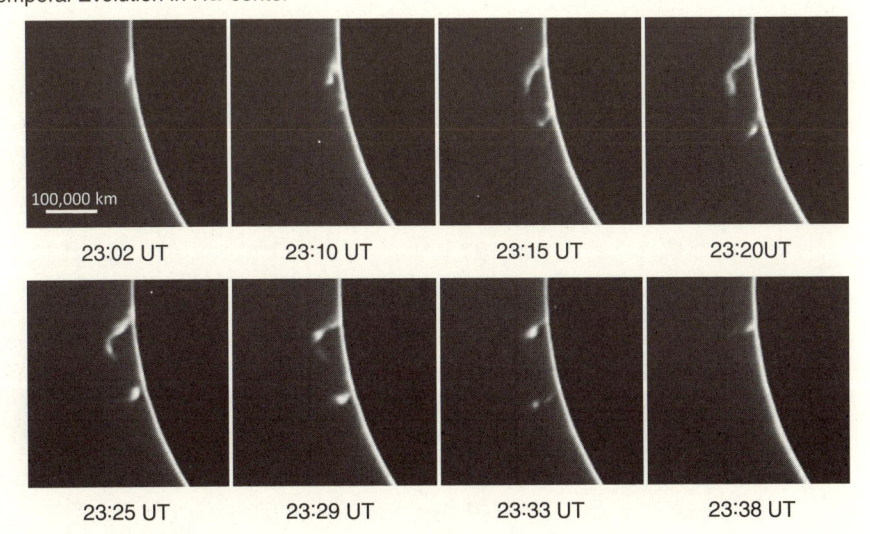

100,000 km

23:02 UT   23:10 UT   23:15 UT   23:20 UT

23:25 UT   23:29 UT   23:33 UT   23:38 UT

Nov 18, 2002 23:37−03:38 UT Prominence Eruption

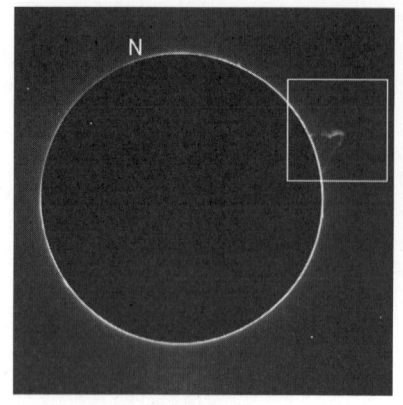

02:04 UT

WN 0-10
P = 20.6 [deg]   B0 = 2.5 [deg]

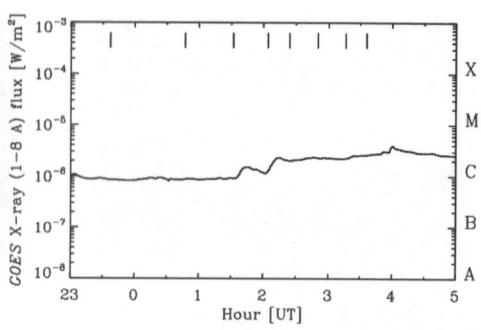

A prominence loop rose up slowly with a speed of 4 km/s, was accelerated to 25 km/s and exploded and faded away. In corona, a bubble structured CME was observed to ascend with a speed of 580 km/s by SOHO/LASCO after the prominence eruption.

プロミネンスが 4 km/s でゆっくりと上昇し，25 km/s まで加速された後消滅した．この後コロナでは，泡構造をした CME が 580 km/s の速さで上昇しているのが SOHO/LASCO で観測された．

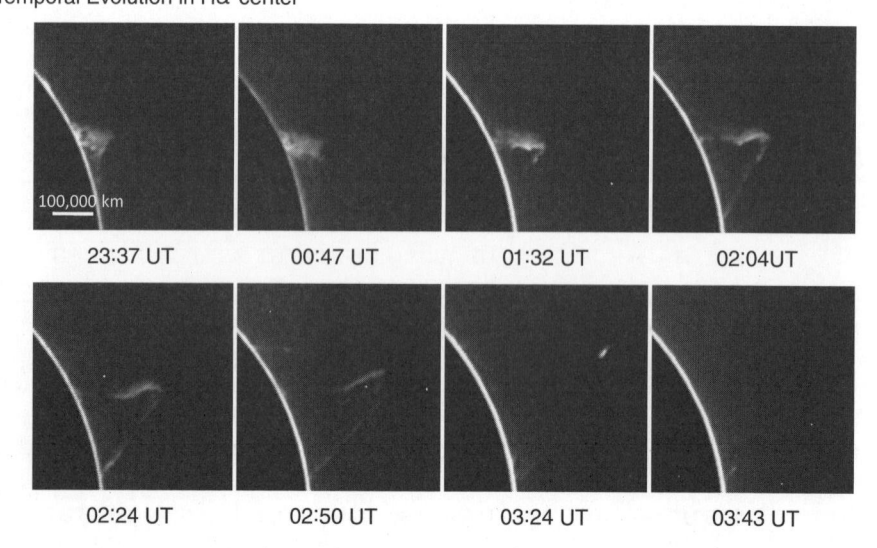

Temporal Evolution in Hα center

100,000 km

23:37 UT   00:47 UT   01:32 UT   02:04 UT

02:24 UT   02:50 UT   03:24 UT   03:43 UT

Nov 23, 2002 01:25−02:00 UT Surge Prominence

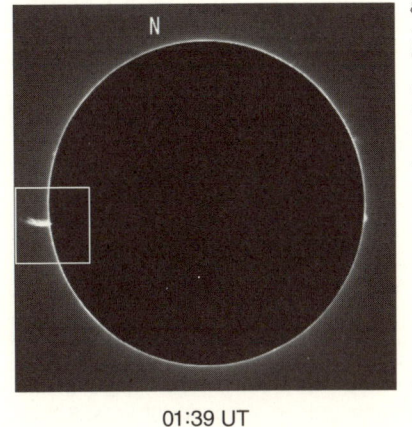

01:39 UT

EN 10-15
P = 19.0 [deg]   B0 = 1.9 [deg]

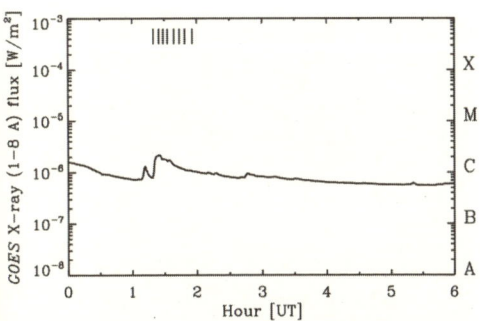

A surge prominence was ejected with a speed of 120 km/s from the NOAA10202 just behind the east limb. No prominent CME was associated with the event.

東のリムの裏側に位置していたNOAA10202から，120 km/sの速さで放出されたサージプロミネンスである．CMEは付随していなかった．

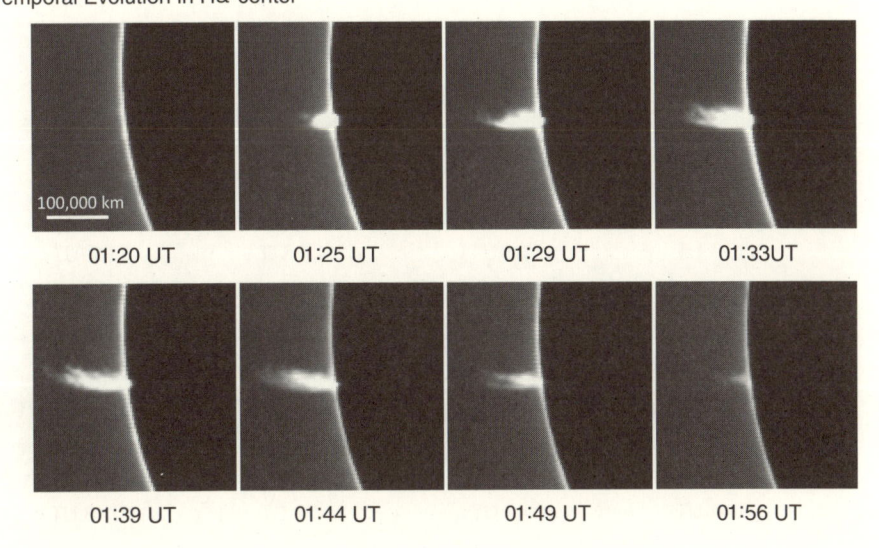

Temporal Evolution in Hα center

01:20 UT   01:25 UT   01:29 UT   01:33 UT

01:39 UT   01:44 UT   01:49 UT   01:56 UT

Apr 09, 2003 23:28−00:19 UT Prominence Eruption

23:39 UT

WS 0-10
P = -26.3 [deg]   B0= -6.1 [deg]

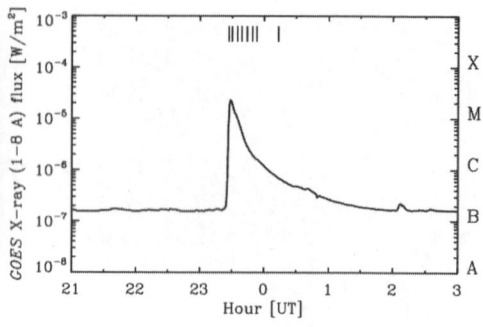

In association with an M2.5 flare in NOAA10326 located near the west limb, the prominence erupted with a speed of around 300 km/s. SOHO/LASCO detected an associated CME.

西のリム近傍の活動領域 NOAA10326 で発生した M2.5 クラスのフレアに伴って，約 300 km/s の速度でプロミネンスが噴出した．SOHO/LASCO 観測によりこれに応じた CME が発生していることも確認された．

Temporal Evolution in Hα center

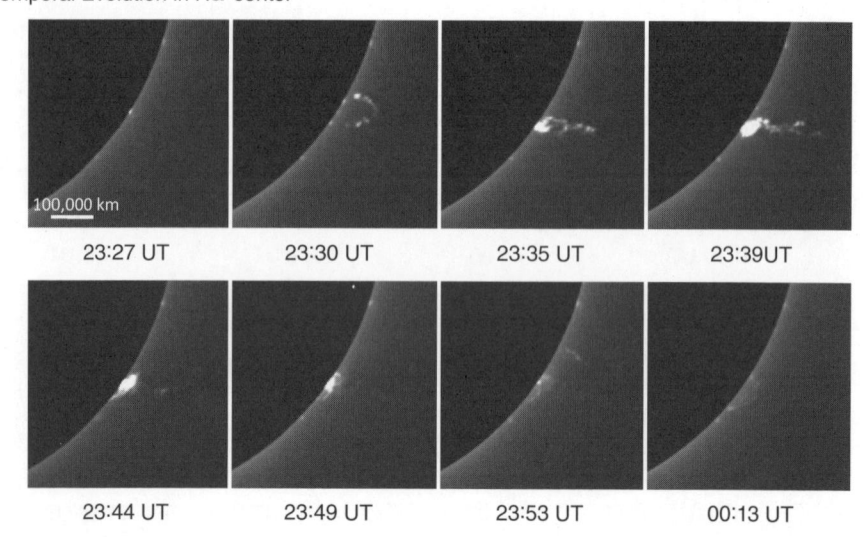

100,000 km

23:27 UT   23:30 UT   23:35 UT   23:39 UT

23:44 UT   23:49 UT   23:53 UT   00:13 UT

Apr 27, 2003 01:02−03:29 UT Prominence Eruption

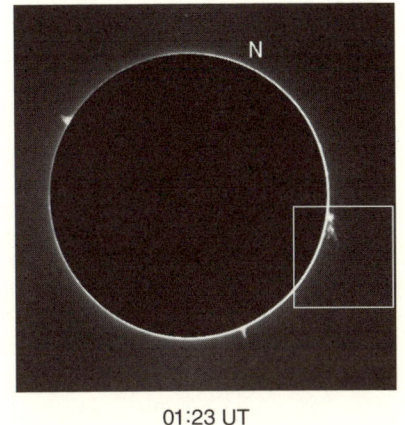

01:23 UT

WN0-15
P = -24.9 [deg]   B0= -4.6 [deg]

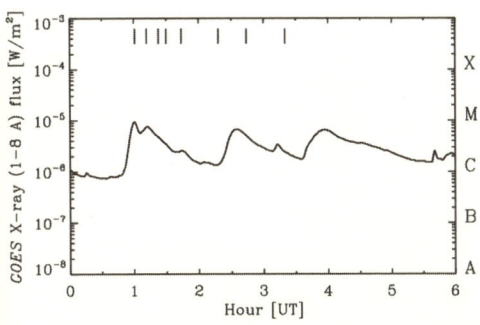

The filament eruption was driven by a C9.3 flare in a flare-productive δ-type region NOAA10338. A CME was observed to expand with a speed of 470 km/s by SOHO/LASCO.

このフィラメント爆発は，フレアを活発に起こしたδ型活動領域 NOAA10338 で発生した C9.3 クラスのフレアで引き起こされた．このとき，470 km/s の速さで膨張する CME が SOHO/LASCO で観測されている．

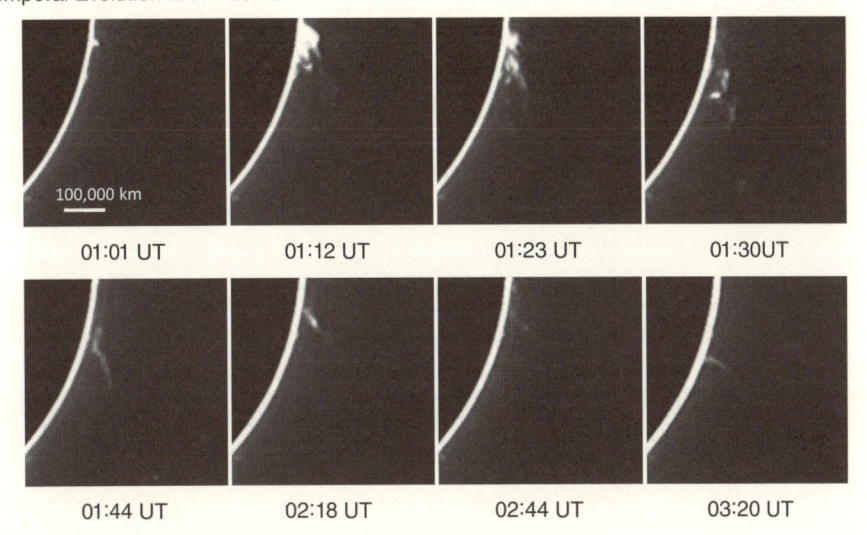

Temporal Evolution in Hα center

100,000 km

01:01 UT   01:12 UT   01:23 UT   01:30 UT

01:44 UT   02:18 UT   02:44 UT   03:20 UT

参考文献／References

Akiyama, S., Takeuchi, T., Mizuno, Y., Shibata, K. and Morimoto, T.
"The Relationship between CME Interactions and Complex IP Disturbances"
SHINE (Solar Heliospheric and Interplanetary Environment) Meeting at Banff/Canada, 2002

Asai, A., Ishii, T.T., Kurokawa, H., Yokoyama, T., and Shimojo, M.
"EVOLUTION OF CONJUGATE FOOTPOINTS INSIDE FLARE RIBBONS DURING A GREAT TWO-RIBBON FLARE ON 2001 APRIL 10"
ApJ, Vol586, pp624–629, 2003

Asai, A., Yokoyama, T., Shimojo, M., Masuda, S., Kurokawa, H., and Shibata, K.
"FLARE RIBBON EXPANSION AND ENERGY RELEASE RATE"
Ap J, Vol611, pp557–567, 2004

Belkina, I. L., Akimov, L. A., Beletsky, S. A.
"Solar flare 27 august 1999 from the 1083 nm helium line observations"
Kinematika i Fizika Nebesnykh Tel, vol. 18, no. 3, p. 217-226, 2002

Chae, J., Moon, Y., Park, Y.
"Determination of magnetic helicity content of solar active regions from SOHO/MDI magnetograms"
Solar Physics, Volume 223, Issue 1-2, pp. 39-55, 2004

Chandra, R., Schmieder, B., Aulanier, G., Malherbe, J. M.
"Evidence of Magnetic Helicity in Emerging Flux and Associated Flare"
Solar Physics, Volume 258, Issue 1, pp.53-67, 2009

Eto, S., Isobe, H., Narukage, N., Asai, A., Morimoto, T., Thompson, B., Yashiro, S., Wang, T., Kitai, R., Kurokawa, H. and Shibata, K.
"Relation between a Moreton Wave and an EIT Wave Observed on 1997 November 4"
PASJ, Vol.54, No.3, pp.481-491, 2002

Hanaoka, Y., Kurokawa, H., Enome, S., Nakajima, H., Shibahashi K., Nishio, M., Takano T., Torii, C., Sekiguchi, H., Kawashima, S., Bishimata, T., Shinohara, N., Irimajiri, Y., Koshiishi, H., Shiomi, Nakai, Y., Funakoshi, Y., Kitai, R., Ishiura, K. and Kimura, G.
"Simultaneous observations of a prominence eruption followed by a coronal arcade formation in radio, soft X-rays, and H-alpha"
PASJ, vol.46, pp.205-216, 1994

Hanaoka, Y. and Shinkawa, T.
"Heating of Erupting Prominences Observed at 17Ghz"
ApJ, vol.510, pp.466-473, 1999

Ishii, T.T., Asai, A., Kurokawa, H., and Takeuchi, T. T.
"Magnetic Neutral Line Rotations in Flare-Productive Regions"
Highlights of Astronomy, Vol. 13, as presented at the XXVth General Assembly of the IAU - 2003 [Sydney, Australia, 13 - 26 July 2003]. Edited by O. Engvold. San Francisco, CA: Astronomical Society of the Pacific, ISBN 1-58381-189-3. XXIX + 1085 pp., p.138, 2005

Isobe, H., Yokoyama, T., Shimojo, M., Morimoto, T., Kozu, H., Eto, S., Narukage, N., Shibata, K.
"Reconnection Rate in the Decay Phase of a Long Duration Event Flare on 1997 May 12"
ApJ, Volume 566, pp. 528-538, 2002.

Kardapolova, N. N., Borisevich, T. P., Peterova, N. G., Lesovoĭ, S. V.
"Coronal mass ejection of April 27, 2003, and evolution of the active region NOAA 10338 in the radio"
Astronomy Reports, Volume 52, Issue 5, pp.409-418, 2008

Koi, T., Terasawa, T., et al.
"The Observation of low Energy Ions Accelerated by an Interplanetary Shock on February 21st 1994 with the Geotail HEP Experiment"
24th International Cosmic Ray Conference, Vol. 4, held August 28-September 8, 1995 in Rome, Italy. Edited by N. Iucci and E. Lamanna. International Union of Pure and Applied Physics, p.381, 1995

Kundu, M. R., White, S. M., Garaimov, V. I., Manoharan, P. K., Subramanian, P., Ananthakrishnan, S., Janardhan, P.
"Radio Observations of Rapid Acceleration in a Slow Filament Eruption/Fast CME Event"
AGU, Fall Meeting 2003, abstract #SH21A-06, 2003.

Kurokawa, H., Ishiura, K., Kimura, G., Nakai, Y., Kitai, R., Funakoshi, Y. and Shinkawa, T.
"Observations of Solar H alpha Filament Disappearances with a New Solar Flare-Monitoring-Telescope at Hida Observatory"
J.Geomag.Geoelectr., 47, 1043-1052, 1995

Kurokawa, H., Wang, T., Ishii, T. T.
"Emergence and Drastic Breakdown of a Twisted Flux Rope to Trigger Strong Solar Flares in NOAA Active Region 9026"
ApJ, Volume 572, pp. 598-608, 2002

Lee, J., Gallagher, P. T., Gary, D. E., Harra, L. K.
"Radio and X ray Observations of a Limb Flare during the Max Millennium Campaign"
AGU Spring Meeting 2001, abstract #SP51A-03, 2001

Liu, Y., Kurokawa, H., and Shibata, K.
"PRODUCTION OF FILAMENTS BY SURGES"
ApJ, Vol631, L93–L96, 2005

McAllister, A.H., Kurokawa, H., Shibata, K. and Nitta, N.
"A Filament Eruption and Accompanying Coronal Field Changes on November 5, 1992"
Solar Physics, vol.169, pp.123-149, 1996

Morimoto, T. and Kurokawa, H.
"A Method for the Determination of 3-D Velocity Fields of Disappearing Solar Filaments"
PASJ vol.55, pp.503-518, 2003

Morimoto, T. and Kurokawa, H.
"Eruptive and Quasi-Eruptive Disappearing Solar Filaments and Their Relationship with Coronal Activities"
PASJ vol.55, pp.1141-1151, 2003

Mosalam Shaltout, M. A.
"The First Two Solar Proton Flares in the Onset of Solar Cycle 23"
The Sun and Space Weather, 24th meeting of the IAU, Joint Discussion 7, August 2000, Manchester, England, meeting abstract, 2000

Narukage, N. and Shibata, K.
"Moreton waves observed at Hida Observatory"
Multi-Wavelength Investigations of Solar Activity, IAU Symposium, vol.223, pp.367-370, 2004

Narukage, N., Hudson, H.S., Morimoto, T., Akiyama, S., Kitai, R., Kurokawa, H. and Shibata, K.
"Simultaneous Observation of a Moreton Wave on 1997 November 3 in H-alpha and

Soft X-Rays"
ApJ, Volume 572, Issue 1, pp. L109-L112., 2002

Narukage, N., Morimoto, T., Kadota, M., Kitai, R., Kurokawa, H., Shibata, K.
"X-Ray Expanding Features Associated with a Moreton Wave"
PAS J, Vol.56, No.2, pp. L5-L8, 2004

Narukage, N., Ishii, T. T., Nagata, S., UeNo, S., Kitai, R., Kurokawa, H., Akioka, M., Shibata, K.:
"Three Successive and Interacting Shock Waves Generated by a Solar Flare"
ApJL, 684, pp. L45-L49, 2008.

Okamoto, T. J., Nakai, H., Keiyama, A., Narukage, N., Ueno, S., Kitai, R., Kurokawa, H., and Shibata, K.
"Filament Oscillations and Moreton Waves Associated with EIT Waves"
ApJ, vol.608, pp.1124-1132, 2004

Park, S., Lee, J., Choe, G. S., Chae, J., Jeong, H., Yang, G., Jing, J., Wang, H.
"The Variation of Relative Magnetic Helicity around Major Flares"
ApJ., Volume 686, Issue 2, pp. 1397-1403, 2008

Shibata, K.
"Evidence of Magnetic Reconnection in Solar Flares and a Unified Model of Flares"
Ap&SS, Vol264, 129, 1999

Shibata, K., Eto, S., Narukage, N., Isobe, H., Morimoto, T., Kozu, H., Asai, A., Ishii, T., Akiyama, S., Ueno, S., Kitai, R., Kurokawa, H., Yashiro, S., Thompson, B. J., Wang, T., and Hudson,H.S.
"Observations of Moreton Waves and EIT Waves"
COSPAR Colloquia Series 2002, p.279, 2002

Takasaki, H., Asai, A., Kiyohara, J., Shimojo, M., Terasawa, T., Takei, Y., Shibata, K.
A Quantitative Study of the Homologous Flares on 2000 November 24
ApJ, Volume 613, Issue 1, pp. 592-599, 2004

Terasawa, T., Oka, M.;,Nakata, K., Keika, K., Nosé, M., McEntire, R. W., Saito, Y., Mukai, T.
" 'Cosmic-ray-mediated' interplanetary shocks in 1994 and 2003
Advances in Space Research, Volume 37, Issue 8, p. 1408-1412, 2006

Tian, L., Liu, Y., Yang, J., Alexander, D.

"The Role of the Kink Instability of a Long-Lived Active Region AR 9604"
Solar Physics, Volume 229, Issue 2, pp.237-253, 2005

Tian, L., Alexander, D., Nightingale, R.
"Origins of Coronal Energy and Helicity in NOAA 10030"
ApJ, Volume 684, Issue 1, pp. 747-756, 2008

Warmuth, A., Vršnak,B., and Hanslmeier, A.
"Flare waves revisited"
Hvar Observatory Bulletin, vol.27, no.1, pp.139-149, 2003

Warmuth, A., Vršnak, B., Magdalenić, J., Hanslmeier, A., and Otruba, W.
"A multiwavelength study of solar flare waves. I. Observations and basic properties"
As & Ap, vol.418, pp.1101-1115, 2004

Warmuth, A., Vršnak, B., Magdalenić, J., Hanslmeier, A., and Otruba, W.
"A multiwavelength study of solar flare waves. II. Perturbation characteristics and physical interpretation"
As & Ap, vol.418, pp.1117-1129, 2004

付録
Appendix

A FMT関係論文
A Selected Papers Relating to FMT

B 関連論文リスト
B FMTrelated Papers

C 太陽地球系エネルギー国際共同研究(STEP) シンポジウム報告よりFMT関連報告集
C FMT-Related Reports in Proceedings of the Symposiums of Solar-Terrestrial Energy Program (STEP)

付録 A
FMT 関係論文

Appendix A
Selected Papers Relating to FMT

A-1

太陽フィラメント消失現象の三次元速度場決定法
A Method for the Determination of 3-D Velocity Fields of Disappearing Solar Filaments

森本太郎, 黒河宏企
京都大学大学院理学研究科附属天文台

(原論文) PASJ: Publ. Astron. Soc. Japan 55, 503-518, 2003 April 25

Taro MORIMOTO and Hiroki KUROKAWA
Kwasan and Hida Observatories, Kyoto University

概要

本論文では，太陽面上におけるフィラメント消失現象の三次元速度場について詳細な研究を行った．飛騨天文台のフレア監視望遠鏡（FMT）によってHα線中心および±0.8Åで観測された3波長の単色像で観測された現象にBeckersのクラウドモデルを当てはめることによって，フィラメント消失現象の三次元速度場の導出法を開発し，5つのフィラメント消失現象を解析

Abstract

The 3-D velocity fields of disappearing filaments (Disparition Brusques: DBs) on the solar disk were extensively studied in order to determine their 3-D velocity fields. Applying Beckers' cloud model to 5 DB events observed in the Hα line center and Hα±0.8Å with the Flare Monitoring Telescope (FMT) at Hida Observatory, we developed a method to derive the complete 3-D velocity field of DBs. The line-of-sight velocity is obtained (i) by

した．視線速度の時間変化は(i)フィラメントのHα線プロファイルを計算し，(ii)観測されたフィラメントのコントラストの時間変化に最もよく一致するドップラーシフトを計測することによって求められた．視線垂直速度は，時間的に連続した画像間でフィラメントの内部構造を追尾することによって得た．今回の手法では，光学系の有効バンドパス，散乱光およびドップラー増光現象の補正を行った．また，フィラメント活動中におけるプラズマの状態変化や彩層からの背景光強度の変動，散乱光の推定のための点広がり関数の形状に起因する速度誤差についても議論した．この手法によってフィラメント消失現象の三次元軌道の計算を行うことで，宇宙天気予報の精度がより高まることを強調しておく．

1. 序論

太陽表面上のフィラメント消失現象はしばしば地磁気嵐を引き起こす (Wright, McNamara 1983)．太陽縁でのプロミネンス噴出とコロナ質量放出現象とはよい相関をもつ (Munro et al. 1979) が，フィラメント消失現象とコロナ質量放出現象および地磁気嵐の関係はいまだよく理解されていない．これは，Hα線中心のみによる観測ではフィラメント消失現象における視線速度を計測することができずに，フィラメントが惑星間空間に噴出したのか判断できないことによる．それゆえに，フィラメント消失現象のドップラーシフト量を計測して三次元速度場を求めることが宇宙天気予報にとって重要である．

多くの研究者がフィラメント消失現象について論文を発表している (Webb et al. 1976; McAllister et al. 1992, 1996; Khan et al. 1998) が，彼らの速度場診断はフィラメントの見た目の運動のみによるものであった．Kubota et al. (1992) は飛騨天文台ドームレス太陽望遠鏡を用いてフィラメント消失現象のスペクトル観測を行い，

calculating the Hα line profile of the filament, and (ii) by measuring the Doppler shift which best fits to the observed temporal variations of contrasts of the filament. The tangential velocity is obtained by tracing the motions of the internal structures on the successive images. In this method, corrections for the effective filter bandwidths of the instrument, stray lights, and Doppler brightening effect are performed. We also discuss the velocity errors which arise from the intrinsic variations of the filament plasma during its activation, the fluctuations in intensities of the background chromosphere, and the choice of different forms of the spread function for estimating the stray light. It is emphasized that the calculation of three-dimensional vector trajectories of a disappearing filament with our method can enhance the quality of a space weather forecast with better certainty.

Key words: Sun: chromosphere — Sun: filament — Sun: filament disappearances — Sun: prominences

1. Introduction

Disappearing filaments (DBs) on the solar disk often cause geomagnetic storms (Wright, McNamara 1983). Although prominence eruptions at the solar limb have a good correlation with coronal mass ejections (CMEs) (Munro et al. 1979), the relationship among the DBs, CMEs, and geomagnetic storms is not yet well understood. This is because one cannot measure the line-of-sight velocity of a disappearing filament and determine whether it is ejected into interplanetary space or not with observations in the Hα line center alone. Thus, it is very important to measure the Doppler shift and three-dimensional velocity field of a disappearing filament for a space weather forecast.

Although many authors have published papers on DBs (Webb et al. 1976; McAllister et al. 1992, 1996; Khan et al. 1998), their velocity diagnostics are limited to the filament's apparent motion. Kubota et al. (1992) made spectroscopic observations of a DB with Domeless Solar Telescope at Hida Observatory, and measured its ascending motion at several sections along the axis of the filament. A

フィラメントの軸に沿った数点における上昇運動を計測した．そもそも，スペクトル観測は視線速度を計測するための便利な方法であるが，視野はスリット上のみに制限される．Maltby (1976) と Schmieder et al. (2000) はフィラメントの視線速度を，「クラウドモデル」(Beckers 1964) を用いて計測した．しかし彼らの研究の主な目的は乱流や電子温度などの内部の物理パラメータの解明に重点を置いており，速度場の導出ではなかった．それゆえ，フィラメント消失現象における三次元速度場計測の体系的な研究が強く求められていた．FMT は Hα 線中心のみならず両ウイングをも用いてフィラメント消失現象の同時観測が行えるため，我々は Beckers のクラウドモデルを用いてフィラメント消失現象の視線速度を導出することが可能である．この論文では，速度診断の手法を定式化し，それを飛騨天文台のフレア監視望遠鏡 (FMT: Kurokawa et al. 1995) で観測された5つのフィラメント消失現象に適用した結果を述べる．また，この手法で発生する誤差を求め，それらの補正についても詳細に議論する．

2節では，本研究に用いられた観測データとイベント選択について述べる．消失するフィラメントの視線速度とそれに垂直方向の速度の導出方法，および誤差の補正は1999年2月16日のイベントを例にして3節にて紹介する．4節では，イベント選択された5つの典型例について詳細な結果を述べる．最後に，速度計測のまとめを5節にて行う．

2. 観測および観測データ
2.1. Hα観測
2.1.1. 飛騨フレア監視望遠鏡(FMT)

京都大学飛騨天文台のフレア監視望遠鏡 (FMT：Kurokawa et al. 1995) は 1992年5月より定常観測を行っており，Hα線中心，青色側および赤色側ウイング，連続光そしてプロミネンス観測（掩蔽円盤を用いた Hα線中心観測）により5種類の太陽全面像を観測している．フィラメント消失現象を研究するにあたって，我々は Hα 線中

spectroscopic observation is a useful tool to measure the line-of-sight velocity precisely, but the field of view is limited by the slit. Maltby (1976) and Schmieder et al. (2000) measured the line-of-sight velocities of filaments by using the "cloud model" (Beckers 1964). However, in these studies, the main goals were not to derive the velocity field, but to investigate the internal plasma parameters, such as the microturbulent velocities or electron temperatures. Therefore, a systematic study to measure the 3-D velocity fields of DBs has been highly required. Since FMT can simultaneously observe DBs not only in the Hα line center, but also in its two wings, we are able to derive the line-of-sight velocities of DBs by using the Beckers' cloud model. In this paper, we present the method of velocity diagnostics and its application to five DB events observed by the Flare Monitoring Telescope (FMT: Kurokawa et al. 1995) at Hida Observatory. We also estimate the errors involved in this method and discuss the corrections for them in detail.

In section 2, we describe the data used in this study and the selection of events. Methods to derive the line-of-sight and tangential velocities of a disappearing filament, and the estimations and corrections for errors are proposed in section 3 with an example of the 1999 February 16 event. In section 4, we present the results of the five selected typical events in detail. Finally, we summarize the velocity measurements in section 5.

2. Observations and Data
2.1. Hα Observations
2.1.1. The Hida Flare Monitoring Telescope (FMT)

The Flare Monitoring Telescope (FMT: Kurokawa et al. 1995) at Hida Observatory, Kyoto University has had a routine observational program since 1992 May, and provides 5 full disk images in the Hα line center, Hα blue wing, Hα red wing, continuum, and prominence mode (Hα line center with occulting disk to detect prominences). To study DBs, we used Hα line center, blue wing, and red

Appendix A Selected Papers Relating to FMT | 333

表1 フレア監視望遠鏡（FMT）のフィルター特性.

望遠鏡名	フィルター		
	中心波長	透過幅	フィルタータイプ
Hα 中心望遠鏡	6562.8Å	0.5Å	ファブリペロ
Hα 赤色側ウイング望遠鏡	6563.6Å	0.6Å	ファブリペロ
Hα 赤色側ウイング望遠鏡	6562.0Å	0.6Å	ファブリペロ

Table 1. Filter parameters of the three telescopes of the Flare Monitoring Telescope(FMT)system.

Telescope name	Filter		
	Central wavelength	Passband	Filter type
Hα center telescope	6562.8Å	0.5Å	A Fabry-Perot
Hα red wing telescope	6563.6Å	0.6Å	A Fabry-Perot
Hα blue wing telescope	6562.0Å	0.6Å	A Fabry-Perot

心，青色側および赤色側ウイングによる観測データを用いた．これらの観測で使用したフィルターの特性を表1に示す．それぞれの画像は2秒間隔で撮影され，間欠式ビデオレコーダーに記録される．1996年9月以降は，観測画像は1分間隔でデジタル記録もされている．観測データのピクセルサイズは4.2秒角である．FMTは活動領域やフィラメントの長期間発展のみならずフレアやサージ，フィラメント消失現象やプロミネンス噴出など様々な現象を観測してきた．

FMTはフィラメント消失現象の速度場を解明するのに最適な機器の1つである．通常のHα線中心画像による観測とは異なり，FMTは長短両ウイングの2波長でも観測することが可能であり，このことによりフィラメント消失現象の視線速度を得ることができる．図1に1999年2月16日のフィラメント消失現象の例を示す．フィラメント消失前（02:00 UT）には，Hα線中心画像ではダークフィラメントが活動領域 NOAA 8458の南側に見られる．このフィラメントの時間発展は以下の通りである．Hα線中心画像でフィラメントが消失し始

wing images. The filter parameters of these three telescopes are presented in table 1. Each image was obtained with 2-s cadence and recorded with a timelapse video-recorder. Images were also recorded in digital form with 1-min cadence from 1996 September. The pixel size of the data was 4″.2. The FMT system has recorded many phenomena, such as flares, surges, DBs, prominence eruptions as well as long-term evolution of active regions and filaments.

FMT is one of the best instruments used to investigate the velocity field of a DB. Unlike ordinary Hα filtergrams at line center, FMT provides additional images in the 2 wings, which enable us to derive the line-of-sight velocity of a DB. One example of a DB observed on 1999 February 16 is displayed in figure 1. The dark filament is seen at the south of the active region NOAA8458 in the Hα line center image before the start of disappearance (02:00 UT). Its dynamical evolution can be summarized as follows. The central part of the filament appears on the blue wing image as the filament begins to fade in the Hα line center image (02:45 UT). After 02:45 UT, the filament is accelerated toward the south-west and

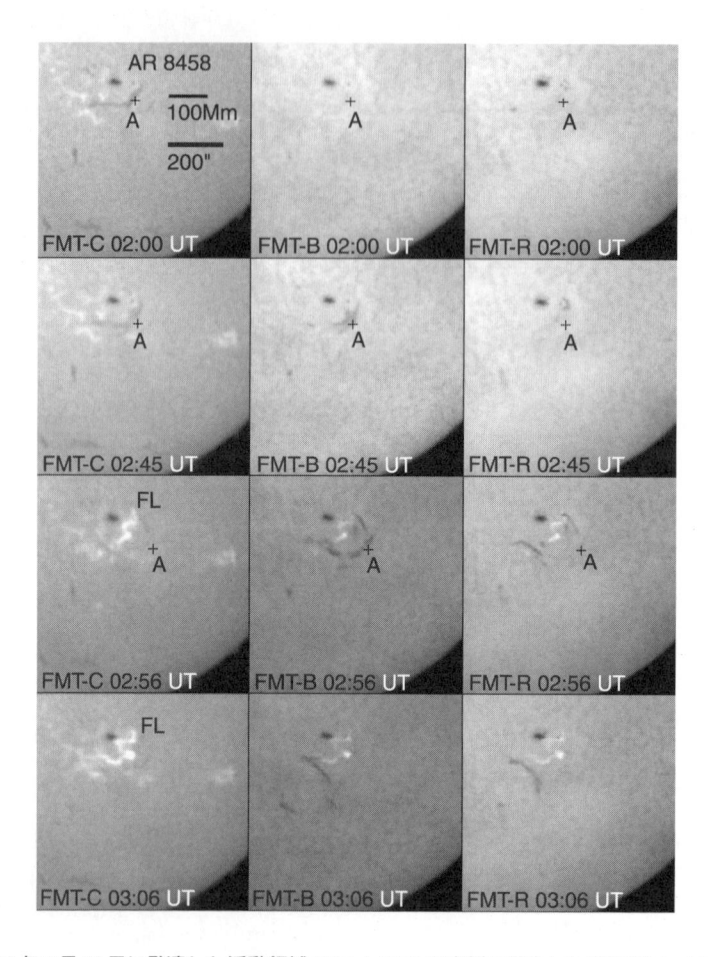

図1 1999年2月16日に発達した活動領域 NOAA 8458 の南側で発生した噴出型フィラメントの FMT の部分像（左列：FMT Hα線中心，中列：青色側ウイング，右列：赤色側ウイング）．フィラメントは最初 Hα線中心画像でのみ観測される（左上図）が，青色側ウイング画像で暗く観測され南西方向に伸びる．Hα線中心画像でフィラメントが完全に消失するとツーリボンフレア（FL）がフィラメントの位置で発達する（02:49 UT 以降）．青色側および赤色側ウイング画像では後半段階においても足元の構造が観測される（02:56 UT 以降）．A 点はフィラメントの中央部分に位置しており，視線速度の診断がされた位置である（本文参照）．この論文を通じて，太陽の北は上，東は左である．

Fig. 1. Series of partial FMT images showing the eruptive filament at the south of the well developed active region NOAA8458 on 1999 February 16 (left column: FMT center, middle: blue wing, and right: red wing images). The filament first appearing dark in only the Hα line center image (top-left panel), becomes dark in the blue wing and expands toward the southwest. As the main part vanishes in the Hα line center, a two-ribbon flare (FL) develops at the filament site (from 02:49 UT). Blue- and red-shifted structures are seen at the footpoints in the later phase (02:56 UT-). Point A is located at the middle of the filament and is used to show the line-of-sight velocity diagnosis (see main text). The solar north is up and the east is left, which holds for all the other solar images in this article.

めると同時に，青色側ウイング画像ではフィラメントの中央部分が観測された (02:45 UT)．02:45 UT 以降，フィラメントは南西方向に加速され，大きなΩ型の構造となった．このとき，フィラメントは青色側ウイング画像で最も暗く観測された．フィラメント消失直後にツーリボンフレアが発生し，02:56 UT に Hα 線中心で観測された．フィラメントの大半が消失した後，その足元が青色および赤色側ウイング画像の両方で暗くなった．このことは，多くのプロミネンス噴出現象で観測される，ねじれが解ける運動を示している可能性がある．

2.1.2. イベント選択

FMT の活動現象データベース[1] を用いて，我々は 1992 年 6 月から 2000 年 6 月までの期間で発生した 2747 個のフィラメ

shows a large Ω structure. At that moment, the filament is most prominent on the blue wing image. A two-ribbon flare occurs just after the disappearance, and it is seen on the Hα line center image at 02:56 UT. After the disappearance of the main part of the filament, its footpoints become dark in both blue and red wing images, which may indicate untwisting motions commonly observed in many prominence eruptions.

2.1.2. Event selection

Making use of the FMT data base[1] of active phenomena, we selected 35 out of 2747 DB events from 1992 June to 2000 June. The criteria for the selection were (i) the size of the event (filament) is

表 2　フィラメント消失現象のリスト．

番号	時刻 (UT)		場所/NOAA
1	1992 年 11 月 05 日	00:15–02:15	S20W17/
2	1999 年 02 月 16 日	01:42–04:15	S27W18/8458
3	2000 年 01 月 19 日	00:28–01:47	N08W18/8829
4	2000 年 01 月 28 日	05:35–06:20	S28W20/8841
5	2000 年 05 月 08 日	04:19–07:40	S21W03

注　各列はそれぞれ，本論文において本論文において解析された 5 イベントのイベント番号，時刻，および場所（消失フィラメントの位置する活動領域番号）である．

Table 2.　List of DB events.

No.	Time (UT)		Location/NOAA
1	1992 November 05	00:15–02:15	S20W17/
2	1999 February 16	01:42–04:15	S27W18/8458
3	2000 January 19	00:28–01:47	N08W18/8829
4	2000 January 28	05:35–06:20	S28W20/8841
5	2000 May 08	04:19–07:40	S21W03

Note.　Each column gives the event number, time, and location (active region number to which the disappearing filament belongs) for the 5 events studied in this paper, respectively.

1　http://www.kwasan.kyoto-u.ac.jp/Hida/FMT/obs-report.html

ト消失現象から35個のイベントを選択した．選択の条件は，(i)SXT や EIT との画像比較のために，イベント（フィラメント）のサイズが60 Mm 以上であること，(ii)イベント全体を通じて，観測が雲や日の入りなどによって中断されていないことであった．本研究では，これら35個のイベントの中から，三次元速度場の導出に適した5つの典型的なフィラメント消失現象を選択した．選択された5つのイベントは表2に表されている．最初の列はイベント番号を示し，時刻および場所は2番目および3番目の列にそれぞれ示されている．イベントが発生した活動領域の NOAA 番号も3番目の列に示されている．

FMT のデジタル/ビデオ画像は60秒間隔で選択された．フィラメントの運動速度が大きい場合，画像間隔は30秒とした．ビデオテープのデータに対しては，京都大学花山天文台の AD コンバータシステムによりデジタル化を行った．

3. 解析

典型的なフィラメントは，長さ60-600 Mm，幅 0.4-1.5 Mm の長大な構造をもつ(Tandberg-Hanssen 1995) ため，速度の大きさや運動方向はフィラメント中の場所によって大きく変化する．プロミネンス噴出現象についての多くの観測的研究では，しばしばプロミネンスの先行部分の運動のみを解析しているが，これは噴出現象全体の理解には不十分である．したがって，フィラメント消失現象の運動の理解には，フィラメント各点における三次元速度の計測が求められる．本論文では，FMT の Hα 画像の各ピクセルに対して視線速度と視線垂直速度を求める手法を紹介する．

手法の紹介の前に，本論文で用いられている速度に関する用語の説明を行う．視線速度とは文字通り視線方向に沿った速度成分であり，負が我々に向かってくる方向を表す．視線垂直速度とは，FMT の画像上の見かけの運動の速度である．視線垂直速度には x および y の2成分があるため，視線速度と合わせて，「上昇」速度を求めることができる．上昇速度とは，フィラメ

larger than 60 Mm to make the comparison with SXT and EIT images easier; (ii) throughout the event, the observation is not interrupted due to clouds or sunset. In this study, we selected five typical events which were suitable to present the details of the new method to obtain 3-D velocity fields of disappearing filaments from these 35 events. These five events are listed in table 2. The first column gives the event number; the date and location are in the second and third columns, respectively. The NOAA numbers of active regions where the events took place are also presented in the third column.

We selected digital or video images from the FMT telescopes, at a 60-s cadence. When the apparent motion was rapid, we opted for the 30-s cadence. In the case of video-tape data, the signals were digitized with an AD converter system at Kwasan Observatory, Kyoto University.

3. Analysis

Since a typical filament has an elongated structure with a length about 60-600 Mm and a width of 0.4-1.5 Mm (Tandberg-Hanssen 1995), the amplitude of the velocity and the direction of motion considerably differ at different positions when it is activated. Many observational studies on prominence eruptions often consider only the motion of the leading part of a prominence, which is insufficient to thoroughly understand its evolution. Hence, it is required to measure the 3-D velocity field at each portion of a filament to understand the motion of a DB. We propose here a method to obtain line-of-sight and tangential velocities at all of the pixels on the FMT Hα images.

Before describing the method, we briefly explain the velocity terms which are used in this paper. The line-of-sight velocity is, literally, the speed of motion along the line-of-sight, and is negative when it is toward us. The tangential velocity means the speed of an apparent motion on the FMT images. Since the tangential velocity has two components, namely x and y components, we can convert the line-of-sight and tangential velocities to obtain the "upward" velocity. The upward velocity means the speed of

ントの位置における太陽表面に鉛直な速度成分である．これに関して，もしフィラメントが太陽表面に向かって運動している場合は，「上昇」の替わりに「下降」という言葉を用いる．最後に，全速度という言葉は，運動方向に依らない速度の絶対値を表す．

3.1. 消失するフィラメントの視線速度
3.1.1. 基本原理

フィラメントは通常，周囲の彩層に対して暗い構造として観測されるが，これはフィラメントが下部の太陽表面からの放射を散乱させるためである．図1の左上のパネルには，1999年2月16日02:00 UTにFMTで観測されたHα線中心画像の例が示されている．我々は，フィラメント上の点 (x) と観測波長 ($\Delta\lambda$，Hα線中心からの波長差) に対して，フィラメントの光度と周囲の背景光度の比をコントラスト [$C(\Delta\lambda)$] として定義した．コントラストは式(1)のように記述できる．

$$C(x,\Delta\lambda) = [I_P(x,\Delta\lambda) - I_{R0}(\Delta\lambda)]/I_{R0}(\Delta\lambda). \quad (1)$$

この式の中で，I_pはフィラメントの光度，I_{R0}は周囲の彩層の光度である．

コントラストの波長依存性はフィラメントとそれを取り巻く周囲からのHα強度プロファイルの相違に起因する．フィラメントのHα線プロファイルの変化は，主としてフィラメント全体の運動すなわちドップラーシフト量によるため，FMTで観測されたコントラスト値からドップラーシフト量を推測することが可能である．したがって，ドップラーシフト量を求めるには，フィラメントと周囲の彩層のHα線プロファイルを正しく知る必要がある．静穏な彩層プロファイルとして，我々は標準Hα線プロファイル (Kurucz et al. 1984) を用いた．このことから，視線速度の導出において重要な手順は(i) FMTの画像からフィラメントのHα線プロファイルを得ること，そして(ii)観測されたコントラストプロファイルに最もよく一致するドップラーシフト量を計算することに帰結する．FMTが画像観測機器かつ3波長のみでの観測であった

motion of a filament in the radial direction at the filament site. In this case, if the motion is toward the solar surface, we use the term "downward" instead of "upward". The final velocity term is the total velocity, and is the magnitude of the speed of a motion regardless to the direction of its propagation.

3.1. Line-of-Sight Velocity of a Disappearing Filament
3.1.1. Basic idea

A filament usually appears as a dark feature against the surrounding chromosphere since it scatters lights from the chromosphere underneath. One example is the Hα line center image obtained by FMT at 02:00 UT on 1999 February 16 in the top-left panel of figure 1. We define the contrast [$C(x,\Delta\lambda)$] at a specified position (x) on a filament, and in a specified wavelength ($\Delta\lambda$: the difference between the observational wavelength and Hα line center) as a ratio in intensities between filament and surrounding chromosphere. It then can be written as

$$C(x,\Delta\lambda) = [I_P(x,\Delta\lambda) - I_{R0}(\Delta\lambda)]/I_{R0}(\Delta\lambda). \quad (1)$$

In this equation, I_P is the intensity from the filament, and I_{R0} is that from the surrounding chromosphere.

Contrasts arise from the difference between the Hα intensities from a filament and the chromosphere which surrounds it. Considering that a change of a filament's Hα line profile is primary due to its bulk motion, viz. Doppler shift, we can estimate the Doppler shift according to the contrast values observed by FMT. Thus, it requires us to know the Hα line profile of a filament and the surrounding chromosphere to measure the Doppler shift. For the static chromospheric Hα line profile, we made use of the reference Hα line profile (Kurucz et al. 1984). The important processes in deriving the line-of-sight velocities, therefore, are (i) to obtain the Hα line profile of a filament from FMT images, and (ii) to calculate the Doppler shift which best fits to the observed contrast values. Since FMT is an imaging instrument and has observations in only three bands, we need some assumptions to obtain an Hα line profile. In the next subsection, we introduce the

め，Hα線プロファイルを得るためには多少の仮定を必要とする．以降の小節では視線速度計算のための仮定と具体的な方法について述べる．

本研究では，すべてのイベントにおいて，彩層からの背景光のHα線プロファイルとしてKurucz et al. (1984) による参照Hα線プロファイルを用いた．表2に示されている通り，今回選択された5つのイベントは太陽面中心では発生していないため，観測されたコントラストは太陽面中心からリムにかけての光度変化やHα線プロファイルの形状変化の影響を受けている．しかし，フィラメントのHα線プロファイルが太陽面中心からリムにかけてどのように変化するのか，観測データがないためにこれらの効果を推定するのは難しい．White (1962) によるHα線の周辺減光に関する研究結果を用いると，5イベントが発生した地点での周辺減光の効果は，最大でも太陽面中心光度の2%であった．それゆえ，本研究ではHα線の太陽面中心－リムへの変化は視線速度決定に際して十分無視できると判断した．

3.1.2. 基本手法

我々の手法はBeckersのクラウドモデル (Beckers 1964) を基本としている．クラウドモデルは太陽面上の「雲状」の現象に対して，物理量や運動を導出するために幅広く用いられてきた (Grossmann-Doerth, von Uexküll, 1973, 1977; Loughhead 1973; Bray 1974; Alissandrakis et al. 1990; Tsiropoula et al. 1992; Heinzel et al. 1995, 1999; Paletou 1997)．フィラメント真下の彩層光度 $[I_{R0}(\Delta\lambda)]$ を直接観測することは不可能なので，替わりにフィラメント周辺の平均光度 $[I_C(x, \Delta\lambda)]$ を用いた．

$$I_{R0}(\Delta\lambda) = \langle I_C(x, \Delta\lambda)\rangle_x. \quad (2)$$

本研究では，フィラメント周囲50 Mm以内の領域を平均した．この50 Mm領域は，フィラメントの典型的な高度 (10 Mm) よりも大きく，フィラメントを照射する平均光度を求めるのに十分広い領域である．

assumptions and concrete measures to calculate the line-of-sight velocities.

In this study, we used the reference Hα line profile provided by Kurucz et al. (1984) as the Hα line profile of the background chromosphere for all of the events. As shown in table 2, the 5 events did not occur at the center of the disk, the observed contrast is affected by the center-to-limb variations of intensity and profile of Hα line. It is, however, difficult to estimate the effect because we have no observational data on the center-to-limb variations of the Hα line from the filaments. By making use of the result of White (1962), who studied the limb darkening in Hα line, the calculated decrease of the Hα line center intensity from the regions of the 5 filaments compared to the disk center is, at most, 2%. In this study, therefore, we assumed that the effect of the center-to-limb variation of the Hα line intensity is insignificant for the determination of the line-of-sight velocity.

3.1.2. Basic method

Our method is based on Beckers' cloud model (Beckers 1964). This method has been extensively used to derive the physical parameters, or motions, of "cloud-like" solar structures (Grossmann-Doerth, von Uexküll, 1973, 1977; Loughhead 1973; Bray 1974; Alissandrakis et al. 1990; Tsiropoula et al. 1992; Heinzel et al. 1995, 1999; Paletou 1997).

Since we can not observe the intensity of chromosphere beneath the filament $[I_{R0}(\Delta\lambda)]$ directly, the mean intensity of surrounding chromosphere $[I_C(x, \Delta\lambda)]$ over a large area is used, instead of $I_{R0}(\Delta\lambda)$, as

$$I_{R0}(\Delta\lambda) = \langle I_C(x, \Delta\lambda)\rangle_x. \quad (2)$$

In this study, a region within 50 Mm distant from the filament is selected as the surrounding chromosphere. 50 Mm is sufficiently larger than the typical height of filaments (10 Mm) and covers a sufficient area, which accounts for the irradiation.

If the source function (S) is constant

もし源泉関数 (S) が視線方向に沿ってすべての波長 $\Delta\lambda$ で一定であるならば，コントラストは以下のように書ける．

$$C(\Delta\lambda) = \left[\frac{S}{I_{R0}(\Delta\lambda)} - 1\right]\{1 - \exp[-\tau(\Delta\lambda)]\}, \quad (3)$$

ここで，$\tau(\Delta\lambda)$ は光学的厚みである．源泉関数 (S) と彩層からの背景放射光度は近傍の連続光光度で規格化されている．ドップラー幅 ($\Delta\lambda_D$) とドップラーシフト量 ($\Delta\lambda_S$) も視線方向で一定とみなすと，光学的厚みは $\tau(\Delta\lambda) = \tau_0 \exp\{-[(\Delta\lambda - \Delta\lambda_S)/\Delta\lambda_D]^2\}$ で表される．ここで，τ_0 は波長中心における光学的厚みである．ドップラー幅 ($\Delta\lambda_D$) は温度 T と微小乱流 ξ_t を用いて，以下のように表される．

$$\Delta\lambda_D = (\lambda_0/c)\sqrt{\xi_t^2 + 2kT/m}. \quad (4)$$

以上より，最低4波長のコントラスト値が与えられれば，4つの不定値 S, τ_0, $\Delta\lambda_D$, および $\Delta\lambda_S$ の値を決定することが可能となる．しかしながら，FMTで得られるデータは3波長のみなので，このままでは物理量を求めることが不可能である．さらに，FMTの観測波長は波長中心に対して対称 (± 0.8Å) であり，Hα 線プロファイルが通常波長中心に対して対称であることを考慮すると，ドップラーシフトしていないプロファイルに対しては正味のところ2つのコントラスト値しか利用できない．

これらのデータの不足を補うため，我々は以下の2つの仮定を設けた．(i) フィラメント消失現象が発生する時刻 ($t = 0$) より以前では，フィラメントは全体的に静止しており ($\Delta\lambda_S = 0$)，その他のパラメータ (S, τ_0, $\Delta\lambda_D$) は一様である．ここで，τ_0 と $\Delta\lambda_D$ はフィラメント消失開始後も変化しないものとする．(ii) 両ウイングにおけるコントラスト値が最小となった時点のドップラーシフト量を 0.8Å とする．
$\Delta\lambda_S(t = 0) = 0$ とする仮定は，フィラメントが活性化する以前は静穏であると見なしてよいことに基づく．仮定(i)と式(3)により，フィラメント消失以前のHα線中心と青色側ウイングのコントラスト (C_1, C_2) は以下のように書ける．

along the line-of-sight for all $\Delta\lambda$, we can write

$$C(\Delta\lambda) = \left[\frac{S}{I_{R0}(\Delta\lambda)} - 1\right]\{1 - \exp[-\tau(\Delta\lambda)]\}, \quad (3)$$

where $\tau(\Delta\lambda)$ is the optical thickness. The source function (S) and the intensity of the background chromosphere are normalized by the intensity of the nearby continuum. Assuming that the Doppler width ($\Delta\lambda_D$) as well as the Doppler shift ($\Delta\lambda_S$) is also constant along the line-of-sight, the optical thickness can be obtained via $\tau(\Delta\lambda) = \tau_0 \exp\{-[(\Delta\lambda - \Delta\lambda_S)/\Delta\lambda_D]^2\}$, where τ_0 is the optical thickness at the line center with a Gaussian form. The Doppler width ($\Delta\lambda_D$) can be expressed in terms of the temperature T and the microturbulence ξ_t as

$$\Delta\lambda_D = (\lambda_0/c)\sqrt{\xi_t^2 + 2kT/m}. \quad (4)$$

Finally, given at least four contrast values, we can derive the four unknown parameters, namely S, τ_0, $\Delta\lambda_D$, and $\Delta\lambda_S$. It is impossible, however, for us to derive these parameters with FMT data alone, since we have only three data points along the wavelength. Moreover, the wavelength differences of the central wavelength of the 2 wings' filters against Hα line center are the same (0.8Å) and the Hα line profile is normally symmetric against Hα line center, we have substantially only two available data points if the profile is not Doppler shifted.

In order to make up for the insufficient observational data, we made two assumptions: (i) before the onset of a disappearance ($t = 0$), the filament has no bulk motion ($\Delta\lambda_S = 0$), and all the other parameters ($S, \tau_0, \Delta\lambda_D$) are uniform. Here, τ_0 and $\Delta\lambda_D$ do not change even after the onset of the disappearance. (ii) The amount of the Doppler shift is 0.8Å when and where the contrast in the blue or red wing is its minimum.

The assumption $\Delta\lambda_S(t=0) = 0$ is made since the filament can be regarded as being a stationary structure before its activation. With assumption (i) and equation (3), we can write the Hα line center and blue wing's contrasts before the onset (C_1 and C_2, respectively) as

$$C_1 = \left[\frac{S}{I_{R0}(0)} - 1\right][1 - \exp(-\tau_0)], \quad (5)$$

$$C_1 = \left[\frac{S}{I_{R0}(0)} - 1\right][1 - \exp(-\tau_0)], \quad (5)$$

$$C_2 = \left[\frac{S}{I_{R0}(-0.8)} - 1\right]$$
$$\times \left[\left[1 - \exp\left\{-\tau_0 \exp\left[-\left(\frac{0.8}{\Delta\lambda_D}\right)^2\right]\right\}\right]\right] \quad (6)$$

仮定(ii)は，フィラメントのブルーシフトの最大値（$|\Delta\lambda_{Smax}|$）が 0.8Å 以下の時には不適である．その場合，別の仮定をおく必要がある．ドップラーシフト量以外の 3 つのパラメータ（S, τ_0, $\Delta\lambda_D$）のうち，ドップラー幅はほかの 2 つよりも静穏フィラメントにおいては値の変動が少ないことが知られている．Kubota (1968) は静穏フィラメントのスペクトル観測を行い，9 個の Hα プロファイルについてドップラー幅を計測した．Jefferies and Orrall (1958) は静穏フィラメントの 36 個の Hα プロファイルからドップラー幅を計測した．これらの研究によると，ドップラー幅の平均値は 0.26Å，標準偏差は 0.02Å である．これより，ドップラー幅の変動はほかの 2 パラメータと比較して非常に小さいので，$|\Delta\lambda_{Smax}| < 0.8$Å のイベントに対しては，固定したドップラー幅を用いた．それゆえ，仮定(ii)は以下のように書ける．

$$C_3 = \left[\frac{S}{I_0(-0.8)} - 1\right][1 - \exp(-\tau_0)],$$
$$(\text{for } |\Delta\lambda_{Smax}| \geq 0.8\text{Å}) \quad (7)$$

または，

$$\Delta\lambda_D = 0.26 \pm 0.02, \quad (\text{for } |\Delta\lambda_{Smax}| < 0.8\text{Å}) \quad (8)$$

ここで，C_3 は青色側ウイングにおける最小コントラストである．

上述の式と仮定より導出された 3 つの物理パラメータ（S, τ_0, $\Delta\lambda_D$）と $\Delta\lambda_S = 0$ より，最終的にフィラメントの Hα 線プロファイルを求めることができる．しかし，今回の研究では(i)フィルターの透過幅，(ii)散乱光効果，および(iii)ドップラー増光効果の補正を行う必要がある．なぜならばこれらの効果がコントラストに大きな影響を与えるからである．

表 1 に示されているように，各望遠鏡には決まった有効透過幅があり，このことにより得られる観測光度は実際の光度に透過

$$C_2 = \left[\frac{S}{I_{R0}(-0.8)} - 1\right]$$
$$\times \left[\left[1 - \exp\left\{-\tau_0 \exp\left[-\left(\frac{0.8}{\Delta\lambda_D}\right)^2\right]\right\}\right]\right] \quad (6)$$

respectively.

Assumption (ii) is invalid when the maximum amount of blue shift of a filament ($|\Delta\lambda_{Smax}|$) is less than 0.8Å. In this case, we need another assumption. Among the other three parameters, we notice the Doppler width for a static quiescent filament, which shows less variation than do the other two parameters. Kubota (1968) made a spectroscopic observation of a quiescent filament, and measured the amount of Doppler widths of 9 Hα line profiles. Jefferies and Orrall (1958) also measured it with 36 Hα line profiles of quiescent filaments. According to their studies, the mean value of the derived Doppler widths is 0.26Å with a standard deviation of 0.02 Å.

Since this shows a very small variation compared to the variations made by the other two parameters, we used a fixed Doppler width for those events with $|\Delta\lambda_{Smax}| < 0.8$Å. So far, assumption (ii) can be expressed as

$$C_3 = \left[\frac{S}{I_0(-0.8)} - 1\right][1 - \exp(-\tau_0)],$$
$$(\text{for } |\Delta\lambda_{Smax}| \geq 0.8\text{Å}) \quad (7)$$

or

$$\Delta\lambda_D = 0.26 \pm 0.02, \quad (\text{for } |\Delta\lambda_{Smax}| < 0.8\text{Å}) \quad (8)$$

where C_3 is the minimum contrast value of the blue wing.

With the three physical parameters (S, τ_0, and $\Delta\lambda_D$) derived using the equations and the assumptions above together with $\Delta\lambda_S = 0$, we can finally get the Hα line profile of the filament. In this study, however, we need to make corrections of (i) the effect of the filter's band width, (ii) the effect of stray light, and (iii) the Doppler brightening effect, since they may change the observed contrasts notably.

As shown in table 1, each telescope has a filter with a finite effective band width, which makes the observed intensity to be the sum of the intensities within its band multiplied by the transmitting function of the filter. In order to correct this effect,

関数を畳み込み積分したものとなる．この効果を補正するために，透過関数をガウス型とみなして式(3)を以下のように書き換える．

$$C_{\text{FMT}}(\Delta\lambda_{\text{obs}}) = \int_{-\infty}^{\infty} C(\Delta\lambda) \exp\left[-\left(\frac{\Delta\lambda - \Delta\lambda_{\text{obs}}}{\Delta\lambda_{\text{w}}}\right)^2\right] d\Delta\lambda \Big/ \int_{-\infty}^{\infty} \exp\left[-\left(\frac{\Delta\lambda - \Delta\lambda_{\text{obs}}}{\Delta\lambda_{\text{w}}}\right)^2\right] d\Delta\lambda. \quad (9)$$

式中の $\Delta\lambda_{\text{w}}$ は有効透過幅，$\Delta\lambda_{\text{obs}}$ はフィルターの中心波長である．散乱光の影響も考慮にいれた．ダークフィラメントの光度は近傍の彩層からの散乱光によって増加されるために，FMTの生画像のコントラストは実際のものより小さくなる．我々は散乱の効果は常に一定であるとし，また点広がり関数はガウス型である (Kubota 1968：詳細は付録1) として補正を行った．

有効透過幅［式(9)］と散乱の効果を補正することで，観測されたコントラスト (C_1, C_2, C_3) を実際のコントラスト (C'_1, C'_2, C'_3) に変換することができる．この変換は，3つの物理パラメータを先行研究 (Grossmann-Doerth, von Uexküll 1971, 1973, 1977; Loughhead 1973; Bray 1974; Alissandrakis et al. 1990; Tsiropoula et al. 1992; Heinzel et al. 1995, 1999; Paletou 1997) で得られた範囲内で少しずつ繰り返し変化させることによって行った．各反復毎に，フィルターの透過幅と散乱光の関数を用いて，観測されたコントラストを最もよく再現するパラメータの組み合わせを求める．次に C'_1, C'_2, C'_3 の値を式(5)-(7)に代入することでフィラメントの物理パラメータ (S, τ_0, $\Delta\lambda_{\text{D}}$) とHαプロファイルを求める．このプロファイルを波長シフトさせることによって，様々なドップラーシフト量におけるフィラメントのコントラストプロファイルを計算する．コントラストプロファイルによって，観測されたコントラスト値を直接フィラメントのドップラーシフト量に対応することができる．しかし，計算されたコントラストプロファイルはFMTで観測されたコントラスト値と同等ではない．なぜなら，上記の手法で求めら

we rewrite equation (3) by assuming that the transmitting function has a Gaussian form, and used

$$C_{\text{FMT}}(\Delta\lambda_{\text{obs}}) = \int_{-\infty}^{\infty} C(\Delta\lambda) \exp\left[-\left(\frac{\Delta\lambda - \Delta\lambda_{\text{obs}}}{\Delta\lambda_{\text{w}}}\right)^2\right] d\Delta\lambda \Big/ \int_{-\infty}^{\infty} \exp\left[-\left(\frac{\Delta\lambda - \Delta\lambda_{\text{obs}}}{\Delta\lambda_{\text{w}}}\right)^2\right] d\Delta\lambda. \quad (9)$$

In this equation, $\Delta\lambda_{\text{w}}$ is the effective band width and $\Delta\lambda_{\text{obs}}$ is the central wavelength of the filter, respectively. The effect of stray light is also taken into account. Since the intensity from the dark filament is intensified by the stray light of the nearby chromosphere, the contrast on the raw FMT images is weaker than the real case. We corrected this effect, assuming that it is always constant, and the scattering function has a Gaussian form (Kubota 1968: see appendix 1 for more detailed discussion).

Using the functions for the effective band width [equation (9)] and the scattering effect, we converted the observed contrasts (C'_1, C'_2, and C'_3) to the real ones (C'_1, $C'_{2,}$, and C'_3). This conversion was performed by changing the three physical parameters iteratively with slight steps within ranges consistent with previous studies (Grossmann-Doerth, von Uexküll 1971, 1973, 1977; Loughhead 1973; Bray 1974; Alissandrakis et al. 1990; Tsiropoula et al. 1992; Heinzel et al. 1995, 1999; Paletou 1997). In each iteration, the functions of the filterbandwidths and stray light are used to find the parameters which best fit to the observed contrasts. We then substituted C'_1, C'_2, and C'_3 to equations (5)-(7) and obtain the physical parameters (S, τ_0, and $\Delta\lambda_{\text{D}}$) and hence the Hα line profile of the filament. By shifting this line profile, we then calculated the contrast profile for various values of the Doppler shift of the filament. The contrast profile directly relates the observed contrast values to the Doppler shift of the structure. The calculated contrast profile, however, is not equivalent to the observed contrasts with FMT, since the Hα line profile derived with the method described above is now free from the effects of filters' band widths and stray light. We, therefore, using the

れたHα線プロファイルはフィルターの有効透過幅や散乱光の影響を考慮していないためである．それゆえ，補正式[式(9)および点広がり関数]を用いてFMTで観測されたコントラスト値に合致するようなコントラストプロファイルを計算した．次に，新しく計算したコントラストプロファイルと観測されたコントラスト値をフィラメント全域のピクセルで比較を行い，ドップラーシフト量を求めた．

我々はまた，ドップラー増光現象の補正も行った．ドップラー増光現象は，運動するフィラメントからのHα線プロファイルを増光させる働きをする (Rompolt 1967a,b)．今回の研究では，ドップラー増光現象を補正するために源泉関数 (S) が時間依存変数であるとして扱った．この補正に関する詳細な議論は，本論文の末尾の付録2を参照されたい．

3.1.3. 1999年2月16日のイベントの視線速度

この小節では，1999年2月16日のイベントを例にして，我々の手法についてより詳しく解説を行う．イベントの時間発展を追うために，フィラメント消失以前の02:00 UT におけるHα線中心画像で観測されたフィラメントの中心点をA点と定義した．FMTの画像上でのA点の初期位置およびその時間発展の様子は図1において「+」で示されている．これらの位置は，後（小節3.2）でその導出法を議論する視線垂直速度に基づいて求められた．A点におけるコントラストの時間変化は図2の左上のパネルに示されている．02:40 UTより以前では，Hα線中心（∗）でフィラメントは暗く，またコントラスト値は負である．また青色側ウイング（◇）と赤色側ウイング（△）ではコントラスト値はほとんど0に近い値である．02:40 UT，フィラメントが南西の方向に膨張を開始すると，A点におけるHα線中心のコントラストは増加して0に近付く．また，青色側ウイングではコントラストの急激な減少が見られる．青色側コントラストは 02:48 UT に最小値となり，A点のドップラーシフト量が青色

correction functions [equation (9) and the scattering function], calculated the contrast profile to fit the observed contrast with FMT. We then derived the Doppler shift by comparing the newly calculated contrast profile and the observed contrasts at all of the pixels on the filament.

We also corrected the Doppler brightening effect (DBE). The DBE acts to brighten the Hα line profile from a moving filament (Rompolt 1967a,b). In this study, the source function (S) was handled as a time-dependent variable in order to correct the DBE. See appendix 2 in the end of this paper for a detailed discussion of this effect and the correction.

3.1.3. The line-of-sight velocity of 1999 February 16 event

In this subsection, we explain our method in more detail by using the 1999 February 16 event as an example. To see an overview of its evolution, we defined a point (point A) at the middle of the filament on the Hα line center image at 02:00 UT before the start of its disappearance. Its initial and subsequent locations on the FMT images are displayed with "+" marks in figure 1. These positions were derived according to the tangential velocity obtained by a method which we discuss later (subsection 3.2). The time-dependent contrast for the trajectory of point A is plotted in the top-left panel of figure 2. Before 02:40 UT, the filament is dark, or the contrast values are negative, in the Hα line center (asterisk marks) and no or little contrast is detected in the blue (diamond marks) and red (triangle marks) wings. As the filament starts to expand toward the south-west at 02:40 UT, the contrast in Hα line center at point A gradually increases to zero and the contrast in the blue wing shows a sudden depletion. The blue wing's contrast reaches its minimum at 02:48 UT, indicating that the amount of Doppler shift for point A reaches -0.8 Å

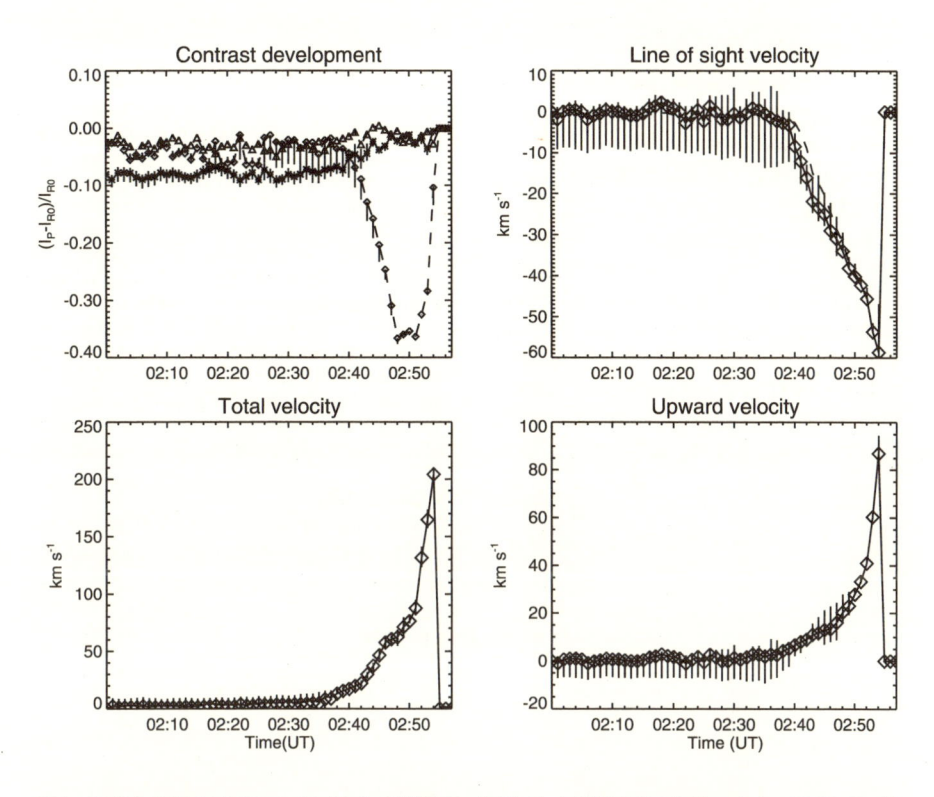

図2 1999年2月16日のフィラメントの観測コントラストと導出された速度. 左上: FMT Hα線中心 (*), 青色側ウイング (◇), および赤色側ウイング (△) で観測されたコントラスト変化. 右上: 導出された視線速度 (正: 前進運動, 負: 後退運動), ドップラー増光現象の補正の有無 (実線および破線). 左下および右下: A点における全速度および上昇速度 (正: 上昇, 負: 下降). 上昇速度はA点における視線速度と視線垂直速度を変換して得た. A点に対応する部分は強く加速されて消失している.

Fig. 2. Observed contrasts and derived velocities of the filament of the 1999 February 16 event. Top-left: The development of the contrasts in the FMT Hα line center (asterisks), blue wing (diamond), and red wing (triangle). Top-right: The derived line-of-sight velocity (+ : receding motion, − : approaching motion) with and without the correction of the Doppler brightening effect (solid and dash lines, respectively). Bottom-left and right: The total and upward velocity of point A (+ : upward, − : downward). The upward velocity was obtained by applying a coordinate conversion to the line-of-sight and tangential velocities of point A. The column corresponding to point A vanishes with its upward velocity being strongly accelerated.

側フィルターの中心波長とHα線中心との波長差にあたる−0.8Å (=−35 km/s) に達したことを示している. 02:50 UTから青色側ウイングのコントラストが急激に増加した後, A点は消滅している. このイベ

($=-35\,\mathrm{km\,s^{-1}}$), which is equal to the wavelength difference between the central wavelength of blue wing's filter and Hα line center. Point A disappears after a sudden increase in the contrast of the blue wing from 02:50 UT. Throughout

ントを通して，赤色側ウイングにおけるコントラストはほぼ0であった．このことは，太陽表面に向かう運動がなかったことを意味する．

これらの観測されたコントラストと式(5)-(7) を用いて，フィラメントの3つの物理パラメータを求めた．これらのパラメータ $[S, \tau_0, \Delta\lambda_D]$ はそれぞれ $[0.12, 3.1, 0.26]$ となった．これらのパラメータを元に計算された Hα 線プロファイルを図3に示す．フィラメントの光度は参照 Hα 線プロファイル (Kurucz et al. 1984) と比較して $\Delta\lambda = \pm 0.5$Å の範囲では小さい．このためにフィラメントは Hα 線中心では太陽表面に対して暗い構造として観測される．線中心での平らなプロファイル部分は静穏フィラメントで通常観測される特徴 (e.g.,

the event, the contrast of the red wing remained almost zero, which means no receding motion.

Using these observational contrasts with equations (5)-(7), we derived the three physical parameters of this filament. These parameters $[S, \tau_0, \Delta\lambda_D]$ are $[0.12, 3.1, 0.26]$, respectively. The calculated Hα line profile with these parameters is plotted in figure 3. The intensity of the filament is weaker than the Hα reference line profile (Kurucz et al. 1984) within the wavelength range of $\Delta\lambda = \pm 0.5$Å so that the filament is observed as a dark feature against the disk near the Hα line center. The flat bottom at the line center is one of the characteristics which are often observed in quiescent filaments (e.g., Kubota 1968).

The contrast profiles were calculated by giving various amounts of Doppler

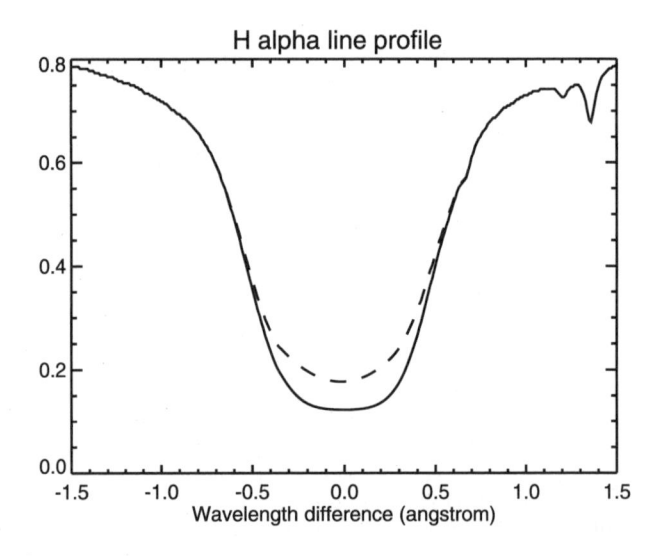

図3 計算から求められた1999年2月16日のダークフィラメントのHα線プロファイル（実線）．破線は参照Hα線プロファイル（Kurucz et al. 1984）であり，本研究では彩層の背景Hα線プロファイルとして用いた．横軸はHα線中心から計測した観測波長，縦軸は近傍の連続光光度で規格化した吸収線光度．

Fig. 3. Calculated Hα line profile of the dark filament of the 1999 February 16 event (solid line). The dash line is a reference Hα line profile (Kurucz et al. 1984), which was used as the Hα line profile of the background chromosphere in this study. The abscissa represents the observational wavelength measured from Hα line center, and the ordinate does the line intensity normalized by the nearby continuum.

Kubota 1968) の1つである.

コントラストプロファイルは図3のHα線プロファイルに様々なドップラー速度を与えて計算され,その結果は図4に示されている. $\Delta\lambda_S = 0.0$ のとき,コントラストは±0.2Å近傍で2つの最小値を取る.このことはフィラメントのプロファイルが線中心で平らな底をもつことに起因する.ドップラーシフト量が増大するにつれ,コントラストは強められる.特定の波長に注目してみると,フィラメントのドップラーシフト量が増加するにつれてコントラスト値が連続的に変化することが見てとれる.例として,計算されたコントラストの-0.8Åでの値はドップラーシフト量が-0.40Åから-0.60Åと変化するに従い-0.20から-0.66に減少している.

velocities to the filament Hα line profile of figure 3, and are shown in figure 4. When $\Delta\lambda_S = 0.0$, it shows two minima near ±0.2 Å due to the flat bottom of the filament line profile near the line center. As the Doppler shift increases, the contrast strengthens. Notice that the contrast successively changes at a fixed wavelength as the Doppler shift of the filament increases. For example, the calculated contrast at -0.8 Å increases from -0.20 to -0.66 with an increase of the Doppler shift from -0.40 Å and -0.60 Å.

As described in the previous subsection, this contrast profile was corrected for the effects from the filter's effective band width and stray light. Thus, we can not directly compare this contrast profile with the observed contrasts to obtain the Doppler shift of the observed filament.

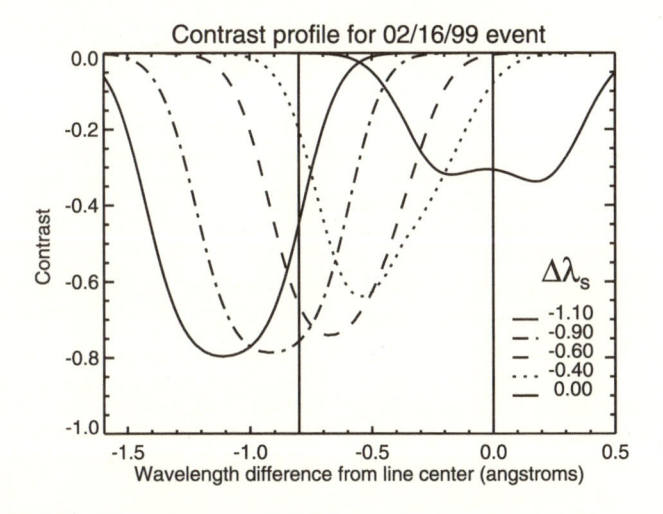

図4 計算から求められた1999年2月16日のイベントのコントラストプロファイル.コントラストは図3のフィラメントのHα線プロファイルを用いてドップラーシフト ($\Delta\lambda_S$) を 0.0Å (実線),−0.4Å (破線),−0.9Å (一点鎖線),および−1.1Å (三点鎖線) として計算された.横軸はHα線中心から計測した観測波長,縦軸はコントラスト.2本の縦線はFMTのHα中心フィルターと青色側フィルターの中心波長を示す.

Fig. 4. Calculated contrast profile for the 1999 February 16 event. The contrasts were calculated for Doppler shifts ($\Delta\lambda_S$) of 0.0Å (solid line), −0.4Å (dot line), −0.6Å (dash line), −0.9Å (dot-dash line), and −1.1Å (dot-dot-dot-dash line) with the filament Hα line profile of figure 3. The abscissa represents the observational wavelength measured from the Hα line center and the ordinate is the contrast. The two vertical solid lines show the central wavelength of the FMT center and the blue wing filters.

Calculated contrast

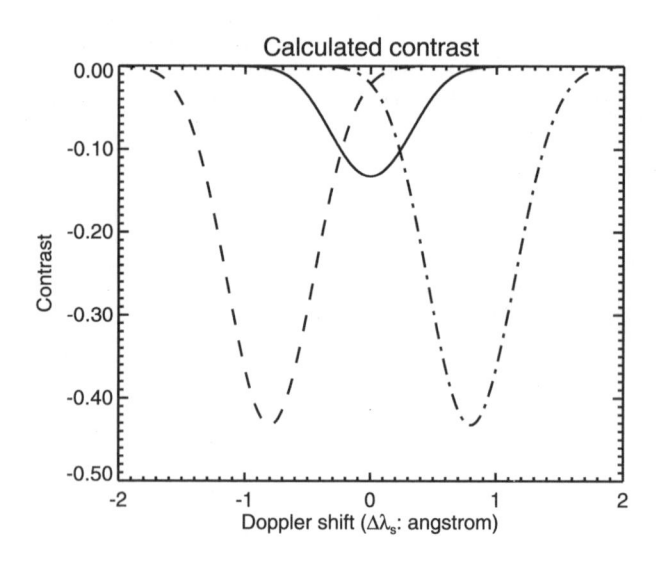

図5 1999年2月16日のイベントのコントラストプロファイル．フィルターの有効透過幅と散乱光の影響が考慮されている．横軸はドップラーシフト量，縦軸はコントラスト．実線はFMTのHα線中心画像，破線は青色側ウイング画像，および一点鎖線は赤色側ウイング画像．

Fig. 5. Another kind of contrast profile for the 1999 February 16 event. The effects of filters' band widths and stray light are taken into account. The abscissa is the Doppler shift and the ordinate is the contrast. The solid line is for FMT Hα line center images, the dashed line for the blue wing, and dot-dashed line for the red wing.

前小節で述べたように，このコントラストプロファイルはフィルターの有効透過幅と散乱光の影響による誤差が補正されている．このため，このコントラストプロファイルを観測されたコントラスト値と直接比較してフィラメントのドップラーシフト量を求めることはできない．有効透過幅と散乱光の影響を考慮して計算されたコントラストプロファイルを図5に示す．ここで，横軸は観測波長ではなく，フィラメントのドップラーシフト量である．縦軸は3つのフィルター：Hα線中心（実線），Hα-0.8 Å（破線），Hα$+0.8$ Å（一点鎖線）で観測されたコントラスト値を示す．図4と5のコントラスト値の絶対値を比較すると，図5におけるコントラスト値のほうが減少していることがわかる．これは，有効透過幅と散乱光の影響が常にコントラストを弱める方向に働くためである．

The contrast profile calculated with the two effects taken into account is given in figure 5, where the abscissa represents not the observational wavelength, but the Doppler shift of the filament. The ordinate gives contrasts observed with the three different filters: Hα center (solid line), Hα-0.8 Å (dash line), and Hα$+0.8$ Å (dot-dash line). Comparing the absolute values of contrasts in figures 4 and 5, one can find that they are reduced in figure 5. This is because the two effects always act to weaken contrasts. For example, in figure 4, the contrast with $\Delta\lambda_S = 0.0$ (solid line) at the line center is -0.31, while in figure 5 it is -0.13. Another difference is the broadening of the profiles due to the finite band widths of the filters. For example, in figure 4, no contrast is detectable in $\Delta\lambda = -0.6$ Å when the Doppler shift is zero, while we can find a weak contrast even with the ± 0.8 Å filters at $\Delta\lambda_S = 0$ in figure 5.

実際，図4では$\Delta\lambda_S = 0.0$（実線）における線中心でのコントラスト値は-0.31であるが，図5では-0.13となっている．また，フィルターの透過幅が有限の値であることによってプロファイルの幅が増大していることも差異として認められる．例えば，図4においてドップラーシフト量がなしの場合，$\Delta\lambda = -0.6$ÅではコントラストᏯは0であるが，一方図5では$\Delta\lambda_S = 0$の場合± 0.8Åにおいても弱いコントラストが残存している．

図5を用いて1999年2月16日のイベントでのA点における視線速度の時間変化が求められ，図2の右上のパネルに示されている．この視線速度の変化をFMTで観測されたコントラスト（左上のパネル）と比較すると，視線速度の時間発展がコントラストの変化と一致していることがわかる．02:40 UT より以前では，A点はHα線中心では一定のコントラスト値を示し，両ウイングではほぼコントラストがない．このことは視線方向の運動をほとんどしていないことを意味している．青色側ウイングのコントラストは02:40 UT に突然強まり始め，一方Hα線中心のコントラストは0まで増加する．このコントラストの変化にしたがって，視線速度は一様に加速されて02:48 UT には~ 38 km/s に達する．このとき，青色側ウイングのコントラストは最小値を取る．38 km/s のドップラー速度は-0.8Åのドップラーシフト量に対応する．このシフト量は青色側ウイングの中心波長とHα線中心との波長差に相当する．02:48 UT に青色側ウイングのコントラストが上昇に転じた後もフィラメントは加速し続け，A点が完全に消失する02:55 UT の直前には~ 59 km/s まで達する．

3.1.4. 誤差の評価

3.1.4.1. プラズマパラメータの変化による効果

運動するフィラメントのHα線光度が増大する原因として考えられるものには，フィラメントの内部で物理状態が変化していることがあげられる．この効果を無視し得る場合（e.g., Hyder, Lites 1970）もある

The time-dependent line-of-sight velocity of point A of the 1999 February 16 event was derived with figure 5, and is shown in the top-right panel of figure 2. Comparing this with the observed contrasts by FMT (top-left panel), we can see the evolution of the line-of-sight velocity is in good agreement with the evolution of its contrast. Point A shows stationary contrast in the Hα line center, and almost no contrast in the 2 wings, indicating little line-of-sight motion until 02:40 UT. The contrast in the blue wing suddenly starts to strengthen from 02:40 UT, while it decreases to zero in the Hα line center. Following this change in the contrasts, the line-of-sight velocity is monotonically accelerated and reaches ~ 38 kms^{-1} at 02:48 UT when the contrast in the blue wing reaches its minimum. The line-of-sight velocity of 38 kms^{-1} corresponds to a Doppler shift of -0.8Å, which is equal to the wavelength difference between the blue wing's central wavelength and Hα line center. After a reversal of the blue wing's contrast at 02:48 UT, the filament continues to be accelerated, and it reaches ~ 59 km s^{-1} just before the complete disappearance of point A at 02:55 UT.

3.1.4. Error estimations

3.1.4.1. Effect of variations of plasma parameters

Another origin to which we can attribute the brightening of a moving filament of Hα line is intrinsic variations of the physical state of the filament. Although there are some cases in which this effect can be ignored (e.g. Hyder,

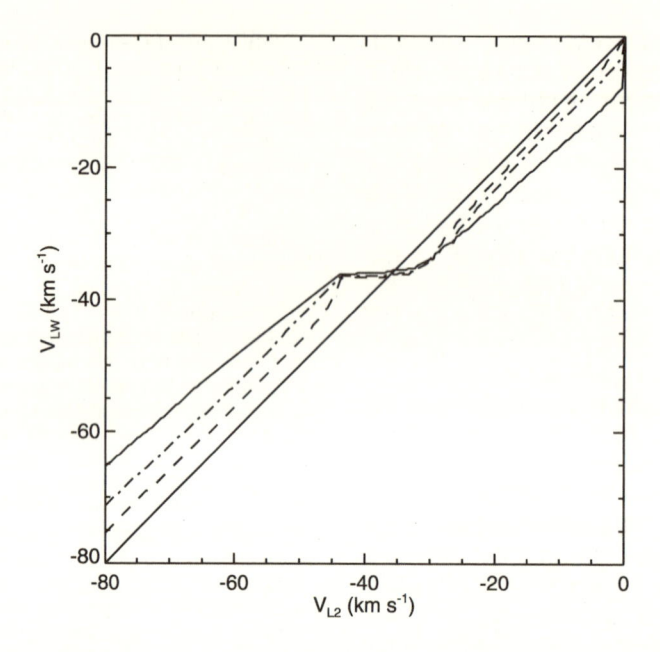

図6 ドップラー幅を 0.26Å（ケース1），0.32Å（ケース2），0.43Å（ケース3），および 0.65Å（ケース4）と変化させた場合のフィラメントの視線速度（km/s）．横軸は運動するフィラメントの真の視線速度（V_{L2}）．縦軸はドップラー幅を 0.26Å と固定した場合に我々の手法で求められた速度（V_{LW}）．実線はケース1，破線はケース2，一点鎖線はケース3，および三点鎖線はケース4．

Fig. 6. Line-of-sight velocity (km s^{-1}) of moving filaments with Doppler widths of 0.26Å (case 1), 0.32Å (case 2), 0.43Å (case 3), and 0.65Å (case 4). The abscissa represents the real line-of-sight velocity (V_{L2}) of the moving filaments. The ordinate is the velocity (V_{LW}) which is measured by our method with a fixed Doppler width of 0.26Å. The solid line is for case 1, dashed line for case 2, dot-dashed line for case 3, and dot-dot-dot-dashed line for case 4.

が，特にプロミネンス噴出のようにドップラー幅が増大している場合には，観測されるコントラストに一定の影響を及ぼす．我々は簡単な実験を行い，ドップラー幅が変動するに従い視線速度がどのように影響を受けるのかを調査した．この実験において，運動するフィラメントの視線速度（V_{L2}），源泉関数（S），および光学的厚み（τ_0）は1999年2月16日のイベントのフィラメントのものを使用した．そしてドップラー幅（$\Delta\lambda_D$）を 0.32Å（ケース2），0.43Å（ケース3），0.65Å（ケース4），および 0.26Å（変化させず：ケース1）と変化させた．0.32Å，0.43Å，および 0.65Åのドッ

Lites 1970), it has a certain effect on the observed contrast, especially when the Doppler width of a filament increases, as is often observed in prominence eruptions. We performed a simple experiment to see how the line-of-sight velocity might be affected by the variation of the Doppler width. In this experiment, we assumed a moving filament with its line-of-sight velocity (V_{L2}), its source function (S), and optical thickness (τ_0) being equal to the original values of 1999 February 16 filament. We then changed its Doppler width ($\Delta\lambda_D$) to 0.32 Å (case 2), 0.43 Å (case 3), 0.65 Å (case 4), and 0.26 Å (unchanged: case 1). The Doppler widths of 0.32 Å, 0.43 Å, and 0.65 Å

ブラー幅はそれぞれ、フィラメントの温度を 8000 K とした時の微小乱流が 15, 20, 30 km/s の場合に相当する. 1999 年 2 月 16 日のイベントのフィラメントにおける S, τ_0 および $\Delta\lambda_D$ と同じ値を用いて視線速度 (V_{LW}) を計算した. その結果を図 6 に示す.

ケース 2, 3, および 4 では、求められた視線速度 (V_{LW}) は $V_{L2} = -37$ km/s ($\Delta\lambda_D = -0.8$ Å：青色側ウイングの中心波長と Hα 線中心との波長差) 近傍でほぼ一定となる. これは広がった Hα 線プロファイルの底が平らになることにより、ウイング部分と比較して線中心近傍における光度変化が小さくなることによる. 青色側ウイング部分はブルーシフト量 ($V_{L2} < 0$ km/s) の導出に最も影響を与えるため、狭いドップラー幅 (0.26 Å) から計算した速度は青色側フィルターの中心波長 (-0.8 Å) に近付く傾向がある. このことにより計算された速度の大きさは、$V_{L2} > -37$ km/s の範囲で実際の速度よりも大きく、$V_{L2} < -37$ km/s の範囲では小さくなる. 誤差の最大値は、視線速度が -60 km/s の場合、ケース 2 において 5 km/s, ケース 3 では 7 km/s, およびケース 4 で 10 km/s となった.

3.1.4.2. 背景彩層光度の変動

上記以外に誤差の原因となるものとして、背景光度の変動があげられる. 式 (3) から観測されたコントラストを計算する際に、フィラメント直下の光度として、フィラメント周辺の広範囲の彩層光度の平均値を用いた. これは静穏領域のように周辺の彩層光度がほぼ一様であった場合には有効である. しかし、もしプラージュのような明るい活動領域が周辺に存在した場合、平均光度は実際にフィラメントを照らしている彩層光度とは異なる可能性がある. 更に、フィラメントは長く伸びた構造をもつために、平均光度はダークフィラメント下部の各ピクセルでの彩層光度とは異なりうる.

そこで、背景光度の平均値 (I_{R0}) と標準偏差 (σ) を用いて $I_{R0} + \sigma$ のようにコントラスト値に上下限を設けた上で、視線速度の上下限を計算した. 視線速度を示した図

correspond to the microturbulence of 15, 20, and 30 km s^{-1}, respectively, with a fixed temperature of 8000 K in the filament. We then calculated their line-of-sight velocities (V_{LW}) with our method, fixing S, τ_0, and $\Delta\lambda_D$ equal to the original values of 1999 February 16 event's filament. The results are shown in figure 6.

In situational cases 2, 3, and 4, the derived line-of-sight velocities (V_{LW}) are almost constant near $V_{L2} = -37$ km s^{-1} ($\Delta\lambda_S = -0.8$ Å：which corresponds to the wavelength difference between the blue wing filter's central wavelength and Hα line center). This is due to the broadened bottom of the Hα line profile where the intensity variation is smaller compared to that in the line wings. Since the blue wing is most responsible for the derivation of blue shift values ($V_{L2} < 0$ km s^{-1}), calculated velocities with narrower line width (0.26 Å) tend to be closer to the central wavelength of the blue wing's filter (-0.8 Å). This causes the amplitude of the calculated velocity to be larger when $V_{L2} > -37$ km s^{-1} and smaller when $V_{L2} < -37$ km s^{-1} than the real ones. The maximum magnitudes of errors are 5 km s^{-1} for case 2, 7 km s^{-1} for case 3, and 10 km s^{-1} for case 4 when the line-of-sight velocity is -60 km s^{-1}.

3.1.4.2. Fluctuations in intensities of background chromosphere

Another significant error arises from the fluctuations in the background intensities. Since in the calculation of the observed contrast values with equation (3), we used the averaged chromospheric intensities over a large area surrounding the filament as the intensity of the chromosphere beneath the filament. This is valid if the intensities of nearby chromosphere are almost uniform, like in quiet regions. If there is a bright active region or plage, however, the averaged intensity may differ from the intensity of chromosphere which irradiates the filament. Moreover, since the filament is a long structure, the averaged intensity may also be different from the intensity of chromosphere under the dark filament at each pixel.

Thus, we set the upper and lower limit of contrast values by using the averaged (I_{R0}) and standard deviation (σ) of

中のエラーバーは，背景光度変動と式（8）中のドップラー幅変動の上下限から求められた誤差の和となっている．これらのエラーバーの平均値を計算した結果，5 km/s と求められた．

3.2. 消失するフィラメントの視線垂直速度

3.2.1. 手法

視線垂直速度の導出は，FMT の連続画像中のフィラメントの運動を追うことによって行った．FMT の画像中でフィラメント内の小さなガス塊（blob，ブロブ）を数多く同定し，その動きをそれぞれ追跡した．これによって運動の速度と方向を得た．図 7 は 1999 年 2 月 16 日にフレアモニターで観測された Hα 線中心画像である．02:54 UT から 02:55 UT までの間におけるブロブの運動速度と方向が矢印にて示されている．フィラメント上に平均距離 15 Mm の間隔で同定されたブロブは，フィラメント中央部分では最大速度 400 km/s で南西方向に運動している．これらのブロブの速度を補間することにより，フィラメント上の全てのピクセルにおける視線垂直速度を求めた．

導出された視線垂直速度はフィラメントの各部分の軌道を求めるためにも用いられた．ピクセル位置 $[x(t), y(t)]$ で速度 $[v_x, v_y]$ が検出された場合，このピクセル上の物体は，時刻 $t+\delta t$ の画像では $[x(t+\delta t), y(t+\delta t)] = [x, y] + \delta t \times [v_x, v_y]$ の位置に移動する．このことを用いて，イベント全体におけるフィラメントの各部分の軌道を決定した．この手法を用いて求められた A 点の位置変化を図 1 に示す．A 点のコントラストおよび視線速度の時間発展は，上記の手法で得られた各時刻における A 点の位置を用いて得られた．

3.2.2. 誤差の推定

フィラメント中のブロブの位置測定に際して，パソコンを用いたカーソルによる

background intensities as $I_{R0} \pm \sigma$ and calculated the upper and lower limit of the line-of-sight velocity. All of the error bars in the figures of the line-of-sight velocity are given as the sum of the errors due to the intensity fluctuation and those obtained from the upper and lower limit of the Doppler width variations in equation (8). The mean value of these error bars was calculated and found to be $5 \mathrm{~km\,s^{-1}}$.

3.2. Tangential Velocity of Disappearing Filaments

3.2.1. Method

We derived the tangential velocity of a disappearing filament by tracing the apparent motions of filament blobs on the successive FMT images. We identified as many blobs on the FMT images as possible, and traced them on the subsequent images, we thus obtained the magnitudes of velocities and directions of the motions. Figure 7 shows an FMT center image on which the velocities of identified blobs between 02:54 UT and 02:55 UT of the 1999 February 16 event are shown with arrows. The blobs are distributed with a mean distance of 15 Mm, and are moving toward the southwest with a maximum velocity of nearly $400 \mathrm{~km\,s^{-1}}$ at the middle part of the filament. By interpolating the blob velocities, we derived the tangential velocity at all of the pixels on the filament pixels.

The derived tangential velocity was also used to obtain the trajectory of each element of the filament. Since a pixel at $[x(t), y(t)]$ with a tangential velocity of $[v_x, v_y]$ would shift to $[x(t+\delta t), y(t+\delta t)] = [x, y] + \delta t \times [v_x, v_y]$ on the next image at $t+\delta t$, we can easily find its trajectory throughout the event. The apparent shifts in the positions of point A shown in figure 1 were derived with this method. The evolutions of the contrasts and line-of-sight velocity of point A were also obtained by measuring these values at the derived positions.

3.2.2. Error estimations

In measuring the positions of the blobs the images were expanded by a factor of three to reduce the error involved in the

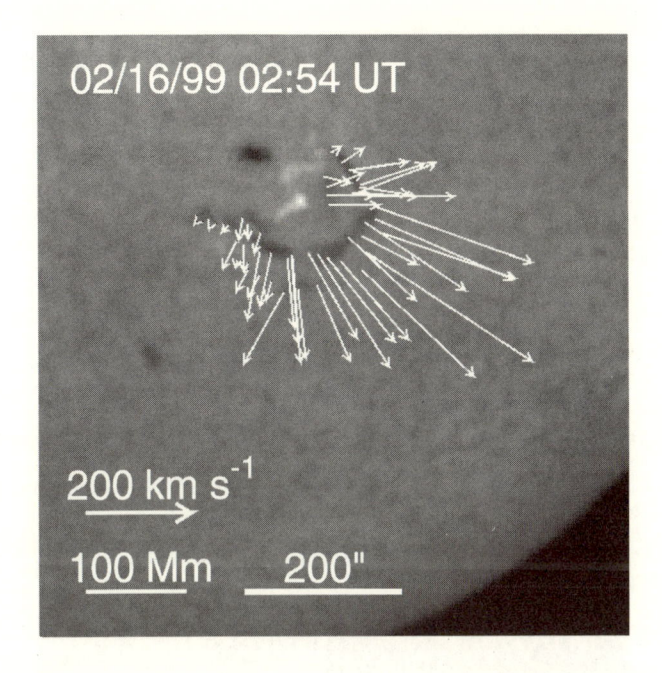

Appendix A Selected Papers Relating to FMT | 351

図7 1999年2月16日の02:45 UTにFMTで観測されたHα線中心画像. フィラメントが南西方向に膨張している. フィラメント上のブロッブの分布およびその視線垂直速度が矢印で示されている. 視線垂直速度はブロッブの位置変化を02:54 UTから02:55 UTの間で追跡して求めた.

Fig. 7. FMT Hα line center image at 02:54 UT on 1999 February 16, showing the filament expanding toward the south-west. The distribution of defined blobs on the filament and their tangential velocities are shown with arrows. The tangential velocities were calculated according to the apparent motions of the blobs at 02:54 UT and 02:55 UT.

位置指定に伴う誤差を減らすために画像は3倍に拡大して用いた. それゆえ, 計測の確度は $U_m = 1000$ km (拡大画像におけるピクセルサイズ) となる. したがって, 位置変化の不定性は $U_d = \sqrt{2U_m^2} = 1400$ km である. フィラメントがT秒の間に1ピクセル以上移動した際における速度計算の不定性は $U_v = U_d/T$ となる. 補間された点における速度の不定性はブロッブと同じ方法を用いて行った. 5つのイベントにおいて, 計算された視線垂直速度の平均誤差は8 km/sである.

cursor positioning on the workstation. The measurement accuracy is therefore estimated as $U_m = 1000$ km (one pixel on the enlarged image). The uncertainty for a change in position is then $U_d = \sqrt{2U_m^2} = 1400$ km. In calculating the velocities, the uncertainty is $U_v = U_d/T$ when the filament shifts its position more than 1 pixel in a time interval of T seconds. The uncertainty in the velocities of interpolated pixels was estimated in the same manner as that for the blobs. The calculated mean value of errors in the tangential velocity of the 5 events is $8 \mathrm{km s}^{-1}$.

4. 結果と議論

4.1. 消失するフィラメントの例

この節では，5つのフィラメント消失現象（1999年2月16日，1992年11月5日，2000年1月19日，2000年1月28日，および2000年4月6日のイベント）について，速度の時間発展の詳細な解析結果を述べる．

4.1.1. 1999年2月16日のイベント

導出された速度場から判断すると，フィラメントは2つの足元は太陽表面に残したまま中央部分から噴出した．フィラメントの中央部分は南西の方向に伝播し，全速度200 km/s で FMT の視野から消失した．

4. Results and Discussions

4.1. Examples of Hα Disappearing Filaments

In this section, we first present an analysis of five DB events (1999 February 16, 1992 November 05, 2000 January 19, 2000 January 28, and 2000 April 6 events) to see some details of their velocity evolutions.

4.1.1. The 1999 February 16 event

According to the derived velocity fields, the filament was ejected from the middle of it with its two legs remaining anchored to the photosphere. The middle part of the filament propagated toward the southwest to disappear from the FMT field of view with a total velocity of over 200 km s^{-1}.

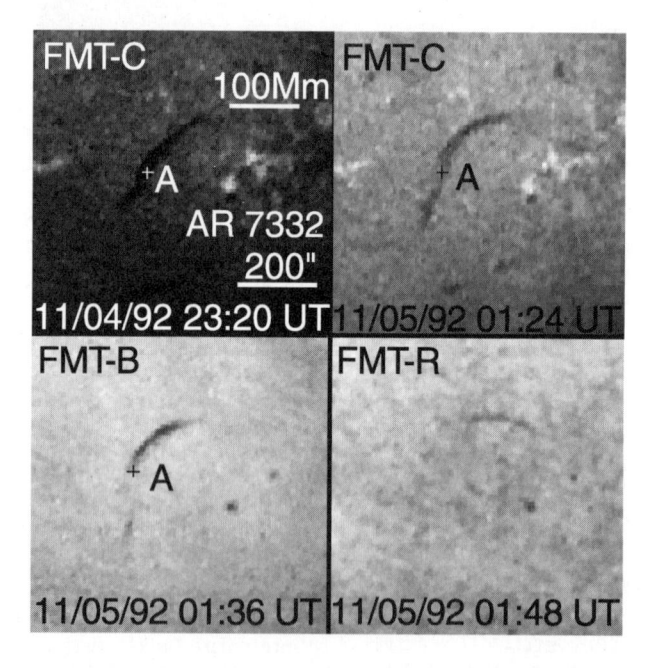

図8　1992年11月5日に発生した静穏フィラメント消失現象の FMT 部分像の時間変化．A 点の位置は「+」で示されている．A 点は 01:38 UT に消失したため，右下の画像には示されてはいない．

Fig. 8. Series of partial FMT images showing the disappearance of a quiescent filament on 1992 November 05. The positions of the point A are shown with '+' marks. Since point A disappears at 01:38 UT, its position is not displayed on the bottom-right panel.

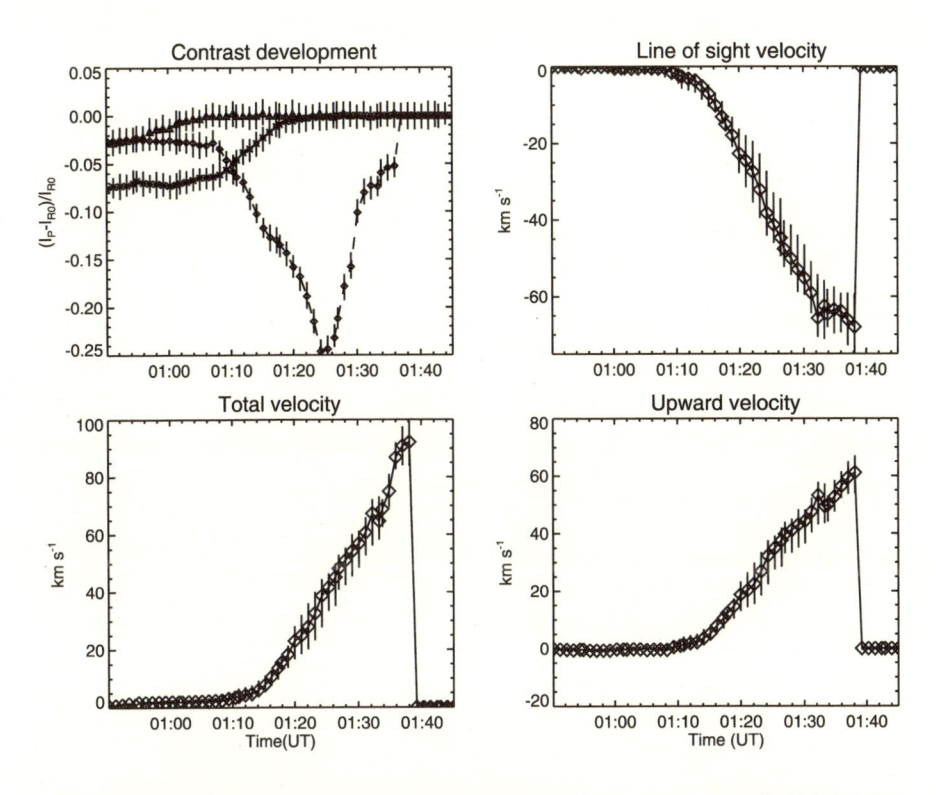

Fig. 9. Time-dependent evolutions of the observed contrasts and line-of-sight velocity (top panels) of point A of 1992 November 05 event. Its total and upward velocities are also presented in the bottom panels.

4.1.2. 1992年11月5日のイベント

1992年11月5日，活動領域NOAA 7332の東側（S20W17）においてHαフィラメント噴出現象とそれに伴うコロナ形状の変化が発生した．McAllister et al. (1996)がFMTのデータを用いてこのイベントの研究を行った．彼らの研究では，消失するフィラメントの形状変化と運動について解析し，「ようこう」のSXTで観測されたコロナにおける巨大アーケード形成との比較を行った．彼らはまた，ドップラーシフト量ではなくフィラメントの上昇角度を想定する事で視線速度を導出した．このイベントのFMTの画像は彼らの論文の図2に示

4.1.2. The 1992 November 05 event

On 1992 November 5, an Hα filament eruption and related coronal restructurings took place to the east (S20W17) of an active region of NOAA7332. McAllister et al. (1996) conducted a study on this event using FMT data. They investigated the morphology and motions of the disappearing filament and compared it with a large arcade formation observed by Yohkoh SXT. They derived the line-of-sight velocity by assuming the rise angle of the filament and not by measuring the Doppler shift. FMT images of this event can be referred to in figure 2 of their article.

されている.

フィラメントの三次元速度を示すために，我々は図8に示されているフィラメント上の1点(A点)を指定した.フィラメントは足元を固定したまま北東に向けて膨張した.A点における3波長でのコントラスト値と視線速度は図9の上段のパネルに示されている.A点は01:00 UT頃に徐々に上昇を開始した.この時にHα線中心のコントラストが弱まり始め，青色側ウイングではコントラストの減少が見られる.01:19 UTにA点の視線速度は-36 km/s(~-0.8 Å)に達した時に，青色側が最小値を取っている.その後青色側ウイングのコントラストが上昇に転じてからも視線速度は加速を続け，01:38 UTには-65 km/sに達する.これは，ほかの2波長のコントラストがほぼ0であり続けたためである.図9の下のパネルには全速度と上昇速度が示されている.このイベントは太陽面中心近傍で発生し，またフィラメントの視線垂直速度が小さかったために，上昇速度は視線速度の時間発展をよく反映している.上昇速度はA点が消失するまで増加し続けた.

4.1.3. 2000年1月19日のイベント

2000年1月19日にFMTでフィラメント消失現象が観測された.図10には，Hαフィラメント消失現象の連続画像が表示されている.この図では，活動領域NOAA 8829の南側に位置するフィラメントが2つの足元を残したまま中央部分から消失していく様子が見て取れる.

参照点としてA点をフィラメント中央部分に設定し，その位置を図10中に表示してある.

図11はA点のコントラストと視線速度を示したものである.コントラストは青色側ウイングでは次第に強くなる一方で，Hα線中心では検出されない.01:10 UTから01:24 UTにかけて，青色側ウイングのコントラストは最小値を取り，視線速度は-38 km/sでほぼ一定となる.その後視線速度は再び加速を始めて-67 km/sに達し，青色側ウイングのコントラストが上昇するに応じて検出されなくなる.このイベントは

In order to demonstrate the 3-D velocity of the filament, we selected a point (point A) on the filament in figure 8. The filament expands to the north-east, fixing its both footpoints. The contrast values on the three bands images and corresponding line-of-sight velocity derived for the point is plotted in the top panels of figure 9. Point A starts to ascend slowly around 01:00 UT, when its contrast in the Hα center starts to become weak and that in the blue wing begins to be stronger. The line of-sight velocity of this point reaches -36 km s^{-1} (~-0.8 Å) at 01:19 UT when the blue wing contrast attains its minimum. Though the blue wing's contrast again decreases to zero after that, the line-of-sight velocity continues to be accelerated earthwards up to -65 km s^{-1} at 01:38 UT because the other two contrasts remains almost zero. The bottom panels of figure 9 are its total and upward velocities. Since the event took place near the disk center and the filament exhibited little apparent motion, the upward velocity well reflects the line-of-sight velocity. The upward velocity increases until the disappearance of point A.

4.1.3. The 2000 January 19 event

On 2000 January 19, FMT observed a filament disappearance. Sequential images of this Hα DB are displayed in figure 10 which shows a filament located at the south of the active region of NOAA8829 starts to disappear from the middle part with its eastern and western footpoints visible throughout the event.

Sample point A is selected at the middle of the filament, and its positions are shown at three different times in figure 10. Figure 11 illustrates the evolution of the contrast and line-of-sight velocity of point A. The contrast develops gradually in the blue wing while it disappears in Hα line center. From 01:10 UT to 01:24 UT, the blue wing contrast reaches its minimum and the line-of-sight velocity becomes almost constant about -38 km s^{-1}. After that, the line-of-sight velocity again starts to increase up to -67 km s^{-1} and disappears in accordance with the increase of the blue wing's contrast values. Since this event occurred near the disk center (N08W18) with less tangential

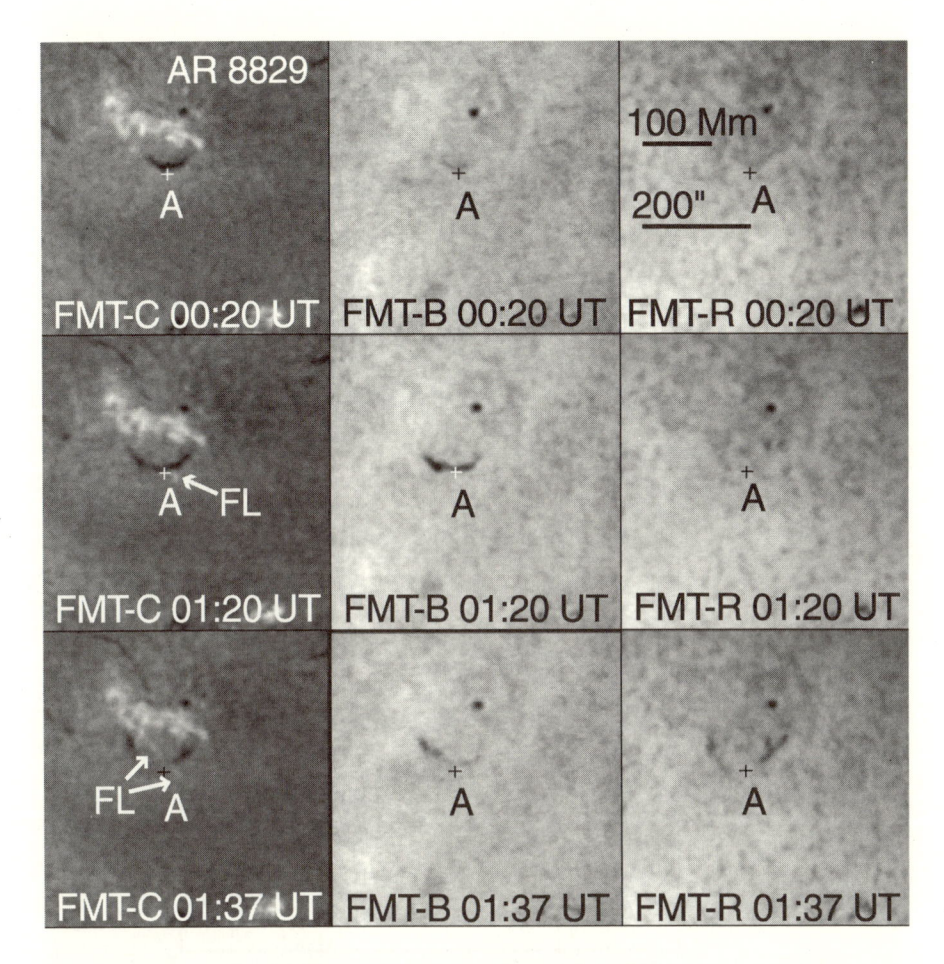

Appendix A Selected Papers Relating to FMT | 355

図10 2000年1月19日に活動領域 NOAA 8829 の南側にて発生したフィラメント消失現象の FMT 画像. 各段はそれぞれ, 00:20 UT, 01:20 UT, および01:37 UT における画像. フィラメントは中央部分から消失し, 01:20 UT 以降は H α 線中心画像で小規模なツーリボンフレアが発生した. 本文で説明された A 点はフィラメントの中央部分に位置する. 左列は FMT の H α 線中心画像, 中列は青色側ウイング画像, および右列は赤色側ウイング画像である.

Fig. 10. FMT images showing the DB at the south of the active region NOAA 8829 on 2000 January 19. From the top to the bottom, images are at 00:20 UT, 01:20 UT, and 01:37 UT, respectively. The filament disappears from the middle and a faint ribbon flare (FL) is seen on the Hα line center image after 01:20 UT. The point A described in the main text is located at the middle of the filament. The left column is the FMT Hα line center images, the middle is the blue wing images, and the right column is the red wing images.

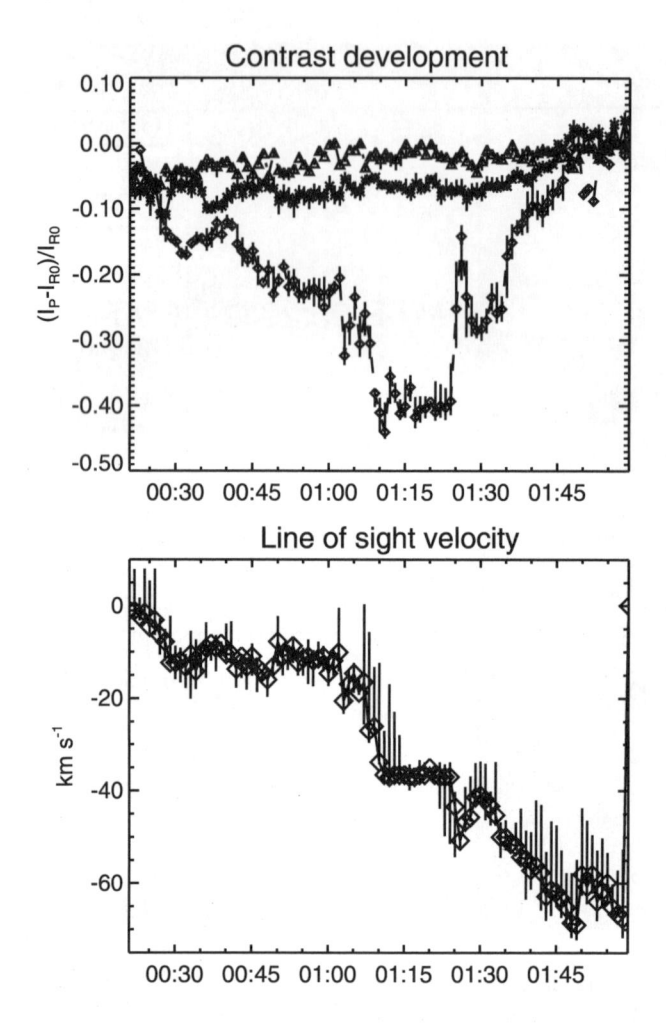

図 11　2000 年 1 月 19 日のイベントにおけるコントラスト（上段）および視線速度（下段）の時間変化．

Fig. 11. Contrasts (top panel) and line-of-sight velocity (bottom panel) of point A of 2000 January 19 event are shown as a function of time.

ほぼ太陽面中心（N08W18）で発生し視線垂直速度が小さかったために，上昇速度は視線速度の形状をよく反映している．

4.1.4. 2000 年 4 月 6 日のイベント

2000 年 4 月 6 日，活動領域 NOAA 8945 の南西 100 Mm の位置でフィラメント消失

motion, the upward motion can be well represented by the line-of-sight velocity.

4.1.4. The 2000 April 06 event

A filament disappearance took place 100 Mm distant to the south-west of the

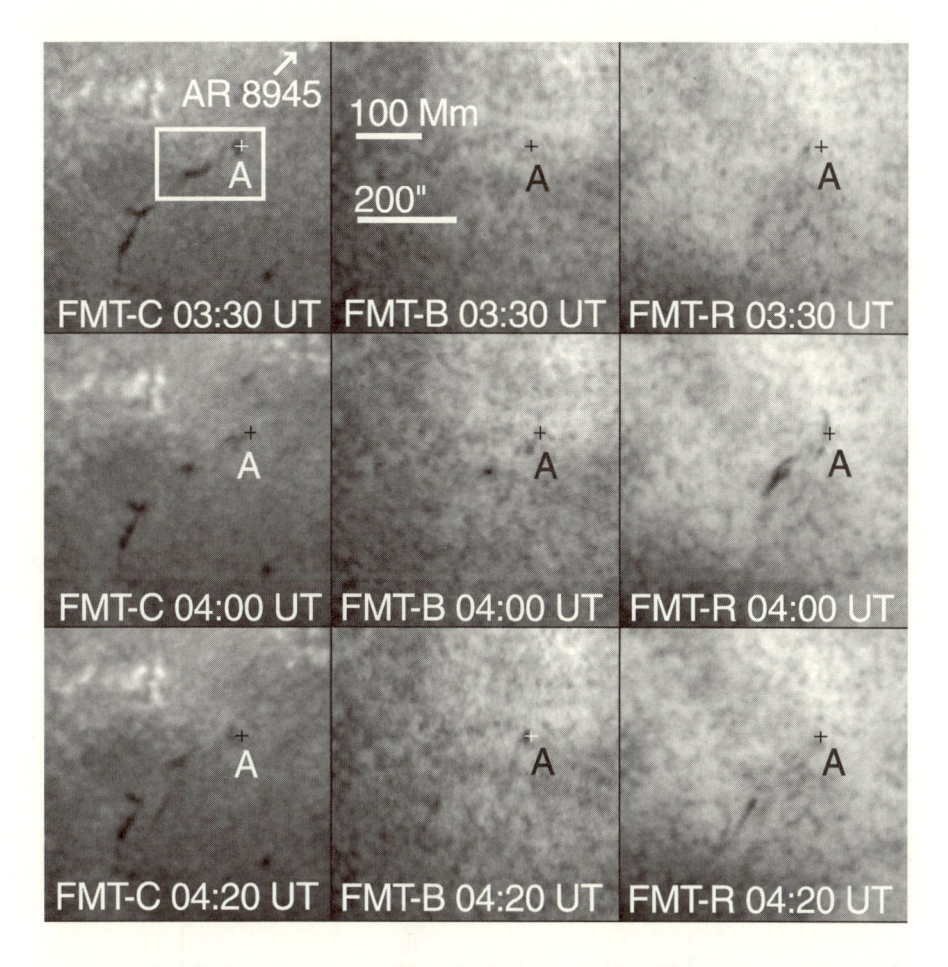

Appendix A Selected Papers Relating to FMT | 357

図 12 2000 年 4 月 6 日のフィラメント消失現象における FMT 画像の時間変化．白い枠で囲われたフィラメントの北側半分が南東方向に移動した後に消失した．参照点として A 点が北側の足元に位置している．

Fig. 12. Series of partial FMT images of the 2000 April 6 DB event. The northern half of the filament, indicated by the white box, shifted its position toward the south-east and disappeared. The sample point A is located at the northern footpoint.

現象が発生した．消失は図 12 の左上のパネル中にある白枠で囲われたフィラメント北側半分の領域で起こった．Hα 線中心画像でのフィラメント消失と同時に，赤色側ウイング画像においてフィラメントが青色側ウイング画像よりも顕著に観測された．これはこのイベントを通じてレッドシフトが優勢であることを示す．

active region of NOAA 8945 on 2000 April 6. The disappearance occurred at the northern half of this filament, which is indicated by the white box in the top-left panel of figure 12. With the disappearance in the Hα line center, the filament appears to be more prominent on the red-wing images than the blue-wing images, indicating that a red shift is dominant

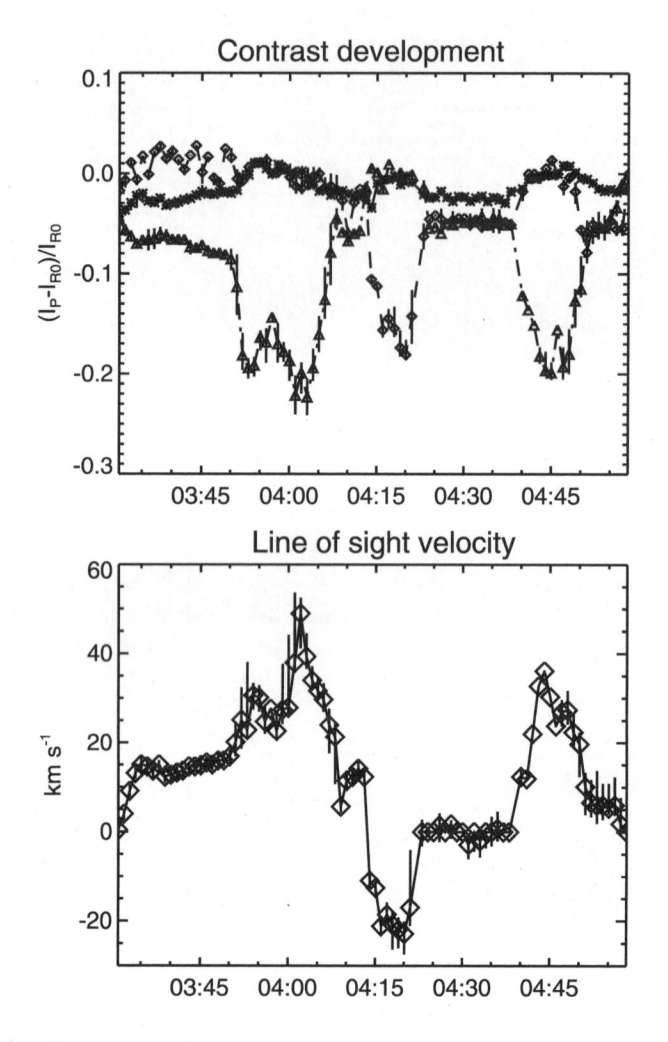

図13 2000年4月6日のイベントにおけるコントラスト（上段）および視線速度（下段）の時間変化．

Fig. 13. Time-dependence of the contrast and the line-of-sight velocity of point A of the 2000 April 6 event shown in the top and bottom panels, respectively.

参照点（A点）とその位置変化の様子を図12に示す．図13はA点における観測されたコントラストと視線速度のグラフである．このグラフの時間発展は複雑であり，4つの段階に分けられる．ステージ1:フィラメント消失開始の03:30 UTから04:03 UT．フィラメントはHα線中心画像で消

throughout the event.

A sample point (point A) and its apparent motion toward the south-east are also shown in figure 12. Figure 13 illustrates the development of the observed contrast and the line-of-sight velocity of point A. Their evolutions are complicated and can be divided into four phases. Stage 1: from the onset at 03:30 UT to 04:03 UT. It fades on FMT center images and becomes visible on red-wing images while no signature is found in the blue wing, which means a receding motion. Stage 2: 04:03 UT–C04:23 UT. It is lost in the red wing and appears in the blue wing. The line-of-sight motion accelerated earthward up to -22 km s^{-1}. Stage 3: 04:23 UT–C04:37 UT. It disappears in the 2 wings and loses its velocity. Stage 4:04:37 UT–. It shows a receding motion, again. The whole filament shows little upward motion and a significant receding motion in the later phase.

4.1.5. The 2000 January 28 event

Another example is the 2000 January 28 event, which took place near the center of the south-west quadrant of the solar disk (S20W28). A series of partial FMT images are shown in figure 14. A dark filament in the west of an active region NOAA 8841 disrupts toward the north-west, first appearing dark in the blue wing from 05:42 UT to 05:54 UT, and in the Hα line center and red wing in the later stage from 06:20 UT. A compact two-ribbon flare occurred near the eastern footpoint of the filament, which can be seen in the FMT Hα line center image at 05:54 UT, and lasted until the end of this event.

A sample point A and its tangential motions are shown in figure 14. The time-dependent observed contrast and the line-of-sight velocity for this point are displayed in figure 15. Their development can be divided into the following consecutive stages: (i) contrast in the Hα line center is dominant and point A is stationary until 05:35 UT, (ii) an increase in the blue wing's contrast values, which is associated with a steep earthward acceleration ($\sim$ 06:09 UT), (iii) as the red wing's contrast becomes stronger, the direction of the line-of-sight motion reverses and point A is accelerated up to 18 kms^{-1}. As the velocity of point A indicates, the filament is lifted off at first with a tangential motion toward the north-west. It, however, is decelerated after 05:49 UT and begins to flow back onto the solar surface some 120 Mm away from its original position.

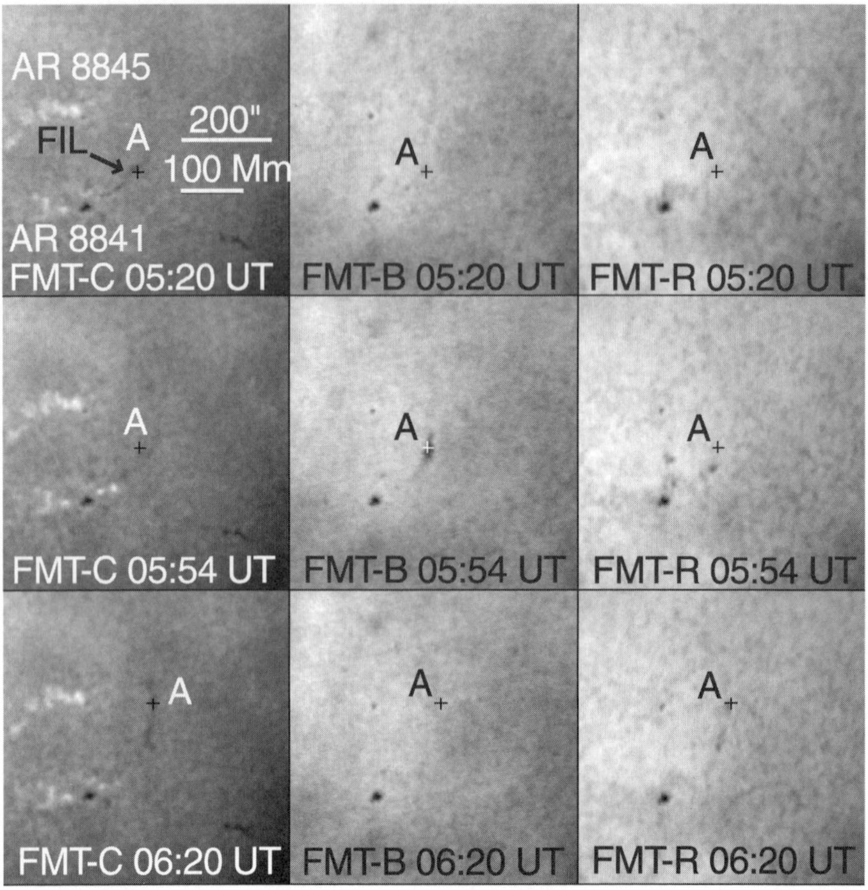

図14 2000年1月28日にFMTで観測されたフィラメント消失現象. 活動領域NOAA 8841の西側に位置するフィラメント (FIL) が北西方向に放出された後減速されている. コンパクトなHαフレア (FL) がフィラメント消失現象と同時に発生した. 参照点 (A点) は「+」で示されている.

Fig. 14. FMT observations of the DB on 2000 January 28. A filament (FIL) at the west of the active region NOAA 8841 is ejected toward the north-west, but decelerated afterwards. A compact Hα flare (FL) occurred coincident with the DB. The positions of the sample point (point A) are shown with '+' marks.

UT以降は減速し始め, 初期位置からおよそ120 Mm離れた太陽表面上に落下している.

5. 要約と結論

本研究では, 飛騨天文台のフレア監視望

5. Summary and Conclusions

Using the Hα line center, blue-wing and

Contrast development

Fig. 15. Contrast and line-of-sight velocity of sample point A of the 2000 January 28 event plotted as functions of time.

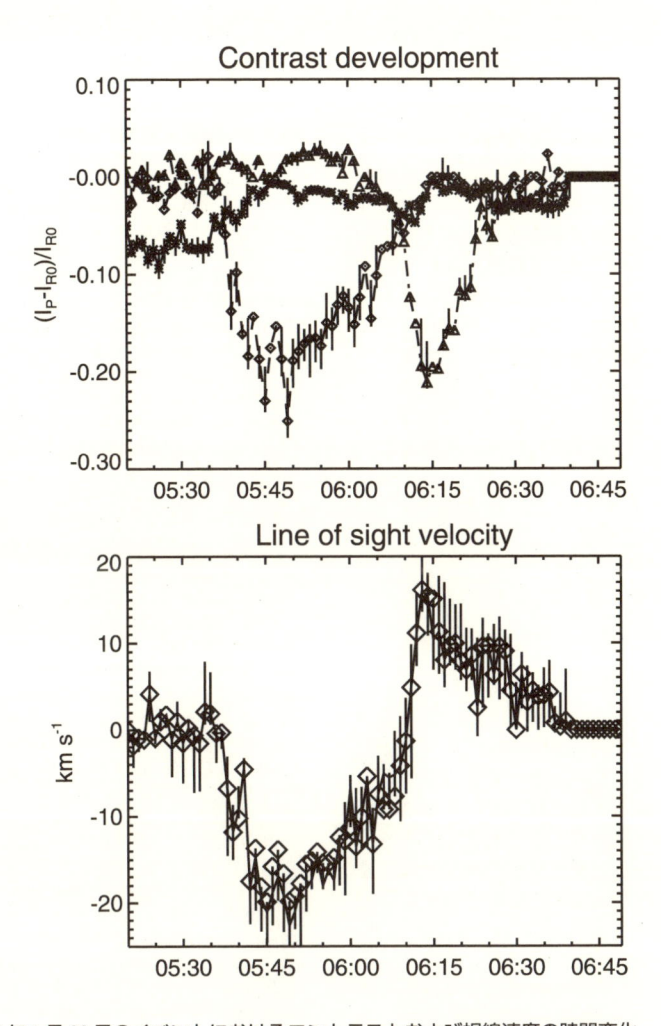

図 15　2000 年 1 月 28 日のイベントにおけるコントラストおよび視線速度の時間変化.

遠鏡（FMT）の Hα 線中心画像，青色および赤色側ウイング画像を用いて，フィラメント上の各ピクセルにおいて視線速度と視線垂直速度の両方を導出する方法を初めて確立した．Beckers のクラウドモデルを元に 3 節で紹介した仮定とあわせてフィラメントの Hα 線プロファイルを計算し，観測されたコントラスト値と再現されたコン

red-wing images of the Flare Monitoring Telescope (FMT) at Hida Observatory, we first established a method to derive both the line-of-sight and tangential velocities at all of the pixels on disappearing filaments. By calculating the Hα line profile of a filament based on the Beckers' cloud model with the assumptions described in section 3, we compared the

トラストプロファイルを比較してドップラーシフト量を求めた．視線垂直速度はフィラメント中のブロブの運動を追うことにより求められた．本研究では上記の手法を用いて 5 つのイベントに対して視線および視線垂直速度を計算した．得られた速度場は $H\alpha$ フィラメントのコントラストの時間発展および FMT で観測された連続画像上のフィラメントの運動の様子とよい一致を見せた．また，こうして得られた視線速度および視線垂直速度を組み合わせることで，消失フィラメントの三次元速度場を求めた．さらに我々は，計算された速度の誤差となりうる要素についての考察も行った．視線速度計算に対しては以下の問題についての調査を行った：(i) 散乱光，(ii) FMT の有効透過幅，(iii) $H\alpha$ 線プロファイルのドップラー幅増加現象，(iv) ドップラー増光現象，および (v) 背景光度の変動の 5 つである．

散乱光はフィラメントと周囲の彩層からの $H\alpha$ 線光度の真の値を求める際に最も問題となる点である．本研究では点広がり関数をガウス型［式 (A2)：Kubota (1968)］と仮定して補正を行った．先行研究の大半は散乱光の問題を扱う際にガウス型ではなくローレンツ型の点広がり関数［式 (A3)］を適用している．我々は点広がり関数の違いが計算された速度でどのような差を発生するかについて調査した．1999 年 2 月 16 日のイベントに対し，点広がり関数の形状を変えた結果得られる視線速度の差を見積もったところ -5 km/s 以下であった．それゆえ，視線速度の決定には点広がり関数の形状選択は重要ではないといえる．なお，有効透過幅についてもガウス型の関数を仮定して補正を行った．

フィラメントが活性化した際の $H\alpha$ 線プロファイルが幅広くなることはよく知られている．今回の手法では，FMT による観測だけでは $H\alpha$ 線のドップラー幅増大は計測が困難であるため，我々は先行研究で求められたドップラー幅の上限 0.65Å までの様々なドップラー幅に対して取りうる誤差を見積もった．結果として得られた視線速度に対する誤差は最大 10 km/s であると推

observed contrast values and computed contrast profiles in order to obtain the Doppler shift. The tangential velocity was obtained by tracing the internal structures (blobs). In this work, we calculated these velocities for five selected events with the method. The obtained velocity fields show fairly good agreement with the temporal variations of the contrasts of the $H\alpha$ filaments and their apparent motions on successive FMT images. The derived line-of-sight and tangential velocities were combined to yield the 3-D velocity field of the disappearing filament. We investigated the effects of several factors that could result in errors in the calculated velocity. For the line-of-sight velocity, we examined the following issues: (i) the stray light, (ii) the effective filter-bandwidths of the FMT, (iii) the broadening of the $H\alpha$ line profile, (iv) the Doppler brightening effect (DBE), and (v) the fluctuations in the background intensity.

The stray light presents the greatest difficulty when attempting to obtain the true $H\alpha$ line intensities of the filaments and the surrounding chromosphere. In this study, we assumed a Gaussian shape for the spread function [equation (A2): Kubota (1968)] and performed correction. Most previous authors have not considered Gaussian spread functions, but have adopted Lorentzian spread functions in their studies of stray light [equation (A3)]. We also investigated the difference in the calculated velocity as the result of using these two different forms of the spread function. We estimated that the difference in the line-of-sight velocity for the 1999 February 16 event, as a result of the choice of the spread function, was less than 5 km s^{-1}. Therefore, we conclude that the choice of the spread function in this study is insignificant for determining the lineof- sight velocity. The effective filter-bandwidths were also corrected by assuming that their functional form is Gaussian.

It is well known that the $H\alpha$ line of a filament becomes wider when it is activated. In our method, it is difficult to measure the broadening of the $H\alpha$ line with FMT observations alone. Therefore, we estimated the possible errors due to a line broadening of up to 0.65 Å, which is expected from previous studies. We

定された．ドップラー増光現象は視線速度の誤差となる重要な要素の1つである．我々は1999年2月16日のイベントに対し，ドップラー増光現象が視線速度に及ぼす誤差を見積もるために簡単な実験を行い，誤差が20 km/sに及ぶことを明らかにした．ドップラー増光現象はフィラメント全体の運動によって引き起こされるため，我々の手法を用いて計算された速度からこの効果の補正を行った．我々の手法では，背景光度は一様であると仮定しているために背景光度の変動はフィラメントのコントラストの誤差となりうる．この影響による誤差は，視線速度のグラフ中に線分でその範囲が示されている．この誤差の平均値は5 km/sとなった．

視線垂直速度の誤差はパソコン画面上において位置指定する際の誤差に起因する．フィラメントが長時間静穏である場合には視線垂直速度誤差は小さく，平均8 km/s程度である．

しかし，フィラメントが連続する2枚の画像上で1ピクセル以上移動した場合，この誤差は23 km/sまで大きくなりうる．この誤差は背景光度変動による誤差と合計されて，上昇速度と全速度のグラフ中に誤差範囲を示す線分として表示されている．したがって，今回の手法で補正されずに速度誤差として扱われるのは，Hα線のドップラー幅増大現象と背景光度変動による誤差，および視線垂直速度誤差の合計となる．前節の議論から得られる速度誤差の平均値はおよそ20 km/sとなる．正確な速度導出のためにはこの値は十分に小さいとはいえないが，今回の手法はフィラメント消失現象の運動の時間発展の研究には適していると考えられる．それゆえ，消失するフィラメントの速度や加速度の時間変化だけではなく，フィラメントが惑星間空間に放出されたか太陽大気中に留まったかを判断することも可能である．このことは宇宙天気予報にとってきわめて重要であり，我々の手法の意義を示している．

京都大学飛騨花山天文台の方々による協力，また充実した意見や議論に深く感謝する．

estimated the errors in the calculated line-of-sight velocity as a result, and found that it is, at most, 10 kms^{-1}. The DBE is also one of the biggest factors which produces considerable errors in the line-of-sight velocity. We performed a simple experiment for the 1999 February 16 event to estimate the error in the line-of-sight velocity, and found that it could become as large as 20 kms^{-1}. Since the DBE is produced by the bulk motion of a filament, we performed a correction for this effect with the calculated velocity fields from our method. The fluctuations in the background intensity may also result in errors in the observed contrasts of the filament, because the background chromospheric intensity is assumed to be uniform in our method. The errors in the calculated line-of-sight velocities due to this effect are presented with error bars in all of the figures showing the temporal variations of the line-of-sight velocities. The mean value was found to be 5 kms^{-1}.

The error in the tangential velocity is due to the error involved in positioning the cursor on the workstation screen. This error is small when the filament is stationary for a long time and has a mean value of 8 kms^{-1}. However, when the filament shifts its position by more than one pixel on successive FMT images, it can become as large as 23 kms^{-1}. This error, combined with the error due to the fluctuations in the background intensities, is shown as an error bar in all of the figures showing upward or total velocities. Therefore, the errors in the velocities which are not corrected for in the method are the sum of the errors due to the Hα line broadening, the fluctuations in the background intensities, and the tangential velocity errors. From the discussion above, this yields a mean value for the errors of about 20 kms^{-1}. Although this is not sufficiently small to obtain an accurate magnitude for the velocity, our method is suitable for studying the temporal evolution of the motion of DBs. It is therefore possible not only to study the temporal variation of the velocity and acceleration of the disappearing filament, but also to judge whether it is ejected into interplanetary space or remains in the solar atmosphere. This is very important for space weather forecasts, and

る．FMT のデータベースは M. Kadota, Y. Nasuji, M. Kamobe, そして S. Ueno の人々によって作成，運営されている．またレフェリーには参考となるコメントや指摘，そして短時間に審査をして頂いたことに対してお礼を申し上げる．P. F. Chen と D. B. Brooks の両博士による議論や励ましの言葉にも感謝する．

付録1 点広がり関数

位置 (x_0, y_0) にある点 P で観測される光度を求めるには，太陽面上の各点 Q からの散乱光を積分する必要があり，以下の式によって求められる．

$$I(x_0, y_0) = \int_{\psi_1}^{\psi_2} \int_{\rho_1}^{\rho_2} F(R^2)\Psi(\varrho)\varrho\,d\varrho\,d\psi, \quad (A1)$$

ここで，$F(R^2)$ を Q における真の光度，$\Psi(\varrho)$ は点広がり関数（Zwaan 1965）である．今回の解析では，我々は Kubota (1968) で紹介されたガウス型の点広がり関数を用いた．

$$\Psi(\varrho) = \frac{a}{\pi} \exp(-a\varrho^2), \quad (A2)$$

ここで a は散乱パラメータである．このガウス型の点広がり関数は，以下の式で表される良く知られた Zwaan のローレンツ型の点広がり関数とは異なる．

$$\Psi(\varrho) = (1-\epsilon)\frac{1-m}{\pi b_1^2}\exp[-(\varrho/b_1)^2] \\ + \frac{m}{\pi b_2^2}\exp[-(\varrho/b_2)^2] + \frac{\epsilon A}{B^2+\varrho^2}, \quad (A3)$$

ここで ϵ, B, は散乱パラメータ，b_1, b_2 および m は像滲みパラメータである．また A は規格化因子（e.g., Brahde 1972 参照）である．Mullan (1973) は，ローレンツ型とガウス型の点広がり関数がどのように黒点の光度の測定精度に影響を及ぼすかについて調査した．その結果，ローレンツ型の点広がり関数は積分の際に発散して規格化が困難であり，ガウス型の点広がり関数の方が散乱光の決定により適している事が明らかとなった．

demonstrates the significance of our method.

The authors wish to thank all members at Hida and Kwasan Observatories, Kyoto University for the operation of telescopes and fruitful comments and discussions. The database of FMT is compiled and maintained by M. Kadota, Y. Nasuji, M. Kamobe, and S. Ueno. The authors thank an anonymous referee for his/her helpful comments, suggestions, and hard work to examine this paper in a limited time. We are also grateful to Dr. P. F. Chen and Dr. D. H. Brooks for thorough discussions and encouragements.

Appendix 1. The Spread Function

In order to find the observable intensity in point P with coordinates (x_0, y_0), one must integrate the stray light from every point Q on the disk, according to the formula

$$I(x_0, y_0) = \int_{\psi_1}^{\psi_2} \int_{\rho_1}^{\rho_2} F(R^2)\Psi(\varrho)\varrho\,d\varrho\,d\psi, \quad (A1)$$

where $F(R_2)$ is the true intensity in Q, and $\Psi(\varrho)$ is the stray light function (Zwaan 1965). In this study, we used a spread function with a Gaussian form, which was originally introduced by Kubota (1968) as

$$\Psi(\varrho) = \frac{a}{\pi}\exp(-a\varrho^2), \quad (A2)$$

where a is the scattering parameter. This Gaussian form of the spread function is different from the well known Lorentzian function or the spread function of Zwaan, which is given by

$$\Psi(\varrho) = (1-\epsilon)\frac{1-m}{\pi b_1^2}\exp[-(\varrho/b_1)^2] \\ + \frac{m}{\pi b_2^2}\exp[-(\varrho/b_2)^2] + \frac{\epsilon A}{B^2+\varrho^2}, \quad (A3)$$

where ϵ and B are the scattering parameters, b_1, b_2 and m are the blurring parameters, and A is the normalization factor (see, e.g., Brahde 1972). Mullan (1973) examined the contribution to the inaccuracies in the intensity of a sunspot arising from the choice of a Lorentzian or Gaussian spread function. As a result, they concluded that Lorentzians lead to divergence and normalization difficulties, and that Gaussian functions are more suitable for the determination of scattered light.

In order to investigate the difference in

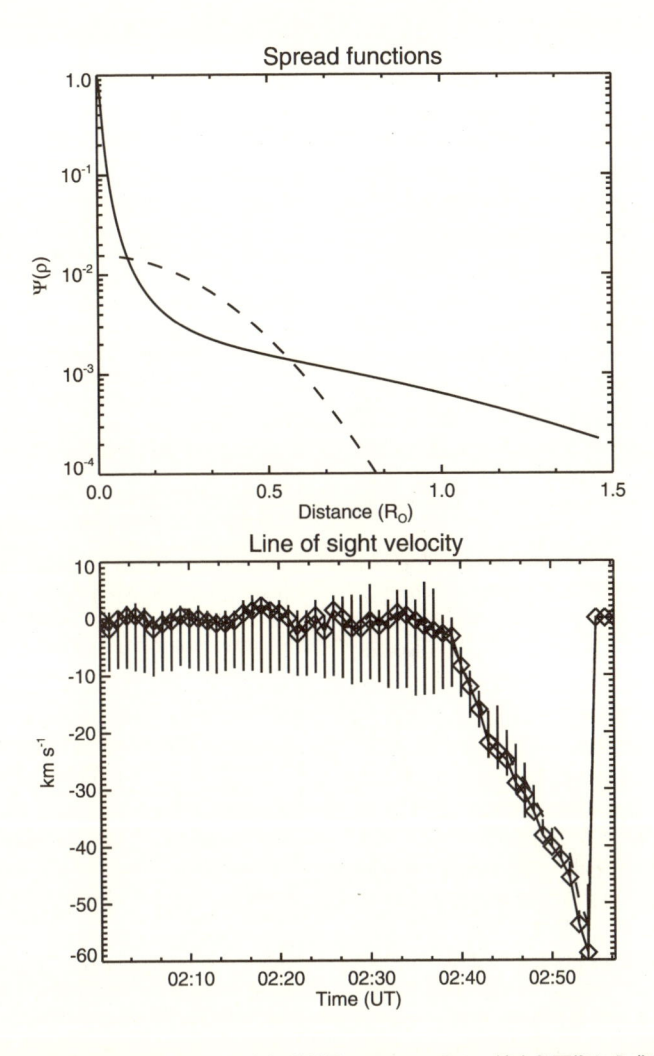

図16 上段は1999年2月16日のイベントにおける02:00 UTのHα線中心画像から求められたローレンツ型点広がり関数（実線）．今回の研究で用いられたガウス型点広がり関数を破線で示す．横軸は太陽半径で規格化した距離，縦軸は任意の単位における点広がり関数の大きさ．1999年2月16日のイベントの視線速度を，ガウス型（誤差付きの実線）およびローレンツ型（破線）の点広がり関数を用いて計算した結果を下段に示す．

Fig. 16. The top panel shows the Lorentzian spread function derived for the Hα line center image of the 1999 February 16 event at 02:00 UT (solid line). The Gaussian spread function used in this study is shown by the dashed line. The abscissa represents the distance in units of solar radius and the ordinate is the scale of the spread functions in arbitrary units. The calculated line-of-sight velocity of the 1999 February 16 event with the Gaussian (solid line with error bars) and the Lorentzian (dashed line) spread functions are plotted as functions of time in the bottom panel.

点広がり関数の形状の違いによる視線速度への影響を調べるため，1999年2月16日のイベントにおける視線速度を計算する際に，本研究で用いた式（A2）のガウス型点広がり関数の替わりに式（A3）で表されるローレンツ型の点広がり関数を用いた実験を行った．Hα線中心と両ウイングにおける周辺減光効果はWhite（1962）のものと同じにした．White（1962）の周辺減光の効果はFMTに3つあるフィルターの波長域に完全には対応していないため，我々の観測波長に応じたHα線近傍での周辺減光効果を得るべくWhite（1962）の周辺減光パラメータを線形補間した．1999年2月16日の02:00 UTの時点におけるFMTのデータから求められた点広がり関数を図16の上のパネルに示す．

2つの点広がり関数の大きな違いは，注目する点 (x_0, y_0) から遠く離れた領域からの光度の寄与である．この寄与はガウス型よりもローレンツ型の点広がり関数で大きい．これは，大気中のエアロゾルやダスト，光学系の汚れや傷，設計不良による0°から90°の散乱光を表す散乱項［式（A3）の第3項］に起因するものである．我々は式（A2）の替わりに式（A3）を用いて，視線速度の再計算を行った．ローレンツ型の点広がり関数を用いて計算された視線速度は図16の下のグラフに破線で表されている．実線はガウス型の点広がり関数を用いて計算した視線速度で，図2の右上のグラフと同じものである．これらの2本のグラフにおける速度差は5 km/s以下であり，これより視線速度の決定には点広がり関数の形状は重要ではないと結論付けられる．しかし散乱光の効果は日によって変動し，別の日のデータでは速度差が大きくなる可能性もあるために，この現象には留意しておく必要がある．

付録2　ドップラー増光現象（DBE）

一般的にドップラー増光現象（DBE）とは，狭帯域フィルター（Kawaguchi et

the calculated lineof- sight velocity due to the choice of these different forms of spread function, we calculated the line-of-sight velocity of the 1999 February 16 event by replacing the Gaussian spread function of equation (A2), which is used in this study, by the Lorentzian function of equation (A3). The limb darkening in the Hα line center and wings is assumed to be the same as in White (1962). Since the limb darkening in White (1962) does not completely cover the wavelength range of the 3 band filters of the FMT, we performed a linear interpolation for the limb-darkening parameters presented in White (1962) to yield the limb darkening around the Hα line for our calculation. We then obtained the spread function derived from the FMT data of the 1999 February 16 event at 02:00 UT; this is shown in the top panel of figure 16.

The main difference between these two forms of the spread function is in the contribution from intensities far away from the point of interest (x_0, y_0). This contribution is found to be larger for the Lorentzian function than for the Gaussian function. This is because of the scattering term [the third term of equation (A3)] which arises as a result of the light scattered at angles from 0. to 90. by aerosols, dust in the atmosphere, and by dirty, scratched, and generally imperfect optics. Using equation (A3) instead of equation (A2), we recalculated the line-of-sight velocity. The obtained line-of-sight velocity with a Lorentzian spread function is shown in the bottom panel of figure 16 by a dashed line. The solid line represents the line-of-sight velocity obtained with a Gaussian spread function and is identical to that in the top-right panel of figure 2. The difference in the velocities in this figure is found to be less than 5 km s^{-1}, and hence we conclude that the choice of the form of the spread function is not crucial for determining the line-of-sight velocity. However, we must pay attention to the amplitude of the velocity difference because the effect of stray light changes from day to day; thus, it may be enhanced in different data sets.

Appendix 2. The Doppler Brightening Effect (DBE)

In general, the Doppler brightening

al. 1984) を用いてHα線中心で運動するフィラメントやプロミネンスを観測した際の光度変化の1つの要因である．ドップラー増光現象は，源泉関数が主に彩層の散乱光によって支配されている場合に太陽の動径方向速度によって源泉関数が変動することに起因するものである．この現象は太陽表面とフィラメントとの相対運動によってフィラメントのHα線吸収係数プロファイルの波長範囲内における放射量が増加し，放射による励起率が上昇することによって発生する．Rompolt (1967a,b) によるプロミネンスのドップラー増光現象に関する研究の後，多くの著者によってその重要性が証明された (Beckers 1968; Roy, Tang 1975; Labonte 1979;Rompolt 1980)．Hyder and Lites (1970) はプロミネンスの大きな光度を解釈する際のドップラー増光現象の重要性を強調した．彼らは1969年3月1日に発生したプロミネンス噴出現象で観測された光度変化は，ドップラー増光現象によって電子温度や電子の個数密度などの物理パラメータを大きく変えることなしにすべてのプロミネンスの光度変化を説明できることを示した．Rompolt (1980) は光学的に薄いプロミネンスに対するドップラー増光現象をプロミネンスの高度 (H)，全速度 (V) および太陽表面に垂直方向から測った運動方向の傾き (D) の関数として表した．彼は運動するプロミネンスと静止したプロミネンスの源泉関数の比をグラフにして表した．彼の結果によると，全速度100 km/sで運動するプロミネンスの源泉関数は太陽表面に静止するプロミネンスの250%以上に達し得ることが分かった．

それゆえ，ドップラー増光現象を無視した場合，視線速度に相当な誤差が発生する可能性がある．この誤差を見積もるために簡単な実験を行った．運動するフィラメントの速度と高度を $[V(\mathrm{kms}^{-1}), H(\mathrm{Mm})]$ = [25, 20], [50, 60], [75, 90]（それぞれ源泉関数を1.2倍，2.2倍，および2.4倍した場合に相当）とおき，各々のコントラストプロファイルを計算した．そして源泉関数を実際の値にした上で視線速度 (V_LD)

effect (DBE) is one of the origins of variations in the intensity from moving filaments/ prominences seen at the center of the Hα line with a narrow band monochromatic filter (Kawaguchi et al. 1984). The DBE is due to the variation of the source function resulting from the radial velocities, if the source function is mainly dominated by scattered chromospheric light. It occurs when the amount of radiation incident in the wavelength range of the line absorption coefficient profile of a filament increases due to the relative motion between the Sun's surface and the scattering filament, thus increasing the radiative excitation rate for a given line transition. After Rompolt (1967a,b) conducted studies on DBE of prominences, many authors approved of its importance (Beckers 1968; Roy, Tang 1975; Labonte 1979; Rompolt 1980). Hyder and Lites (1970) emphasized the importance of DBE in interpreting the high brightness of prominences. They applied their analysis to the observed brightness variation of the 1969 March 1 eruptive prominence, and found that all of the brightness variation could be attributed to Doppler brightening with no significant variations of the physical parameters, such as the electron temperature and the electron number density. Rompolt (1980) computed the DBE for an optically thin prominence as a function of the prominence's height (H), its total velocity (V) and the value of the inclination of the direction of the velocity to the vertical line (D). He then made a figure showing the ratio of the source function of a moving prominence to a stationary prominence. According to his result, the source function of a moving prominence can be more than 250% of that of a stationary prominence on the solar surface, when its total velocity is 100 km s^{-1}.

Therefore, it might cause a considerable error in the line-of-sight velocity if we neglect DBE. In order to estimate this error, we performed a simple experiment. Setting the velocity and height of the moving filament to be $[V(\mathrm{kms}^{-1}), H(\mathrm{Mm})]$ = [25, 20], [50, 60], [75, 90] (which correspond to the source functions multiplied by a factor of 1.2, 2.2, and 2.4, respectively), we computed the

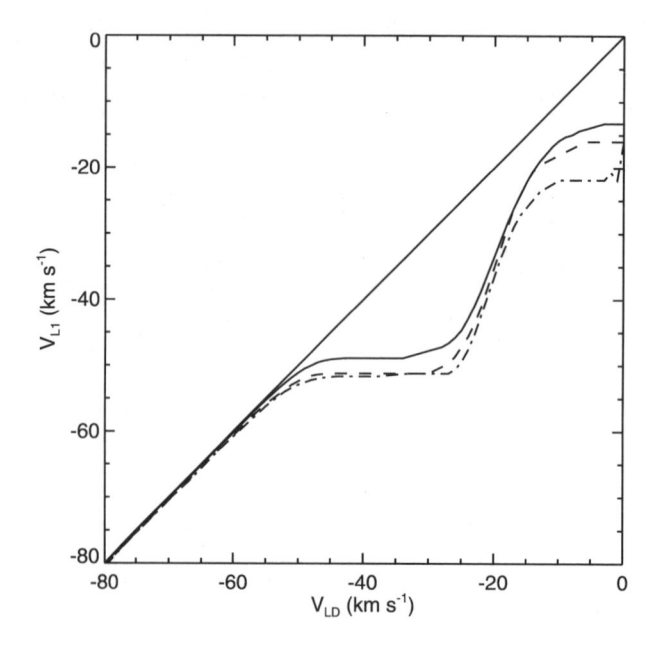

図17 ドップラー増光現象が視線速度に及ぼす影響の試算結果. 横軸は我々の手法で導出した視線速度 (V_{LD}：km/s), 縦軸は真の視線速度 (V_{L1}).

Fig. 17. Result of an experiment to investigate the effect of DBE on the derived line-of-sight velocity. The abscissa represents the line-of-sight velocity derived with our method (V_{LD}: kms^{-1}) while the ordinate is the real line-of-sight velocity (V_{L1}).

を計算した. この計算において, フィラメントは Rompolt (1980) と同じく光学的に薄く, 運動方向は真に動径方向 ($D = 0$) であると想定した. 図17はそれぞれの速度と高度の組み合わせに対して, 横軸に計算で求められた視線速度 (V_{LD}), 縦軸に実際の視線速度 (V_{L1}) を取ったグラフである. 図中に比較のために源泉関数を変えなかった場合の視線速度を示す. 視線速度が50 km/s 以下の場合, 計算で求められた速度は常に実際の速度よりも小さい. これは源泉関数を増大させることで本来の源泉関数を用いた場合よりもコントラストが弱くなることが原因である. ブルーシフトが -0.8 Å (青色側フィルターの中心波長) を超えると, ドップラー増光現象が最も顕著に現れる波長帯であるフィラメントのHα線プロファイルの中央部がFMTの機器の観

contrast profile for each case. We then calculated the line-of-sight velocity (V_{LD}) while setting the source function to be the same as the original one. In this calculation, we assumed the filament is optically thin, as did Rompolt (1980), and that the direction of the motion is purely radial ($D = 0$). Figure 17 is the result for the case in which the abscissa is the derived line-of-sight velocity (V_{LD}) and the ordinate is the real line-of-sight velocity (V_{L1}). In this figure, for a comparison, we plot the line-of-sight velocity with an unaltered source function. When the line-of-sight velocity is less than 50 kms^{-1}, the derived velocity is always slower than the real value. This is due to the enlarged source function, which makes a weaker contrast than that made with the original source function. When the amount of blue shift exceeds -0.8 Å (the central wavelength of blue

測波長領域から外れてしまう．その結果青色側ウイングのコントラストにはその影響が現れなくなり，誤差は無視できるようになる．

　ドップラー増光現象はフィラメントの全体の運動によって発生する．フィラメントの高度，速度，および運動方向を指定する事でドップラー増光現象の補正をすることができる．我々の手法における補正方法は以下の通りである．時刻 t_1 でドップラー増光現象を考慮して源泉関数 S_1 を定めると，次の時刻では源泉関数は大きく変化しないと考えられる．これより S_1 を用いて t_2 における暫定的な視線速度 (V'_2) を導出する．この速度 (V'_2) を用いて，t_2 におけるフィラメントの高度，全速度，および運動方向を求め，これから適切な源泉関数 (S_2) を計算する．最後に S_2 を用いる事で真の視線速度 (V_2) を得る．この方法は $t=0$ における源泉関数が既知の場合には帰納的に適用できる．上記の議論では，Rompolt (1980) の場合と同じくフィラメントは光学的に薄いものとして扱っている．一般的にドップラー増光現象の効果はフィラメントが光学的に厚い場合には小さくなる．Heinzel and Rompolt (1987) は光学的に薄い場合と厚い場合においてプロミネンスからの放射を非局所熱平衡の計算を用いて行い，源泉関数の変化を評価した．その結果によると，全速度が 150 km/s 以下の条件では光学的に薄い場合と厚い場合で源泉関数の差はおよそ7%以下である．源泉関数が 250% 変動する場合もあり，これは7%よりも遙かに大きいので，我々は光学的に薄いフィラメントを仮定して補正を行った．

wing's filter), the central part of the filament's Hα line, on which DBE is most effective, is out of the wavelength regime of FMT instruments, and no longer responsible for the contrast on the blue-wing images, so that the error becomes negligible.

　The DBE is produced by the bulk motion of a filament. Given the height, velocity, and direction of a moving filament, we can correct it. The procedure for this correction in our method is as follows. Given a source function S_1 with DBE at time t_1, we assume that the source function at the next time t_2 is altered little, and derive a provisional line-of-sight velocity (V'_2) at t_2 with S_1. With the velocity (V'_2), we obtain the height, total velocity, and direction of the filament at t_2, and also obtain the appropriate source function (S_2). With S_2, we finally obtain the true line-of-sight velocity (V_2). This step can be inductively applied, since the source function at $t=0$ is known. In the above discussions, we assume that the filament is optically thin, as did Rompolt (1980). In general, the DBE is reduced if a filament is optically thick. Heinzel and Rompolt (1987) computed the radiations from both optically thin and thick prominences using non-LTE techniques, and evaluated the variations of the source functions. According to their results, the difference of the source functions between these two cases are generally less than 7%, provided that the total velocity is less than 150 km s^{-1}. Since the source function may vary by more than 250%, which is considerably larger than 7%, we performed a correction with an assumption of optically-thin filaments.

参考文献／References

Alissandrakis, C. E., Tsiropoula, G., & Mein, P. 1990, A&A, 230, 200
Beckers, J. M. 1964, PhD Thesis, University of Utrecht
Beckers, J. M. 1968, Sol. Phys., 3, 367
Brahde, R. 1972, Sol. Phys., 26, 318
Bray, R. J. 1974, Sol. Phys., 38, 377
Grossmann-Doerth, U., & von Uexküll, M. 1971, Sol. Phys., 20, 31
Grossmann-Doerth, U., & von Uexküll, M. 1973, Sol. Phys., 28, 319

Grossmann-Doerth, U., & von Uexküll, M. 1977, Sol. Phys., 55, 321
Heinzel, P., Kotrc, P., Mouradian, Z. & Buyukliev, G. T. 1995, Sol. Phys., 160, 19
Heinzel, P., Mein, N., & Mein, P. 1999, A&A, 346, 322
Heinzel, P., & Rompolt, B. 1987, Sol. Phys., 110, 171
Hyder, C. L., & Lites, B. W. 1970, Sol. Phys., 14, 147
Jefferies, J. T., & Orrall, F. Q. 1958, ApJ, 127, 714
Kawaguchi, I., Nakai, Y., Funakoshi, Y., & Kim, K.-S. 1984, Sol. Phys., 91, 87
Khan, J. I., Uchida, Y., McAllister, A. H., Mouradian, Z., Soru-Escaut, I., & Hiei, E. 1998, A&A, 336, 753
Kubota, J. 1968, PASJ, 20, 317
Kubota, J., Kitai, R., Tohmura, I., & Uesugi, A. 1992, Sol. Phys., 139, 65
Kurokawa, H., Ishiura, K., Kimura, G., Nakai, Y., Kitai, R., Funakoshi, Y., & Shinkawa, T. 1995, J. Geomag. Geoelectr., 47, 1043
Kurucz, R. L., Furenlid, I., Brault, J., & Testerman, L. 1984, Solar Flux Atlas from 296 to 1300 nm (Sunspot, New Mexico: National Solar Observatory)
Labonte, B. J. 1979, Sol. Phys., 61, 283
Loughhead, R. E. 1973, Sol. Phys., 29, 327
Maltby, P. 1976, Sol. Phys., 46, 149
McAllister, A. H., et al. 1992, PASJ, 44, L205
McAllister, A. H., Kurokawa, H., Shibata, K., & Nitta, N. 1996, Sol. Phys., 169, 123
Mullan, D. J. 1973, Sol. Phys., 32, 65
Munro, R. H., Gosling, J. T., Hildner, E., MacQueen, R. M., Poland, A. I., & Ross, C. L. 1979, Sol. Phys., 61, 201
Paletou, F. 1997, A&A, 317, 244
Rompolt, B. 1967a, Acta Astron., 17, 329
Rompolt, B. 1967b, Acta Universitatis Wratislaviensis, 75
Rompolt, B. 1980, Hvar Observatory Bulletin, 4, 39
Roy, J.-R., & Tang, F. 1975, Sol. Phys., 42, 425
Schmieder, B., Delannée, C., Yong, D. Y., Vial, J. C., & Madjarska, M. 2000, A&A, 358, 728
Tandberg-Hanssen, E. 1995, The Nature of Solar Prominences (Dordrecht: Kluwer Academic Publishers), 15
Tsiropoula, G., Georgakilas, A. A., Alissandrakis, C. E., & Mein, P. 1992, A&A, 262, 587
Webb, D. F., Krieger, A. S., & Rust, D. M. 1976, Sol. Phys., 48, 159
White, O. R. 1962, ApJS, 7, 333
Wright, C. S., & McNamara, L. F. 1983, Sol. Phys., 87, 401
Zwaan, C. 1965, Rech. Astron. Obs. Utrecht, 17, No. 41

A-2

噴出型および準噴出型太陽フィラメント消失現象と
コロナ活動との関係

Eruptive and Quasi-Eruptive Disappearing Solar Filaments and Their Relationship with Coronal Activities

森本太郎, 黒河宏企
京都大学大学院理学研究科附属天文台

(原論文) PASJ: Publ. Astron. Soc. Japan 55, 1141-1151, 2003 December 25

Taro MORIMOTO and Hiroki KUROKAWA
Kwasan and Hida Observatories, Kyoto University

概要

太陽面上における35個のフィラメント消失現象の三次元速度場を計測することにより、Hαフィラメント消失現象の運動とそれに関連したコロナ活動現象の因果関係を研究した。フィラメント消失現象の三次元速度場を求めることで、フィラメントが惑星間空間に放出されたか、太陽大気中に留まったかを判定する手法を発展させた。この方法を用いてフィラメント消失現象の分類を行い、コロナ質量放出現象(CME)などのコロナ活動の有無との比較を行った。フィラメント噴出の発生後には必ずアーケードが形成され、準噴出型フィラメントの場合には軟X線や極紫外線における局所的な変化が見られた。噴出型フィラメントとCMEの密接な関係も発見された。太陽観測機SOHOの高視野分光コロナグラフで観測された15例のフィラメント消失現象中、8例の噴出型イベントはすべてCMEが観測された一方で、7例の準噴出型イベントでは見られなかった。これらの観測的結果より、Hα消失フィラメントの運動はコロナ活動やCMEと深く関連し、その三次元速度場の詳細な解明は加速や加熱のメカニズムの理解のみならず、CMEや地磁気嵐の予報にとっても重要である。

Abstract

By measuring the 3-D velocity fields of 35 disappearing filaments (Disparition Brusques: DBs) on the solar disk, we studied the causal relationship between the motions of Hα DBs and the associated coronal phenomena. Using the derived 3-D velocity fields of the DBs, we developed a method to judge whether a DB is ejected into interplanetary space or whether it remains in the solar atmosphere. We compared the DB type thus obtained with the presence of coronal mass ejections (CMEs) and other associated coronal activities. It is inferred that eruptive filaments are always followed by the formation of arcades, while most quasi-eruptive events are followed by localized changes in soft X-rays and the EUV. A close causal relation between eruptive filaments and CMEs was also found: of 15 DBs for which Solar and Heliospheric Observatory Large Angle and Spectrometric Coronagraph data were available, all eight of the eruptive ones were associated with CMEs, while no CMEs were found following any of the 7 quasieruptive ones. These observational results indicate that the motions of Hα disappearing filaments are causally related to the associated coronal activities and also to the appearance of CMEs, and that an accurate analysis of

1. 序論

太陽面におけるフィラメント消失現象はしばしば地磁気嵐を引き起こす（Wright, McNamara 1983）。太陽面の縁におけるプロミネンス噴出現象と CME とは強い相関関係にある（Munro et al. 1979）が，フィラメント消失現象と CME，および地磁気嵐の関係はいまだよくわかっていない。これは，Hα 線中心のみの観測では，消失したフィラメントが惑星間空間に放出されたのか否かの判別がつかないことによる。Mouradian, Soru-Escaut, and Pojoga（1995）はフィラメント消失現象を動的フィラメント消失現象（Démoulin, Vial 1992）と熱的フィラメント消失現象の2つのカテゴリに分類した。熱的フィラメント消失現象は，エネルギー入射量が増加してプラズマが加熱され，水素が電離することにより Hα でフィラメントが消失する現象である。この場合，加熱されたプロミネンスは極紫外線で観測され，その後冷却されると再び Hα で観測されるようになる。動的フィラメント消失現象は3つの形態に分類される（Kiepenheuer 1953）。「噴出型」では，フィラメント全体が対称的に上昇し，少なくともその一部は Hα 線で観測されなくなるまで上昇し続ける。残りの部分は太陽面に向かって流れ落ちる。「準噴出型」は上昇した後に断片に分割され，彩層の複数の地点に落下する。「低速溶解型」では，フィラメント中のプラズマ質量流出速度が供給速度よりも大きい結果，フィラメントが消失する（Martin 1973）。

Hα 線中心で観測したこれらのフィラメントの形態学的な特徴にはいくつかの共通点がある。準噴出型フィラメントは，噴出した後にプラズマが元いた場所に落下する際にも再び観測される。しかし，初期の加速段階ではフィラメント本体の大半は Hα 線中心では消失し，その形状は激しく変化

their 3-D velocities is important not only for a better understanding of their acceleration and heating mechanisms, but also for predicting the occurrence of CMEs and geomagnetic storms.

Key words: Sun: coronal mass ejections (CMEs) — Sun: filaments. — Sun: flares

1. Introduction

Disappearing filaments (DBs) on the solar disk often cause geomagnetic storms (Wright, McNamara 1983). Although there is a strong correlation between prominence eruptions at the solar limb and coronal mass ejections (CMEs) (Munro et al. 1979), the relationship among the DBs, CMEs, and geomagnetic storms is not yet well understood. This is because one cannot determine whether a disappearing filament is ejected into interplanetary space or not, with observations in the Hα line center alone. Mouradian, Soru-Escaut, and Pojoga (1995) defined two categories of DB: dynamic DB (DBd) (Démoulin, Vial 1992) and thermal DB (DBt). DBt is the disappearance of a filament in Hα due to an increase in the energy input which, as it heats the plasma, ionizes the hydrogen. In this case, the heated prominence appears in extreme ultraviolet lines, and once cooled, reappears in Hα. There have been three forms of DBd processes recognized (Kiepenheuer 1953). In the "eruptive" one, the whole filament ascends symmetrically and at least a part of it continues to ascend until it becomes invisible. Other parts become streamers flowing back to the disk. The "quasi-eruptive" filament ascends, breaks into fragments, and then its mass flows into the chromosphere at several locations. In the "slow dissolution", the rate of mass loss to the chromosphere exceeds the rate of new accumulation of mass and consequently the filament disappears (Martin 1973).

There are some similarities in the morphological features of DBs in Hα line center filtergrams. A quasi-eruptive filament reappears when its material falls back to the initial location. During the first acceleration phase, however, most of the filament body disappears and its shape is changed dramatically. These signatures are similar to those of an

する．これらの特徴は噴出型フィラメントと同様である．熱的フィラメント消失現象では，フィラメントは大きな上昇運動は示さず，急速に消失した数時間後に再び現れる．フィラメントの熱的消失段階のHα線中心画像は準噴出型のものと似ている（Gilbert et al. 2000）．低速溶解型フィラメントでは，フィラメント消失に至るまで形状や位置の激しい変化は見られず，熱的フィラメント消失現象との共通点がある．

噴出型フィラメントはしばしば，それに引き続くCMEのコアとして観測される（Illing, Hundhausen 1985）．コロナ噴出現象が太陽面中心付近，特に西半球面から発生した場合は，太陽面の縁で発生した場合に比べて地磁気に対する影響が最も大きくなる（Zhang et al. 2003）ために，動的フィラメント消失現象は太陽辺縁部のプロミネンス噴出現象よりも地磁気に影響を及ぼす．それゆえ，フィラメント消失現象が噴出型であるか否かの判定は宇宙天気予報にとって非常に重要となる．このためには，消失フィラメントのドップラーシフト量と三次元速度場を計測してフィラメント消失現象のタイプを決定する必要がある．我々の以前の研究（Morimoto, Kurokawa 2003; 論文 I）では，フィラメント消失現象の三次元速度場を得る手法を開発した．この手法によって，フィラメント消失現象の速度場全体の時間発展を明らかにし，関連するコロナ活動との比較を行うことができる．この手法を使用して，飛騨天文台のフレア監視望遠鏡（FMT：Kurokawa et al. 1995）で観測された35個のフィラメント消失現象について解析を行い，体系的な三次元速度場の計測を行った．

本研究では，フィラメント消失現象のタイプ（噴出型か準噴出型か）を判定する別の手法を紹介する．次に，この手法で得られたフィラメント消失現象のタイプを，Yohkoh/SXT, GOES, および SOHO/EIT, LASCO によって観測されたコロナ活動と比較し，フィラメント消失現象の運動の特徴とコロナ活動，およびCMEとの因果関係について詳細に議論する．

2節では，本研究に用いられたデータと

eruptive filament. A filament in the DBt process exhibits little upward motion, but shows significant disappearance, followed by a reappearance of the filament within a few hours. Also, its disappearance from Hα line center filtergrams resembles a quasi-eruptive filament (Gilbert et al. 2000). A filament in slow-dissolution does not show any violent changes in its shape or location until its disappearance, and there are similarities with DBt.

An eruptive prominence is often observed as a core of a subsequent CME (Illing, Hundhausen 1985). Since a CME that originates from the solar disk, especially from the western hemisphere, is known to be more geoeffective (Zhang et al. 2003) than one from the solar limb, DBd must be more geoeffective than a prominence eruption. It is, therefore, very important for space weather forecasts to investigate whether a DB is eruptive or not. Therefore, we need to measure the Doppler shift and three-dimensional (3-D) velocity field of a disappearing filament in order to determine the type of DB. In our previous study (Morimoto, Kurokawa 2003; hereafter Paper I), we developed a new method to obtain the complete 3-D velocity field of a DB. This method enables us to see the overall evolution of the velocity field of a DB and to compare it with the associated coronal activities. Making use of this method, we analyzed 35 DB events observed by the Flare Monitoring Telescope (FMT: Kurokawa et al. 1995) at Hida Observatory and systematically measured their 3-D velocity fields. In this paper, we present another method to judge the type (eruptive or quasi-eruptive) of DB. We next compare the obtained types of our DB events with the coronal activities observed by the Yohkoh/SXT, GOES, and SOHO/EIT and LASCO instruments and discuss the causal relationship among the characteristics of DB motions, coronal activities and CMEs in detail.

In section 2, we describe the data used in this study and the selection of events. The methods, to obtain the 3-D velocity field of a disappearing filament and to judge the type of DB, are proposed in section 3. We also describe the data from Yohkoh/SXT, GOES, and SOHO/EIT and LASCO in the same section. In section 4,

イベント選択について述べる．消失フィラメントの三次元速度場を得る手法とフィラメント消失現象のタイプを判定する手法は 3 節にて紹介する．Yohkoh/SXT，GOES および LASCO のデータも 3 節にて紹介する．4 節では，35 個のフィラメント消失現象から 5 つの典型例を選択し，その結果について詳細に述べる．最後に，全イベントの結果をまとめた上での結論を 5 節にて述べる．

2. 観測および観測データ
2.1. Hα観測
2.1.1. 飛騨フレア監視望遠鏡（FMT）

FMT は 1992 年 5 月より定常観測を行っている．この望遠鏡により，Hα線中心のみならず±0.8Åによる計 3 種類の Hα 太陽全面像が観測されている．FMT の 3 波長による観測では，フィラメントの視線方向の速度と視線に垂直な方向の速度を計測することが可能であり，これよりフィラメントの三次元速度場を得ることができる．

論文 I では，FMT で観測された 5 つの典型的なフィラメント消失現象についての詳細な解析結果が示されている．そのうちの 1 つが 1999 年 2 月 16 日に発生したイベントであり，この現象の FMT による観測を図 1 に示す．消失現象の開始前（02:00 UT）には，Hα線中心画像では活動領域 NOAA 8458 の南側にダークフィラメントが観測される．このフィラメントの動的な進化は以下のように記述される．Hα線中心画像でフィラメントが消失し始めると同時に，青色側ウイング画像ではフィラメントの中央部分が観測される（02:45 UT）．02:45 UT 以降，フィラメントは南西方向に加速され，大きな Ω 型の構造となる（02:56 UT）．この時，青色側ウイング画像ではフィラメントが最も顕著に現れる．02:56 UT にフィラメント消失直後にツーリボンフレアが発生し，Hα線中心で観測される．論文 I の図 1 に，FMT で観測されたこのフィラメント消失現象の詳細が示されている．

we present the results for five typical events selected from our 35 DB events in detail. Finally, summarizing the results of all events, we draw our conclusions in section 5.

2. Observations and Data
2.1. Hα Observations
2.1.1. The Hida flare monitoring telescope (FMT)

FMT has a routine observational program, which has been run since 1992 May. It provides 3 full disk Hα images, not only in its line center, but also in Hα±0.8 Å. The FMT observations in the 3 bands enable us to measure the line-of-sight and tangential velocities of filaments, and hence their complete 3-D velocity fields.

Five detailed examples of DB events observed by the FMT are illustrated in Paper I. One of them is the 1999 February 16 event and the FMT observation of this event is also displayed in figure 1. The dark filament is seen south of the active region NOAA 8458 in the Hα line center image before the start of its disappearance (02:00 UT). Its dynamical evolution can be summarized as follows. The central part of the filament appears in the blue wing image (02:45 UT) as the filament begins to fade in the Hα line center image. After 02:45 UT, the filament is accelerated toward the south-west and shows a large Ω structure (02:56 UT). At that moment, the filament is most prominent in the blue wing image. A two-ribbon flare occurs just after its disappearance, and it is seen in the Hα line center image at 02:56 UT. We refer to figure 1 of Paper I, for more detailed FMT observations of this filament disappearance.

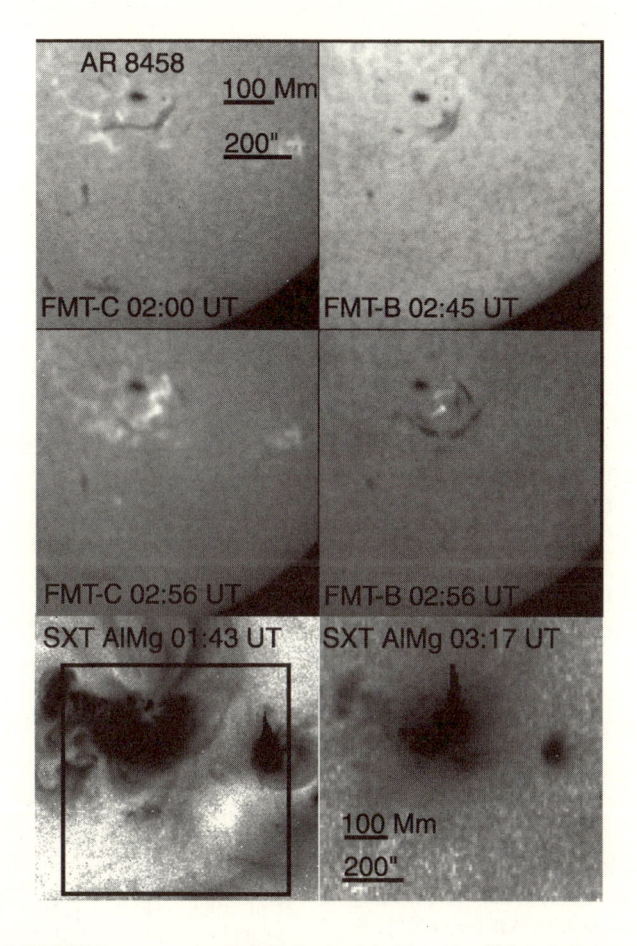

図1 発達した活動領域 NOAA 8458 の南側で 1999 年 2 月 16 日に発生した噴出フィラメントの FMT 画像の時間変化（上側 4 枚）．観測時間および波長（C：Hα線中心画像，B：Hα青色側ウイング画像）は各画像の下部に示されている．フィラメントは最初 Hα線中心画像で暗く観測された（左上図）後，青色側ウイング画像で暗く現れて南西方向に膨張した．このイベントに関連する軟 X 線フレアの発生前後における SXT 太陽全面像の一部分を示す（下側 2 枚）．FMT の視野は左下の図中に枠にて示されている．FMT と SXT の空間サイズは異なることに注意．各図中の黒線を参照．この論文を通じて，太陽の北は上，東は左である．

Fig. 1. Series of partial FMT images showing the eruptive filament south of the well-developed active region NOAA 8458 on 1999 February 16 (the four top panels). The time and the band (C: Hα line center image, B: Hα blue wing image) are shown in the bottom of each of the four panels. The filament first appeared dark in the Hα line center image (top-left panel), became dark in the blue wing and expanded toward the south-west. Two partial full-frame SXT images (the two bottom panels) before and after the onset of the associated soft X-ray flare are shown. The FMT field of view is indicated by the box in the bottom-left panel. Note that the spatial scales for the FMT and SXT images are different, and are indicated by the bars in the figure. The solar north is up and east is left, which holds for all other solar images in this article.

2.1.2. イベント選択

FMT の活動現象データベース[1] を用いて，我々は 1992 年 6 月から 2000 年 6 月までの期間で 2747 個のフィラメント消失現象から 35 個のイベントを選択した．選択の条件は，(i)SXT や EIT との画像比較のために，イベント（フィラメント）のサイズが 60 Mm 以上であること，(ii) イベント全体を通じて，観測が雲や日の入りなどによって中断されていないことであった．表 1 に選択された 35 個のイベントを示す．この中には論文 I で紹介された 5 つの典型例も含まれる．最初の列はイベント番号，2 番目は日付，3 番目の列は場所を示す．イベントが発生した活動領域の NOAA 番号も同じく 3 番目の列に記されている．

FMT のデジタル/ビデオ画像は 60 秒（フィラメントの運動速度が大きい場合は 30 秒）ごとに選択された．ビデオテープのデータに対しては，京都大学花山天文台の AD コンバータシステムによりデジタル化を行った．

2.2. コロナ観測

Hα フィラメント消失現象やプロミネンス噴出現象に伴い，コロナアーケードや一時的な減光現象，および CME がしばしば観測されることはよく知られている（e.g. Sheeley et al. 1975; Webb et al. 1976; Hanaoka et al. 1994; McAllister et al. 1992, 1996; Khan et al. 1998）．

しかし，噴出の有無によるコロナ変動の差はいまだよくわかっていない．本研究では，このことを Yohkoh/SXT, GOES, および SOHO/EIT, LASCO のデータを用いて解明を行った．

2.2.1.「ようこう」および GOES 観測

「ようこう」の軟 X 線望遠鏡（SXT）(Ogawara et al. 1991; Tsuneta et al. 1991) は太陽コロナの軟 X 線による全面像および部分像を 1991 年より観測している．太陽の全面像および部分像は通常 50 秒間隔

2.1.2. Event selection

Making use of the FMT data base[1] of active phenomena, we selected 35 out of 2747 DB events from 1992 June to 2000 June. The criteria for the selection were (i) the size of the event (filament) is larger than 60 Mm to make the comparison with SXT and EIT images easier, and (ii) throughout the event, the observation is not interrupted due to clouds or sunset. The 35 events are listed in table 1, and the five events that we discussed in Paper I are included in these 35 events. The first column gives the event No., and the date and location are in the second and third columns, respectively. The NOAA numbers of the active regions where the events took place are also presented in the third column.

We used an FMT image every 60 seconds (30 seconds when the apparent motion is fast) from video-tape or digital form data. In the case of the video-tape data, the signals were digitized with an AD converter system at Kwasan Observatory, Kyoto University.

2.2. Observations of the Corona

It is well known that coronal arcades, transient dimmings and CMEs are often observed to be associated with Hα DBs and prominence eruptions (e.g., Sheeley et al. 1975; Webb et al. 1976; Hanaoka et al. 1994; McAllister et al. 1992, 1996; Khan et al. 1998). However, the difference in coronal changes between events with and without an eruption is still not well understood. We investigated this using data from Yohkoh/SXT, GOES, and SOHO/EIT and LASCO.

2.2.1. Yohkoh and GOES observations

The Yohkoh Soft X-ray Telescope (Ogawara et al. 1991; Tsuneta et al. 1991) has provided full and partial images of the Sun's corona in soft X-rays from 1991. A full and partial frame image is obtained, usually, every 50 seconds with a pixel size

---

[1] http://www.kwasan.kyoto-u.ac.jp/Hida/FMT/obs-report.html

表1 全35個のフィラメント消失現象イベントのリスト.*

番号	時刻（UT）	場所/NOAA	タイプ（γ）	SXT	EIT	GOES（分）	CME
1	1992/11/05 00:15 – 02:15	S20W17	E (43)	A		-	
2	1993/01/23 05:55 – 06:39	N30W32	E (27)	A		B1.5	
3	1993/04/20 03:45 – 06:10	S22E70/7480	E (60)	A		B7.4(10)	
4	1993/05/15 22:29 – 24:27	S28E30	E (38)	A		B3.6	
5	1993/05/15 22:55 – 24:01	S32E38	E (64)	A		-	
6	1993/07/29 00:21 – 01:22	N10E38/7555	E (60)	-		B3.1	
7	1993/08/12 01:25 – 02:44	N13W43/7562	Q (0)	LB		B5.1	
8	1993/10/21 02:51 – 05:19	S15W08	E (75)	A		B2.2	
9	1993/10/21 22:24 – 22:54	N20W63	E (27)	A		B5.4	
10	1994/01/05 06:04 – 07:05	S10W22/7647	E (27)	A		M1.0	
11	1994/01/05 06:06 – 06:36	S15W08/7646	E (40)	A		B8.6	
12	1994/02/20 00:17 – 01:24	N01W00/7671	E (59)	A		M4.0	
13	1994/09/05 06:16 – 08:00	S05W03/7773	E (35)	A			
14	1994/09/14 01:30 – 04:37	S10W70	Q (0)			-	
15	1995/03/20 23:54 – 26:50	S19W08/7854	Q (0)	LB		B3.4	
16	1995/10/20 01:23 – 02:13	S18W52/7912	Q (0)	-		B1.1	
17	1996/05/13 05:56 – 06:39	S08W07/7962	Q (0)	LB	-	B2.9	無
18	1997/05/12 04:27 – 05:26	N20W09/8038	E (29)	A	A/EW/D	C1.3	有
19	1997/10/20 22:30 – 24:56	S30E43	E (33)	A	A/FE/D	B1.2	有
20	1998/04/11 02:09 – 08:00	S25E12/8194	Q (0)	LB	-	B8.5	無
21	1998/04/11 04:15 – 05:41	S37E35/8195	E (50)	A		C3.0	有
22	1998/04/29 05:11 – 05:40	S10E32/8210	E (29)	-	A/FE/D	B7.7	有
23	1998/05/20 01:40 – 03:30	N31W75	Q (0)	-	-	-	無
24	1998/09/05 00:08 – 01:37	S36W37	Q (0)	A		C1.5	
25	1998/09/20 02:00 – 05:28	N20E70/8340	E (63)	A		M1.8	
26	1998/10/27 23:34 – 36:39	N18E40/8369	Q (0)	-	LB	C1.6	無
27	1999/01/30 00:00 – 01:50	S34E20	E (84)	A		B3.3	
28	1999/02/09 03:07 – 05:22	S27W39/8453	E (13)	A	A/EW/D†	C2.3	有
29	1999/02/16 01:42 – 04:15	S27W18/8458	E (53)	A		M3.2	
30	1999/06/01 06:29 – 07:08	S23E17/8557	Q (0)	LB		C6.2	無
31	2000/01/19 00:28 – 01:47	N08W18/8829	E (36)	A	A/D	C1.4	有
32	2000/01/28 05:35 – 06:20	S28W20/8841	Q (0)	LB	LB	B4.4	無
33	2000/04/06 03:48 – 05:48	S27W02	Q (0)		LB	C1.4	無
34	2000/04/25 01:05 – 01:47	N23W27/8972	E (51)	A	A/FE	C1.1	有
35	2000/05/08 04:19 – 07:40	S21W03	E (45)	A	A/FE/D	B6.8	有

*左列より，イベント番号，時刻(UT)，フィラメントが位置する場所および活動領域番号，フィラメント消失現象のタイプ(E：噴出型，Q：準噴出型)および噴出率(γ)，Yohkoh/SXT 観測によるコロナ活動タイプ(A：ループやアーケード形成等の大規模コロナ再配置，LB：大規模な変化を伴わない既存ループ増光現象や小規模局所的フレアループ形成，空欄：データなし，-：データ不足で判別できず)，SOHO/EIT 観測によるコロナ活動タイプ(A：アーケード形成，LB：大規模な変化を伴わない既存ループ増光現象や小規模局所的フレアループ形成，EW：EIT波，D：減光現象，FE：ダークフィラメント噴出現象，空欄：データなし，-：データ不足で判別できず，†：λ 195 画像の替わりにλ 284 画像を使用)，軟 X 線イベントにおける GOES の最大強度(空白：データなし，-：X 線強度ピーク検出されず)，SOHO/LASCO 観測による CME との関連性(空白：データなし)．活動領域番号はイベントが活動領域内で発生した場合にのみ表示．

Table 1. Complete list of our 35 DB events.*

No.	Time	Location/NOAA	Type (γ)	SXT	EIT	GOES (min)	CME
1	11/05/92 00:15 – 02:15	S20W17	E (43)	A		-	
2	01/23/93 05:55 – 06:39	N30W32	E (27)	A		B1.5	
3	04/20/93 03:45 – 06:10	S22E70/7480	E (60)	A		B7.4(10)	
4	05/15/93 22:29 – 24:27	S28E30	E (38)	A		B3.6	
5	05/15/93 22:55 – 24:01	S32E38	E (64)	A		-	
6	07/29/93 00:21 – 01:22	N10E38/7555	E (60)	-		B3.1	
7	08/12/93 01:25 – 02:44	N13W43/7562	Q (0)	LB		B5.1	
8	10/21/93 02:51 – 05:19	S15W08	E (75)	A		B2.2	
9	10/21/93 22:24 – 22:54	N20W63	E (27)	A		B5.4	
10	01/05/94 06:04 – 07:05	S10W22/7647	E (27)	A		M1.0	
11	01/05/94 06:06 – 06:36	S15W08/7646	E (40)	A		B8.6	
12	02/20/94 00:17 – 01:24	N01W00/7671	E (59)	A		M4.0	
13	09/05/94 06:16 – 08:00	S05W03/7773	E (35)	A			
14	09/14/94 01:30 – 04:37	S10W70	Q (0)			-	
15	03/20/95 23:54 – 26:50	S19W08/7854	Q (0)	LB		B3.4	
16	10/20/95 01:23 – 02:13	S18W52/7912	Q (0)	-		B1.1	
17	05/13/96 05:56 – 06:39	S08W07/7962	Q (0)	LB	-	B2.9	no
18	05/12/97 04:27 – 05:26	N20W09/8038	E (29)	A	A/EW/D	C1.3	yes
19	10/20/97 22:30 – 24:56	S30E43	E (33)	A	A/FE/D	B1.2	yes
20	04/11/98 02:09 – 08:00	S25E12/8194	Q (0)	LB	-	B8.5	no
21	04/11/98 04:15 – 05:41	S37E35/8195	E (50)	A	-	C3.0	yes
22	04/29/98 05:11 – 05:40	S10E32/8210	E (29)	-	A/FE/D	B7.7	yes
23	05/20/98 01:40 – 03:30	N31W75	Q (0)	-	-	-	no
24	09/05/98 00:08 – 01:37	S36W37	Q (0)	A		C1.5	
25	09/20/98 02:00 – 05:28	N20E70/8340	E (63)	A		M1.8	
26	10/27/98 23:34 – 36:39	N18E40/8369	Q (0)	-	LB	C1.6	no
27	01/30/99 00:00 – 01:50	S34E20	E (84)	A		B3.3	
28	02/09/99 03:07 – 05:22	S27W39/8453	E (13)	A	A/EW/D†	C2.3	yes
29	02/16/99 01:42 – 04:15	S27W18/8458	E (53)	A		M3.2	
30	06/01/99 06:29 – 07:08	S23E17/8557	Q (0)	LB		C6.2	no
31	01/19/00 00:28 – 01:47	N08W18/8829	E (36)	A	A/D	C1.4	yes
32	01/28/00 05:35 – 06:20	S28W20/8841	Q (0)	LB	LB	B4.4	no
33	04/06/00 03:48 – 05:48	S27W02	Q (0)		LB	C1.4	no
34	04/25/00 01:05 – 01:47	N23W27/8972	E (51)	A	A/FE	C1.1	yes
35	05/08/00 04:19 – 07:40	S21W03	E (45)	A	A/FE/D	B6.8	yes

* From the left column, event No., time (UT), location and the AR number to which the filament belongs, the type of DB (E: Eruptive, Q: Quasi-eruptive) with the ejection rate (γ), the class of coronal activities observed by Yohkoh/SXT (A: Global restructuring of corona, such as formations of new loops of arcade. LB: Only brightenings of pre-existing coronal loops and/or small and localized flare loop formation without global changes. Blank: No data, and -: not enough data to distinguish the coronal activity), the type of coronal activities observed by SOHO/EIT (A : Arcade formation. LB: The same abbreviation for that of SXT. EW: EIT wave. D: Dimming. FE: Dark filament ejections. Blank: No data, and -: not enough data.
† : We used λ 284 images instead of λ 195), peak flux of accompanying soft X-ray events from GOES satellite (Blank: No data, and -: no peak of X-ray flux was found), the CME association with SOHO/LASCO data (Blank: No data), respectively. The AR number is presented only if the event occurred in an active region.

of 4."9 (half resolution) or 9."8 (quarter resolution). Yohkoh observations correspond to a temperature regime greater than 1MK, and we can see the distribution of hot plasma and changes of magnetic structures associated with a DB. We used full-frame images taken with the two thinnest soft X-ray filters: one consisting of a thin layer of Al, abbreviated as the Al.1 filter (sensitive to plasmas with temperature from 10^6 to several times 10^6 K), and the other a composite filter consisting of Al, Mg, Mn, and C, abbreviated as the AlMg filter (sensitive to plasmas with temperature from 1.5×10^6 to a few times 10^7 K) (Tsuneta et al. 1991). We also used the soft X-ray flux data obtained by Geostationary Operational Environmental Satellite (GOES) to investigate whether the peak flux of soft X-ray events is coincident with the Hα DBs. Note that the magnitude of GOES X-ray event is expressed in units of watts per meter^2 and classified into classes designated by letters B, C, and M, with flux starting at 10^{-7}, 10^{-6}, and 10^{-5} Wm^{-2}, respectively.

2.2.2. SOHO observations

The Extreme Ultraviolet Imaging Telescope (EIT) (Delaboudinière et al. 1995) on board the Solar and Heliospheric Observatory (SOHO) spacecraft provides images in four extreme ultraviolet lines (Fe IX λ 171, Fe XII λ 195, Fe XV λ 284, He II λ 304). For the observations reported here, we mainly used λ 195 Å data. However, for the 1999 February 9 event λ 195 Å data was not available; hence, we used the λ 284 Å data for that event. The EIT 195 Å line is most sensitive to coronal plasma with the temperature peaked at 1.4 MK, which is lower than that of SXT. It obtains three full-disk images every hour and its CCD pixel size is 2".6 in the full-resolution mode. Since Yohkoh/SXT has a periodic data gap of about 40 minutes every 100 minutes, called *Yohkoh night*, some fast coronal changes can possibly be missed by SXT. Hence, we made use of EIT data together with SXT data to check the coronal activities for events after 1996.

To check the CME association with DBs, we also used data from the Large Angle and Spectrometric Coronagraph (LASCO)

イベントには高視野分光コロナグラフ (LASCO) (Brueckner et al. 1995) のデータを使用した．LASCO は 1.1-30 $R_\odot$ の範囲の太陽コロナを3つの異なるコロナグラフで観測できる．今回の研究では，CME との関連を調べるために C2 白色光コロナグラフ (2-6 $R_\odot$) のみを使用した．

3. 解析

3.1. 消失フィラメントの三次元速度場

　消失フィラメントの三次元速度場の導出法を説明する前に，この論文で用いられる速度の用語について述べる．視線速度とは視線方向に沿った速度成分である．「前進」および「後退」とは，それぞれ視線に沿って我々に近付く運動と離れる運動を示す．視線垂直速度とは，フレア監視望遠鏡の連続画像上における運動の速度である．視線垂直速度は空の平面上で定義されるために，x および y の 2 成分がある．x と y の方向は太陽の西と北に相当する．これらから座標変換を行うことで，「上昇」速度を求めることができる．上昇速度は太陽表面からフィラメントが鉛直方向に外側に向かう運動の速度である．この逆の用語として，「下降」は鉛直方向に太陽表面に向かう運動を示す．最後に，「全」速度は進行方向を無視した際のフィラメントの速さである．

　論文 I では，消失フィラメントの視線速度と視線垂直速度を導出する手法を紹介した．Hα フィラメント消失現象の速度の詳細な計算方法は論文 I にて示されているので，この論文では簡潔に述べるにとどめる．この新しい手法では，Beckers のクラウドモデル (Beckers 1964) のコントラストプロファイルを用いて，視線速度の時間発展を Hα 線中心，Hα-0.8Å，および Hα+0.8Å の 3 波長画像におけるコントラストの変化から求める．フィラメントの各ピクセルに対してコントラスト計算を行うことで，フィラメントのどの部分でも視線速度を計測することができる．この手法では，視線速度の精度を上げるために FMT の有効透過幅，散乱光およびドップラー増光効果 (DBE) の補正も行った．視線垂直方向の速さおよび方向は，FMT の連続画像

(Brueckner et al. 1995) for 15 events after 1996. LASCO images the solar corona from 1.1-$30R_\odot$ with three different coronagraphs. In this study, we used only the C2 white-light coronagraph (2-$6R_\odot$) data to check the CME association.

3. Analysis

3.1. The 3-D Velocity Field of a Disappearing Filament

Before explaining the method for deriving the 3-D velocity field of a disappearing filament, we describe the velocity terms used in this paper. The "line-of-sight" velocity refers to the speed of the filament motion along the line-of-sight. We use "ascending" and "receding" for the motions toward and away from us along the line-of-sight, respectively. The "tangential" velocity is the speed of the apparent motion seen on successive images. The tangential velocity consists of two components of velocity, namely x and y, because it is defined on the plane of the sky. The directions x and y correspond to solar west and north. Then, by converting the coordinate, we can obtain the "upward" velocity. The upward velocity is the speed of filament motion radially outward from the solar surface. The opposite term "downward" represents a motion radially toward the solar surface. Finally, "total" velocity is the speed of the filament regardless of its direction of propagation.

In Paper I, we presented a method to derive both the line-of-sight and tangential velocities of a disappearing filament. Since the detailed calculation procedures of velocity in disappearing Hα filaments are presented in Paper I, we explain it only briefly in this paper. In this new method, the time variation of the line-of-sight velocity is derived by interpreting the temporal variations of the contrasts observed in the three filtergrams in the Hα line center, Hα-0.8 Å, and Hα+0.8 Å in terms of a contrast profile based on Beckers' cloud model (Beckers 1964). Since we calculated the contrast at each pixel on the filament, the line-of-sight velocity can be measured everywhere on the filament. In this method, we also make corrections for the effective band widths

上でフィラメントのガス塊（blob, ブロッブ）を追跡することにより得られた．個々のブロッブの速度を周辺のピクセルに対して補間することにより，フィラメント上の各ピクセルにおける視線垂直速度を求めた．最後に視線速度と視線垂直速度を組み合わせることにより，消失フィラメントの三次元速度場が得られる．

求められた視線垂直速度はフィラメント上の各部分の軌道を決定するためにも用いられた．位置 $[x(t), y(t)]$ で速度が $[v_x, v_y]$ のピクセル上の物体は時刻 $t + \delta t$ の画像では $[x(t+\delta t), y(t+\delta t)] = [x, y] + \delta t \times [v_x, v_y]$ の位置に移動することを用いて，イベント全体におけるフィラメントの各部分の軌道を容易に求めることができる．このことにより，フィラメントの各部分における三次元速度の時間発展を得ることができる．このプロセスはフィラメント上の各ピクセルに対して個別に行い，完全な三次元速度場を求める．

3.2. フィラメント消失現象の分類

今回選択した35個のフィラメント消失現象の中には，「低速溶解型」と呼ばれ下降運動のみで上昇運動を示さないタイプや，また熱的フィラメント消失現象のようにほとんど運動を見せないタイプは見出されなかった．それゆえ，ここではフィラメントの三次元速度場を用いて「噴出型」と「準噴出型」とを区別する手法について議論する．

FMTの画像から見えなくなった後では，フィラメント中のブロッブが惑星間空間に放出されたか否かを判定することは不可能なため，ブロッブの上昇加速度の特徴から噴出型と準噴出型の区別を行った．フィラメントのブロッブが惑星間空間に噴出したか否かを判定するために，「上昇速度が正で，FMTの視野から消えた時点でも加速を続けている」という基準を満たすものを噴出型とみなした．これは，太陽の重力から逃れるためには，フィラメントはHα線

of the FMT filters, the stray light and the Doppler brightening effect (DBE) in order to accurately determine the line-of-sight velocity. The tangential speeds and directions are obtained by tracing the motions of the internal structures (blobs) of the filament on the successive FMT images. The velocities of the identified blobs are then interpolated to the surrounding pixels to yield the tangential velocity at each pixel on the filament. When combined with the line-of-sight velocity, it then gives us the complete 3-D velocity field of the disappearing filament.

The derived tangential velocity is also used to obtain a trajectory for each element of the filament. Since a pixel at $[x(t), y(t)]$ with a tangential velocity of $[v_x, v_y]$ would shift to $[x(t+\delta t), y(t+\delta t)] = [x, y] + \delta t \times [v_x, v_y]$ on the next image at $t + \delta t$, we can easily find the trajectory of an element of the filament represented by each pixel throughout the event. This enables us to obtain the time-dependent evolution of the 3-D velocity of each element. This procedure is performed for all the pixels on the filament individually to reconstruct its complete 3-D velocity field.

3.2. Types of DBs

In our 35 samples of DBs, we found neither the "slowdissolution" type, in which we see only downward motion without any significant upward motions, nor the DBt with little or no motions. We, therefore, discuss a method used to distinguish "eruptive" and "quasi-eruptive" types of DBs from their derived 3-D velocity fields here.

Since we cannot determine whether a blob of a filament is ejected into interplanetary space or not after they disappear from FMT images, we distinguished eruptive and quasieruptive filaments by examining the acceleration characteristics of the upward velocity of a blob. In order to judge whether the blob is erupted into interplanetary space or not, we employed a criterion for an eruptive blob as follows: "the upward motion must be positive and must be accelerating when it disappears from the FMT field of view". This is because the Hα filaments should be further accelerated after its

で観測されなくなった以降も加速する必要があるためである．

この基準は同一の視線上に異なる運動を示す部分が複数存在する場合には十分ではない．この場合には，例え一部が加速されFMTの画像から消失したとしてもほかの部分が視線上で観測されることとなる．その場合，上昇速度の時間変化は急激な減少を示し，この視線上の物体は放出されなかったと判定されてしまう．フィラメントはしばしば噴出中に変形するために，上記のような現象は時々起こりうる．この種の誤判定を避けるために，「上昇速度が太陽の重力加速度（～ 0.27 km/s^2）よりも大きな減速を示し，その後の上昇速度が0以上であった場合は，視線上の複数の物体の一部が放出されたとみなす」という基準を設けた．この基準は，最大減速度は自由落下から求められるという根拠に基づいている．この2つの基準が満たされない場合，視線上の物体は放出されなかったと判定される．例として，論文Ⅰにおいて1999年2月16日に発生したイベントで参照点としてA点を定義しその速度変化を計測した．求められたA点の上昇速度は論文Ⅰの図2に示している．この図によると，A点における上昇速度と全速度は02:45 UTにA点が消失するまで単調に増加していることから，A点は惑星間空間に放出されたと判定した．

図2はFMTで1999年2月16日のイベントで02:00 UTに観測されたHα線中心画像であり，白い斜線の網掛けはその後の消失現象において惑星間空間に放出されたとみなされるピクセル位置の分布を示したものである．フィラメントの中央部と西側部分の大半が放出された．付け加えると，この解析により放出したフィラメントの量を見積もることができる．投影効果を除くためにフィラメントを太陽面中心まで仮想的に移動させ，「放出率（γ）」を$\gamma = N_E/N_P$として計算した．ここでN_PとN_Eはそれぞれ移動後の画像においてフィラメント上のピクセルの数と「放出された」ピクセルの数である．計算された放出率は表1の4列目の括弧内にパーセントで示されてい

disappearance from the Hα line to escape from the solar gravity.

This criterion is not sufficient when there are more than two structures which exhibit different motions on the same line-of-sight. In this case, even if one of them is accelerated and ejected so that it disappears from FMT images, the other structure remains visible on the line-of-sight. In this case the evolution of its upward velocity would show a sudden drop to a smaller velocity, leading us to conclude that this column is not ejected. Since a filament is often distorted in its eruption, such a case sometimes occurs. In order to avoid this kind of misjudgment, we employed a second criterion as follows: "if the upward velocity shows a sudden deceleration that is more than the solar gravitational acceleration value (~ 0.27 km s^{-2}) and the subsequent upward velocity is not zero, one of the structures on the same line-of-sight has been ejected". This is based on a reasonable assumption that the maximum deceleration is due to the free-fall motion. Unless these two criteria are satisfied, the column is regarded as unejected. For example, in Paper I, we defined a sample point (point A) on the FMT images of the 02/16/99 event and obtained its velocity evolution. The derived upward velocity of point A is shown in figure 2 of Paper I. According to the figure, the upward velocity and the total velocity of the element represented by point A is monotonically accelerated until its disappearance at 02:54 UT, and we judged that point A was ejected into interplanetary space.

Figure 2 illustrates the distribution of the pixels that were judged to be ejected into interplanetary space on their subsequent disappearance, with the white shadow region on the FMT Hα line center image of the 02/16/99 event at 02:00 UT. The central part and most of the western part of the filament were ejected. Furthermore, this analysis can be used to estimate the amount of the ejected part of the filament. In order to remove projection effects, we rotated the filament to the disk center, and calculated the "ejection rate (γ)" which is defined as $\gamma = N_E/N_P$, where N_P is the number of pixels occupied by the filament and N_E is the number of "ejected"

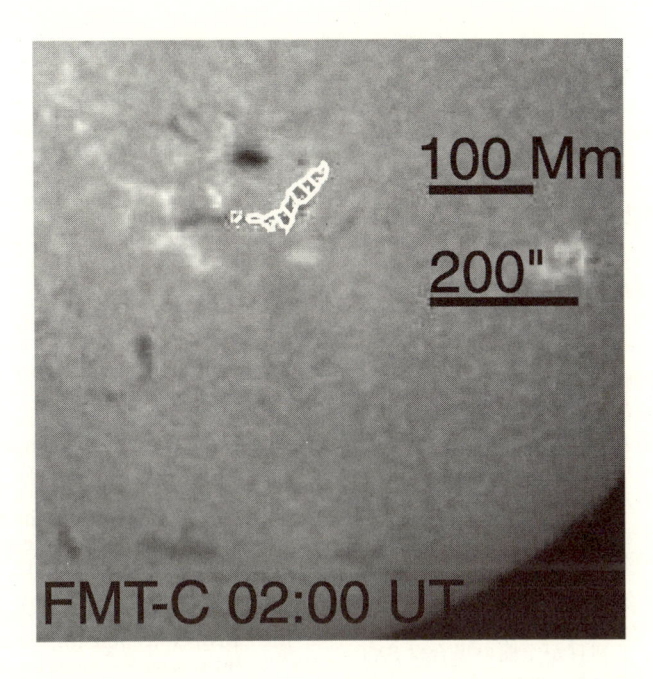

Appendix A Selected Papers Relating to FMT | 383

Fig. 2. 噴出前の 02:00 UT における FMT の Hα 線中心画像．イベント発生後に加速して消失する部分の元の状態を示す．消失する部分はフィラメントの中央部分から西側の足元までの網掛けの領域に分布している．

Fig. 2. Elements which disappeared with increasing upward velocities traced back to the first image of FMT Hα line center at 02:00 UT before the onset of this event. They are distributed in the shaded region from the middle to the western footpoint of the filament.

る．「放出率」が 100 % はフィラメント全体が惑星間空間に放出されたことを示し，0 % はフィラメント全体がコロナ中に留まるか太陽表面に落下したことを意味する．放出率は単なるピクセル数の比率であり，体積比や質量費などの物理量ではないことに注意する必要があるが，それらについての大まかな見積もりを立てることができる．噴出型フィラメントの放出率は $\gamma > 0$ %，準噴出型フィラメントの放出率は $\gamma = 0$ % とした．フィラメント消失現象のタイプ（「E」が噴出型，「Q」が準噴出型）は表1の4列目に表示されている．1999年2月16日のイベントでは，放出率は $\gamma = 53$ % であり，フィラメントの約半分が放出され

pixels on the rotated image. The ejection rate was calculated, and is listed in the parenthesis as a percentage value in the fourth column of table 1. An "ejection rate" of 100% means that the whole filament is ejected into interplanetary space and 0% means that the whole filament remains in the corona or falls back to the surface. One must keep in mind that the ejection rate is merely a ratio of pixel numbers, and is not a physically meaningful quantity, such as volume or mass ratio, though it can give us a rough estimation of them. We defined eruptive filaments as those with $\gamma > 0$% and quasi-eruptive filaments as events with $\gamma = 0$%. The types of DBs ("E" for eruptive filament, "Q" for quasi-eruptive filament) are also listed in the fourth column of table 1. For the 02/16/99 event, the ejection rate is $\gamma = 53\%$ and nearly half of the filament is estimated to be ejected. As a result, we found that 23 out of 35 events were eruptive and that the other 12 events were quasi-eruptive DBs.

3.3. Coronal Changes in Soft X-Rays and the EUV

3.3.1. Yohkoh/SXT data

We investigated the coronal activities, such as arcade formations and transient loop brightenings, which are associated with Hα DBs, using Yohkoh/SXT data. We classified the coronal activities into two categories: one case is when a large-scale rearrangement of coronal structures, such as arcade formation, takes place, and is abbreviated as "A" in the fifth column of table 1. The other is the case in which the coronal activity is limited to a localized brightening of a preexisting loop or a very small flare loop formation, designated as "LB". Blanks are left for events with no SXT data, and "-" marks for events without enough data to judge the coronal activities well.

The two bottom panels of figure 1 are the Yohkoh/SXT images obtained before and after the onset of the soft X-ray event which is related to the 02/16/99 DB. An X-ray class M3.2 flare took place starting at 02:48 UT when the Hα filament started to be accelerated abruptly. This flare appears as an intense brightening of the flare loops (see SXT image at 03:16:40 UT) followed by large arcade formation.

3.3.2. SOHO/EIT data

We also used SOHO/EIT data to investigate the coronal activities. Since EIT 195 is most sensitive to a plasma with a temperature of 1.4 MK, which is sufficiently lower than that of SXT, a few flare-related phenomena that SXT can not observe are detectable. The abbreviation for the types of EUV activities are as follows: "A" for arcade formations, "LB" for brightenings of small localized structures or preexisting loops, "FE" for dark filament eruptions, "EW" for EIT waves and "D" for dimmings. EIT waves

である．EIT波はEITによって発見され，コロナの増光に引き続いて減光領域が広がる一時的な波状の現象である（Thompson et al. 1999, 2000）．EIT波はしばしばフレアやCMEに伴って発生する．減光現象は一般的にはCMEによってコロナ中の物質が引き抜かれた結果発生すると考えられ，それゆえ減光領域はCMEの発生源の重要な証拠であると考えられる（Harra, Sterling 2001）．したがって，EIT波と減光現象はアーケード形成と同様にコロナ構造の大規模な再配置であると見なすことができる．EITのデータによって分類されたコロナ活動のタイプは表1の6列目に表示されている．ここでもSXTデータによる分類と同様に，空白はEITデータが存在しなかった場合であり，「-」はデータ不足を示す．

3.3.3. GOESデータ

GOESの低エネルギー帯（1-8Å）による軟X線強度の積算データを用いてX線イベントの最大強度を求めた結果を表1の7列目に表示する．列中の空白はデータの欠落，「-」はHαフィラメント消失現象に関連する軟X線の増光が背景レベルに隠れている場合である．

3.4. CME

噴出型フィラメントはCMEを伴う一方で，準噴出型は伴わないと考えられる．しかし，状況によってはX線プラズモイド放出現象（Shibata et al. 1995; Ohyama, Shibata 1997, 1998）や，放出された磁束管が濃いプラズマで満たされていない場合のように，Hα線では観測されない放出現象が生じることもありうる．この場合，Hαフィラメント消失現象のタイプは準噴出型であってもCMEが観測されると考えられる．一方で，準噴出型フィラメントがほぼ視線に沿って90 km/s以上の速度で運動する場合，消失現象の初期段階では加速を続けたままFMTの視野からは消える．この場合，消失現象の後半における減速段階を観測できずに，CMEを伴わない噴出型フィラメントと判定しうる．フィラメント消失

were discovered by the EIT as transient wave-like structures with enhanced coronal emission that are followed by an expanding dimming region (Thompson et al. 1999, 2000). They are often accompanied by flares and CMEs. Dimmings have generally been assumed to be due to a discharge of CME material from the corona, and thus the dimming regions are thought to be an important signature of the sources of CMEs (Harra, Sterling 2001). EIT waves and dimmings, therefore, can be regarded as large-scale rearrangements of coronal structures, like arcade formations. The types of coronal activities found with EIT data are shown in the sixth column of table 1. Again, blanks are left for events without EIT data and "-" marks for events with inadequate data, as in the case of the SXT classification.

3.3.3. GOES data

The integrated soft X-ray flux data from the GOES satellite low-energy channel (1.8Å) is used to check the peak flux of X-ray events; the results are listed in the seventh column of table 1. We left blanks for events with inadequate data, and "-" marks for events in which soft X-ray enhancements related to Hα DBs are buried in the background flux.

3.4. CME

It is expected that eruptive filaments are accompanied by CMEs and quasi-eruptive filaments are not. A discrepancy, however, may arise from some situations in which there are ejections invisible in the Hα line, such as X-ray plasmoid ejection (Shibata et al. 1995; Ohyama, Shibata 1997,1998), and ejection of a magnetic flux rope which is not filled with dense plasma. In these cases, although the Hα DB type is quasieruptive, we would have a CME. Besides, if a quasi-eruptive filament's motion is nearly along the line-of-sight at more than 90 km s^{-1}, the filament disappears from the FMT field of view in the early stage of its disappearance with an increasing upward velocity. In this case, we may miss its evolution and its deceleration in the later phase, and hence we may have no CME. In order to investigate more meaningful

現象と CME，またはコロナ活動と CME とのより有意な相関関係を得るために，1996 年 5 月以降のイベントに対しては SOHO/LASCO の白色光コロナグラフの画像を使用し，フィラメント消失現象に伴って CME が観測されるかを調べた．結果を表 1 の最終列にあげる．

4. 結果と議論

4.1. Hαフィラメント消失現象の例

この節では 5 つのフィラメント消失現象（1999 年 2 月 16 日，1992 年 11 月 5 日，2000 年 1 月 19 日，2000 年 4 月 6 日，および 2000 年 1 月 28 日）について，速度の時間発展の詳細や軟 X 線や極紫外線，白色光コロナグラフで観測されたコロナ活動の特徴についての解析結果を述べる．次に 35 個のイベントすべてに対する解析結果のまとめを踏まえて，Hαフィラメント消失現象とそれに伴う現象の関連性について議論する．

4.1.1. 1999 年 2 月 16 日のイベント（噴出型）

3 節で述べたように，1999 年 2 月 16 日のイベントにおけるフィラメントは噴出型であった．参照点（A 点）に代表されるフィラメントの速度変化は論文 I の図 1 に示されている．求められた速度場によると，フィラメントは中央部分から放出される一方で，2 つの足元は光球面に固定されていた．放出率は 53 ％ であり，中央部分と西側部分の大半は放出された．噴出に引き続いて同じ領域で M3.2 クラスのフレアが発生し，巨大なアーケードが形成された．

4.1.2. 1992 年 11 月 5 日のイベント（噴出型）

1992 年 11 月 5 日，活動領域 NOAA 7332 の東側（S20W17）の静穏領域にて Hα フィラメント噴出が発生し，それに伴いコロナ構造の再配置が起こった．McAllister et al. (1996) は FMT のデータを用いたこのイベントの研究を行った．彼らは消失フィラメントをその形態と運動の観点から研究し，Yohkoh/SXT で観測された巨大

relationships between DBs and CMEs, and also coronal activities and CMEs, we made use of white-light coronagraph images from SOHO/LASCO for events after 1996 May. We checked if CMEs were observed associated with the DB events. The results of the CME association are listed in the last column of table 1.

4. Results and Discussions

4.1. Examples of Hα Disappearing Filaments

In this section, we first present an analysis of five DB events (02/16/99, 11/05/92, 01/19/00, 04/06/00, and 01/28/00 events) to see some details of their velocity evolutions and some typical signatures of coronal activities in soft X-rays, EUV and white light coronagraphs. We then discuss the relationships between Hα DBs and their associated phenomena based on a summary of the results for all 35 events.

4.1.1. The 1999 February 16 event (eruptive)

As we have previously discussed in section 3, the filament of the 02/16/99 event was eruptive. The detailed evolution of the velocity represented by a sample point (point A) of this filament is shown in figure 1 of Paper I. According to the derived velocity fields, the filament was ejected from the middle while its two legs remained anchored to the photosphere. The ejection rate was 53% and the central part and most of the western part were ejected. Subsequently, an M3.2 flare occurred in the same region and a large arcade formation took place.

4.1.2. The 1992 November 5 event (eruptive)

On 1992 November 5, an Hα filament eruption and related coronal restructuring took place in a quiet region to the east (S20W17) of active region NOAA 7332. McAllister et al. (1996) have conducted a study of this event using FMT data. They investigated the morphology and motions of the disappearing filament and compared them with the large arcade

アーケード形成と比較した．彼らはドップラーシフト量の計測ではなくフィラメントの上昇角度を推定することにより視線速度を導出した．このイベントに関するFMT画像は彼らの論文の図2および論文Iの図8にて参照できる．

論文Iでは，このイベントの参照点（A点）を定義し，その速度変化を同論文の図9に示した．三次元速度場からは，A点が消失するまで加速を続け上昇速度が増加し続けることがわかった．フィラメント全体の三次元速度場を見ると，1999年2月16日のイベントと同様にフィラメントが中央部分から放出されたことが示されている．このイベントの放出率は43％であった．

噴出に引き続いて，Hαでは弱いツーリボンフレアが観測され，軟X線観測では巨大アーケードが形成された．このアーケード形成に関するSXTの連続部分像はMcAllister et al. (1996)の図8に示されている．01:20 UTからの「ようこうの夜」(*訳注　日陰) に伴う観測中断後，02:54 UTには発達した軟X線アーケードが初めて観測された．

4.1.3. 2000年1月19日のイベント（噴出型）

2000年1月19日，FMTによってフィラメント噴出が観測された．図3の上側4枚の図はFMTで得られたHαフィラメント消失現象の連続画像であり，活動領域NOAA 8829の南側に位置するフィラメントがその中央部分から消失していく一方で，両方の足元はイベントを通じて観測されている様子が示されている．導出されたこの消失フィラメントの三次元速度場からは，フィラメントの両足部分が光球面に固定されながら中央部分は上方に加速されている様子が見て取れる．このイベントの放出率を計算したところ36％と求められた．

図3の下側2枚の図は00:13:26 UTと01:47:12 UTにSXTで観測された全面像から一部を切り出したものである．これらの画像中からはフィラメントの初期位置であるNOAA 8829の南側でアーケード形成が発生していることがわかる．もう1つの現

Appendix A　Selected Papers Relating to FMT　|　387

formation observed by Yohkoh/SXT. They derived the line-of-sight velocity by assuming a rise angle of the filament, but not by measuring the Doppler shift. The FMT images of this event can be referred to in figure 2 of their article and figure 8 of Paper I.

We also defined a sample point (point A) of this filament in Paper I and showed its velocity evolution in figure 9 therein. The 3-D velocity field of point A shows that it was accelerated and its upward velocity increased until its disappearance. The complete 3-D velocity field of this filament also shows that the filament was ejected from its middle, similarly to the 02/16/99 event. The calculated ejection rate for this event was $\gamma = 43\%$.

Subsequently, a faint two-ribbon flare in Hα and a large arcade formation in soft X-rays took place. A series of SXT partial frame images of this arcade formation are shown in figure 8 of McAllister et al. (1996). After a data gap due to Yohkoh night, from 01:20 UT, new well-developed soft X-ray arcade structures were first observed at 02:54 UT.

4.1.3. The 2000 January 19 event (eruptive)

On 2000 January 19, FMT observed an eruptive filament. Sequential FMT images of this Hα DB are displayed in the four top panels of figure 3, which show a filament located south of the active region NOAA 8829 that started to disappear from the middle part with its eastern and western footpoints visible throughout the event. The derived 3-D velocity field of this disappearing filament reveals that the filament was accelerated upward from its middle, leaving both footpoints tied to the solar surface. The calculated ejection rate for this event is 36%.

The two bottom panels of figure 3 are cutouts of two fullframe images from SXT at 00:13:26 UT and 01:47:12 UT. As can be seen in these images, arcade formation takes place south of NOAA 8829 where the filament was initially located. Another signature is a brightening of the interconnecting loop (ICL) between NOAA 8829 and NOAA 8831 in the

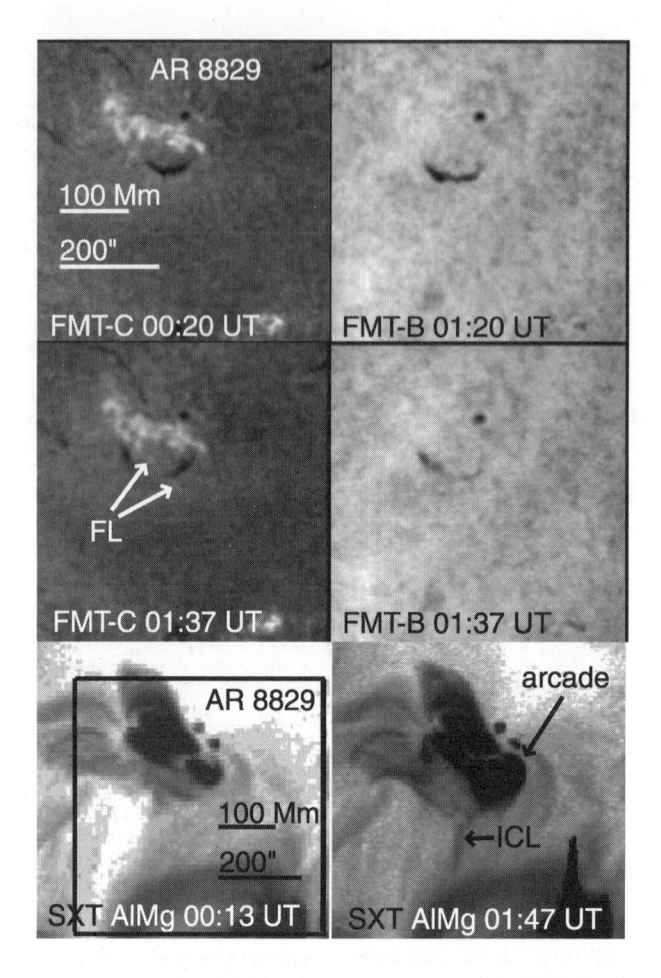

図3 2000年1月19日のイベント．上側4枚の図：FMT観測による活動領域 NOAA 8829 の南側で発生したフィラメント消失現象．フィラメントは中央部分から消失し，01:37 UT には Hα 線中心画像では微弱なリボンフレア（FL）が見られる．下側2枚の図：フィラメント噴出現象に伴う軟X線イベントの前後（00:13 UT および 01:47 UT）における SXT 太陽全面像の一部分．「arcade」は新しく形成されたループが連なっている部分を示し，相互連結ループは「ICL」で示されている．左下の図中には上記 FMT 画像の視野を表す枠が示されている．

Fig. 3. 2000 January 19 event. Four top panels: FMT images showing the DB south of the active region NOAA 8829. The filament disappears from the middle and a faint ribbon flare (FL) is seen on the Hα line center image at 01:37 UT. The two bottom panels: Two partial SXT full-frame images before (00:13 UT) and after (01:47 UT) the onset of the associated soft X-ray event. The "arcade" indicates a newly formed arcade of loops and the interconnecting loop is denoted by the abbreviation "ICL" in the figure. The box in the left-bottom panel shows the field of view of the FMT images.

象として，NOAA 8829 と南半球にある NOAA 8831 とを繋ぐ相互連結ループ（ICL）の増光があげられる．これらの現象は EIT によっても NOAA 8829 の南東側の 2 つの減光構造として観測された．このコロナ活動は「A」タイプと判定された．Hα フィラメントが完全に消失した 1 時間後の 02:54 UT には，LASCO C2 コロナグラフで不規則な波面をもつ拡散型 CME が観測された．CME の放出角度は北極軸から 166° であり，波面の広がりは 68° であった．CME の投影速度は約 800 km/s であり，C2 の視野内を南向きに進行した．

4.1.4. 2000 年 4 月 6 日のイベント（準噴出型）

2000 年 4 月 6 日に NOAA 8945 の南西側 100 Mm の静穏領域にてフィラメント消失現象が発生した．この Hα フィラメントの FMT による観測は図 4 の上側 4 枚に示されている．このフィラメントの北側半分が Hα 線中心画像で薄れるに従い，青色側ウイング画像よりも赤色側ウイング画像においてフィラメントが顕著に観測された．このことはイベントを通じてレッドシフトが優勢であることを示す．フィラメント全体では上昇運動はほとんど観測されず，後半の段階では強い下降運動が見られる．放出率は 0 ％であった．

フィラメントの北側部分は上昇運動を示すことなく南東方向へ移動した．フィラメントの北西から南東への移動は 03:59 UT に NOAA 8945 で発生した C1.4 フレアによるものである．フレアが開始すると，EIT 画像ではフィラメントに沿って微かで細長い増光現象が観測された（図 4 の下側 2 枚）．細長い増光部分は NOAA 8945 から南東方向に伸び 04:47 UT までにはフィラメント全体を覆った．この増光部分の見かけの移動はフィラメントの運動と一致しているため，フィラメントがエネルギー入射によって活性化させられ，極紫外線でもプラズマを増光させた可能性がある．それゆえ，この現象は NOAA 8945 からのエネルギー入射によってフィラメントが加熱された証拠とも解釈できる．この考えは熱的

southern hemisphere. These signatures were also observed by EIT with two dimming structures south-east of NOAA 8829. We classified this coronal event as type "A". A diffuse CME with a ragged front occurred an hour after a complete disappearance of the Hα filament, first appearing at 02:54 UT, in the LASCO C2 coronagraph. Its central polar angle was 166° from the solar north and the angular width was 68°. The projected speed of the CME was about 800 km s^{-1} and it traveled southward through the C2 field of view.

4.1.4. The 2000 April 6 event (quasi-eruptive)

A filament disappearance took place in a quiet region 100 Mm to the south-west of active region NOAA 8945 on 2000 April 6. The FMT observations of this Hα filament are displayed in the four top panels of figure 4. As the northern half of the filament fades in the Hα line center, the filament appears to be more prominent on the red wing images than the blue wing images, indicating that the red shift is dominant throughout the event. The whole filament shows little upward motion, but a significant receding motion in the later phase. The ejection rate for this event is 0%.

The northern part of the filament travels toward the south-east without showing any significant rising motion. The motion of the filament from the north-west to the south-east can be attributed to the C1.4 flare activity, which peaked at 03:59 UT in NOAA 8945. Following the start of this flare, a faint and long brightening along the dark filament is observed in the EIT images (the two bottom panels of figure 4). This long and slender brightening proceeds from NOAA 8945 toward the south-east and covers the entire filament by 04:47 UT. Since this apparent motion of the brightening is consistent with the motion of the filament, the filament may be activated by the energy input, which also brightens the plasma in EUV. This, therefore, can be interpreted as being a signature of heating of the filament due to the energy input

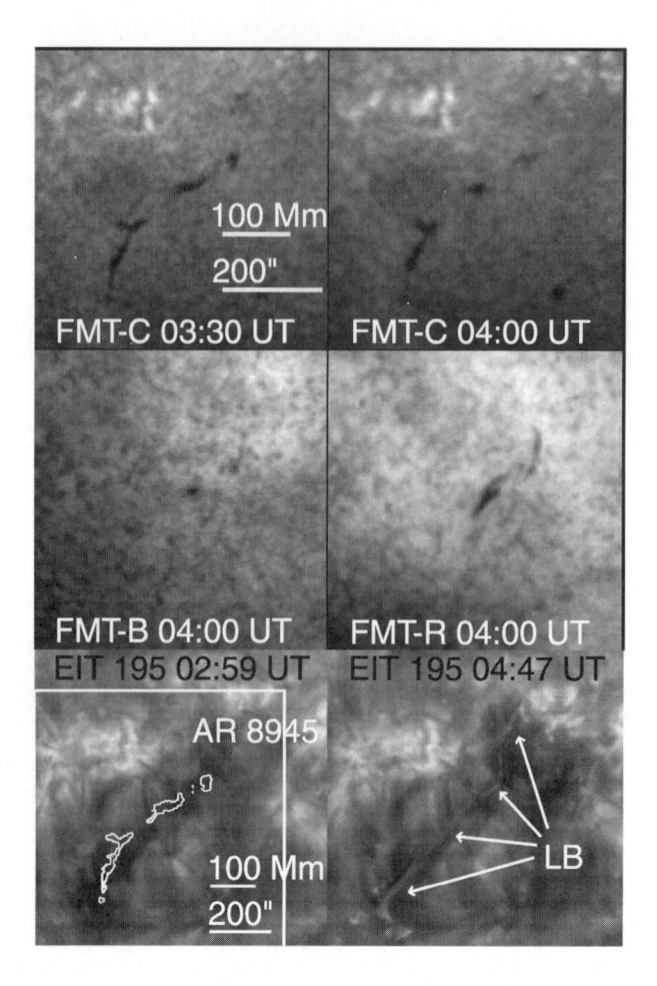

図4 2000年4月6日のイベント．上側4枚の図：FMT観測による消失前のフィラメント活動の様子．フィラメントの北側半分は南東方向に移動した後消失した．噴出型イベントとは異なり，フィラメントは青色側ウイング画像よりも赤色側ウイング画像において顕著に見られる．下側2枚の図：EIT 195画像ではHαフィラメント消失現象から引き続くコロナ活動が見られる．微弱な細長い増光現象がフィラメントの位置をすべて覆う形で発生している（04:47 UT）．白い枠は上記FMT画像の視野を表し，Hαフィラメントの初期位置は白い輪郭で示されている．

Fig. 4. 2000 April 6 event. Four top panels: Series of partial FMT images showing the filament activities before its disappearance. The northern half of the filament shifted its position toward the south-east and disappeared. Unlike eruptive events, the filament appears darker in the red wing images than in the blue wing images. The two bottom panels: EIT 195 images showing the coronal activities which followed the Hα DB event. The faint and filamentary brightening is seen to cover the entire filament region (04:47 UT). The white box and contour plot indicate the field of view of the FMT images and the initial location of the Hα filament, respectively.

フィラメント消失現象と密接に関連しているが，我々はこの消失現象は熱によるものではないと結論付けた．今回のフィラメントの大部分は極紫外線を発さないのに加え，Hαで観測され続けたためである．このタイプのコロナ活動では，Hαフィラメントの領域では新たなアーケード形成や大規模なコロナ構造の再配置が観測されなかったために「LB」と分類された．このイベントに関してはSXTとLASCOのデータは利用できなかった．

4.1.5. 2000年1月28日のイベント（準噴出型）

準噴出型フィラメントのほかの例には2000年1月28日に太陽面の南西部分のほぼ中心（S20W28）で発生したイベントがある．図5の上側4枚にフレア監視望遠鏡で観測された太陽の部分像の連続画像を示す．活動領域NOAA8841の西側に位置するダークフィラメントが北西に向けて分裂していく様子が，最初は青色側ウイング画像で05:42 UTに観測される．後半段階の06:20 UT以降ではHα線中心画像と赤色側ウイング画像にて再度観測される．コンパクトなツーリボンフレアがフィラメントの東側の足元近傍で発生している様子が極紫外線観測によっても確認できる（図5の下側2枚）．導出された三次元速度場によると，フィラメントは最初上昇すると同時に北西方向に移動している．しかしその後05:49 UT以降は減速が始まり，初期位置からおよそ120 Mm離れた太陽表面上に落下している．

このイベントはGOESの軟X線強度データで最大B4.4クラスの非常に小規模な軟X線フレアを伴う．EIT 195による2枚の太陽部分像が図5の下側2枚に示されている．噴出前にはフィラメントは2つの活動領域NOAA 8841と8845に挟まれた暗い領域の南側に存在している．05:57 UTにフィラメントが上昇運動を示した際にHαにおける小規模フレアが発生し，EIT 195の画像では関連してコンパクトな増光現象と微弱なループ増光現象（L1およびL2）が現れた．L1はフレアの発生場所から南

from NOAA 8945. Although this idea is closely related to that of DBt, we conclude that this disappearance is not of thermal origin; most of the filament does not light up in EUV, and remains visible in Hα. We classified this type of coronal activity as "LB", since no formation of a new arcade of loops and no global restructuring of the corona took place at the site of the Hα filament. We have no SXT and LASCO data for this event.

4.1.5. The 2000 January 28 event (quasi-eruptive)

Another example of a quasi-eruptive filament is the 01/28/00 event, which took place near to the center of the south-west quadrant of the solar disk (S28W20). Series of partial FMT images are shown in the four top panels of figure 5. The dark filament to the west of active region NOAA 8841 disrupts toward the north-west and appears first in the blue wing at 05:42 UT. It subsequently reappears in the Hα line center and red wing, in the later stages, from 06:20 UT. A compact two-ribbon flare occurred near the eastern footpoint of the filament which is also seen in the EUV images (the two bottom panels of the same figure). According to the derived 3-D velocity field of this filament, the filament initially lifts off with a tangential motion toward the north-west. However, it is decelerated after 05:49 UT and begins to flow back onto the solar surface some 120 Mm away from its original position.

This event is accompanied by a very small and weak soft X-ray event with a peak of B4.4 in the integrated soft X-ray flux data from GOES. Two EIT 195 partial frame images are displayed in the bottom panels of figure 5. Before the eruption, the filament is lying to the south of a dark region between two active regions, NOAA 8841 and 8845. The EIT 195 image at 05:57 UT, when the filament is showing ascending motion, shows a compact brightening corresponding to the compact flare in Hα and some faint loop brightenings (L1 and L2). L1 is an Ω shaped structure extending southward

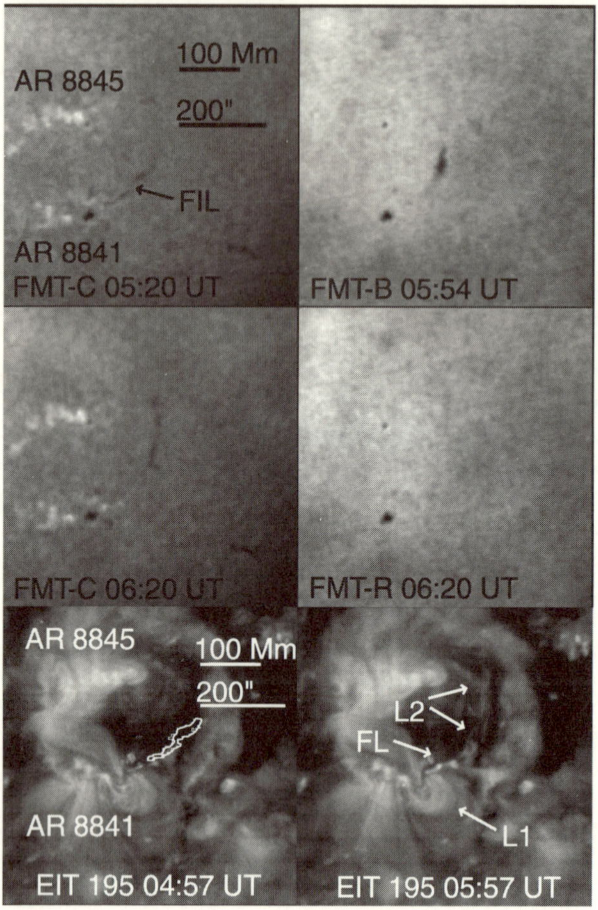

図5 2000年1月28日のイベント．上側4枚の図：FMTによる2000年1月28日のフィラメント消失現象の観測結果．活動領域 NOAA 8841 の西側に位置するフィラメント（FIL）が北西方向に放出されるが，その後減速される．B4.4 クラスの X 線強度を示すコンパクトな Hα フレアがフィラメント消失現象に伴って発生した．下側2枚の図：2000年1月28日のEIT 195 画像には，NOAA 8841 におけるコンパクトなフレア（FL）の発生後にループの増光現象（L1 および L2）が現れる．L1 は既存のループの増光現象であり，L2 は消失する Hα フィラメントとほぼ一致する．Hα フィラメントの初期位置は白い輪郭で示されている．

Fig. 5. 2000 January 28 event. Four top panels: FMT observations of the 01/28/00 DB event. A filament (FIL) west of the active region NOAA 8841 is ejected toward the north-west, but decelerated afterwards. A compact Hα flare with B4.1 X-ray flux occurred coincident with the DB. The two bottom panels: EIT 195 images of the 01/28/00 event showing loop brightenings (L1 and L2), which took place after the compact flare (FL) at NOAA 8841. L1 consists of brightenings of preexisting loops, and L2 is almost coincident with the disappearing Hα filament. The white contour indicates the initial location of the Hα filament.

方へ伸びるΩ型の構造をもち，膨張は見られなかったことからこれは既存の活動領域ループの増光によるものと考えられる。L2 は新しく増光したループであり，分裂したフィラメントの移動経路と一致する。この現象は 2000 年 4 月 6 日のイベントにおける EIT 195 の微弱なフィラメント型の増光現象と非常によく似ているために，フレアの発生場所からのエネルギー入射によってフィラメント中の磁気ループの一部分が増光したものとみなすことができる。これらのコロナ活動は EIT や SXT 画像において大規模なコロナ構造の再配置が見られなかったことから「LB」タイプと分類された。LASCO C2 のデータからはこのフィラメント消失現象後に CME が発生しなかったことが確認された。

以上の 5 つの例で示されているように，フィラメント消失現象に伴ってフィラメントが惑星間空間に放出されるか否かを判定する際には，消失直前の速度の時間変化がきわめて重要である。今回のような解析では，消失フィラメントの運動方向と測定可能な速度の最大値の関係に注意する必要がある。FMT 機器の観測波長域は有限なため，フィラメントが視線方向に沿って 60-90 km/s 以上の速度で運動した場合にはフィラメントは消失する。それゆえ，60-90 km/s よりも大きな視線速度をもつ運動は観測することができない。一方，空の平面上における運動では 90 km/s 以上の速度をもつフィラメントであっても FMT を用いて観測することが可能である。この場合，フィラメントの速度が 90 km/s を超過した後に再び減速した場合でも正しく準噴出型と判定される。しかし，視線方向の運動であった場合は，減速段階が観測されないためにこのフィラメントは噴出型であると判定される。上記のような誤判定は，フィラメントが活性化して分裂した場合にしばしば極めて変形するために起こりうる。それゆえ，より精度の高い分類を可能にするためには，FMT の観測波長域を現在の ±0.8 Å から，少なくとも CME の速度の下限（〜 100 km/s：e.g., Gosling et al. 1976）より十分に大きい 150 km/s の視線速度を計測可能

from the flare site, which shows little expansion, and thus may be due to brightenings of the preexisting AR loops. L2 is a newly brightened loop, which coincides with the path of the disrupted filament. This can be interpreted as a brightening of a part of the magnetic loops in the filament due to the energy input from the flare site, since it is very similar to the faint filamentary brightening in the EIT 195 images of the 04/06/00 event. We classified these coronal signatures as type "LB", because no global restructuring occurred in the EIT and SXT images. LASCO C2 data confirm the absence of a CME following this DB.

As shown in the five examples above, in judging whether the filament is ejected into interplanetary space during a DB event or not, the evolution of its velocity just before the disappearance is crucial. In this type of analysis, we need to pay attention to the following relation between the direction of motion and the detectable maximum velocity of a disappearing filament. Due to the limited wavelength coverage of the FMT instruments, a filament would disappear if its line-ofsight velocity exceeds some 60-90 $km s^{-1}$. We, therefore, can not observe its evolution with its line-of-sight velocity larger than 60-90 $km s^{-1}$. For example, if the motion of a filament element is on the plane of the sky, we can observe it with FMT even its velocity exceeds 90 $km s^{-1}$. In this case, when the element is decelerated after its velocity reaches more than 90 $km s^{-1}$, this element is judged as quasi-eruptive. However, if its motion were along the line of sight, this element would be judged as being eruptive because one could not observe the deceleration. This kind of misclassification possibly happens because the motion of disrupted filaments is often significantly altered in the early stages of their activation. Therefore, for more accurate classification, it is highly desirable to extend the wavelength coverage of the filters of the FMT from ±0.8Å to, at least, ±2.0Å in the future in order to measure the line-ofsight velocity up to 150 $km s^{-1}$, which is well above the lowest speed of CMEs ($\sim$ 100 $km s^{-1}$: e.g., Gosling et al. 1976).

である±2.0Åまで広げることが強く望まれる．

4.2. フィラメント消失現象とコロナ変動の関係

上記の5例のフィラメント消失現象より，噴出型フィラメント消失現象イベントは以下のように特徴付けられる．(i) 噴出型フィラメントは赤色側ウイングよりも青色側ウイングで顕著に観測される．(ii) Hαではフレアリボンがフィラメントの真下に広がる．(iii) 軟X線や極紫外線ではアーケード形成および減光現象が観測され，コロナ構造の再配置が発生する．一方，準噴出型フィラメント消失現象イベントは以下のように特徴付けられる．(i) 低速の一時的な上昇運動が見られ，その後太陽表面に向かう下降運動が発生する．(ii) Hαでのフレアや増光現象はコンパクトであるか，フィラメントから離れた位置に発生する．(iii) 軟X線や極紫外線では局所的なフレアおよびループ増光現象が見られるが，コロナ磁場は開放しない．(iv) EIT 195 画像ではフィラメントチャンネルに沿って微弱で細長い増光が見られる．これらの特徴を元に，Hαフィラメント消失現象のタイプとコロナ活動およびCMEとの関連性を，35個のフィラメント消失現象の解析結果を用いて示す．

4.2. The Relationship between DBs and Coronal Changes

From these five examples of DBs shown above, we notice that the characteristics of eruptive DB events are as follows: (i) an eruptive filament appears strongly in the blue wing with fewer signatures in the red wing, (ii) the flare ribbons in Hα spread widely beneath the filament, and (iii) the formation of arcades and/or dimmings, indicating global restructuring of the corona, are observed in soft X-rays and the EUV. Conversely, quasi-eruptive DB events can be characterized by (i) a temporary rising motion with small velocities, followed by a downward motion to the surface; (ii) Hα flares or brightenings, which are compact or far from the filament's site; (iii) confined flare and/or loop brightenings without any signatures of the opening up of coronal magnetic fields in soft X-rays or the EUV; and (iv) faint filamentary brightenings along the filament channel in EIT 195 images. With these signatures in mind, we now take a look at the relationships between the types of Hα DBs, coronal activities and CME associations using the results for all 35 DB events.

4.2.1. フィラメント消失現象と SXT および EIT で観測されたコロナ変動

Yohkoh/SXTによるコロナ活動の観測結果を分類した結果，22個の「A」タイプおよび6個の「LB」タイプのイベントが確認された．SOHO/EITを用いた10個のフィラメント消失現象の観測では，7個が「A」タイプ，および3個が「LB」タイプのイベントに分類された．EIT観測による「A」タイプのイベント中，4イベントがフィラメント噴出現象（「FE」），6イベントが減光現象（「D」），および2イベントがEIT波（「EW」）と関連している．SXTとEITの両方の観測データを利用できた7イベント中では，両者の結果に相違は存在しなかった．これらの結果はまとめて表1に示

4.2.1. DBs vs. coronal changes observed by SXT and EIT

As a result of the classification of coronal activities observed by Yohkoh/SXT, we have 22 "A" and 6 "LB" type events. Also, SOHO/EIT observations of 10 DB events reveal that 7 events are "A" type and 3 events are "LB" type. Among the EIT "A" type events, 4 events are associated with filament eruptions ("FE"), 6 events with dimmings ("D") and 2 events with EIT waves ("EW"). There are 7 events in which we have both SXT and EIT data, and their results show no disagreement in their types of activity. These results are collected and listed in table 1. "LB" type events can be categorized into two groups: (i) intense but very compact (less than 50 Mm) AR

されている．「LB」タイプのイベントは2つのグループに分類することができる：(i) Green et al. (2002) で研究された局所的なフレアと同様の，光度が高くコンパクト (50 Mm 以下) な活動領域フレア，および (ii) Hαフィラメントの運動と一致して進行する一時的で微弱なフィラメント状の増光現象（2000年4月6日のイベント等）．2番目のケースの増光現象はフィラメントに沿った磁気ループ中のプラズマの加熱によるものと判断される．エネルギーやトリガーとなる不安定性はフィラメント近傍におけるフレア活動によって与えられると考えられる．例として，2000年4月6日の微弱なフィラメント状増光現象は近傍の活動領域 NOAA 8945 で発生した C1.4 クラスのフレアと直接結合している．

フィラメント消失現象のタイプと SXT および EIT で観測されたコロナ活動の関係は表2にまとめられている．31個のイベントが SXT または EIT で観測された．噴出型フィラメントと判定された22個のイベントすべてがコロナアーケード形成（「A」）と関連している．9個の準噴出型フィ

flare brightenings which have the same properties as the confined flares studied by Green et al. (2002), and (ii) transient and faint filamentary brightenings which proceed coincidently with the position and apparent motions of the Hα filaments (e.g., the 04/06/00 event). In the second case, the brightening is considered to occur due to a heating of plasma in the magnetic loop along the filament. The energy or a triggering instability seems to be supplied by flare activity near the filament site. For example, the faint filamentary brightening of the 04/06/00 event is directly connected to a C1.4 flare in the nearby active region NOAA 8945.

The relationship between types of DBs and the coronal activities observed by both SXT and EIT is summarized in table 2. We have 31 events which were well observed by SXT and/or EIT. All 22 eruptive filaments are associated with coronal arcade formations (abbreviated to "A"). Among the 9 quasi-eruptive filaments, 1 event is accompanied by arcade formation, while the other 8 events are followed by localized brightenings, or confined flares without any global changes in the coronal structures (abbreviated to

表2 フィラメント消失現象のタイプと SXT および EIT で観測されたコロナ活動の関係．

フィラメント消失現象のタイプ	A	LB	計
噴出型	22 (71%)	0 (0%)	22
準噴出型	1 (3%)	8 (26%)	9
	23	8	31

Table 2. Relation between the types of DBs and coronal signatures observed by SXT and EIT.

DB type	A	LB	Total
Eruptive	22 (71%)	0 (0%)	22
Quasi-eruptive	1 (3%)	8 (26%)	9
	23	8	31

ラメント中，1 イベントがアーケード形成を伴ったが，残りの 8 イベントはコロナ構造を変化させない局所的な増光現象や小規模フレア（「LB」）を伴った．これらの結果から，噴出型フィラメントは常に軟 X 線や極端紫外線における明るいアーケード形成と関連する一方で，コロナ磁場を開放させないフレアや微弱なフィラメント状増光現象は噴出型フィラメントには見られないことがわかる．31 個中 30 個（97 ％）のイベントがこのシナリオを支持している．

1 つの例外が 1998 年 9 月 5 日のイベントであり，これは Hα 観測では準噴出型フィラメントであったが，軟 X 線では大規模なアーケード形成が観測された．Hα データの解析から考察すると，このフィラメントは中央部分から上昇し 35 Mm 前後の高度に達した後にフィラメント全体が 2 つに分裂し，太陽表面へと流れ落ちた．FMT 画像からは 200 Mm 近くのツーリボンフレアが発生したことが確認された．このツーリボンフレアに伴って，軟 X 線では巨大なアーケード（≥ 350 Mm）が形成された．これらの Hα によるツーリボンフレアと軟 X 線によるアーケードのサイズはフィラメントのサイズ（〜 70 Mm）と比較して極めて大きいため，Hα 線では観測されない磁束管の噴出が起こった可能性がある．この仮説を支持する観測的証拠として，フィラメントチャンネルが 450 Mm 以上の長さに渡って伸び，太陽面中心付近に存在する活動領域 NOAA 8323 の南側に位置するフィラメントと今回噴出したフィラメントとを繋いでいることがあげられる．このイベントに関しては EIT や LASCO のデータは利用できず，残念ながら放出現象が発生したのか確認することはできない．しかし，噴出型フィラメントとコロナアーケード形成および準噴出型フィラメントと局所的増光現象との間には強い一対一対応（97 ％）があると結論付けられる．

4.2.2. X 線の最大強度

30 個のフィラメント消失現象について GOES データから軟 X 線（1-8Å）の積算強度を調べた結果，得られた最大強度を表

"LB"). These results imply a scenario in which eruptive filaments are always associated with bright arcade formations in soft X-rays and the EUV, and that flares and faint filamentary brightenings, as they do not show any evidence of field opening in soft X-rays and EUV, are unlikely to be associated with filament eruptions. Thirty out of the 31 events (97%) are in positive correlation with this scenario.

One exception is the 09/05/98 event, which is quasi-eruptive in Hα, but with a large arcade formation in soft X-rays. According to our analysis of Hα data, after this filament lifted off from its middle part and reached some 35 Mm in height, the whole filament broke into fragments and became streamers flowing back to the surface. FMT images revealed a large two-ribbon flare with a length of nearly 200 Mm. Corresponding to this two-ribbon flare, a large arcade (≥ 350 Mm) formed in soft X-rays. Since the sizes of the Hα two-ribbon flare and the arcade in soft X-rays were so large compared to the length of the filament (∼ 70 Mm), it is possible that the eruption of a flux rope, which was invisible in the Hα line, might have taken place. One observational piece of evidence to support this idea is the presence of a long filament channel that extended over more than 450 Mm, connecting the filament and another filament south of the active region NOAA 8323 near to the disk center. Since we have no EIT and LASCO data for this event, we cannot check if any ejections were seen, unfortunately. Nonetheless, we can conclude that there is a strong one-to-one correspondence (97%) between eruptive filaments and coronal arcade formations, and between quasi-eruptive filaments and localized brightenings.

4.2.2. X-ray peak flux

We checked the integrated soft X-ray flux (1-8Å) from GOES data and list the peak flux for 30 events in the seventh

1の7列目に表示する．軟Ｘ線のピーク強度とフィラメント消失現象のタイプの関係は以下のようにまとめられる．(i) M1.0クラス以上のピーク強度をもつイベントはすべて噴出型である．(ii) 噴出型および準噴出型の平均最大強度はそれぞれ C6.0 および C1.3 である．(iii) 噴出型イベントにおけるピーク強度の最小値は B1.2（1997年10月20日のイベント）であり，準噴出型イベントにおけるピーク強度の最大値は C6.2（1999年6月1日のイベント）であった．これらの3つの結果より，先行研究 (e.g., Munro et al.1979) にあるように準噴出型よりも噴出型イベントの方がより大きなＸ線強度イベントを伴う傾向にあることが窺える．しかし，噴出型イベントであっても常に準噴出型よりも大きな軟Ｘ線活動を示すとは限らない．我々は比較的大きな軟Ｘ線強度のイベントを伴う準噴出型イベントの共通点について調べた．準噴出型イベントのうち，軟Ｘ線強度がC1.0を超えたイベントは4個あり，その中で3イベントは活動領域にて局所的なフレアを伴い，残りの1イベントは2000年4月6日のもので，すでに解説したように (4.1.4節参照) フィラメントから 100 Mm ほど離れた領域において長寿命フレアとアーケード形成が発生した．このことより，たとえ大きな軟Ｘ線強度がフィラメントの場所で観測された場合でもフレアが底部コロナに限定されている場合は，このイベントは準噴出型であると見なせるといえる．

4.2.3. CME との関連性

SOHO/LASCO のデータが利用できた15個のイベントに対して CME との関連性を調べたものを表3に示す．関連性の精度を高めるために，LASCO C2 の視野における CME の速度を求めた上で CME の発生時刻を計算し，それがＨαにおけるフィラメント消失現象の開始時刻から90分以内であるか否かの判定を行った．90分の間に複数のイベントが発生したために関連性が不確定である場合には，EIT 195 の差分画像を用いて真に関連性があるかを確認した．結果を表3に示す．8例の噴出型イベ

column of table 1. The relation between the peak flux in soft X-rays and the types of DB can be summarized as follows: (i) all events with a peak flux above M1.0 are eruptive, (ii) the averaged flux for eruptive and quasi-eruptive events are C6.0 and C1.3, respectively, (iii) the lowest peak flux for eruptive events is B1.2 (10/20/97 event) while the highest flux for quasieruptive events is C6.2 (06/01/99 event). These three signatures indicate that eruptive events tend to be associated with higher X-ray intensity events than quasi-eruptive events, as previous observational studies have shown (e.g.,Munro et al. 1979). Eruptive events, however, are not always associated with higher soft X-ray activities than quasi-eruptive events. We investigated a common signature in quasi-eruptive events with rather high soft X-ray intensity events. There are four quasi-eruptive events with soft X-ray flux greater than C1.0, and we found that three events correspond to confined flares in active regions, and one is the 04/06/00 event in which a long-duration flare with an arcade formation took place some 100 Mm away from the filament site as described above (see subsubsection 4.1.4). This implies to us that even an event with high-intensity soft X-ray activity at the filament site is likely to be quasi-eruptive if the flare is confined to the lower corona.

4.2.3. CME associations

The CME association results are listed for 15 events, for which we give SOHO/LASCO data in table 3. In order to clearly establish this association, we calculated the onset time of a CME according to its speed in the LASCO C2 field of view, and checked if it is within a time range of 90 minutes from the start of the DB in Hα. Furthermore, if the association is still unclear due to more than two solar events occurring in the specified time range, we made use of EIT 195 running-differential images to ensure that the association was real. The results are summarized in table

表3 CMEとフィラメント消失現象のタイプとSXTおよびEITで観測されたコロナ活動の関連性.*

フィラメント消失現象のタイプ（SXTおよびEIT）	有	無	計
噴出型（A）	8(53%)	0(0%)	8
準噴出型（LB）	0(0%)	7(47%)	7
	8	7	15

*すべての噴出型および準噴出型フィラメントはそれぞれコロナ活動の「A」タイプおよび「LB」タイプにて示されている．

Table 3. Associations of CMEs to eruptive and quasi-eruptive DBs.*

DB type (SXT & EIT)	Yes	No	Total
Eruptive (A)	8 (53%)	0 (0%)	8
Quasi-eruptive (LB)	0 (0%)	7 (47%)	7
	8	7	15

* All eruptive and quasi-eruptive filaments listed here are of "A" and "LB" type in coronal signatures, respectively.

ントはすべてCMEと関連しているのに対し，7例の準噴出型イベントはいずれもCMEとの関連はない．さらに，軟X線および極紫外線により観測される高いコロナ活動は，CMEが随伴する場合と完全に一致している．一方，フィラメント消失に伴う軟X線のピーク強度と，CMEの随伴の有無とは有意な関連性は見られない．

5. 要約と結論

飛騨天文台のフレア監視望遠鏡（FMT）のHα線中心，青色側ウイングおよび赤色側ウイング画像を用いて35個の太陽フィラメント消失現象の三次元速度場を導出し，フィラメント消失現象とそれに伴うコロナ活動の因果関係を調べた．フィラメント消失現象の三次元速度場の計測方法と5つのフィラメント消失現象に適用して得られた速度場は我々の先行研究（論文I）に詳細に述べられている．

速度場のデータの解析結果，35イベント中「低速溶解型フィラメント消失現象」

3. All 8 eruptive events are associated with CMEs and all 7 quasi-eruptive events have no association with CMEs. The coronal activities in soft X-ray and EUV are also in perfect agreement with the CME association. There is, however, no meaningful correlation between the peak soft X-ray flux and the CME association.

5. Summary and Conclusions

Using the Hα line center, blue-wing and red-wing images of the Flare Monitoring Telescope (FMT) at Hida Observatory, we derived the 3-D velocity field of 35 disappearing solar filaments (DBs) and studied the causal relationship between DBs and the associated coronal activities. The methods used to obtain the 3-D velocity field of a DB and the derived velocity fields of the five events, which are also shown in this paper, are presented in detail in our previous study (Paper I).

Using the velocity data, we found no "slow-dissolution" and "thermal DB" in

および「熱的フィラメント消失現象」は観測されなかった.「低速溶解型」はフィラメントのプラズマの供給と損失のバランスが崩れた際に発生する. このタイムスケールは通常, 今回観測された35イベントのタイムスケール（最大6時間）よりも大きく（数時間から数日）, このため「低速溶解型」はフィラメント選択から漏れた可能性がある. 熱的フィラメント消失現象に関していえば, 静穏でありながらHαでフィラメントが消失した後に極紫外線で観測されるようなイベントは見られなかった. このことより, 熱的フィラメント消失現象は稀な現象であると結論付けられる. しかし一方で, 以下のような現象が発生したことに注意する必要がある. まず, フレアを伴った準噴出型フィラメント消失現象の発生後に, 微弱なフィラメント状の増光現象が極紫外線および軟X線で観測されたが, これはフィラメントのプラズマの加熱現象の可能性がある. 次に, FMTのピクセルサイズよりも小さいスケールで加熱が起こった結果フィラメントが消失するという現象はFMTでは観測されなかった可能性がある.

噴出型と準噴出型フィラメントの速度は底部コロナでは互いに同程度である（Gilbert et al. 2000）. それゆえ, 両者を判別する手法を確立させることが重要である. 本研究では, フィラメント中のブロブが惑星間空間に放出されたことを示すために, 「上昇速度が正で, FMTの-0.8Åの画像視野から消えるまで加速を続ける」という基準を設けた. これは, 太陽の重力から逃れるためには, 噴出されたフィラメントは-0.8Åの画像で観測されなくなった以降も加速する必要があるためである. しかし, この基準は視線上に異なる運動を示す部分が複数存在する場合には十分ではない. この場合には, たとえ一部が加速されFMTの画像から消失したとしてもほかの部分が視線上で観測されることとなる. その場合, 上昇速度の時間変化は急激な減少を示し, この視線上の物体は放出されなかったと判定される. フィラメントはしばしば噴出中には変形するために, 上記のような現象が

the 35 DB events. The "slow-dissolution" is caused because of the imbalance between the supply and loss of the filament's plasma. Since the time scale is usually long (from several hours to days) compared to the time scales of our 35 events (six hours, at most), "slow-dissolution" events might not be selected in this data set. As for DBt events, we have no event which exhibits little motion and appearance in EUV following its disappearance in Hα. This leads us to the conclusion that DBt is a rare phenomenon. At the same time, however, we also have to keep in mind the following issues. First, faint filamentary brightenings in EUV and soft X-rays found after some quasi-eruptive DBs with flare activities may be one of the signatures of heating of the filament plasma. Second, it is possible that filament disappearances due to heating on spatial scales smaller than the pixel size of the FMT data are not observed.

The amplitudes of the velocities of eruptive and quasieruptive filaments are similar to each other in the low corona (Gilbert et al. 2000). It is, therefore, important to establish a method to distinguish these two different types. In this study, we employed the following criterion: "the upward velocity must be positive and must increase before disappearance from the -0.8 Å filter passband" for a blob to be ejected into interplanetary space. This is because eruptive filaments should be further accelerated after their disappearance from the -0.8Å filter passband to escape from the solar gravity. This criterion, however, is not sufficient when there are more than two structures that exhibit different motions on the same line-of-sight. In this case, even if one of them is accelerated and ejected so that it disappears from the FMT images, the other structure remains visible on the line-of-sight. Therefore, the evolution of its upward velocity would show a sudden drop to a smaller velocity, implying that this column is not ejected. Since a filament is often distorted during its eruption, such a case sometimes occurs. In order to avoid this kind of misclassification, we employed another criterion, as follows: "if the upward velocity shows a sudden deceleration which is more than

時々起こりうる．この種の誤判定を避けるために，「上昇速度が太陽の重力加速度（～0.27 km/s²）よりも大きな減速を示し，その後の上昇速度が0以上であった場合は，視線上の複数の物体の一部分が放出されたとみなす」という基準を設けた．この基準は，最大減速度は自由落下から求められるという合理的な根拠に基づいている．これらの基準を用いて，フィラメントの各部分が惑星間空間に放出されたか否かを判定した．また，フィラメント上のピクセル数と「放出された」ピクセル数の比である放出率（γ）を計算した．12イベントは放出率が0であり，「準噴出型」フィラメント消失現象と分類され，残り23イベントは「噴出型」と判定された．噴出型フィラメント消失現象はすべて$\gamma \geq 13\%$と有意な放出率を示した．

太陽面上では視線方向に沿った構造の重なりが少ないため，軟X線および極紫外線によるコロナ構造とHαフィラメント活動とを比較してフィラメント噴出現象に特徴的な状況を解明することが容易にかつ正確に行える．Yohkoh/SXTおよびSOHO/EITのデータを用いて，22個のHα噴出型イベントと9個の準噴出型イベントがそれぞれコロナにどのような影響を及ぼしたか調べた．噴出型イベントに関しては，新しいアーケード形成や減光現象など，軟X線および極紫外線においてコロナの大規模な再配置が付随することが明らかになった．9個の準噴出型イベントのうち8個は，コロナの大規模な変動は見られなかった．準噴出型フィラメントに対するコロナ活動の共通点として，(i)活動領域中の局地的なフレア，および(ii)近傍のフレア活動によってエネルギーやトリガーとなる不安定性を与えられた結果発生したと考えられるフィラメントの位置や運動方向に一致する微弱なフィラメント状増光現象，があげられる．

Hαフィラメント消失現象とCMEとの関連性も調査された．SOHO/LASCOのデータが利用できる期間中には8個の噴出型および7個の準噴出型イベントが観測された．8個の噴出型イベントすべてが

the solar gravitational acceleration value (~ 0.27 km s^{-2}), and the subsequent upward velocity is not zero, then one of the structures on the same line-of-sight has been ejected". This is based on the reasonable assumption that the maximum deceleration is due to the free-fall motion. We judged whether each element of the filament was ejected into interplanetary space or remained in the solar atmosphere, with these criteria. We also calculated the ejection rate (γ), which is the ratio of the number of "ejected" pixels to the total number of pixels occupied by the filament. Twelve events have a zero "ejection rate" and are classified as "quasi-eruptive" DBs; the other 23 events are classified as "eruptive" DBs. All of the eruptive DBs have significant ejection rates of $\gamma \geq 13\%$.

Since many structures do not overlap along the line of sight on the solar disk, we are able to easily and precisely compare the coronal structures in soft X-rays and/or EUV with an Hα filament activity and investigate the preferential conditions for filament eruption. Using Yohkoh/SXT and SOHO/EIT data, we examined the coronal response to 22 Hα eruptive events and 9 quasi-eruptive events. For eruptive events, we found that they were all associated with global restructuring of nearby coronal structures, such as new arcade formation and dimmings in soft X-rays and EUV. We found no global coronal changes in 8 of the 9 quasi-eruptive events. The common signatures of the coronal activities for quasi-eruptive filaments are: (i) confined flare in an active region, and/or (ii) faint filamentary brightenings coincident with the location and direction of filament motion probably due to the energy or a triggering instability which seems to be supplied by nearby flare activity.

The CME association with Hα DBs was also studied. We had 8 eruptive DB events and 7 quasi-eruptive events, for which SOHO/LASCO data were available. We found that all 8 eruptive DBs were accompanied by CMEs, while no CMEs followed any of the 7 quasi-eruptive events. In addition, all of the coronal activities corresponding to the 8 eruptive events are "A" type and are associated with CMEs, while those associated with

CMEを伴い，7個の準噴出型イベントはすべてCMEとは関連性がなかった．さらに，8個の噴出型イベントに関連するコロナ活動はすべて「A」タイプでCMEを伴った一方で，7個の準噴出型イベントではコロナ中に開いた磁場構造を作らない「LB」タイプであり，CMEも発生しなかった．

本研究にて明らかになった結果より，Hαフィラメントや CME の加速機構を明らかにするためには，フィラメント消失現象の三次元速度場を計測した上で軟X線や極紫外線におけるアーケードや減光現象などのコロナ活動との比較を行うことが重要であるといえる．また，Hαフィラメント消失現象のタイプと CME のタイプとが良好な一対一対応を示したことを踏まえると，我々の手法によるフィラメント消失現象の三次元速度場導出法およびその分類基準は，CME や地磁気嵐の予報に有効であるといえる．より正確な三次元速度場の導出およびHαフィラメントやCMEの加速機構の解明，ひいては宇宙天気予報には，将来FMTの空間分解能の向上や観測波長域の拡大等の改良が必要とされる．

京都大学飛騨天文台および花山天文台の方々による協力，また充実した意見や議論に深く感謝する．FMTのデータベースはM. Kadota, Y. Nasuji, M. Kamobe, そしてS. Ueno の人々によって作成，運営されている．「ようこう」のデータは宇宙科学研究本部（ISAS）の主導の元，日本－米国－英国の共同プロジェクトの厚意によって提供された．GOESデータは World Data Center A for Solar-Terrestrial Physics, NGDC, NOAAの厚意による．SOHOデータは欧州宇宙機関（ESA）およびNASAの国際共同ミッションの厚意による．P. F. ChenとD. B. Brooksの両博士による議論や励ましの言葉にも感謝する．

the 7 quasi-eruptive events are "LB" type without any signature of the opening of the coronal magnetic fields, and none are followed by CMEs.

From the clear results obtained in this study, we can conclude that it is important to measure the 3-D velocity fields of DBs and to compare them with coronal activities, such as soft X-ray and EUV arcades and dimmings in order to study the acceleration mechanism of Hα filaments and CMEs. At the same time, based on the good one-to-one correspondence between the types of Hα DBs and CMEs, it should be emphasized that our methods of deriving the 3-D velocity field of a DB and our criteria for classifying the DB types are effective for predicting CMEs and geomagnetic storms. Future development of the FMT system to improve its spatial resolution and wavelength coverage is necessary for a more precise 3-D velocity measurement and for studying the acceleration mechanism of Hα filaments and CMEs and space weather forecasting.

The authors wish to thank all members at Hida and Kwasan Observatories, Kyoto University, for the operation of telescopes and fruitful comments and discussions. The database of FMT is compiled and maintained by M. Kadota, Y. Nasuji, M. Kamobe, and S. Ueno. The Yohkoh data is provided by courtesy of a joint Japan-US-UK project managed by the Institute of Space & Astronautical Science (ISAS) of Japan. GOES data are courtesy of the World Data Center A for Solar-Terrestrial Physics, NGDC, NOAA. SOHO data is courtesy of a mission of international cooperation between European Space Agency (ESA) and NASA. We are also grateful to Dr. P. F. Chen, and Dr. D. H. Brooks for thorough discussions and encouragement.

参考文献／References

Beckers, J. M. 1964, PhD Thesis, University of Utrecht
Brueckner, G. E., et al. 1995, Sol. Phys., 162, 357
Delaboudinière, J.-P., et al. 1995, Sol. Phys., 162, 291
Démoulin, P., & Vial, J. C. 1992, Sol. Phys., 141, 289
Gilbert, H. R., Holzer, T. E., Burkepile, J. T., & Hundhausen, A. J. 2000, ApJ, 537, 503
Gosling, J. T., Hildner, E., MacQueen, R. M., Munro, R. H., Poland, A. I., & Ross, C. L. 1976, Sol. Phys., 48, 389
Green, L. M., Matthews, S. A., van Driel-Gesztelyi, L., Harra, L. K., & Culhane, J. L. 2002, Sol. Phys., 205, 325
Hanaoka, Y., et al. 1994, PASJ, 46, 205
Harra, L. K., & Sterling, A. C. 2001, ApJ, 561, L215
Illing, R. M. E., & Hundhausen, A. J. 1985, J. Geophys. Res., 90, 275
Khan, J. I., Uchida, Y., McAllister, A. H., Mouradian, Z., Soru-Escaut, I., & Hiei, E. 1998, A&A, 336, 753
Kiepenheuer, K. O. 1953, in The Sun, The Solar System I, ed. G. P. Kuiper (Chicago: University of Chicago Press), 322
Kurokawa, H., Ishiura, K., Kimura, G., Nakai, Y., Kitai, R., Funakoshi, Y., & Shinkawa, T. 1995, J. Geomag. Geoelectr., 47, 1043
Martin, S. F. 1973, Sol. Phys., 31, 3
McAllister, A. H., et al. 1992, PASJ, 44, L205
McAllister, A. H., Kurokawa, H., Shibata, K., & Nitta, N. 1996, Sol. Phys., 169, 123
Morimoto, T., & Kurokawa, H. 2003, PASJ, 55, 503 (Paper I)
Mouradian, Z., Soru-Escaut, I., & Pojoga, S. 1995, Sol. Phys., 158, 269
Munro, R. H., Gosling, J. T., Hildner, E., MacQueen, R. M., Poland, A. I., & Ross, C. L. 1979, Sol. Phys., 61, 201
Ogawara, Y., Takano, T., Kato, T., Kosugi, T., Tsuneta, S., Watanabe, T., Kondo, I., & Uchida, Y. 1991, Sol. Phys., 136, 1
Ohyama, M., & Shibata, K. 1997, PASJ, 49, 249
Ohyama, M., & Shibata, K. 1998, ApJ, 499, 934
Sheeley, N. R. Jr., et al. 1975, Sol. Phys., 45, 377
Shibata, K., Masuda, S., Shimojo, M., Hara, H., Yokoyama, T., Tsuneta, S., Kosugi, T., & Ogawara, Y. 1995, ApJ, 451, L83
Thompson, B. J., et al. 1999, ApJ, 517, L151
Thompson, B. J., Reynolds, B., Aurass, H., Gopalswamy, N., Gurman, J. B., Hudson, H. S., Martin, S. F., & St. Cyr, O. C. 2000, Sol. Phys., 193, 161
Tsuneta, S., et al. 1991, Sol. Phys., 136, 37
Webb, D. F., Krieger, A. S., & Rust, D. M. 1976, Sol. Phys., 48, 159 Wright, C. S., & McNamara, L. F. 1983, Sol. Phys., 87, 401
Zhang, J., Dere, K. P., Howard, R. A., & Bothmer, V. 2003, ApJ, 582, 520

1997年11月4日に観測されたモートン波とEIT波の関係

Relation between a Moreton Wave and an EIT Wave Observed on 1997 November 4

衛藤茂[1], 磯部洋明[2], 成影典之[1], 浅井歩[2], 森本太郎[2], Barbara THOMPSON[3],
八代誠司[3], Tongjiang WANG[4], 北井礼三郎[2], 黒河宏企[2], 柴田一成[2]

[1] 京都大学宇宙物理学教室, [2] 京都大学大学院理学研究科附属天文台, [3] NASA Goddard Space Flight Center, [4] Max-Plank-Institute für Aeronomie

(原論文)PASJ: Publ. Astron. Soc. Japan 54, 481-491, 2002 June 25

Shigeru ETO[1], Hiroaki ISOBE[2], Noriyuki NARUKAGE[1], Ayumi ASAI[2], Taro MORIMOTO[2],
Barbara THOMPSON[3], Seiji YASHIRO[3], Tongjiang WANG[4], Reizaburo KITAI[2],
Hiroaki KUROKAWA[2], and Kazunari SHIBATA[2]

[1] Department of Astronomy, Kyoto University, [2] Kwasan and Hida Observatories, Kyoto University,
[3] NASA Goddard Space Flight Center, [4] Max-Plank-Institute für Aeronomie

概要

1997年11月4日に活動領域8100で起こったXクラスフレアイベントを解析し、フレアに付随する波動現象である，彩層中のモートン波とコロナ中のEIT波の関係を調べた．モートン波の観測には，飛騨天文台のフレア監視望遠鏡 (FMT) で得られた，Hα，Hα+0.8Å，Hα-0.8Å像を用いた．EIT波の観測には，SOHO衛星搭載の極紫外線撮像望遠鏡 (EIT) で得られた，極紫外線像を用いた．モートン波とEIT波の伝播速度はそれぞれ，715 km/s, 202 km/s 程度であった．モートン波が観測された時間にEIT波が観測されることはなかったが，フィラメントの振動からモートン波が継続していたであろうことは示唆されている．今回の観測に関していえば，伝播速度と波面位置の情報から，モートン波とEIT波の物理的実態が異なることが示された．モートン波がEIT波に先行して起こることから，EIT波がMHDファストモード衝撃波ではないことが示唆された．

Abstract

We consider the relationship between two flare-associated waves, a chromospheric Moreton wave and a coronal EIT wave, based on an analysis of an X-class flare event in AR 8100 on 1997 November 4. A Moreton wave was observed in Hα, Hα+0.8 Å, and Hα-0.8 Å with the Flare-Monitoring Telescope (FMT) at the Hida Observatory. An EIT wave was observed in EUV with the Extreme ultraviolet Imaging Telescope (EIT) on board SOHO. The propagation speeds of the Moreton wave and the EIT wave were approximately 715 km s^{-1} and 202 km s^{-1}, respectively. The times of visibility for the Moreton wave did not overlap those of the EIT wave, but the continuation of the former is indicated by a filament oscillation. Data on the speed and location clearly show that the Moreton wave differed physically from the EIT wave in this case. The Moreton wave preceded the EIT wave, which is inconsistent with an identification of the EIT wave with a fast-mode MHD shock.

Key words: Sun: chromosphere — Sun: corona — Sun: flares — Sun: MHD

1. 序論

　モートン波は太陽フレアに伴って生じる波動で，Hα 画像，特にウイング画像でよく観測されている (Moreton 1960,1961; Moreton, Ramsey 1960; Athay, Moreton 1961; Smith, Harvey 1971)．伝播速度は特定の方向性をもち，その大きさはおよそ 400-1100 km/s である．アーク状の波面をもち，しばしば Type-II 電波バーストを伴うことも報告されている (Kai 1970)．また，Uchida (1968, 1970, 1974) は，コロナ中を伝播する MHD ファストモード衝撃波が，彩層中でモートン波として観測されるという説を確立した．しかしながら，モートン波の発生機構については未解明のままである．

　近年，SOHO 衛星搭載の極紫外線撮像望遠鏡 (EIT) は EIT 波と呼ばれる，モートン波とは別の波動的な現象を，コロナ中で観測した (Moses et al. 1997; Thompson et al. 1998)．EIT 波はフレアやコロナ質量放出 (CMEs) に伴い，ほぼ等方的に 200-300 km/s の速度で伝播する．また，EIT 波は Type-II 電波バーストを伴うことも知られている (Klassen et al. 2000)．Thompson et al. (1999, 2000) は，EIT 波がモートン波のコロナ中での対応物であると主張したが，両者の間の伝播速度やパターンの著しい違いは大きな謎とされており，EIT 波がモートン波と同じ現象かどうかは議論の余地がある．

　この論文では上述の議論に答えるため，1997 年 11 月 4 日に起こった X クラスフレアに関して，飛騨天文台のフレア監視望遠鏡で撮像されたモートン波の Hα 像と，SOHO/EIT で撮像された EIT 波の EUV 像を用いて解析を行った．一連のデータは，今回のモートン波は EIT 波とは異なることを示している．

　第 2 節では観測方法と観測機器についてまとめた．第 3 節ではモートン波と EIT 波に関する主要な結果を記述した．最後に第 4 節では，まとめと議論を載せた．

1. Introduction

Moreton waves are flare-associated waves seen in Hα, especially in the wing of Hα (Moreton 1960, 1961; Moreton, Ramsey 1960; Athay, Moreton 1961; Smith, Harvey 1971). They propagate within a somewhat restricted solid angle at speeds of ~ 400-1100 km s^{-1}. They have arclike fronts, and are often associated with type-II radio burst (Kai 1970). Uchida (1968, 1970, 1974) established that Moreton waves are a chromospheric manifestation of MHD fast-mode shocks propagating in the corona. However, the generation mechanism of Moreton waves has not yet been clarified.

Recently, the Extreme ultraviolet Imaging Telescope (EIT) aboard the Solar and Heliospheric Observatory (SOHO) discovered another wave-like phenomenon in the solar corona, which is now called EIT waves (Moses et al. 1997; Thompson et al. 1998). They are associated with flares or coronal mass ejections (CMEs), and propagate at speeds of 200-300 km s^{-1}, sometimes nearly isotropic in direction. It is also known that EIT waves are well associated with type-II radio bursts (Klassen et al. 2000). Thompson et al. (1999, 2000) suggested that EIT waves are coronal manifestation of Moreton waves. However, marked differences in the propagation pattern and speed are a big puzzle, which raises the question whether EIT waves are the same as Moreton waves.

The purpose of this paper is to examine this question based on Hα images taken with the Flare-Monitoring Telescope (FMT) at the Hida Observatory of the Moreton wave associated with an X-class flare on 1997 November 4 and the EUV images of the EIT wave taken with the SOHO/EIT of the same flare. These data show that the Moreton wave is different from the EIT wave.

In section 2, the observing methods and instrumentations are summarized. In section 3, the main results of an analysis of the Moreton wave and the EIT wave are described; in section 4, a summary and a discussion are given.

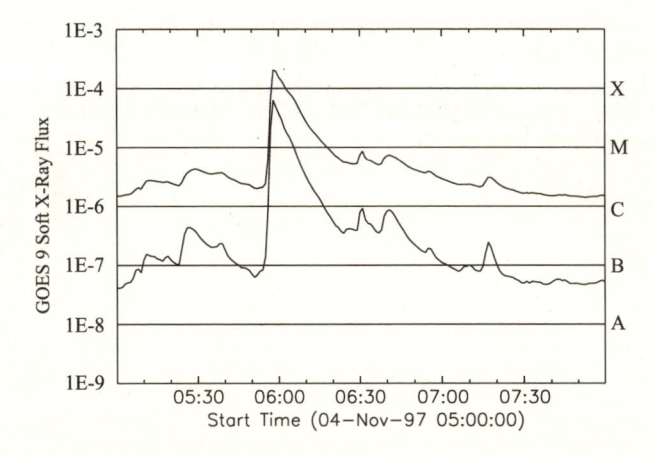

図1　1997年11月4日の 05:00 UT から 08:00 UT に GOES で観測された X 線フラックス．上側の曲線は 1-8Å 帯のもので，下側の曲線は 0.5-4Å 帯のものである．フレアは 05:52 UT から始まって，05:58 UT にピークを迎えた．

Fig. 1. X-ray emission for 05:00-08:00 UT on 1997 November 4 as recorded by GOES at the 1-8Å channel (higher flux curve) and at the 0.5-4Å channel (lower flux curve). The flare started at 05:52 UT and the time of its peak was 05:58 UT.

2. 観測

本論文では 1997年11月4日 05:52 UT に，S14°，W34° の活動領域 NOAA8100 で起こった X クラスフレアに伴う，彩層中のモートン波とコロナ中の EIT 波に関して考察する．図1に GOES 軟 X 線のライトカーブを示した．このフレアは 05:52 UT に始まり，05:58 UT にピークを迎えた．本研究ではこのフレアに伴う CME や Type-II 電波バースト，プロミネンス噴出についても考察するが,「ようこう」のデータは蝕 (日陰) の時間であったため用いていない．

モートン波とプロミネンス噴出は，京都大学飛騨天文台フレア監視望遠鏡 (FMT) (Kurokawa et al. 1995) の Hα 像で観測した．FMT では Hα 線中心，Hα+0.8Å，Hα-0.8Å と連続光の太陽全面像と，Hα 線中心でのリム上空の画像を取得する．今回の解析には Hα 線中心，Hα+0.8Å，Hα-0.8Å の太陽全面像を用いた．図2には 6:00 UT

2. Observations

In this paper, we consider two kinds of flare-associated waves, a chromospheric Moreton wave and a coronal EIT wave accompanied by an X-class flare, which occurred in NOAA 8100 at S14°, W34° at 05:52 UT on 1997 November 4. A GOES plot of the flare is shown in figure 1. The flare started at 05:52 UT, and its peak was at 05:58 UT. We also consider the CME, the type-II radio burst and the prominence eruption associated with the flare. Unfortunately, Yohkoh missed this event because of its night.

The Moreton wave and the prominence eruption were observed in Hα with the Flare-Monitoring Telescope (FMT) (Kurokawa et al. 1995) at the Hida Observatory of Kyoto University. The FMT observes four full disk images in Hα center, Hα+0.8 Å, Hα-0.8 Å, and continuum, and one solar limb image in Hα center. We used three full disk images in Hα center, Hα+0.8 Å, and Hα-0.8 Å. Figure 2 shows the image in Hα center at 6:00:00 UT. The flare occurred in the

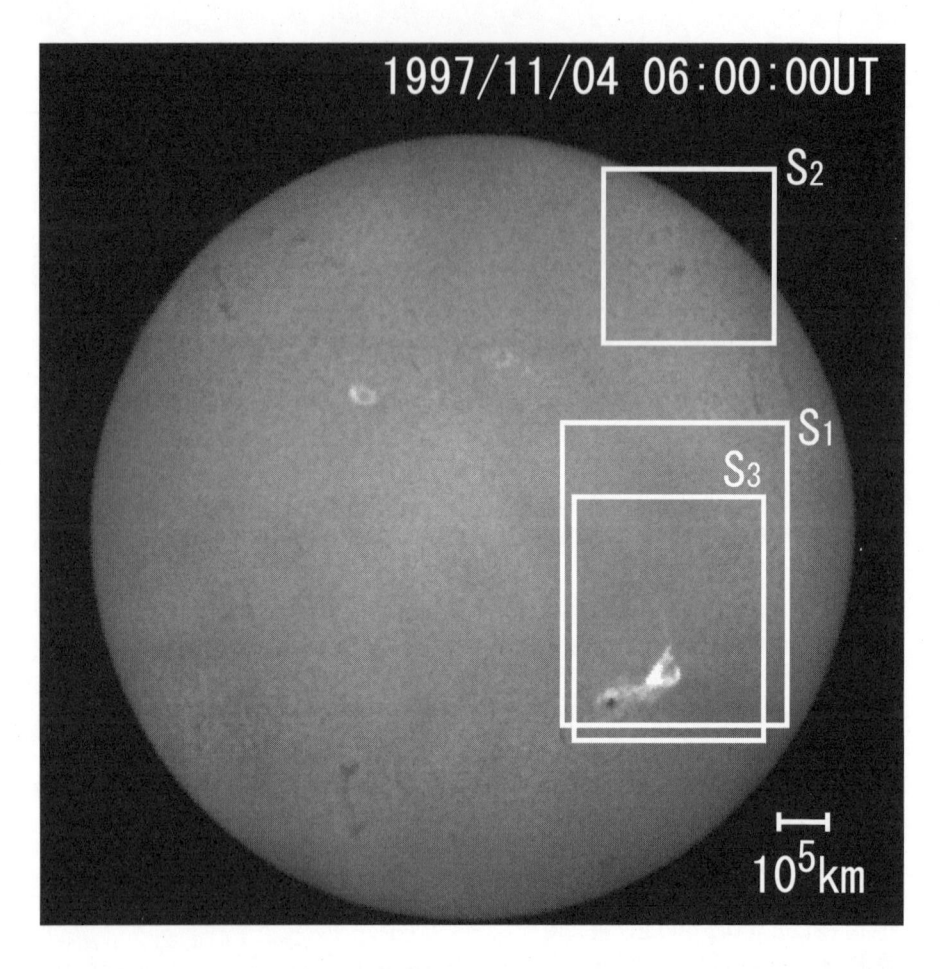

図2 1997年11月4日の06:00 UT に飛騨天文台のFMTで撮像されたHα線中心での太陽全面像．矩形領域 S3 に見られる Hα で明るい領域は，05:52 UT から始まった X クラスフレアによるものである．図 3, 図 4 の視野は矩形領域 S1 であり，S2, S3 はそれぞれ図 5, 図 10 の視野に対応している．

Fig. 2. Full Sun Hα (center) image at 06:00:00 UT on 1997 November 4, taken with FMT at Hida Observatory. An Hα brightening is seen in box S3, which is due to an X-class flare which started at 05:52 UT. The field of view of figures 3 and 4 is box S1. Boxes S2 and S3 correspond to the fields of view shown in figures 5 and 10, respectively.

におけるHα線中心画像を示した．フレアは南西にある明るい領域で発生した．今回の観測ではHα撮像の時間間隔は1分であるが，FMTはさらに高速の撮像が可能である．Hα線中心，Hα+0.8Å, Hα−0.8Å の順に撮像を行い，異なる波長間のタイム bright region in the southwest. The time resolution of the Hα images used in this paper is 1 min, though the FMT operates at a higher time resolution. The exact times of Hα center, Hα+0.8 Å, and Hα −0.8 Å were 0 s, 1 s, and 2 s after each minute during the observed time interval.

Hα−0.8Å Hα center Hα+0.8Å

Original Image

1997/11/04
06:00 UT

Running
Difference
(6:00 − 5:59)

Wavefront

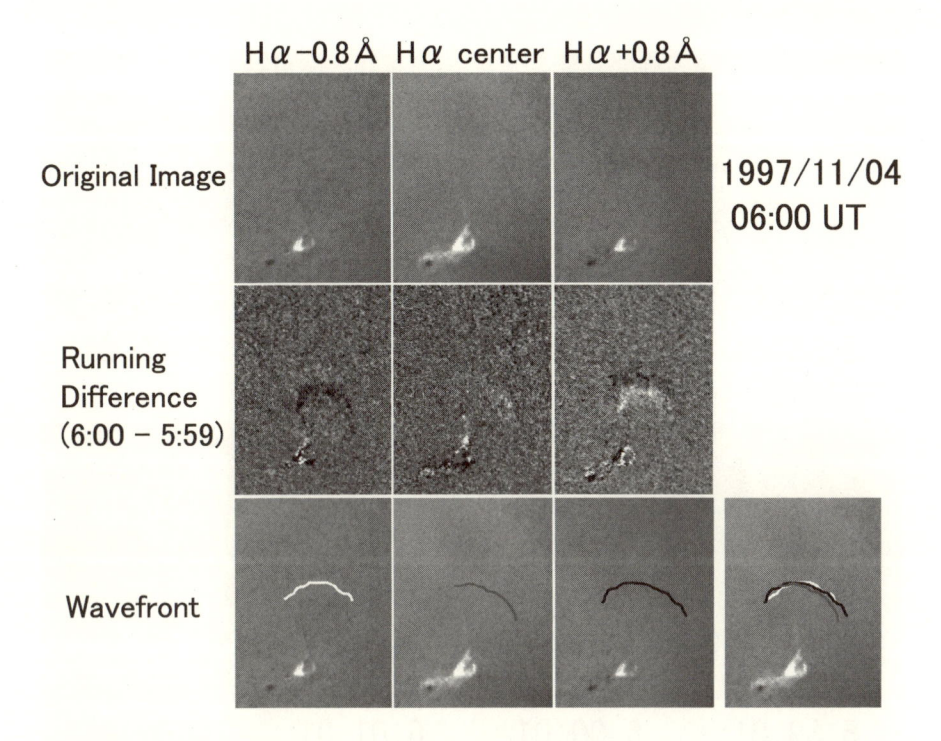

図3 1997年11月4日の06:00 UTにおけるモートン波の位置（左：Hα−0.8Å, 中：Hα線中心, 右：Hα+0.8Å）。視野は図2のS1と対応している．上段には元画像を示した．中段の図は，06:00 UTの画像から05:59 UTの画像を差し引いた差分画像である．下段には各波長から求めたモートン波の波面位置が示されている．Hα−0.8Å, Hα線中心，Hα+0.8Åから求められた波面はそれぞれ，白，灰色，黒の曲線で描いている．右の画像には全波長の波面を重ねて示している．

Fig. 3. Locations of the Moreton wave in Hα-0.8Å (left), Hα center (center), and Hα+0.8Å (right) at 06:00 UT on 1997 November 4. The field is box S1 in figure 2. The top panels are original images. The middle panels are the "Running difference" images which are images at 06:00 UT subtracted by a previous image at 05:59 UT. The bottom panels show the location of the Moreton wave identified with each wavelength image. The wave fronts of Hα-0.8Å, Hα center, and Hα+0.8Å images are shown by the white, gray, and black curves, respectively. The right image shows the wave-fronts at all wavelengths.

ラグは1秒である．1000 km/sで2秒進むと2000 kmになるが，この距離はCCDのピクセルサイズより小さいので，波長間のタイムラグは無視できるほど小さいものといえる．我々の取り扱う波動の速度は1000 km/s以下のものである．

EUV像はSOHO衛星搭載のEIT (Delaboudinière et al. 1995) で取得した．本研究では120万度，コロナでの典型的な

These differences in the observed times were negligible in this study, because a wave with a velocity of 1000 km s^{-1} propagates 2000 km in 2 s, and this distance is within the pixel scale. The velocity of the wave which we will discuss was less than 1000 km s^{-1}.

The EUV images were observed with the EIT (Delaboudinière et al. 1995) on board SOHO. The observation in the

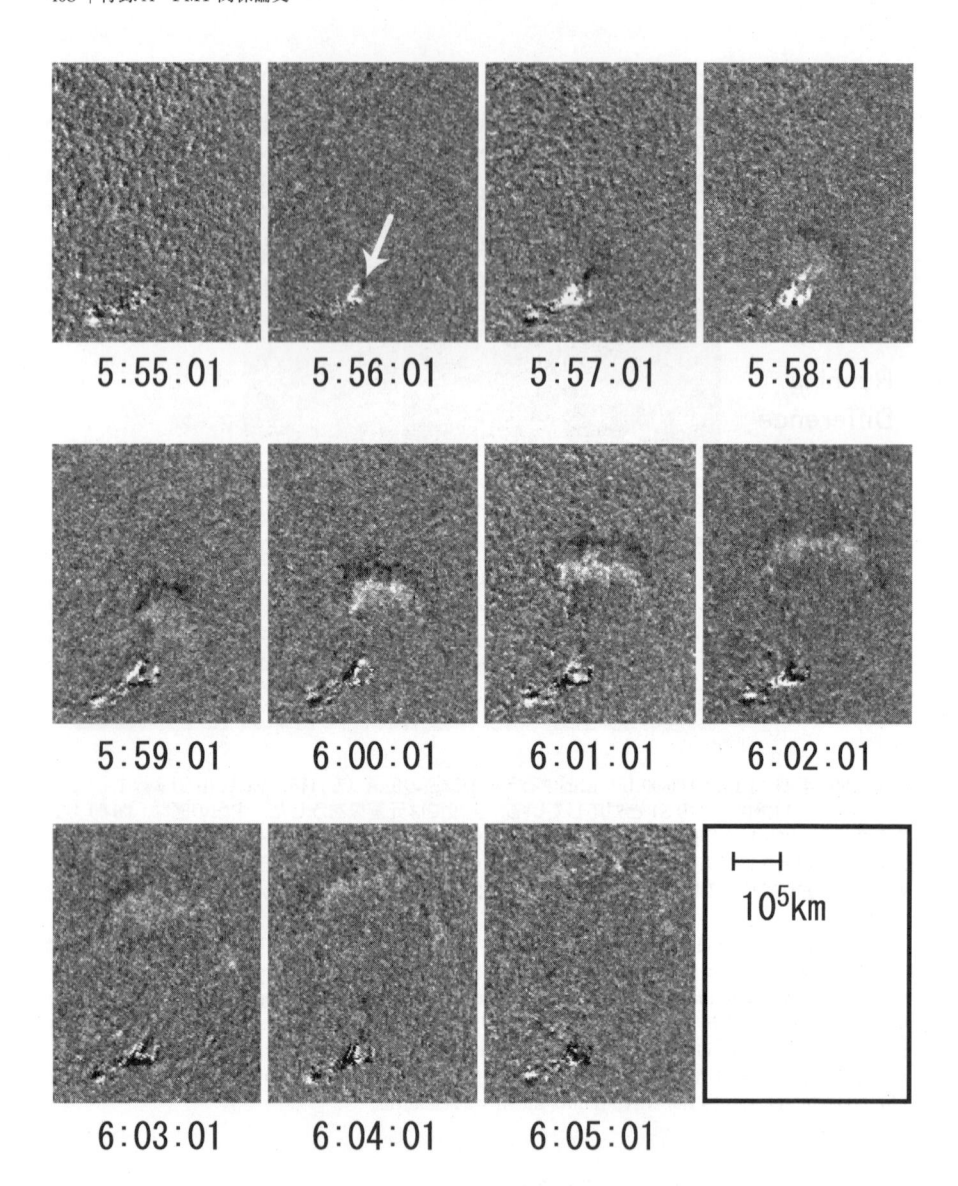

図4 05:57:01 UT から 06:08:01 UT に 1 分ごとに撮られた Hα+0.8Å の差分画像．視野は図 2 の領域 S1 であり，モートン波の波面が白黒の縞模様として鮮明に浮かび上がっている．

Fig. 4. Running differences of Hα+0.8Å images taken at every minute from 05:57:01 UT to 06:08:01 UT. The field of view is box S1 in figure 2. The wave-fronts of Moreton wave can be clearly seen as dark and white strips.

密度をもつプラズマから放出されるFe XII 線からなる，195Å付近の波長域を用いた．時間分解能は約17分である．

CMEの観測にはSOHO衛星搭載のLarge Angle and Spectrometric Coronagraph (LASCO, Brueckner et al. 1995) を用いた．Type-II 電波バーストはHiraiso Radio Spectrograph (HiRAS, Kondo et al. 1995) で記録された．

3. 結果

図3には，FMTで取得した，図2の領域S1に対応するHα−0.8Å（左），Hα線中心（中），Hα+0.8Å（右）の画像を載せた．上段には06:00 UTにおける元画像，中段には差分画像，つまり06:00 UTの画像から05:59 UTの画像を引いたものを載せた．差分画像ではモートン波の波面がくっきりと捉えられている．下段では差分画像から得られた波面を，元画像の上に重ねて表示した．白，灰色，黒の線はそれぞれ，Hα−0.8Å，Hα線中心，Hα+0.8Åの波面を表している．右下の画像に示される通り，波面の場所は波長が違ってもほぼ同じであるので，波面が最も鮮明に観測されるHα+0.8Åを用いて，以下の解析を行った．

Hα+0.8Åの差分画像を図4に示した．視野は図2の領域S1である．図4中の05:56 UTにおける，矢印で示された小さく暗い領域はモートン波の発生点と推測される．発生点は図4の05:57:01 UTでの白く見える領域であるフレア領域とは異なっている．ここではフレア領域は最も明るい領域と定義している．モートン波の波面は05:58 UTから06:04 UTにおいて同定された．06:05 UT以降の画像では，モートン波は波面を検出できないほど減衰している．このことは，06:04 UT以降モートン波の伝播が止まったのを意味するのではなく，波面の検出は不能になったものの，この時間以降もモートン波は伝播を続けたことを示唆している．後述するように，モートン波が北西のリムまで伝播した間接的な証拠もある．

図5はEIT画像の差分画像であり，擾

present study consisted of a wavelength range centered on 195 Å, comprising a Fe XII line emitted at 1.2×10^6 K at typical coronal density. The time resolution was about 17 min.

A CME was observed with the Large Angle and Spectrometric Coronagraph (LASCO, Brueckner et al. 1995) on board SOHO. A type-II radio burst was recorded with the Hiraiso Radio Spectrograph (HiRAS, Kondo et al. 1995).

3. Results

Figure 3 shows the FMT images observed in Hα − 0.8 Å (left column), Hα center (middle column), and Hα + 0.8 Å (right column) in box S1 indicated in figure 2. The top panels show the original images at 06:00 UT, and the middle panels show "running difference" images, or the images at 06:00 UT, which were subtracted by a previous image at 05:59 UT. In the running difference images the wave-fronts of the Moreton wave can be clearly seen. In the bottom panels the wave-fronts identified in the running difference image in each wavelength are overlaid on the original images. The white, gray, and black lines illustrate the wave-fronts in Hα − 0.8 Å, Hα center, and Hα + 0.8 Å, respectively. As shown in the right-bottom panel, the locations of the wave-fronts are not much different in the three wavelengths images. Thus, in the following analyses we use Hα + 0.8 Å images, in which the wave-fronts are most clearly seen.

The running difference images of Hα + 0.8 Å are shown in figure 4. The field of view is illustrated by box S1 in figure 2. The dark small region indicated by the arrow in the image at 05:56:01 UT in figure 4 seems to be an origination point of the Moreton wave. It is different from the brightest region during the flare at Hα, which is the white region in the image at 05:57:01 UT in figure 4. We define the brightest region as a flare site. The wave-fronts of the Moreton wave can be identified in the images from 05:58 UT to 06:04 UT. In the images after 06:05 UT, the Moreton wave is too diffuse to identify its wave-front. Naturally, this does not mean that the Moreton wave stopped propagating at 06:04 UT. It probably propagated farther after 06:04 UT, though its wave-front can not be seen. As

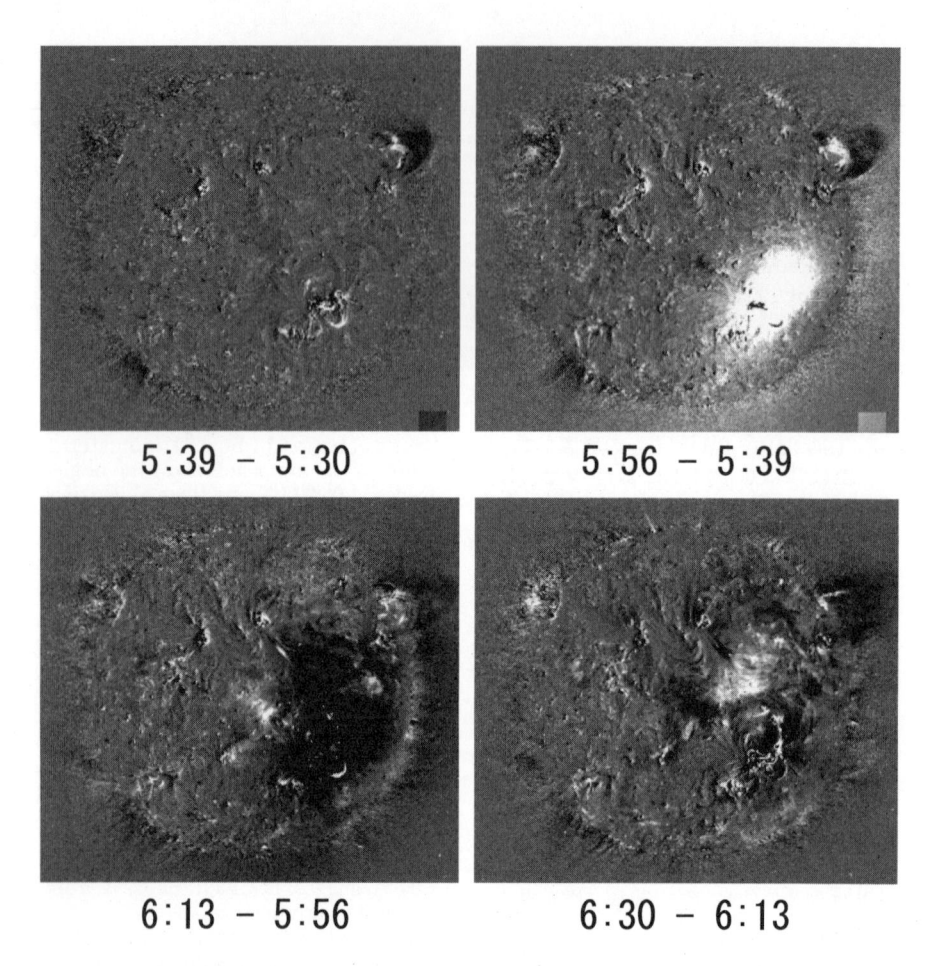

図5 EIT の画像から合成した差分画像．例えば，左上段は，05:39 UT の画像から 05:30 UT の画像を差し引いたものである．右上段の楕円状に白く見える領域は主として望遠鏡の散乱光によるものである．下段にはフレア領域から擾乱が伝播していく様子が窺える．

Fig. 5. Running differences of EIT images. The top-left panel shows the image at 05:39 UT subtracted by that of 05:30 UT, and so on. The white oval region in the top-right panel is mainly due to scattered light in the telescope. In the lower panels the disturbance seems to propagate away from the flare site.

乱がフレア領域から伝わる様子が見て取れる．05:56 UT における楕円状の明るい構造（図5右上）は主に望遠鏡の散乱光によるもので，おそらく本当の波面ではない．したがって，EIT 画像による波面同定は，06:13 UT から 06:30 UT の間のみ可能で

reported later, there is indirect evidence that the Moreton wave propagated farther near the northwest limb, though its wavefront is not seen in the region.

Figure 5 shows the running differences of the EIT images. A disturbance is seen to propagate from the flare site. The

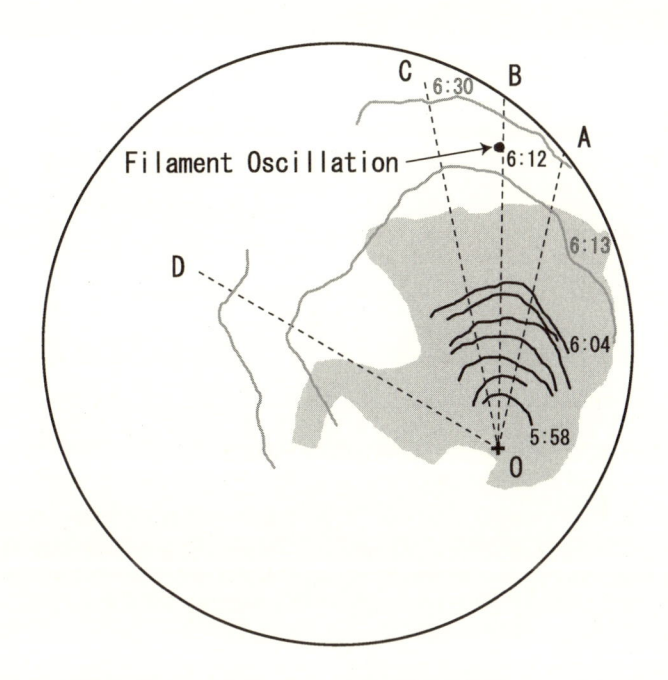

図6 観測された EIT 波の波面（灰色線）とモートン波の波面（黒線），そしてコロナ減光領域（灰色領域）．EIT 波の波面はそれぞれ，06:13 UT と 06:30 UT のもので，モートン波の波面はそれぞれ，05:58:01 UT から 06:04:01 UT の間に1分ごとに観測されたものである．06:12:00 UT 頃に振動し始めたフィラメントの位置は黒丸で示されている．コロナ減光領域は図5中に示された，06:13 UT と 05:56 UT の差分画像で見られる暗い領域に対応しており，時間と共に減少していく傾向にある．図7中の，フレア領域 O から波面までの距離は，直線 OA, OB, OC, OD に沿って計られている．

Fig. 6. Observed wave-fronts of the EIT wave (gray lines) and the Moreton wave (black lines), and the dimming region (light gray region). The EIT wave-fronts are those at 06:13 UT and 06:30 UT, and the Moreton wave-fronts are those from 05:58:01 UT to 06:04:01 UT with interval of 1min. The location of the filament which started to oscillate from about 06:12:00 UT is also indicated by the black filled circle. The dimming region corresponds to the dark region in a running difference image at 6:13-5:56 UT in figure 5, which is a depletion region during this period. The distances of wave-fronts from the flare site O along the lines O-A, O-B, O-C, and O-D are plotted in figure 7.

あった．一方，モートン波は 05:58 UT から 06:04 UT の間同定されているので，モートン波と EIT 波の波面の位置を同時刻に比較することはできなかった．

図6では，モートン波の波面（黒線）を 05:58 UT から 06:04 UT まで1分ごとに描き，さらに 06:13 UT と 06:30 UT における EIT 波の波面（灰色線）を描いた．図の灰色領域はコロナ減光領域であり，点 O は

bright oval structure in the 05:56 UT image (upper-right panel) is mainly due to scattered light in the telescope, and is probably not a real wave-front. Hence, the wave-fronts in the EIT images are identified only at 06:13 UT and 06:30 UT. On the other hand, the wave-fronts of the Moreton wave are identified from 05:58 UT to 06:04 UT. Thus, we can not directly compare the locations of theMoreton wave and the EIT wave at the same time.

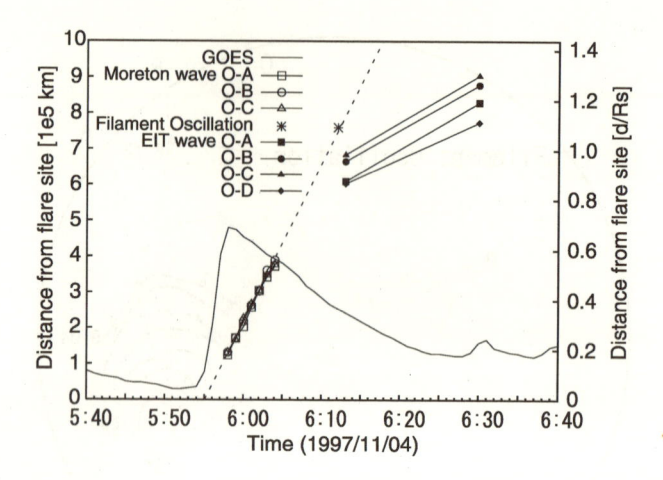

図7 モートン波(中抜きの記号)とEIT波(黒塗りの記号)の距離の時間変化. 距離の定義は図6キャプション参考のこと. 06:12:00 UTのアスタリスクはフィラメントの振動開始時刻とフレア領域からの距離を表している. フレアの開始時刻とピーク時刻を示すために, GOES軟X線(1-8Å帯)フラックスのライトカーブを載せた.「Rs」は太陽半径(695800 km)を表す.

Fig. 7. Time evolutions of the distance of the Moreton wave (open symbols) and the EIT wave (filled symbols) from the flare site (point O in figure 6) along the lines O-A, O-B, O-C, and O-D in figure 6. An asterisk at 06:12:00 UT represents the start of the filament oscillation and the distance of the filament from the flare site. To illustrate the flare start and its peak, GOES X-ray flux data at the 1-8Å channel (higher flux curve in figure 1) are overplotted. 'Rs' represents the solar radius (695800 km).

フレア領域を示している. EIT波とモートン波はどちらも北に伝播したが, 伝播の角度が異なっている. モートン波の伝播角度は60°と狭い角度であるのに対し, EIT波の伝播角度はより広い角度であった. 図7には直線OA, OB, OC, ODに沿った, フレア領域Oからの波面の距離をプロットした. 距離は光球から計り, 曲面の効果も考慮した. モートン波の伝播速度は, ほぼ一定で715 km/s程度であった. 一方EIT波の伝播速度は, 伝播方向によって多少異なるものの, 平均的には202 km/s程度であった. 得られた伝播速度と伝播角度はすべて典型的なものであった.

FMTで観測された静穏型フィラメントの振動は, 06:04 UT以降もモートン波が伝播していた間接的証拠といえる. このフィラメントは図6の黒丸で示されている

Figure 6 shows the wave-fronts of the Moreton wave (black lines) from 05:58:01 UT to 06:04:01 UT every minute, those of the EIT wave (gray lines) at 06:13 UT and 06:30 UT, and the coronal dimming region observed in the EIT images. Point O indicates the flare site. Although both the EIT wave and the Moreton wave propagate northward, the angles of their propagation are different. While the Moreton propagates in a narrow angle of about 60°, the EIT wave propagates in a relatively wide angle. The distances of the wave-fronts from the flare site O along the lines O-A, O-B, O-C, and O-D are plotted in figure 7. The distances are measured along the photosphere. The curvature of the solar surface is taken into account. The propagation velocity of the Moreton wave is almost constant at 715 km s^{-1}. The velocities of the EIT wave are slightly different in different directions. The

Appendix A　Selected Papers Relating to FMT | 413

Hα-0.8Å　　Hα center　　Hα+0.8Å

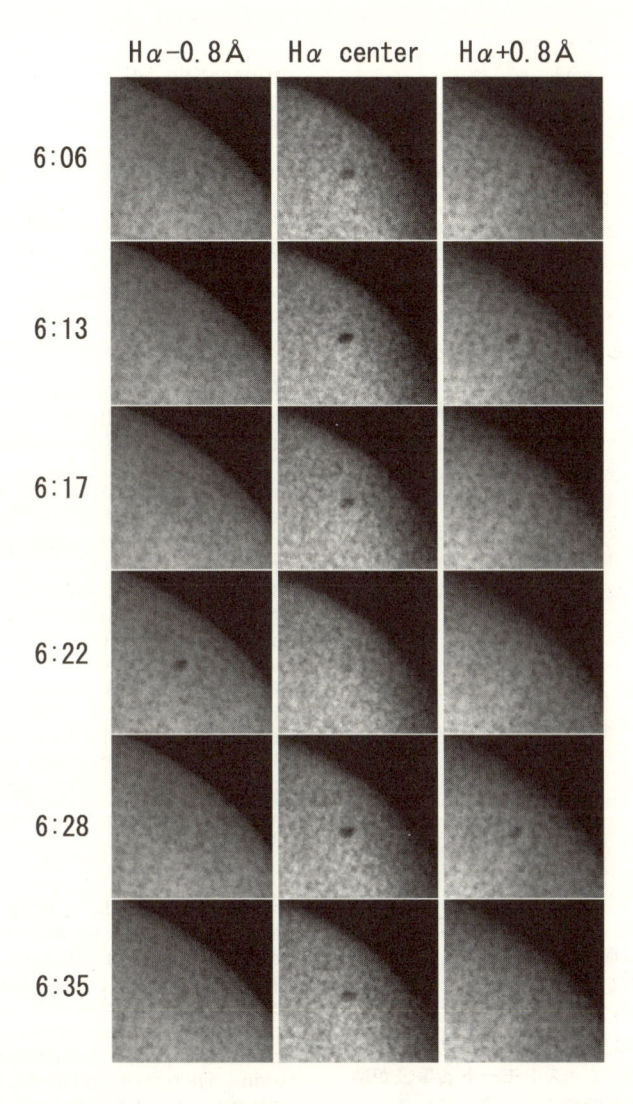

6:06

6:13

6:17

6:22

6:28

6:35

図8　振動するフィラメントのHα画像（Hα-0.8Å;左，Hα線中心;中，Hα+0.8Å;右）視野は図2の領域S2である．このフィラメントは06:12:00 UT頃から振動し始めた．06:13:00 UTには，ダークフィラメントがHα+0.8Åでかすかに見え，鉛直下向きに運動していることが推測される．また，06:22 UTにはフィラメントがHα線中心で見えなくなり，Hα-0.8Åで見えることから，06:22 UTにはフィラメントが上昇運動していることが示唆される．

Fig. 8. Hα images (Hα-0.8Å ; left, Hα center; center, and Hα+0.8Å ; right) of an oscillating dark filament. The field of view is box S2 in figure 2. The filament started to oscillate at about 06:12:00 UT. The faint dark filament was seen in the Hα+0.8Å at 06:13 UT, which means that the filament was undergoing downward motion, whereas the filament disappeared in the Hα center and appeared in the Hα-0.8Å image at 06:22 UT, indicating that the filament was undergoing upward motion at 06:22 UT.

ように，北西のリムに位置していた．図8にFMTのHα−0.8Å，Hα線中心，Hα+0.8Å画像で捉えられた，フィラメント振動の様子を示す．視野は図2の領域S2に対応している．このフィラメント振動は06:12:00 UTから始まっており，図7中の対応する時刻と距離にアスタリスクを付けたところ，モートン波の計測点を外挿した直線上に乗った．このことは，フィラメント振動が「不可視」のモートン波によって引き起こされる，という予想を支持している (Smith, Harvey 1971)．もしこの解釈が正しいのであれば，モートン波は06:12 UTには北西のリムに位置するこのフィラメント近傍に達していたに違いないと考えられる．一方，EIT波は06:13 UTにおいてもこのフィラメントの位置にまで達しておらず，モートン波とEIT波は同時刻に同一の位置にはいなかったと結論できる．

4. 議論

我々はまず，フレアに付随して発生した波である，彩層のモートン波とコロナのEIT波の同定について議論する．広く受け入れられているUchidaモデル (Uchida 1968, 1970) は，モートン波がコロナ中を伝播するMHDファストモード衝撃波が彩層と接する境界であると主張している．したがって，このモデルが正しいならば，コロナ中にモートン波と同一位置に同一速度で伝播する対応物が存在しているはずである．Thompson et al. (2000) は，EIT波には波面輪郭が明るくくっきりしたものと，ぼんやりと拡散したものの2種類に分かれることを報告した．そして，典型的なのは後者である．くっきりしたEIT波とモートン波の位置は同一であるのに対して，拡散型EIT波とモートン波の関係はいまだ

average velocity is 202 kms^{-1}. These propagation velocities and angles are typical for both Moreton waves and EIT waves.

The FMT observed an oscillation of a quiescent filament, which is the indirect evidence mentioned above that the Moreton wave probably propagated farther after 06:04 UT. The filament is located near the northwest limb, which is indicated by the filled circle in figure 6. Figure 8 shows the FMT images of the oscillating filament at the wavelengths of Hα−0.8 Å, Hα center, and Hα+0.8 Å. The field of view is shown by box S2 in figure 2. The oscillation of the filament started at 06:12:00 UT. We plot an asterisk in figure 7 corresponding to the starting time of the oscillation and the distance from the flare site of the filament. The asterisk almost lies on the extrapolated line of the Moreton wave. It is thus reasonable to conjecture that the filament oscillation was caused by an 'invisible' Moreton wave (Smith, Harvey 1971). If this interpretation is correct, the Moreton wave must have reached the filament near the northwest limb at 06:12:00 UT. On the other hand, the EIT wave did not reach the filament even at 06:13 UT. Consequently, we can say that the Moreton wave and the EIT wave are not co-spatial.

4. Discussion

We first discuss the identification of two flare-waves, a chromospheric Moreton wave and a coronal EIT wave. In Uchida's model (1968, 1970), which is most widely accepted, the Moreton wave is a sweeping skirt on the chromosphere of the MHD fast-mode shock which propagates in the corona. Therefore, this model predicts the existence of a coronal counterpart of the chromospheric Moreton wave at the same place and with the same velocity as that of the Moreton wave. Thompson et al. (2000) reported that there are two components in EIT waves, i.e. bright and sharp, and 'traditional' diffuse EIT waves. The sharp EIT wave and Hα Moreton wave are cospatial, whereas the relationship between the diffuse EIT and Moreton wave was not clear. In the event discussed in this paper, only the diffuse EIT wave was observed. From similar

不明瞭である．この論文で議論する現象では，拡散型 EIT 波にのみが観測された．Thompson et al.(2000) や Warmuth et al.(2001) による同様の観測が主張するところによると，拡散型 EIT 波は減速されたモートン波であり，結局のところどちらの種類の EIT 波もモートン波のコロナ中での対応物である，とのことである．しかしながら前節で述べた通り，我々はモートン波と EIT 波の位置は異なっており，伝播速度も大きく異なることを明らかにした．したがって我々は，少なくともこの現象については，EIT 波はモートン波の対応物ではないと結論する．

注意しておきたいのは，この観測事実は Uchida モデルを否定するものではないということである．この事実はただ，Uchida モデルが主張しているコロナでの擾乱が，

observations reported by Thompson et al. (2000), Warmuth et al. (2001) suggested that the diffuse EIT waves are decelerated Moreton wave, namely that not only the sharp EIT waves but also the diffuse EIT waves are coronal counterparts of the chromospheric Moreton waves. In the previous section, however, we revealed that the Moreton wave and the diffuse EIT wave are not co-spatial and that they have very different velocities. Therefore, we conclude that at least in this event, the diffuse EIT wave is not the coronal counterpart of the Moreton wave.

It is worth noting that this observed fact does not deny Uchida's model. It only suggests that an EIT wave is not the coronal disturbance predicted by Uchida's model. Uchida argues that one of the conditions that Moreton wave occurs is that the flare (the source of the MHD fast-mode shock) occurs near the edge of the

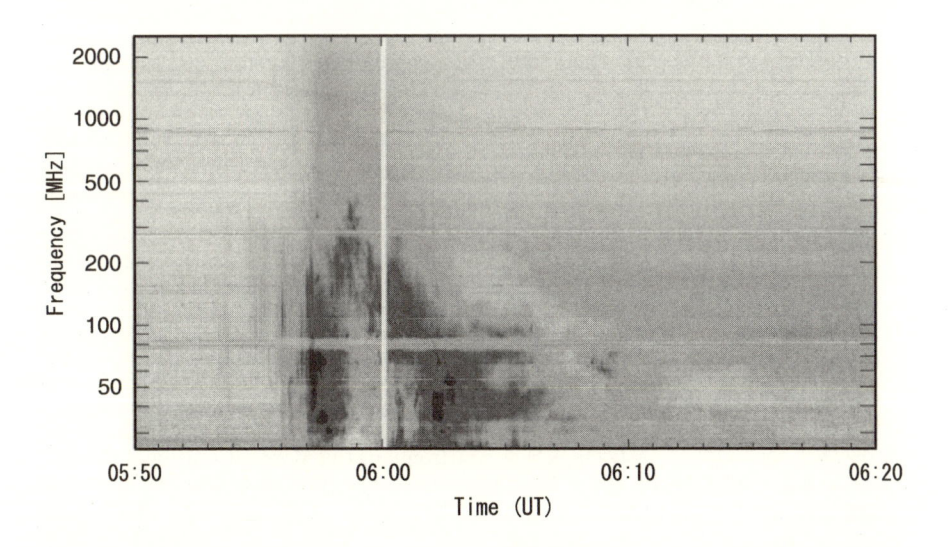

図9　1997年11月4日の 05:50-06:20 UT に Hiraiso Radio Spectrograph（HiRAS）で観測されたメーター波電波のデータ．Type-II と Type-III の電波バーストを示している．05:58 UT のほぼ垂直な線は Type-III 電波バーストを示しており，05:59 UT から 06:06 UT にかけてのバーストは Type-II バーストを示している．

Fig. 9. Metric radio data for 05:50-06:20 UT on 1997 November 4 observed with the Hiraiso Radio Spectrograph (HiRAS), showing type-II and type-III radio bursts. The almost vertical line at about 05:58 UT represents the type-III radio burst, and the burst about from 05:59 UT to 06:06 UT is the type-II radio burst.

EIT 波として観測されるわけではないことを主張するにすぎない．Uchida が主張しているモートン波発生の条件は，MHD ファストモード衝撃波の源であるフレアが，活動領域の縁近くで発生することであり，その伝播方向は活動領域から遠ざかる方向となる．この論文では上記の条件を満たしており，これから議論する Type-II 電波バーストの観測もまた Uchida モデルと整合性のあるものになっている．

上述の通り，Type-II 電波バーストがこの現象に伴って発生した．図 9 は HiRAS で観測された電波スペクトログラフである．05:58 UT 付近のほぼ垂直な線は Type-III バーストであり，05:59 UT から 06:06 UT 付近の曲線は Type-II バーストである．コロナ中の電子密度をモデル (Mann et al. 1999) から評価することで，Type-II 電波バーストの光球からの距離と速度を計算することができる．Type-II 電波バーストの平均速度は約 890 km/s であり，EIT 波の速度 (202 km/s) よりはモートン波の速度 (715 km/s) に近いものであった．さらに図 13 から分かるように，バーストの光球からの距離はモートン波のものと酷似している．Uchida (1974) はモートン波と Type-II バーストが，コロナ中を伝播する MHD ファストモード衝撃波が見せる 2 つの側面に過ぎないことを主張した．速度と距離に関する，モートン波と Type-II 電波バーストの対応関係はこのアイデアを支持するものである．

次に考察するのは，モートン波，EIT 波とコロナ磁場の関係である．図 10 に SOHO/MDI (Scherrer et al. 1995) で得られた光球磁場を用いて外挿したポテンシャル磁場 (Yan et al. 1991, 1993, Wang et al. 2002) と，モートン波，EIT 波の波面を描いた．モートン波の伝播方向と開き角は，一見，フレア領域の北端につながるループに沿っているように見え，その周辺ではコロナ減光が EIT 画像から観測された．しかし，フィラメント振動が観測されていることから，モートン波は閉じた磁気ループを横切って伝播したとも考えられる．モートン波が観測できる条件は磁場形状とコロ

active region, and the direction of propagation is opposite to the active region. The event reported in this paper satisfies these conditions. The type-II burst observations, which we now discuss, are also in accordance with Uchida's model.

As mentioned above, a type-II radio burst was associated with this event. Figure 9 shows a radio spectrograph taken with HiRAS. An almost vertical line at about 05:58 UT and a curve from about 05:59 UT to 06:06 UT in figure 9 are type-III and II radio bursts, respectively. Based on the coronal electron density model (Mann et al. 1999), we can calculate the distance from the photosphere and the velocity of the type-II radio burst. The average velocity of the type-II radio burst is about 890 $km s^{-1}$, which is close to that of the Moreton wave (about 715 $km s^{-1}$) compared with that of the EIT wave (about 202 $km s^{-1}$). Furthermore, the distances of the type-II radio burst are very similar to those of the Moreton wave, as shown in figure 13. Uchida (1974) suggests that the Moreton wave and the type-II radio burst are two aspects of one agent, or the MHD fast-mode shock propagating in the corona. The similarity of the velocity and the distances of the Moreton wave and the type-II radio burst supports this idea.

Next, we consider the relationship between the coronal magnetic field, the Moreton wave, and the EIT wave. Figure 10 shows the extrapolated coronal magnetic-field lines based on the potential field calculation (Yan et al. 1991, 1993; Wang et al. 2002) using a photospheric magnetogram obtained with the SOHO/MDI (Scherrer et al. 1995) and the wavefronts of the Moreton wave and the EIT wave. The direction of propagation and the angular range of the wave front of the Moreton wave, at first sight, seem to be guided by the loops connected to the northern edge of the flare site where the coronal dimming was observed in the EIT images, though the filament oscillation suggests that the Moreton wave propagates across the closed magnetic loops. The visibility of the Moreton wave may be related to the magnetic field configuration and coronal dimming. On the other hand, the direction of propagation of the EIT

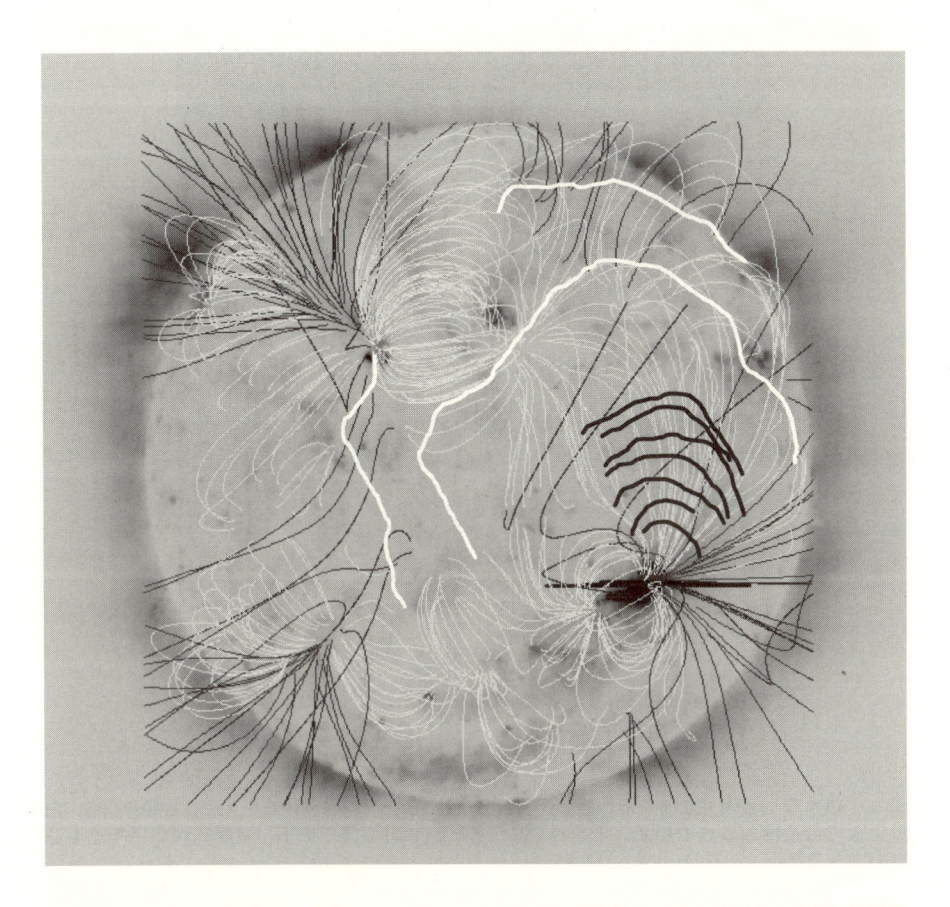

図10 ポテンシャル磁場の上に描かれた，モートン波（太い黒線）と EIT 波（太い白線）の波面．細い黒線と白線はそれぞれ開いた磁力線と閉じた磁力線を表している．波面は図6のものと同様である．背景は 05:56 UT の EIT の画像である．

Fig. 10. Wave fronts of the Moreton wave (thick black lines) and the EIT wave (thick white lines) overlaid on the potential magnetic field lines. The thin black and white lines show the open and closed field lines, respectively. The wave fronts are the same as that in figure 6. The background is the EIT image at 05:56 UT.

ナ減光に関係があるようである．その一方で，EIT 波の伝播方向はループ方向に収束しておらず，ループの外側，つまりコロナ減光領域の外側にまで伝播している．波が閉じた磁気ループを横切って伝播するという点では，フィラメント周辺まで伝播したであろうモートン波は EIT 波と似た性質を示している．しかしながら上述の通り，フィラメント振動を根拠に，EIT 波とモー

wave dose not seem to be confined by the loops, and the EIT wave is observed outside them, namely outside the dimming region. Considering that the waves propagated across the closed magnetic field, the Moreton wave around the filament may be similar to the EIT wave. However, as we have mentioned above, the EIT wave-front is not co-spatial with the Moreton wave front suggested by the filament oscillation and the propagation

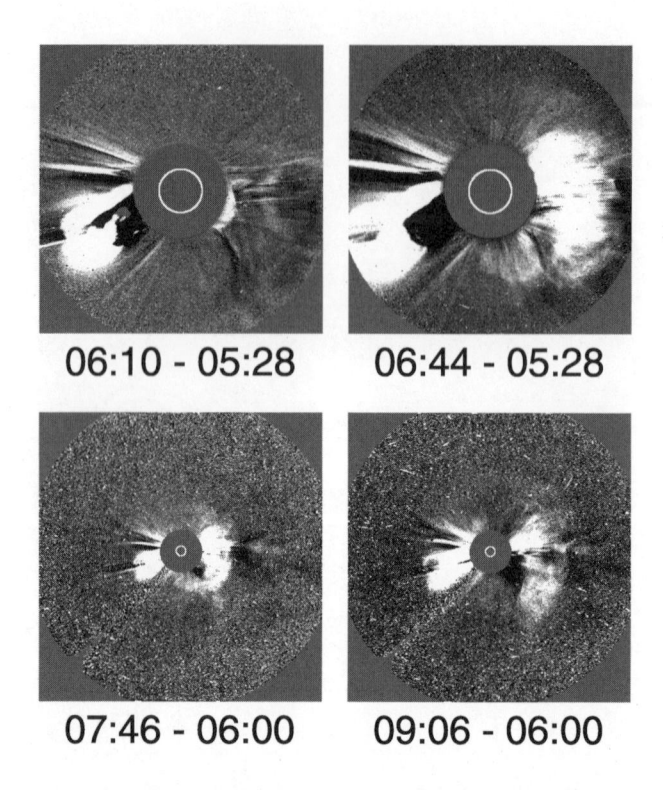

図11 1997年11月4日のモートン波に伴うハローCMEを示したSOHO/LASCOの差分画像．白丸は太陽の位置を表している．上段はC2，下段はC3の画像である．波面先端の平均速度は790 km/sである．

Fig. 11. SOHO/LASCO difference images showing the halo CME associated with the Moreton wave on 1997 November 4. The white circle represents the optical disk of the Sun. The upper and lower panels are C2 and C3 images, respectively. The average velocity of the leading edge is about 790 km s^{-1}.

トン波の波面がずれていることや，両波の伝播速度が大きくことなることが推測される．このことは，モートン波はその伝播速度がアルフベン速度を大きく超える衝撃波(Wang 2000)であり，EIT波はモートン波によって引き起こされた通常のファストモード波であるという可能性を示唆しているかもしれない．別の可能性として，EIT波は長波長の閉じ込められたMHD波であり，水平方向の群速度がファストモードの速度よりもずっと小さいことも考えられる(Uchida et al. 2001)．

speeds of both waves are clearly different. It may be possible that the Moreton wave is a super-Alfvenic shock (Wang 2000) whose propagation speed is larger than that of the MHD fast-mode wave, whereas the EIT wave is a usual fast-mode MHD wave which is excited by the Moreton wave. An alternative possibility is that the EIT wave is a trapped MHD wave with a long wavelength, whose horizontal group-velocity is much less than the fast-mode speed (Uchida et al. 2001).

A halo CME was observed in association with this event. Figure 11 shows SOHO/

Hα−0.8Å　Hα center　Hα+0.8Å

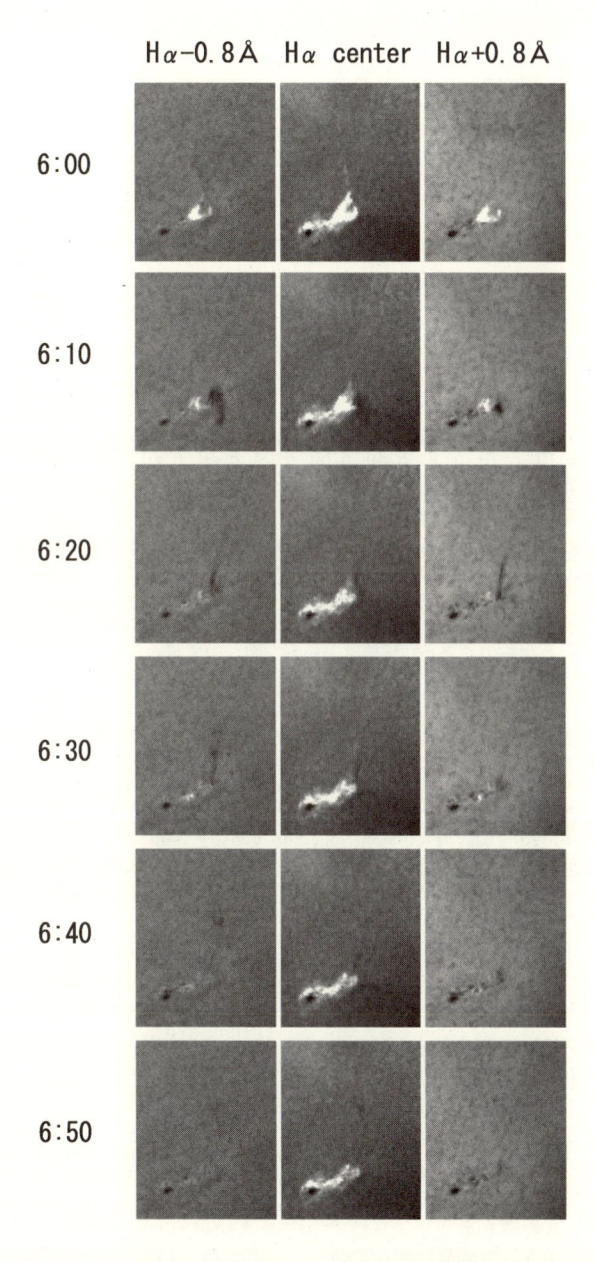

図12　1997年11月4日の06:00 UTから始まったプロミネンス噴出を，Hα−0.8Å（左），Hα線中心（中），Hα+0.8Å（右）を用いてFMTで撮像したもの．視野は図2の領域S3である．

Fig. 12. Images of the prominence eruption starting from 06:00 UT on 1997 November 4 obtained in Hα−0.8Å (left), Hα (center), and Hα+0.8Å (right) with FMT. The field of view is box S3 in figure 2.

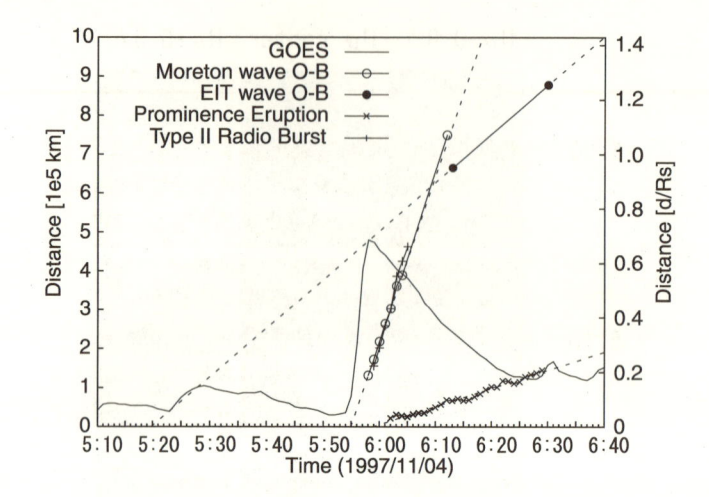

図 13 モートン波と EIT 波，Type-II 電波バースト，プロミネンス噴出，の距離と時刻の関係に，GOES 軟 X 線（1-8Å 帯）フラックスのライトカーブを重ねて描いたグラフ．縦軸はそれぞれ，モートン波と EIT 波については図 6 中の直線 OB に沿って計ったもの，Type-II 電波バーストについてはコロナ中の電子密度モデルをもとに光球から計ったもの，プロミネンス噴出については太陽表面に射影した距離を計ったものである．破線はモートン波，EIT 波，プロミネンス噴出の計測点を外挿した線である．「Rs」は太陽半径（695800 km）を表す．

Fig. 13. Time evolutions of the distances of the Moreton wave, the EIT wave, the type-II radio burst, and the prominence eruption with GOES X-ray flux data at the 1-8Å channel (higher flux curve in figure 1). The vertical axis represents the distances from the flare site along the line **O–B** in figure 6 for the Moreton wave and the EIT wave, the distance from the photosphere calculated with the coronal electron density model for the type-II radio burst, and the projected distance for the prominence eruption. The dashed lines are extrapolated lines for the Moreton wave, the EIT wave, and the prominence eruption. 'Rs' represents the solar radius (695800 km).

ハロー CME もこの現象に伴って観測された．図 11 に SOHO/LASCO の差分画像を載せた．現象が起こる前の画像を引いたものである．06:10 UT に明るい先端が C2 コロナグラフの遮蔽板の西側に現れ，06:44 UT にはぼんやりした構造が遮蔽板の周りを取り囲んでいる．先端の平均速度は約 790 km/s であり，これはモートン波の速度と一致するものでもある．

プロミネンス噴出もこの現象に伴って発生した．図 12 は Hα − 0.8Å，Hα + 0.8Å，Hα 線中心の波長で捉えたプロミネンス噴出の様子で，06:00 UT，06:10 UT，06:20 UT，06:30 UT，06:40 UT，06:50 UT のものである．見かけの速度は約 77 km/s で

LASCO difference images with a suitable base image in the pre-event stage subtracted. A bright leading edge appeared above the western occulting disk of the C2 coronagraph at 06:10 UT, and faint extensions surrounded the occulting disk in the second image at 06:44 UT. The average velocity of the leading edge is about 790 km s^{-1}. This velocity is again similar to that of the Moreton wave.

A prominence eruption also occurred in association with the same event. Figure 12 shows the prominence eruption at the wavelengths of Hα − 0.8 Å, Hα center, Hα + 0.8 Å at 06:00 UT, 06:10 UT, 06:20 UT, 06:30 UT, 06:40 UT, and 06:50 UT. The projected erupting velocity is about 77 km s^{-1}. The original site of the erupting

あった．このプロミネンス噴出の発生点は図4の矢印で示された小さく黒い領域とは異なる．噴出方向はモートン波の伝播方向とほぼ一致している．

最後に，本現象で生じたすべての現象，フレア，モートン波，フィラメント振動，EIT波，Type-II 電波バースト，CME，プロミネンス噴出，の関係について議論する．図13にモートン波，EIT波，Type-II 電波バースト，プロミネンス噴出の距離と時間の関係を，GOESのX線フラックス（1-8Å帯）のライトカーブとともに示す．ここで示したのは，図6の直線 OB 上のフレア領域からのモートン波，EIT波の距離，Type-II 電波バーストの光球からの距離をコロナの電子密度モデルから推定したもの，プロミネンス噴出を投影した距離である．破線はモートン波と EIT 波，プロミネンス噴出の距離を外挿したものである．フレアが始まったのは 05:52 UT であるが，軟X線フラックスが上昇したのは 05:55 UT からである．モートン波，EIT波，プロミネンス噴出の発生時刻はそれぞれ，05:55 UT, 05:20 UT（図6のすべての線から評価したもの），05:59 UT と推定される．上で議論した通り，モートン波の距離は Type-II 電波バーストの距離とほぼ同じである．モートン波と Type-II 電波バーストの発生時刻は軟X線フラックスの立ち上がりと同時刻である一方，プロミネンス噴出の発生時刻はフレアのピーク時刻に対応するので，これらの現象はフレアと強く関連していると考えられる．もし，EIT波が実際に 05:20 UT に発生し，一定速度で伝播したとすると，擾乱は 05:30 UT と 05:39 UT の画像で確認されるはずである．しかしながら，05:30 UT の画像を 05:39 UT の画像から差し引いた差分画像（図5左上）には擾乱は見られなかった．この EIT 波の特性は Thompson et al. (1999) でも報告されている．彼らは，拡散型 EIT 波は，コロナ減光領域の外側でしか観測されないので，フレア領域よりもむしろ，コロナ減光領域から始まっているのかもしれないと主張した．今回の現象についても同様のことがいえる（図6参照のこと）．特筆すべきことは，

prominence is different from the small black region indicated by the arrow in figure 4. The direction of the eruption is approximately the same as the direction of propagation of the Moreton wave.

Finally, we discuss the relationship between all phenomena in this event: flare, Moreton wave (including filament oscillation), EIT wave, type-II radio burst, CME, and prominence eruption. Figure 13 shows the time evolutions of the distances of the Moreton wave, the EIT wave, the type-II radio burst, and the prominence eruption with GOES X-ray flux data at the 1-8Å channel (higher flux curve in figure 1). We plot the distances from the flare site along line **O-B** in figure 6 for the Moreton wave and the EIT wave, the distance from the photosphere calculated with the coronal electron density model for the type-II radio burst, and the projected distance for the prominence eruption. The dashed lines are extrapolated lines for the Moreton wave, the EIT wave, and the prominence eruption. The rapid increase in the soft X-ray flux started at 05:55 UT, though the flare started at 05:52 UT. The extrapolated start time of the Moreton wave, the EIT wave, and the prominence eruption are 05:55 UT, 05:20 UT (estimated with all lines' data in figure 6), and 05:59 UT, respectively. As discussed above, the distance of the Moreton wave was similar to that of the type-II radio burst. The start times of the Moreton wave and the type-II radio burst were similar to that of the rapid increase in the soft X-ray flux, and the start time of the prominence eruption was nearly simultaneous with the flare peak time. These phenomena seem to be closely related to the flare. If the EIT wave actually started at 05:20 UT and propagated with a constant velocity, the disturbance should be seen in the images at even 05:30 UT and 05:39 UT. However, there is no disturbance in the running difference image at 05:39 UT subtracted by the image at 05:30 UT (upper-left panel in figure 5). This characteristic of EIT waves is also reported by Thompson et al. (1999). They suggest that the diffuse EIT waves may have started not from the flare site, but from the boundary of the coronal dimming region, since they were observed only outside the dimming region. This is

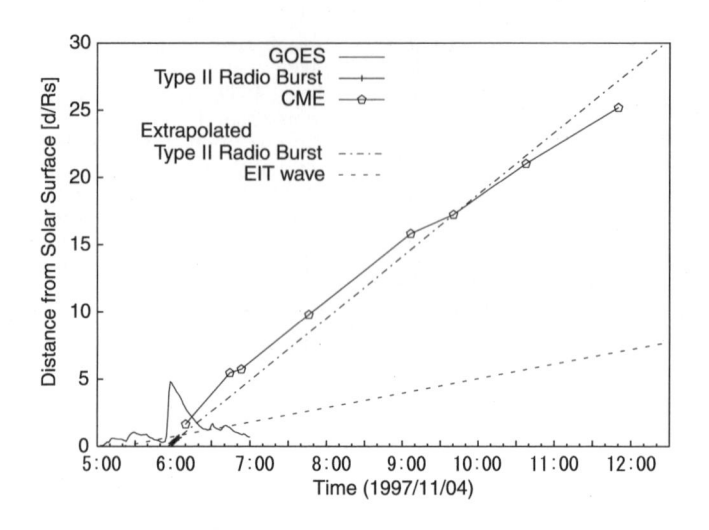

図14 Type-II 電波バーストと CME の，距離と時刻の関係に，GOES 軟 X 線（1-8Å 帯）フラックスのライトカーブを重ねて描いたグラフ．Type-II 電波バーストについては，コロナ中の電子密度モデルを元に光球から計ったものである．CME については，位置角 243°に沿って計測したものである．2種類の破線は，Type-II 電波バーストと EIT 波の計測点を外挿したものである．「Rs」は太陽半径（695800 km）を表す．

Fig. 14. Time evolutions of the distance from the photosphere of the type-II radio burst and the CME with GOES soft X-ray data at the 1-8Å channel (higher flux curve in figure 1). The distance of the type-II radio burst was calculated using the coronal electron density model. The distance of the CME was measured at a position angle of 243°. Two kinds of dashed lines represent the extrapolated lines of the type-II radio burst and the EIT wave, respectively. 'Rs' represents the solar radius (695800 km).

Hα 線中心で見られる振動フィラメントの輝度の急激な変化は 06:22 UT に起こっており，これは EIT 波がフィラメントを通過した時刻に対応していることである．

図14 の縦軸には光球からの距離を，Type-II 電波バーストと CME について描かせたもので，1-8Å の GOES 軟 X 線フラックスのライトカーブを重ねてプロットしたものである．ここで，CME の距離は位置角 243°の方向に沿って計測した．CME は Type-II 電波バーストのすぐ後に生じたように見えることから，共通の領域で発生したことが推測できる．EIT 波とは違って，モートン波，Type-II 電波バースト，CME は同程度の速度を示した．図13, 図14 から，EIT 波以外の現象は非常に関連性の高い現

also true in this event (see figure 6). It is interesting to note that the drastic change in the brightness of the oscillating dark filament in Hα center occurred at about 6:22 UT (see figure 8), which corresponds to the time when the EIT wave passed the filament.

Figure 14 shows the time evolutions of the distances from the photosphere of the type-II radio burst and the CME with GOES soft X-ray data at the 1-8 Å channel (higher flux curve in figure 1). The distance of the CME was measured at a position angle of 243°. The CME seems to follow the type-II radio burst, implying a common origin. In contrast with the EIT wave, the Moreton wave, the type-II radio burst, and the CME have similar velocities. From figures 13 and 14, we can say that

象であるということができる。これらの現象の関連性は今後の論文でより詳細に議論していく予定である。

著者一同は有益なコメントを頂いたH. S. Hudson, M. Kadota 両氏に対して、また、Type-II 電波バーストのデータを提供して頂いた Hiraiso Solar Terrestrial Research Center に対して深く感謝している。SOHO は European Space Agency（ESA）と National Aeronautics and Space Administration（NASA）の国際共同ミッションである。本研究は JSPS Japan–US Cooperation Science Program（主要研究者：K.Shibata, K.I.Nishikawa）により、部分的にサポートされたものである。

these phenomena, except for the EIT wave, seem to be closely involved. The relationship between these phenomena will be discussed in more detail in our future paper.

The authors wish to thank H. S. Hudson and M. Kadota for their useful comments, and the Hiraiso Solar Terrestrial Research Center for providing us with the data of the type-II radio burst. SOHO is a mission of international cooperation between European Space Agency (ESA) and National Aeronautics and Space Administration (NASA). This work was supported in part by the JSPS Japan-US Cooperation Science Program (Pricipal Investigators: K. Shibata and K. I. Nishikawa).

参考文献／References

Athay, R. G., & Moreton, G. E. 1961, ApJ, 133, 935
Brueckner, G. E., Howard, R. A., Koomen, M. J., Korendyke, C. M., Michels, D. J.,Moses, J. D., Socker, D. G., Dere, K. P., et al. 1995, Sol. Phys., 162, 357
Delaboudinière, J.-P., Artzner, G. E., Brunaud, J., Gabriel, A. H., Hochedez, J. F., Millier, F., Song, X. Y., Au, B., et al. 1995, Sol. Phys., 162, 291
Kai, K. 1970, Sol. Phys., 11, 310
Klassen, A., Aurass, H., Mann, G., & Thompson, B. J. 2000, A&A, 141, 357
Kondo, T., Isobe, T., Igi, S., Watari, S., & Tokumaru, M. 1995, J. Commun. Res. Lab., 42, 111
Kurokawa, H., Ishiura, K., Kimura, G., Nakai, Y., Kitai, R., Funakoshi, Y., & Shinkawa, T. 1995, J. Geomag. Geoelectr., 47, 1043
Mann,G., Jansen, F., MacDowall, R. J., Kaiser, M. L., & Stone, R.G. 1999, A&A, 348, 614
Moreton, G. E. 1960, AJ, 65, 494
Moreton, G. E. 1961, Sky and Telescope, 21, 145
Moreton, G. E., & Ramsey, H. E. 1960, PASP, 72, 357
Moses, D., Clette, F., Delaboudinière, J.-P., Artzner, G. E., Bougnet, M., Brunaud, J., Carabetian, C., Gabriel, A. H., et al. 1997, Sol. Phys., 175, 571
Scherrer, P. H., Bogart, R. S., Bush, R. I., Hoeksema, J. T., Kosovichev, A. G., Schou, J., Rosenberg, W., Springer, L., et al. 1995, Sol. Phys., 162, 129
Smith, S. F., & Harvey, K. L. 1971, in Physics of the Solar Corona, ed. C. J. Macris (Dordrecht: Reidel), 156
Thompson, B. J., Gurman, J. B., Neupert, W. M., Newmark, J. S., Delaboudinière, J.-P., St. Cyr, O. C., Stezelberger, S., Dere, K. P., Howard, R. A., & Michels, D. J. 1999, ApJ, 517, 151
Thompson, B. J., Plunkett, S. P., Gurman, J. B., Newmark, J. S., St. Cyr, O. C., & Michels, D. J. 1998, Geophys. Res. Lett., 25, 2465
Thompson, B. J., Reynolds, B., Aurass, H., Gopalswamy, N., Gurman, J. B., Hudson, H. S., Martin, S. F., & St. Cyr, O. C. 2000, Sol. Phys., 193, 161
Uchida, Y. 1968, Sol. Phys., 4, 30
Uchida, Y. 1970, PASJ, 22, 341
Uchida, Y. 1974, Sol. Phys., 39, 431
Uchida, Y., Tanaka, T., Hata, M., & Cameron, R. 2001, Publ. Astron. Soc. Australia, 18, 345
Wang, T., Yan, Y., Wang, J., Kurokawa, H., & Shibata, K. 2002, ApJ, in press
Wang, Y.-M. 2000, ApJ, 543, L89

Warmuth, A., Vrsnak, B., Aurass, H., & Hanslmeier, A. 2001, ApJ, 560, L105
Yan, Y., Yu, Q., & Kang, F. 1991, Sol. Phys., 136, 195
Yan, Y., Yu, Q., & Shi, H. 1993, in Advances in Boundary Element Techniques, ed. J. H. Kane, G. Maier, N. Tosaka, & S. N. Atluri (New York: Springer-Verlag), 447

A-4

1997年11月3日のモートン波における
Hαと軟X線の同時観測

SIMULTANEOUS OBSERVATION OF A MORETON WAVE ON 1997 NOVEMBER 3 IN Hα AND SOFT X-RAYS

成影典之[1], Hugh S. Hudson[2], 森本太郎[1], 秋山幸子[1,3],
北井礼三郎[1], 黒河宏企[1], 柴田一成[1]

[1] 京都大学大学院理学研究科附属天文台, [2] Solar Physics Research Corporation,
[3] Naval Research Laboratory

(原論文) The Astrophysical Journal, 572:L109-L112, 2002 June 10

N. Narukage[1], H. S. Hudson[2], T. Morimoto[1], S. Akiyama[1,3], R. Kitai[1], H. Kurokawa[1], and K. Shibata[1]

[1] Kwasan and Hida Observatories, Kyoto University
[2] Solar Physics Research Corporation
[3] Naval Research Laboratory

概要

1997年11月3日の04:36 UTから04:41 UTの間，京都大学附属飛騨天文台にあるフレア監視望遠鏡において，Hα線（中心と±0.8Å）の観測でモートン波が観測された．モートン波が観測された活動領域（NOAA活動領域8100）は，「ようこう」衛星に搭載されている軟X線望遠鏡でも同時刻に観測されており，軟X線の画像でも波のような擾乱（X線波）が見つかった．X線波の波面の位置や方向は，モートン波のそれとほぼ一致する．モートン波とX線波の伝播速度はそれぞれ，490±40 km/sと630±100 km/sであった．X線波がMHDファストモード衝撃波（MHD fast mode shock）であるとすると，MHDの衝撃波理論と波面の前面側および後面側で観測される軟X線光度とに基づいて，その伝播速度を見積もることができる．ファストモード衝撃波の推定速度は400から760 km/sで，観測されたX線波の速度とほぼ一致する．X線波のファストモードマッハ数は1.15から1.25くらいであることもわかった．これらの結果から，X線波はコロ

ABSTRACT

We report the observation of a Moreton wave in Hα (line center and ±0.8 Å) with the Flare Monitoring Telescope at the Hida Observatory of Kyoto University at 4:36-4:41 UT on 1997 November 3. The same region (NOAA Active Region 8100) was simultaneously observed in soft X-rays with the soft X-ray telescope on board *Yohkoh*, and a wavelike disturbance ("X-ray wave") was also found. The position of the wave front as well as the direction of propagation of the X-ray wave roughly agree with those of the Moreton wave. The propagation speeds of the Moreton wave and the X-ray wave are about 490 ± 40 and 630 ± 100 km s^{-1}, respectively. Assuming that the X-ray wave is an MHD fast-mode shock, we can estimate the propagation speed of the shock, on the basis of MHD shock theory and the observed soft X-ray intensities ahead of and behind the X-ray wave front. The estimated fast shock speed is 400-760 km s^{-1}, which is in rough agreement with the observed propagation speed of the X-ray wave. The fast-mode Mach number of the X-ray wave is also estimated to be about 1.15-1.25. These results suggest

ナを通過する弱い MHD ファストモード衝撃波であり，すなわちモートン波のコロナにおける対応物である，と考えられる．

1. 序論

モートン波とは，フレアに伴って Hα 線（特に Hα 線のウィング）で観測される，太陽面上を横切って伝播する波である（Moreton 1960; Smith & Harvey 1971）．モートン波の伝播速度は 500-1500km/s で，限られた角度方向にアーチ状の波面が伝わって行く．Type-II 電波バースト（Kai 1969）や EIT 波（Thompson et al. 2000; Klassen et al. 2000; Warmuth et al. 2001）と対応していることも多い．モートン波は，コロナにおける MHD ファストモード衝撃波と彩層の交わった所であると解明された（Uchida 1968, 1970; Uchida, Altschuler, & Newkirk 1973）．しかしながら，モートン波の出現メカニズムはまだ明らかになっていない．

近年，「ようこう」衛星に搭載されている軟 X 線望遠鏡（SXT）がフレアに伴ったコロナにおける波のような擾乱を発見した（Khan & Hudson 2000）．この軟 X 線で見られる波を，この論文では X 線波と呼ぶ．モートン波と X 線波の初めての同時観測は，Khan & Aurass（2002）によって報告された．この論文では，2 番目の観測例を報告する．興味深いことに，我々のイベントと Khan-Aurass のイベントは同じ日（1997 年 11 月 3 日）に同じ活動領域（NOAA 活動領域 8100）で観測されているが，時間帯が異なる（それぞれ 04:32 UT と 09:04 UT）．次の日には，同じ活動領域で別のモートン波が観測された（Eto et al. 2002）．

この論文の目的は，X 線波はモートン波のコロナにおける対応物，すなわち MHD ファストモード衝撃波なのか，という基本的な疑問を，MHD 衝撃波理論と「ようこう」衛星の軟 X 線画像に基づいて解明することである．解析の結果，観測された X 線波の性質は，コロナ MHD ファストモード

that the X-ray wave is a weak MHD fast-mode shock propagating through the corona and hence is the coronal counterpart of the Moreton wave.

Subject headings: shock waves—Sun: chromosphere—Sun: corona—Sun: flares—Sun: magnetic fields

1. INTRODUCTION

Moreton waves are flare-associated waves observed to propagate across the solar disk in Hα, especially in the wing of Hα (Moreton 1960; Smith & Harvey 1971). They propagate at speeds of 500-1500 km s^{-1} with arclike fronts in somewhat restricted angles and are often associated with type II radio bursts (Kai 1969) and coronal EIT waves (Thompson et al. 2000; Klassen et al. 2000; Warmuth et al. 2001). The Moreton wave has been identified as the intersection of a coronal MHD fast-mode weak shock wave and the chromosphere (Uchida 1968, 1970; Uchida, Altschuler, & Newkirk 1973). However, the generation mechanism of a Moreton wave has not been made clear yet.

Recently, the soft X-ray telescope (SXT) on board *Yohkoh* discovered wavelike disturbances in the solar corona associated with flares (Khan & Hudson 2000), which we call "X-ray waves" in this Letter. The first simultaneous observation of a Moreton wave and an X-ray wave was reported by Khan & Aurass (2002). In this Letter, we report a second example. Interestingly, our event and the Khan-Aurass event were observed on the same day (1997 November 3) in the same region (NOAA Active Region 8100) but at different times (04:32 and 09:04, respectively). The next day, another Moreton wave occurred in this region (Eto et al. 2002).

The purpose of this Letter is to examine the basic question of whether the X-ray wave is the coronal counterpart of the Moreton wave, i.e., a coronal MHD fast-mode shock, based on MHD shock theory and *Yohkoh*/SXT soft X-ray images. The result shows that the observed properties of the X-ray wave are consistent with the wave interpreted as a coronal MHD fast-mode weak shock.

In § 2, the instrumentation and observing methods are summarized. In § 3, the results of the analysis of the Moreton

衝撃波であると解釈して矛盾しないということがわかった.

2節では,装置と観測方法に関してまとめている.3節ではモートン波とX線波解析結果を述べ,最後に4節にはまとめと議論が述べられている.

2. 観測

この論文では,1997年11月3日にNOAA活動領域8100 (S20°, W13°) で起こったCクラスフレアに伴って観測された,2種類の波を研究する.1つは彩層のモートン波,もう1つはコロナのX線波である.フレアは04:32 UTから増光を始め,04:38 UTに最も明るくなった.

モートン波は,京都大学飛騨天文台にあるフレアモニター望遠鏡 (FMT; Kurokawa et al. 1995) でHα線 (中心と±0.8Å) を用いて観測された.フレア監視望遠鏡は4種類 (Hα線中心と±0.8Å と連続光) の全面像と,1種類の太陽縁像 (Hα線中心) を撮影する.我々は,全面像をHα+0.8Åで撮影した画像を使用した.というのは,モートン波の波面はHα の赤色側ウィングで最も鮮明に見ることができるからである.これは,波面における質量が,彩層とぶつかるときに下向きに動いているからである.この論文で使われている画像の時間分解能は1分である (ただし,FMTはもっと速い時間分解能で動作している).ピクセルサイズは4.2秒角である.

X線波は「ようこう」衛星のSXT (Tsuneta et al. 1991) で軟X線の波長を用いて観測された.SXTは0.28-4 keV (波長3-45Åに対応) のエネルギー領域にある軟X線光子に対して感度がある.この論文では,AlMgフィルターを用いて観測された部分像を使用した.観測の時間分解能はだいたい50秒である.半分の分解能と1/4の分解能をもつ画像のピクセルサイズはそれぞれ,4.91秒角と9.82秒角である.

光球における磁場は,Solar and Heliospheric Observatory (SOHO) に搭載されているMichelson Doppler Imager (MDI; Scherrer et al. 1995) で観測された.時間分解能とピクセルサイズはそれぞれ,

wave and the X-ray wave are described. In §4, a summary and a discussion are given.

2. OBSERVATIONS

In this Letter, we study two kinds of flare-associated waves: the chromospheric Moreton wave and coronal X-ray wave associated with a C-class flare in NOAA AR 8100 at S20°, W13° on 1997 November 3. The flare started at 04:32 UT and peaked at 04:38 (*GOES* times).

The Moreton wave was observed in Hα (line center and ±0.8 Å) with the Flare Monitoring Telescope (FMT; Kurokawa et al. 1995) at the Hida Observatory of Kyoto University. The FMT observes four full-disk images (in the Hα line center and ±0.8 Å and continuum) and one solar limb image (in the Hα line center). We use the full-disk images in Hα+0.8 Å, because a Moreton wave front is best seen in the red wing of the Hα line (Hα+0.8 Å). This is because the mass motion at the wave front is downward when the front enters the chromosphere. The time resolution of the images used in this Letter is 1 minute, although the FMT operates at a higher time resolution. The pixel size is $4''.2$.

The X-ray wave was observed in soft X-rays with *Yohkoh*/SXT (Tsuneta et al. 1991). The SXT is sensitive to soft X-ray photons in the energy range ~ 0.28-4 keV, which corresponds to wavelengths ~ 3-45 Å. In this Letter, we use partial frame images observed with the AlMg filter. The observing time cadence is about 50 s. The pixel sizes of the half- and quarter-resolution images are $4''.91$ and $9''.82$, respectively.

The magnetic field of the photosphere was observed with the Michelson Doppler Imager (MDI; Scherrer et al. 1995) on board the *Solar and Heliospheric Observatory* (*SOHO*). The time resolution and pixel sizes are about 90 minutes and $2''$, respectively.

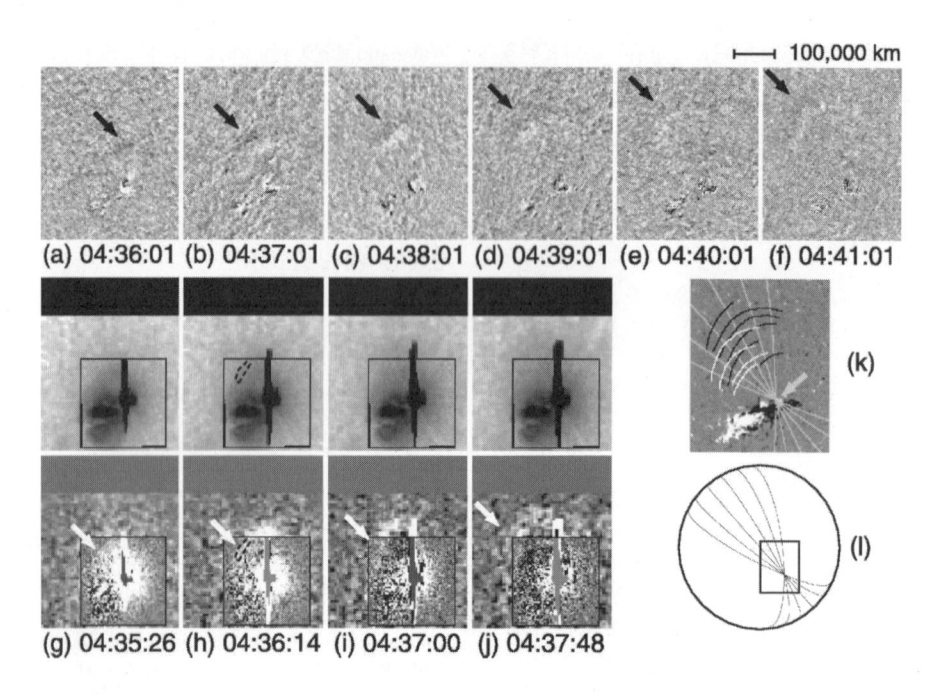

図1　1997年11月3日に撮影されたNOAA活動領域8100. (a)-(f) Hα+0.8Åでのモートン波（黒矢印）の差分画像. (g)-(j) X線波（白矢印）のAlMgフィルター軟X線像（中央のパネル）とその差分画像（下のパネル）. 四角で囲われている内側は半分の空間分解能（4.91秒角），外側は1/4の空間分解能（9.82秒角）にそれぞれ対応する. (k) 04:36:01 UTから04:41:01 UTまでの1分ごとにおけるモートン波の波面（黒）と04:35:26, 04:36:14, 04:37:00, 04:37:48 UTにおけるX線波の波面（白）. 背景は04:51:04 UTに撮影された光球磁場. 灰色の線はフレアの位置（灰色矢印）を通る大円を示している. (l)に示されている長方形，円，線はそれぞれ(a)-(k)の視野，太陽の縁，大円をそれぞれ意味している.

Fig. 1.—Observed images on 1997 November 3 at NOAA AR 8100. (*a–f*) Hα+0.8Å running difference images of a Moreton wave (*black arrows*). (*g–j*) Soft X-ray (*middle panels*) and running difference (*bottom panels*) images of an X-ray wave (*white arrows*) taken with the AlMg filter. The images inside and outside the boxes are half-resolution (4″.91) and quarter-resolution (9″.82) images, respectively. (*k*) Wave fronts of the Moreton wave at every minute from 04:36:01 to 04:41:01 UT (*black lines*) and the X-ray wave at 04:35:26, 04:36:14, 04:37:00, and 04:37:48 UT (*white lines*) overlaid on the photospheric magnetic field observed at 04:51:04. Gray lines show the great circles through the flare site (*gray arrow*). The rectangle, circle, and lines shown in (*l*) are the field of view of (*a*)-(*k*), the limb of the Sun, and the great circles, respectively.

約90分と2秒角である.

3. 結果

図1はNOAA活動領域の1997年11月3日に観測された像を示している. 図1の

3. RESULTS

Figure 1 shows the observed images on 1997 November 3 of NOAA AR 8100. Figures 1*a*-1*f* are "running difference"

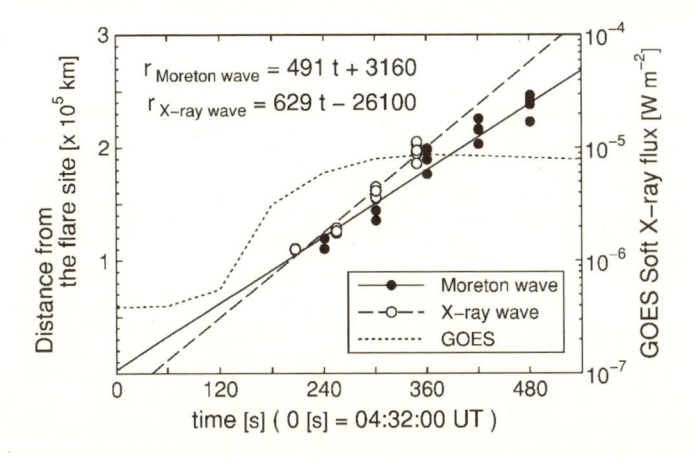

図2 モートン波（黒丸）とX線波（白丸）の伝播距離．1次の多項式近似がそれぞれ実線（モートン波）と破線（X線波）で示されている．伝播速度はそれぞれ，モートン波は 490±40 km/s，X線波は 630±100 km/s である．点線は GOES 9 衛星で観測された軟 X 線のフラックスを示している．

Fig. 2. — Propagation of the Moreton wave (*filled circles*) and the X-ray wave (*open circles*). First-degree polynomial fits of the Moreton wave (*solid line*) and the X-ray wave (*dashed line*) are shown. The propagation speeds of the Moreton wave and the X-ray wave are 490 ± 40 and 630 ± 100 km s^{-1}, respectively. The dotted line shows the soft X-ray flux observed by the *GOES 9* satellite.

(a)-(f)は，Hα+0.8Åで観測された差分画像（running difference：各画像から直前の画像を差し引きしたもの）である．差分画像の手法により，モートン波を明確に見ることができる．図1の(g)-(j)は，軟X線像（中央のパネル）とその差分画像（下のパネル）である．図1(k)は MDI の視線方向磁場の上に波面の見かけの位置を重ねたものである．差分画像を使って，モートン波（黒線）は 04:36:01 UT から 04:41:01 UT まで，X 線波（白線）は 04:35:26 UT から 04:37:48 UT まで，視覚的に見出すことができる．両方とも大体同じ方向に伝播し，位置もほぼ一致している．視野の大きさは図1(l)で説明している．

図2はフレア位置から波面までの大円に沿った距離をプロットしたものである．フレア位置は，最も明るい領域と定義した．モートン波とX線波の伝播速度を4つの大円上における点の分散を用いて求めると，

images (where each image has the previous image subtracted from it) observed in Hα+0.8 Å. The running difference method clearly shows the motion of the Moreton wave. Figures 1g-1j are the direct soft X-ray images (*middle panels*) and running difference images (*bottom panels*). In Figure 1k, the apparent positions of the wave fronts are overlaid on the longitudinal magnetogram from MDI. Using the running difference images, the Moreton wave (*black lines*) can be identified visually from 04:36:01 to 04:41:01 UT, and the X-ray wave (*white lines*) from 04:35:26 to 04:37:48 UT. Both propagate roughly in the same direction and agree approximately in location. The field of view is illustrated by the box in Figure 1l.

In Figure 2, the distances of the wave fronts from the flare site along the great circles are plotted. We assume the brightest region is the flare site. The propagation speeds of the Moreton wave

それぞれ 490±40 と 630±100 km/s でほぼ一定である．実際は，モートン波は少し減速しているように見える．我々が測定した位置は，ピーク位置というよりむしろ，波面の先端を検出している．

4. 議論

Uchida モデル (1968, 1970; Uchida et al. 1973) では，モートン波は弱い MHD ファストモード衝撃波がコロナ中を伝播する時に彩層と交差する裾の部分であると考えている．すなわちこのモデルは，長波長の Type-II 電波バーストもコロナにおける伝播の徴候として説明することができ，彩層におけるモートン波の対応物がコロナにおいて存在することを予測している．実際，我々が解析したイベントでは，X 線波の伝播方向だけでなく波面の位置も，モートン波のそれとほぼ一致している [図1(k)]．

ここで，X 線波が本当に MHD ファストモード衝撃波（ここからは単に高速衝撃波と呼ぶ）なのかどうか，すなわちモートン波のコロナにおける対応物であるのかどうかを調べてみよう．X 線波を高速衝撃波と仮定すると，MHD の衝撃波理論と波面の前面側および後面側で観測される軟 X 線光度とに基づいて，その伝播速度を推定することができる．もしこの衝撃波の推定速度が観測された X 線波の伝播速度と一致すれば，X 線波は高速衝撃波であると結論づけることができる．

MHD 衝撃波理論において密度比が急に変化する条件 (Priest 1982) は，

$$X \equiv \frac{\rho_2}{\rho_1}, \quad (1)$$

に対して

$$(v_1^2 - Xv_{A1}^2)^2 \{Xc_{s1}^2 + \frac{1}{2}v_1^2 \cos^2\theta_1 [X(\gamma-1)-(\gamma+1)]\}$$

$$+ \frac{1}{2} v_{A1}^2 v_1^2 \sin^2\theta_1 X$$

$$\times \{[\gamma + X(2-\gamma)]v_1^2 - Xv_{A1}^2[(\gamma+1)-X(\gamma-1)]\} = 0,$$

$$(2)$$

$$\frac{v_{2x}}{v_{1x}} = X^{-1}, \quad (3)$$

and the X-ray wave are roughly constant at 490±40 and 630±100 km s^{-1}, respectively, using the scatter of points from the four great circles as a measure. In fact, the Moreton wave seems to be somewhat decelerating. The positions refer to the leading edge of the wave front rather than to the peak.

4. DISCUSSION

In the model of Uchida (1968, 1970; Uchida et al. 1973), a Moreton wave is the sweeping skirt on the chromosphere of a weak MHD fast-mode shock that propagates in the corona. This model thus also explains the meter-wave type II radio burst as a signature of the coronal propagation and predicts the existence of a coronal counterpart of the chromospheric Moreton wave. In fact, in our event, the position of the wave front as well as the direction of propagation of the X-ray wave roughly agree with those of the Moreton wave (Fig. 1k).

Let us now examine whether the X-ray wave is an MHD fast-mode shock (hereafter simply called a fast shock), i.e., the coronal counterpart of the Moreton wave. Assuming that the X-ray wave is a fast shock, we can estimate its propagation speed, based on MHD shock theory and the observed soft X-ray intensities ahead of and behind the X-ray wave front. If this estimated speed of the shock is consistent with the observed propagation speed of the X-ray wave, we can conclude that the X-ray wave is a fast shock.

The jump condition of MHD shock theory (Priest 1982), for

$$X \equiv \frac{\rho_2}{\rho_1}, \quad (1)$$

gives

$$(v_1^2 - Xv_{A1}^2)^2 \{Xc_{s1}^2 + \frac{1}{2}v_1^2\cos^2\theta_1[X(\gamma-1)-(\gamma+1)]\}$$

$$+ \frac{1}{2}v_{A1}^2 v_1^2 \sin^2\theta_1 X$$

$$\times \{[\gamma+X(2-\gamma)]v_1^2 - Xv_{A1}^2[(\gamma+1)-X(\gamma-1)]\}=0,$$

$$(2)$$

$$\frac{v_{2x}}{v_{1x}} = X^{-1}, \quad (3)$$

$$\frac{v_{2y}}{v_{1y}} = \frac{v_1^2 - v_{A1}^2}{v_1^2 - Xv_{A1}^2}, \quad (4)$$

$$\frac{v_{2y}}{v_{1y}} = \frac{v_1^2 - v_{A1}^2}{v_1^2 - Xv_{A1}^2}, \tag{4}$$

$$\frac{p_2}{p_1} = X + \frac{(\gamma-1)Xv_1^2}{2c_{s1}^2}\left(1 - \frac{v_2^2}{v_1^2}\right), \tag{5}$$

$$\frac{T_2}{T_1} = \frac{p_2/p_1}{\rho_2/\rho_1} = \frac{p_2/p_1}{X}. \tag{6}$$

である.ここでは,衝撃波の前面側の量を添字 1,後面側の量を添字 2 で表している.また,座標系は衝撃波を基準とした運動座標系である.式(2)で, v_{A1} はアルフベン速度 $[=\beta_1/(\mu_0\rho_1)^{1/2}]$ を示し, c_{s1} は音速を意味し, $(\gamma\rho_1/\rho_1)^{1/2} = (\gamma RT_1/\tilde{\mu})^{1/2}$ で与えられる.ここで,$\tilde{\mu}$ は平均原子量(この論文では Priest 1982 に従って 0.6 としてある).ガンマは比熱比(この論文では 5/3), θ_1 は衝撃波の法線(x軸)に対する上流の磁場の傾き角を表す.y軸は衝撃波の法線に対して垂直であり,速度ベクトルがxy平面に乗るように取っている.式(2)からは v_1 に対して 3 つの解が得られるが,そのうち最も値の大きいものがファストモード衝撃波に対応する. θ_1 が 90°に近づく限界では,衝撃波は垂直になる.これら6つの方程式を用いて,前面の量 $(\rho_1, T_1, B_1, \theta_1, v_1)$ から後面の量 $(\rho_2, T_2, B_2, \theta_2, v_2)$ を決定する.

最も単純な近似では,軟X線の明るさはプラズマ密度の二乗と装置特有の温度関数に比例する.

$$I_X = n^2 lf(T) \propto \rho^2 f(T), \tag{7}$$

ここで I_X は軟X線の明るさ, n はプラズマの数密度, l は視線方向のプラズマスケール長, $f(T)$ はプラズマの放射率とフィルター感度(Tsuneta et al. 1991)を表す.つまり,上記方程式は $(I_{X1}, T_1, B_1, \theta_1, v_1)$ が $(I_{X2}, T_2, B_2, \theta_2, v_2)$ を決める,と書き換えることができる.いい換えれば,もし $(I_{X1}, T_1, B_1, \theta_1, I_{X2})$ がわかれば, $(v_1, T_2, B_2, \theta_2, v_2)$ が求められる(別のアプローチについては,Appendix と図 3 を参照).衝撃波の前面側(領域 1)での温度 (T_1) や密度 (ρ_1) といった物理基本量は,フレア前の画像とフィルター比手法(Tsuneta et al. 1991)を用いて計算することができる.コロナの磁場強度 (B_1) は MDI の画

$$\frac{p_2}{p_1} = X + \frac{(\gamma-1)Xv_1^2}{2c_{s1}^2}\left(1 - \frac{v_2^2}{v_1^2}\right), \tag{5}$$

$$\frac{T_2}{T_1} = \frac{p_2/p_1}{\rho_2/\rho_1} = \frac{p_2/p_1}{X}. \tag{6}$$

Here the quantities ahead of the shock are denoted by 1 and those behind by 2, in a frame of reference moving with the shock. In equation (2), v_{A1} is the Alfvén velocity $[=\beta_1/(\mu_0\rho_1)^{1/2}]$, c_{s1} the sound speed, given as $(\gamma\rho_1/\rho_1)^{1/2} = (\gamma RT_1/\tilde{\mu})^{1/2}$, where $\tilde{\mu}$ is the mean atomic weight (taken as 0.6 in this Letter; Priest 1982), γ the ratio of specific heats (= 5/3 in this Letter), and θ_1 the inclination of the upstream magnetic field to the shock normal (the x-axis). The y-axis is taken to be perpendicular to the shock normal, such that the magnetic field and velocity vectors are in the x-y plane. There are three solutions for v_1 from equation (2), among which the largest corresponds to the fast-mode shock. In the limit of $\theta_1 \rightarrow 90°$, the shock becomes perpendicular. Using these six equations, the quantities ahead of the shock $(\rho_1, T_1, B_1, \theta_1, v_1)$ determine those behind $(\rho_2, T_2, B_2, \theta_2, v_2)$.

In the simplest approximation, the soft X-ray intensity is proportional to the square of the plasma density and an instrument-specific function of the temperature,

$$I_X = n^2 lf(T) \propto \rho^2 f(T), \tag{7}$$

where I_X is the soft X-ray intensity, n is the plasma number X density, l is the line-of-sight plasma scale, and $f(T)$ represents the plasma emissivity and the filter response (Tsuneta et al. 1991). Hence, the above can be rewritten such that $(I_{X1}, T_1, B_1, \theta_1, v_1)$ determine $(I_{X2}, T_2, B_2, \theta_2, v_2)$. In other words, if we know $(I_{X1}, T_1, B_1, \theta_1, I_{X2})$, we can find $(v_1, T_2, B_2, \theta_2, v_2)$. (See the Appendix and Fig. 3 for a different approach.) The physical parameters ahead of the shock (i.e., in region 1), such as temperature (T_1) and density (ρ_1) are obtained from the preflare images and the filter ratio method (Tsuneta et al. 1991). The coronal magnetic field strength (B_1) is derived from the MDI images.

We have four images of the X-ray wave (Figs. 1g–1j), but three of them are inappropriate for analysis. The wave front in Figure 1g is too near the flare site and too greatly influenced by scattered light

像から導いた．

我々は 4 枚の X 線波の画像をもっているが，そのうち 3 枚は解析には適していない．図 1(g) における波面の位置は，フレアの位置に近すぎてフレアからの散乱光に強く影響を受ける．図 1(i) と (j) の波面は暗すぎて，ノイズと実際の信号を区別することができない．そこで我々は 04:36:14UT に観測された，図 1(h) の 1 枚のみを解析する．

我々は，I_{X1} と I_{X2} を測定し，$I_{X2}/I_{X1} = 3.27$ を得た．X 線波の明るさは，図 1(h) で破線で囲われている領域で測定した．T_1 値はフレア前に SXT で薄膜 Al フィルターと AlMg フィルターを用いて撮影された全面像を用いて計算し，T_1 は 2.25 ± 0.25 MK と見積もられた．この T_1 を SXT の感応関数（response function）へ代入すると，放射尺度（emission measure）が得られる．密度（ρ_1）の計算は，観測された領域の視線方向スケール（l）をコロナの圧力スケールハイト [2.00-2.50 MK では $(1.00-1.26) \times 10^5$ km] であると仮定して行なった．光球の磁場（B_{p1}）は MDI によって観測されており，コロナの磁場（B_1）は B_{p1} の 1/3 もしくは半分くらいであると仮定した．この仮定は Appendix や図 3 において議論されている異なるアプローチによって正当化されている．$\theta_1 = 90°$ のとき衝撃波は垂直であり，$\theta_1 = 60°$ のときは斜め衝撃波に対応する．式 (1) から (7) を用いると，これらの観測量から表 1 に示されているように衝撃波の伝播速度（$v_{sh} = v_{1x} = v_1 \cos\theta_1$）が求められる．

高速衝撃波の推定速度は 400-760 km/s で，観測された X 線波の伝播速度 630 ± 100 km/s とほぼ一致する．X 線波のファストモードマッハ数は 1.15-1.25 の範囲にあり，矛盾しない．これらの結果から，X 線波はコロナを伝わる高速衝撃波であり，すなわちモートン波のコロナにおける対応物であるといえる．

図 1(k) から，X 線波の伝播方向だけでなく波面の位置も，モートン波のそれと大体合っていることがわかる．しかし，正確な位置や速度は異なる．我々は，これらの

from the flare. The intensities of the wave fronts in Figures 1i and 1j are too low, so we cannot distinguish the real signal from noise. We analyze a single image, Figure 1h, observed at 04:36:14.

We measure I_{X1} and I_{X2} and obtain $I_{X2}/I_{X1} = 3.27$, where the intensities of the X-ray wave are measured in the area encircled by a dashed line in Figure 1h. The value of T_1 is calculated using preflare full-frame images taken with the thin Al and AlMg filters of SXT as $T_1 = 2.25 \pm 0.25$ MK. Substituting this T_1 into the SXT response function gives the emission measure. We calculate the density (ρ_1) assuming that the line-of-sight distance (l) of the observed region is the coronal pressure scale height [$(1.00-1.26) \times 10^5$ km for 2.00-2.50 MK]. The photospheric magnetic field strength (B_{p1}) was observed by MDI, and we assume that the coronal magnetic field strength (B_1) is one-third or one-half of B_{p1}. This assumption is justified from a different approach discussed in the Appendix and Figure 3. When $\theta_1 = 90°$, the shock is perpendicular, and $\theta_1 = 60°$ corresponds to an oblique shock. Using equations (1)-(7), these observational properties give the propagation speed of the shock ($v_{sh} = v_{1x} = v_1 \cos\theta_1$) as shown in Table 1.

The estimated fast-shock speed, 400-760 km s^{-1}, is in rough agreement with the observed propagation speed of the X-ray wave, 630 ± 100 km s^{-1}. The fast-mode Mach number of the X-ray wave is consistent with values in the range of 1.15-1.25. These results suggest that the X-ray wave is a fast shock propagating through the corona and hence is the coronal counterpart of the Moreton wave.

From Figure 1k, we see that the position of the wave front as well as the direction of propagation of the X-ray wave roughly agree with those of the Moreton wave, although the exact positions and speeds are different. We think these differences may be explained by the three-dimensional distribution of the Alfvén speed. We will study the three-dimensional structure of the propagation of the coronal shock with more observed data and will simulate the flare-associated coronal wave in our future work.

From the above analysis, we can estimate the coronal plasma velocity just

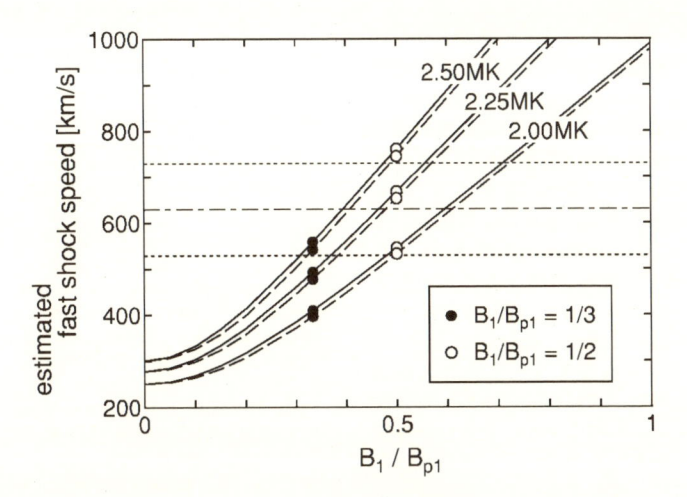

図3 垂直方向（実線）と斜め方向（破線）の衝撃波の推定速度．B1とBp1はそれぞれ，コロナ磁場強度と光球磁場強度を表す．鎖線と点線はそれぞれX線波の観測された伝播速度（630 km/s）とエラー範囲（±100 km/s）を示す．

Fig. 3. — Estimated perpendicular (*solid lines*) and oblique (*dashed lines*) shock speeds. Band B_1 are B_{p1} the coronal and photospheric magnetic field strengths, respectively. The dot-dashed and dotted lines show the observed propagation speeds of the X-ray wave (630 km s^{-1}) and the error bar (±100 km s^{-1}), respectively.

違いはアルフベン速度の三次元分布によって説明できるのではないか，と考えた．将来の研究として，より大量の観測データでコロナ衝撃波の伝播の三次元構造を調べることや，フレアにともなうコロナ波をコンピューターシミュレーションすることを考えている．

上記の解析によって，モートン波（ファストモードのMHD衝撃波）のすぐ後面側にあるコロナプラズマ速度v_1-v_2を100-200 km/sと見積もることができる．これは，SOHO/Coronal Diagnostics Spectrometer や Solar-B/EUV Imaging Spectrometer で観測した場合の視線方向速度50-100 km/sに対応する．

我々の結果では，M_f = 1.15-1.25 であり，モートン波の伝播速度はMHDファストモード速度と同程度であり，弱いMHDファストモード衝撃波であるという Uchidaモデルと矛盾しない．その一方で，Wang (2000)はEIT波の伝播速度はコロ

behind the Moreton wave (fast-mode MHD shock), v_1-v_2, to be about 100-200 km s^{-1}, which would be observed at 50-100 km s^{-1} along the line of sight with the *SOHO*/Coronal Diagnostics Spectrometer or the *Solar-B*/EUV Imaging Spectrometer. Our result, M_f = 1.15-1.25, indicates that the propagation speed of the Moreton wave is comparable to the MHD fast-mode speed and consistent with Uchida's model of a weak MHD fast-mode shock. On the other hand, Wang (2000) showed that the propagation speed of an EIT wave is comparable to the coronal MHD fast-mode speed and suggested that the Moreton wave may be associated with a highly super-Alfvénic shock. This seems inconsistent with our result. But if we assume that the Alfvén (and the fast-mode) velocity is increasing with height in the low corona (Mann et al. 1999), and that X-ray waves and EIT waves are propagating at two different heights, Wang's suggestion may not be in contradiction to our results.

ナにおけるファストモードの速度と同程度であることを示し，モートン波はその伝播速度がアルフベン速度を大きく超えている衝撃波と関係しているという可能性を提示した．これは我々の結果と矛盾するように思える．しかし，アルフベン速度（ファストモード速度）がコロナ底部では高さとともに増加していて（Mann et al. 1999），さらにX線波とEIT波は異なる高さの層を伝播していると考えると，Wangのアイデアは我々の結果と矛盾しない．

　図2にあるようなモートン波とX線波への近似直線は，これらの波の開始時間はフレアよりもいくらか早かったという可能性を示している．しかし，モートン波は普通フレアに伴って起こるので，それはおかしいということになる．これは，図2に見て取れるモートン波の減速によって説明できるかもしれない．モートン波の減速は，Warmuth et al. (2001)の結果と一致するし，モートン波（おそらくX線波も）の初速度が実際は，我々の観測した平均速度よりも速かった，つまり，衝撃波は初期にはもっと強かった，ということを暗示している．

　最後に，EIT波とType-II電波バーストが今回解析したモートン波と同時に観測されていた，ということにも触れておく．EIT波とモートン波の間の関係については，次の論文（Narukage et al. 2002, in preparation）で研究する予定である．

　S. Eto, H. Isobe, A. Asai, H. Kozu, T. T. Ishii, and M. Kadotaとの実りのある議論に感謝する．「ようこう」衛星は日本の国家プロジェクトであり，ISAS，関係国内機関，米国および英国との相互国際協力によって打ち上げられ運用されている．SOHOはEuropean Space Agency (ESA)とNASAの国際協力ミッションである．この研究の一部はJSPS Japan-US Cooperation Science Programの助成を受けて行われたものである（研究責任者：K. ShibataおよびK. I. Nishikawa）．

　The fitting lines for the Moreton wave and X-ray wave in Figure 2 suggest that the start time of these waves may be somewhat earlier than the flare. This is curious since Moreton waves usually occur with flares. It may be explained by the deceleration of the Moreton wave as seen in Figure 2. This is in accordance with the results of Warmuth et al. (2001) and implies that the initial velocities of the Moreton wave and probably also the X-ray wave were actually higher than the mean values we found; i.e., the shock may be stronger in the early phase.

　Finally, we note that an EIT wave and a type II radio burst were also observed simultaneously with this Moreton wave. The relationship between the EIT wave and the Moreton wave will be studied in our future paper (N. Narukage et al. 2002, in preparation).

　The authors thank S. Eto, H. Isobe, A. Asai, H. Kozu, T. T. Ishii, and M. Kadota for useful discussions. The *Yohkoh* satellite is a Japanese national project, launched and operated by ISAS, and involving many domestic institutions, with multilateral international collaboration with the US and the UK. *SOHO* is a mission of international cooperation between the European Space Agency and NASA. This work was supported in part by the JSPS Japan-US Cooperation Science Program (principal investigators: K. Shibata and K. I. Nishikawa).

表1 高速衝撃波の推定速度 v_{sh}

Parameter	$B1/B_{p1}=1/3$ [a]			$B1/B_{p1}=1/2$		
$B1$(G)	2.66	2.66	2.66	3.99	3.99	3.99
$T1$(MK)	2.00	2.25	2.50	2.00	2.25	2.50
EM1 [b] (cm^{-5})	$10^{27.3}$	$10^{27.0}$	$10^{26.9}$	$10^{27.3}$	$10^{27.0}$	$10^{26.9}$
n_1(cm^{-3})	$10^{8.7}$	$10^{8.5}$	$10^{8.4}$	$10^{8.7}$	$10^{8.5}$	$10^{8.4}$
v_{A1}(km/s)	274	328	369	411	492	553
c_{s1}(km/s)	215	228	240	215	228	240
$\theta_1=90°$						
T_2(MK)	2.33	2.75	3.16	2.34	2.77	3.19
X	1.25	1.32	1.36	1.24	1.31	1.35
v_{sh}(km/s)	410	493	559	546	668	761
M_f [c]	1.18	1.23	1.27	1.18	1.23	1.26
$\theta_1=60°$						
T_2(MK)	2.33	2.75	3.16	2.34	2.78	3.20
X	1.24	1.32	1.36	1.24	1.30	1.34
v_{sh}(km/s)	397	478	542	533	653	745
M_f [c]	1.14	1.20	1.23	1.15	1.20	1.24

注：$\theta_1=90°$は垂直衝撃波，$\theta_1=60°$は斜め衝撃波を意味する．
a B_{p1}は光球の磁場強度．
b EM$_1$ $=n_1^2 l$ で，l は視線方向のスケール（コロナの圧力スケールハイトと同等としている）．
c M_fはファストモードマッハ数．

付録

この論文では，コロナ磁場強度 B_1 が光球磁場強度 B_{p1} の 1/2 または 1/3 であると仮定してファストモード衝撃波の速度（$v_{sh} = v_{1x} = v_1\cos\theta_1$）を推定し，$v_{sh}$ が X 線波の速度の観測値とほぼ等しいことを見出した．しかし，別のアプローチも可能である．つまり，$(I_{X1}, T_1, \theta_1, v_1, I_{X2})$ がわかれば $(B_1, T_2, B_2, \theta_2, v_2)$ が決まるから，X 線波の伝播速度の観測値と MHD 衝撃波理論［式(1)-(7)］からコロナ磁場強度を求めることができる．図3は式(1)-(7)に基づいて高速衝撃波の推定速度と B_1/B_{p1} の関係を求めたものである．この図と X 線波の伝播速度の観測値（630±100 km/s）から B_1/B_{p1} は 0.3-0.7 の範囲にあると推定できる．この値は，源泉表面場外挿法（source-surface field extrapolations）により求められたコロナ磁場強度（Dere 1996; Wang 2000 など）とよく一致していると思われる．したがって，このアプローチでも X 線波が MHD

APPENDIX

In this Letter, we estimated the fast-shock speed ($v_{sh} = v_{1x} = v_1\cos\theta_1$) by assuming the coronal magnetic field strength, $B_1 = (\frac{1}{3}$ or $\frac{1}{2})B_{p1}$, and found that v_{sh} is comparable to the observed X-ray wave speed, confirming that the X-ray wave is an MHD fast-mode shock. However, we can take a different approach; it is possible to estimate the coronal magnetic field strength, if we use the observed X-ray wave propagation speed and the MHD shock theory (eqs. [1]-[7]), since $(I_{X1}, T_1, \theta_1, v_1, I_{X2})$ determines $(B_1, T_2, B_2, \theta_2, v_2)$. Figure 3 shows the estimated fast shock speed as a function of B_1/B_{p1} on the basis of equations (1)-(7). From this figure and the observed propagation speed of the X-ray wave, 630±100 km s^{-1}, we can estimate B_1/B_{p1} to be 0.3-0.7. This value seems to be consistent with the calculated coronal magnetic field strength using source-surface field extrapolations (e.g., Dere 1996; Wang 2000). Hence, this approach

TABLE 1
Estimated Fast-Shock Speed v_{sh}

Parameter	$B_1/B_{p1} = 1/3$[a]			$B_1/B_{p1} = 1/2$		
B_1 (G)	2.66	2.66	2.66	3.99	3.99	3.99
T_1 (MK)	2.00	2.25	2.50	2.00	2.25	2.50
EM_1[b] (cm^{-5})	$10^{27.3}$	$10^{27.0}$	$10^{26.9}$	$10^{27.3}$	$10^{27.0}$	$10^{26.9}$
n_1 (cm^{-3})	$10^{8.7}$	$10^{8.5}$	$10^{8.4}$	$10^{8.7}$	$10^{8.5}$	$10^{8.4}$
v_{A1} (km s^{-1})	274	328	369	411	492	553
c_{s1} (km s^{-1})	215	228	240	215	228	240
$\theta_1 = 90°$						
T_2 (MK)	2.33	2.75	3.16	2.34	2.77	3.19
X	1.25	1.32	1.36	1.24	1.31	1.35
v_{sh} (km s^{-1})	410	493	559	546	668	761
M_f[c]	1.18	1.23	1.27	1.18	1.23	1.26
$\theta_1 = 60°$						
T_2 (MK)	2.33	2.75	3.16	2.34	2.78	3.20
X	1.24	1.32	1.36	1.24	1.30	1.34
v_{sh} (km s^{-1})	397	478	542	533	653	745
M_f	1.14	1.20	1.23	1.15	1.20	1.24

Note. — $\theta_1 = 90°$ means perpendicular shock, and $\theta_1 = 60°$ oblique.
[a] B_{p1} is the photospheric magnetic field strength.
[b] $EM_1 = n_1^2 l$, where l is the line-of-sight length, which is assumed equal to the pressure scale height of the corona.
[c] M_f is fast-mode Mach number.

ファストモード衝撃波であることが裏付けられる.

also confirms that the X-ray wave is an MHD fast-mode shock.

参考文献／References

Dere, K. P. 1996, ApJ, 472, 864
Eto, S., et al. 2002, PASJ, in press
Kai, K. 1969, Sol. Phys., 10, 460
Khan, J. I., & Aurass, H. 2002, A&A, 383, 1018
Khan, J. I., & Hudson, H. S. 2000, Geophys. Res. Lett., 27, 1083
Klassen, A., Aurass, H., Mann, G., & Thompson, B. J. 2000, A&AS, 141, 357
Kurokawa, H., et al. 1995, Geomag. Geoelectr., 47, 1043
Mann, G., Klassen, A., Estel, C., & Thompson B. J. 1999, in Proc. Eighth *SOHO* Workshop, Plasma Dynamics and Diagnostics in the Solar Transition Region and Corona, ed. J.-C. Vial & B. Kaldeich-Schümann (ESA SP-446; Noordwijk: ESA), 477
Moreton, G. E. 1960, AJ, 65, 494
Priest, E. R. 1982, Solar Magnetohydrodynamics (Dordrecht: Reidel), chap. 5.4
Scherrer, P. H., et al. 1995, Sol. Phys., 162, 129
Smith, S. F., & Harvey, K. L. 1971, in Physics of the Solar Corona, ed. C. J. Macris (Dordrecht: Reidel), 156
Thompson, B. J., et al. 2000, Sol. Phys., 193, 161
Tsuneta, S., et al. 1991, Sol. Phys., 136, 37
Uchida, Y. 1968, Sol. Phys., 4, 30
—— 1970, PASJ, 22, 341
Uchida, Y., Altschuler, M. D., & Newkirk, G., Jr. 1973, Sol. Phys., 28, 495 Wang, Y. M. 2000, ApJ, 543, L89
Warmuth, A., Vrsnak, B., Aurass, H., & Hanslmeier, A. 2001, ApJ, 560, L105

モートン波に伴うX線で見られた爆発現象

X-Ray Expanding Features Associated with a Moreton Wave

成影典之，森本太郎，門田三和子，北井礼三郎，黒河宏企，柴田一成
京都大学大学院理学研究科附属天文台

(原論文) PASJ: Publ. Astron. Soc. Japan 56, L5-L8, 2004 April 25

Noriyuki NARUKAGE, Taro MORIMOTO, Miwako KADOTA, Reizaburo KITAI,
Hiroki KUROKAWA, and Kazunari SHIBATA
Kwasan and Hida Observatories, Kyoto University

概要

太陽外縁近傍で発生したHα線モートン波と軟X線で見られたコロナ中の2種の爆発噴出現象の同時観測について報告する。X線現象のうち速い爆発現象はX線波動であり，遅いものは放出ガスであると考えられる。彩層内のモートン波は太陽面上を1040±100 km/sの速度で伝播した。一方，X線波動は太陽表面から上空のコロナに向かって1400±250 km/sで伝播した。このX線波動は電磁流体力学波の高速伝播モードの衝撃波であると同定した。そのマッハ数 (M_f) は1.13-1.31であり，伝播するにつれて減少した。ちょうど M_f が1になった時点でモートン波が見えなくなった。この結果に加えて，衝撃波の三次元構造および衝撃波と噴出ガス（さらにはHαフィラメント上昇）の相互関係についてこの論文では議論する。

1. 序論

モートン波は，フレア発生時に太陽表面上を伝播する波であってHα単色像で観測されるものである (Moreton 1960; Smith and Harvey 1971)。この波は，その伝播速度が500-1500 km/sで，波面は一定の角度の円弧状であり，しばしばType-II電波バーストと同時に観測される (Kai 1969)。こ

Abstract

We report on the simultaneous observation of a Moreton wave in Hα and two kinds of coronal expanding features in soft X-rays near the solar limb. We consider the faster X-ray feature and the slower one as being an "X-ray wave" and "ejecta", respectively. The chromospheric Moreton wave propagated on the solar disk at a speed of 1040 ± 100 km s^{-1}, whereas the coronal X-ray wave propagated outside of the disk toward the outer corona at 1400 ± 250 kms^{-1}. We identified the X-ray wave as MHD fast-mode shock. The fast-mode Mach number (M_f) of the X-ray wave was also estimated to be about 1.13-1.31, which decreases during propagation. The timing when the M_f became "1" is consistent with that of the disappearance of the Moreton wave. Moreover, we discuss the 3-dimensional structure of the shock wave and the relation between the shock wave and the ejecta (and Hα filament eruptions).

Key words: shock waves—Sun: chromosphere—Sun: corona—Sun: flares

1. Introduction

Moreton waves are flare-associated waves observed to propagate across the solar disk in Hα (Moreton 1960; Smith, Harvey 1971). They propagate at speeds of 500-1500 kms^{-1} with arc-like fronts in somewhat restricted angles, and are often

のモートン波は，上空のコロナ中を伝播する電磁流体力学的（MHD）高速モードの弱い衝撃波が上空から彩層に突入することによって彩層を励起することにより形成されると考えられている（Uchida 1968, 1974）．しかし，波源での波の発生機構はいまだはっきりとはわかっていない．

近年，大規模で過渡的なコロナ擾乱が Solar and Heliospheric Observatory (SOHO：太陽・太陽圏天文台) 衛星に搭載された Extreme ultraviolet Imaging Telescope（EIT：極端紫外線撮像望遠鏡）によって多数発見されている．現在では，この現象は通常 EIT 波といわれている（Thompson et al. 1998; Klassen et al. 2000; Eto et al. 2002）．

また，「ようこう」衛星搭載の軟 X 線望遠鏡（SXT）は，太陽フレアに付随する波動のような擾乱現象を発見した（Kahn, Hudson 2000; Hudson et al. 2003）．この論文ではこの波動を X 線波ということにする．モートン波と X 線波の同時観測はこれまでに 2 例報告されている（Kahn, Aurass 2002; Narukage et al. 2002）．本論文は，2000 年 3 月 3 日に太陽外縁付近で発生したモートン波と X 線波について報告する．同時観測に成功したのは今回で 3 例目となった．

本論文の目的は，太陽外縁近傍で発生したモートン波と 2 種類の X 線爆発噴出現象の同時観測を基にフレアに付随する MHD 衝撃波の特性を明らかにすることである．速い方の X 線現象は X 線波，遅い方の X 線現象はガス噴出現象であると考えられる．そして，X 線波の研究からそれが MHD 高速モード衝撃波であることがわかった．その衝撃波の高速モードマッハ数（M_f）は衝撃波が伝播するにつれて減少した．M_f の変化とモートン波の消失との間に相関があることを発見した．さらに，モートン波と X 線波の研究からこの現象での衝撃波の三次元構造が詳細に明らかになった．衝撃波と噴出ガス（Hα フィラメントの噴出についても）との関係についてもまた議論する．

2 節は，観測装置と観測方法についての

associated with type-II radio bursts (Kai 1969). The Moreton wave has been identified as the intersection of a coronal MHD fast-mode weak shock wave and the chromosphere (Uchida 1968, 1974). However, the generation mechanism of a Moreton wave has not yet been cleared.

Recently, many large-scale coronal transients have been discovered using the Extreme ultraviolet Imaging Telescope (EIT) on board Solar and Heliospheric Observatory (SOHO). These features are now commonly called "EIT waves" (Thompson et al. 1998; Klassen et al. 2000; Eto et al. 2002).

Moreover, the Soft X-ray Telescope (SXT) on board Yohkoh discovered wave-like disturbances in the solar corona associated with flares (Khan, Hudson 2000; Hudson et al. 2003), which we call "X-ray waves" in this paper. Two simultaneous observations of Moreton waves and X-ray waves were reported (Khan, Aurass 2002; Narukage et al. 2002). In this paper, we report on a third example, which was observed near the solar limb on 2000 March 3.

The purpose of this paper is to consider the properties of a flare-associated MHD shock wave based on the simultaneous observation of a Moreton wave and two kinds of X-ray expanding features observed near the solar limb. We consider the faster X-ray feature and the slower one as being an "X-ray wave" and "ejecta", respectively. Therefore, we examine the X-ray wave and identify it as an MHD fast-mode shock. The fast-mode Mach number (M_f) of the shock wave decreased during the propagation. We found a relationship between M_f and the disappearance of the Moreton wave. Moreover, in this event the 3-dimensional structure of the shock wave was clearly revealed by the Moreton wave and the X-ray wave. The relation between the shock wave and the ejecta (and Hα filament eruptions) is also discussed.

In section 2, the instrumentation and observing methods are summarized. In section 3, the results of analyses of the Moreton wave and the X-ray wave are described. In section 4, a summary and a discussion are given.

2. 観測

2000年3月3日 S15, W60 にある NOAA 活動領域 8882 で発生した M クラスフレアに付随したモートン波, X 線現象を観測した. そのフレアは 02:08 UT に始まり, 02:14 UT に極大を迎えたものである (時刻は GOES 衛星で測定された X 線強度変化から求められたもの).

モートン波は京都大学飛騨天文台のフレア監視望遠鏡 (Flare Monitoring Telescope; FMT) (Kurokawa et al. 1995) で Hα線 (中心波長と ±0.8Å) で観測された. FMT は4つの太陽全面画像 (Hα中心波長と ±0.8Åおよび連続光) と1つの太陽外縁画像 (Hα中心波長) を観測しており, これらの5種の画像はすべて同時に取得されたものである. FMT では時間分解能の高い観測を行っているが, 本研究では時間分解能 14s の画像を用いて解析した. ピクセルサイズは4.2秒角である.

X 線波とガス噴出は Yohkoh/SXT (Tsuneta et al. 1991) により軟 X 線で観測された. SXT は 0.28-4 keV のエネルギー帯 (波長 3-45Å に相当する) の軟 X 線光子に感度をもつ. 私たちは Al-Mg フィルターで観測された部分画像を用いた. 観測時間間隔は 14 秒である, 1/2 分解能画像と 1/4 分解能画像のピクセルサイズはそれぞれ 4.91, 9.82 秒角である.

光球磁場は SOHO 衛星に搭載されている Michelson Doppler Imager (MDI) 装置 (Scherrer et al. 1995) で観測された. 時間分解能, ピクセルサイズはそれぞれ 90 分, 2 秒角である. メーター波 Type-II 電波バーストは Hiraiso Radio Spectrograph (HiRAS) によって記録されたものを用いた (Kondo et al. 1995).

3. 結果

図1は 2000 年 3 月 3 日に観測された NOAA AR 8882 の像である. 図(a)-(c) の上段の図は Hα+0.8Å で作成した差分画

2. Observations

We observed a Moreton wave and X-ray features associated with an M-class flare in NOAA Active Region 8882 at S15, W60 on 2000 March 3. The flare started at 02:08 UT and peaked at 02:14 (GOES times).

The Moreton wave was observed in Hα (line center and +/−0.8Å) with the Flare Monitoring Telescope (FMT) (Kurokawa et al. 1995) at Hida Observatory of Kyoto University. The FMT observed four full-disk images (in the Hα line center and +/−0.8 Å, and continuum) and one solar limb image (in the Hα line center). These images were observed co-temporal. The time resolution of the images used in the present work was 14 s, though the FMT operated at a higher time resolution. The pixel size was 4″.2.

The X-ray wave and ejecta were observed in soft X-rays with Yohkoh/SXT (Tsuneta et al. 1991). The SXT is sensitive to soft X-ray photons in the energy range ∼0.28-4 keV, which corresponds to wavelengths ∼3-45 Å. We used partial frame images observed with an Al-Mg filter. The observing time cadence was 14 s. The pixel sizes of the half and quarter-resolution images were 4″.91 and 9″.82, respectively.

The magnetic field of the photosphere was observed with the Michelson Doppler Imager (MDI) (Scherrer et al. 1995) on board SOHO. The time resolution and pixel sizes are about 90 min and 2″, respectively. A metric type-II radio burst was recorded with the Hiraiso Radio Spectrograph (HiRAS) (Kondo et al. 1995).

3. Results

Figure 1 shows the observed images on 2000 March 3 of NOAA AR 8882. The top panels of images (a)-(c) are "running

Appendix A Selected Papers Relating to FMT | 441

⊢━━⊣ 100,000 km

(a) 02:13:01 (b) 02:13:43 (c) 02:14:11

(d)

(e)

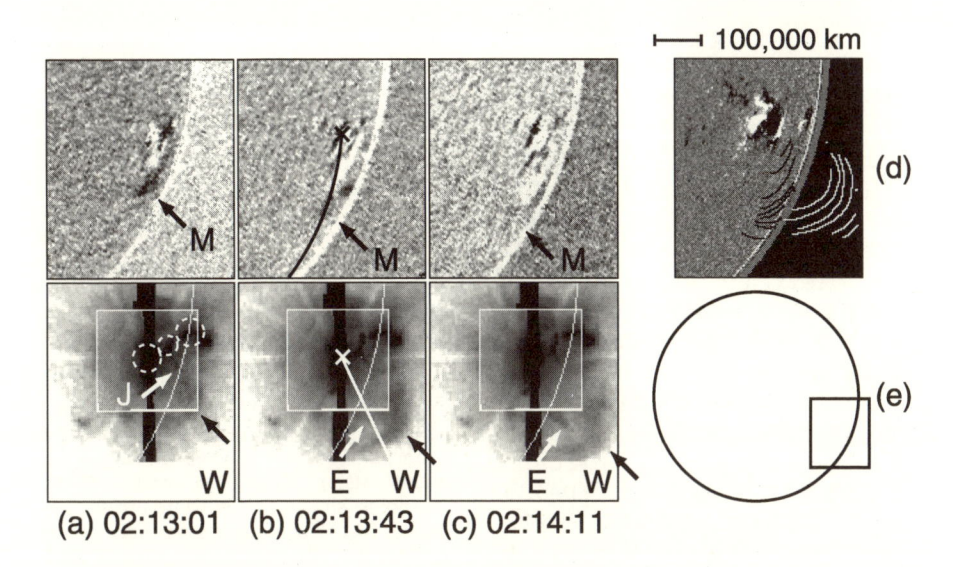

図1 2000年3月3日のNOAA AR 8882の観測画像．(a)-(c)の上段の画像はモートン波（黒い矢印M）のHα+0.8Å差分画像である．(a)-(c)の下段の画像はAl-MgフィルターでX線撮影された X 線波（黒い矢印W），噴出ガス（白い矢印E），ジェット（白い矢印J）の白黒反転軟 X 線画像である．画像の内側，外側の四角はそれぞれ1/2分解能（4".91），1/4分解能（9".82）の画像である．(a)の白い破線の円は軟X線で明るい領域を示す．(b)の×印と線はそれぞれフレアの発生した位置と波を計測した位置を示す．Hα線で見たときに最も明るい場所をフレア発生点としている．(d)は02:12:01から02:15:01 UTまでのモートン波（黒い線），02:13:01から02:14:11 UTまでのX線波（白い線）の波面をそれぞれ01:35:02 UTに観測された光球磁場の画像に重ねたものである．これらの波面位置は眼で定めた．(e)の四角と円はそれぞれ(a)-(d)の視野と太陽外縁を示す．

Fig. 1. Observed images on 2000 March 3, at NOAA AR 8882. Top panels of images (a)-(c) are Hα+0.8Å "running difference" images of a Moreton wave (black arrows "M"). Bottom panels of images (a)-(c) are negative soft X-ray images of an X-ray wave (black arrows "W"), ejecta (white arrows "E") and a jet (white arrow "J") taken with the Al-Mg filter. The images inside and outside the boxes are half-resolution (4".91) and quarter-resolution (9".82) images, respectively. The white dashed circles in image (a) show the bright regions in soft X-rays. The crosses and lines in image (b) are the flare site and measures of the wave, respectively. We assume that the brightest region in Hα is the flare site. Image (d) shows the wave fronts of the Moreton wave from 02:12:01 to 02:15:01 UT (black lines) and the X-ray wave from 02:13:01 to 02:14:11 UT (white lines) overlaid on the photospheric magnetic field observed at 01:35:02 UT. These wave fronts were determined visually. The box and circle shown in image (e) are the field of view of images (a)-(d) and the limb of the Sun, respectively.

像（running difference）である（差分画像とはある画像から1分前の画像を引いたものであり，それぞれの画像の変化した成分を示す）．この方法はモートン波の動きを明瞭に示すことができる．下段は軟X線画像を白黒反転して示したものである．

difference" images (where each image has the image before one minute subtracted from it) observed in Hα+0.8 Å. The running-difference method clearly shows the motion of the Moreton wave. The bottom panels are the negative soft X-ray images. There are two kinds of expanding

2種類の爆発噴出現象があり黒い矢印W, 白い矢印Eで示す. フレア中の軟X線で明るい領域を(a)の3つの白い破線の円で示す. フレアが発生したとき, 左と右の領域は同時に明るくなる. そしてWの現象は左の領域から広がる. 数10秒後, 領域の中心部で輝点とジェットのような現象Jを観測した.

Wの現象は「ループの上昇」あるいは「X線波」という2つの可能性が考えられる. もしWが「ループの上昇」ならば, ループの端 (いい換えれば足下) は太陽表面に固定されていなければならない. しかしながら, Wの南側の端は南方に動き, 伝播するモートン波のちょうど上に位置している (図1および3より). 一方で白い矢印Jによって示されている北側の足下は固定されているように見える. WとJのX線で明るくなる時刻の時間差から, 私たちはWとJが同じループのものではないと考える. したがってWは「ループの上昇」現象ではなく「X線波」であり, EとJはそれぞれ「噴出ガス」と「ジェット」と考えることができる.

(d)は波面の見かけの位置をMDIの視線方向磁場マップに重ね書きしたものである. モートン波 (黒線) は02:12:01 UTから02:15:01 UTまで, X線波 (白線) は02:13:01 UTから02:14:11 UTまでの間見られた. 彩層モートン波は太陽表面上を伝播し, コロナX線波はコロナ上空に向かって伝播した. 観測領域は図1(e)の四角で示している.

図2(f), (g)はそれぞれ大円に沿ってのモートン波位置の時間変化, 天球面上のX線増光位置の時間変化を示す. この位置の時間変化から求まるモートン波, X線波の伝播速度はほぼ等速でそれぞれ 1040 ± 100 km/s, 1400 ± 250 km/s である. 実際はモートン波はいくらか減速しているようである. 噴出ガスEは490 km/sで動いた.

4. 議論

Uchidaモデル (1968, 1974) では, モートン波は弱いMHD高速モード衝撃波面が

features shown by the black arrows, "W", and the white arrows, "E". In the flare, the three regions indicated by the white dashed circles in image (a) brightened in soft X-rays. When the flare occurred, the left and right regions became bright at the same time and feature "W" expanded from the left region. After several tens of seconds, we observed the brightening and a jet-like feature, "J", at the center region.

There are two possibilities: either feature "W" is an "erupting loop" or an "X-ray wave". If "W" is an "erupting loop", the edges (i.e., footpoints) should be fixed on the solar disk. However, the southern edge of "W" moved southward and was located just above the propagating Moreton wave (see figures 1 and 3). Meanwhile, the apparent northern footpoint, shown by the white arrow, "J", seems to be fixed. However, because of the time lag between the X-ray brightenings related to "W" and "J", we think "W" and "J" are not parts of the same loop. Hence we consider "W" is not an "erupting loop" but instead an "X-ray wave". "E" and "J" are considered to be "ejecta" and a "jet", respectively.

In image (d), the apparent positions of the wave fronts are overlaid on the longitudinal magnetogram from MDI. The Moreton wave (black lines) can be identified visually from 02:12:01 to 02:15:01 UT, and the X-ray wave (white lines) from 02:13:01 to 02:14:11 UT. The chromospheric Moreton wave propagated on the solar disk, whereas the coronal X-ray wave propagated toward the outer corona. The field of view is illustrated by the box in figure 1 (e).

Figures 2 (f) and (g) show timeslice images of the Moreton wave along the great circle and the X-ray expanding features along the plane-of-sky, respectively. The propagation speeds of the Moreton wave and the X-ray wave derived from the timeslice images are roughly constant at 1040 ± 100 km s^{-1} and 1400 ± 250 km s^{-1}, respectively. In fact, the Moreton wave seems to be somewhat decelerating. The ejecta, "E", moved at speeds of 490 km s^{-1}.

4. Discussion

In Uchida's model (1968, 1974), a Moreton wave is a sweeping skirt on the

(f) Timeslice image of the Moreton wave

(g) Timeslice image of the X-ray wave

(h) Estimated fast-mode Mach number

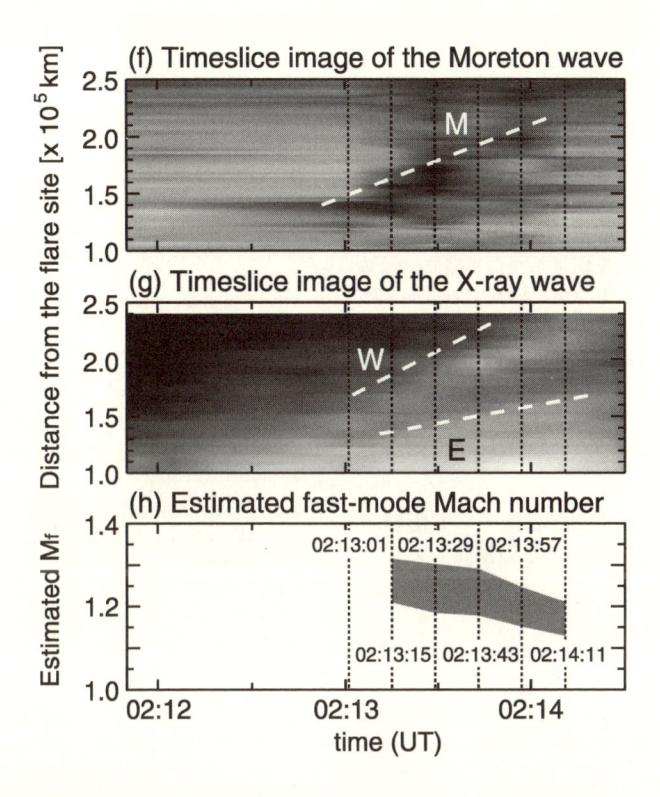

図2 画像の時間変化と高速モード衝撃波マッハ数の時間変化．(f)は大きな円弧[図1(b)の黒い線]に沿ったHα画像の時間発展を示したものである．(g)は天球面[図1(b)の白線]に沿った軟X線画像の時間発展を示したものである．時間変化画像から導かれるモートン波 "M"，X線波 "W"，噴出ガス "E" の速度はそれぞれ1040 km/s，1400 km/s，490 km/sである．(h)の灰色部分は高速モード衝撃波のマッハ数の変化を示したものである．

Fig. 2. Timeslice images and estimated fast-mode Mach number. Image (f) is the timeslice image of the Hα images along the great circle [black line in figure 1 (b)]. Image (g) shows the timeslice image of the soft X-ray images along the plane-of-sky [white line in figure 1 (b)]. The speeds of the Moreton wave, "M", the X-ray wave, "W", and the ejecta, "E", derived from the timeslice images are 1040 km s^{-1}, 1400 km s^{-1}, and 490 km s^{-1}, respectively. The gray area of image (h) is the range of the estimated fast-mode Mach number.

コロナ中をスカート上に拡がり，それが彩層面を掃くときに彩層に引き起こす擾乱である．このモデルはメーター波 Type-II 電波バーストがコロナ中を伝播する波によるということも説明し，さらに彩層モートン波に対応してコロナ中の擾乱があることを予言する．このX線波がモートン波に対応するコロナ中の擾乱である可能性がある．

chromosphere of a weak MHD fast-mode shock that propagates in the corona. This model thus also explains the meter-wave type-II radio burst as a signature of the coronal propagation, and predicts the existence of a coronal counterpart of the chromospheric Moreton wave. There is the possibility that the X-ray wave is a coronal counterpart of the Moreton wave.

実際，X 線波の１つの例では，MHD 理論を用いて MHD 高速モード衝撃波であることが確認されている (Narukage et al. 2002)．

今回のイベントでは，フレアは太陽外縁近傍で発生した．従って X 線波がコロナ上空に向かって伝播する一方，彩層モートン波は太陽表面を伝播した．X 線波の伝播する方向はモートン波の伝播する方向とは異なっている［図 1(d)］．しかしながら，伝播速度はほぼ等しい．三次元構造を考えると，この結果は X 線波が彩層モートン波に対応するコロナ中の現象であるという考えと矛盾していないということができる．

4.1 X 線波が MHD 高速モード衝撃波かどうかについての検討

X 線波が観測されたとき，Hiraiso Radio Spectrograph（HiRAS）はメーター波 Type-II 電波バーストを記録した．Type-II 電波バーストは MHD 衝撃波の存在を示す．Mann のコロナ電子密度モデル (Mann et al. 1999) を基に光球からの電波源高度を計算すると 02:14 UT では 223000 km，02:16 UT では 417000 km であった．したがって衝撃波の速度は 1600 ± 300 km/s である．02:14 UT の電波源高度は X 線現象と一致する［図 2(g)］．Mancuso et al. (2002) は別の密度モデルを用いて同じ衝撃波の速度は 1100 km/s であると報告している．これらの値は X 線波の伝播速度 1400 ± 250 km/s とほぼ一致している．したがって，X 線波がコロナの MHD 衝撃波である可能性がある．

X 線波が高速モード衝撃波であると仮定すると MHD 衝撃波の理論（ランキン – ユゴニオの関係式）と波面の前方（領域 1）と後方（領域 2）の軟 X 線光度（I_X）を用いて衝撃波の伝播速度を推定することができる．もしこの推定した衝撃波の速度が観測された X 線波の伝播速度と一致したならば，私たちは X 線波を高速モード衝撃波であると結論づけることができる．

$(I_{X1}, T_1, B_1, \theta_1, I_{X2})$ がわかれば，$(v_1, T_2, B_2, \theta_2, v_2)$ がわかる (Narukage et al.

One case of the X-ray wave was examined using MHD theory, and was identified as an MHD fast-mode shock (Narukage et al. 2002). In this event, the flare occurred near the solar limb. Hence the chromospheric Moreton wave propagated on the solar disk, whereas the coronal X-ray wave propagated toward the outer corona. The direction of propagation of the X-ray waves disagrees with those of the Moreton wave [figure 1 (d)]. However, the propagation speeds are comparable. Considering the 3-dimensional structure, we can say that these results are not inconsistent with the assumption that the X-ray wave is the coronal counterpart of the Moreton wave.

4.1. Examination Whether the X-Ray Wave is an MHD Fast-Mode Shock

The Hiraiso Radio Spectrograph (HiRAS) recorded a metric type-II radio burst when the X-ray wave was observed. The type-II radio burst indicates the existence of an MHD shock wave. Based on Mann's coronal electron density model (Mann et al. 1999), we can calculate the source height from the photosphere as being 223000 km at 02:14 and 417000 km at 02:16, and hence the speed of the shock wave as being 1600 ± 300 km s^{-1}. The source height at 02:14 is consistent with the X-ray features [see figure 2 (g)]. Mancuso et al. (2002) also reported the speed of the same shock wave as being 1100 km s^{-1} using another density model. These are roughly consistent with the propagation speed of the X-ray wave, 1400 ± 250 km s^{-1}. Hence, there is the possibility that the X-ray wave is a coronal MHD shock wave.

Assuming that the X-ray wave is fast shock, we can estimate its propagation speed, based on MHD shock theory (Rankine–Hugoniot relation) and the observed soft X-ray intensities (I_X) ahead of (i.e., in region 1) and behind (i.e., in region 2) the X-ray wave front. If this estimated speed of the shock is consistent with the observed propagation speed of the X-ray wave, we can conclude that the X-ray wave is a fast shock.

If we know $(I_{X1}, T_1, B_1, \theta_1, I_{X2})$, we can find $(v_1, T_2, B_2, \theta_2, v_2)$ (see Narukage et al.

2002).我々は,02:13:15,02:13:29,02:13:43,02:13:57,02:14:11 UT の I_{X1} と I_{X2} を測定して I_{X2}/I_{X1} = 3.74, 4.18, 4.02, 3.35, 2.83 を得た.温度 T_1 は SXT (Tsuneta et al.1991)の薄膜 Al フィルターと Al-Mg フィルターで撮ったフレア発生前の全面像を用いて 02:13:15 UT では T_1 = 2.25-2.75MK, 02:13:29-02:14:11 UT では T_1 = 2.00-2.50MK であると求められた.SXT の感応関数 (response function) にこの T_1 を代入すると放射尺度 (emission measure) が得られる.観測領域の視線方向の厚さ (l) をコロナの圧力スケールハイト(2.00-2.75MK では 1.00-1.38 × 10^5 km)とすると,コロナ密度 (ρ_1) は $10^{8.3}$-$10^{8.6}$ cm^{-3} である.l は衝撃波の空間的大きさとほぼ同じであり,推定された ρ_1 は実際のコロナ密度と一致する.光球磁場強度 (B_{p1}) は MDI によって観測されており,コロナ磁場強度 B_1 は B_{p1} の 1/3 または 1/2 であると考えられる (Dere 1996).θ_1 が 90°のとき,衝撃波は垂直衝撃波であり,θ_1 が 60°のときは斜め衝撃波となる.観測から衝撃波の伝播速度 $(v_{sh} = v_1 \cos\theta_1)$ が求められる.高速モード衝撃波の推定された速度 900-1900 km/s は,観測された X 線波の伝播速度 1400 ± 250 km/s とほぼ一致する.

上記の考察によって X 線波がコロナ中を伝播する高速モード衝撃波であることが示唆される.衝撃波の速度から X 線波の高速モードマッハ数も 1.13-1.31 であることがわかった.この値はモートン波の伝播速度は MHD 高速モードと同程度であることを意味しており,モートン波が弱い MHD 高速モード衝撃波起源のものであるという Uchida モデルと一致する.モートン波の減速があることを図 2(f) で見ることができる.これは Warmuth et al. (2001) の結果と一致する.X 線波面(高速モード衝撃波)が時間と共に巾広くなる様子,あるいは明るさが淡くなる様子が図 2(g) で確認できる.推定された高速モードマッハ数も図 2(h) のように減少している.減少率は -0.0016 /s である.02:14:11 UT の後,X 線波は視野から外に出てしまう.もしマッハ数が同じ割合で減少していくと,

2002). We measure I_{X1} and I_{X2} observed at 02:13:15, 02:13:29, 02:13:43, 02:13:57, and 02:14:11 UT, and obtain I_{X2}/I_{X1} = 3.74, 4.18, 4.02, 3.35, and 2.83, respectively. The temperature (T_1) was calculated using pre-flare full frame images taken with the thin Al and Al-Mg filters of SXT (Tsuneta et al. 1991) as T_1 = 2.25-2.75MK at 02:13:15 UT and T_1 = 2.00-2.50MK at 02:13:29-02:14:11 UT. Substituting this T_1 into the SXT response function gave the emission measure. We calculated the coronal density (ρ_1) as $10^{8.3}$-$10^{8.6}$ cm^{-3}, assuming that the line-of-sight thickness (l) of the observed region was the coronal pressure scale-height (1.00-1.38 × 10^5 km for 2.00-2.75 MK). The l is comparable to the scale of the waves, and the estimated ρ_1 is consistent with the real coronal density. The photospheric magnetic field strength (B_{p1}) was observed by MDI, and we assumed that the coronal magnetic field strength (B_1) was 1/3 or 1/2 of B_{p1} (Dere 1996). When θ_1 = 90°, the shock is perpendicular, and θ_1 = 60° corresponds to an oblique shock. These observational properties give the propagation speed of the shock $(v_{sh} = v_1 \cos\theta_1)$. The estimated fast shock speed, 900-1900 km s^{-1}, is in rough agreement with the observed propagation speed of the X-ray wave, 1400 ± 250 km s^{-1}.

The above examinations suggest that the X-ray wave is a fast shock propagating through the corona. The fast-mode Mach number of the X-ray wave was also estimated to be 1.13-1.31 using the estimated shock speeds. These values indicate that the propagation speed of the Moreton wave is comparable to the MHD fast-mode speed, and consistent with Uchida's model of a weak MHD fast-mode shock. Deceleration of the Moreton wave can be seen in figure 2 (f). This is in accordance with the results of Warmuth et al. (2001). We can see that the X-ray wave (fast shock) becomes wider, i.e., diffuses, in figure 2 (g), and that the estimated fast-mode Mach number decreases in figure 2 (h). The decrease rate is -0.0016 s^{-1}. After 02:14:11 UT, the X-ray wave propagated outside of the field of view. If the Mach number decreased at the same rate, at 02:16:10 UT the Mach number would become "1". The timing roughly agrees with the timing of the

02:16:10 UT にマッハ数は1になっていたであろう．このタイミングはモートン波が消失するタイミングとほぼ一致する．Uchida モデルでは，モートン波面はコロナ衝撃波によって引き起こされた彩層での下向き運動領域と考えられている．私たちの結果は Uchida モデルを支持する明瞭な証拠である．

4.2 フレアに付随する衝撃波の三次元構造

今回のフレアは太陽外縁近傍で発生した．したがって，コロナ X 線波がコロナ上空に向かって伝播した一方で，彩層モートン波は太陽表面上を伝播した（図1）．これら2種類の波は異なるように見える．しかし，上記の結果から2つの波はフレアに付随する同一の波であることがわかった．

図3は 02:13:01 と 02:13:43 UT でのモートン波（左図），X線波（中図）の様子を示す．図3(i)と(j)に描かれている球面は，その中心がフレア領域にあり，半径がそれぞれ 150000 km，200000 km のものである．概略図（右図）で，X線波（赤い線）が球面に沿っているのに対してモートン波（青い線）の波面は球面と太陽表面が交わっているところである．したがって，観測された2つの波の伝播の様子から判断して，フレアに付随する衝撃波球面の三次元構造は，最も単純な近似（等方的に衝撃波が伝播するとしたもの）で予想されるような形状（球面）をしていることが分る．

4.3 X線波と噴出ガスとの関係

このイベントでは，噴出ガス E は X 線波 W の後ろ側に観測され 490 km/s の速度で動いていた．衝撃波が観測された初期 (02:13:15 UT) では，$X \equiv \rho_2/\rho_1 = v_1/v_2$ は 1.3-1.4 である．v_{ejecta} を噴出ガスの速度とする．衝撃波近傍の構造は一次元近似できると仮定すると，$v_{\text{sh}}/v_{\text{ejecta}} = v_1/(v_1 - v_2) = X/(X-1)$ から v_{ejecta} は 330-400 km/s となる．この値は噴出ガス E の観測された速度とほぼ一致する．図2(g)で破線 E は破線 W と 02:12:20 UT に交差する．この時刻にはすでにモートン波は観測され

disappearance of the Moreton wave. In the Uchida model, the front of the Moreton wave is identified as the region of downward motion in the chromosphere caused by the coronal shock wave. Our result is clear evidence for the Uchida model.

4.2. 3-D Structure of the Flare-Associated Shock Wave

The flare occurred near to the solar limb. Hence, the chromospheric Moreton wave propagated on the solar disk, whereas the coronal X-ray wave propagated toward the outer corona (figure 1). These two kinds of waves seem to be different. However, the above result suggests that these two waves are the same flare-associated waves.

Figure 3 shows the Moreton wave (left panels) and the X-ray wave (center panels) at 02:13:01 and 02:13:43 UT. The spheres in figures 3 (i) and (j) are centered on the flare site and have radii of 150000 km and 200000 km, respectively. In the outline images (right panels), the wavefront of the Moreton waves (blue lines) corresponds to the intersection of the sphere and the solar disk, whereas the X-ray waves (red lines) seem to be along the sphere. Hence, in the simplest approximation (symmetrical shock expansion), the propagation of these two waves indicates the 3-dimensional structure of the flareassociated shock wave (spheres).

4.3. Relation between the X-Ray Wave and the Ejecta

In this event, the ejecta, "E", were observed behind the X-ray wave, "W", and moved at a speed of 490 km s^{-1}. In the early phase of the shock observation (at 02:13:15 UT), $X \equiv \rho_2/\rho_1 = v_1/v_2$ is 1.3-1.4. If a one-dimensional approximation is assumed around the shock front, $v_{\text{sh}}/v_{\text{ejecta}} = v_1/(v_1-v_2) = X/(X-1)$ and v_{ejecta} is estimated to be 330-400 km s^{-1}, where v_{ejecta} is the speed of the ejecta. This value is roughly consistent with the observed speed of the ejecta, "E". In figure 2 (g), the dashed line "E" intersects the dashed line "W" at about 02:12:20 UT. At this

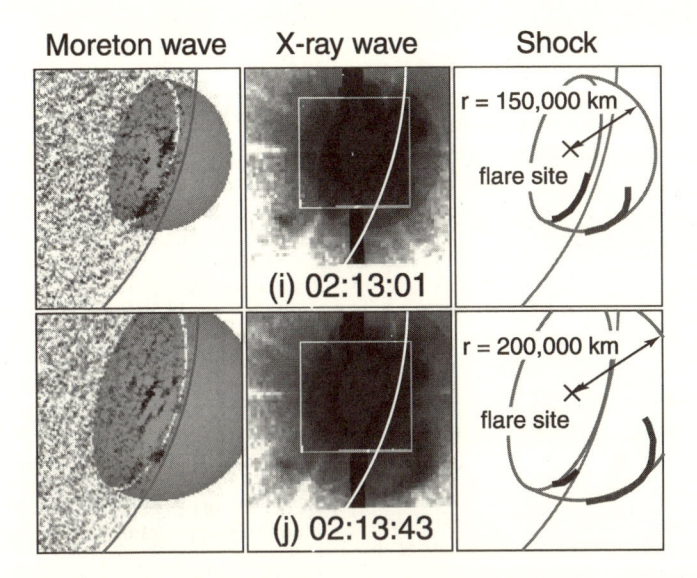

図3 フレアに付随する衝撃波の三次元構造. 左, 中央の図はそれぞれモートン波のHα+0.8Å 差分画像, X 線波の白黒反転軟 X 線画像である. (i), (j)は 02:13:01, 02:13:43 UT に観測されたものである. 球の中心はフレア発生地点であり, (i), (j)での半径はそれぞれ 150000 km, 200000 km である. 右図は衝撃波の輪郭を示したものであり, モートン波の波面を青線, X 線波の波面を赤線で示す.

Fig. 3. Observation of the 3-dimensional structure of the flareassociated shock wave. The left and center panels show Hα+0.8Å "running difference" images of the Moreton wave and negative soft X-ray images of the X-ray wave, respectively. Images (i) and (j) were observed at 02:13:01 and 02:13:43 UT. The center of the spheres is located at the flare site and the radii in images (i) and (j) are 150000 km and 200000 km, respectively. The right panels show outline images of the shock wave. The lines are the wavefronts of the Moreton wave (blue lines) and the X-ray wave (red lines).

ていた. これらの結果をもとにすると, ガス噴出によって衝撃波が引き起こされたのではなく, 衝撃波によって噴出ガスが加速されたのだろうと考えられる. しかしながら, その他の可能性も捨てきることはできない. たとえば波を発生させた機構と同じ機構によってガス噴出が始まったのか, あるいは別の密接に関連した機構(例えば磁気平衡の崩壊)で発生したのかという可能性がある.

最後に, モートン波とフィラメント噴出の関係について考えてみよう. 飛騨天文台で観測されたすべてのモートン波は, 波が伝播したあと同じ方向にフィラメント噴出

time, the Moreton wave was already observed. Based on these results, we suggest that the shock may not have been driven by the ejecta, but instead the ejecta may have been accelerated by the shock. However, we cannot really distinguish this possibility from the other scenario, where the ejecta is launched either by the same process that launches the wave, or by another closely related mechanism (e.g., loss of magnetic equilibrium).

Finally, we discuss the relation between Moreton waves and filament eruptions. In all of the Moreton wave events observed at Hida observatory, Hα filaments erupted in the same direction after the waves propagated (e.g., Eto et al. 2002). Before

を伴っていた (Eto et al. 2002). 噴出の前は, フィラメントはHα単色像では確認できていなかった. 今回のイベントでは, フィラメント噴出はHα－0.8Å単色像で02:22 UTにおいてはじめて確認できるようになった. 我々はフィラメント噴出が噴出ガスEばかりでなくモートン波と密接に関係していると考えている.

花山, 飛騨天文台の全構成員に感謝する. 「ようこう」衛星は日本の国家プロジェクトであり, ISAS, 関係国内機関, 米国および英国との相互国際協力によって打ち上げられ運用されている. SOHOはEuropean Space Agency (ESA) とNASAとの国際協力ミッションである. Hiraiso Radio Spectrograph (HiRAS) はHiraiso Solar Terrestrial Research Centerによって運用され, 太陽電波バーストを監視している.

私たちは故内田豊博士に深く感謝するとともに安らかに眠られることを切に祈ります. この論文を故人への追悼論文といたします.

the eruption, the filaments were invisible in Hα. In this event, the filament eruption became visible at 02:22 UT in Hα − 0.8 Å. We consider these filament eruptions to be closely related to the Moreton waves as well as the ejecta, "E".

The authors thank all members of the Kwasan and Hida Observatories for useful comments. The Yohkoh satellite is a Japanese national project, launched and operated by ISAS, and involving many domestic institutions, with multilateral international collaboration with the US and the UK. SOHO is a mission of international cooperation between European Space Agency (ESA) and NASA. The Hiraiso Radio Spectrograph (HiRAS) is operated by Hiraiso Solar Terrestrial Research Center, and monitoring solar radio bursts.

We are deeply grateful to the late Dr. Yutaka Uchida and pray for the repose of his soul. We dedicate this paper to his memory.

参考文献／References

Dere, K. P. 1996, ApJ, 472, 864
Eto, S., et al. 2002, PASJ, 54, 481
Hudson, H. S., Khan, J. I., Lemen, J. R., Nitta, N. V., & Uchida, Y. 2003, Sol. Phys., 212, 121
Kai, K. 1969, Sol. Phys., 10, 460
Khan, J. I., & Aurass, H. 2002, A&A, 383, 1018
Khan, J. I., & Hudson, H. S. 2000, Geophys. Res. Lett., 27, 1083
Klassen, A., Aurass, H., Mann, G., & Thompson, B. J. 2000, A&AS, 141, 357
Kondo, T., Isobe, T., Igi, S., Watari, S., & Tokumaru, M. 1995, J. Commun. Res. Lab., 42, 111
Kurokawa, H., Ishiura, K., Kimura, G., Nakai, Y., Kitai, R., Funakoshi, Y., & Shinkawa, T. 1995, J. Geomag. Geoelectr., 47, 1043
Mancuso, S., Raymond, J. C., Kohl, J., Ko, Y.-K., Uzzo, M., & Wu, R. 2002, A&A, 383, 267
Mann, G., Jansen, F., MacDowall, R. J., Kaiser, M. L., & Stone, R. G. 1999, A&A, 348, 614
Moreton, G. E. 1960, AJ, 65, 494
Narukage, N., Hudson, H. S., Morimoto, T., Akiyama, S., Kitani, R., Kurokawa, H., & Shibata, K. 2002, ApJ, 572, L109
Scherrer, P. H., et al. 1995, Sol. Phys., 162, 129
Smith, S. F., & Harvey, K. L. 1971, in Physics of the Solar Corona, ed. C. J. Macris (Dordrecht: Reidel), 156
Thompson, B. J., Plunkett, S. P., Gurman, J. B., Newmark, J. S., St. Cyr, O. C., Michels, D. J., & Delaboudinière, J.-P. 1998, Geophys. Res. Lett., 25, 2465
Tsuneta, S., et al. 1991, Sol. Phys., 136, 37
Uchida, Y. 1968, Sol. Phys., 4, 30

Uchida, Y. 1974, Sol. Phys., 39, 431
Warmuth, A., Vršnak, B., Aurass, H., & Hanslmeier, A. 2001, ApJ, 560, L105

太陽フィラメント消失とそれに随伴する フレアアーケードの間のエネルギー関係について

Energetic Relations between the Disappearing Solar Filaments and the Associated Flare Arcades

森本太郎, 黒河宏企, 柴田一成, 石井貴子
京都大学大学院理学研究科附属花山飛騨天文台

(原論文) Publ. Astron. Soc. Japan 62, 939-949, 2010 August 25

Taro MORIMOTO, Hiroki KUROKAWA, Kazunari SHIBATA, Takako T. ISHII
Kwasan and Hida Observatories, Kyoto University

概要

太陽の消失フィラメントの持つ力学的なエネルギーと, 軟 X 線で観測される随伴フレアアーケードの熱エネルギー間の関係を時間的発展の観点と, 統計的な相関関係から調査した結果を報告する. 10 個の噴出型フィラメントの三次元的な速度場を求め, その単位体積当たりの力学的エネルギー増加率 ϵ_{mc} を評価し, Yohkoh/SXT データから得られる単位体積当たりの熱エネルギー解放率と比較した. 統計解析から, 近似的に $\epsilon_{th} \propto \epsilon_{mc}^{1.9}$ という関係式が得られた. この関係式は, 磁気リコネクション過程によるポインティング流束によりアーケードにエネルギーが供給されること, およびフィラメントはローレンツ力によって加速されることによって解釈すれば説明可能である. また, この解釈は, 熱的および力学的エネルギー上昇率が, フィラメント消失領域の平均的な磁場強度に強く依存することからも支持できるものである. 我々は, 二種のエネルギー増加率の時間変化の比較から, いくつかの例では, 力学的エネルギーの増加が, 熱エネルギーの増加より先行することを見出した. この研究で見出された関係は次のことを意味する. フィラメント加速と高温プラズマをつくる機構は

Abstract

We present the temporal and statistical relations between the mechanical energies of disappearing solar filaments and the thermal energies of the associated flare arcades in soft X-rays. Measuring the 3-D velocity fields of 10 eruptive filaments, we calculated their mechanical energy gain rate, ϵ_{mc}, per unit volume and compared it to the thermal energy release rate per unit volume, ϵ_{th}, derived with Yohkoh/SXT data. For the statistical relation, we found a relation that can be approximated as $\epsilon_{th} \propto \epsilon_{mc}^{1.9}$. This relation can be explained by interpreting the energy input to an arcade via the Poynting flux in the magnetic reconnection process and the acceleration of a filament by the Lorentz force. This explanation is also supported by the strong dependence of the observed increase rates of both the thermal and mechanical energy densities on the mean magnetic field strength of the source region. We also investigated their temporal variations, and found that the start time of increase in the mechanical energy of a filament preceded that of the thermal energy of the coronal arcade in some cases. These relations imply that the basic mechanisms that accelerate a filament and create a hot plasma are different, and both energy increase rates are determined primary by the magnetic

異なっていることである．また，両方のエネルギー増加率を決めているのは，その領域の磁場強度であることである．

1. 序論

プロミネンス噴出やフィラメント消失に伴って，Hα線ツーリボンフレアが発生し，軟X線やEUVでは明るいアーケード構造が形成されることはよく知られている (Sheeley et al. 1975; Webb et al. 1976; McAllister et al. 1992, 1996; Hanaoka et al. 1994; Khan et al. 1998; Tripathi et al. 2004)．フレアと，プロミネンス・フィラメント噴出やコロナ質量放出（CME）などの質量放出現象との相互関係は，これまでも大事な研究テーマであった．この相互関係の研究が，広範囲にかつ詳細になされた結果，2つの現象には強い相関があることが明らかになった (Munro et al. 1979; Webb & Hundhausen 1987; Sheeley et al. 1983; St.Cyr & Webb 1991; Harrison 1995)．人工衛星Yohkohが打ち上げられフレア研究が進んだことによって，フレアと質量放出現象は共に1つの磁気リコネクション過程に含まれるものであると広く認識されるようになった (Shibata et al. 1996)．この考えは，Hirayama (1974) が提唱したフィラメント噴出がフレアループの上空の鉛直な電流シートでの磁気リコネクションを駆動するモデルと，CSHKPモデルとして知られている同種の磁場構造についてのその他の研究に基づいている (Carmichael 1964; Sturrock 1966, Hirayama 1974; Kopp & Pneuman 1976)．

フィラメント噴出と軟X線アーケード形成が磁場と強く関係していることも知られていた (Pevtsov 2002)．強烈なフレアはしばしば磁場の強い活動領域で起こるのに対して，弱いフレアは活動領域から離れたところで起こる傾向にある．また，活動領域で発生したプロミネンス噴出は大きな加速を受けて，ときには高速のスプレー型プロミネンス（Warwick 1957）と呼ばれる一方，弱い磁場領域で発生したものは数時間もの間ゆっくりと加速されることがわかっている (Valnicek 1964, Tandberg-

field strengths.

Key words: Sun: coronal mass ejections (CMEs) — Sun: filaments — Sun: flares

1. Introduction

It is well known that prominence eruptions and filament disappearances are often followed by two ribbon flares in the Hα line, and bright arcade formations in soft X-rays and EUV (e.g., Sheeley et al. 1975; Webb et al. 1976; McAllister et al. 1992, 1996; Hanaoka et al. 1994; Khan et al. 1998; Tripathi et al. 2004). The relation between flares and mass ejections, including filament/prominence eruptions as well as coronal mass ejections (CMEs), has been a major research subject. Every broad study that extensively investigated this relation has also revealed that these two phenomena have strong association (Munro et al. 1979; Webb & Hundhausen 1987; Sheeley et al. 1983; St. Cyr & Webb 1991; Harrison 1995). Owing to detailed studies on solar flares after the launch of Yohkoh satellite, it became commonly recognized that a flare and mass ejection are the processes involved in a magnetic reconnection process (Shibata et al. 1996), based on a model which Hirayama (1974) proposed that a filament eruption triggers magnetic reconnection in the vertical current sheet above the flare loop and other studies on similar magnetic configurations, known as the CSHKP model (Carmichael 1964; Sturrock 1966; Hirayama 1974; Kopp & Pneuman 1976).

It is also known that the origins of filament eruptions and the formations of soft X-ray arcades are magnetic (e.g., Pevtsov 2002). High-intensity flares often occur in active regions where the magnetic field is strong, and low-intensity flares are likely taking place away from active regions. At the same time, a prominence eruption that originates from an active region tends to undergo strong acceleration, and is often called "a spray prominence" (Warwick 1957), while that from a region with a weak magnetic field strength takes much more time (up to several hours) to be accelerated (Valnicek 1964; Tandberg-Hanssen et al. 1980; Rompolt 1990), although it is subject to some exceptions (Sterling et al. 2001; Shakhovskaya et al. 2002).

These facts give us an insight that a

Hanssen et al. 1980, Rompolt 1990). もちろん, 一般的な傾向に反する例もある (Sterling et al. 2001, Shakhovskaya et al. 2002).

以上のことから, フィラメント噴出とX線フレアアーケードやフレアループとの間には, 物理的な相互関連があるかもしれないと予想される. ところが, これまで精力的に研究されてきたのもかかわらず, 両者の物理的なつながりは明らかにされなかった. Hundhausen (1997) は, Solar Maximum Mission (SMM) 搭載のコロナグラフで撮影されたコロナ質量放出 (CMEs) を解析して, CMEsと軟X線現象と比較して次のような結論をえている. (ⅰ) 強いX線フレアは, コロナ質量放出の発生にとって必要条件でも十分条件でもない. (ⅱ) コロナの加熱を意味する軟X線フレアの強さは, それに伴う質量放出現象の特質(速さ, 質量, およびエネルギー)とは強い相関が見られない. Hundhausen (1997) は, 質量放出に必要なエネルギーと, 軟X線フレア加熱に必要なエネルギーとの間に簡単な関係があると考えるべき物理的な理由はないとも述べている.

Zhang et al. (2001) は, Solar and Heliospheric Observatory (SOHO) 衛星に搭載された Extreme Ultraviolet Imaging Telescope (EIT) (Delaboudiniere et al. 1995) と Large Angle and Spectrometric Coronagraph (LASCO) (Brueckner et al. 1995)で観測された4つのCMEの運動の時間発展について, Geostationary Operational Environmental Satellite (GOES) による軟X線観測と比較した. 彼らは, CMEの急激な加速は, 軟X線の増光とときを合わせていることを明らかにした. ただし, 両者のエネルギー(CMEの場合は速さ, フレアの場合は軟X線ピーク強度)の間には, 相関が見られないと報告している. Moon et al. (2002) は, X線強度がC1クラス以上のリムフレアに伴うCMEの解析から, CMEの速さとフレアの軟X線強度とは弱い相関があることを見出している.

両者の相関についてのこれまでの研究で

filament in its eruption and an arcade or a flare loop formation observed in soft X-rays should have some physical connection to each other. Extensive studies based on observations, however, have failed to figure out certain physical links between them. Hundhausen (1997), analyzing the images of coronal mass ejection (CMEs) taken by the Solar Maximum Mission (SMM) coronagraph, compared the observed signatures of CMEs to those of associated soft X-ray events, and concluded that (i) intense soft X-ray flares are neither a necessary nor a sufficient condition for the occurrence of coronal mass ejections, (ii) the intensity of any soft X-ray flare, which implies heating of the corona, that accompanies a mass ejection is not closely related to the characteristics (such as speed, mass, and energy) of the ejection. Hundhausen (1997) also mentioned that there is no physical reason why the energy available for mass ejection should have any simple relation to the energy available for a soft X-ray flare.

Zhang et al. (2001) carefully investigated the temporal evolutions of the motions of four CMEs observed by Extreme Ultraviolet Imaging Telescope (EIT) (Delaboudiniere et al. 1995) and Large Angle and Spectrometric Coronagraph (LASCO) (Brueckner et al. 1995) instruments aboard the Solar and Heliospheric Observatory (SOHO) spacecraft, and compared it to the Geostationary Operational Environmental Satellite (GOES) soft X-ray flux observations. They demonstrated that the impulsive acceleration phase of CMEs coincides very well with the rising phase of the associated soft X-ray flares. They, however, reported that there is no correlation between the energetics of CMEs (in terms of speeds) and flares (in terms of peak flux). Moon et al. (2002) analyzed CMEs with limb flares greater than C1 class, and found a weak correlation between the speed of CMEs and the associated flare X-ray flux.

In these studies, however, they only focused on the temporal relation, or compared the peak intensities of soft X-ray events with the velocities of the mass ejections. We therefore, in order to seek physically more meaningful temporal

は，軟X線のピーク強度と質量放出現象の速さの比較あるいは時間変化の相関に注視されてきた．我々は，物理的により有意な相関を調べるために，噴出フィラメントの運動エネルギーと重力ポテンシャルエネルギーの2つを求め，それらとフレアアーケードやループの熱エネルギーと比較した．更に，力学的エネルギー密度と熱エネルギー密度の増加率と対象領域の平均磁場強度との関係を調査した．これは考えている2つの現象が磁場に関係しているものであり，磁場強度との相関を調べることが適切であると考えられるためである．

我々の過去の論文（Morimoto & Kurokawa 2003a，以降 論文1と参照）では，太陽表面上の5個の消失フィラメント（Disparition Brusques：DB）の三次元速度場をもとめた．これは飛騨天文台のフレア監視望遠鏡で観測されたHα線中心波長とHα±0.8Åの三波長像にBeckersのクラウドモデル解析を施したものである．次に35個のDB現象について，三次元速度場を測定し，HαDBの運動と随伴するコロナ現象の因果関係を研究した（Morimoto & Kurokawa 2003b，以降 論文2と参照）．本論文では，噴出フィラメントの力学的エネルギー増加率と軟X線でみられるフレアアーケードの熱エネルギー増加率の2つの量について，時間的発展の相互の関係，および強度の統計的な相関を調べた結果を報告する．

第2節では，本研究で用いたデータとイベントの選択法について述べる．消失フィラメントの視線方向速度および視線に垂直な方向の速度の導出法の簡単な説明を第3節で言及する．第4節の前半では，2つのイベントについてフィラメントの力学的エネルギー増加率とコロナアーケードの熱エネルギー増加率の時間変化の相互関係について述べ，後半では，選択されたすべてのイベントのデータを用いて，両者の強度の統計的な相関についての結果を示す．第5節では，すべてのイベントのまとめをした後，結論を述べる．

and statistical relations, estimated both the kinetic and potential energies of eruptive filaments, and compared them directly to the released thermal energies of the associated arcades or flare loops. Furthermore, we investigated the relation between the measured increase rates of the mechanical energy density and the thermal energy density with the mean photospheric magnetic field strength of the source region, because both processes are magnetic, and it may be appropriate to relate them in terms of the magnetic field strength.

In our previous paper (Morimoto & Kurokawa 2003a, hereafter Paper I), we studied the 3-D velocity fields of disappearing filaments (disparitions brusques: DBs) on the solar disk, applying Beckers' cloud model to 5DB events observed in Hα line center and Hα±0.8Å with the Flare Monitoring Telescope (FMT) at Hida Observatory. By measuring the 3-D velocity fields of 35 DBs on the solar disk, we studied the causal relation between the motions of Hα DBs and the associated coronal phenomena (Morimoto& Kurokawa 2003b, hereafter Paper II). In this paper, we present the temporal and statistical relations between the mechanical energies of eruptive filaments and the thermal energies of the associated flare arcades in soft X-rays.

In section 2, we describe the data used in this study and the selection of events. A brief explanation of the methods used to derive the line-of-sight and transversal velocities of a disappearing filament is given in section 3. We then show the results of two events to see the temporal relation of the energy of a disappearing filament and the thermal energy of the associated arcades in the first half of section 4. We also present their statistical relation using all of the selected events. Finally, summarizing the results of all events, we give a conclusion in section 5.

2. 観測とデータ
2.1. Hα線観測

　消失フィラメントの速度やエネルギーを求めるためには，京都大学飛騨天文台のフレア監視望遠鏡（FMT：Kurokawa et al. 1995）で取得したHα像を用いた．この望遠鏡は，太陽全面のHα線中心波長，Hα−0.8Å（青色側ウィング）およびHα+0.8Å（赤色側ウィング）での単色像を観測している．3つの波長帯での観測を用いると，視線方向および視線に垂直な方向の速度が得られることになり，三次元速度が求められることになる．

　FMTで観測されたフィラメント消失の一例を図1に示す．このイベントは，2000年5月8日に起こったもので，太陽面の中央経度線上の南半球側に位置していた．消失の前は，フィラメントはHα線中心波長像（左側パネル）でのみ見えていた．04:17 UTに，フィラメントの南側の部分が上昇を始め，青色側ウィング（中央パネル）で黒い模様として現れた．Hα線中心波長ではこの部分は徐々に見えなくなった．この南の部分がすべて見えなくなった05:36 UTには，図1の下左パネルに示すように，消失フィラメントの両脇にツーリボンフレアが発生した．このとき，フィラメント中央部および北側の部分は，北西の方角に移動続けており，赤色側ウィング像（右パネル）で顕著であった．これは，現象の後半では，後退運動が主となったことを意味している．

2.2. ようこう観測

　ようこう軟X線望遠鏡（SXT：Ogawara et al. 1991; Tsuneta et al. 1991）は，1991年以来，太陽コロナの軟X線全面画像と部分画像（それぞれFFIとPFI）を取得している．FFIは，通常50秒ごとに，ピクセルサイズ4.9秒角（1/2分解能）あるいは9.8秒角（1/4分解能）で得られている．PFIは，ピクセルサイズ2.5秒角（最高分解能）で，2, 3秒から30秒までの時間間隔で得られている．この望遠鏡には5枚のX線フィルターがあり，フィルター比法

2. Observations and Data
2.1. H. Observations

We used Hα images obtained by the Flare Monitoring Telescope (FMT: Kurokawa et al. 1995) at Hida Observatory, Kyoto University to investigate the velocities and energetics of disappearing filaments. This telescope provides full-disk solar Hα images not only in its line center, but also in blue and red wings, whose central wavelength are Hα−0.8Å and Hα+0.8Å, respectively. The FMT observations in the 3 bands enable us to measure the line-of-sight and transversal velocities of the filaments, and hence their complete 3-D velocity fields.

One example of FMT observations of a filament disappearance is displayed in figure 1. This event took place on 2000-May-08 near to the central meridian in the southern hemisphere. Before the onset of disappearance, the filament is visible in only the Hα line center image (left panels). From 04:17 UT, the southern part of the filament starts to ascend from its southern leg, and appears as a dark feature in the blue wing (middle panels), and then fades in the Hα line center images. At 05:36 UT, when the southern part vanishes completely, a two-ribbon flare appears at both sides of the disappeared filament, as shown in the bottom-left panel of figure 1. At this moment, the middle and northern parts of the filament still continue to move towards the north-west, and it becomes prominent in the red wing (right panels), which indicates that the receding motion is dominant in the later phase.

2.2. Yohkoh Observations

Yohkoh Soft X-ray Telescope (SXT: Ogawara et al. 1991; Tsuneta et al. 1991) provides full and partial frame images (FFIs and PFIs, respectively) of the solar corona in soft X-rays from 1991. An FFI is obtained, usually, per 50 seconds with a pixel size of $4''\!.9$ (half resolution) or $9''\!.8$ (quarter resolution). A PFI is obtained with a pixel size of $2''\!.5$ (full resolution), and the range of time of the cadence is from a few seconds to 30 seconds. This telescope is equipped with five X-ray analysis filters, which enables us to derive

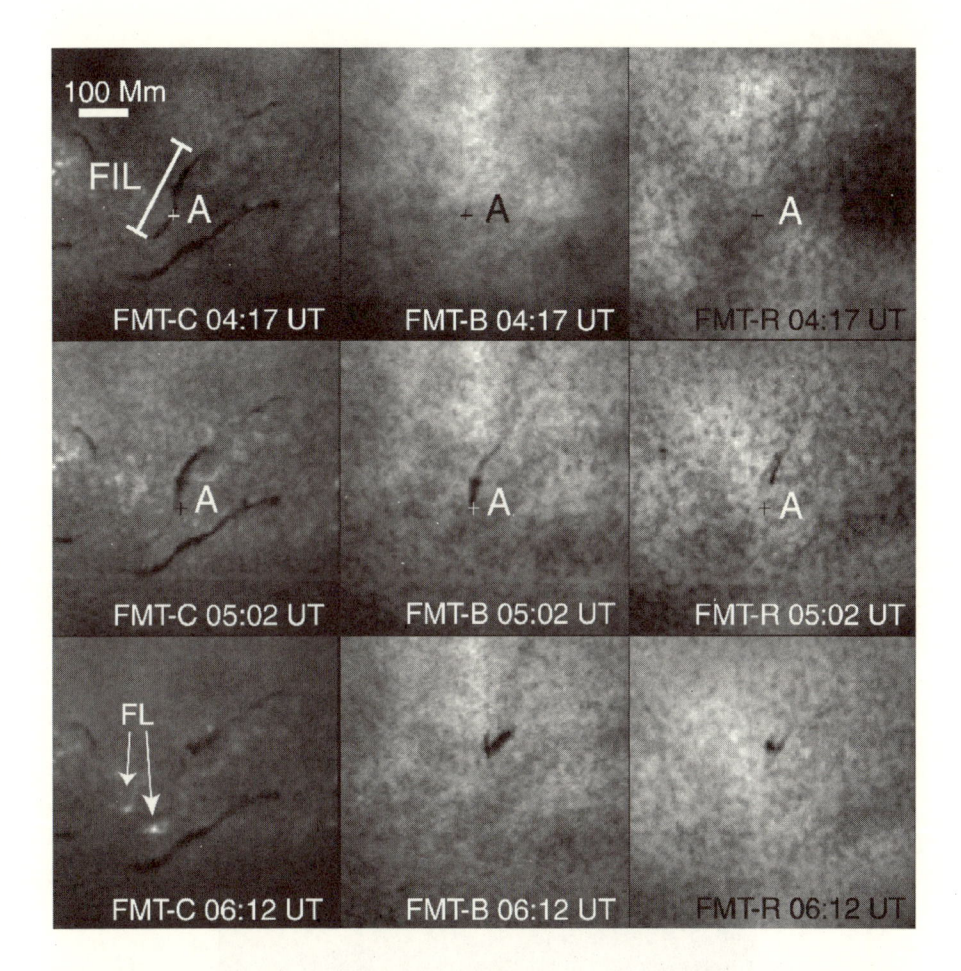

図1 2000年5月8日イベントのFMT画像．観測時刻とフィルターの種類は（Hα線中心像はC，Hα青色側ウィング像はB，Hα赤色側ウィング像はR）各パネルの下部に示されている．最上部左のパネルにFILという文字が付されている棒状の印でフィラメントを示す．イベント開始前の04:17 UTには，Hα線中心画像でのみ見えており，05:02 UTからフィラメントの南側が消え始めた．フィラメントの南側が完全に消えたちょうどそのとき，Hα線中心像で，ツーリボンフレア（FL）が現れている．太陽の北が上方向で，東は左側である．本論文では，図内の方角は以降これにしたがって示されている．

Fig. 1. FMT images of the 2000-May-08 event. The observational time and filter passband (C for Hα line center, B for Hα blue wing, and R for Hα red wing) are presented in the bottom of each panel. The filament (denoted by the bar shown in the left-top panel with "FIL"), which can be seen only in Hα line center before onset of the event (04:17 UT), starts to disappear from the southern part (05:02 UT). Just after the southern part of the filament vanishes, a two-ribbon flare (FL) appears on the Hα center images. The solar north is up and east is left, which holds for all of the solar images in this paper.

図2 全面画像から切り出した2000年5月8日の軟X線イベント像．上のパネルには，Hαフィラメントの初期位置とFMTの視野が白い輪郭線と四角で示されている．強い加速を受けて惑星間空間に噴出したフィラメントの南部分上空に 07:46:13 UT カスプ形のループ（CL）が見えている．

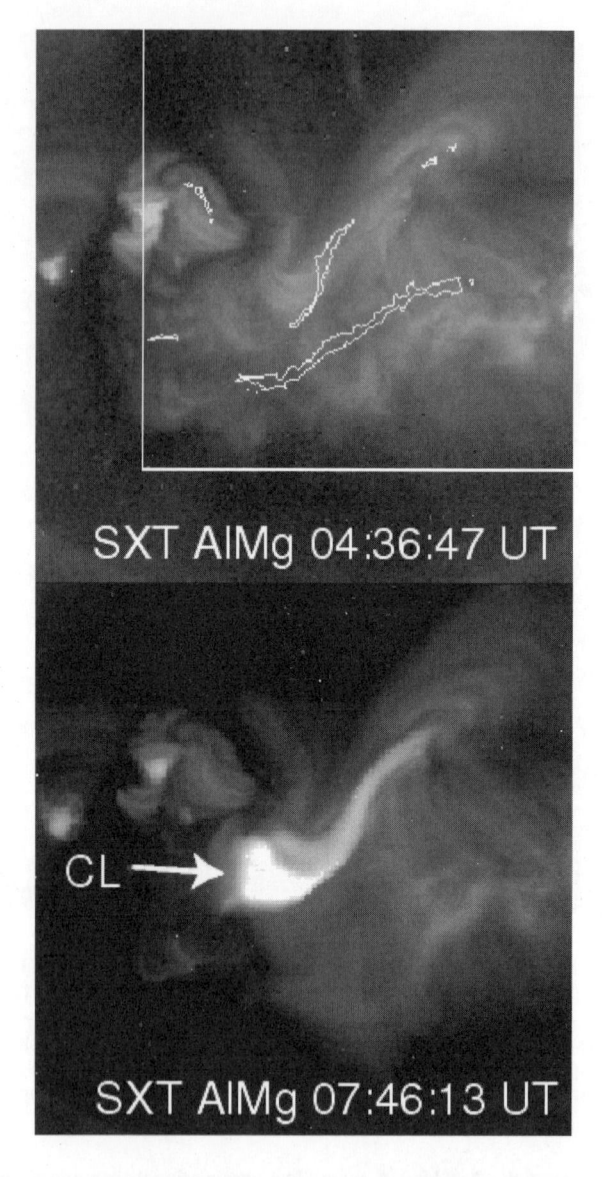

Fig. 2. Series of SXT partial full frame images showing the soft X-ray event on 2000-May-08. The initial locations of the Hα filaments and the FMT field of view of figure 1 are indicated by the white contour plots and the box, respectively in the top panel. A cusp-shaped loop (CL) is seen at 07:46:13 UT over the southern part of the filament, which underwent a strong acceleration and was ejected into the interplanetary space.

(Hara et al. 1992) でデータを解析することで，アーケードの温度，放射尺度を求めることができる．

2000年8月8日のイベントのYohkoh/SXT像が図2に示されている．このイベントの発生前の04:36:37 UTのHαフィラメントの様子（図1の上左）が，図2に白い輪郭線で示されている．ようこうの日陰が終わった直後の05:57:27 UTには，軟X線では，長大で淡いシグモイド構造の南の部分が明るくなっていた．07:46:13 UTには，明るいカスプ形ループのアーケードが観測された．これらのループの位置は，FMTでその消失が観測されたHαフィラメントの南の部分に対応する（図1）．軟X線シグモイドの中央部および北側の部分も少し明るくなっていた．10個のイベントを解析するときには，空間および時間分解能がよいPFIを主として用いた．適当なPFIがないときには，FFIを用いた．

2.3. その他のデータ

軟X線イベントに伴うそのピーク強度は，Geostationary Operational Environmental Satellite（GOES）の観測を利用した．フィラメントの加速やコロナの加熱は磁場の変化によって引き起こされると思われるので，各領域のコロナの磁場強度Bの指標として光球磁場Bfを採用して調査した．光球磁場は，National Solar Observatory/Kitt Peak Vacuum Telescopeのデータを用いた．符号を無視して平均化したBfを求めて，表1の第6列に挙げた．2000年5月8日のイベントについては，LASCOの白色光コロナグラフ像を用いてCMEの速度変化を調べた．このイベントは後ほど詳述する．

2.4. イベントの選択

論文2では，35個のDBについて三次元速度場の詳細解析を行い，それらを噴出

the temperatures and emission measures of arcades with filter ratio method (Hara et al. 1992).

The Yohkoh/SXT images of the 2000-May-08 event are illustrated in figure 2. At 04:36:37 UT, before the onset of this event (top panel), the Hα filaments are indicated by the white contour. The soft X-ray corona was found to become bright at the southern part of a large faint sigmoid-like structure at 05:57:27 UT, or just after the data gap due to Yohkoh night. An arcade of bright cusp-shaped loops is seen in the image at 07:46:13 UT. These locations of the cusp-shaped loops correspond to the southern part of the Hα filament, which disappeared in the FMT Hα center images (figure 1). The rest part of the soft X-ray sigmoid-like structure, which corresponds to the middle and northern parts of the Hα filament, is also seen to slightly brighten. In an analysis of 10 events, we used FFIs only in the cases in which we did not have appropriate PFIs, because the spatial and temporal resolutions of PFIs were better than those of FFIs.

2.3. Other Data

The soft X-ray flux data obtained by Geostationary Operational Environmental Satellite (GOES) are used to see the peak fluxes associated with the soft X-ray events. Since the acceleration of a filament and heating of coronal plasma appear to be the consequences of the magnetic field changes, we examined the photospheric magnetic field strength, B_f, of the region as an indicator of the coronal magnetic field strength, B. As the photospheric magnetogram, we used data from the National Solar Observatory/Kitt Peak Vacuum Telescope. The values of B_f are listed in the sixth column of table 1, by taking the unsigned average over the arcade region. We also used whitelight coronagraph images from the LASCO instrument to see the velocity evolution of the CME accompanied with the event on 2000-May-08, which will be discussed in some detail later.

2.4. Event Selection

In Paper II, we extensively analyzed 3-D velocity fields of 35DBs, and classified them into eruptive and quasi-eruptive

表1 本研究で解析されたイベントの一覧リスト
Table 1. Complete list of the events analyzed in this study.*

No	Time	Location/NOAA	ϵ_{mc}	ϵ_{th}	B_f	GOES
1	1993-Apr-20 03:45–06:10	S22E70/7480	3.8	8.3	6.7	B7.4
2	1993-May-15 22:29–24:27	S28E30	2.7	3.8	5.2	B3.6
3	1993-Oct-21 02:51–05:19	S15W08/(7605)	8.1	4.4	6.0	B2.2
4	1994-Jan-05 06:04–07:05	S10W22/7647	25	250	46	M1.0
5	1994-Feb-20 00:17–01:24	N01W00/7671	14	370	22	M4.0
6	1994-Sep-05 06:16–08:00	S05W03/7773	8.7	19	16	
7	1997-Oct-20 22:30–24:56	S30E43	5.0	19	12	B1.2
8	1998-Sep-20 02:00–05:28	N20E70/8340	11	570	41	M1.8
9	2000-Jan-19 00:28–01:47	N08W18/8829	2.8	4.2	8.1	C1.4
10	2000-May-08 04:19–07:40	S21W03	4.0	9.0	9.8	B6.8

左から順に，イベント番号，フィラメント消失現象の時間，フィラメントの概略位置と（フィラメントが属しているのが明らかなとき）NOAA活動領域番号，単位体積当たりの力学的および熱エネルギー解放率（×10^{-4} ergs cm^{-3} s^{-1}），軟X線イベント下の平均光球磁場強度（G），GOES衛星によるアーケード構造形成時の軟Xピーク強度を示している．1994年9月5日イベントについては，GOES強度は公表されていない．

*From the left column, event number, time of Hα filament disappearance, the approximate location of the filament with the NOAA number for a case in which the filament is belonging to an active region, the mechanical energy and thermal energy release rates per unit volume (×10^{-4} erg cm^{-3} s^{-1}), the mean photospheric magnetic field strength below the soft X-ray events (G), the peak soft X-ray flux of the associated arcade formation obtained by GOES satellite, respectively. We left a blank for the soft X-ray flux of the 1994-Sep-05 event because we have no GOES data for this event.

型と準噴出型に分類した．噴出型というのは，消失フィラメントの全体あるいは少なくともその一部が惑星間空間に放出されるものをいう．準噴出型とは，フィラメントが上昇をし始めるが，バラバラに分裂し下層の彩層に帰還するものや，太陽大気中にとどまるものをいう．本論文では，この35イベントの中から次の条件を満たすものを10イベント選び出した．（i）本論文ではフィラメント加速の様子を研究するのが目的であるので，噴出型であること．（ii）軟X線イベントの増光時の熱エネルギー解放率を正しく求められる程度に，多種のフィルター，高い時間分解能および適正な露出時間で観測されたSXT像が十分あって，コロナ温度や放射尺度を導出できるイベントであること．本論文では，消失フィラメント加速に対応する軟X線増光時に注目して解析する．選択された10イベントは，その発生時刻，場所データとともに表1にあげられている．

types. "Eruptive" means that all or, at least a part of the disappearing filament is ejected into interplanetary space. The "quasi-eruptive" filament ascends, breaks into fragments, and then flows down into the chromosphere at several places, or remains in the solar atmosphere. We selected 10 events out of these 35 events with the following criteria: (i) The filament is judged to be eruptive, since we are concerned with the problem of how a filament is accelerated. (ii) We have sufficient SXT images taken with different filters and good temporal cadence with a reasonable exposure time to obtain the temporal variations of coronal temperatures and emission measures, and to determine the thermal energy release rate in the rise phase of soft X-ray events. We focused on the rise phase because the energetics of disappearing filaments in their acceleration phase is considered to be related to the rise phase of the soft X-ray events. The selected 10 events are listed in table 1 along with their occurrence time and locations.

Appendix A Selected Papers Relating to FMT | 459

3. 解析
3.1. Hαフィラメントのエネルギー
3.1.1. 消失フィラメントの速度

　消失フィラメントの速度場の導出は論文1で詳述されているので，ここでは簡単に紹介する．消失フィラメントの三次元速度場は，フィラメントの視線方向速度と視線に垂直な方向の速度からなる．我々の方法では，視線速度場はHα線中心波長，Hα−0.8ÅおよびHα+0.8Åの3枚の単色像データにBeckersのクラウドモデル解析(Beckers 1964)を適用して求める．視線速度を正確に導出するために，FMTのフィルターの有効透過帯，散乱光のレベル，ドップラー効果による源泉関数の上昇(Doppler Brightening)ということも考慮に入れて補正している．視線に垂直な面内での横方向の速さと方向は，FMT時系列像上でフィラメントの内部構造（ブロブ，Blobs）を追跡することで求められる．この横方向の速度を基にすると，フィラメント内部の各点が動いた軌跡を定めることができる．それぞれの軌跡上での三次元速度場の値から，フィラメント各部分の上昇速度を導くことができる．この横方向の速度場は，フィラメントの各ピクセルごとの運動の軌跡をつくることにも使用される．これを用いると各ピクセルごとの三次元速度場が求まり，フィラメントの動きの詳細な時間変化を導くことが可能となる．

　三次元速度場の時間変化を，2000年5月8日のイベントについて例示してみよう．イベントが始まる前の04:17 UTにおいて，フィラメント内にある点を指定し（図1での点A)，その点が時々刻々移動してゆく様子を追尾する．図1では，点Aの位置が+印で示されている．図3の上左パネルには，点Aの視線方向速度の時間変化が示されている．点Aは消失の開始時から加速され始め，完全に消失する05:18 UTの直前には−48 km/sに達した．図3上右には，点Aの上昇速度の時間変化を示す．点Aの動きは，ほぼ視線方向沿いであったため，上昇速度は視線速度とほぼ同じ値になっている．

3. Analysis
3.1. Energetics of the H. Filaments
3.1.1. The velocities of disappearing filaments

　The detailed calculation procedures of the velocity in the disappearing Hα filaments are presented in Paper I, so we only briefly explain them in this paper. The 3-D velocity field of a disappearing filament is obtained by calculating the line-of-sight and transversal velocities of the filament. In this new method, the time variation of the line-of-sight velocity is derived using the three filtergrams obtained in Hα center, Hα−0.8Å, and Hα+0.8Å on the basis of Beckers' cloud model (Beckers 1964). We also make corrections for the effective band widths of the FMT filters, the scattered light and the Doppler brightening effect (DBE) in order to accurately determine the line-of-sight velocity. The transversal speeds and directions are measured by tracing the internal structures (blobs) of a filament on successive FMT images. This transversal velocity enables us to determine the trajectory of each element inside the filament. With the line-of-sight and transversal velocities of each trajectory, we calculate the corresponding upward velocity. The transversal velocity is next used to derive the trajectory of each pixel on the filament. This enables us to obtain the time-dependent evolution of the 3-D velocity for each pixel. This procedure is performed for all the pixels on the filament individually to reconstruct the complete 3-D velocity field.

　In order to see an example of the 3-D velocity evolution, we define a sample point (point A in figure 1) on the filament of the 2000-May-08 event at 04:17 UT before its onset. The defined and subsequent positions of point A are shown on the FMT images with "+" marks in figure 1. The top-left panel of figure 3 shows the temporal evolution of the line-of-sight velocity of the point. The point A is accelerated from the onset of the disappearance, and attains −48 km s^{-1} at 05:18 UT just before its disappearance. The top-right panel in figure 3 shows the upward velocity of point A. Since the motion of point A is almost along the line-of-sight, the upward velocity is quite similar to the line-of-sight velocity.

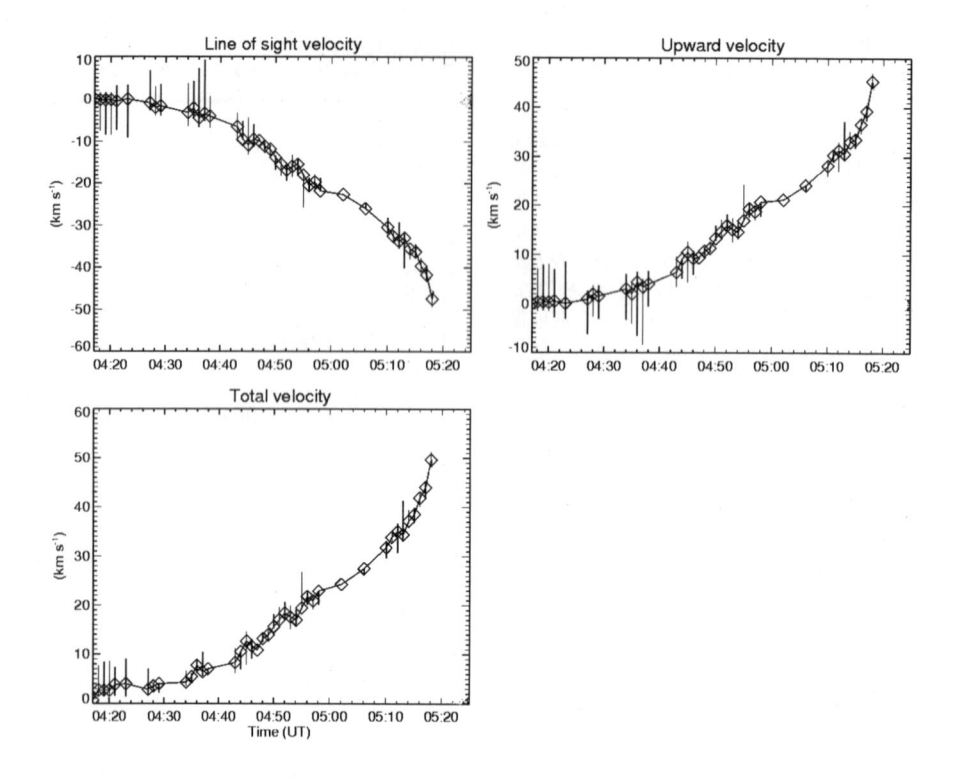

Fig. 3. Line-of-sight velocity of the point A of the 2000-May-08 event plotted as a function of time at the top-left panel in which negative values correspond to approaching motion. The upward and the total velocities of the same point are given at the top-right and bottom-left panels, respectively.

3.1.2. The mechanical energy gain rate

The mechanical energy gained by an element x with unit volume inside a filament $e_{\mathrm{mc}}(x,t)$ at an arbitrary time t is the sum of its kinetic and potential energies; it can be written as

$$e_{\mathrm{mc}}(x,t) = \frac{1}{2}\rho_p v(x,t)^2 + \int_{R_\odot}^{H(x,t)+R_\odot} G \frac{M_\odot \cdot \rho_p}{r^2} dr, \quad (1)$$

where $R_\odot$, $M_\odot$, and G are the solar radius, solar mass, and gravitational constant, respectively. Since the density of a

び重力定数を表す．フィラメントのガス密度ρ_pは，FMT観測だけでは求まらないので，通常のプロミネンスの平均密度を採用して$\rho_p = 1.7 \times 10^{-13}$ g/cm^{-3}とした．この値は，10^{11} cm^{-3}の粒子数密度に対応する（Jefferies & Orrall 1963）．また，単位体積当たりのフィラメントガス充満率を0.1とした（Engvold 1976; Simon et al. 1986）．フィラメントガスの3次元速度の大きさを$v(x, t)$とする．この値は，上で述べた論文1の方法で導き，高さ$H(x, t)$は，上昇速度$v_u(x, t)$を用いて以下の式で求めた．

$$H(x,t) = \int_{t_0}^{t} v_u(x, t')dt', \quad (2)$$

ここで，t_0はフィラメント要素の加速開始時刻である．式(1)と(2)を用いると，求められた軌跡に沿って噴出するフィラメントの要素がFMTの視野から消える時刻t_fでの最終的な$e_{mc}(x, t_f)$を求めることができる．この量を時間差$t_f - t_0$で割ることにより，力学的エネルギー増加率$\epsilon_{mc}(x) = e_{mc}(x, t_f)/(t_f - t_0)$が得られる．最後に，このエネルギー増加率をすべての上昇要素について平均化して，単位体積当たりの平均力学的エネルギー上昇率$\epsilon_{mc} = \langle \epsilon_{mc}(x) \rangle_x$を求めることができる．

　この節の計算の一例を以下に示す．2000年5月8日のイベントの点Aは，完全消失前の05:18 UTにおける高さは4.9×10^9 cmで，その重力ポテンシャルエネルギーは2.1 ergs cm^{-3}であった．運動エネルギーは2.1×10^{-1} ergs cm^{-3}であったので，全力学的エネルギー増加率は$e_{mc}(A, t_f) = 2.3$ ergs cm^{-3}であった．加速開始，消失時刻はそれぞれ$t_0 = $ 04:17 UTと$t_f = $ 05:18 UTであったので，力学的エネルギー増加率は$\epsilon_{mc}(A) = 6.3 \times 10^{-4}$ ergs cm^{-3} s^{-1}となった．フィラメントが移動したすべての軌道のうち，惑星間空間に放出された要素が存在したものについて平均すると，噴出した要素についての平均的な力学的エネルギー増加率が得られることになる．すべてのイベントについて上記の計算をした結果が，表1の第4列にあげられている．

filament, ρ_p, can not be obtained with FMT observations alone, we adopted an average value of an ordinary prominence, or $\rho_p = 1.7 \times 10^{-13}$ g cm^{-3}, which corresponds to a number density of 10^{11} cm^{-3} (Jefferies & Orrall 1963) with a filling factor of 0.1 (Engvold 1976; Simon et al. 1986). The total velocity, $v(x, t)$, is obtained by the velocity diagnostics described above and in Paper I, and the height $H(x, t)$ is given by the upward velocity, $v_u(x, t)$, via an equation,

$$H(x,t) = \int_{t_0}^{t} v_u(x, t')dt', \quad (2)$$

where t_0 is the onset time of the acceleration of the element. Using equation (1) and equation (2), we obtain the final mechanical energy, $e_{mc}(x, t_f)$, of the eruptive element on the trajectory at time t_f when the element disappeared from the FMT field of view. We then divide it by the time difference between t_0 and t_f to obtain the mechanical energy gain rate, $\epsilon_{mc}(x) = e_{mc}(x, t)/(t_f - t_0)$. Finally, we average the derived mechanical energy gain rates over all the eruptive elements and obtain the mean mechanical energy gain rate per unit volume $\epsilon_{mc} = \langle \epsilon_{mc}(x) \rangle_x$.

For example, the height of the element that corresponds to point A of the 2000-May-08 event at 05:18 UT just before its disappearance was 4.9×10^9 cm, and its potential energy is therefore 2.1 erg cm^{-3}. The kinetic energy of the same element is 2.1×10^{-1} erg cm^{-3} which, combined with the potential energy, gives a total mechanical energy of $e_{mc}(A, t_f) = 2.3$ erg cm^{-3}. Since the start time of acceleration and the time of its disappearance of the element are $t_0 = $ 04:17 UT and $t_f = $ 05:18 UT, respectively, we obtained its mechanical energy gain rate to be $\epsilon_{mc}(A) = 6.3 \times 10^{-4}$ erg cm^{-3} s^{-1}. This value is used together with those of the other trajectories which are judged to be ejected into the interplanetary space to calculate the mean value of the mechanical energy gain rate, ϵ_{mc}, of all eruptive elements. The calculated results for all events are listed in the fourth column of table 1.

3.2. 熱エネルギー解放率

単位体積当たりの熱エネルギーの上昇率 ($\epsilon_{\rm th}$) は, Isobe et al. (2002) の方法に則って計算した. (ⅰ) コロナの熱エネルギーを求めたいある時刻 $t = t_{S0}$ を指定する. (ⅱ) 時刻 $t = t_{S0}$ より以前の時刻 $t = t_{S1} < t_{S0}$ に2種のフィルターを通して撮影された2枚のSXT像, および時刻 $t = t_{S0}$ より以後の時刻 $t = t_{S2} > t_{S0}$ に同様に撮影された2枚のSXT像を選択する. (ⅲ) 選択された画像内のアーケード領域の軟X線強度の平均を求める. (ⅳ) 2種のフィルター画像の平均値にフィルター比法を適用して, 時刻 t_{S1}, t_{S2} のときの温度 (T_{S1}, T_{S2}), 放射尺度 (EM_{S1}, EM_{S2}) を求め, これらの値から次の数式のように直線内挿で $t = t_{S0}$ 時の温度 T, 放射尺度 EM を求める.

$$T = \frac{T_{S2} - T_{S1}}{t_{S2} - t_{S1}}(t_{S0} - t_{S1}) + T_{S1}. \tag{3}$$

次に, 電子密度 n と単位体積当たりの平均熱エネルギー $e_{\rm th}(t)$ を次式で計算する.

$$n(t) = \sqrt{\frac{EM(t)}{V(t)}}, \tag{4}$$

$$e_{\rm th}(t) = 3n(t)k_{\rm b}T(t), \tag{5}$$

ここで, 体積 V はSXT画像の1ピクセルの面積に, 視線方向のアーケードの長さ l を乗算したものである. l の値は, アーケードの高さあるいは幅に等しいと仮定した. 2000年5月8日のイベントについて, 熱エネルギー密度 $e_{\rm th}(t)$ の時間変化を図4にダイアモンド印で示した. このイベントのときにはPFI画像が利用できたので, 30秒以下の時間間隔で変化を追うことができ, 熱エネルギー密度が上昇しているのが見てとれる. 解放エネルギーをより正確に見積もるために, 放射および伝導冷却の効果も考慮に入れることが必要である. このとき, 熱エネルギー解放率は次の式になる.

$$\epsilon_{\rm th} = \frac{de_{\rm th}}{dt} + l_{\rm r} + l_{\rm c}, \tag{6}$$

ここで, $l_{\rm r}$, $l_{\rm c}$ は単位体積当たりの平均放射冷却率, 平均伝導冷却率である. これらは, 次式で与えられる.

3.2. Thermal Energy Release Rate

The mean increasing rate of thermal energy per unit time ($\epsilon_{\rm th}$) is calculated in the following procedure based on Isobe et al. (2002). (i) We select an SXT image (obtained at $t = t_{S0}$) without saturated pixels. (ii) We select two SXT images taken with a different filter and without saturated pixels before ($t = t_{S1} < t_{S0}$) and after the first image ($t = t_{S2} > t_{S0}$). (iii) We average the intensity of the arcade, and (iv) we derived two temperatures (T_{S1}, T_{S2}) and emission measures (EM_{S1}, EM_{S2}) from the averaged intensities of the two pairs of images. Finally, we obtain T and EM at $t = t_{S0}$ by a linear interpolation, i.e.,

$$T = \frac{T_{S2} - T_{S1}}{t_{S2} - t_{S1}}(t_{S0} - t_{S1}) + T_{S1}. \tag{3}$$

Then, we calculate the electron number density, n, and the mean thermal energy per unit volume, $e_{\rm th}(t)$, with the following equations:

$$n(t) = \sqrt{\frac{EM(t)}{V(t)}}, \tag{4}$$

$$e_{\rm th}(t) = 3n(t)k_{\rm b}T(t), \tag{5}$$

where the volume V is the area of one pixel on the SXT images multiplied by the line-of-sight length, l, which we assume to be equal to the height or the width of the arcade. The calculated temporal variation of the thermal energy density $e_{\rm th}(t)$ of the arcade of the 2000-May-08 event is plotted in figure 4 with diamonds. Since we have PFIs for this event, the time cadence of the images is less than 30 seconds, and we can see the increase in the thermal energy density. We also take the radiative and conductive coolings into account for a more accurate measurement of the released energy. The thermal energy release rate per unit volume, $\epsilon_{\rm th}$, is therefore given by

$$\epsilon_{\rm th} = \frac{de_{\rm th}}{dt} + l_{\rm r} + l_{\rm c}, \tag{6}$$

where $l_{\rm r}$ and $l_{\rm c}$ are the mean radiative and conductive loss rates per unit volume, respectively. They are written as

$$l_{\rm r} = n^2 Q(T) \simeq 10^{-17.73} T^{-2/3} n^2 \ ({\rm erg\ s^{-1}\,cm^{-3}}), \tag{7}$$

$$l_{\rm c} = \frac{d}{ds}\left(\kappa \frac{d}{ds}T\right) \simeq 9.0 \times 10^{-7} \frac{T^{7/2}}{s^2}$$

$$({\rm erg\ s^{-1}\,cm^{-3}}). \tag{8}$$

Appendix A Selected Papers Relating to FMT | 463

Energy gain rate

[Figure: Plot showing e_mc (squares), e_th (diamonds), CME velocity (asterisks), and GOES X-ray flux (solid line) versus Time [05/08/00] (UT) from 05:00 to 08:00. Left axis: e_{mc} & e_{th} (ergs cm^{-3}) from 10^{-4} to 10.0. Right axis: CME velocity (km s^{-1}) from 200 to 450. GOES class: B, C, M marked near right axis.]

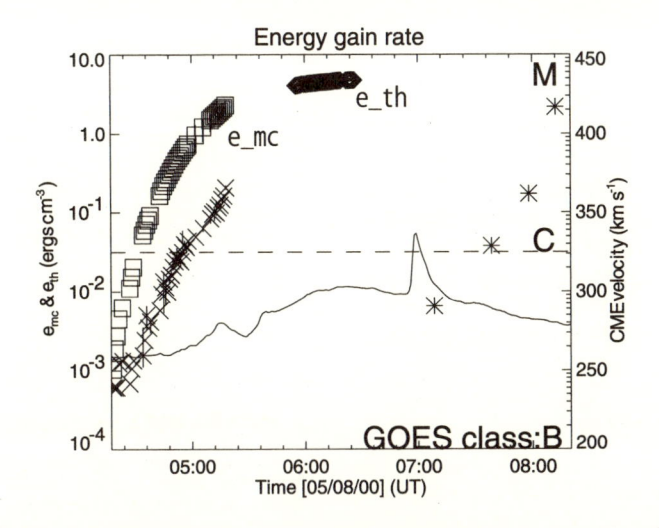

図4 FMT, SXT, GOES, および LASCO で観測された様々なパラメーターの時間発展. □印は単位体積当たりの平均力学的エネルギー (e_{mc}), ×印は単位体積当たりの平均運動エネルギーを示す. 熱エネルギー密度 (e_{th}) の時間変化はダイアモンド印で, CME の速さは∗印で示されている. エネルギー密度の目盛は左側の縦軸に記されており, 速度の目盛は右側の縦軸にある. 実線は, GOES 衛星による太陽全面積分 X 線強度変化を示す. 右側の縦軸近くのそのクラスを B, C, M で示している.

Fig. 4. Temporal evolutions of various parameters obtained with FMT, SXT, GOES, and LASCO instruments. The squares represent the mean mechanical (e_{mc}) and the crosses show the mean kinetic energies of the filament per unit volume. The temporal variation of the thermal energy density (e_{th}) and the CME velocity are plotted by diamonds and asterisks, respectively. The amount of energy densities and CME velocity are indicated by the lefthand and righthand axes, respectively. The solid line is for the integrated soft X-ray flux data from GOES, and its class is denoted by "B", "C", and "M" beside the righthand axis.

$$l_r = n^2 Q(T) \simeq 10^{-17.73} T^{-2/3} n^2 \text{ (erg s}^{-1}\text{cm}^{-3}\text{)}. \quad (7)$$

$$l_c = \frac{d}{ds}\left(\kappa \frac{d}{ds}T\right) \simeq 9.0 \times 10^{-7} \frac{T^{7/2}}{s^2}$$
$$(\text{erg s}^{-1}\text{cm}^{-3}). \quad (8)$$

上式で $Q(T)$ は $10^{6.3} < T < 10^7$ の温度範囲での放射冷却関数 (Rosner et al. 1978), $\kappa = 9.0 \times 10^7 T^{5/2}$ は Spitzer 熱伝導率 (Spitzer 1956) であり, s はアーケードループ全長の半分の長さである. 熱エネルギー密度の時間微分 de_{th}/dt は, 観測された熱エネルギー密度 e_{th} の時間変化を最小二乗法で滑らかな関数に近似して求めた. 本論文では, 式 (6) – (8) により単位体積当た

In the above equations, $Q(T)$ is the radiative loss function for the temperature range of $10^{6.3} < T < 10^7$ (Rosner et al. 1978), $\kappa = 9.0 \times 10^7 T^{5/2}$ is the Spitzer thermal conductivity (Spitzer 1956), and s is the half-length of an arcade loop. The time derivative of the observed thermal energy density, de_{th}/dt, is obtained by applying the least-squares fitting method to the temporal evolution of the observed thermal energy density, e_{th}. Using equations (6)–(8), we then calculated the thermal energy release rate, ϵ_{th}, per unit volume.

In analyzing an event for which we only have PFIs, we make another assumption to obtain a more appropriate energy

りの熱エネルギー解放率 $\epsilon_{\rm th}$ を算出した．

すべて PFI 画像しかない場合のイベントについては，もう1つの仮定をおくことで，より適切なエネルギー解放率の算出をすることができる．このような場合，軟 X 線イベントの開始時刻は FMT で観測される Hα フレアリボンの開始時刻と同時であり，すべての追加の熱エネルギーは Hα フレア開始後になされたと仮定して，解放率を算出した．単位体積当たりの熱エネルギー解放率は，表1の第5列にあげられている．

4. 結果とその議論

この節では，まず，2000年5月8日と1998年9月20日の2つのイベントを取り上げ，消失フィラメントの力学的エネルギー密度 $e_{\rm mc}$ の時間発展と軟 X 線アーケードの熱エネルギー密度 $e_{\rm th}$ の時間発展の相互関係を示す．この2つのイベントについては，時間分解能の高い PFI 画像が得られており，アーケードの熱エネルギー密度の時間発展を詳しく研究することができた．次に，10個のイベントについて，2つのエネルギー増加率の間の関係を統計的に研究した結果を議論する．

4.1. 時間発展の相互関係

4.1.1. 2000年5月8日のイベント

フィラメントガスの平均力学的エネルギー密度 $e_{\rm mc}$ の時間変化は図4に示されている．×印は平均運動エネルギー密度を示し，□印は運動エネルギー密度と重力ポテンシャルエネルギーを合わせた平均力学エネルギー密度を示している．$e_{\rm mc}$ の平均増加率 $\epsilon_{\rm mc}$ は 6.4×10^{-4} ergs cm^{-3} s^{-1} であった．力学的エネルギー $e_{\rm mc}$ の変化は単純な直線で近似できない振る舞いであったため，全速度が 15 km/s 以上の時間帯を数個の時間帯に分割して，それぞれの時間帯での増加率を求めることによって，その上下限値を評価した．ある速度以上を示した時間帯を選んだのは，論文1で述べたように，速度の測定精度が 15 km/s 未満であるためである．このイベントの $\epsilon_{\rm mc}$ の上下限値はそれぞれ 1.8×10^{-3}, 3.1×10^{-4} ergs cm^{-3} s^{-1}

release rate. In this case, to derive the energy release rate, we assume that the start time of the soft X-ray event is equal to the time when an Hα ribbon flare first appears on FMT images, and that all of the surplus thermal energy is supplied after onset of the Hα flare. The thermal energy release rates per unit volume are listed the fifth column of table 1.

4. Results and Discussions

In this section, we first present the temporal relation between the evolution of the mechanical energy density, $e_{\rm mc}$, of disappearing filaments and that of the thermal energy density, $e_{\rm th}$, of the associated soft X-ray arcades, using two events that occurred on 2000-May-08 and 1998-Sep-20. In these events, we fortunately have SXT PFIs of the associated arcades with a high time cadence, which is suitable for a detailed study of the temporal variation of the thermal energy densities of the arcades. We then discuss the statistical relation between these two energy increase rates, using the results of all 10 events.

4.1. The Temporal Relation

4.1.1. The 2000-May-08 event

The temporal variation of the mean mechanical energy density, $e_{\rm mc}$, of elements inside the filament is plotted in figure 4. The crosses show the mean kinetic and the squares the mechanical energy density, which is the sum of the kinetic and potential energy densities. It is clear that the filament possesses much more potential energy than the kinetic energy. The mean increase rate of $e_{\rm mc}$ is $\epsilon_{\rm mc} = 6.4 \; 10^{-4}$ erg cm^{-3} s^{-1}. Since the increase of the mechanical energy density, $e_{\rm mc}$, can be approximated not by a linear function, we measured the upper and lower limits by dividing the event into several time ranges in which the increase of the total velocity is more than 15 km s^{-1}. This is because, as we mentioned in Paper I, the magnitude of errors of the measured velocities is less than 15 km s^{-1}. The upper and lower limits

であった．

このイベントに伴う軟X線増光現象は，長時間にわたって継続する LDE イベントというもので，図 4 の実線で描かれた GOES 衛星が観測したX線強度変化が示すように，05:01 UT からゆっくりと増光をはじめ，06:23 UT にピーク強度に達した．07:00 UT の鋭いピークは，この現象とは関係のない領域で起こったものである．軟X線の増光は，フィラメントの加速が起こってから 30 分以上も後に発生したことになる．このことは，軟X線の増光は物質放出の加速開始時刻より遅れて始まるという Hundhausen (1997) が見出したことと一致する．このイベントに伴うアーケード形成（図 2）は，ようこうの日陰が終わった 05:55 UT から Yohkoh/SXT 観測が始まったため，Hα フィラメント加速の全期間にわたる SXT 像はなく，その後半の時間帯のみ解析が可能であった．図 4 には，この限られた時間帯での熱エネルギー密度 e_{th} の時間変化がダイアモンド印で示されている．このとき，平均増加率は 9.1×10^{-4} ergs cm^{-3} s^{-1} であった．

SOHO/LASCO の観測によると，時刻 06:50 UT のとき LASCO C2 の視野内に，不規則な形の波面を持つコロナ放出現象（CME）が現れた．この時刻は，FMT 像で Hα フィラメントが消えて (5:20 UT) から 1 時間半後であった．LASCO 像上で時々刻々 CME の先端部の位置を測定してその速さをもとめて，結果を図 4 に星印で示している．この CME は LASCO C2 の視野内では，08:20 UT まで加速され続けた．LASCO C2 の視野内での CME の加速があるからといって，CME あるいはプロミネンスが発生時から同じような加速を受けていたとは必ずしも言えない（Tripathi et al. 2006）．従って，04:17 UT に始まったフィラメント加速が，この時間帯まで継続していたかどうかは定かではない．05:55 UT から 06:26 UT までの SXT 観測でとらえられた熱エネルギー増加率が，フィラメントおよび CME の加速に関与していた可能性はある．

of ϵ_{mc} of this event are 1.8×10^{-3} and 3.1×10^{-4} erg cm^{-3} s^{-1}, respectively.

The soft X-ray event associated with this event is a long duration event (LDE), very gradually starting at 05:01 UT and reaching its peak at 06:23 UT according to GOES soft X-ray flux data, shown by the solid line in figure 4. The sharp peak around 07:00 UT is not related to the event in the problem. The start time of the soft X-ray event is more than a half hour after the onset time of the acceleration of the filament, which is consistent with the results by Hundhausen (1997), who found that the onset times of X-ray events commonly lag behind the start times of the acceleration of mass ejections. Since the Yohkoh/SXT observations of the arcade formation (see figure 2) associated with the Hα filament eruption start at 05:55 UT after the Yohkoh night, we do not have enough SXT data that covers the acceleration phase of the Hα filament. The thermal energy density, e_{th}, is plotted in figure 4 with diamonds. It increases within the specified time range, and their mean increase rate is found to be 9.1×10^{-4} erg cm^{-3} s^{-1}.

According to the CME observation by SOHO/LASCO, a CME with a ragged front appeared at 06:50 UT in the LASCO C2 field of view. This time was one and a half hours after disappearance of the Hα filament in FMT images (5:20 UT). The CME velocity obtained by measuring the position of its leading edge on the time-sequenced LASCO images is also shown in the same panel with asterisk marks. This CME is continuously accelerated within the LASCO C2 field of view until 08:20 UT. The acceleration of a CME in LASCO/C2 FOV does not necessarily mean that the CME or the associated prominence has followed the same profile since its onset (Tripathi et al. 2006). Thus, we cannot distinguish whether the acceleration of the filament from its onset of eruption at 04:17 UT continued or not. The increase in the thermal energy release rate during SXT observations from 05:55 UT to 06:26 UT, may be related to the accelerations of both the filament and the CME.

4.1.2. 1998年9月20日イベント

1998年9月20日に太陽の東の縁近傍でフィラメント噴出が発生した．FMT像から部分的に切り取って時間順に並べたものを図5に示す．消失現象が始まる前の02:00 UT には，活動領域 NOAA 8340 の東側に1つのフィラメントが存在し，Hα線中心画像でのみ確認できた．02:40 UT に，フィラメント本体の西側でツーリボンフレアが発生した．FMT像では，03:09 UT にフレアが明瞭に見えており，3種のHαバンドすべての像でフィラメントが活発化しているのが観測された．このフレアの際，フィラメントは最大 27 km/s の上昇速度で 30-40 Mm の高さまで持ち上げられた．

図6の実線で示されているように，このフィラメント噴出の後，02:33 UT から始まり GOES クラス M1.8 のピーク強度を持つ軟X線アーケードが形成された．ようこうによる観測は 02:40 UT から始まっており，このフレアの PFI 画像の一例が図5の下右パネルに示されている．図6には，力学的エネルギー密度 e_{mc} と熱エネルギー密度 e_{th} の時間変化も示されている．ダイアモンド印の熱エネルギー密度 e_{th} は，実線で描かれた軟X線の変化と歩調を合わせて，増加の後減少に転じている．(02:40 UT から 02:56 UT の間の) 増光時の，単位体積当たりの熱エネルギー解放率は ϵ_{th} = 0.57 ergs cm^{-3} s^{-1} であった．フィラメントの平均力学的エネルギー増加率 ϵ_{mc} も，02:33 UT から始まった軟X線の増光時に，最大 ϵ_{mc} = 2.7 × 10^{-4} ergs cm^{-3} s^{-1} の値となる顕著な増加を示した．

しかし，熱エネルギー密度と力学的エネルギー密度の増加の様子は，02:42 UT 以降異なっている．熱エネルギー密度 e_{th} は 02:56 UT で最大になるまで増加しているのに対して，力学的エネルギー密度 e_{mc} は 03:08 UT まで増加していない．これは，02:42 UT 以降フィラメント運動が一時的に停止したことによる．熱エネルギー密度 e_{th} が減少しつつある 03:08 UT からフィラメントの噴出部が再び上空に向かって加速されたのち消失した．このことが，力学的エネルギー密度 e_{mc} の更なる増加の原因と

4.1.2. The 1998-Sep-20 event

A filament eruption occurred near the solar east limb on 1998-Sep-20. A series of partial FMT images are displayed in figure 5. The dark filament is initially located to the east of an active region, NOAA 8340, and is visible only in the Hα line center image at 02:00 UT before its disappearance. From 02:40 UT, a two-ribbon flare takes place at the west side of the main body of the filament. This flare is clearly seen on FMT images at 03:09 UT, and we can see that the filament is activated, but still visible, in all images. The filament was lifted up to a height of 30-40 Mm with a maximum upward velocity of 27 km s^{-1} when the flare occurred.

This filament eruption was followed by soft X-ray arcade formation from 02:33 UT with a peak integrated flux of M1.8 class, observed by GOES, which is shown in figure 6 by the solid line. The Yohkoh observation started from 02:40 UT, and one of the PFIs of this flare is shown in the bottom-right panel of figure 5. The temporal variations of the mean mechanical energy density, e_{mc}, of the filament and the mean thermal energy density, e_{th}, of the arcade are also shown in figure 6. The thermal energy density, e_{th} (diamonds), rises and then falls almost consistently with the soft X-ray flux (the solid line). The mean release rate of the thermal energy per unit volume, ϵ_{mc}, during its rise phase (from 02:40 UT to 02:56 UT) was calculated to be ϵ_{th} = 0.57 erg cm^{-3} s^{-1}. Similar to the thermal energy density, the mean mechanical energy of the filament, ϵ_{mc}, shows a notable increase with a maximum rate of ϵ_{mc} = 2.7 × 10^{-4} erg cm^{-3} s^{-1} during the rise phase of the soft X-ray event from 02:33 UT.

The growth of both thermal and mechanical energy densities, however, are not in coincidence after 02:42 UT. Although the thermal energy density, e_{th}, increases until its peak at 02:56 UT, the mean mechanical energy density of the filament, e_{mc}, does not show any significant enhancements until 03:08 UT. This corresponds to a temporary suspension of filament motion from 02:42 UT. Then, from 03:08 UT when the thermal energy density was slowly decaying, the "eruptive" part of the filament started to be accelerated

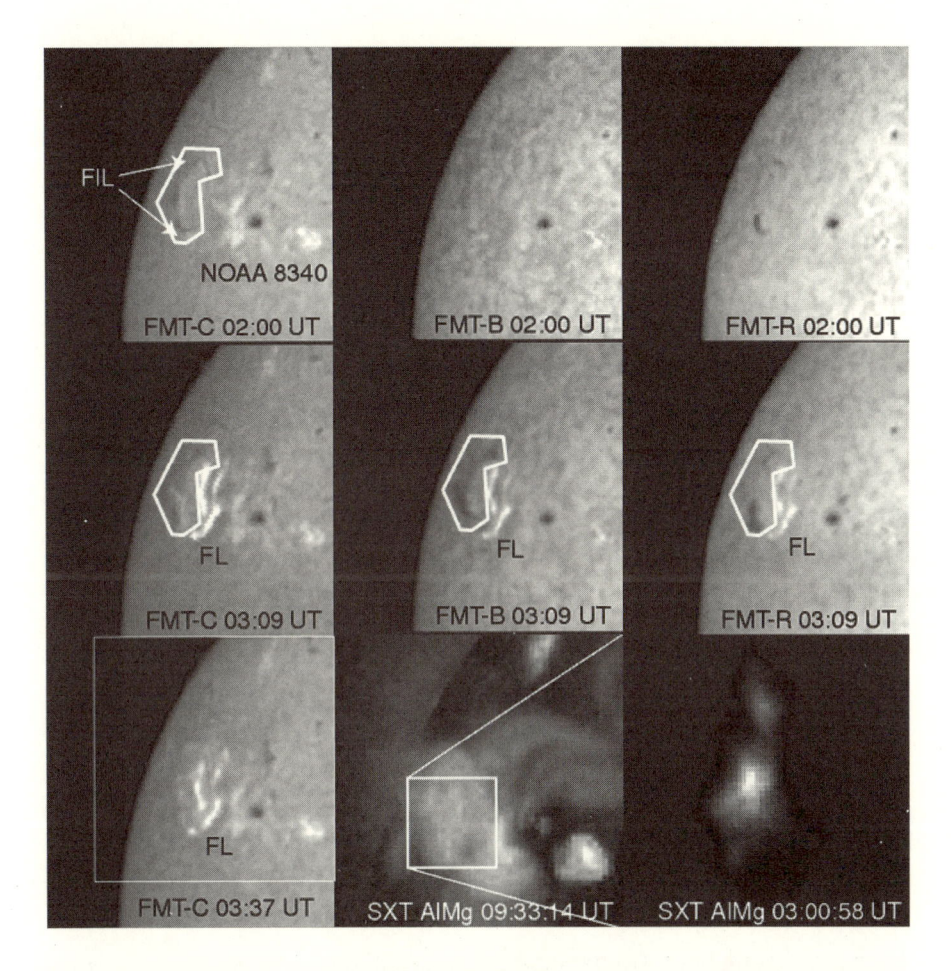

図 5 1998 年 9 月 20 日イベントの FMT, SXT 観測画像. このイベントは東の縁近くで発生して, その後強いフレアが Hα (FL で示したもの) および軟 X 線で観測された. 下右のパネルは, 図 6 の熱エネルギー密度を求める際に使用した SXT の PFI 画像の例である. フィラメントは, 上左パネルに FIL という文字が付された 2 本の矢印間に位置していた. 下中パネルの視野は, 下左パネル内に白い四角で示されている. また, 下中パネルの白い四角は SXT PFI 画像の視野を示す.

Fig. 5. FMT and SXT observations of the event on 1998-Sep-20. This event occurred near the east limb, followed by an intense Hα (FL) and soft X-ray flares. The bottom-right panel is one of the SXT PFIs which are used to derive the thermal energy densities shown in figure 6. The filament is located between the arrows shown in the top-left panel with an abbreviation of "FIL". The SXT field of view of the bottom-middle panel is indicated by the white box in the left-bottom panel and the white box in the bottom-middle panel stands for the SXT PFI field of view.

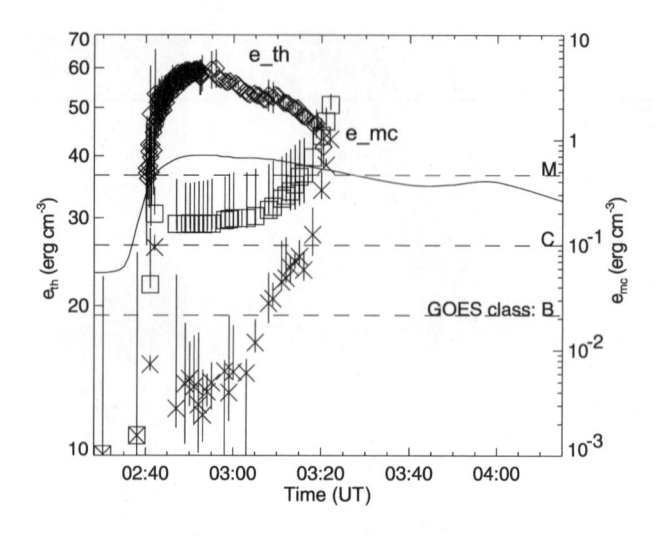

図6 1998年9月20日イベント．アーケードの熱エネルギー密度（e_{th}：ダイアモンド印）とフィラメントの単位体積当たりの力学的エネルギー密度（e_{mc}：□印），運動エネルギー密度（×印）の時間変化が示されている．実線は，GOES軟X線強度の時間変化を示す．力学的エネルギーと運動エネルギーの目盛は右側の縦軸に示されている．また，軟X線強度のクラス（B，C，M）は，右の縦軸のそばに示されている．

Fig. 6. Thermal energy density (e_{th}: diamond) of the arcade and the mean mechanical energy of the disappearing filament per unit volume (e_{mc}: square) with its kinetic energy density (cross) are plotted as functions of time. The solid line represents the soft X-ray flux observed by GOES. The magnitudes of mechanical and kinetic energy densities are indicated by the righthand axis and the level of class (B, C, M) of soft X-ray flux is given beside the righthand axis.

なっている．この後半の加速時の力学的エネルギーの増加率は $\epsilon_{mc} = 1.9 \times 10^{-3}$ ergs cm^{-3} s^{-1} にも達しており，初期の加速時の最大増加率に比べてほぼ一桁程度大きくなっていることは注目に値する．02:42 UTまでの初期の加速のときは，運動エネルギーが力学的エネルギーの大部分を占めていたがフィラメントの動きが停止した02:42 UT以降は重力ポテンシャルに比べて無視できる程度になっている．この結果は，2000年5月8日のイベントの場合とも一致しており，重力ポテンシャルエネルギーがフィラメント噴出の加速時には重要であることを示している．

軟X線イベントの増光時に，e_{mc} と e_{th} が時間的に歩調を合わせて変化していたが，

upward again and disappeared, thus causing an increase in the mechanical energy density, e_{mc}. One should notice that the increase rate of the mechanical energy density in the later phase is up to $\epsilon_{mc} = 1.9 \times 10^{-3}$ erg cm^{-3} s^{-1}, and this is roughly an order of magnitude larger than the maximum increase rate in the initial phase of this event. The contribution of kinetic energy of the filament to the mechanical energy is very large in the initial acceleration phase until 02:42 UT, though it then becomes negligible compared to the potential energy, since the filament ceased its motion at 02:42 UT. This result agrees with the result of the 2000-May-08 event, and indicates that the potential energy is very important during the acceleration phase of a filament

03:08 UT から始まる急激な（2回目の）力学的エネルギー密度 e_{mc} の増加時には，熱エネルギー密度は減少していた．GOES 衛星の X 線観測によると，軟 X 線強度は 03:45 UT から再び上昇を始め 03:57 UT に C8.0 の2回目のピークに達している．FMT では，フィラメントが視野から消えた後，軟 X 線の増光と共に最初と同じ場所で2回目のツーリボンフレアが起こっていることが観測されている．02:32 UT および 03:08 UT からそれぞれ始まった力学的エネルギー増加は，02:33 UT から始まる1回目の軟 X 線イベントばかりでなく，03:45 UT から始まる2回目のイベントに関係している可能性は否定できない．

4.2. 統計的な関係

10 個のイベントについて，Hαフィラメントの単位体積当たりの力学的エネルギー増加率 ϵ_{mc} と，SXT データから得られる単位体積当たりの熱エネルギー増加率 ϵ_{th} との間の相関関係を図7に示す．この図には，アーケード下の光球平均磁場強度の概略値も記している．この相関関係から，(ⅰ) 熱エネルギー解放率と力学的エネルギーとの間は正の相関関係にあること，(ⅱ) 磁場強度が強い場所で起こるイベントにはより多くのエネルギーが供給されることがわかる．この相関関係を表示するとき，噴出フィラメントの力学的エネルギー増加率としては，それぞれのフィラメントの代表的な一部分を取って計算したものを用いている．観測によると，噴出の間はフィラメントの各部分は異なる速さで運動している．この部分ごとの運動エネルギーの相違，ひいては力学的エネルギーの相違はありうるが，その相違は速度測定誤差範囲内（論文1で述べられたように 15 km/s）にあると思われる．

図7には，最小二乗法で求まる両者間の直線的な関係 $\epsilon_{\mathrm{th}} \propto \epsilon_{\mathrm{mc}}^{1.9}$ を破線で示している．この関係式が成り立つ理由を，蓄積された磁気エネルギーがポインティングエネルギー流速により磁気リコネクション領域に運ばれ，熱エネルギーに変換されるとい

eruption.

Though the evolutions of e_{mc} and e_{th} in the rise phase of the soft X-ray event is temporally in coincidence, the temporal variations of the thermal energy density decreases during an abrupt increase of the mechanical energy density, e_{mc}, from 03:08 UT. According to the GOES X-ray flux, however, the soft X-ray flux increases again from 03:45 UT, and reaches a second peak of C8.0 at 03:57 UT. The FMT observations also show another Hα two ribbon flare at the same site associated with an enhancement in the soft X-ray flux just after the filament disappeared from the FMT field of view. Increases in its mechanical energy from 02:32 UT and 03:08 UT may correspond to not only the soft X-ray events from 02:33 UT, but also to events from 03:45 UT.

4.2. The Statistical Relation

The relation between the derived mean mechanical energy gain rates of the Hα filaments per unit volume, ϵ_{mc}, and the thermal energy release rate per unit volume, ϵ_{th}, obtained with SXT data for all 10 events are displayed in figure 7. We also give rough values of the mean magnitude of photospheric magnetic field beneath the arcade B_{f} in the figure. This relation clarifies that (i) the thermal energy release rates and the mechanical energy gain rates have a positive correlation and (ii) more energy is supplied for events with stronger photospheric magnetic field strengths. To calculate the mean value of the mechanical energy gain rate of all the eruptive elements, we used the value of one representative element. From observations it appears that different parts of a filament move with different speeds during an eruption. The velocity differences of each element would affect the kinetic energy of the filament, and in turn the mean mechanical energy, but we believe that such effects are included in the errors of the measured velocities (15 km s^{-1}, mentioned in Paper I).

The most probable correlation between the increase rates of the thermal energy of the coronal arcades and the mechanical energy of filaments derived by the least-squares fitting method is shown by the dashed line in figure 7, which represents

うことと，フィラメントはローレンツ力で加速されるということをもとに，オーダー計算で考えてみる．熱エネルギー増加率は式(6)より

$$\epsilon_{\mathrm{th}} = 2\frac{B^2}{4\pi L}v_{\mathrm{in}}, \qquad (9)$$

と書ける．ここで，B, v_{in}, L は磁気リコネクション領域のコロナ磁場強度，流入速度，長さを示す（Isobe et al. 2002）．流入速度は，アルフベン速度 V_{A} と X 形の磁気中性点を磁力線が通り抜ける率を示す磁気リコネクション率 M_{A} を用いて $v_{\mathrm{in}} = M_{\mathrm{A}}V_{\mathrm{A}}$ と書き表わせる．磁気リコネクション率 M_{A} の磁場強度に対する依存性は，リコネクション機構によって異なる．低速リコネクションと高速リコネクション機構があり，前者の例としては Sweet-Parker 機構 (Sweet 1958; Parker 1957) を取り上げ，後者の例としては Petchek 機構 (Petschek 1964) を取り上げる．圧力勾配の効果を無視してオーダー解析をすると，磁気レイノルズ数 R_{m} を用いて，それぞれのリコネクション率は $M_{\mathrm{A}} \cong 1/R_{\mathrm{m}}^{1/2}$, $\cong \pi/8 \log R_{\mathrm{m}}$ と表わされる（Priest & Forbes 2000）．$\log R_{\mathrm{m}}$ はゆっくりと変わる関数であるので，単位体積当たりの熱エネルギーの解放率の磁場強度 B に対する依存性は，Sweet-Parker 機構のとき $\epsilon_{\mathrm{th}} \propto B^{2.5}$, Petchek 機構のとき $\epsilon_{\mathrm{th}} \propto B^3$ となる．

フィラメントの力学的エネルギー増加率は，フィラメント内部の単位体積が単位時間に受けるローレンツ力による仕事 $[j \times B/c](v + 1/2\alpha_{\mathrm{p}})$ から評価できる．ここで，v はフィラメントの速度，α_{p} は加速度を表す．通常，α_{p} は加速初期を除いて速度 v より遥かに小さいので，フィラメントの力学的エネルギー増加率の磁場強度 B に対する依存性は，近似的に $\epsilon_{\mathrm{mc}} \propto B^2$ となる．以上のことから，2つのエネルギー解放率は $\epsilon_{\mathrm{th}} \propto \epsilon_{\mathrm{mc}}^{\gamma}$ と書ける．Sweet-Parker 機構のとき $\gamma = 1.3$, Petchek 機構のとき $\gamma = 1.5$ となる．以上のように，ポインティングエネルギー流束が熱エネルギーの源であり，ローレンツ力がフィラメント加速の駆動力であるという仮説のもとでのオーダー計算によって，$\epsilon_{\mathrm{th}} \propto \epsilon_{\mathrm{mc}}^{1.9}$ という関係

the relation $\epsilon_{\mathrm{th}} \propto \epsilon_{\mathrm{mc}}^{1.9}$. In order to interpret this relation, we compile an order-of-magnitude analysis of both the energy increase rates by assuming that the stored magnetic energy is carried by the Poynting flux to the magnetic reconnection region, and is converted into the observed thermal energy, and a filament is accelerated by the Lorentz force. Then, for the increase rate of thermal energy, equation (6) can be rewritten as

$$\epsilon_{\mathrm{th}} = 2\frac{B^2}{4\pi L}v_{\mathrm{in}}, \qquad (9)$$

where B, v_{in}, and L are the coronal magnetic field strength, the inflow speed, and the length scale of the reconnection region, respectively (Isobe et al. 2002). The inflow speed is expressed as $v_{\mathrm{in}} = M_{\mathrm{A}} V_{\mathrm{A}}$ with the Alfvén velocity, V_{A}, and the reconnection rate, M_{A}, which characterizes the rate at which the field lines move through the X-type neutral point. The dependence of the reconnection rate, M_{A}, on the magnetic field strength, B, differs from slow to fast reconnection mechanisms. For slow and fast reconnections, we consider the Sweet-Parker (Sweet 1958; Parker 1957) and the Petschek (Petschek 1964) mechanisms in which the reconnection rates, M_{A}, are given by the magnetic Reynolds number, R_{m}, as $M_{\mathrm{A}} \approx 1/R_{\mathrm{m}}^{1/2}$ and $M_{\mathrm{A}} \approx (\pi/8) \log R_{\mathrm{m}}$, respectively, by an order-of-magnitude analysis while ignoring the effect of pressure gradients (Priest & Forbes 2000). Since $\log R_{\mathrm{m}}$ is slowly varying, this yields the dependence of the thermal-energy release rate per unit volume, ϵ_{th}, on the magnetic field strength, B, as $\epsilon_{\mathrm{th}} \propto B^{2.5}$ and $\epsilon_{\mathrm{th}} \propto B^3$ for the Sweet-Parker and the Petschek mechanisms, respectively.

For the increase rate of the mechanical energy of a filament, we find that the work done by the Lorentz force per unit time for an element with unit volume inside a filament is $[j \times B/c](v + 1/2\alpha_{\mathrm{p}})$, where v is the velocity of the element and α_{p} is its acceleration value. Since α_{p} is usually much smaller than the velocity, v, except for the very beginning of activation, the dependence of the mechanical energy gain rate of the element on the magnetic field strength, B, is approximated as $\epsilon_{\mathrm{mc}} \propto B^2$. This then yields the relations between two energy release rates as $\epsilon_{\mathrm{th}} \propto \epsilon_{\mathrm{mc}}^{\gamma}$, where $\gamma = 1.3$ for the Sweet-Parker

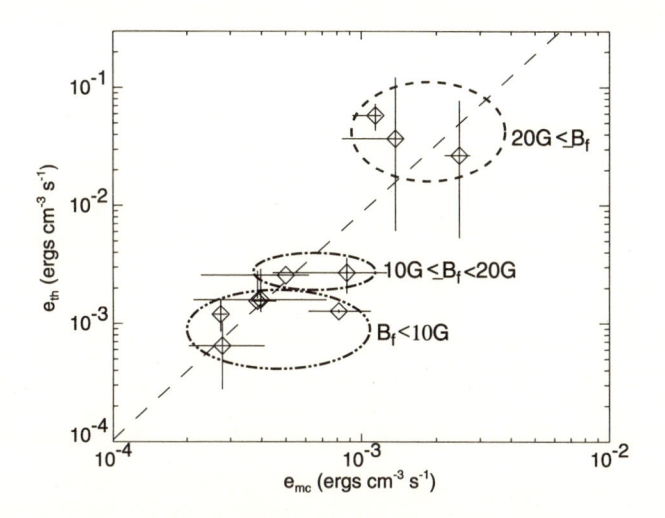

図7 10個のイベントについて，単位体積当たりの力学的エネルギー増加率 ϵ_{mc} と単位体積当たりの熱エネルギー解放率 ϵ_{th} の関係が示されている．破線は，最小二乗法で求められた両者の関係式 $\epsilon_{th} \propto \epsilon_{mc}^{1.9}$ を描いたものである．アーケード下の平均光球磁場強度 B_f の概略値を楕円で囲って示している．

Fig. 7. Relation between the mechanical energy gain rate per unit volume, ϵ_{mc}, and the thermal energy release rate per unit volume, ϵ_{th}, for the selected 10 events. The dashed line represents the relation $\epsilon_{th} \propto \epsilon_{mc}^{1.9}$ obtained by the least-squares fitting method. The rough magnitudes of the mean photospheric magnetic field strength beneath the arcade, B_f are indicated by the circles.

が説明できるし，また，図7に概略が示されている2つのエネルギー増加率の磁場強度 B_f に対する依存性も説明ができる．

5. まとめと結論

京都大学飛騨天文台のフレア監視望遠鏡（FMT）の Hα 線中心および Hα±0.8Å 像を用いて，10個の消失フィラメントの視線速度とそれに垂直な横方向の速度を測定して，三次元速度場を導いた．視線方向速度は，3つの波長像をみたとき，運動があるときにはドップラー効果によって，フィラメントの各部分が周りの彩層に比較して

type reconnection and $\gamma = 1.5$ for the Petschek type. These two order-of-magnitude analyses based on an assumption of the Poynting flux as a source of thermal energy and the Lorentz force as a driving force for the filament, can explain the relation $\epsilon_{th} \propto \epsilon_{mc}^{1.9}$ as well as the dependence of the two energy increase rates on the magnetic field strengths, B_f, which is roughly indicated in figure 7.

5. Summary and Conclusions

Using the Hα line center, Hα±0.8Å images obtained by the Flare Monitoring Telescope (FMT) at Hida Observatory, we derived the complete 3-D velocity field of 10 solar disappearing filaments by measuring both the line-of-sight and transversal velocities. The line-of-sight velocity is obtained by interpreting the temporal variation of contrasts that are created by the intensity difference

コントラストが変化することを利用して求めた．横方向の速度は，フィラメントの内部構造の移動を追うことで求めた．この2つの速度成分を合成して，フィラメントの三次元速度場を求めた．

運動エネルギーおよび重力ポテンシャルエネルギーの時間変化を測定することは，論文1で詳しく記述されているように，我々のこの新しい方法によって可能となったものであって，コロナグラフ観測やHα線中心波長のみの従来の観測ではできなかったことである．また，本論文では，Yohkoh/SXTデータにフィルター比法 (Hara et al. 1992) を適用して，随伴して発生するフレアアーケードで解放されている熱エネルギー密度も求めた．熱エネルギー解放率を求める際には，放射および伝導冷却効果も考慮に入れた．従来の質量放出現象およびフレアアーケード形成の研究では質量放出の速さやX線強度を元にして議論されていたが，本論文で初めて実際のエネルギーを求めて両者の相関関係が研究された．

力学的エネルギー密度 e_{mc} と熱エネルギー密度 e_{th} の時間変化の比較は，SXT PFI画像がある2つのイベント (2000年5月8日と1998年9月20日のイベント) について行われた．両イベントにおいて，力学的エネルギー密度 e_{mc} の増加は，常に熱エネルギー密度 e_{th} の増加に対応していた．力学的エネルギー密度 e_{mc} の増加は，熱エネルギー密度 e_{th} の増加より時間的に先行していることがしばしばであった．この時間変化のずれは，X線イベントが質量加速より時間的に遅れることが普通であるという従来の研究結果 (Hundhausen 1997) とも一致する．本論文で研究した10イベントのうち，X線イベントの開始が不明であった2例を除いて他の8例では，この時間ずれが確かめられた．

次に，力学的エネルギー密度の平均増加率 ϵ_{mc} および熱エネルギー密度の平均増加率 ϵ_{th} を，それぞれフィラメント加速時の時間平均，X線の急激な増光時の時間平均で求めた．この両者のエネルギー増加率には，図7のように強い相関があることがわ

between the filament and the surrounding chromosphere in the Hα line center and H$\alpha \pm 0.8$Å at every pixel on the filament due to the Doppler shift of the Hα line profile of the filament. The transversal velocity is derived by tracing the internal structures of the filament. These two components of the velocity are combined to yield the complete 3-D velocity field of the filament.

This new method is described in more detail in our previous study (Paper I), and enables us to measure the temporal variations of the kinetic and potential energies, which are difficult to obtain with coronagraph observations or ordinary Hα observations in the Hα line center alone. We also calculated the released thermal energy density of the associated flare arcades by applying the filter ratio method (Hara et al. 1992) to Yohkoh/SXT data. In order to make the more accurate measurement of the thermal energy release rate, we took the radiative and conductive coolings into account. Since previous studies on the energetics of mass ejections and flare/arcade formations have been done in terms of the speed of mass ejections and the intensity of the associated soft X-ray events, this study is the first one that presents the real energetic relation between them.

The temporal variations of the mechanical energy density, e_{mc}, and the thermal energy density, e_{th}, were compared for two events (the 2000-May-08 and the 1998-Sep-20 events) in which we had SXT PFIs. In both events, an increase of the mechanical energy density, e_{mc}, seems to always correspond to an increase of the thermal energy density, e_{th}, though the increase of the mechanical energy density, e_{mc}, often temporally precedes the corresponding thermal energy density, e_{th}. This temporal disagreement is consistent with our previous study in which the onset times of X-ray events commonly lag behind the start times of the acceleration of mass ejections (Hundhausen 1997). In our 10 data sets, this was also confirmed in 8 events, while in the remaining 2 events, we could not identify the start time of their soft X-ray events.

We next calculated the mean increase rate of the mechanical energy density, ϵ_{mc},

かった．従来の研究では，CMEの運動とフレア強度との間に弱い相関があることが見つかっていた（Moon et al. 2002）のに対して，本論文で質量放出エネルギーとフレアエネルギーとの間に強い相関があることが初めて明らかになった．このことは，両者の関係を探るには，熱エネルギー評価ばかりでなく，質量放出現象について射影効果などの影響を受けずに速度場を求めることの重要性を示している．

図7で示されている2つのエネルギー増加率の関係は，$\epsilon_{th} \propto \epsilon_{mc}^{1.9}$という式で表される．この関係は，蓄積された磁気エネルギーがポインティングエネルギー流速により運ばれ，磁気リコネクションにより熱エネルギーに変換されるということ，およびフィラメントの加速はローレンツ力によってなされることを仮定する簡略モデルで説明することができる．両エネルギー密度増加率が，アーケード周辺の平均光球磁場強度に強く依存していることもこの考え方を支持することがらである．

上述の結果は，時間的な変化は同時ではないけれども，フィラメントのエネルギー増加率とフレアアーケードのエネルギー増加率が互いに関係していることを示している．フィラメント加速機構とプラズマ加熱機構は直接的な関係はなく，磁場強度が両機構に関係しているだけかもしれない．このことから，フィラメント加速はローレンツ力でなされるものの，アーケードプラズマは磁気リコネクション過程の何らかの機構によって加熱されていて，蓄積された磁気エネルギーが一方では加速に使われ，他方では加熱に使われているということを示すのかもしれない．

有益なコメントによって本論文を整えることに役立った査読者に感謝する．また，京都大学花山飛騨天文台の構成員にも，その望遠鏡運用および実りのあるコメントや議論に対して謝意を表する．FMTデータベースは，M. Kadota, Y. Nasuji, M. KamobeおよびS.UeNoの方々によって管理運営されている．ようこうのデータは，宇宙航空科学研究所（ISAS）の管理のもと，日本-米国-英国の共同プロジェクトから提供さ

and that of the thermal energy, ϵ_{th}, by taking the average of the amounts of increase at both energies over the time during the acceleration and impulsive phases, respectively. We then found a strong correlation between these two energy increase rates, which is shown in figure 7. Since previous studies on the energetic relation between the CME kinematics and the flare strength found only a weak correlation (e.g., Moon et al. 2002), our result is the first evidence of a strong correlation between the mass ejection energy and the flare energy. This also indicates that it is very important to accurately measure the velocity fields, which are free from any projection effect of mass ejections as well as the released thermal energy to seek the actual relation between them.

The correlation between these two energy increase rates shown in figure 7 is best represented by the relation $\epsilon_{th} \propto \epsilon_{mc}^{1.9}$. This relation can be explained by a simple model in which the stored magnetic energy is carried by the Poynting flux to be converted into thermal energy via magnetic reconnection; also, the Lorentz force plays the primary role in accelerating a filament. The strong dependence of both the energy increase rates on the mean photospheric magnetic field strength, B_f, of the arcade region also supports this interpretation.

These above results mean that the energy increase rates of the filaments and the associated flare arcades are related to each other, though they are temporally not coincident. There may be no direct connection between the mechanisms of the accelerating mass ejections and heating the coronal plasma; they merely have a connection in terms of the magnetic field strength. This supports the idea that the filament is accelerated by the outward Lorentz force, while the arcade plasma is heated by some mechanisms closely involved in magnetic reconnection; also, they are two different manifestations of the process that dissipates the stored magnetic energy.

We would like to thank the anonymous referee for useful comments that improved our paper. We wish to thank all members at Hida and Kwasan

れている．GOES データは，World Data Center A for Solar-Terrestrial Physics, NGDC, NOAA によるものである．SOHO データは，European Space Agency（ESA）と NASA 国際共同ミッションによって提供されている．本研究は，日本文科省学術創成研究科研費「宇宙天気予報の基礎研究」（17GS0208　代表研究者　柴田一成）から補助を受けている．また，本研究は文科省グローバル COE プログラム「普遍性と創発性から紡ぐ次世代物理学」から補助を受けている．

Observatories, Kyoto University, for operation of the telescopes as well as fruitful comments and discussions. The database of FMT is compiled and maintained by M. Kadota, Y. Nasuji, M. Kamobe, and S. UeNo. The Yohkoh data is provided by courtesy of a joint Japan-US-UK project managed by the Institute of Space & Astronautical Science (ISAS) of Japan. GOES data are courtesy of the World Data Center A for Solar-Terrestrial Physics, NGDC, NOAA. SOHO data is courtesy of a mission of international cooperation between European Space Agency (ESA) and NASA. This work was supported by a Grant-in-Aid for Creative Scientific Research "The Basic Study of Space Weather Prediction" (17GS0208, Head Investigator: K. Shibata) from the Ministry of Education, Culture, Sports, Science and Technology (MEXT) of Japan. This work was supported by a Grant-in-Aid for the Global COE Program "The Next Generation of Physics, Spun from Universality and Emergence" from the MEXT.

参考文献／References

Beckers, J. M. 1964, PhD thesis, University of Utrecht
Brueckner, G. E., et al. 1995, Sol. Phys., 162, 357
Carmichael, H. 1964, in Proc. AAS-NASA Symp., The Physics of Solar Flares, ed. W. N. Hess, NASA-SP50 (Washington D.C.: NASA), 451
Delaboudiniére, J.-P., et al. 1995, Sol. Phys., 162, 291
Engvold, O. 1976, Sol. Phys., 49, 283
Hanaoka, Y., et al. 1994, PASJ, 46, 205
Hara, H., Tsuneta, S., Lemen, J. R., Acton, L.W., & McTiernan, J. M. 1992, PASJ, 44, L135
Harrison, R. A. 1995, A&A, 304, 585
Hirayama, T. 1974, Sol. Phys., 34, 323
Hundhausen, A. 1997, in the Many Faces of the Sun: A Summary of the Results from NASA's Solar Maximum Mission, ed. K. T. Strong et al. (New York: Springer-Verlag), Ch. 5
Isobe, H., Yokoyama, T., Shimojo, M., Morimoto, T., Kozu, H., Eto, S., Narukage, N., & Shibata, K. 2002, ApJ, 566, 528
Jefferies, J. T., & Orrall, F. Q. 1963, ApJ, 137, 1232
Khan, J. I., Uchida. Y., McAllister, A. H., Mouradian, Z., Soru-Escaut, I., & Hiei, E. 1998, A&A, 336, 753
Kopp, R. A., & Pneuman, G. W. 1976, Sol. Phys., 50, 85
Kurokawa, H., Ishiura, K., Kimura, G., Nakai, Y., Kitai, R., Funakoshi, Y., & Shinkawa, T. 1995, J. Geomagn. Geoelectr., 47, 1043
McAllister, A., et al. 1992, PASJ, 44, L205
McAllister, A. H., Kurokawa, H., Shibata, K., & Nitta, N. 1996, Sol. Phys., 169, 123
Moon, Y.-J., Choe, G. S., Wang, H., Park, Y. D., Gopalswamy, N., Yang, G., & Yashiro, S. 2002, ApJ, 581, 694

Morimoto, T., & Kurokawa, H. 2003a, PASJ, 55, 503 (Paper I)
Morimoto, T., & Kurokawa, H. 2003b, PASJ, 55, 1141 (Paper II)
Munro, R. H., Gosling, J. T., Hildner, E., MacQueen, R. M., Poland, A. I., & Ross, C. L. 1979, Sol. Phys., 61, 201
Ogawara, Y., Takano, T., Kato, T., Kosugi, T., Tsuneta, S., Watanabe, T., Kondo, I., & Uchida, Y. 1991, Sol. Phys., 136, 1
Parker, E. N. 1957, J. Geophys. Res., 62, 509
Petschek, H. E. 1964, in Proc. AAS-NASA Symp., The Physics of Solar Flares, ed. W. N. Hess, NASA SP-50, (Washington D.C.: NASA), 425
Pevtsov, A. A. 2002, Sol. Phys., 207, 111
Priest, E., & Forbes, T. 2000, Magnetic Reconnection: MHD Theory and Applications (New York: Cambridge Univ. Press), Ch.4
Rompolt, B. 1990, Hvar Obs. Bull., 14, 37
Rosner, R., Tucker, W. H., & Vaiana, G. S. 1978, ApJ, 220, 643
Shakhovskaya, A. N., Abramenko, V. I., & Yurchyshyn, V. B. 2002, Sol. Phys., 207, 369
Sheeley, N. R., Jr., et al. 1975, Sol. Phys., 45, 377
Sheeley, N. R., Jr., Howard, R. A., Koomen, M. J., & Michels, D. J. 1983, ApJ, 272, 349
Shibata, K., Yokoyama, T., & Shimojo, M. 1996, Adv. Space Res., 17, 197
Simon, G., Schmieder, B., Démoulin, P., & Poland, A. I. 1986, A&A, 166, 319
Spitzer, L., Jr. 1956, Physics of Fully Ionized Gases (New York: Interscience)
St. Cyr, O. C., & Webb, D. F. 1991, Sol. Phys., 136, 379
Sterling, A. C., Moore, R. L., & Thompson, B. J. 2001, ApJ, 561, L219
Sturrock, P. A. 1966, Nature, 211, 695
Sweet, P. A. 1958, in IAU Symp. 6, Electromagnetic Phenomena in Cosmical Physics, ed. B. Lehnert (Cambridge: Canbridge Univ. Press), 123
Tandberg-Hanssen, E.,Martin, S. F.,& Hansen, R. T. 1980, Sol. Phys., 65, 357
Tripathi, D., Bothmer, V., & Cremades, H. 2004, A&A, 422, 337
Tripathi, D., Solanki, S. K., Schwenn, R., Bothmer, V., Mierla, M., & Stenborg, G. 2006, A&A, 449, 369
Tsuneta, S., et al. 1991, Sol. Phys., 136, 37
Valniček, B. 1964, Bull. Astron. Inst. Czech., 14, 207
Warwick, J. W. 1957, ApJ, 125, 811
Webb, D. F., & Hundhausen, A. J. 1987, Sol. Phys., 108, 383
Webb, D. F., Krieger, A. S., & Rust, D. M. 1976, Sol. Phys., 48, 159
Zhang, J., Dere, K. P., Howard, R. A., Kundu, M. R., & White, S. M. 2001, ApJ, 559, 452

付録 B
フレア監視望遠鏡関連論文リスト

Appendix B

FMT-Related Papers

Akiyama, S., Takeuchi, T., Mizuno, Y., Shibata, K. and Morimoto, T. "The Relationship between CME Interactions and Complex IP Disturbances" SHINE (Solar Heliospheric and Interplanetary Environment) Meeting at Banff/Canada, 2002.

Eto, S., Isobe, H., Narukage, N., Asai, A., Morimoto, T., Thompson, B., Yashiro, S., Wang, T., Kitai, R., Kurokawa, H. and Shibata, K. "Relation between a Moreton Wave and an EIT Wave Observed on 1997 November 4" PASJ, Vol.54, No.3, pp.481-491, 2002

Hanaoka, Y., Kurokawa, H., Enome, S., Nakajima, H., Shibahashi K., Nishio, M., Takano T., Torii, C., Sekiguchi, H., Kawashima, S., Bishimata, T., Shinohara, N., Irimajiri, Y., Koshiishi, H., Shiomi, Nakai, Y., Funakoshi, Y., Kitai, R., Ishiura, K. and Kimura, G. "Simultaneous observations of a prominence eruption followed by a coronal arcade formation in radio, soft X-rays, and H-alpha" PASJ, vol.46, pp.205-216, 1994

Hanaoka, Y. and Shinkawa, T. "Heating of Erupting Prominences Observed at 17Ghz" ApJ, vol.510, pp.466-473, 1999

Hori, K., Glover, A., Akioka, M., and Ueno, S. "Tether Cutting Action in Two Sigmoidal Filaments" COSPAR Colloquia Series 2002, p.139, 2002

Ishitsuka, J.K., Ishitsuka, M., Aviles, H.T., Sakurai, T., Nishino, Y., Miyazaki, H., Shibata, K., Ueno, S., Yumoto, K. and Maeda, G. "A Solar Station for Education and Research on Solar Activity at a National University in Peru" Bull. Astr. Soc. India, vol.35, pp.709-712, 2007

Isobe, H. and Tripathi, D. "Large amplitude oscillation of a polar crown filament in the pre-eruption phase" As & Ap, vol.449, pp.L17-20, 2006

Isobe, H., Tripathi, D., Asai, A., and Jain, R. "Large-Amplitude Oscillation of an Erupting Filament as Seen in EUV, H-alpha and Microwave Observations" Solar Physics, vol.246, pp.89-99, 2007

Isobe, H., Yokoyama, T., Shimojo, M., Morimoto, T., Kozu, H., Eto, S., Narukage, N., Shibata, K. "Reconnection Rate in the Decay Phase of a Long Duration Event Flare on 1997 May 12" ApJ, Volume 566, pp. 528-538, 2002.

Kimura, G., Ueno, S., Kitai, R., Nagata, S., Shibata, K. "Renewal of CCD cameras for the flare monitoring telescope (FMT) of the Hida Observatory, Kyoto University" (in Japanese) Proceedings of Symposium on Techniques in Astronomy 2006, pp.91-93, 2006

Kurokawa, H., Ishiura, K., Kimura, G., Nakai, Y., Kitai, R., Funakoshi, Y. and Shinkawa, T. "Observations of Solar H alpha Filament Disappearances with a New Solar Flare-Monitoring-Telescope at Hida Observatory" J.Geomag.Geoelectr., 47, 1043-1052, 1995

Li, H., Sakurai, T., Ichimoto, K. and Ueno, S. "Magnetic Field Evolution Leading to Solar Flares I. Cases with Low Magnetic Shear and Flux Emergence" PASJ, Vol.52, pp.465-481, 2000

Li, H., Sakurai, T., Ichimoto, K. and Ueno, S. "Magnetic Field Evolution Leading to Solar Flares II. Cases with High Magnetic Shear and Flare-Related Shear Change" PASJ, Vol.52, pp.483-497, 2000

Liu Y. and Kurokawa, H. "On a Surge: Properties of an Emerging Flux region" Ap. J, 610,1136, 2004

Liu Y., Kurokawa, H., Kitai, R., Ueno, S., and Su, J.T. "Study of Surges : I. Automatic Detection of Dynamic Halpha Dark Features from High-Cadence Full-disk Observations" Solar Physics, vol.228, pp.149-164, 2005

Liu Y., Su, J.T., Morimoto, T., Kurokawa, H., and Shibata, K. "Observations of an Emerging Flux Region Surge: Implications for Coronal Mass Ejections Triggered by Emerging Flux" ApJ, vol.628, pp.1056-1060, 2005

Liu Y., Kurokawa, H. and Shibata, K. "Production of Filaments by Surges" ApJ, vol.631, pp.L93-96, 2005

McAllister, A.H., Kurokawa, H., Shibata, K. and Nitta, N. "A Filament Eruption and Accompanying Coronal Field Changes on November 5, 1992" Solar Physics, vol.169, pp.123-149, 1996

Morimoto, T. and Kurokawa, H. "Acceleration Time Scales of Solar Disappearing Filaments" COSPAR Colloquia Series 2002, p.291, 2002

Morimoto, T. and Kurokawa, H. "A Method for the Determination of 3-D Velocity Fields of Disappearing Solar Filaments" PASJ vol.55, pp.503-518, 2003

Morimoto, T. and Kurokawa, H. "Eruptive and Quasi-Eruptive Disappearing Solar Filaments and Their Relationship with Coronal Activities" PASJ vol.55, pp.1141-1151, 2003

Narukage, N., Hudson, H.S., Morimoto, T., Akiyama, S., Kitai, R., Kurokawa, H. and Shibata, K. "Simultaneous Observation of a Moreton Wave on 1997 November 3 in H-alpha and Soft X-Rays" ApJ, Volume 572, Issue 1, pp. L109-L112., 2002

Narukage, N., Morimoto, T., Kadota, M., Kitai, R., Kurokawa, H., Shibata, K. "X-Ray Expanding Features Associated with a Moreton Wave" PAS J, Vol.56, No.2, pp. L5-L8, 2004.

Narukage, N., Morimoto, T., Kitai, R., Kurokawa, H. and Shibata, K. "Multi-wavelength Observations of a Moreton Wave on 2000 March 3" The Proceedings of the IAU 8th Asian-Pacific Regional Meeting, Volume II, pp.449-450, 2002

Narukage, N. Shibata, K. "Moreton waves observed at Hida Observatory" Multi-Wavelength Investigations of Solar Activity, IAU Symposium, vol.223, pp.367-370, 2004

Okamoto, T. J., Nakai, H., Keiyama, A., Narukage, N., Ueno, S., Kitai, R., Kurokawa, H., and Shibata, K. "Filament Oscillations and Moreton Waves Associated with EIT Waves" ApJ, vol.608, pp.1124-1132, 2004

Shibata, K., Eto, S., Narukage, N., Isobe, H., Morimoto, T., Kozu, H., Asai, A., Ishii, T., Akiyama, S., Ueno, S., Kitai, R., Kurokawa, H., Yashiro, S., Thompson, B. J., Wang, T., and Hudson,H.S. "Observations of Moreton Waves and EIT Waves" COSPAR Colloquia Series 2002, p.279, 2002

Su, J.T., Liu Y., Zhang, H. Q., Kurokawa, H., Yurchyshyn, V., Shibata, K., Bao, X. M., Wang, G. P., and Li, C. "Evolution of Barb Angle and Filament Eruption" ApJ, vol.630, pp.L101-104, 2005

Ueno, S., Nagata, S., Kitai, R. and Kurokawa, H. "Features of Solar Telescopes at the Hida Observatory and the Possibilities of Coordinated Observations with SolarB" ASP Conference Series, vol.32, pp.319-324, 2004

Ueno, S., Shibata, K., Kimura, G., Nakatani, Y., Kitai, R., Nagata, S. et al. "Chain-Project and Installation of the Flare Monitoring Telescopes in Developing Countries" Bull. Astr. Soc. India, vol.35, pp.697-704, 2007

Warmuth, A., Vršnak,B., and Hanslmeier, A. "Flare waves revisited" Hvar Observatory Bulletin, vol.27, no.1, pp.139-149, 2003

Warmuth, A., Vršnak, B., Magdalenić, J., Hanslmeier, A., and Otruba, W. "A multiwavelength study of solar flare waves. I. Observations and basic properties" As & Ap, vol.418, pp.1101-1115, 2004

Warmuth, A., Vršnak, B., Magdalenić, J., Hanslmeier, A., and Otruba, W. "A multiwavelength study of solar flare waves. II. Perturbation characteristics and physical interpretation" As & Ap, vol.418, pp.1117-1129, 2004

付録 C

太陽地球系エネルギー国際共同研究 (STEP) シンポジウム報告より FMT 関連報告集

Appendix C

FMT-Related Reports in Proceedings of the Symposiums of Solar-Terrestrial Energy Program (STEP)

第 2 回シンポジウム報告（平成 3 年（1991 年）5 月 29, 30, 31 日）pp.7-13
黒河宏企，中井善寛，船越康宏，北井礼三郎
「太陽フレアおよびプロミネンス噴出の観測」
Proceedings of the second symposium (1991, May 29-31), pp.7-13
Hiroki Kurokawa, Yoshihiro Nakai, Yasuhiro Funakoshi and Reizaburo Kitai
"Observations of Solar Flares and Prominence Eruptions"

第 3 回シンポジウム報告（平成 4 年（1992 年）5 月 28, 29, 30 日）pp.21-28
黒河宏企，中井善寛，船越康宏，北井礼三郎，石浦清美，木村剛一
「飛騨天文台の新太陽フレア監視望遠鏡」
Proceedings of the third symposium (1992, May 28-30), pp.21-28
Hiroki Kurokawa, Yoshihiro Nakai, Yasuhiro Funakoshi, Reizaburo Kitai, Kiyomi Ishiura and Gouichi Kimura
"New Flare-Monitoring Telescope of Hida Observatory"

第 4 回シンポジウム報告（平成 5 年（1993 年）6 月 21, 22, 23, 24 日）pp.61-71
黒河宏企，石浦清美，木村剛一，北井礼三郎，船越康宏，中井善寛，新川雄彦，谷田貝宇
「太陽プロミネンス活動現象の観測」
Proceedings of the fourth symposium (1993, June 21-24), pp.61-71
Hiroki Kurokawa, Kiyomi Ishiura, Gouichi Kimura, Reizaburo Kitai, Yasuhiro Funakoshi, Yoshihiro Nakai, Takehiko Shinkawa and Hiroshi Yatagai
"Observations of Solar Prominence Activities with the New Flare-Monitoring Telescope"

第 5 回シンポジウム報告（平成 6 年（1994 年）6 月 5, 6, 7, 8, 9, 10 日）p.6

黒河宏企，石浦清美，木村剛一，中井善寛，新川雄彦，北井礼三郎，船越康宏
「飛騨天文台の新太陽フレア監視望遠鏡によるHαフィラメント消失の観測」
Proceedings of the fifth symposium (1994, June 5-10), p.6
Hiroki kurokawa, Kiyomi Ishiura, Gouichi Kimura, Yoshihiro Nakai, Takehiko Shinkawa, Reizaburo Kitai, Yasuhiro Funakoshi
"Observations of Hα Filament Disappearance with the New Flare-Monitoring Telescope at Hide Observatory"

第6回シンポジウム報告（平成7年（1995年）11月15, 16, 17, 18日）pp.195-204
黒河宏企，石浦清美，木村剛一，新川雄彦，北井礼三郎，船越康宏，越石英樹，鰕目信三
「太陽Hαフィラメント消失の観測」
Proceedings of the sixth symposium (1995, July 15-18), pp.195-204
Hiroki kurokawa, Kiyomi Ishiura, Gohichi Kimura, Takehiko Shinkawa, Reizaburo Kitai, Yasuhiro Funakoshi, Hideki Koshiishi, Shindo Enome
"Observations of Disappearing Hα Filaments on the Solar Disk"

STEP 第2回シンポジウム報告

Observations of Solar Flares and Prominence Eruptions

Hiroki Kurokawa, Yoshihiro Nakai, Yasuhiro Funakoshi and Reizaburo Kitai

Kwasan and Hida Kyoto University
Kamitakara, Gifu 506-13, Japan
Yamashina, Kyoto 607, Japan

ABSTRACT

In the period of Solar Terrestrial Energy Program (STEP), we study the energy build-up and release processes of solar flares and eruptive prominences which are considered as the sources of the interplanetary and geomagnetic storms.

The observations are made to study the following subjects by using the 60 cm Domeless Solar Telescope and a new Flare Patrol Telescope which will be installed in this financial year at Kwasan and Hida 0bservatories.

(1) Evolutional characteristics and dynamical features of Hα solar flares and prominence eruptions.
(2) Causal relations among prominence motions, Hα flares, X-ray flares and coronal mass ejections (CME).
(3) Active region evolution and processes of magnetic shear development

The morphological study of disparition brusque (DB), flare spray and eruptive prominence is important for us to understand the explosive energy release processes from the solar surface into the interplanetary space. There have been, however, only a few published observations which can demonstrate a detailed process of a filament eruption projected on the solar disk due to the difficulty in obtaining Hα images of high-speed phenomena on the solar surface.

The new Flare Patrol Telescope will be able to monitor the solar active phenomena simultaneously in Hα line center and Hα line wings and enable us to study the dynamical features of prominence eruptions and solar flares in more details. The fundamental characteristics of the telescope are (A) to get simultaneously full disc images of the Sun and enlarged images of active regions through several Hα monochromatic filters, (B) to get Hα monochromatic images simultaneously in multiple wavelengths such as Hα center, Hα+0.8Å, Hα-0.8Å, Hα+1.2Å and Hα-1.2Å. and (C) to record all Hα images with CCD-VTR systems.

STEP 第2回シンポジウム報告

太陽フレアーおよびプロミネンス噴出の観測

京都大学理学部附属天文台
黒河宏企
中井善寛
船越康宏
北井礼三郎

1. 研究目的

　地磁気嵐や惑星間空間擾乱が，コロナ爆発（CME），フレア，プロミネンス噴出等の太陽表面現象に起因する事は，色々な観測からその傍証が得られており，これにともなって，(1) これらの太陽表面現象のエネルギー蓄積および放出機構，(2) 惑星間空間における輸送機構，(3) 惑星磁気圏衝突流入機構の具体的研究が，STEP 計画の重要課題のひとつとなっている．本研究は，太陽活動領域の進化，太陽フレアーの発達過程，プロミネンス噴出等の観測を行なうことによって，(1) の課題に取り組もうとするものである．

　具体的な研究テーマは，次の二点に重点を置いている．
(A) 強い磁気嵐を起す太陽面爆発について，フレア及びプロミネンス運動噴出の発達過程を調べ，互いの因果関係を明らかにする．
(B) 活動領域の誕生から複合化の過程を調べ，どのような形の磁気シアーがどの様な過程で発達すれば，大フレアーの発生に到るかを明らかにする．

　観測は，飛騨天文台のドームレス太陽望遠鏡を用いて高分解 H α 単色像及び分光観測を実施すると同時に，広視野で多波長同時撮影可能なフレアー監視望遠鏡を新設することによって，太陽面爆発現象の動的な姿を明らかにすることを狙っている．また，花山天文台でも H α 単色像による観測を実施する．

　プロミネンスの上昇あるいは噴出がフレアーの引金となることが，多くの人々によって指摘されており（Smith & Ramsey (1964), Hirayama (1974))，またプロミネンス噴出とコロナ爆発（CME）との密接な関係も明らかにされている（Munro et al. (1984)).しかし，フレアから CME に到る（あるいは Harrison (1986) いうように，CME がフレアを起している場合もある）一連の太陽面爆発現象の中でのプロミネンスの役割がいま一つ明かでない．このことについては，太陽面上におけるプロミネンス（フィラメント）の噴出過程の観測の困難さにその一因を押しつけることができる．すなわち，たいていのフレアモニター像は，H α 線中心のみで撮影されているので，約 20 km/s を越え

STEP 第2回シンポジウム報告

る視線方向の運動を検出できないわけである．

　今回 STEP 計画で新設するフレア監視望遠鏡では，Hα 中心に加えて，H$\alpha \pm 0.8$Å，H$\alpha \pm 1.2$Å の両翼像も常時記録することを狙っており，約 100 km/s を越える上昇下降運動の形態をより詳しく連続的に観測することができるものと期待される．

2. フレアの発達過程とプロミネンス噴出との因果関係

　プロミネンスの上昇が引金となってフレアが発生するというシナリオがある．さらに具体的には，上昇プロミネンスの下で，引き伸ばされた磁力線の再結合によってフレアエネルギーが解放されるというモデルがある（Hirayama（1974）．Kopp & pneuman（1976））．また一方では，プロミネンス消失と地磁気嵐（Wright & McNamara 1983），及び CME（Munro et al. 1984）との高い相関が報告されているし，CME が太陽コロナ中を出発した数十分後に，CME アーチの片足で，フレアーが発生するという観測も提出されている（Harrison1986）．このように，フレアーから CME にいたる一連の太陽面爆発現象による，惑星間空間への突発的エネルギー放出機構を理解する上で，プロミネンスの運動と噴出過程の観測が一つのヒントを与えるのではないかと考えられる．

　フレアの発生とプロミネンス消失の関係について，最もまとまった観測は Smith & Ramsey（1964）のものであるが，重要度2以上の Hα フレア 71 個の中，約 1/3 について，フレアの近辺または上で，フレアの最盛期の前に，フィラメントの消失が見られたと報告している．しかし彼女達の観測は，H$\alpha \pm 0.5$Å の off-band に留まっているので，フィラメント消失が実際に噴出によるものかどうか，不明な場合が多いと考えられる．また，Martre et al.（1977）は H$\alpha \pm 3/4$Å の単色像を使って preflare 活動の観測をしており，Soru-Escaut & Mouradian（1990）は H$\alpha \pm 0.5$Å 像でフィラメントの熱的消失について調べているが，いずれもフィラメント噴出の運動形態を調べることのできるデータまでは示していない．要するに，この程度の Hα off-band 像では，上昇後すぐに 100 km/s を越える高速度となるフレアスプレイや活動領域フィラメントの，噴出初期の形態ですら記録することが難しいのである．実際，太陽表面上でのフィラメント噴出形態変化について詳しい観測を示して，解析した論文は非常に少ない（Tandberg-Hansen et al（1980））．

　図1にその数少ない一例を示している．この場合，フレア発生後浮き上がった（001916UT）Hα フィラメントが，004334UT の Hα 線中心像では消失しているが，004449UT（H$\alpha +1.0$Å）及び 004539UT（H$\alpha -1.0$Å）の Hα off-band 像ではその回転噴出している様子がはっきりと捕らえられている（Kurokawa et al.1987）．太陽周縁上でのプロミネンス噴出についての観測はこれまでにも数多く出版され，上昇速度の変化について貴重なデータを提供しているが，これらの場合，その下の磁場構造やフレアの

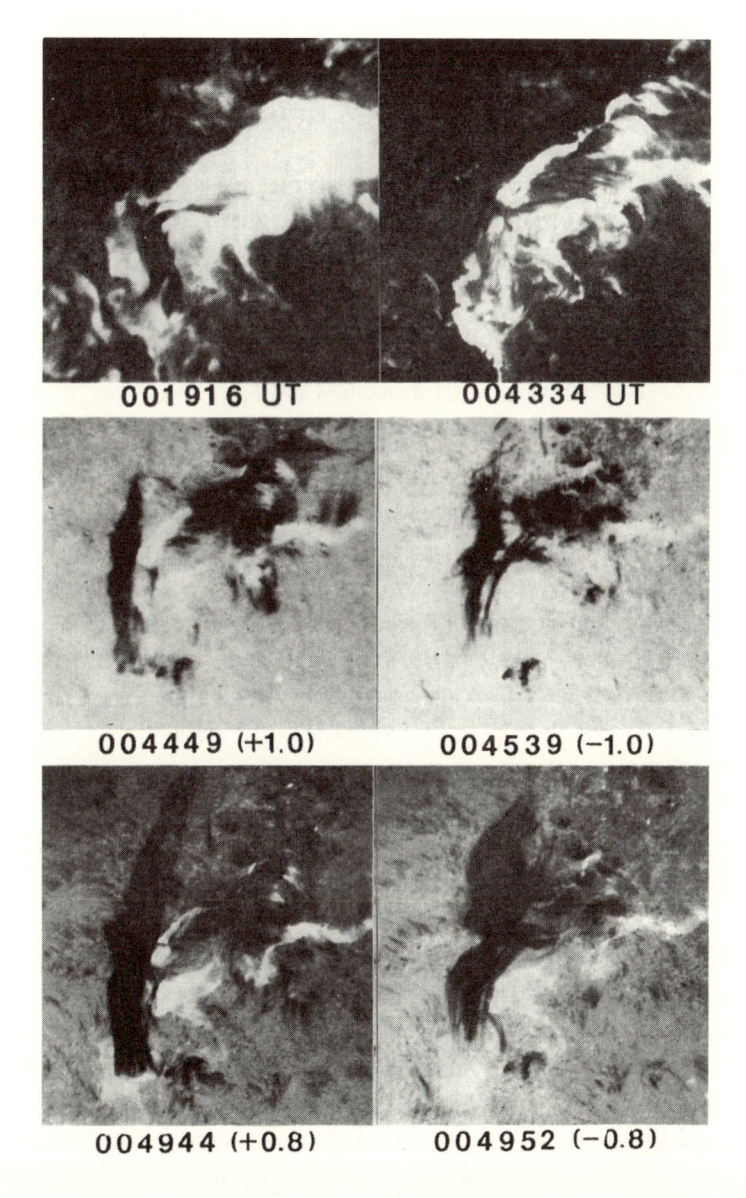

STEP 第2回シンポジウム報告

図1 1984 April 25 フレアに伴ったフィラメントの噴出. 004334UT の Hα 中心像で, フィラメントが消失しているが, Hα+1.0Å, Hα-1.0Å 等ではその運動がみられる

STEP 第 2 回シンポジウム報告

実態が見えないので，最も知りたい噴出前の磁場構造との位置関係や，フレア発生との因果関係を調べることが困難である．従って，図1の様な太陽面上での活動領域フィラメント噴出の詳しい観測がさらに必要とされる訳である．

その為，新設されるフレアー監視望遠鏡は，100 km/s を越えるプロミネンス噴出の初期の形態変化を，複数の $H\alpha$ off-band フィルターで同時に連続的に記録できるという，従来にない特長を持つ予定である．新望遠鏡の基本性能（目標値）は次の通り．

(A) 太陽全面像と拡大像を同時に撮影できる．
(B) 撮像はすべて，CCD カメラービデオ録画方式である．
(C) 全面像は，周縁の紅炎活動をも撮影できる．
(D) 拡大像の視野は 500 arcsec x 500 arcsec 程度を標準とする．
(E) 全面像は，$H\alpha \pm 0.0\text{Å}$ (0.5Å幅), $H\alpha + 0.8\text{Å}$ (0.5Å幅), $H\alpha - 0.8\text{Å}$ (0.5Å幅), $H\alpha \pm 0.0\text{Å}$ (3.0Å幅) の4波長で，順次撮影できる．同一波長の撮影頻度は，4コマ/4秒程度とする．
(F) 拡大像は，$H\alpha \pm 0.0\text{Å}$ (0.5Å幅), $H\alpha + 0.8\text{Å}$ (0.5Å幅), $H\alpha - 0.8\text{Å}$ (0.5Å幅), $H\alpha + 1.2\text{Å}$ (0.5Å幅), $H\alpha - 1.2\text{Å}$ (0.5Å幅), 及び $H\alpha \pm 0.0\text{Å}$ (10.0Å幅) の6波長で撮影できる．同一波長の撮影頻度は，3コマ/6秒程度（スローモード）から10コマ/秒程度（フレアーモード）へ敏速に切り替えられる．
(G) 拡大像の視点移動にかかわらず，全面像が常時監視できるオフセット機構を持っている．
(H) 空間分解能
　　全面像系： 4 seconds of arc
　　拡大像系： 1 second of arc
(I) 追尾精度： ± 1 arcsec／1分間, ± 5 arcsec／10分間, ± 10 arcsec／1時間．

3. 強いフレアを起す活動領域の発達過程の特徴

強いフレア発生に必要な環境がいかにして構築されるかの問題（Flare Build-up Problem）は，フレア物理のみならず，地球惑星間空間へのエネルギー放出の研究即ち STEP の課題の一つに対して本質的に重要な鍵を内包していると思われる．この問題について最近の観測結果をまとめると，New Emerging Flux Region（新浮上磁場領域）と Magnetic Shear（磁気シアー）の二つのキーワードが浮かび上がってくる．

しかし，一方では，次の事実も忘れてはならない．すなわち，多くの Emerging Flux Region (EFR) は，さしたるフレア活動を起さずに浮上してくるが，一部の限られた特別の EFR のみが強いフレアー活動を引き起こすこと．磁気シアーも一般的には安定

STEP 第2回シンポジウム報告

DEVELOPMENT OF MAGNETIC SHEAR IN NOAA
6233   (25 AUG.- 2 SEP., 1990)

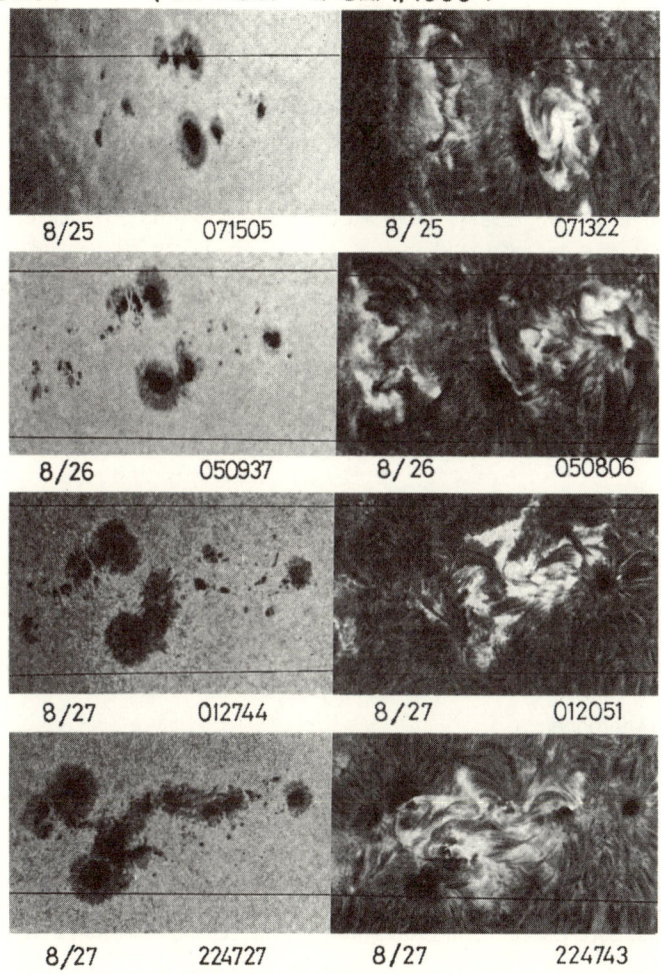

図2　フレアー活動の原因となる磁気シアーの発達例．左側：$H\alpha-5.0Å$ 像　右側：$H\alpha$ 中心像．8/27 日に急激な磁気シアーの発達がみられる．

STEP 第 2 回シンポジウム報告

して存在することが多く，これに何か別の要因が加わって初めて，強いフレア発生にいたることである．

従って STEP 期間中の観測課題として，(1) 強いフレア活動を引き起こす浮上磁場領域の特徴は何か？(2) どの程度の磁気シアーがどの様な過程で発達したときに強いフレアが発生するか？の問題に焦点を合わせる必要がある．この課題は将来，強い惑星間擾乱を引き起こすフレア発生の予報の研究につながるものである．

図 2 にこの観測の一例としてドームレス太陽望遠鏡で撮影された NOAA6233 領域の発達過程を示している．左側は，$H\alpha - 5.0$Å の波長で撮影された黒点群の成長過程を示している．右側は $H\alpha$ 線中心像である．この中で黒い筋模様は，黒点間を結ぶ transverse 磁場の方向を示しているので，磁気シアー構造の発達過程を調べるのに非常に有効な手段を与えてくれる．この NOAA6233 領域の場合，8/27 後半における急速な黒点群の発達と磁気シアーの発達がフレア活動の前ぶれとなっている．

参考文献／References

Harrison,R.A.:1986, Astron. Astrophys. 162, 283-291

Hirayama,T.:1974, Solar Phys. 34, 323.

Kurokawa,H., Hanaoka,Y., Shibata,K., and Uchida,Y.:1987, Solar Phys. 108, 251-264.

Martres,M.-J., Soru-Escaut,l., and Nakagawa,Y.:1977. Astron. Astrophys. 59, 255.

Munro,R.H., Gosling,J.T., Hildner,E., MacQueen,R.M., Poland,A.I., and Ross,C.L.:1979. Solar Phys. 61, 201-215.

Smith,S.F., and Ramsey,H.E.:1964, Z. Astrophys.60, 1-18

Tandberg-Hanssen,E., Martin,S.F., and Hansen,R.T.:1980, Solar Phys. 65, 357.

Wright,C.S., and McNamara,L.F.:1983, Solar Phys. 87, 401.

STEP 第3回シンポジウム報告

New Flare-Monitoring Telescope of Hida Observatory

Hiroki Kurokawa, Yoshihiro Nakai, Yasuhiro Funakoshi,
Reizaburo Kitai, Kiyomi Ishiura and Gouichi Kimura

Kwasan and Hida Observatories, Kyoto University
Kamitakara, Gifu 506-13

ABSTRACT

A new Flare-Monitoring Telescope was built at the Hida Observatory of Kyoto University in the first financial year of the Japan STEP project. The overview of the telescope is given in this paper.

We have been studying the energy build-up and release processes of solar flares and eruptive prominences by using the 60 cm Domeless Solar Telescope at the Hida Observatory. This subject is important not only in the study of solar physics itself, but also for the solar-terrestrial energy transport, because the solar flares and eruptive prominences are considered as the sources of the interplanetary and geomagnetic storms. In order to make a further progress in this study during the STEP period, we constructed the new Flare-Monitoring Telescope.

The Flare-Monitoring Telescope consists of six small telescopes, namely, four Full Disc Telescopes, one Prominence Telescope and one Photoelectric Guiding Telescope. They are assembled in a large fork arm which is driven with a telescope control unit. The objective lens of each telescope is 64 mm in diameter. The four Full Disc Telescopes observe full solar images in four different wavelengths, those are, Hα line-center, Hα+0.8Å, Hα−0.8Å and continuum. The Prominence Telescope observes solar prominences outside the solar limb in Hα line-center by using two occulting cones which reflect the light of the solar disc. These five solar images are continuously observed with five CCD cameras and recorded with five time-lapse video-tape-recorders.

The new telescope covers up the narrow field of view of the high resolution Domeless Solar Telescope in studying various types of solar active phenomena. In fact, it has already observed many flares, surges and eruptive prominences since April, 1992. The Hα+0.8Å and Hα−0.8Å full disk images and the solar limb image are found to be especially useful for studying the dynamical features of Hα prominences and filaments. Two examples of filament disappearance and prominence eruption observed with the new telescope are demonstrated.

STEP 第3回シンポジウム報告

飛騨天文台の新太陽フレアー監視望遠鏡

京都大学理学部附属天文台
黒河　宏企　　中井　善寛
船越　康宏　　北井　礼三郎
石浦　清美　　木村　剛一

1. はじめに

　STEP 計画における我々の研究課題については，すでに第1回及び第2回 STEP シンポジウムで述べられているが，その要点は次の通りである．
(1) フレア及びプロミネンス爆発現象のエネルギー蓄積機構，解放機構を調べること．
(2) フレア及びプロミネンス爆発とコロナ物質放出現象（CME），惑星間空間擾乱，地磁気嵐等との因果関係を調べること．
　以前から我々は，高分解能の60cmドームレス太陽望遠鏡を用いて，これらの研究を行ってきたが（Kurokawa et al. 1991），STEP 期間中に更にこれを推し進めるために，その初年度（平成3年度）において，太陽フレア監視望遠鏡を新設した．この望遠鏡は波長の異なる4個の太陽全面像と，1個の太陽周縁プロミネンス像の合計5個の太陽像を同時に連続観測できるものであって，高分解太陽望遠鏡（ドームレス太陽望遠鏡）の宿命である狭視野の欠点を補って，フレア及びプロミネンス消失の観測に威力を発揮することが期待されている．実際すでに，4月からのテスト観測を含む3カ月の間に，多くのフレアやプロミネンス活動現象が観測されている．
　以下に，新太陽フレア監視望遠鏡の概要と観測例を紹介する．

2. 望遠鏡の構成

　新太陽フレア監視望遠鏡は，4本の太陽全体像望遠鏡と1本の太陽周縁像（プロミネンス像）望遠鏡及び，1本の光電追尾望遠鏡の合計6本の小望遠鏡（対物レンズ口径:64mm）で構成されている．これらは，1台のフォーク式架台に同架されて，1体として運転制御されている．図1に，新設された3mドームと望遠鏡の外観が示されている．右上に見られるのは，ドームレス太陽望遠鏡である．また，図2には，高さ2mのコンクリート台の上に設置された望遠鏡本体が示されている．望遠鏡は西村製作所製である．

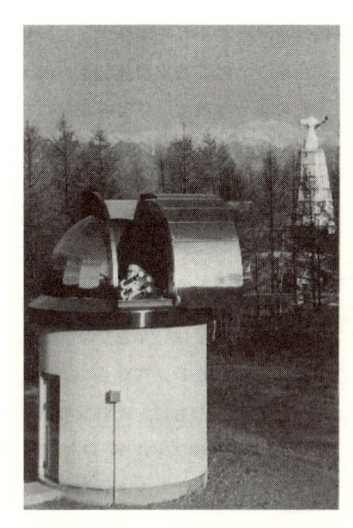

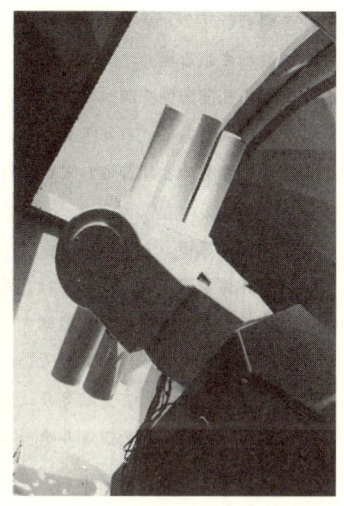

STEP 第3回シンポジウム報告

図1　飛騨天文台に新設されたドーム及び太陽フレア監視望遠鏡．右後方にはドームレス太陽望遠鏡が見られる．

図2　太陽フレア監視望遠鏡の外観．

　これまで，世界の各観測所で用いられている，フレアモニター望遠鏡は，Hα線中心のみで連続撮影しているのに対して，飛騨天文台の太陽フレア監視望遠鏡は，次の新しい特徴を持つように，設計製作されている．
(1) Hα線中心，Hα＋0.8Å，Hα－0.8Å，連続光の4波長で太陽全体像を同時に連続観測できる．特に，Hα＋0.8ÅとHα－0.8ÅのようなHα線翼での全体像の連続撮影は，これまでに例を見ないものである．
(2) 太陽周縁のプロミネンス像を透過幅の広い（3Å）のHαフィルターで連続観測できる．
(3) 上記5個の太陽像を5台のCCD-VTRシステムで同時連続記録できる．
　これらによって，フレアだけでなく，太陽面上のダークフィラメント（暗條）や太陽周縁のプロミネンス（紅炎）の運動状態をより詳しく観測できるものと考えている．

　HαフィラメントやHαプロミネンスの消失現象の原因には，(1) 上昇した後太陽面上に落下する場合，(2) 高温になって，Hα線では見えなくなる場合，(3) 噴出して，惑星間空間に放出される場合，等を考えることができる．惑星間空間擾乱や地磁気嵐とプロミネンス消失の相関関係を調べるためには，上記のような消失の原因を正確に区別

STEP 第3回シンポジウム報告

できる観測が必要不可欠である．Hα+0.8Å と Hα-0.8Å 全体像は，この点について重要な役割を果たすものと期待されている．

3. 太陽全体像望遠鏡

4本の太陽全体像望遠鏡は，それぞれ，Hα 線中心像，Hα+0.8Å 像，Hα-0.8Å 像および連続光像を観測する．これらの光学系は，単色フィルターを除いてすべて同一であって，図3に示されている通りである．対物レンズからの光束は，テレセントリックレンズ系によって，F30 に引き伸ばされるとともに，その主光線がすべて光軸に平行になるように，単色フィルター（ファブリペロ干渉フィルター）に導入される．これによって太陽像全面にわたって，同一波長が得られる．縮小レンズ系は第1焦点の太陽像を約1/3に縮小して，CCD 受光面上に，約6 mm の太陽像を結像する．これら2組のレンズ系に使用されているレンズは，ミノルタカメラ技術センターカメラ開発部の協力を得て選択されたものである．

各望遠鏡に使用されている単色フィルターのデータは表1にまとめられている．

4. プロミネンス像望遠鏡

太陽周縁プロミネンスの観測効率を上げるために設計されたものである．基本的な光学系は前述の太陽全体像望遠鏡のものと同一としながら，しかも太陽面本体を隠すように，遮光円錐が組み込まれている．すなわち，テレセントリックレンズ系の後方に置かれた第1遮円錐によって，まず大部分の太陽本体光を Hα フィルター入射前にカットし，次に第1焦点面に置いた第2遮光円錐によって完全に太陽本体を除いている．また，Hα フィルターとして通常より透過幅の広いもの（3Å）を用いており，高速で運動しているプロミネンスでも撮影できるようにしている．

表1

Telescope Name	Monochromatic Filter		
	Central Wavelength	Passband	Filter Type
Hα center Tele.	6562.8 Å	0.42Å	Fabry-Pero (DayStar)
Hα+0.8Å Tele.	6563.6 Å	0.5Å	Fabry-Pero (DayStar)
Hα-0.8Å Tele.	6562.0 Å	0.5Å	Fabry-Pero (DayStar)
Continuum Tele.	6100 Å	60Å	multi-layer coating
Prominence Tele.	6562.8 Å	3Å	Fabry-Pero (DayStar)

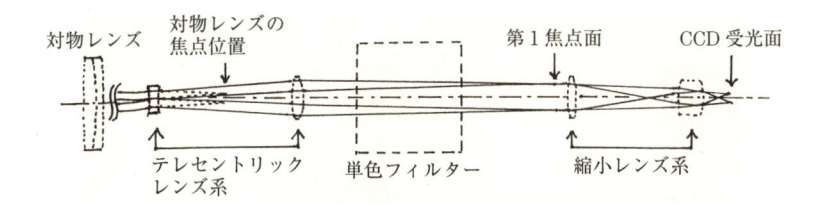

STEP 第3回シンポジウム報告

図3 太陽全面像望遠鏡の光学系.

5. 録画システムおよび観測例

5台のカメラで撮影された5太陽像は5台のタイムラプスVTRによって，連続録画されている．図4は5台のモニターに出画された5太陽像を示している．上段左がHα中心像，右がHα+0.8Å像，下段左からプロミネンス像，Hα-0.8Å像，連続光像である．

図5には，フィラメント消失の1例が示されている．(a)のHα中心像で，西半球（右側）中程に見られるフィラメントが(b)の02h02mUTでは消失している．この間のHα-0.8Å像とHα+0.8Å像の1例が(c)及び(d)に示されている．(c)で黒い部分はフィラメントの上昇運動を示しており，(d)はフィラメントの両端の落下運動を示している．

図6はプロミネンス像望遠鏡で捕らえられたプロミネンス爆発の1例である．

このように，新太陽フレア監視望遠鏡は，フレアのみならず，プロミネンス活動の観

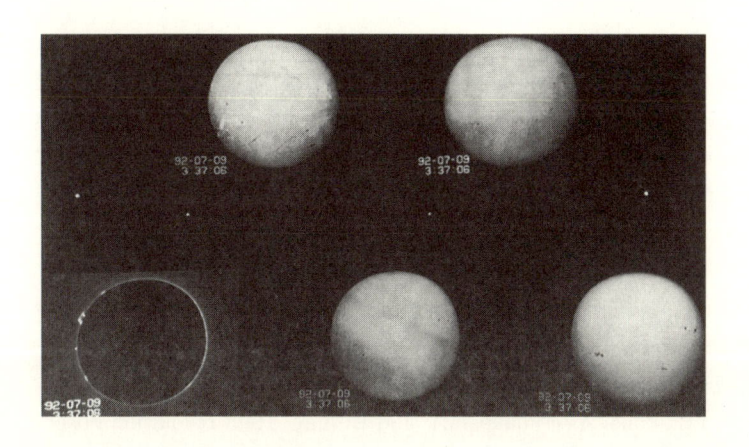

図4 5台のモニターに出画された5太陽像．上段左:Hα中心像，右:Hα+0.8Å像，下段左:プロミネンス像，中:Hα-0.8Å像，右:連続光像

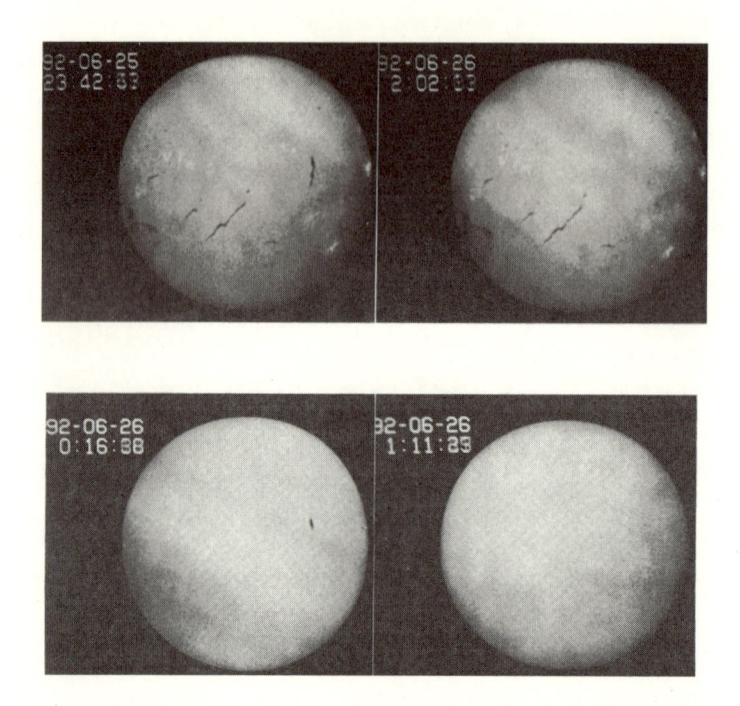

STEP 第3回シンポジウム報告

図5 June 26, 1992 のフィラメント消失. (a) Hα中心像：西半球（右側）中ほどのフィラメントが (b) 02h02mUT の Hα中心像では消失している. (c) Hα−0.8Å 像：黒い部分はフィラメントの上昇運動を示している. (d) Hα+0.8Å 像：フィラメント両端の薄黒い部分が落下している.

測に大変有効であることが判る．今後ドームレス太陽望遠鏡の高分解能観測と組み合わせることによって，はじめに述べた研究課題についての視野を広げる事ができるとともに，Yohkoh 衛星の軟 X 線画像や野辺山の電波ヘリオグラフ像との同時観測による成果も期待される．

参考文献／References

Kurokawa,H., Nakai,Y., Funakoshi,Y. and Kitai,R.: High resolution observations of active phenomena obtained at Hida Observatory, Adv. Space Research Vol.11 No.5 pp.233-240.

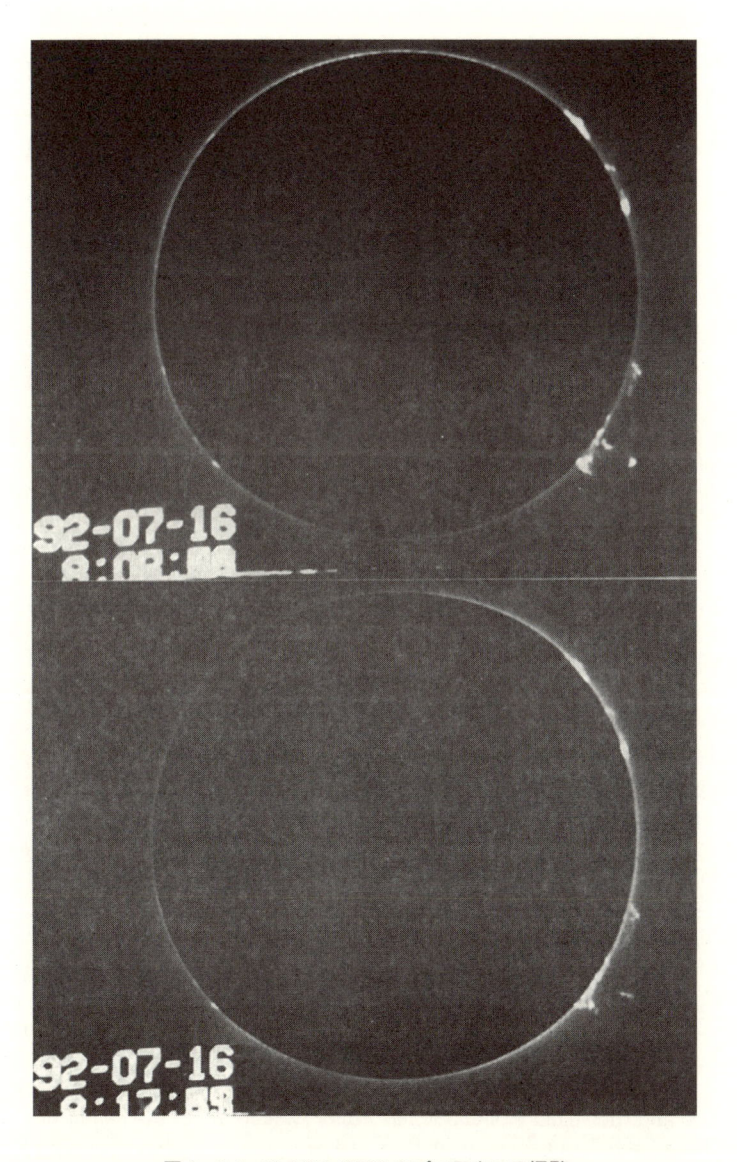

STEP 第3回シンポジウム報告

図6　July 16, 1992 08UT のプロミネンス爆発

STEP 第4回シンポジウム報告

OBSERVATIONS OF SOLAR PROMINENCE ACTIVITIES WITH THE NEW FLARE-MONITORING TELESCOPE

Hiroki Kurokawa[1], Kiyomi Ishiura[1], Gouichi Kimura[1], Reizaburo Kitai[1], Yasuhiro Funakoshi[1], Yoshihiro Nakai[2], Takehiko Shinkawa[2] and Hiroshi Yatagai[3]

[1]*Hida Observatory, Kyoto University, Kamitakara, Gifu 506-13*
[2]*Kwasan Observatory, Kyoto University, Yamashina, Kyoto 607*
[3]*Dept. of Geophys., Facul. of Scien., Kyoto Univ., Kyoto 606*

ABSTRACT

The new solar Flare-Monitoring Telescope (FMT) of Hida Observatory is unique among currently-operating flare patrol telescopes in the world, because it simultaneously records four full-solar-disc images in Hα line center, Hα+0.8Å, Hα−0.8Å and continuum, and one solar-limb (prominence) image in Hα line-center with five time-lapse video-tape-recorders (Kurokawa et al. 1992:*STEP GBRSC NEWS vol.*2, 6-8, Kurosawa *et al.* 1992:*Proceedings of Third STEP Symposium pp*21-28).

After its first year operation, it was confirmed that the FMT system is very effective to record and study the dynamical features of various types of prominence activities. The Hα−0.8Å and Hα+0.8Å images are especially useful for detecting the motions of activating prominences on the solar disc. They enable us to discriminate the dynamical disappearance from the thermal one of Hα dark filaments. In fact, we analyzed the time-lapse movies of the large filament disappearance of 5 November, 1992 obtained with the FMT, and concluded that this filament disappeared because it erupted out into the interplanetary space.

Many limb events, which provide us with more detailed informations of the vertical structures of prominence eruptions, sprays, surges and post-flare-loops, were also observed with the Prominence Telescope of the FMT. The prominence eruption of 31 July, 1992 is a typical example of them obtained with the FMT. In this event, a large quiescent prominence was found to start its eruption when its northernmost leg was detached from the solar limb. The ascending motion of the prominence was gradually accelerated and the velocity of about 100 kms^{-1} was found at the height of about 1×10^5 km. The ascending motion of this prominence was also observed with the Nobeyama Radioheliograph, and a bright soft x-ray arcade was found to develop along the prominence channel after the prominence eruption with the Soft X-ray Telescope aboard Yohkoh Satellite. The detailed comparison among the Hα, 17 GHz and soft x-ray images of this event was made by Hanaoka *et al.*(to be published in *Publ. Astron. Soc. Japan*, 1993).

STEP 第4回シンポジウム報告

In conclusion, more observations of dynamical features of the prominence eruptions and filament disappearances with the new Flare-Monitoring Telescope of Hida Observatory will shed more light upon the relationships among prominence eruptions, coronal mass ejections, and interplanetary and geomagnetic disturbances.

STEP 第4回シンポジウム報告

太陽プロミネンス活動現象の観測

京都大学理学部
黒河宏企, 石浦清美, 木村剛一
北井礼三郎, 船越康宏, 中井善寛
新川雄彦, 谷田貝宇

1. はじめに

　京都大学飛騨天文台に新設された太陽フレア監視望遠鏡は昨年の6月から定常観測に入り, その後順調に稼動して, 太陽フレアやプロミネンス爆発等の活動現象の監視観測を行っている.

　世界の各観測所で用いられている従来のフレアパトロール望遠鏡に対して, この望遠鏡の新しい点は, $H\alpha$線中心だけでなく, $H\alpha + 0.8\text{Å}$ 及び $H\alpha - 0.8\text{Å}$ の $H\alpha$ 線両翼の波長でも同時に, 太陽全面を連続監視していることである (Kurokawa et al. 1992a, 黒河他 1992b). これによって, 太陽面上のフィラメント消失をはじめ, 種々のプロミネンス活動の運動状況を太陽全面にわたって常時監視して, 記録できるようになった.

　また, 遮光円錐を装備した太陽周縁望遠鏡も同架しているので, プロミネンス爆発, サージ, スプレイ, ループプロミネンス等の高さ方向の運動を細大漏らさず監視記録することが出来る.

　以下に, 前年度における観測システムの整備状況を簡単に述べるとともに, 1年間にわたって得られた観測結果の中から, 代表的な現象の例を紹介する.

2. 新太陽フレア監視望遠鏡観測システム

　第1図に観測システム全体の系統図を示している. 監視望遠鏡本体は, 4本の太陽全体像望遠鏡と, 1本の太陽周縁像望遠鏡 (プロミネンス望遠鏡) 及び1本の光電追尾望遠鏡の合計6本の小望遠鏡で構成されており, 1台の赤道儀に同架されて, 一体として運転駆動されている (Kurokawa et al. 1992a, 黒河他 1992b). これらの小型望遠鏡群の作る太陽像は, CCDカメラからビデオ信号ケーブルによって観測コントロール室とドームレス太陽望遠鏡棟に転送されている.

　観測コントロール室では, 5台の14インチモニター上に出画されるとともに, 5台のタイムラプスVTRによって, 2秒に1フレーム (連続光像は4秒に1フレーム) の割合で同時連続録画されている.

STEP 第4回シンポジウム報告

太陽フレアー監視望遠鏡システム　概念図

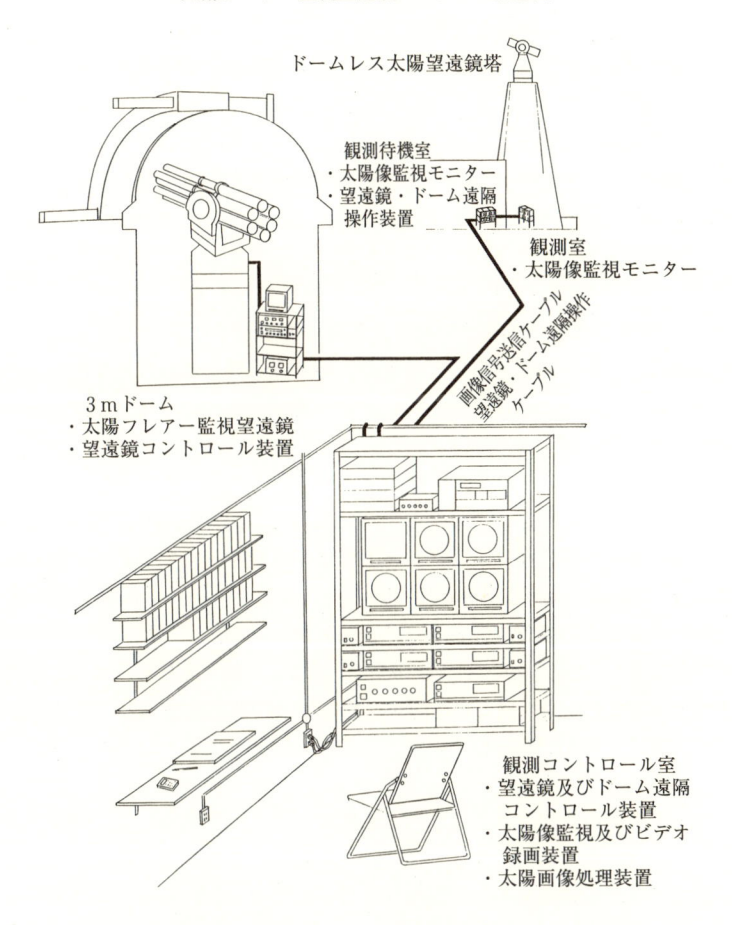

図1　飛騨天文台の太陽フレア監視望遠鏡システム系統図

　ドームレス太陽望遠鏡棟においては，一階観測室前ロビー，一階垂直分光観測室内，及び二階水平分光器観測室内の3箇所に設置されたビデオモニター群に出画されている．太陽フレア監視望遠鏡からドームレス太陽望遠鏡棟まで約250m間の画像転送ケーブルの配管工事は前年度に新たに行われたものである．この時同時に，制御用信号ケーブルも敷設されたので，現在では観測コントロール及びドームレス太陽望遠鏡からも，太陽フレア監視望遠鏡駆動及びドーム回転スリット開閉等のリモートコントロールを行うことができる．ドームレス太陽望遠鏡では，その高空間分解機能の故に，通常は，太陽全

STEP 第4回シンポジウム報告

面の約 1/30 の部分像しか観測できないが，フレア監視望遠鏡から太陽全体像の転送を受けて，常時太陽全面の活動状況を監視できるようになったので，フレアやプロミネンス活動等の突発現象に際して，観測領域の切り換えを機敏に行えるようになった．

画像データ記録方式に関しては，ディジタル記録方式の導入が現在進められている．もともと，この太陽フレアー監視望遠鏡においては，フレアー及びプロミネンス爆発等の発生を監視すること，及びそれらの運動を映画的手法によって解析することを第一目的としているので，ビデオ録画方式を採用している．1コマごとの画像解析を行う為には，ビデオ画像の必要コマを後でディジタル化することによって対応している．この方式は大量データを高時間分解で連続して容易に記録蓄積出来る利点があるが，欠点はVTR 記録時に画像劣化を伴うことである．最初から直接ディジタイズするディジタル記録方式を併用させれば，倍ぐらい分解能の良い画像が得られるはずである．

3. フィラメント消失の観測

Hαダークフィラメント（以下では単にフィラメントと呼ぶ）はその一生の間に，何回かその一部あるいは全体が，数分から1時間位の間に消失したり，再現したりすることがある．地磁気嵐の発生がフィラメント消失と相関を持っているという統計結果が報告されており (Wright and McNamara 1983)，CMEとプロミネンス爆発が相伴って起こることが多いことも指摘されている (Munro et al. 1984)．CMEに伴って惑星間衝撃波が検出されてもいる (Sheeley et al. 1985)．これらの事実を考え併せると，フィラメント消失←→CME→地磁気嵐という図式を想定するのは自然なことであろう．しかし，個々のイベントについてこれらの間の具体的な因果関係を実証した例は未だほとんどない．

フィラメント消失の要因には3通り考えられる．(1) 上昇加速して，惑星空間に抜け出ていくもの，(2) 光球に落下するもの，(3) 高温となりHα線では見えなくなるもの，である．地磁気嵐との相関を調べる為には，(1) によるものに限るべきであろう．

飛騨天文台の太陽フレア監視望遠鏡のHα−0.8Å像とHα+0.8Å像を用いれば，フィラメント消失時の運動状況を調べることができるので，上記3通りの区別に役立てることができるものと期待できる．実際，Hα−0.8Å像で見ると，表1にイベントリストの抜粋を示しているように，色々な種類のプロミネンスの運動が観測されている．以下に，代表的な例を紹介する．

3-1. 1992年11月5日のフィラメント消失

これは，大きなフィラメント全体の消失を，その始まりから最後まで連続的に，Hα線中心及び両翼で同時観測したものとしては，最初の例である．

図2の左側には，Hα中心線，右側にはHα−0.8Å像によって，この消失の時間変化

STEP 第4回シンポジウム報告

表1 Hα−0.8Å 像で観測されたフィラメント活動現象の一覧表(一部)
(飛騨天文台太陽フレアー監視望遠鏡イベントリストよりの抜粋)

Date	Time(UT) start	Time(UT) end	Location	Features in Hα−0.8Å	Features in Hα+0.8Å	Note
1993/05/15	22 40	24 00	S20 E35	Moving Dark Filament	Fil. Activ.	
05/16	00 00	00 20	S20 E35	Moving Dark Filament	Fil. Disap.	
05/19	04 45	05 15	S05 E55	Moving Dark Fibril	Fil. Disap. Flare	small
05/20	(2055)	21 15	N25 W15	Moving Dark Fibril	Fil. Disap.	Partial
05/20	21 43	22 25	N32 W10	Moving Dark Fibril	Fil. Disap.	Partial
05/20	21 50	22 05	N25 E02	Moving Dark Fibril	Fil. Activ.	small
05/21	06 54	08 00	N08 E45	Moving Dark Fibril	Fil. Disap.	Partial
05/25	03 25	04 08	N10 E60	Moving Dark Fibril	Fil. Disap.	small
05/25	04 00	04 30	N25 E65	Faint Dark Filament	Fil. Disap.	
05/25-26	20 57	03 00	S05 E25	Recurrent Surging	EFR-Surges	EFR-Surges
05/27	02 22	02 35	N25 E65	Moving Dark Filament	Flare-Surge	
05/27	02 45	03 35	S05 E40	Moving Dark Fibril	Fil. Activ.	Partial
05/27	03 30	04 10	S25 W08	Moving Dark Fibril	Fil. Disap.	small
05/27	21 00	21 57	N28 E60	Moving Dark Fibril	Fil. Disap.	small
05/28	02 20	03 19	N22 E42	Fil. Disap., Flare	Fil. Disap. Flare	
05/29	01 08	01 21	N12 E08	Moving Dark Fibril	Flare-Ejection	Fast Eject
05/29	03 12	03 27	N15 E05	Moving Dark Filament	Surge	
05/31	23 25	23 42	S02 W40	Moving Dark Fibril	Flare, Fil. Disap.	small
06/06	(0126)	01 38	S10 W15	Moving Dark Fibrils		Ejection?
06/06	01 25	07 00	S10 W15	Recurrent Surging	Surge Activity	
06/06	07 20	07 40	S10 W15	Big Dark Ejection	Flare-Ejection	Big Eject.
06/07	05 42	06 18	S10 W40	Flare and Ejection	Flare-Ejection	Eruptive

を示している.矢印で示されたフィラメントがHα中心では,01h32m45sUTですでに消えかかっており,02h14mUTでは完全に消失して,代わりにその場所に明るい筋模様(faint two ribbon flare)が見られる.Hα−0.8Å像では,逆に最初の0h52m03sUTでは,ほとんどこのフィラメントは見えないが,01h32m48s UT以後では,黒く現れ,その後その中心部が太陽面中心方向に移動して,湾曲した形に変化して行く様子が見られる.

図3には,このフィラメント周辺のみを拡大して,Hα中心像(±0.0Å),Hα−0.8Å

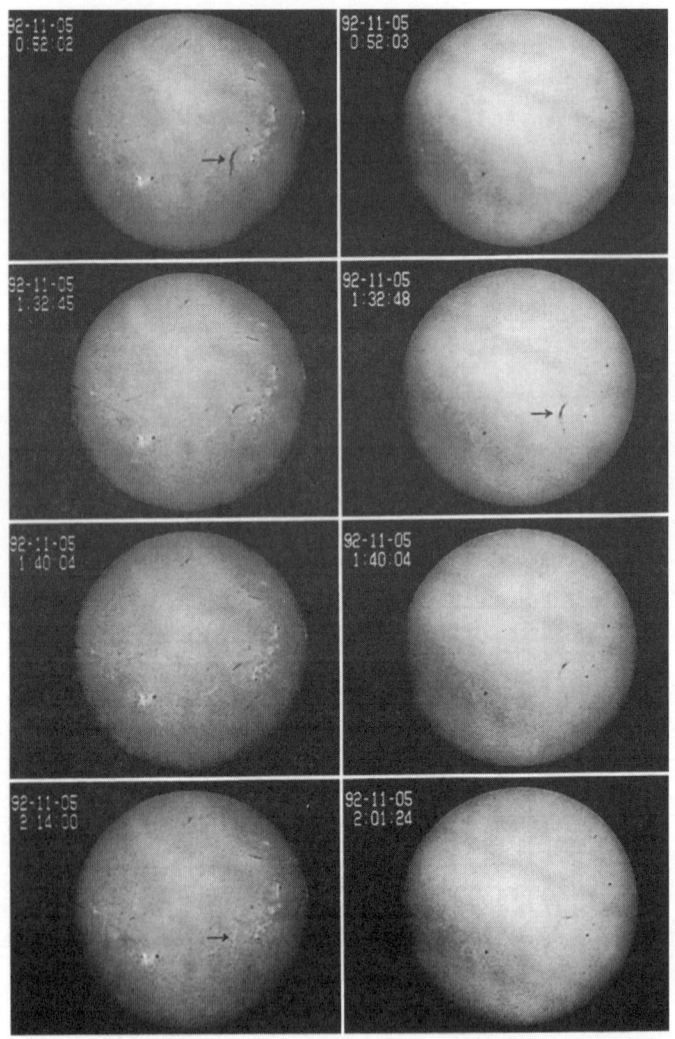

STEP 第4回シンポジウム報告

図2 1992年11月5日のフィラメント消失現象の時間変化. 矢印のフィラメントが消失した. 左側: Hα中心像, 右側: Hα − 0.8Å 像.

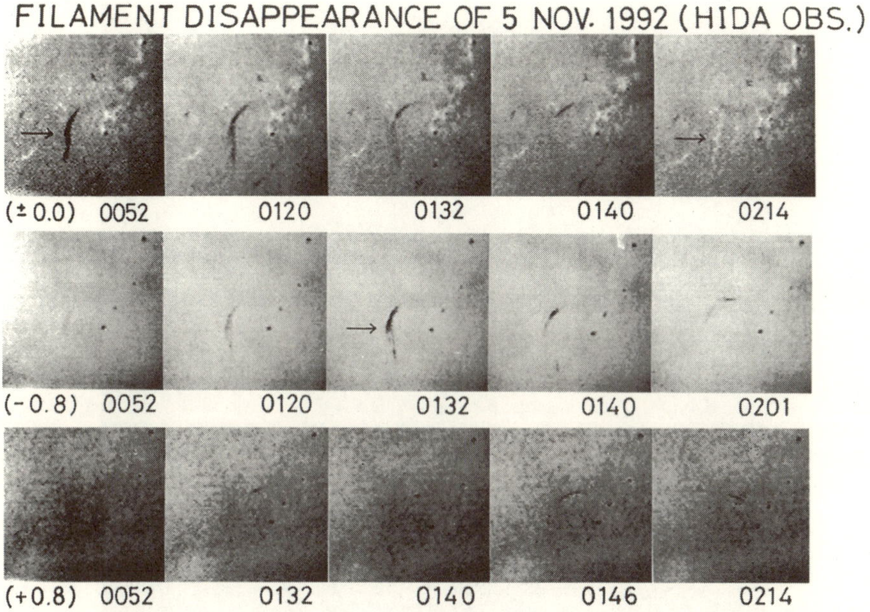

STEP 第4回シンポジウム報告

FILAMENT DISAPPEARANCE OF 5 NOV. 1992 (HIDA OBS.)

図3 図2のフィラメント消失の拡大写真. 上段:Hα中心像, 中段:Hα − 0.8Å 像, 下段:Hα + 0.8Å 像.

像 (−0.8), Hα+0.8Å 像 (+0.8) の順に示している. Hα+0.8Å 像では, フィラメントの中央部分はほとんど見られないが, 0140UT 頃から, 両端 (特に上方の端) に落下している部分が見られる. これらの像の比較から, このフィラメントは, 左上方に上昇して, その大半は, 惑星間空間へと抜け出て行ったものと判定する事ことができる. なおこのフィラント消失に際して, Yohkoh 軟X線望遠鏡では, 明るいX線アーケードが観測されており, 互いの関係について詳細な解析が行われている (McAllister et al. 1994).

3-2. フレアに伴うプロミネンス噴出

前述のフィラメント消失は, 静穏領域プロミネンス噴出の典型的な例であるが, 活動領域のプロミネンスが, フレアの前後に噴出する場合は, より激しい運動を示すことが多い. 図4にはこの様なものとして, 今年の観測の中から3例を示している. 上から NOAA7448 領域に発生したX線重要度 M1.8, Hα重要度 1N フレア, NOAA7500 領域に発生したX線重要度 4.4, Hα重要度 2B フレア, NOAA7518 領域に発生した C1.8, SN フレアである. それぞれに対応する右側の Hα−0.8Å 像では, 黒いプロミネンス噴

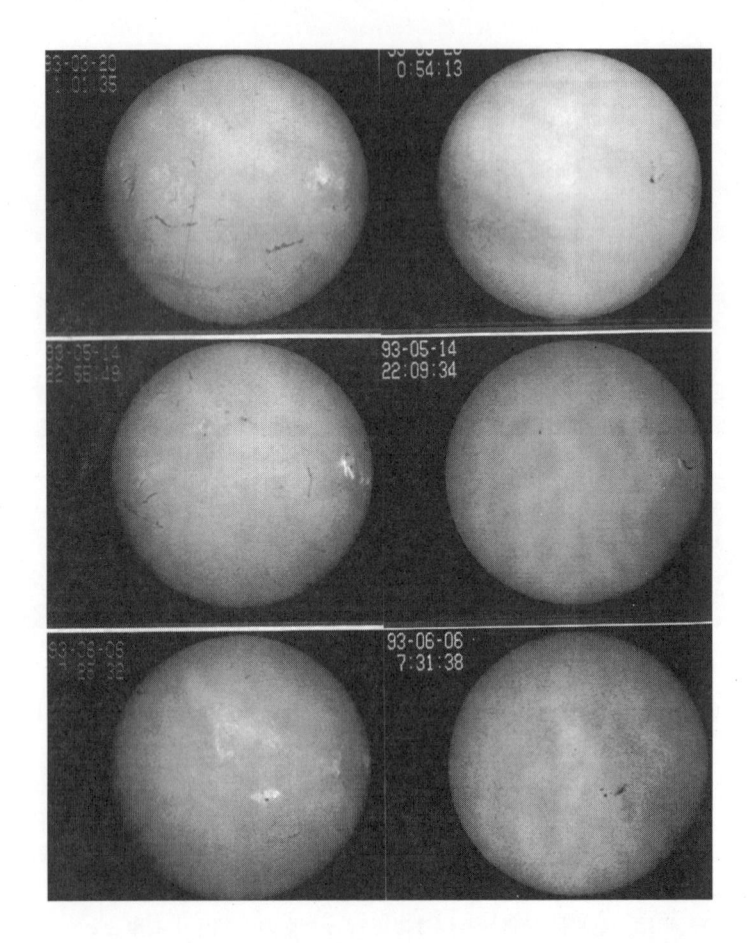

STEP 第4回シンポジウム報告

図4 フレアに伴ったプロミネンス噴出現象. 上段：20 March,1993（NOAA7448），中段：14 May,1993（NOAA7500），下段：6 June,1993（NOAA7518）. 左列：Hα中心像，右列：Hα−0.8Å像.

出が見られる. これらの噴出に伴って，何らかの惑星間空間擾乱が発生していることが予想される.

4. 太陽縁におけるプロミネンス噴出

太陽周縁でプロミネンス活動が発生すれば，その高さ方向の形態変化や運動を詳しく調べる事が出来る. 図5に与えられている1992年7月31日のプロミネンス噴出は，大規模な静止型紅炎の上昇噴出の典型的なものであった. 紅炎の足が北（上）の方から順

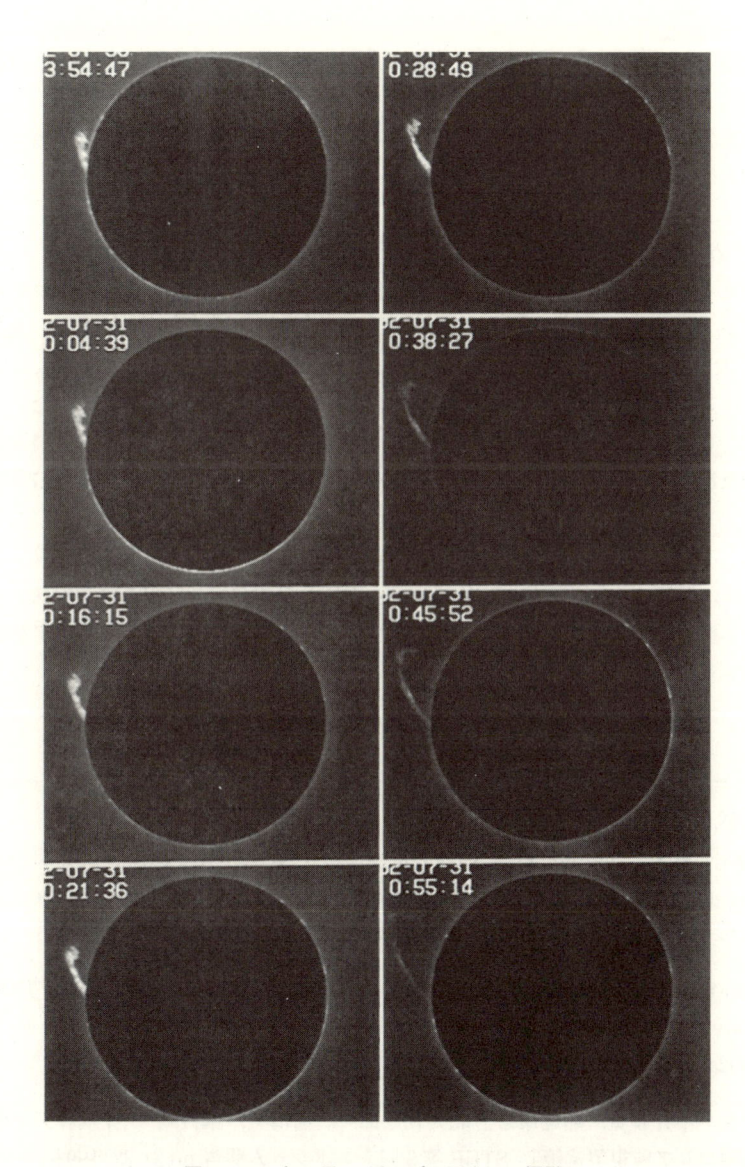

STEP 第 4 回シンポジウム報告

図 5　1992 年 7 月 31 日のプロミネンス爆発.

STEP 第 4 回シンポジウム報告

番に太陽面を離れるとともに，ゆっくりと加速して惑星間空間へ噴出してゆく運動が見事に捕らえられている．

この現象は，野辺山電波ヘリオグラフと Yohkoh 軟 X 線像でも観測されており，これ等と Hα 像との詳しい比較解析が Hanaoka et al. (1993) によって行われた．

5. まとめ

飛騨天文台の新太陽フレア望遠鏡は，稼動以来 1 年余りの間に，多くのフレア及びプロミネンス活動現象を観測した．この中で，特に，Hα 中心，Hα−0.8Å 像，及び Hα＋0.8Å 像の同時連続観測は従来のパトロール望遠鏡には無いものであり，太陽面上のフィラメント活動を解析するものとして，非常に有用であることが確認された．これによって，フィラメント消失時の運動を詳しく分析できるので，フィラメント消失と惑星間空間及び地磁気嵐との相関についての解析の精度を上げることが出来るものと期待できる．

また，プロミネンス活動及び消失時の形態変化を，Yohkoh 軟 X 線像と比較することによって，プロミネンス周辺部の磁場構造とそのリコネクションのモデルの検証，プロミネンス噴出の発生機構の研究及びそれと CME との因果関係の研究等にも手がかりを与えることが期待される．

参考文献／References

Hanaoka,Y, et al.; "Simultaneous Observations of a Prominence Eruption followed by a Coronal Arcade Formation in Radio, Soft X-ray and Hα", Publ. Astron. Soc. Japan, to be published, 1993.

Kurokawa,H., Nakai,Y., Funakoshi,Y., Kitai,R., Ishiura,K. and Kimura,G.; "The New Flare-Monitoring Telescope at Hida Observatory", STEP GBRSC NEWS vol.2, 6-8. 1992a.

黒河宏企，中井善寛，船越康宏，北井礼三郎，石浦清美，木村剛一;「飛騨天文台の新太陽フレア監視望遠鏡」，STEP 第 3 回シンポジウム報告 pp.21-28, 1992b.

McAllister,A., 黒河宏企，柴田一成，「1992 年 11 月 5 日のフィラメント消失」，STEP 第 4 回シンポジウム報告，本集録，1994.

Munro,R.H., Gosling,J.T., Hildner,E., MacQueen,R.M., Poland,A.I., and Ross,C.L.; "The

STEP 第4回シンポジウム報告

Association of Coronal Mass Ejection Transients with Other Forms of Solar Activity", Solar Phys. 61, 201-215.

Sheeley,Jr.N.R., Howard,R.A., Koomen,M.J., Michels,D.J., Schween.R., Muhlhaiser, K.H., Rosenbauer,H.; "Associations between Mass Ejections and Interplanetary Shocks", J. Geophys. Res., 90, 163, 1984.

Wright.C.S. and McNamara,L.F.; "The Relations between Disappearing Solar Filaments, Coronal Mass Ejections, and Geomagnetic Activity", Solar Phys. 87, 401-417, 1983.

STEP 第5回シンポジウム報告

飛騨天文台の新太陽フレア監視望遠鏡による Hαフィラメント消失の観測

京大理学部 附属天文台　　黒河宏企, 石浦清美, 木村剛一, 中井善寛,
新川雄彦, 北井礼三郎, 船越康宏

（論文要約）

　STEP 計画の一年目において, 京都大学理学部附属天文台では, 太陽フレア監視望遠鏡を飛騨天文台に新設した. この論文では, この新型望遠鏡システムの概要を述べ, これによって得られた成果の例を紹介している.

　現在世界各地で稼動しているフレアパトロール望遠鏡に較べて, この望遠鏡の新しい特長は, $H\alpha$ 線中心のみならず, $H\alpha-0.8Å$ と $H\alpha+0.8Å$ の $H\alpha$ 線両翼でも同時に, 太陽全面像を連続して記録できることである. その他に, 連続光による太陽全面像と太陽周縁のプロミネンス活動を監視できるプロミネンス望遠鏡を加えて, 計5個の太陽像を同時に5台のビデオテープレコーダーで2秒に1コマの割合で連続撮影している. この望遠鏡の稼動以来, 多くのプロミネンス活動が太陽面上及び周縁で観測されているが, 中でも $H\alpha-0.8Å$ 像及び $H\alpha+0.8Å$ 像が消失フィラメントの運動特性を調べる上で, 非常に有用であることが明らかになった. 即ち, これらの像を比較することによって, フィラメント消失がフィラメントの惑星空間への噴出によるものか, 太陽面への落下によるものか, またはそのどちらでもないのかを判定することがある程度可能になった.

　一方, Hαフィラメントが消失した後, 軟X線で明るいアーケード構造が形成される例が, YOHKOH の軟X線望遠鏡で数多く観測されているが, 我々のフレア監視望遠鏡での観測との比較から, より運動の激しいフィラメント消失に対応して, より顕著な軟X線アーケード構造が形成されるらしいことが明らかになってきた. またこれらは地磁気嵐発生の有無あるいはその強弱とも関係しているようである. 実例として, 1992年11月5日と1993年10月21日のフィラメント消失における Hα像と軟X線像の変化を比較して示している. 消失フィラメントの運動は前者の方がよりダイナミックであり, これに伴って, 軟X線アーケードが形成されたが, 後者のゆっくりとした消失の後には, 紐状の長いループがフィラメント軸に沿ってゆっくりと明るくなるのが観測されたのみであった.

STEP 第6回シンポジウム報告

OBSERVATIONS OF DISAPPEARING Hα FILAMENTS ON THE SOLAR DISK

Hiroki Kurokawa[1], Kiyomi Ishiura[1], Gohichi Kimiura[1],
Takehiko Shinkawa[2], Reizaburo Kitai[1], Yasuhiro Funakoshi[1]
Hideki Koshiishi[3], Shindo Enome[3]

[1]*Kwasan and Hida Observatories, Kyoto University, Kamitakara, Gifu 506-13*
[2]*Kwasan and Hida Observatories, Kyoto University, Yamashina, Kyoto 607*
[3]*Nobeyama radio Observatory, NAOJ, Minamisaku, Nagano 384-13*

ABSTRACT

The new solar Flare Monitoring Telescope (FMT) installed at Hida Observatory during the STEP period successively obtained time-lapse video movies of disappearing Hα filaments simultaneously in Hα line center, Hα−0.8Å and Hα+0.8Å. They enable us to study morphological and evolutional features of disappearing filaments. The Hα −0.8Å and Hα+0.8Å movies are found to be especially useful for the study of their dynamical characteristics. The dynamical features of three typical filament disappearances are comparatively examined and their relations to geomagnetic and interplanetary disturbances are discussed in this paper. The 20 Feb. 1994 event is an eruption of an active region filament followed by a 3B flare. It was most dynamic and energetic event and rapidly disappeared from the passband of Hα line center and Hα −0.8Å filters in about 5 and 10 minutes, respectively. This was also observed by Nobeyama Radio Heliograph in 17 GHz. By analyzing the motions of the disappearing filament in Hα−0.8Å and 17 GHz images, we conclude that the filament erupted to wide directions from the southeast through the southwest. Such an expanding eruption of the filament in a wide angle resulted in the interplanetary shocks detected by Ulysses even at S54 on 27 Feb. (Gosling et al. 1994) as well as the strong geomagnetic storm of 21 Feb. The 5 Nov. 1992 filament disappearance is a typical one in the quiet region. It slowly rose up and erupted into the interplanetary space. It gradually faded out from the passband Ha-0.8Å filter in about an hour. After its disappearance, a very faint Hα two-ribbon flare emission was observed, and the Soft X-ray Telescope aboard Yohkoh observed an arcade of bright x-ray loops (McAllister et al. 1996). This event caused the geomagnetic storm of 9 Nov. The 21 Oct. 1993 filament disappearance in the quiet region is of nearly the same size as but less dynamical than the 5 Nov. 1992 event. It was neither followed by Hα flare emission nor by x-ray arcade formation. Instead, several bright string-like x-ray loops were observed to form along nearly the same direction as the filament over about 10 hours

STEP 第 6 回シンポジウム報告

after the filament disappearance. The identification of a geomagnetic storm caused by this event is not conclusive. The dynamical characteristics of the three disappearing filaments and their relations to Hα flare, soft x-ray bright structure and geomagnetic storm are summarized in the following table.

Event	filament vol. cm^3	filament mass (g)	speed km/s	kinetic ener. (erg)	Hα flare	X-ray emiss.	geoma storm
5/11/92	2.5×10^{28}	4.2×10^{16}	~ 120	$\sim 3.0 \times 10^{30}$	faint	arcade	○
20/2/94	6.1×10^{28}	1.0×10^{17}	~ 520	$\sim 1.3 \times 10^{32}$	3B	bright arcade	◎
21/10/93	1.0×10^{28}	1.7×10^{16}	~ 100	$\sim 8.5 \times 10^{29}$	none	string	△

STEP 第6回シンポジウム報告

太陽 Hα フィラメンント消失の観測

京大理附属天文台
黒河宏企，石浦清美，木村剛一，
新川雄彦，北井礼三郎，船越康宏

国立天文台野辺山
越石英樹，鮫目信三

1. はじめに

太陽面上で発生する Hα フィラメント消失と，地磁気嵐の発生との間の，因果関係を調べようとする試みが，これまでいくつか成されているが，未だはっきりとした結論は得られていない．その中で注目されるのは，多くのフィラメントについて統計的に調べた Wright and McNamara (1983) の仕事である．彼等は，大きな Hα フィラメントが消失すると，地磁気擾乱が発生し，その擾乱の強さは，消失フィラメントのサイズや黒み，及び日面上の位置や太陽活動周期の位相等に依存することを見い出した．また，フィラメント消失から地磁気嵐発生までの時間間隔は，典型的なものについては 3 日～6 日であり，大きく黒いフィラメントの方が小さく淡いものより短いという結果を出している．

しかし，一言でフィラメント消失といっても，その大半が太陽面上に落下するものと惑星間空間へ飛び出すものがあり，また，温度が上昇して Hα 線では見えなくなるいわゆる熱的消失（thermal DB）も考えられる．惑星間空間に飛び出すものの中にも，急激に加速されるものと，ゆっくり加速されるもの，一方向に集中的に噴出するもの，大きな立体角へ膨張的に放出されるもの等，いくつかの運動特性に分類出来そうである．

さらに，最近の Yohkoh SXT による観測から，フィラメント消失後に明るい軟 X 線アーケード構造やフィラメント軸に沿った明るいループ構造が形成されることが明らかになった．また一方では，フィラメント消失の無い場合にも，軟 X 線アーケード構造が形成されたという観測もある．この様なことから，フィラメント消失過程とコロナ磁場の変動再結合過程機構がいかに互いに関わっているか，またそれらが地磁気嵐や惑星空間擾乱の強さや特徴とどのように関わっているか等，個々のフィラメント消失について具体的に且つ総合的に調べることが重要であると考えられ，これがまさに STEP 計画における我々の主たる研究テーマであった．以下にこれまでに得られた観測成果の例を紹介する．

STEP 第6回シンポジウム報告

2. Hαフィラメント噴出の典型例

飛騨天文台に新設されたフレア監視望遠鏡(Flare Monitoring Telescope: FMT)によって撮影された多くのHαフィラメント消失の中から，典型的な3例を選んでその特性を比較する．

2-1　20 Feb. 1994 の活動領域フィラメント消失

活動領域 NOAA7671 領域を取り囲んでいた大きな Hαフィラメントが 20 Feb.1994 0050UT から 0100UT にかけて消失し，それに引き続いて Hα 重要度 3B, X線クラス M4 の大フレアが発生した（Kurokawa & Shinkawa 1995）．図1にはこの現象の太陽面上の位置が判るように，Hα中心像（上側）と Hα−0.8Å像（下側）によって消失前後の形の変化を示している．図2には，このフィラメントの周辺部分だけを切り取って，さらに詳しい時間変化を示している．これによると，フィラメントの西側半分は Hα−0.8Å像において，0043UT 頃からゆっくり見えはじめ，0058UT 以後急速に消えて 0101UT ではすでに見えない．Hα−0.8Å像の透過半値幅は 0.6Å であるので，視線方向速度に換算すると，約 20〜50 km/s の範囲が見えていることになる．従って，ゆっくり見えはじめて急速に消えるという上述の事実は，フィラメントが加速上昇運動しており，速度 20 km/s 前後における加速度より 50 km/s 前後における加速度の方が大きくなっていることを示している．

また，図2に見られるように，フィラメントの西側部分は，その場でほとんど位置を変えずに消失しているので，その上昇運動はおおむね視線方向に平行であったことが判る．それに対して，フィラメントの東側部分は，南東方向へ移動しながら消えていく成分が顕著に見られる．この南東方向に移動していくフィラメントの先端を追跡することによって，フィラメント噴出の視線方向に垂直な速度成分を求めることができる．その結果は，0102UT から 0107UT 間の平均速度が約 75 km/s であった．この時，フィラメントの視線方向速度は上述したように，Hα−0.8Å像で見えているということより，約 20 km/s から 50 km/s の範囲にあるはずであるから，このフィラメント東側部分の噴出方向角度は，視線方向に対して約 56°〜73°，噴出全速度は 78〜90 km/s と計算される．

このフィラメント消失とそれに伴うフレアーは，国立天文台野辺山の 17 GHz 電波ヘリオグラフでも観測された（Koshiishi et al. 1995）．図3に，この 17 GHz 像の変化を示している．フレアの始まる前（0103UT 以前）から明るい二領域は黒点とプラージュの上であるが，その間で南東方向に広がった暗い領域が問題の Hαフィラメントに相当する部分である．なお，Hα−0.8Å像で見たその Hαフィラメントの位置が白い線で囲んで示されている．

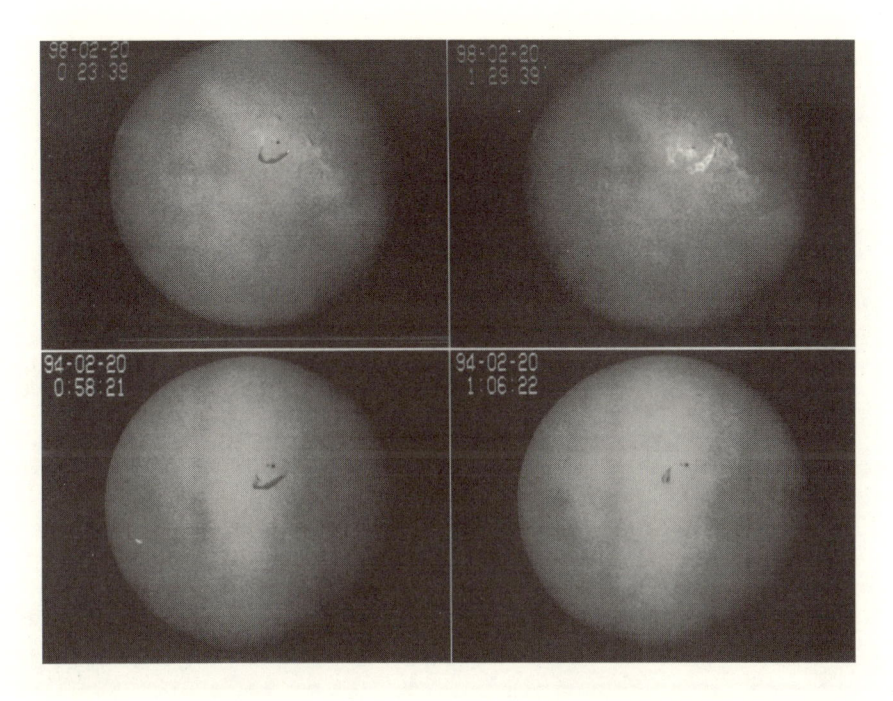

STEP 第6回シンポジウム報告

図1 20 Feb. 1994 の活動領域フィラメントの消失とそれに伴って発生した 3B フレア．上段：Hα 線中心像．下段：Hα－0.8 像．（飛騨天文台太陽フレア監視望遠鏡撮影）

　図3から判るように，噴出初期の 0057UT では，フィラメントの Hα－0.8Å 像は 17 GHz 像の暗い部分のほぼ中心付近にまだ重なっているが，噴出速度が大きくなった 0100～0103UT では，Hα－0.8Å 像のフィラメントは，17 GHz 像で見えるフィラメントの東端部分に対応している．また，0109UT 以後は，フィラメントは高速となって，Hα－0.8Å 像ではその透過波長範囲を越えてしまって全く見えないが，17 GHz では引き続き南東方向に飛んで行くのが見られる．この 17 GHz フィラメント像の南東先端の見かけの移動速度，すなわち，視線方向に垂直な速度成分は，0106UT～0109UT で約 80 km/s であり，上で求めた Hα－0.8Å 像上での速度成分と良く一致する．更に，0118UT 付近では，370 km/s まで加速していることも 17 GHz 像から求めることが出来る．0106UT 以後，17 GHz の黒いフィラメント領域は白いフレアー領域に反転して，更にそれが急速に膨張発展していくのが見られる．更に 17 GHz 像の中で注目すべきことは，西側（右側）のプラージュ領域上の白く明るい電波源が，0112UT から 0119UT にかけて急速に消えていることである．これは，噴出したフィラメントによって覆い隠された

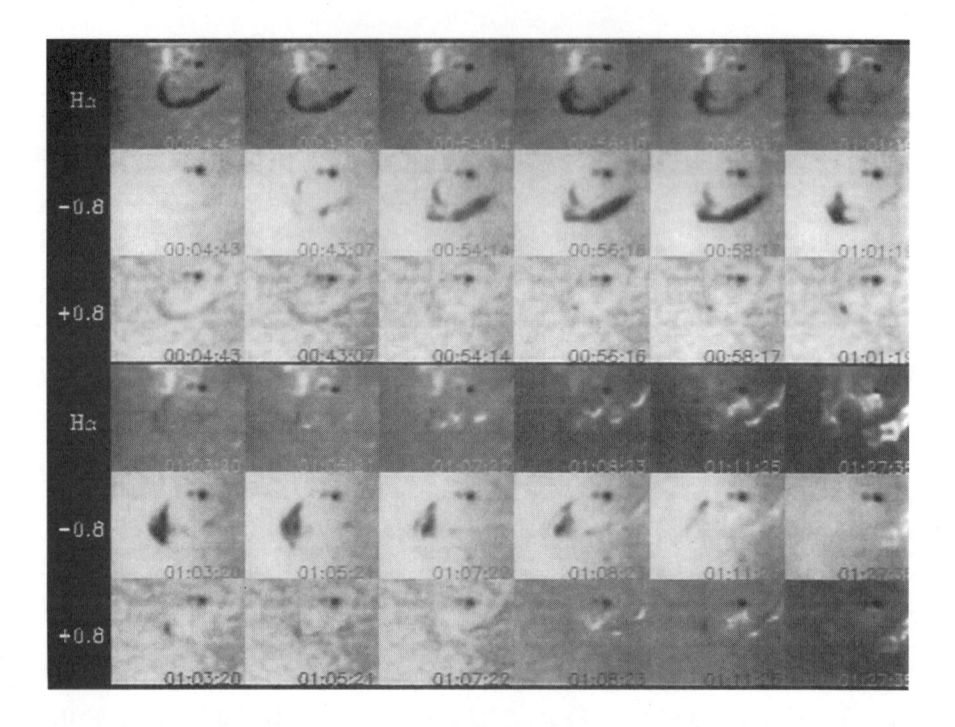

STEP 第6回シンポジウム報告

図2　20 Feb. 1994 の活動領域フィラメントの消失現象の時間変化．上段：Hα線中心像．中段：Hα－0.8Å 像．下段：Hα＋0.8Å 像．（飛騨天文台太陽フレアー監視望遠鏡撮影）

為であると考えられる．Hα−0.8 像では，フィラメント西側部分はほぼ視線方向に噴出した為に，透過波長範囲に見えている時間が短く南西方向への噴出成分が観測されなかったのに対して，17 GHz 像はドップラー効果に関係しないため，この西側部分の噴出移動が捕らえられているわけである．

　この結果，このフィラメントはある限られた方向に集中して噴出したのではなく，東西にかなり幅広い角度で南方向に膨張的に噴出したものと結論することができる．Gosling et al.（1994）によれば，太陽赤道面より54度も南に位置していた惑星空間探査機 Ulysses によって，27 Feb. 1994 にこのフィラメント噴出によると考えられる惑星間空間擾乱が観測された．フィラメント噴出発生の日面緯度は N3 度であり，Ulysses の観測点と57度もの角度差があるが，フィラメントが広い角度で南方向へ膨張噴出したという上記結論を考慮すれば，これも説明できると考えられる．

2-1　5 Nov. 1992 の静穏領域フィラメント消失

　1992年11月5日の0110UTから0210UTにかけて，静穏領域の平均的サイズのHαフィ

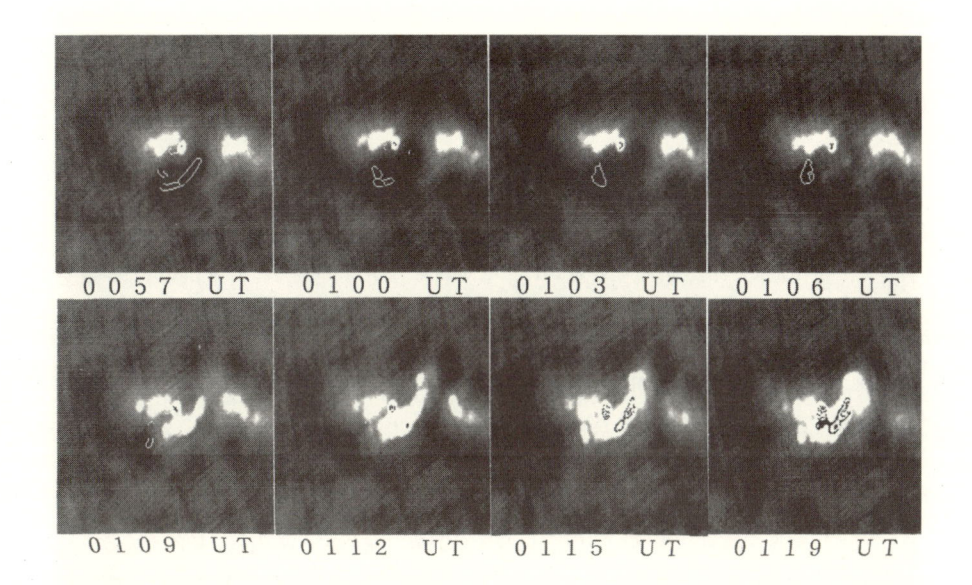

STEP 第6回シンポジウム報告

図3　17 GHz 像で見た 20 Feb. 1994 の活動領域フィラメントの消失現象の時間変化（国立天文台野辺山電波ヘリオグラフ撮影）．白い実線は Hα−0.8Å 像で見た Hα フィラメント（図2）の位置を示す．

ラメント消失噴出現象が観測された．このイベントでは，フィラメント消失後に，元のフィラメントの場所に非常に淡いながら，Hα フレアも観測された．このイベントのフレア監視望遠鏡で撮影された時間変化についての写真は，第4回 STEP シンポジウム集録にすでに掲載されている（黒河他 1995）．また，その詳しい解析は McAllister et al.（1995）によって行われた．

このフィラメント消失はやはり，上昇噴出によるものであり，フィラメントの一部は足元の太陽面に落下したものの，その大部分は惑星空間に抜け出たものと結論される．ただし加速の過程は 20 Feb. 1994 イベントとは対照的にゆっくりしたものであり，最初は約1時間にわたって，約 20 km/s から約 140 km/s まで次第に上昇加速して，噴出していったものと結論される．

これについては，Yohkoh の SXT 像も詳しく解析された（McAllister et al.1995）．フィラメント消失後，フィラメントの場所には軟 X 線の明るいアーケード構造が観測され，アーケードの片側の足元は Hα フレアの場所に対応していることが示された．

2–3　21 Oct. 1993 の静穏領域フィラメント消失

0345UT から 0520UT の約1時間半にわたって，Hα−0.8Å 像で上昇運動が観測され，

STEP 第6回シンポジウム報告

ゆっくりと消失した．Hα＋0.8Å像ではこの間，これに関係した落下物質に相当する構造はほとんど何も観測されなかった．このことから，このフィラメント消失も非常にゆっくりとした上昇運動によるものであると考えられる．ただし，その初期の上昇加速度は5 NOV. 1992よりもさらにゆっくりしており，しかも上記2例のようなフィラメントの膨張噴出を示すような横方向の移動はほとんど見られない．消失の最終段階になって（0500-0520 UT），比較的速い動きが唯一フィラメント南端で軸方向に見られるのみであった．更に上記二例との重要な違いは，このフィラメント消失に関連して，Hαフレアの発生が全く認められなかったこと，及び，いわゆる軟X線アーケード構造が顕著には発達しなかったことである．その代わり，Yohkoh SXTでは，元のフィラメントの軸方向に沿ったいくつかの明るい軟X線ループ構造が，フィラメント消失後，0537UTから1536UTまでという非常に長い間にわたってゆっくりと発達した．

このような特徴を総合すると，このイベントにおいては，ゆっくりとした上昇運動によって消失したフィラメントが，その後その大部分が引き続き加速されて惑星間空間へ抜け出て行ったのかどうかは，前出の2例ほど明らかではない．しかし，Hαフィラメントが上昇した後，再び太陽面に戻る落下運動がHα＋0.8像でなんら観測されなかったこと，またHα中心像で再び見えることのなかったことを考えると，フィラメントの大部分はどこへ消えたのかという大きな疑問が残る．フィラメント噴出によって開いたコロナ磁場が再結合することによって形成されると考えられている顕著な軟X線アーケード構造やHαフレアの発生がなかったことは，このフィラメントの運動が上部コロナ磁場全体を引きずり上げるほどのものではなく，惑星間空間に噴出したにせよ，上部コロナ磁場の一部の箇所からすり抜けるようなものであったと考えるべきかも知れない．また又，軟X線の長時間にわたる明るいループ構造の形成は，このフィラメント消失に伴うゆっくりとしたコロナ磁場の再結合とエネルギー解放が起こっていることを示しており，この具体的な機構の解明も今後の課題である．

3. まとめ

STEP計画によって京都大学飛騨天文台に新設されたフレアー監視望遠鏡は，Hα線中心のみならず，Hα＋0.8Å及びHα－0.8Åの3波長で同時に太陽全面像を連続記録するというこれまでにない特徴を持っている．これによって1992年4月の観測開始以来，太陽縁上のプロミネンス爆発はもとより，太陽面上の多くのHαフィラメント消失についても，それらの運動特性を観測することに成功している（Kurokawa et al. 1996）．解析は未だ途中であるが，ここではそれらの中から典型的な3例のフィラメント消失現象を選んで紹介した．

これらのフィラメント消失の特徴を比較してまとめると表1のようになる．

STEP 第6回シンポジウム報告

表1. 典型的フィラメント消失の特性比較

Event	filament vol. cm^3	filament mass (g)	speed km/s	kinetic ener. (erg)	Hα flare	X-ray emiss.	geoma storm
5/11/92	2.5×10^{28}	4.2×10^{16}	~ 120	$\sim 3.0 \times 10^{30}$	faint	arcade	○
20/2/94	6.1×10^{28}	1.0×10^{17}	~ 520	$\sim 1.3 \times 10^{32}$	3B	bright arcade	◎
21/10/93	1.0×10^{28}	1.7×10^{16}	~ 100	$\sim 8.5 \times 10^{29}$	none	string	△

表中の filament vol. はフィラメントの（長さ）×（平均幅）2で求めた体積である．また，フィラメント中の hydrogen density を 10^{12} cm^{-3} として mass を求めている．speed は，Hα - 0.8Å 像及び 17 GHz 像で求めた視線方向に垂直な速度成分と噴出角度から求めたもので，噴出初期に観測された最大速度を示している．kinetic ener. は $1/2mv^2$ で求めた噴出初期のフィラメントの運動エネルギーである．X-ray emiss. 中の string は，フィラメント消失が関係していると考えられる地磁気擾乱の強さを表しており，京都大学地磁気世界資料センターがまとめた H-component の時間変動グラフを参照した．

表から明らかなように，20 Feb.1994 のフィラメント消失の運動エネルギーは他に較べて2桁大きく，21 Feb. の 0901UT の SC に始まる強い地磁気嵐を引き起こしている．フィラメント消失後わずか 33 時間で擾乱が到達する高速伝搬の典型例であり，活動領域フィラメントの高速噴出（スプレイ）とそれに伴った大フレアによる強い惑星間空間擾乱の典型例であるということが出来る．

5 Nov.1992 のフィラメント消失は，中程度の大きさの静穏領域プロミネンスの噴出であるが，X-ray でアーケード構造の形成が見られ，淡いながら Hα フレアも発生しているので，フィラメント上のコロナ磁場は一度開いた後，再結合したものと考えられる．このことはまた，フィラメントが惑星間空間へ飛び出したことを裏付ける証拠であるとも考えられる．これによる地磁気嵐としては，4日後の 9 Nov. の 4UT 頃から始まるものが対応していると考えられる．

21 Oct.1993 のフィラメント消失はやはり中程度の大きさの静穏領域プロミネンスであるが，さらに低エネルギーのもので，この場合は，X-ray のアーケード構造と呼ぶべきものがほとんど形成されず，Hα フレアも発生しなかった．従って，フィラメントが惑星間空間へ飛び出したかどうかについては疑問が残る．しかし，フィラメントが太陽面に戻ってきたことを示すデータも観測されていないことを考慮すると，やはりこのフィラメントも惑星間空間へ出たものと考えるのが妥当であろう．これに対応する地磁

STEP 第6回シンポジウム報告

気擾乱の候補としては4日後の 25 Oct. の 08UT から始まるものがあるが，これがコロナホールによるものではないかどうかについて，更に詳しく調べる必要がある．

　以上のように，Hα線中心のみならずHα線両翼（Hα line wings）での太陽全面像観測によるフィラメント上昇運動を解析する飛騨天文台の新観測システムによる我々の新しい試みは，フィラメント噴出と惑星間空間擾乱や地磁気嵐との因果関係を詳しく調べる上で，有効であることが立証された．また，これらのHα像データを軟X線像及び電波像と比較することによって，プロミネンス爆発現象の機構とコロナ磁場構造の変動再結合過程及びエネルギー解放機構相互の間の関係を具体的にこれまでよりも詳細に研究できることが示された．

　この観測装置改良の今後の課題としては，フィラメント上昇運動解析の精度を上げるために，現在の3波長による観測を5波長に拡張すること，特にHα−1.5ÅからHα−2.0Å付近のフィルターを増設して，より高速のフィラメント上昇速度を検出できるようにすることが必要である．また，CCDカメラを 1000×1000 素子のものに切り替えて，空間分解能を向上させることも重要である．

参考文献／References

Gosling, J.T., D.J.MccComas, J.L.Phillips, L.A.Weiss, V.J.Pizzo, B.E. Goldstein, and R.J. Forsyth, A new class of forward-reverse shock pairs in the solar wind, Geophys. Res. Lett. 21, 2271, 1994.

Koshiishi, H., K. Fujiki. S. Enome. H. Nakajima. K. Shibasaki. and H. Irie, 17-GHz radio observations of a disparition brusque, a flare and post-flare loops on 1994 February 20, STEP GBRSC vol. 5, NO.1 (May 1995), 5, 1995.

Kurokawa, H., T. Shinkawa, The Hα filament eruption of 20 Feb. 1994 followed by a big flare and geomagnetic and interplanetary disturbances, STEP GBRSC NEWS vol. 5, No.1 (May 1995), 3, 1995.

Kurokawa. H., K. Ishiura, G. Kimura, Y. Nakai, R. Kitai, Y. Funakoshi, and T. Shinkawa, Observations of solar Hα filament disappearances with a new solar flare-monitoring-telescope at Hida Observatory, Geomag. Geoelectr. 47, in press, 1996.

黒河宏企，石浦清美，木村剛一，北井礼三郎，船越康宏，中井善寛，新川雄彦，谷田貝宇，太陽プロミネンス活動現象の観測，太陽地球系エネルギー国際協同研究(STEP)

STEP 第6回シンポジウム報告

第4回シンポジューム報告，p.62-71, 1995.

Wright, C.S. and L.F. McNamara, The relationships between disappearing solar filaments, coronal mass ejections, and geomagnetic activity, Solar Phys. 87, 401, 1983.

あとがき

　フレア監視望遠鏡は，黒河宏企名誉教授が中井善寛元助教授，船越康弘元助手の協力を得て，計画設計設置されたものである．また，石浦清美元技術専門職員，木村剛一技術専門職員，仲谷善一技術職員によるドーム建屋建築，観測システム整備，ドーム自動制御機構導入などがなされ，定常的な観測が行えるようになった．1996年度のデジタル画像取得システムの導入に当たっては，当時院生であった新川雄彦氏の貢献があった．その後も，飛騨天文台職員・ポスドク研究員によってシステムの改良がなされてきたし，日常観測は，歴代の飛騨天文台職員全員があたってきた．そのたゆまぬ努力があって長期の貴重なデータの蓄積が可能となった．本書出版にあたり，これらの方々に感謝したい．

　フレア監視望遠鏡により本書のような素晴らしい成果が得られたのは，まえがきに書いたような望遠鏡の特色もあるが，飛騨天文台技能補佐員門田三和子氏によるモートン波サーベイの努力によるところが大きい．また，門田氏はモートン波だけでなく，フレア，フィラメント噴出，フィラメント消失，フィラメント運動，サージなど，$H\alpha - 0.8Å$の太陽全面像ムービーで観測されるあらゆる太陽活動現象のサーベイリストを作成し，附属天文台ホームページ上で公開してきた．太陽リム画像を用いたプロミネンスデータ解析に関しては，花山天文台技術補佐員であった故名筋容子氏，小林燕氏，磯田安宏氏が行い，これらのデータも同様に公開してきた．門田氏を始めとする，これらの方々の献身的なご努力に深く感謝したい．

　本書の編集は，京都大学大学院理学研究科附属天文台のスタッフの皆さんおよび大学院生の皆さん（松本琢磨，大辻賢一，渡邉皓子，阿南徹）の様々な協力のおかげで完成したものである．とりわけ，編集委員会のメンバー（北井礼三郎氏，上野悟氏，野上大作氏，浅井歩氏，石井貴子氏，門田三和子氏，八木正三氏，鴨部麻衣氏，小森裕之氏，金田直樹氏）の献身的な協力に，心より感謝したい．

　本書は日本学術振興会科学研究費補助金研究成果公開促進費（課題番号225226），および，京都大学教育研究振興財団＜学術研究書刊行助成金＞の援助を受けたことにより，出版が可能となった．（独）日本学術振興会，京都大学研究振興財団，さらに出版・編集で色々助けていただいた京都大学学術出版会編集部の鈴木哲也さん，高垣重和さん，（有）アリカの藤本りおさんに，この場を借りて厚くお礼申し上げるものである．

2010年10月1日

京都大学大学院理学研究科附属天文台 台長 柴田一成

Afterword

The Flare Monitoring Telescope (FMT) was planned, designed, and installed under the supervision of Hiroki Kurokawa (professor emeritus at Kyoto University), with the cooperation of Yoshihiro Nakai (former associate professor) and the late Yasuhiro Funakoshi (former assistant). Kiyomi Ishiura (former technical staff), Goichi Kimura (technical staff), and Yoshikazu Nakatani (technical staff) played important roles in the construction of the dome building, the preparation of the observation system, and the introduction of the automatic dome control system, to facilitate steady observation. Takehiko Shinkawa, who was a graduate student at Kyoto University, contributed to the introduction of a digital image acquisition system in 1996. The FMT system has been upgraded over the years by the staff at Hida Observatory and many postdoctoral researchers. Furthermore, many current and previous members of the staff at Hida Observatory have been engaged in daily observations using the FMT. Thanks to the ceaseless efforts of those members, a large amount of valuable data has been collected over a long period of time, and I am very grateful to all involved.

The FMT, which has superior characteristics as explained in the Introduction, could not have produced the splendid results described in this book without the efforts of Miwako Katoda (assistant technical staff member at Hida Observatory), who has been dedicated to the survey of Moreton waves. Her work has not been limited to Moreton waves; she has also created survey lists of flares, filament eruptions, filament disappearances, filament activities, surges, and other solar activity phenomena observed on video movies of $H\alpha - 0.8$Å solar full disk images. These lists are publicly available on the website of the Kwasan and Hida Observatories. Prominence-data analysis using solar limb images was successively conducted by the late Yoko Nasuji, En Kobayashi, and Yasuhiro Isoda, who all worked as assistant technical staff at Kwasan Observatory. The data they compiled are also available to the public. I sincerely appreciate the dedicated efforts of Miwako Katoda and the other members of staff mentioned above.

This book was completed with various forms of cooperation of staff members and graduate students (Takuma Matsumoto, Ken-ichi Otsuji, Hiroko Watanabe, and Tetsu Anan) at the Kwasan and Hida Observatories, Graduate School of Science, Kyoto University. Special thanks are due to the editorial committee members (Reizaburo

Kitai, Satoru UeNo, Daisaku Nogami, Ayumi Asai, Takako Ishii, Miwako Katoda, Shozo Yagi, Mai Kamobe, Hiroyuki Komori, and Naoki Kaneda) for their dedicated cooperation.

The publication of this book was made possible by financial support through a Grant-in-Aid for Publication of Scientific Research Results (Subject No. 225226) from the Japan Society for the Promotion of Science (JSPS), and a Grant-in-Aid for Publication of Academic Research Book from The Kyoto University Foundation. I am deeply grateful to JSPS and The Kyoto University Foundation. Finally, I would like to thank Mr. Tetsuya Suzuki and Mr. Shigekazu Takagaki at Kyoto University Press and Ms. Rio Fujimoto at Arica Inc. for their professional support in the editing and publishing of this book.

Kazunari Shibata
Director of Kwasan and Hida Observatories, Graduate School of Science, Kyoto University

索引

[あ]

アーケード 18, 87, 93, 95, 103, 107, 115, 139, 147, 153, 247, 249, 279, 309, 353, **371**, 376, 377, 450, 451, 458
ウインキングフィラメント **64**, 127, 295, 299

[か]

カターニア天文台 43
カンツェルヘーエ太陽天文台 43
クラウドモデル 29, 31, 153, **330**, 380, 453
光球 3, 12, 237, 386
光球磁場 115, 416, 428, 440, 445, 458
高視野分光コロナグラフ 380
国立天文台 43
コロナ 3, 8, 42, 45, 48, 64
コロナ減光（減光現象、ディミング）48, 139, 376, 377, 384, 389, 416
コロナ質量放出 4, **8**, **9**, 331, 371, 404, 451
コンパクトフレア（コンパクトなフレア）56, 57, 97, 99, 117, 127, 137, 141, 161, 165, 167, 169, 171, 179, 191, 195, 207, 215, 227, 229, 241, 253, 261, 301, 360, 392

[さ]

サージ 35, 52, **57**, 99, 109, 113, 119, 137, 141, 143, 165, 192, 193, 195, 215, 217, 255, 257, 261, 269, 321
彩層 3, 8, 12, 31, 193, 195, 237, 301, 439
差分画像 65, **77**
磁気ループ 48, 237, 287, 393, 416
磁気シアー 488
磁気リコネクション 9, **45**, 115, 137, 203, 209, 450, 451, 469, 470
スプレー型プロミネンス 451
遷移層 45, 48

[た]

太陽風 3, 19
地磁気嵐（磁気嵐）4, 21, 25, 331, 372, 483, 500
地磁気変動 20, 24, 25, 517
ツーリボンフレア 17, **52**, **53**, 54, 87, 91, 93, 107, 115, 147, 151, 191, 201, 203, 209, 211, 235, 247, 249, 251, 271, 279, 281, 283, 285, 334, 374, 387, 451
ドップラー効果 14
ドップラーシフト 331
ドップラー増光現象 342, 366, 380
ドップラー速度 33

[な]

NOAA 番号 68, 70, 80, 81, 82, 83, 84, 85
野辺山電波観測所 18, 46, 49
野辺山電波ヘリオグラフ（電波ヘリオグラフ）18, **46**, 49, 143, 309, 512

[は]

ハロー型 CME（ハロー CME）**22**, 201, 233, 418, 420
ピック・デュ・ミディ天文台 43
ビッグ・ベア太陽天文台 43
「ひので」衛星 46, 48
平磯太陽観測センター 43
平磯電波分光観測 301
フィラメント 6, 31, 42, 49, 52, 73
フィラメント活動 61, 62
フィラメント消失 18, 32, 33, **61**, 103, 209, 211, 249, 330, 371, 377, 381, 398, 450, 451, 458, 480, 481, 500, 508, 511
フィラメント振動 **64**, 73, 129, 133, 295, 299, 414
フィラメント噴出 19, 35, 52, 117, 139, 161, 191, 227, 229, **371**, 377, 447, 451
浮上磁場領域 488
フレア 4, **8**, 16, 19, 22, 42, 45, 52, 56, 64, 69, 480, 483
フレア監視望遠鏡 27, 480, 490
プロミネンス 5, **6**, 15, 30, 42, 49, 60, 67, 70, 71, 72, 73, 74, 79, 80, 306, 308, 310, 311, 312, 313, 314, 318, 319, 320, 322, 332, 461
プロミネンス活動 52, 480, 498
プロミネンス噴出（プロミネンス爆発、噴出型

プロミネンス）*7, 35, 49, **60**, 71, 309, 331, 405, 419, 420, 451, 480, 483*
ポストフレアループ *52, 72, 147, 209, 315*
ホモロガスフレア *203, 221, 223*
ホワイロウ太陽観測所 *43*

[ま]
マウナロア太陽天文台 *43, 46*
ムードン天文台 *43*
モートン波 *32, 56, **64**, 65, 77, 119, 121, 123, 127, 129, 133, 141, 143, 153, 155, 169, 171, 179, 181, 185, 187, 209, 211, 215, 217, 221, 223, 241, 243, 263, 265, 273, 275, 287, 289, 293, 295, 299, 301, 303, 403, 425, 438*

[や]
ユンナン天文台 *43*
「ようこう」衛星 *18, 32, **45**, 46, 48, 87, 115, 171, 309, 353, 376, 401, 405, 425, 439, 454, 473*

[ら]
ループ型プロミネンス *52, 72*

[C]
CHIAN プロジェクト *37*

[E]
EIT 波 *123, 129, 143, 169, 171, 179, 181, 187, 209, 211, 217, 263, 377, 384, 385, 394, **403**, 404, 420, 426, 433, 439*

[G]
GOES クラス *76*
GOES プロット *75, 76, 79*

[T]
Type-II 電波バースト（II 型の電波バースト） *129, 169, 171, 265, **404**, 415, 420, 426, 438*
Type-III 電波バースト（III 型の電波バースト） *415*

INDEX

[A]
Atmospheric Imaging Assembly (AIA) *46, 47*
Arcade *18, 87, 93, 95, 103, 107, 115, 139, 147, 153, 247, 249, 279, 309, 353,* **371**, *376, 378, 450, 451, 458*

[B]
Big Bear Solar Observatory *44*

[C]
Catania Astrophysical Observatory *44*
CHAIN Project *37*
Chromosphere **3**, *8, 12, 31, 193, 195, 237, 301*
Cloud model *30, 31, 153,* **330**, *380, 453*
Compact flare (Compact ** flare) **56**, *57, 97, 99, 117, 127, 137, 141, 161, 165, 167, 169, 171, 179, 191, 195, 207, 215, 227, 229, 241, 253, 261, 301, 360, 392*
Corona **3**, *8, 45, 48, 60, 64*
Coronal dimming (Dimming) *48, 376, 378, 384, 385, 389, 416*
Coronal mass ejection (CME) *4,* **8**, **9**, *19, 135, 137, 163, 167, 169, 171, 175, 177, 207, 213, 217, 231, 249, 251, 265, 271, 281, 283, 285, 312, 313, 314, 315, 316, 317, 318, 319, 320, 321, 322, 323, 331, 371, 377, 378, 385, 398, 404, 451, 484, 500*
CORONAS-photon *46, 47*

[D]
DB *61, 330, 331, 371, 378, 381, 398, 453*
Doppler brightening effect (DBE) *342, 366, 381*
Doppler effect *14*
Doppler shift *331*
Doppler velocity *30, 33*

[E]
EIT wave *123, 129, 143, 169, 171, 179, 181, 187, 209, 211, 217, 263, 378, 384, 394,* **403**, *404, 420, 426, 433, 439*
Emerging flux region (EFR) *486*

Extreme UltraViolet Imager (EUVI) *46, 47*
Extreme ultraviolet Imaging Telescope (EIT) *46, 47, 48, 139, 175, 177, 197, 199, 201, 207, 227, 231, 237, 239, 247, 249, 251, 259, 261, 279, 281, 312, 313, 314, 319, 336, 373, 377, 378, 385, 398,* **404**, *407, 409, 439, 452*

[F]
Full-disk EUV Telescope (FET) *46, 47*
Filament *6, 31, 53, 73*
Filament activity *61, 88, 148, 196, 198, 238, 258*
Filament disappearance (Disappearance of filament, Disappearing filament) *18, 32, 33,* **61**, *249, 330, 450, 451, 458, 509*
Filament eruption (Erupting filament) *19, 35, 53, 86, 90, 92, 94, 100, 102, 104, 106, 110, 114, 116, 126, 134, 138, 146, 152, 158, 160, 166, 168, 174, 176, 178, 190, 200, 202, 204, 206, 208, 209, 211, 212, 226, 230, 234, 236, 246, 252, 270, 280, 284, 300,* **371**, *447, 451*
Filament oscillation (Oscillatory filament motion, Oscillating filament) **64**, *129, 133, 299, 414*
Flare **8**, *16, 19, 42, 45, 49, 52, 53, 56, 64, 69, 90, 92, 96, 98, 104, 106, 114, 116, 120, 126, 134, 136, 138, 140, 146, 152, 160, 162, 164, 166, 168, 174, 178, 184, 190, 200, 202, 206, 208, 214, 220, 226, 230, 232, 236, 240, 246, 252, 254, 260, 262, 270, 272, 280, 284, 286, 292, 300, 482*
Flare monitoring telescope (FMT) *27, 43, 44, 489*

[G]
Geomagnetic fluctuation *20, 24, 25*
Geomagnetic storm (Geomagnetic disturbance) *4, 21, 25, 331, 372, 482, 489, 509, 510*
Geostationary Operational Environment

528 | INDEX

Satellite (GOES) *46, 47, 376, 377, 378*
GOES Class *76*
GOES Plot *75, 76, 79*

[H]
Halo CME (Halo-type CME) **22,** *201, 233, 418*
Hinode **47,** *48*
Hiraiso Radio Spectrograph (HiRAS) *301, 409, 415, 440*
Hiraiso Solar Observatory *44*
Homologous flare *203, 221, 223*
Huairou Solar Station *44*
H α *6, 10, 11, 13, 29, 41, 42, 43, 44, 64*

[K]
Kanzelhoehe Solar Observatory *44*

[L]
Large Angle Spectrometric Coronagraph (LASCO) *378, 379, 380*
Loop prominence *53, 72*

[M]
Magnetic loop *48, 393, 416*
Magnetic reconnection **9, 45,** *115, 137, 203, 209, 450, 451, 470*
Magnetic shear *482*
Mark-IV K-coronameter *46, 47*
Mauna-Loa Solar Observatory *44, 47*
Meudon Observatory *44*
MgXII Imaging SpectroHeliometer (MISH) *46, 47*
Moreton wave *32, 56,* **64,** *65, 77, 119, 121, 122, 128, 133, 141, 142, 153, 154, 169, 170, 179, 180, 185, 186, 209, 211, 216, 221, 222, 241, 242, 263, 264, 273, 274, 287, 288, 293, 294, 299, 301, 302, 403, 425, 438*

[N]
National Astronomical Observatory of Japan *44*
NOAA 10030 (0030) *65, 84, 254, 256*
NOAA 10069 (0069) *65, 84, 85, 258, 260, 262, 264*
NOAA 10087 *85, 268*

NOAA 10119 *85, 270*
NOAA 10139 (0139) *65, 85, 272, 274*
NOAA 10283 *85, 278*
NOAA 10311 *85, 280*
NOAA 10338 *85, 284*
NOAA 10342 *85, 284*
NOAA 10346 *85, 282*
NOAA 10365 (0365) *65, 85, 286, 288*
NOAA 10412 (0412) *85, 292, 294, 298*
NOAA 10794 (0794) *65, 85, 300, 302*
NOAA 7325 *81, 86*
NOAA 7332 *81, 86*
NOAA 7448 *81, 90*
NOAA 7500 *81, 92*
NOAA 7518 *81, 96, 98*
NOAA 7585 *81, 100*
NOAA 7605 *81, 102, 458*
NOAA 7646 *81, 104, 377, 378*
NOAA 7647 *81, 104*
NOAA 7671 *81, 106, 377, 378, 458*
NOAA 7746 *81, 108*
NOAA 7773 *81, 110, 377, 378, 458*
NOAA 7838 *81, 112*
NOAA 8038 *81, 114, 377, 378*
NOAA 8076 *81, 116*
NOAA 8100 *65, 81, 82, 118, 120, 122, 126, 128, 132, 403, 425, 428*
NOAA 8198 *82, 134*
NOAA 8203 *82, 136*
NOAA 8210 *82, 138, 377, 378*
NOAA 8299 *65, 82, 140, 142*
NOAA 8340 *82, 146, 377, 378, 458, 466*
NOAA 8453 *82, 150, 377, 378*
NOAA 8457 *82, 158*
NOAA 8458 *65, 82, 152, 154, 311, 334, 335, 374, 377*
NOAA 8557 *82, 160*
NOAA 8562 *82, 160*
NOAA 8645 *82, 162*
NOAA 8647 *82, 162*
NOAA 8673 *82, 164*
NOAA 8674 *82, 164*
NOAA 8806 *82, 166*
NOAA 8882 *65, 82, 83, 168, 170, 440, 441*
NOAA 9026 *65, 83, 174, 176*
NOAA 9040 *65, 83, 178, 180*

NOAA 9082 *65, 83, 184, 186*
NOAA 9087 *83, 190*
NOAA 9097 *83, 192*
NOAA 9176 *83, 194*
NOAA 9212 *83, 198*
NOAA 9213 *83, 198*
NOAA 9218 *83, 198*
NOAA 9231 *83, 200*
NOAA 9236 *83, 202*
NOAA 9238 *83, 200*
NOAA 9373 *83, 206*
NOAA 9415 *65, 83, 208, 210*
NOAA 9455 *65, 83, 84, 214, 216, 220, 222*
NOAA 9461 *65, 84, 226, 228*
NOAA 9628 *84, 230*
NOAA 9661 *84, 232*
NOAA 9685 *84, 234*
NOAA 9692 *84, 234*
NOAA 9704 *84, 236*
NOAA 9734 *84, 238*
NOAA 9742 *65, 84, 240, 242*
NOAA 9809 *84, 246*
NOAA 9870 *84, 248*
NOAA 9888 *84, 252*
NOAA number *68, 70, 80, 81, 82, 83, 84, 85*
Nobeyama Radioheliograph *18,* **47,** *49, 143, 309*
Nobeyama Solar Radio Observatory *47, 49*

[P]

Photosphere *3, 12, 237*
Photospheric magnetic field *115, 416, 428, 440, 458*
Pic-du-Midi Observatory *44*
Post flare loop *53, 72, 147, 209, 307, 315*
Prominence *5,* **6,** *15, 42, 49, 60, 67, 70, 71, 72, 73, 79, 80, 306, 332, 461*
Prominence activity *53, 480, 496*
Prominence eruption (Erupting prominence) *35, 49,* **60,** *71, 307, 308, 309, 310, 311, 312, 313, 314, 316, 317, 318, 319, 320, 322, 323, 331, 405, 419, 420, 451, 482*

[R]

Reuven Ramaty High Energy Solar Spectroscopic Imager (RHESSI) *46, 47*

Running difference image *65,* **77**

[S]

Solar Dynamic Observatory (SDO) *46, 47*
Solar EUV Coronagraph (SEC) *46, 47*
Sun Earth Connection Coronal and Heliospheric Investigation (SECCHI) *46, 47*
Solar Magnetic Activity Research Telescope (SMART) *43, 44, 301*
SOlar and Heliospheric Observatory (SOHO) *46, 47*
Solar wind *3, 19*
Spray prominence *451*
Solar TErrestrial RElations Observatory (STEREO) *46, 47*
Surge *35, 53,* **57,** *96, 98, 108, 112, 118, 136, 140, 164, 194, 214, 260, 268, 321*
Solar X-ray Imager (SXI) *46, 47*
Soft X-ray Telescope (SXT) *46, 47, 398*

[T]

TElescope-Spectrometer for Imaging solar Spectroscopy in X-rays (TESIS) *46, 47*
Transition Region And Coronal Explorer (TRACE) *46, 47*
Transition region (TR) *45, 48*
Two-ribbon flare (ribbon flare, Two-ribbon** flare) *17,* **53, 54,** *87, 91, 93, 107, 115, 147, 150, 151, 191, 201, 203, 209, 211, 235, 247, 248, 249, 250, 251, 271, 278, 279, 281, 282, 283, 285, 334, 374, 387, 451*
Type-II radio burst (type II) *129, 169, 171, 265,* **404,** *415, 426, 439*
Type-III radio burst (type III) *415*

[W]

Winking filament **64,** *65, 127, 132, 210, 228, 256, 294, 295, 298, 299*

[X]

X-Ray Telescope (XRT) *46, 47*

[Y]

Yohkoh *18, 32,* **45,** *47, 48, 87, 115, 171, 309, 353, 376, 405, 425, 439, 454, 474*
Yunnan Observatory *44*

執筆者一覧

柴田一成(京都大学大学院理学研究科附属天文台)
Kazunari Shibata (Kwasan & Hida Observatories, Graduate School of Science, Kyoto University)
北井礼三郎(京都大学大学院理学研究科附属天文台)
Reizaburo Kitai (Kwasan & Hida Observatories, Graduate School of Science, Kyoto University)
上野 悟(京都大学大学院理学研究科附属天文台)
Satoru UeNo (Kwasan & Hida Observatories, Graduate School of Science, Kyoto University)
野上大作(京都大学大学院理学研究科附属天文台)
Daisaku Nogami (Kwasan & Hida Observatories, Graduate School of Science, Kyoto University)
石井貴子(京都大学大学院理学研究科附属天文台)
Takako T. Ishii (Kwasan & Hida Observatories, Graduate School of Science, Kyoto University)
門田三和子(京都大学大学院理学研究科附属天文台)
Miwako Katoda (Kwasan & Hida Observatories, Graduate School of Science, Kyoto University)
名筋容子(京都大学大学院理学研究科附属天文台)
Yoko Nasuji (Kwasan & Hida Observatories, Graduate School of Science, Kyoto University)
鴨部麻衣(京都大学大学院理学研究科附属天文台)
Mai Kamobe (Kwasan & Hida Observatories, Graduate School of Science, Kyoto University)
金田直樹(京都大学大学院理学研究科附属天文台)
Naoki Kaneda (Kwasan & Hida Observatories, Graduate School of Science, Kyoto University)
小森裕之(京都大学大学院理学研究科附属天文台)
Hiroyuki Komori (Kwasan & Hida Observatories, Graduate School of Science, Kyoto University)
八木正三(京都大学大学院理学研究科附属天文台)
Shozo Yagi (Kwasan & Hida Observatories, Graduate School of Science, Kyoto University)
浅井 歩(京都大学宇宙総合学研究ユニット)
Ayumi Asai (Unit of Synergetic Studies for Space, Kyoto University)

太陽活動 1992-2003
――フレア監視望遠鏡が捉えたサイクル 23
Solar Activity in 1992-2003
―― Solar Cycle 23 Observed by Flare Monitoring Telescope

2011年2月28日　初版第1刷発行

著　　者　　柴　田　一　成
　　　　　　北　井　礼三郎
　　　　　　上　野　　　悟
　　　　　　野　上　大　作
　　　　　　石　井　貴　子
　　　　　　門　田　三和子
　　　　　　名　筋　容　子
　　　　　　鴨　部　麻　衣
　　　　　　金　田　直　樹
　　　　　　小　森　裕　之
　　　　　　八　木　正　三
　　　　　　浅　井　　　歩

発行者　　　檜　山　爲次郎

発行所　　　京都大学学術出版会
　　　　　606-8315　京都市左京区吉田近衛町69
　　　　　　　　　　京都大学吉田南構内
　　　　　　電話 075(761)6182　FAX 075(761)6190
　　　　　　URL　http://www.kyoto-up.or.jp/

印刷所　　　株式会社太洋社
装　幀　　　鷺草デザイン

ⓒ K. Shibata et al. 2011
Printed in Japan　　　価格はカバーに表示してあります

ISBN978-4-87698-987-4　C3044